Volume 3

GENETIC RESOURCES, CHROMOSOME ENGINEERING, AND CROP IMPROVEMENT

Vegetable Crops

GENETIC RESOURCES, CHROMOSOME ENGINEERING, AND CROP IMPROVEMENT SERIES

Series Editor, Ram J. Singh

Genetic Resources, Chromosome Engineering, and Crop Improvement
Volume 1: Grain Legumes
edited by Ram J. Singh and Prem P. Jauhar

Genetic Resources, Chromosome Engineering, and Crop Improvement
Volume 2: Cereals
edited by Ram J. Singh and Prem P. Jauhar

GENETIC RESOURCES, CHROMOSOME ENGINEERING, AND CROP IMPROVEMENT SERIES

Series Editor, Ram J. Singh

Volume 3

GENETIC RESOURCES, CHROMOSOME ENGINEERING, AND CROP IMPROVEMENT

Vegetable Crops

EDITED BY
RAM J. SINGH

CRC Press
Taylor & Francis Group
Boca Raton   London   New York

CRC Press is an imprint of the
Taylor & Francis Group, an informa business

CRC Press
Taylor & Francis Group
6000 Broken Sound Parkway NW, Suite 300
Boca Raton, FL 33487-2742

First issued in paperback 2019

© 2007 by Taylor & Francis Group, LLC
CRC Press is an imprint of Taylor & Francis Group, an Informa business

No claim to original U.S. Government works

ISBN-13: 978-0-8493-9646-5 (hbk)
ISBN-13: 978-0-367-38995-6 (pbk)

Visit the Taylor & Francis Web site at
http://www.taylorandfrancis.com

and the CRC Press Web site at
http://www.crcpress.com

Dedication

Nicolay Ivanovich Vavilov (November 25, 1887–January 26, 1943)

This book is dedicated to the memory of one of the greatest scientists of his time: Nicolay (also written as Nicolai) Ivanovich Vavilov. His breadth of knowledge included biology, botany, genetics, geography, agronomy, and plant breeding; above all, he was a "plant explorer extraordinaire." He established the law of homologous series in variation, provided theoretical background of plant resistance to infectious diseases and gave us the hypothesis for the centers of origin of cultivated plants. His vision to collect, preserve, and utilize immense genetic diversity for the improvement of the crops of economic importance is now being fulfilled. Thus, the dream of feeding the entire humankind with enough food may come true in the near future; at least, that is our hope.

(Photograph of Professor N. I. Vavilov is a courtesy of Sergey Shuvalov, N. I. Vavilov Research Institute of Plant Industry, St. Petersburg, Russia.)

Preface

The greatest delight the fields and woods minister is the suggestion of an occult relation between man and the vegetable. "I am not alone and unacknowledged." They nod to me and I to them.

Ralph Waldo Emerson (1803—1882)

Major cereals, grain legumes, oilseeds, and vegetable crops have always been an integral part of human civilization since time immemorial; they are also a primary food source for the world population. These crops are a rich source of carbohydrates (cereals), proteins (grain legumes), oils and fats (oilseed crops), and vitamins and minerals (vegetable crops). Vegetable crops have a unique place among domesticated crops because they are enriched in minerals, vitamins and antioxidants. Fresh vegetables are low in fat and sodium. Vegetables are consumed raw as well as cooked by both the vegetarian and non-vegetarian populations of the world. Moderate consumption of vegetables and fruits with cereals and beans reduces the risk of cancer, stress, diabetes, Alzheimer's, heart disease, and stroke (http://www.cncplan.com/vitamins.htm). Lycopene, found in tomato, is a powerful protector of prostrate, breast, digestive tract, cervix, bladder, and skin against cancer. Onion is a rich source of quercetin, a potent antioxidant linked to preventing stomach cancer. Daily consumption of a half raw onion lowers low density lipoprotein (LDL) cholesterol and raises beneficial high density lipoprotein (HDL) cholesterol by as much as 30%. Garlic, consumed raw, has antibacterial, antifungal, and anticancer properties. Garlic also has a blood-thinning quality. It lowers blood cholesterol, reduces high blood pressure, boosts the immune system, elevates mood, and has a calming effect in humans. Brassicas are effective in fighting cancer and are a rich source of vitamins (B, C, and E), iron, folate, zinc, potassium, thiamine, beta carotene, and fiber. Potato contains complex carbohydrates, protein, fiber, and provides vitamins B and C, iron and potassium (http://prostratecanceralternatives.com). Consequently, researchers around the globe are relentlessly examining and breeding vegetable crops for their superior nutritional quality and quantity, increased yield, and disease resistance. Despite their nutritional superiority and profound importance to human health, vegetable crops have received less attention toward needed genetic improvement than they deserve when compared with genetic research on cereals, grain legumes, and oilseeds. However, recently in progress is a new role for vegetable crops: their increasing use as ornamental and decorative plants (e.g. special forms and varieties of amaranth, crucifers, cucurbits, wild alliums, tomato, pepper, and eggplant). These new trends clearly demonstrate that vegetables and related crops may contribute not only to human nutrition but also to the improvement of human environment and quality of life.

Vegetable crops include crucifers (cabbage, cauliflower, kale, and brussels sprouts), tuber bulbs and root crops (potatoes, onions, garlic, turnip, leek, carrot, beet root, and radish), vegetable fruits (tomato, eggplant, okra, and cucurbits), salad crops (spinach, lettuce, amaranth, and celery), and protein-rich grain legumes (common bean, cowpea, faba bean, chickpea, and mungbean). Beans and peas have been excluded from this volume on *Vegetable Crops* because they have been included in the volume *Grain Legumes* of this series.

The majority of the vegetable crops originated in the Old World, particularly in Asia, while potato and tomato originated in the New World—South America. Today, potato and tomato are important vegetables throughout the world. They are considered a miracle gift of the New World and are grown worldwide in a wide range of agro-eco-climatic conditions on a commercial scale as well as on a small scale, often known as "kitchen gardening." Potato production alone accounts for 44% of all the vegetable crops production listed by the Food and Agriculture Organization (FAO) (http://faostat.fao.org). It is ranked as the fourth most important food crop of the world.

Furthermore, more than one-third of all the potatoes are now grown in the developing countries (http://www.cipotato.org).

The intensive varietal improvements of vegetable crops for high yield and improved nutritional quality are primary breeding objectives of various national and international programs. Three international centers for vegetable crops have been established: (1) The International Potato Center [Centro Internacional de la Papa (CIP)] was founded in 1971 in Lima, Peru (http://www.cipotato.org); its primary mandate was to deliver high-yielding potatoes. (2) The Asian Vegetable Research and Development Center (AVRDC) was established in 1971 in Shanhua, Taiwan (www.avrdc.org); its mandate is to improve tomato, brassicas, alliums, cucurbits, peppers, eggplant, okra, and legumes including edamame (edible vegetable soybean). (3) Centro Internacional de Agricultura Tropical (CIAT), Cali, Colombia (www.ciat.cgiar.org) works on beans as one of its primary mandates (described in *Grain Legumes*, Volume 1 of this series), and also works on sweet potatoes. The Indian Institute of Vegetable Research (IIVR) in Varanasi, Uttar Pradesh, India (www.icar.org.in) is a national institute; it concentrates on multidisciplinary research approaches for the improvement of the major vegetable crops grown in India. These and other centers around the world collect, maintain and preserve germplasm resources as potential sources of genes for high yield, resistance to various biotic and abiotic stresses, and for improving nutritional, storage, and shelf life qualities. In Europe, Warwick Horticulture Research Station (HRI) (a part of the University of Warwick, Warwick, UK; formerly known as National Vegetable Research Station (NVRS) and then HRI in Wellesbourne, UK) is a leading world center of vegetable research and breeding. Since its days in Wellesbourne in the middle of the twentieth century, this institution has contributed substantially to progress in several areas (breeding, plant protection, and disease management, plant nutrition and physiology, seed and growing technology, germplasm preservation and characterization, and biotechnology of vegetable sciences).

Most genetic improvements of vegetable crops have been accomplished by conventional breeding assisted by germplasm resources, cytogenetics, plant pathology, entomology, agronomy, cell and tissue cultures, and molecular biology. Because there is no consolidated account of germplasm resources, cytogenetic manipulations, biotechnological approaches, and breeding of vegetable cops, it was important to bring out this book, Volume 3 in the series *Genetic Resources,Chromosome Engineering, and Crop Improvement*. World-renowned scientists were invited to contribute chapters on the vegetable crops of their expertise. This volume consists of 11 chapters dealing with vegetable crops of great economic importance for the developed and the developing countries of the world. These chapters give comprehensive and authoritative accounts of genetic resources and their utilization for improving yields, disease and pest resistance, other agronomic traits, and nutritive quality of the most widely grown and consumed vegetables.

The introductory chapter summarizes the landmark research carried out in ten vegetable crops, giving information on germplasm availability for breeding for high yields and improved nutritional quality. Each of the subsequent chapters (2 through 11) deals, respectively, with one of the ten crops: potato, tomato, brassicas, okra, capsicum, alliums, cucurbits, lettuce, eggplant and carrot. Each chapter provides a comprehensive account of the origin of the crop, its genetic resources in various gene pools, basic and molecular cytogenetics, conventional breeding, and the modern tools of molecular genetics and biotechnology.

Appropriate germplasm collections can be an excellent source for genetic enhancement of various traits in vegetative crops and for broadening their genetic base. The genetic base of major vegetable crops is extremely narrow. The classical example of a narrow genetic base is the Great Irish Potato Famine (1845–1847) caused by a fungus, the potato blight [*Phytophothora infestans* (Mont.) de Bary] that ruined almost 100% of the potato crop of Ireland. Irish farmers were cultivating white potato introduced directly from the Andean Mountains (http://www.american.edu). In view of the narrow genetic base of vegetable crops, three gene pools have now been identified by the scientists: primary (GP-1), secondary (GP-2), and tertiary (GP-3) for each crop. The recommendation is to the use GP-2 and GP-3 resources in producing widely adapted varieties.

Utilization of these resources in producing high-yielding cultivars, resistant to abiotic and biotic stresses, and with improved nutritional qualities is discussed in this book.

Each chapter has been written by one or more experts in the field. I am extremely grateful to all the authors for their outstanding contributions, and to the reviewers of all the chapters. I have been fortunate to know them both professionally and personally, and our communication has been very cordial and friendly. I am particularly indebted to Henry De Jong, Govindjee, Mike Havey, Bob Jarret, Aleš Lebeda, Joseph Nicholas, Carlos Quiros, Phil Simon, and Richard Veilleux for their comments and suggestions on some of the chapters. I am profoundly grateful to Aleš Lebeda and Govindjee for their constructive and invaluable suggestions for the composition of the dedication for this volume. Finally, I thank Steven G. Pueppke, former associate dean and research director at the University of Illinois, Urbana, for his support and encouragement.

This book is intended for scientists, professionals, and graduate students whose interests center upon genetic improvement of crops in general, and major vegetable crops in particular. This book is meant to be a reference for plant breeders, taxonomists, cytogeneticists, agronomists, molecular biologists, food technologists, and biotechnologists. Graduate-level students in these disciplines, with an adequate background in genetics, as well as other researchers interested in biology and agriculture will also find this volume a worthwhile reference. I sincerely hope that the information assembled in this book will help in the much-needed genetic amelioration of vegetables crops to feed the ever-expanding global population. I anticipate that this book will enhance awareness regarding nutritive values of vegetables, thus preventing malnutrition worldwide. I can be reached at ramsingh@uiuc.edu

I end this Preface with a philosophical quotation from Albert Einstein (1879–1955):

> *Human beings, vegetables or cosmic dust, we all dance to a mysterious tune, intoned in the distance by an invisible piper.*

Ram J. Singh
Urbana-Champaign, Illinois

The Editor

Ram J. Singh is an agronomist–plant cytogeneticist in the Department of Crop Sciences, University of Illinois at Urbana-Champaign. He received his PhD in plant cytogenetics under the guidance of the late Professor Takumi Tsuchiya from Colorado State University, Fort Collins, Colorado. He benefited greatly from the cytogenetic expertise of Drs. T. Tsuchiya, G. Röbbelen, and G. S. Khush.

Dr. Singh conceived, planned, and conducted pioneering research related to cytogenetic problems in barley, rice, rye, wheat, and soybean. Thus, he isolated monotelotrisomics and acrotrisomics in barley, identified them by Giemsa C- and N-banding techniques and determined chromosome arm–linkage group relationships. In rice, he established a complete set of primary trisomics and determined chromosome–linkage group relationships. In soybean (*Glycine max*), he established genomic relationships among species of the genus *Glycine* and each species was assigned a genome symbol based on cytogenetics and molecular methods. Dr. Singh constructed, for the first time, a soybean chromosome map based on pachytene chromosome analysis and that laid the foundation for creating a global soybean map. By using fluorescent genomic *in situ* hybridization, he confirmed the tetraploid origin of the soybean.

Dr. Singh has published 69 research papers in reputable international journals including *American Journal of Botany*, *Chromosoma*, *Crop Science*, *Genetics*, *Genome*, *Journal of Heredity*, *Plant Breeding*, and *Theoretical and Applied Genetics*. In addition, he has summarized his research results by writing thirteen book chapters. His book *Plant Cytogenetics* [first edition (1993) and second edition (2003)] is widely used for teaching graduate students. Dr. Singh has presented research findings as an invited speaker at national and international meetings. In 2000, he received the Academic Professional Award for Excellence: Innovation & Creativity from the University of Illinois at Urbana-Champaign. He was invited as visiting professor (October 12, 2004 to January 12, 2005) by Professor Kiichi Fukui, Osaka University, Osaka, Japan. He is the editor of *Genetic Resources, Chromosome Engineering, and Crop Improvement* series and has published *Grain Legumes*, Volume 1 and *Cereals*, Volume 2.

Contributors

K. V. Bhat
NBPGR, Pusa Campus
New Delhi, India

I. S. Bisht
NBPGR, Pusa Campus
New Delhi, India

Hielke ("Henry") De Jong
Potato Research Centre,
    Agriculture and Agri-Food Canada
Fredericton, New Brunswick, Canada

Caroline Djian-Caporalino
UMR INRA 1064/UNSA/CNRS
Centre de Sophia, Unité Interactions Plantes
    Microorganismes et Santé Végétale
Sophia Antipolis, France

I. Doležalová
Department of Botany
Palacký University
Olomouc-Holice, Czech Republic

H. Ezura
Gene Research Center
Graduate School of Life and Environmental
    Sciences
University of Tsukuba
Ibaraki, Japan

Irwin L. Goldman
Department of Horticulture
University of Wisconsin
Madison, Wisconsin

R. Grube
Department of Plant Biology
University of New Hampshire
Durham, New Hampshire

N. Inomata
Biological Laboratory
Konan Women's University
Kobe, Japan

Yuanfu Ji
Gulf Coast Research and Education Center
Institute of Food and Agriculture Sciences
University of Florida
Wimauma, Florida

E. Křístková
Department of Botany
Palacký University
Olomouc-Holice, Czech Republic

Rajesh Kumar
Indian Institute of Vegetable Research
Varanasi, India

Aleš Lebeda
Department of Botany
Palacký University
Olomouc-Holice, Czech Republic

Véronique Lefebvre
INRA, Unité de Génétique et d'Amélioration
    des Fruits et Légumes
Montfavet, France

Alain Palloix
INRA, Unité de Génétique et d'Amélioration
    des Fruits et Légumes
Montfavet, France

E. J. Ryder
U.S. Department of Agriculture
Agricultural Research Service
Salinas, California

Anne-Marie Sage-Daubèze
INRA, Unité de Génétique et d'Amélioration
    des Fruits et Légumes
Montfavet, France

J. W. Scott
Gulf Coast Research and Education Center
Institute of Food and Agriculture Sciences
University of Florida
Wimauma, Florida

Masayoshi Shigyo
Laboratory of Horticultural Science
Department of Biological and
  Environmental Sciences
Yamaguchi University
Yamaguchi, Japan

Philipp W. Simon
USDA-ARS, Vegetable Crops
  Research Unit
Department of Horticulture
University of Wisconsin
Madison, Wisconsin

Major Singh
Indian Institute of Vegetable Research
Varanasi, India

Ram J. Singh
Department of Crop Sciences
National Soybean Research Laboratory
Urbana, Illinois

J. Staub
Department of Horticulture
U.S. Department of Agriculture
Agricultural Research Service
University of Wisconsin
Madison, Wisconsin

Richard E. Veilleux
Department of Horticulture
Virginia Polytechnic Institute and State
  University
Blacksburg, Virginia

M. P. Widrlechner
U.S. Department of Agriculture
Agricultural Research Service
North Central Regional Plant Introduction
  Station
and
Departments of Agronomy and Horticulture
Iowa State University
Ames, Indiana

J. Zalapa
Department of Horticulture
University of Wisconsin
Agricultural Research Service
U.S. Department of Agriculture
Madison, Wisconsin

Contents

CHAPTER 1

Landmark Research in Vegetable Crops

Ram J. Singh and Aleš Lebeda

CONTENTS

1.1 INTRODUCTION

In the nineteenth and twentieth centuries, humans tried to understand in greater detail the rules and processes of evolution, domestication, genetics, and breeding of plants. In the field of evolutionary biology, background was built by C. Darwin (Darwin 1859), domestication by N. I. Vavilov (see Löve 1992) and J. R. Harlan (Harlan 1992), and genetics and breeding by J. G. Mendel (Orel 1996). The Vilmorin-Andrieux family in France pioneered European plant and vegetable breeding in the nineteenth century, as did Luther Burbank in the United States, who gained particular fame for his work with the potato (Stickland 1998). By the 1940s, Darwin's concept of evolution and Mendel's ideas of heredity had been brought together to produce "the modern synthesis" (neo-Darwinism) and this was one of the great intellectual triumphs of the twentieth century (Tudge 2000). These germinal ideas and their combination created the background of modern biology and genetics. Practical application and utilization of this knowledge applied to studies of plant genetic resources has lead to advances in plant breeding.

Many vegetables are ancient cultivated plants that have played an important role in human development. In situ cultivation of plants began more than 10,000 years ago as result of changes in human life style from nomadic and semi-nomadic to sedentary life (Hancock 2004). Sauer (1969) outlined the requirement for plant domestication and early cultivation. As a crucial factor, he considered the status of land (i.e., that the first cultivated lands would have been those that required little preparation). Environmental conditions (rainfall and temperatures favorable for growing plants) were considered important. These conditions played a major role in the development of diverse food plants. N. I. Vavilov postulated at least eight centers of origin and diversity of cultivated plants including vegetable crops (Figure 1.1; Vavilov 1950). All of these centers encompassed modern vegetable crops (Rubatzky and Yamaguchi 1997). The diversity of vegetables is considered an international and world heritage (Stickland 1998).

The current diverse group of modern vegetable crops originated from ancestral species found on all habitable continents, some cultivated for more than 10,000 years and others only recently developed. From their original centers of origin, vegetable crops have been spread by man throughout the world, often to climates far from their natural range. For example, the tomato originated in the tropics of the New World but is now a major vegetable crop in temperate climates. Developing novel traits that facilitate adaptation to alien environments, such as frost-hardiness in the potato, is a goal of modern vegetable breeding (Phillips and Rix 1993). The major objectives and modern practices of vegetable breeding and cultivation have been reviewed in specialized textbooks (Bassett 1986; Kalloo 1988; Kalloo and Bergh 1993; Rubatzky and Yamaguchi 1997; Decoteau 2000).

For future domestication and improvement of useful plants, including vegetables, it is necessary that regional as well as global genetic resources be appropriately collected, maintained, and preserved before they become extinct. The potential of existing biological diversity should be preserved for the benefit of all people (Rubatzky and Yamaguchi 1997). The main aim of this chapter is to summarize recent knowledge and achievement on genetic resources and their taxonomy, diversity, collection, conservation, evaluation, and utilization in breeding of ten major vegetables of economic importance.

1.2 IMPORTANCE OF VEGETABLE CROPS

Cereals are a major source of carbohydrates and fiber, grain legumes supply protein and antioxidants, and oilseeds are an excellent source of dietary fatty acids, whereas vegetable crops are valuable sources of nutrition, including mineral nutrients, antioxidants, and vitamins. The primary vegetable crops included in this volume are potato, tomato, brassicas, okra, pepper, allium, cucurbits, lettuce, eggplant, and carrot. Many legumes, such as common bean, pea, cowpea, pigeonpea, faba bean, chickpea, cowpea, mungbean, azuki bean, and lentil are also used as green vegetables. These legumes are described in *Grain Legumes*, Volume 1 of this series. Green immature soybean pods (known as *Edamame*), described in *Oilseed Crops*, Volume 4, are popular in Japan, Korea, China, and Taiwan. Soybeans are a major source of protein, vitamin A, carbohydrates, and iron, as well as many antioxidants. Maize (sweet corn) is also included in vegetables and is discussed in *Cereals*, Volume 2 of this series.

Vegetable crops belong to the families Alliaceae (onions), Asteraceae (lettuce), Brassicaceae (brassicas), Cucurbitaceae (cucurbits), Malvaceae (okra) and Solanaceae (potato, tomato, pepper, and eggplant) and Apiaceae (carrot, celery), whereas legumes belong to Fabaceae and cereals to

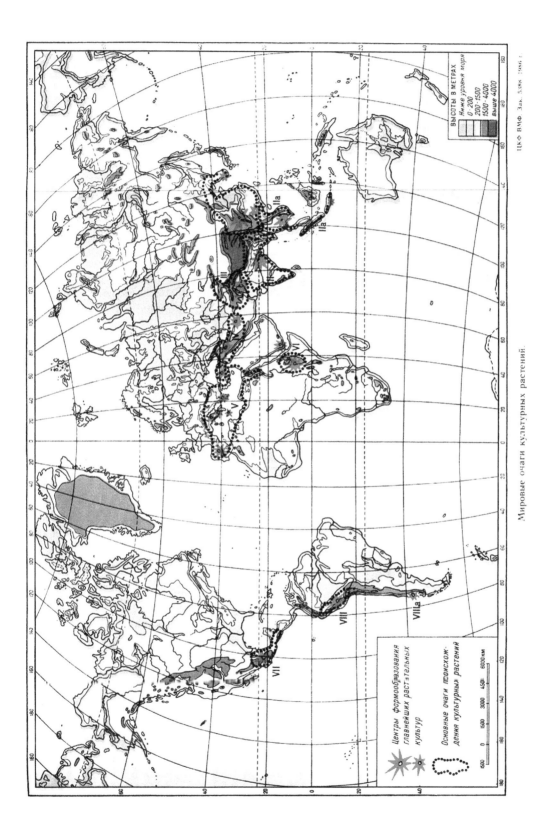

Figure 1.1 **(See color insert following page 304.)** Eight centers of origin and diversity of cultivated plants including vegetable crops. (From Vavilov, N. I., *Origin and Geography of Cultivated Plants*, Nauka, Leningrad, USSR, 1987.)

Poaceae. Vegetable crops are highly diverse in geographical distribution (Chapter 2 through Chapter 11). Vegetables are classified based on edible parts of the plant (root, bulb, seedling, shoot, stem, leaf, immature flower bud, flower, mature and immature fruit, and seed), life cycle (annual, biennial, and perennial), climatic preference (cool season and warm season), and accepted use (cole crops, greens, salad crops, perennial crops, root crops, bulb crops, legumes or pulse crops, sweet corn, solanum crops, and cucurbits) (George 1999; Decoteau 2000).

Vegetables have been divided into five subgroups depending on their use and nutrient content: (1) Dark green vegetables (broccoli, collard green, lettuce, kale, mustard green, spinach, turnip green, and watercress); (2) Orange vegetables (squash, pumpkin, carrot, and sweet potato); (3) Dry beans and peas (common bean, faba bean, chickpea, pea, lentil, and soybean); (4) Starchy vegetables (corn, green pea, green faba bean, and potato); (5) Other vegetables (artichoke, asparagus, bean sprouts, beets, brussels sprouts, cabbage, cauliflower, celery, cucumber, eggplant, green bean, green or red pepper, butterhead lettuce, mushrooms, okra, onion, parsnip, tomato, turnip, and zucchini).

The U.S. Department of Agriculture, Center for Nutrition Policy and Promotion revised the food pyramid guidelines in April 2005. The pyramid recommends eating more dark-green vegetables like broccoli, spinach, and other dark leaf greens, more orange vegetables like carrot and sweet potato, and more grain legumes. A balance between food intake and physical activity and restrictions (limits) on consumption of fat, sugar, and salt are recommended (http://www.mypyramid.gov/pyramid/vegetables.html).

Radish, eggplant, brassicas, soybean, onion, various cucurbits, spinach, mustard, carrot, cabbage, lettuce, muskmelon, and okra originated and were domesticated in the Old World. Potato, tomato, pepper and pumpkin originated in the New World and are now cultivated worldwide. Of all the economically important vegetable crops listed by the Food and Agriculture Organization of the United Nations (FAO), potato production leads (37%) and is followed by cucurbits (19%), tomato (14%), cabbage and cauliflower (10%). Production of other vegetable crops ranges from 1% (okra) to 8% (alliums) (Figure 1.2). Major vegetable crops described in this volume are grown by commercial vegetable industries and total world acreage is given in Figure 1.3 (FAO STAT 2004; http://faostat.fao.org). Potato is the leading crop, covering over 18 million hectares, followed by cucurbits (over 8 million hectares). Tomato, alliums, cabbage and cauliflower occupies 4 million hectares worldwide. Vegetable crops are also produced by small vegetable growers for home consumption, sale in seasonal farmers' markets, or distributed to small businesses. Some vegetable crops, such as gourds (bitter, bottle, sweet, snake, pointed, ash or wax), loofah (smooth, angular), amaranth, mungbean, cowpea, anise, basil, chives, cilantro (coriander),

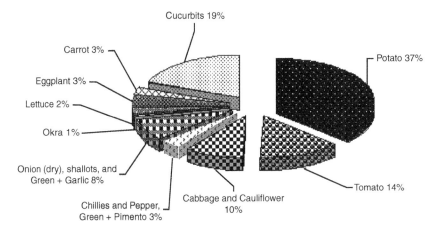

Figure 1.2 Pie diagram showing the world production of the major vegetable in 2004. (From http://faostat.fao.org)

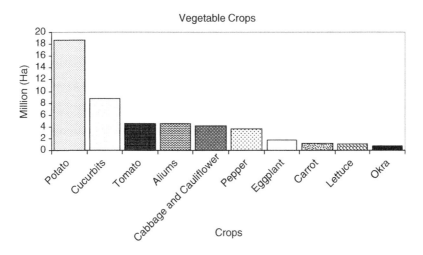

Figure 1.3 Total world area of major vegetable crops in 2004. (From http://faostat.fao.org)

dill, mint, thyme, watercress, parsley roots, alfalfa sprouts, and many others are grown and consumed primarily in restricted regions of the world (Decoteau 2000; http://www.icar.org.in).

Vegetable crops have a unique place among domesticated crops because they are an important source of minerals (Table 1.1), vitamins (Table 1.2), antioxidants, aminoacids, and micronutrients to many populations. A diet rich in vegetables helps to reduce malnutrition in developing and developed countries. Moderate consumption of fresh vegetables along with fruits, cereals, beans, and high quality vegetable oils reduces the risk of cancer, stress, hypertension, obesity, diabetes, Alzheimer's, cholesterol, heart disease, and stroke (http://www.cncplan.com/vitamins. htm; Chapter 10). Vegetables may be consumed raw or cooked in developing countries; vegetable crops are grown seasonally as subsistence crops in kitchen gardens for home consumption. Often, these vegetables are not included in the daily diet during off seasons. Vegetables are particularly essential for those populations that are primarily vegetarian. Varietal improvement of vegetable crops has been neglected, compared to cereals, in the developing countries.

1.3 ESTABLISHMENT OF INTERNATIONAL AND NATIONAL PROGRAMS

The following international and national centers have been established for vegetable research:

1. Asian Vegetable Research & Development Center (AVRDC), Shanhua, Tainan, Taiwan (http://www.avrdc.org): This institute was established in 1971 and is the principal international center for vegetable research and development in the world. Mandated crops are bulb allium, crucifers, legumes, pepper, and tomato. Breeding for improved vegetable soybean, eggplant, gourd, cucumber, and snap bean is also a priority for AVRDC.
2. Centro Internacional de la Papa (CIP; International Potato Center), Lima, Peru (http://www.cipotato.org): This institute was established in 1971. Potato, sweet potato, and other tuber crops are the focus.
3. International Plant Genetics Resources Institute (IPGRI), Rome, Italy (http://www.ipgri. cgiar.org): This institute is involved in the coordination of conservation of germplasm of several vegetable crops.

Table 1.1 Mineral Nutritional Contents of Major Vegetable (Value/100 g of edible portion)

	Calcium (mg)	Iron (mg)	Magnesium (mg)	Phosphorus (mg)	Potassium (mg)	Sodium (mg)	Zinc (mg)	Copper (mg)	Manganese (mg)	Selenium (mcg)	Carbohydrate (g)
Potato	34	7.04	43	101	573	21	0.49	0.817	0.616	0.7	46.07
Tomato	5	0.45	11	24	222	9	0.09	0.07	0.105	0.4	4.64
Broccoli	48	0.879	25	66	326	27	0.4	0.045	0.229	3	5.239
Cabbage	47	0.589	15	23	246	18	0.179	0.022	0.159	0.9	5.43
Cauliflower	22	0.44	15	44	303	30	0.28	0.042	0.156	0.6	5.2
Kale	135	1.7	34	56	447	43	0.44	0.289	0.774	0.9	10.01
Radish	21	0.289	9	18	232	24	0.3	0.039	0.07	0.7	3.539
Okra	81	0.8	57	63	303	8	0.6	0.094	0.99	0.7	7.63
Chili Pepper	18	1.2	25	46	340	7	0.3	0.173	0.237	0.5	9.46
Sweet Pepper	9	0.459	10	19	177	2	0.12	0.065	0.116	0.3	6.43
Onion	20	0.22	10	33	157	3	0.19	0.06	0.137	0.6	8.63
Cucumber	14	0.259	11	20	144	2	0.2	0.032	0.076	0	2.759
Zucchini	15	0.42	22	32	248	3	0.2	0.056	0.127	0.2	2.9
Squash	20	0.46	23	35	195	2	0.26	0.076	0.156	0.2	4.349
Pumpkin	21	0.8	12	44	340	1	0.319	0.126	0.125	0.3	6.5
Lettuce	19	0.5	9	20	158	9	0.22	0.027	0.15	0.2	2.089
Eggplant	7	0.269	14	22	217	3	0.14	0.054	0.13	0.3	6.069
Carrot	27	0.5	15	44	323	35	0.2	0.046	0.142	1.1	10.140

Source: From www.personalhealthzone.com/nutrients/vegetables.html

Table 1.2 Vitamin Contents of Major Vegetables (Value/100 g of edible portion)

	Vitamin C (mg)	Thiamin (mg)	Riboflavin (mg)	Niacin (mg)	Pantothenic Acid (mg)	Vitamin B6 (mg)	Folate Total (mcg)	Vitamin A (IU)	Vitamin E (mg_ATE)
Potato	13.5	0.122	0.106	3.065	0.857	0.614	22	0	0.04
Tomato	19.1	0.058	0.047	0.627	0.247	0.08	15	623	0.38
Broccoli	93.2	0.064	0.119	0.637	0.535	0.159	71	1,542	1.66
Cabbage	32.2	0.05	0.04	0.3	0.14	0.095	43	133	0.104
Cauliflower	46.4	0.057	0.063	0.526	0.652	0.222	57	19	0.04
Kale	120	0.11	1.298	1	0.091	0.271	29	8,900	0.8
Radish	22.8	0.005	0.044	0.3	0.087	0.07	27	7,931	0.001
Okra	21.1	0.2	0.06	1	0.245	0.215	88	660	0.69
Chili Pepper	242.5	0.091	0.091	0.948	0.06	0.27	23	10,750	0.68
Sweet Pepper	190	0.065	0.03	0.508	0.079	0.248	22	5,700	0.689
Onion	6.4	0.041	0.02	0.148	0.106	0.116	19	0	0.31
Cucumber	5.3	0.023	0.021	0.221	0.178	0.042	13	215	0.078
Zucchini	9	0.069	0.03	0.376	0.083	0.009	22	340	0.12
Squash	14.8	0.637	0.037	0.551	0.101	0.108	26	196	0.12
Pumpkin	9	0.05	0.11	0.6	0.298	0.061	16	1,600	1.06
Lettuce	3.9	0.045	0.031	0.187	0.045	0.04	56	330	0.28
Eggplant	1.7	0.052	0.034	0.597	0.252	0.084	19	84	0.03
Carrot	9.3	0.097	0.059	0.928	0.197	0.147	14	28,129	0.459

Source: From www.personalhealthzone.com/nutrients/vegetables.html

4. Centro Internacional de Agricultura Tropical (CIAT), Cali, Colombia (http://www.ciat. cgiar org)· This institute works on beans as one of its primary mandates (described in *Grain Legumes*, Volume 1 of this series) and also works on sweet potatoes.
5. National Programs: National (public) and private industries worldwide have vegetable crop improvement programs. For example, the Indian Institute of Vegetable Research (IIVR), Varanasi, Uttar Pradesh, India (http://www.icar.org.in) collects and maintains germplasm, develops technology to improve vegetable cultivars, and disseminates scientific information to farmers.

Public sectors conduct basic research on crops, develop unique and improved germplasm lines, and release germplasm lines for further improvement by breeders in the private sector (Chapter 9). Private companies release new cultivars and market produce to consumers.

1.4   DEVELOPMENT OF VEGETABLE CROPS INDUSTRIES

Developed and many developing countries have major and specialty vegetable crop industries. Their main objective is to develop improved vegetable crops of economic importance through conventional and molecular methods. They also distribute seeds or clones resistant to abiotic and biotic stress or with improved nutritional quality. Commercial industries are leaders in vegetable breeding (Chapter 6). They usually concentrate on one or two crops grown on large acreages and distribute products to distant supermarkets worldwide. Profit is the bottom line for industry; therefore, they consider only higher-value crops. Vegetables such as beans, beets, cabbage, carrot, sweet corn, cucumber, spinach, tomato, asparagus, broccoli, cauliflower, potato, and others may be processed by freezing, canning, and dehydration before shipping to supermarkets.

1.5   GENE POOLS FOR VEGETABLE CROPS

Based on reviewing literature on hybridization, Harlan and de Wet (1971) proposed three gene pool concepts: primary (GP-1), secondary (GP-2), and tertiary (GP-3) for utilization of germplasm resources for crop improvement. Genetic resources are developed with integrated, multidisciplinary approaches through plant exploration, taxonomy, genetics, cytogenetics, plant breeding, microbiology, plant pathology, entomology, agronomy, physiology, hybridization, and molecular biology, including cell and tissue culture and genetic transformation. These efforts have produced superior vegetable cultivars, rich in nutrients, with resistance to abiotic and biotic stress, high yield, and extended shelf life. The concept of primary, secondary, and tertiary gene pools has played a key role in improving vegetable crops (Chapter 2 through Chapter 11).

1.5.1   Primary Gene Pool

The primary gene pool (GP-1) consists of landraces and biological species and has been identified for vegetable crops described in this volume. Progenitors of cultivated vegetable crops are identified, postulated, and proposed based on geographical distribution, classical taxonomy, cytogenetics, and molecular methods (see Chapter 9). Each chapter describes the primary gene pool in detail. For example, GP-1 for potato includes cultivated diploid ($2n=2x=24$), triploid ($2n=3x=36$), tetraploid ($2n=4x=48$), and pentaploid ($2n=5x=60$) species (Table 2.2). *Solanum tuberosum* is a tetraploid species and is cultivated worldwide. The primary gene pool of pepper includes cultivated species of diverse geographical origin and one wild relative,

C. annuum var. *glabriusculum.* GP-1 of onion has been subdivided into GP-1A and GP-1B. GP-1A of bulb onion includes vegetatively propagated shallot. GP-1B consists of *A. vavilovii* for bulb onion and *A. altaicum* for Japanese bunching onion (Chapter 7). Chapter 8 describes nearly all wild *Cucumis* species that are cross-incompatible with *C. sativus* and *C. melo.* Such wild relatives, including *Cucurbita argyrosperma*, *C. pepo*, and *C. maxima*, grow sympatrically in the Americas, facilitating genetic exchanges between them. The lettuce primary gene pool is represented primarily by *Lactuca sativa*, the wild species *L. serriola*, and seven wild *serriola*-like species that yield fertile hybrids in interspecific crosses (Chapter 9).

1.5.2   Secondary Gene Pool

The secondary gene pool includes species that can yield at least partially fertile F_1 on hybridization with GP-1 (Chapter 2 through Chapter 10). Gene transfer is possible. Many species in GP-2 (Table 2.2) have been used in potato breeding, resulting in cultivars released in North America and Europe (Chapter 2). GP-2 in okra consists of only two species (Chapter 5). Onion is rich with GP-2 species (Figure 7.1). Interspecific crosses among *Cucurbita* species are usually sterile or sparingly fertile; GP-2 for *C. pepo* is scarce (Chapter 8). In lettuce, *L. saligna* belongs to GP-2 (Figure 9.18).

1.5.3   Tertiary Gene Pool

The tertiary gene pool is the outer limit of potential genetic resources for breeding. Prezygotic and postzygotic barriers can cause partial or complete hybridization failure, inhibiting introgression between GP-1 and GP-3 (Singh 2003). GP-3 is available for most vegetable crops (Chapter 2 through Chapter 11). The tertiary gene pool in potato has been divided into those more easily crossed (A) and those that resulted in little or no seed set (B) (Chapter 2). Many interspecific and intergeneric hybrids among cross-incompatible species of *Brassica* have been produced by conventional crossing, embryo rescue, and protoplast fusion (Chapter 4). In pepper, *Capsicum pubescens* is genetically isolated, and breeders have not succeeded with any exchange with the *C. annuum* complex. Extensive exploration of wild *Capsicum* spp. is recommended to identify additional species for GP-3 (Chapter 6). Technology to exploit GP-3 for broadening the genetic base of two *Allium* cultivated crops (Chapter 7), lettuce (Chapter 9), and eggplant (Chapter 10) has yet to be developed.

1.6   GERMPLASM RESOURCES FOR VEGETABLE CROPS

The national and international institutes and private industries for the ten vegetable crops (potato, tomato, brassicas, okra, pepper, onion, cucurbits, lettuce, eggplant, and carrot) presented in this volume collect, maintain, disseminate, and develop breeding lines with resistance to abiotic and biotic stresses. Plant exploration of landraces and wild relatives of the ten vegetables (Chapter 2 through Chapter 11) described in this volume is extensive. They have been characterized based on classical taxonomy, cytogenetics, and molecular methods. It is interesting to note that three solanaceous vegetable crops (potato, tomato, and pepper) originated in the New World, whereas a fourth (eggplant) was domesticated in the Old World. Members of the Cucurbitaceae are predominantly tropical, with 90% of the species found in three main areas: Africa, including Madagascar; Central and South America; and Southeast Asia, including Malaysia (Chapter 8). The center of origin of cultivated lettuce is the Middle East; however, over 90% of accessions of its wild relatives held in genebank collections are represented by only three species (*Lactuca serriola, L. saligna, L. virosa*). They are mostly of European origin, and genebanks do not hold some *Lactuca* species, which are considered progenitors of lettuce (*L. sativa*) (Chapter 9). Like lettuce and *Lactuca,* germplasm collections of carrot and wild *Daucus* species are meager (Chapter 11).

Cytogenetic methods have been used to develop chromosome-linkage group relationships in tomato, potato, alliums, and brassicas. Cytogenetic stocks developed in tomato helped in locating genes and centromeres on linkage maps (Chapter 3). The discovery of the ancestral species of *Brassica napus* (AACC; $2n=38$), *Brassica juncea* (AABB; $2n=36$), and *Brassica carinata* (BBCC; $2n=34$) elucidated the role of allopolyploidization in producing new crops (Chapter 4).

Somatic chromosomes of vegetable crops, except *Allium*, are small. Onion root tips and meiotic pollen mother cells are classic cytological materials used to teach students cell division. Although *Allium* has three basic chromosome numbers ($x=7, 8, 9$) the $x=8$ chromosome number predominates in most of the species (Chapter 7). Chromosome number in okra ranges from $2n=56$ to $2n=199$. Chromosome number in cultivated okra (*Abelmoschus esculentus*) varies from $2n= \pm 66$ to 144. By contrast, basic chromosome number of most solanaceous crops (potato, tomato, pepper, and eggplant) is 12 (Chapter 2, Chapter 3, Chapter 6, and Chapter 10). Pachytene chromosomes of tomato and potato are clearly differentiated by euchromatin and heterochromatin. Carrot ($2n=2x=18$) and most of its relatives are diploids with a relatively low chromosome number, but little is known about chromosome structure and evolution (Chapter 11).

By using aneuploid stocks and pachytene chromosomes, researchers have developed a precise cytogenetic map of tomato (Chapter 3). Primary trisomics have been generated in pepper, and they have been utilized to locate both isozyme loci and monogenic traits. Molecular linkage maps of pepper have been constructed for several mapping populations. Comparative mapping has revealed that tomato and potato genomes differ by only five paracentric inversions on five distinct chromosomes; 18 homologous blocks cover 98% of the tomato genome and 98% of the pepper genome. A minimum of 22 chromosome breaks were responsible for the chromosomal translocations and inversions between pepper and tomato genomes (Chapter 6). Fluorescence in situ hybridization (FISH) and genomic in situ hybridization (GISH) have revolutionized cytogenetic research in potato (Chapter 2), tomato (Chapter 3) and *Allium* (Chapter 7). Molecular linkage maps of potato, tomato, onion, pepper, *Cucumis* species, lettuce, eggplant, carrot, and other vegetables are being constructed. These constructions include classical morphological markers, isozymes, restriction fragment length polymorphism (RFLP) markers, cleaved amplified polymorphic sequences (CAPS), sequence characterized amplified regions (SCARs), mocrosatellites, and single nucleotide polymorphism (SNP) (Chapter 2, Chapter 3, and Chapter 11). Several traits of economic importance have been mapped. These include genes and quantitative trait loci (QTLs) for resistance to pests and pathogens, fruit length, seed cavity size and color, earliness, sex expression, multiple branching, female flowering, and determinate growth (Chapter 2, Chapter 3, Chapter 6, Chapter 8, and Chapter 9). Carbohydrate metabolism genes have also been mapped for potato using CAPs, SCARs, and RFLPs (Chapter 2). QTL for carotenoid genes have been mapped for carrot using amplified fragment length polymorphism (AFLP), and major genes for carbohydrate metabolism and nematode resistance have also been mapped (Chapter 11). Marker-assisted selection (MAS) and studies on synteny are breeders' and cytogeneticists' ultimate objectives. This will enable them to utilize exotic germplasm for breeding better vegetable cultivars (Chapter 2, Chapter 3, Chapter 6, Chapter 8, Chapter 9, Chapter 10, and Chapter 11).

High-yielding vegetable cultivars are a threat to the natural habitat of the allied species and genera. It is important, therefore, that invaluable germplasm resources are collected before they become extinct. International and national institutions as well as private industries are preserving indigenous cultivars, land races, and wild relatives in medium and long-term storages. For example, representative samples of many wild potato species have been collected and maintained in genebanks (ex situ and in situ) around the world (Chapter 2). Germplasm evaluation for variation in root growth, post harvest storage, combining ability for yield, and specific yield components (fruit number and weight) are described in Chapter 8.

1.7   GERMPLASM ENHANCEMENT FOR VEGETABLE CROPS

Varietal improvement programs develop elite lines by multidisciplinary approaches (Figure 1.4). Development of these lines is dependant upon breeding objectives, the type of vegetable crops, and end-use products (long-term versus short-term, reproductive cycle, molecular-aided versus conventional, and loss versus benefit). Other important qualities include: inheritance patterns and heritability of the selected characters (sex expression, flowering date, disease and insect resistance, yield heterosis, modifying plant architecture and quality); horticultural and agronomic traits; and nutritional quality, including processing and consumer preferences from available germplasm (Chapter 2 through Chapter 11). The present genetic base of vegetable crops is narrow because breeders have largely confined their varietal improvement programs to GP-1 (primitive cultivated form, landraces, and wild progenitors). Vegetatively propagated crops fix the desired genes in clones that are maintained for a long times. The high uniformity of modern cultivars makes them susceptible to biotic and abiotic stresses. The potato cv. Russet Burbank was the most important cultivar in Canada in 2000 (Figure 2.4), but it was released in 1872. In the Netherlands, cv. Bintje, still the number two cultivar in 2002, was first commercialized in 1910. The close affinity between the cultivated potato and its wild relatives makes it relatively easy to incorporate related germplasm into the cultivated potato. By using endosperm balance number (EBN) in potato, several traits of economic importance have been introgressed into the cultivated potato from wild species (Chapter 2). In India, okra cv. *Pusa Sawani* was released in 1960, quickly becoming popular

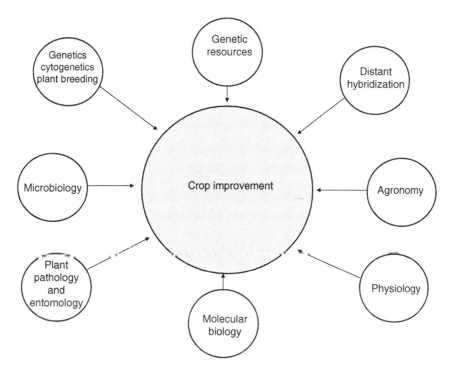

Figure 1.4   Diagrammatic sketch of a multidisciplinary approach for crop improvement. (Modified from Pohelman, J. M. and Sleeper, D. A., *Breeding Field Crop,* 4th ed., Iowa State University Press, Ames, IA, 1995.)

throughout the country and replacing most of the landraces leading to extensive genetic erosion (Chapter 5). The first F_1 pepper hybrid, "Lamuyo," released in 1973 by INRA (France), is still grown commercially (Chapter 6). GP-1A of bulb onion (*Allium cepa*) and Japanese bunching onion (*Allium fistulosum*) consists of local varieties and wild relatives. GP-1B of bulb onion contains a wild relative *A. vavilovii*.

GP-2 may be used to improve many vegetables. Monosomic alien additional lines of *Allium fistulosum* and *A. cepa* have been produced and identified based on morphological features, isozymes, genomic in situ hybridization, ribosomal DNA, and molecular markers (Chapter 7). GP-3 has been exploited only in a few vegetable crops (Chapter 2 through Chapter 10). Wide hybridization and backcrossing to the recurrent parent is necessary in order to transfer a single gene of economic importance (Singh 2003). Genetically diverse, allogamous (although some inbreeding occurs) and economically important cucurbits encompass three genera: *Cucumis*, *Cucurbita*, and *Citrullus*. Other secondary cucurbits include species of *Benincasa*, *Lagenaria*, *Luffa*, *Sechium*, and *Mimordica*. Each genus is geographically isolated, and production of hybrids between genera is difficult due to both pre- and post-hybridization barriers (Chapter 8).

Many exotic vegetable accessions are stored in germplasm banks worldwide (Tanksley and McCouch 1997). However, only a fraction of their valuable genes have been utilized for vegetable crop improvement. The development of new vegetable cultivars is complex and generally takes a longer time compared with breeding new cultivars of cereals or legumes. Conventional breeding (selection from primitive varieties and landraces, pedigree, bulk, back-cross, or single-seed descent methods of selection), mutation breeding, distant hybridization, somaclonal variation, and genetic transformation have helped breeders to select vegetable cultivars resistant to abiotic and biotic stress. Such selections have improved nutritional values, generated higher yield and longer shelf life, and produced unique aromas and flavors (Chapter 11). Photoperiod insensitive cultivars that grow in diverse agro-climatic conditions have been developed (Chapter 2 and Chapter 8).

GP-3 is a rich reservoir of genes of economic importance. This source can be exploited to broaden the genetic base of vegetable crops by wide hybridization (embryo rescue followed by chromosome doubling and subsequent backcrossing to recurrent parent), somaclonal variation, and genetic transformation.

1.7.1 Breeding for High Yield

The ultimate aim of any varietal improvement program is to produce high-yielding cultivars that can be achieved using multidisciplinary approaches. Conventional breeding utilizing GP-1 produced vegetables with high-yield-containing genes for resistance to biotic (fungal, bacterial and virus diseases, and damage by pests) and abiotic (tolerance to heat, cold, adverse soil nutrition, drought and flood, and lodging) stress, improved nutritional quality and quantity, shelf life, and marketability (Chapter 2 through Chapter 11). Conventional breeding methods have produced potato resistant to several diseases, including the most destructive pathogen, *Phytophthora infestans* (Chapter 2). A landrace of *Cucurbita moschata* grown in Nigeria is the only known source of resistance to certain viral diseases. Yield increases in processing cucumber are positively correlated with increased number of fruit-bearing branches. Cultivars with this growth habit have been developed, resulting in yield increases. Conventional breeding has utilized a broad range of germplasm to improve marketable yield of carrot, and hybrid carrot, accounting for the majority of the crop consumed in Europe and North America, relies upon the cytoplasm of wild carrot for stable CMS (Chapter 11).

Haploid technology through anther, irradiated pollen, and ovule culture is being applied for breeding disease-resistant lines (Chapter 2 and Chapter 8). In *Cucumis melo*, somaclonal variants for low temperature germinability and larger fruit were isolated through somatic embryogenesis. The role of transformation in producing abiotic and biotic resistant lines has been summarized in Chapter 3, Chapter 4, and Chapter 8. Monsanto released several transgenic potato cultivars under the trade name NewLeaf®. However, Monsanto withdrew them from the market because they were not acceptable by fast food chains. Hybrid seed production in potato has been developed.

1.7.2 Breeding for Nutritional and Medicinal Value

Some vegetable crops, including many cucurbits (Chapter 8), potato (Chapter 2), *Allium* (Chapter 6), carrot (Chapter 11) and others (Table 1.1 and Table 1.2), have nutritional and medicinal value. In potato, genes encoding human milk proteins (*b-casein* and *lactoferrin*) have been added in hope of providing an inexpensive source of plant-synthesized hypoallergenic human milk for infant formula and baby foods. Human β-amyloid is a peptide that accumulates in plaque during the progression of Alzheimer's disease. A cloned human β-amyloid gene has been transformed into potato with no adverse effect on the plant, but levels of expression were low. This study is a promising step to eventual treatment of Alzheimer's disease. Research is underway to develop inexpensive sources of antiviral pharmaceuticals in transformed potato. Breeding watermelon for an edible oilseed, improved fruit quality, and less bitterness have been successful. Furthermore, the substantial production of cucurbitacins, a class of unusual tetracyclic triterpenoids from *Cucurbita foetidissima*, has potential application in the production of insecticides. A gene from cassava, superoxide dismutase (SOD) for human cosmetic applications (putative anti-aging agent), has been transferred into cucumber fruit by *Agrobacterium*-mediated transformation. Breeding, using wide crosses (Chapter 4), has produced *Brassica* lines with altered erucic acid, glucosinolate content, and fatty acid composition. Pepper has a high concentration of vitamins and antioxidants and helps improve the human diet. Pungency in the red pepper is due to capsaicinoids that have many medicinal applications (Chapter 6).

The leaves of oilseed lettuce are bitter and not eaten. However, its seeds contain 35% oil, which is a source of vitamin E and is used for cooking. Nutrient constituent values are lower in lettuce than in many other vegetables; iceberg lettuce has the lowest nutritional value in many categories compared with the other lettuce types. Plant breeders must consider the growing conditions of lettuce. One of these aspects is the reduction of nitrate content in lettuce grown in glasshouses during periods of low temperatures and low light intensity. Excess nitrate in winter-grown lettuce may lead to health problems, particularly in infants (a condition known as blue baby) (Chapter 9).

Carrot is the single most abundant source of provitamin A carotenoids in the U.S. diet with genetic variation for carotenoid content in orange carrots from 5 to 500 ppm (Chapter 11). Conventional breeding has increased carotene content by over 50% in the last 35 years, and with recent efforts to map QTL, this progress can continue. An interest in purple, red, and yellow carrots opens up an opportunity for breeders to enhance the nutrients responsible for these interesting colors as well.

1.7.3 Breeding for Plant Ideotypes

Early maturing and determinate cultivars have been developed in *Brassica* (Chapter 4), okra (Chapter 5), pepper (Chapter 6), *Cucumis sativus* (Chapter 8), and *Lactuca sativa* (Chapter 9) by conventional and chromosome engineering techniques. Chapter 6 discusses pepper breeding for determinate, semi-determinate and indeterminate growth habits, horticultural type, and fruit quality. Chapter 9 describes seven types of cultivated lettuce classified by morphological traits: butterhead, crisphead, cos, leaf, stalk (asparagus), Latin, and oilseed.

1.7.4 Breeding for Marketability

Fresh and canned vegetables are transported to consumers worldwide. Suitable varieties for the export market are being developed (Chapter 2 through Chapter 11). Pepper dominates the world hot spice trade, and most harvests of fresh pepper take place over 4–12 months. Salsa and hot sauce represent a large market in the United States. The red pigments are widely used in the food industry as an alternative to artificial red food dyes that are known for their toxicity. Pepper is also grown as garden ornamentals (Chapter 6). The convenience of cut and peel or baby carrots has expanded U.S. carrot consumption significantly since the development of this product less than 20 years ago. The value of this product to growers brings major focus for breeding programs to improve marketability aspects of this new product.

1.7.5 Development of Breeding Methods

Breeding methodologies depend upon the objectives of the plant breeder, available germplasm resources, nature of the plant (autogamous to allogamous), agroclimatic conditions, and inheritance of abiotic and biotic stress traits. Chromosome engineering has played a pivotal role in transferring traits of economic importance, such as abiotic and biotic resistance, cytoplasmic male sterility, fertility restoration, earliness, and change in seed color from alien species (Chapter 2 through Chapter 11). Several potato cultivars have originated by spontaneous mutation. Mutation breeding is being used to develop amylase-free starch potato cultivars (Chapter 2). In okra, a mutant EMS 8 carrying resistance to yellow vein mosaic virus (YVMV) and tolerance to fruit borer has been developed by the mutagen ethyl methanesulfonate (Chapter 5). Mutagenesis has generated stable male sterile mutants in pepper, and this trait is commercially exploited for hybrid seed production (Chapter 6).

The impact of somaclonal variation and genetic transformation for the development of better cultivars has been extremely successful for solanaceous and brassicaceous (Chapter 2, Chapter 3, Chapter 4, Chapter 6, and Chapter 10). A somaclonal variant of potato has been released as a variety in Japan. Tubers of this cultivar do not brown after peeling as do the tubers of the source cultivar (Chapter 2). Genetic transformation created FLAVR SAVR™ tomato (Kramer and Redenbaugh 1994). Transgenic tomatoes, containing superior consumer quality, shelf life, and flavor, failed to attract consumers, as they were more expensive than non-transgenic tomato cultivars (Singh 2003). Transgenic *Brassica napus* with herbicide resistance and increased levels of methionine have been produced. Chapter 4 discusses the potential impact of transgenic *Brassica* cultivars on the ecosystem. Transgenic okra, onion, pepper and eggplants have not yet been produced.

1.8 CONCLUSIONS

1. Grain legumes (rich in protein), cereals (rich source of carbohydrate), oilseed crops (rich source of healthy fats), and vegetable crops (an excellent source of minerals, vitamins, fiber, and antioxidants) may be combined to provide an excellent balanced diet for the human population worldwide.
2. Authors recommend extensive plant exploration for wild relatives and local, limited-use cultivars or landraces of the crops before allied genera and species disappear.
3. Classical taxonomy, cytogenetics, and molecular genetics have identified distinct gene pools for the cultivated vegetable crops.
4. Vegetable breeders have confined their efforts to the primary gene pool (GP-1).

Exploitation of and tertiary gene pool (GP-3) is often hampered because of crossability barriers.

5. Vegetable breeders have achieved substantial success in producing vegetable cultivars with resistance to biotic and abiotic stress and increased nutritional quality and quantity by conventional and molecular breeding methods. Genetic research has resulted in high-yielding, nutritional varieties. Commercial companies have developed facilities and the means to transport vegetables to consumers worldwide.

REFERENCES

Bassett, M., Ed., *Breeding Vegetable Crops*, Westport, CT: AVI Publishing, 1986.

Darwin, C., *On the Origin of Species by Means of Natural Selection*, London: John Murray, 1859.

Decoteau, D. R., *Vegetable Crops*, Upper Saddle River, NJ: Prentice Hall, 2000.

FAO STAT, http://faostat.fao.org, 2004 (accessed on June 2005).

George, R. A. T., *Vegetable Seed Production*, 2nd ed., Wallingford, UK: CABI Publishing, 1999.

Hancock, J. F., *Plant Evolution and the Origin of Crop Species*, 2nd ed., Wallingford, UK: CABI Publishing, 2004.

Harlan, J. R., *Crops and Man*, 2nd ed., Madison, WI: American Society of Agronomy and Crop Science Society of America, 1992.

Harlan, J. R. and de Wet, J. M. J., Toward a rational classification of cultivated plants, *Taxon*, 20, 509–517, 1971.

Kalloo, G., *Vegetable Breeding,* Vols I and II, Boca Raton, FL: CRC Press, 1988.

Kalloo, G. and Bergh, B. O., Eds., *Genetic Improvement of Vegetable Crops*, Oxford: Pergamon Press, 1993.

Kramer, M. G. and Redenbaugh, K., Commercialization of a tomato with an antisense polygalacturonase gene: the FLAVR SAVR™ tomato story, *Euphytica*, 79, 293–297, 1994.

Löve, V., *Origin and geography of Cultivated Plants, (English translatin of N. I. Vavilov's Origin and Geography of Cultivated Plants)*, Cambridge: Cambridge University Press, 1992.

Orel, V., *Gregor Mendel: The First Geneticist*, Oxford: Oxford University Press, 1996.

Phillips, R. and Rix, M., *Vegetables*, Singapore: Toppan Printing, 1993.

Rubatzky, V. E. and Yamaguchi, M., *World Vegetables, Principles, Production, and Nutritive Values*, 2nd ed., New York: Chapman & Hall, 1997.

Sauer, C. O., *Agricultural Origins and Dispersals*, Cambridge, MA: MIT Press, 1969.

Singh, R. J., *Plant Cytogenetics*, 2nd ed., Boca Raton, FL: CRC Press Inc, 2003.

Stickland, S., *The Gardener's Guide to Cultivating Diversity*, London: Gaia Books, 1998.

Tanksley, S. D. and McCouch, S. R., Seed banks and molecular maps: unlocking genetic potential from the wild, *Science*, 277, 1063–1066, 1997.

Tudge, C., *In Mendel's Footnotes*, London: Jonathan Cape, 2000.

Vavilov, N. I., *The Origin, Variation, Immunity and Breeding of Cultivated Crops*, Waltham, Massachusetts: Chronica Botanica, 13, 1950.

Vavilov, N. I., *Origin and Geography of Cultivated Plants*, Leningrad, USSR: Nauka, 1987. (In Russian).

CHAPTER 2

Potato

Richard E. Veilleux and Hielke ("Henry") De Jong

CONTENTS

2.1   INTRODUCTION

Although the potato [*Solanum tuberosum* L. $(2n=4x=48)$] is indigenous to the Andean region of South America (see Section 2.3.1), today it is grown in environments that differ vastly with regard to latitude, altitude, daylength, and temperature from where it was first domesticated (Hijmans 2001; Figure 2.1). The adaptability of potato to different climates and cultivation methods has contributed to its ascent over the past four centuries to become the world's fourth major food crop.

During the past 50 years there have been massive changes in the rankings among the major potato-producing countries. In general, the area of potato cultivation has decreased in developed countries and increased in developing countries (Hijmans 2001). For example, the combined production of Germany and Poland dropped from about 29% of the world's total production in 1961–1965 to about 8% in 2001–2005 (Table 2.1; FAOSTAT, 2005). In contrast, potato production of China and India that together during 1961–1965 comprised only about 6% of the world's total production, rose to 29.4% of the total during 2001–2005. On a global basis, between 1961 and 1965 and 2001–2005 the area harvested has decreased by 13%. During this same period, however, global production of potato has increased by 12% and yield by 36%. However, yield per hectare differs considerably among countries, e.g., yields in the United States, Germany, the Netherlands and France were more than double the world average during 2001–2005 (Table 2.1).

In contrast to crops such as cereals and soybean where a large proportion of recent yield increase can be attributed to cultivar improvement, the evaluation of the contribution of breeding to potato improvement is much more complex. One of the reasons for this is that different cultivars are being developed for different end uses (see Item 3 below).

Major factors that have contributed to yield increase of potato include:

1. Seed potato certification. Planting healthy seed potatoes has a significant positive effect on tuber yield. Most developed countries have extensive seed potato certification schemes. Release of propagules routinely from modern tissue culture facilities contributes to efficient seed potato production systems.
2. Improved agronomic practices, including the use of agrochemicals and irrigation. In countries such as the United States, mechanized irrigation (that in turn has brought along fertigation and chemigation) has been a major factor improving potato yield.
3. Breeding. The development of disease-resistant cultivars has had a positive effect on yield. On the other hand, the increasing demand for cultivars with consistently predictable processing traits (especially for french fries and chips) has resulted in greater attention by breeders to improve processing traits while maintaining (but not necessarily increasing) yield (Douches et al. 1996); see also Section 2.5.

As developing countries adapt technologies described under (1), (2), and (3) above, yield increases should result. For example, although the current average yield in China is still slightly below the world average, a relatively small proportional increase there may have a huge positive effect on world potato production.

The potato has been the subject of extensive genetic and cytogenetic research, despite its cumbersome genome, intractable breeding system and inconspicuous chromosomes. Yet, because of the close affinity between the cultivated potato and its wild relatives, it is relatively easy to incorporate related germplasm into cultivated forms. Virtually any desirable trait for breeding purposes appears to be available in the potato germplasm resource. The use of $2n$ gametes in germplasm transfer has been more extensive in potato than for any other crop. Because of its relative facility in tissue culture regeneration and transformation compared to the grasses, potato has been both the object of parasexual protoplast fusions with a range of species as well as the recipient of genes from organisms as alien as man and mouse. Controversy over the production of transgenic potato crops has caused

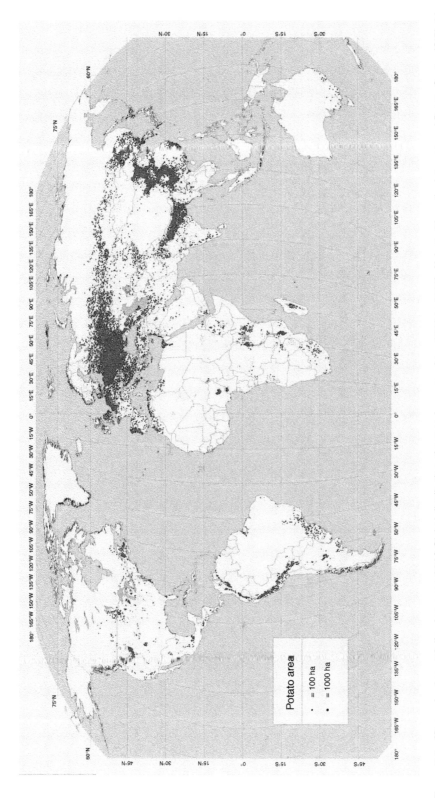

Figure 2.1 (See color insert following page 304.) Global distribution of potato (averages from 2000 to 2002). (Adapted from the International Potato Center (CIP), Lima, Peru. With permission.)

Table 2.1 Changes in Potato Production in the Current Ten Highest Ranking Potato Producing Countries and the World, 1961–1965 Compared with 2001–2005

Country	Production (×1,000,000 Mt)			Area Harvested (×1,000,000 Ha)			Yield (Mt/Ha)		
	1961–1965 Average	2001–2005 Average	Percent Change	1961–1965 Average	2001–2005 Average	Percent Change	1961–1965 Average	2001–2005 Average	Percent Change
China	13.7	69.4	+406	1.52	4.52	+197	9.1	15.4	+69
Russian Federation	—	35.4	—	—	3.20	—	—	11.2	—
India	2.9	24.4	+741	0.40	1.33	+225	7.4	18.4	+148
USA	12.4	20.3	+64	0.55	0.48	−13	22.4	41.8	+87
Ukraine	—	18.5	—	—	1.57	—	—	11.8	—
Poland	43.7	14.7	−67	2.84	0.81	−71	15.4	18.3	+19
Germany	34.3	11.6	−66	1.63	0.28	−83	21.0	40.7	+93
Belarus	—	8.5	—	—	0.55	—	—	15.5	—
Netherlands	3.8	7.1	+86	0.13	0.16	+23	29.2	43.5	+49
France	13.3	6.6	−51	0.77	1.58	+105	17.3	41.7	−141
World	**269.8**	**319.5**	**+12**	**21.88**	**19.02**	**−13**	**12.3**	**16.8**	**+36**

Source: From FAOSTAT. http://faostat.fao.org (accessed on 7/7/2006).

the withdrawal of effectively engineered insect resistant cultivars. However, research into the possibilities of a transgenic crop continues unabated, with the likelihood that new transgenic potatoes will be marketed again in the future. Although potato will never be a model species for genetic research, its importance as a staple in world nutrition guarantees that all of the latest technologies in genomics, proteomics and metabolomics will find applications in potato breeding.

2.2 TAXONOMY

The potato probably has more related wild species than any other crop. The cultivated potato, the wild tuber-bearing species and their allies belong to the genus *Solanum*, subgenus *Potatoe*, and section *Petota*. There are several different taxonomic treatments of the various tuber-bearing *Solanum* species. Hawkes (1990) recognized 235 species (of which seven are cultivated) that he grouped into 19 series (Table 2.2). Spooner and Hijmans (2001) presented a revised list of 206 species and indicated that the series classification of Hawkes (1990) is not well supported by the molecular data set used to date. Of the 176 species for which chromosome counts are recorded, 73% are diploid ($2n = 2x = 24$), 4% are triploid ($2n = 3x = 36$), 15% are tetraploid ($2n = 4x = 48$), 2% are pentaploid ($2n = 5x = 60$), and 6% are hexaploid ($2n = 6x = 72$) (Hawkes 1990). Dodds (1962) differs from other taxonomists in that he included five primitive cultivated species in *S. tuberosum* and then

Table 2.2 Classification of Potato Species and Their Allies

Genus: *Solanum*
  Subgenus: Potatoe
    Section: Petota
      Subsection: *Estolonifera*
        Series I: *Etuberosa*
        Series II: *Juglandifolia*
      Subsection: *Potatoe*
        Superseries: *Stellata*
          Series I: *Morelliformia*
          Series II: *Bulbocastana*
          Series III: *Pinnatisecta*
          Series IV: *Polyadenia*
          Series V: *Commersoniana*
          Series VI: *Circaeifolia*
          Series VII: *Lignicaulia*
          Series VIII: *Olmosiana*
          Series IX: *Yungasensa*
        Superseries: *Rotata*
          Series X: *Megistacroloba*
          Series XI: *Cuneoalata*
          Series XII: *Conicibaccata*
          Series XIII: *Piurana*
          Series XIV: *Ingifolia*
          Series XV: *Maglia*
          Series XVI: *Tuberosa*
          Series XVII: *Acaulia*
          Series XVIII: *Longipedicellata*
          Series XIX: *Demissa*

Source: From Hawkes, J.G. in *The Potato. Evolution, Biodiversity and Genetic Resources*, Smithsonian Institute Press, Washington, DC, 1990. With permission.

further subdivided *S. tuberosum* by groups and ploidy level. Many scientists who are working with primitive cultivated potatoes have adopted this system. More recent taxonomic treatments have drawn from molecular data as well as morphological traits and field studies (Spooner et al. 2004).

2.3 GERMPLASM RESOURCES AND GENE POOLS

2.3.1 Germplasm Resources

The wild potato species are distributed through much of the Americas, from the Southwest US through Central America, along the Andes Mountains from Venezuela through Colombia, Ecuador, Peru, Bolivia, and Northwest Argentina. They also occur in the lowlands of Chile, Argentina, Uruguay, Paraguay, and Southeastern Brazil (Figure 2.2). The habitats are described in detail in several monographs (Hawkes and Hjerting 1989; Ochoa 1990, 1999). The wide range of environments in which wild potatoes are found emphasizes the fact that various species have become adapted to stress conditions and have developed resistances to various pests and diseases. Virtually any desirable trait for breeding purposes appears to be available in the potato germplasm resource (Bamberg et al. 1994a; Hanneman 1989; Hawkes 1990, 1994; Ortiz 1998; Spooner and Bamberg 1994).

The potato was likely domesticated at least 7,000 years ago near Lake Titicaca on the border of what is now Peru and Bolivia (Glave 2001). The greatest diversity of wild potato species is currently found in this area (Hawkes 1990; Simmonds 1995). It is generally accepted that the immediate ancestor of *S. tuberosum* is the primitive cultivated tetraploid *S. andigena,* also known as *S. tuberosum* subsp. *andigena* or *S. tuberosum* Group Andigena that in turn had its origin in the primitive cultivated diploids. However, it is still unclear from which wild diploid species the primitive cultivated diploids were derived (Simmonds 1995). Until recently, most modern potato cultivars have been built on a relatively narrow genetic base (Brown 1993; Love 1999; Pavek and Corsini 2001; Plaisted and Hoopes 1989). However, the building blocks for both broadening the genetic foundation as well as for the transfer of specific traits are readily available in the large germplasm resource.

Several issues regarding the conservation of potato genetic resources, including ownership, collection, classification, and genetic erosion, have been discussed by Bamberg and del Rio (2005). Representative samples of many wild potato species have been collected and are maintained in genebanks around the world. In most cases, the accessions are increased by means of true seed that is generated in the genebank (*ex situ*). In general, this process has not altered the genetic diversity of the *ex situ* germplasm using the current standard techniques that are applied in most major genebanks (del Rio et al. 1997a). However, del Rio et al. (1997b) found significant genetic differences between gene-bank-conserved and re-collected *in situ* populations of several accessions and concluded that *in situ* preservation may be important for the backup of diversity already present in genebanks and for the preservation of new diversity that can be accessed in future re-collections. The major genebanks are organized in a consortium to facilitate their efficient management (Bamberg et al. 1995). Seven potato genebanks collectively contain more than 11,000 wild potato accessions; the inter-genebank potato database contains records of evaluations of several thousand wild potato accessions (Huaman et al. 2000).

2.3.2 Gene Pools

The approximate primary, secondary and tertiary gene pools of potato are shown in Table 2.3. Many species in the secondary gene pool have been used in potato breeding resulting in cultivar releases in North America and Europe over the last century (Love 1999; Plaisted and Hoopes 1989).

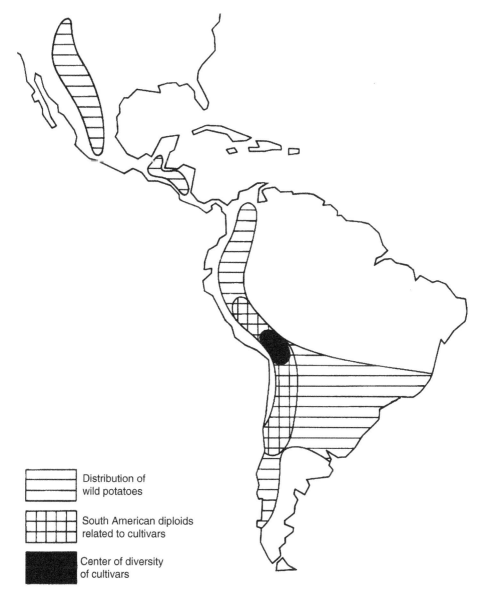

Figure 2.2 Distribution of the tuber-bearing species of *Solanum*. (Adapted from Simmonds, N. W., in *Evolution of Crop Plants*, Simmonds, N. W., Ed., Longman, London, 279–283, 1976.)

Those in the tertiary gene pool were used by Jackson and Hanneman (1999) in an extensive crossability study between a group of 15 tetraploid potato cultivars or breeding lines (used as both male and female parents) and various accessions (522 total) comprising 134 species representing 19 botanical series within the Solanaceae. Crossability was determined by fruit and seed set and, in many cases, the vigor and fertility of the hybrids. Most of the 522 accessions were diploid and therefore would not have resulted in seed set with the tetraploids without the functioning of $2n$ gametes. The tertiary germplasm pool has been divided into those more easily crossed (A) and those that resulted in little or no seed set (B) in the Jackson and Hanneman (1999) study. Crossability at the diploid level with *S. tuberosum* has often yielded different results and several species have been reclassified from the tertiary to the secondary gene pool based on breeding studies at the diploid

Table 2.3 Approximate Primary, Secondary and Tertiary Gene Pools of Potato

Primary Gene Pool (Cultivated Species)	Secondary Gene Pool	Tertiary Gene Pool (A)[a]	Tertiary Gene Pool (B)[a]
Subsection *Potatoe*	**Subsection** *Estolonifera*		
Series *Tuberosa*		**Series I:** *Etuberosa*	**Series I:** *Etuberosa*
S. tuberosum (4*x*)		*S. fernandezianum* (2*x*)	*S. palustre* (2*x*)
S. andigena (4*x*)		*S. etuberosum* (2*x*)	
S. curtilobum (5*x*)		*S. brevidens* (2*x*)	
S. × *chaucha* (3*x*)	**Subsection** *Potatoe* **Superseries**: *Stellata*		
S. × *juzepczukii* (3*x*)	**Series V:** *Commersoniana*	**Series VI:** *Circaeifolia*	**Series II:** *Bulbocastana*
S. phureja (2*x*)	*S. commersonii* (2*x*)	*S. capsicibaccatum* (2*x*)	*S. bulbocastanum* (2*x*)
	Series IX: *Yungasensa*		**Series III:** *Pinnatisecta*
	S. chacoense (2*x*)		*S. brachistotrichum* (2*x*)
	S. tarijense (2*x*)		*S. cardiophyllum* (2*x*)
	S. arnezii (2*x*)		*S. jamesii* (2*x*)
			S. pinnatisectum (2*x*)
			S. tarnii (2*x*)
			S. trifidum (2*x*)
			Series IV: *Polyadenia*
			S. lesteri (2*x*)
			S. polyadenium (2*x*)
			Series VI: *Circaeifolia*
			S. circaeifolium (2*x*)
			Series IX: *Yungasensa*
			S. huancabambense (2*x*)
	Subsection *Potatoe* **Superseries**: *Rotata*		
	Series X: *Megistacroloba*	**Series XII:** *Conicibaccata*	**Series X:** *Megistacroloba*
	S. raphanifolium (2*x*)	*S. moscopanum* (6*x*)	*S. astleyi* (2*x*)
	S. toralapanum (2*x*)	*S. tundalomense* (4*x*)	*S. boliviense* (2*x*)
	Series XVI: *Tuberosa*	**Series XVI:** *Tuberosa*	*S. megistacrolobum* (2*x*)
	S. berthaultii (2*x*)	*S. chancayense* (2*x*)	*S. sanctae-rosae* (2*x*)
	S. bukasovii (2*x*)	*S. hoopesii* (4*x*)	*S. sogarandinum* (2*x*)
	S. canasense (2*x*)	*S. okadae* (2*x*)	**Series XI:** *Cuneoalata*
	S. gourlayi (4*x*)	*S. pampasense* (2*x*)	*S. infundibuliforme* (2*x*)
	S. hondelmannii (2*x*)	*S. ugentii* (4*x*)	**Series XII:** *Conicibaccata*
	S. kurtzianum (2*x*)	**Series XVII:** *Acaulia*	*S. agrimonifolium* (4*x*)
	S. leptophyes (2*x*)	*S. albicans* (6*x*)	*S. chomatophilum* (2*x*)
	S. microdontum (2*x*)	**Series XIX:** *Demissa*	*S. laxissimum* (2*x*)
	S. multidissectum (2*x*)	*S. brachycarpum* (6*x*)	*S. paucijugum* (4*x*)
	S. oplocense (4*x*, 6*x*)	*S. guerreroense* (6*x*)	*S. santolallae* (2*x*)
	S. sparsipilum (2*x*)	*S. iopetalum* (6*x*)	**Series XIII:** *Piurana*
	S. spegazzinii (2*x*)		*S. acroglossum* (2*x*)
	S. × *sucrense* (4*x*)		*S. paucissectum* (2*x*)
	S. vernei (2*x*)		*S. pascoense* (2*x*)
	S. verrucosum (2*x*)		*S. solisii* (4*x*)
	Series XVII: *Acaulia*		*S. tuquerrense* (4*x*)
	S. acaule (4*x*)		**Series XVI:** *Tuberosa*
	Series XVIII: *Longipedicellata*		*S. abancayense* (2*x*)
	S. stoloniferum (4*x*)		*S. acroscopicum* (2*x*)
	Series XIX: *Demissa*		*S. achacachense* (2*x*)
	S. demissum (6*x*)		*S. alandiae* (2*x*)
			S. ambosinum (2*x*)
			S. avilesii (2*x*)
			S. brevicaule (2*x*)
			S. candolleanum (2*x*)
			S. × *doddsii* (2*x*)
			S. dolichocremastrum (2*x*)
			S. gandarillasii (2*x*)

continued

Table 2.3   Continued

Primary Gene Pool (Cultivated Species)	Secondary Gene Pool	Tertiary Gene Pool (A)[a]	Tertiary Gene Pool (B)[a]
			S. incamayoense(2x)
			S. marinasense (2x)
			S. medians (2x)
			S. mochiquense (2x)
			S. multiinterruptum (2x)
			S. neocardenasii (2x)
			S. neorossii (2x)
			S. × *rechei* (3x)
			S. scabrifolium (2x)
			S. venturii (2x)
			S. weberbaueri (2x)
			Series XVIII: *Longipedicellata*
			S. hjertingii (4x)
			Series XIX: *Demissa*
			S. hougasii (6x)
			S. schenckii (6x)

Note: The species are arranged according to the classification by Hawkes (Table 2.1).

[a] The tertiary germplasm pool is divided into those more easily crossed (A) and those that resulted in little or no seed set (B) according to Jackson and Hanneman (1999).

level (Carputo et al. 1997b; Colon et al. 1995; Devine and Jones 2001; Jansky 2000; Novy and Hanneman 1991; Watanabe et al. 1995). Many of the other species in the tertiary gene pool, especially those within Series Tuberosa may likewise be more appropriately classified as secondary gene pool members; however there have not been extensive breeding studies conducted with them. For further discussion on crossability, see also Section 2.6: Germplasm Utilization and Enhancement.

2.4   CYTOGENETICS

2.4.1   Genome Relationships

The cultivated potato species comprise three diploid types (*S. phureja*, *S. stenotomum*, and *S. ajanhuiri*), two triploids (*S. chaucha* and *S. juzepczukii*), tetraploid *S. tuberosum*, in which subsp. *tuberosum* and *andigena* are recognized, and pentaploid S. *curtilobum* (Hawkes 1990). Recent evidence based on both nuclear and chloroplast DNA markers suggests that these cultivated types share a common gene pool that likely derived from *S. stenotomum*. None of the seven species could be uniquely identified (Sukhotu et al. 2004). These results are in agreement with earlier work describing intragenomic and intergenomic relationships in diploid *Solanum* based on crossability of species and chromosome pairing in hybrids (Matsubayashi 1983).

2.4.2   Chromosome Mapping by Cytogenetics and Molecular Methods

The small size of potato chromosomes has been discouraging for cytogeneticists intent on identifying chromosome sets for karyotypic analysis. Most of the chromosomes are similar in size and position of the centromere, making them indistinguishable or nearly so, even with Giemsa staining (Lee and Hanneman 1975; Pijnacker and Ferwerda 1984). Karyotypic analysis of potato has been published, but most potato cytogeneticists have been content to observe mitotic chromosome spreads that are sufficient to provide an accurate count of the chromosome number. The advent of molecular cytogenetics has greatly expanded the possibilities of visualizing whole chromosomes or their constituent parts.

The elegant techniques of genomic *in situ* hybridization (GISH) and fluorescence *in situ* hybridization (FISH) revolutionized research on species with small chromosomes. Initially, the technique was limited to whole chromosome staining (GISH) or to using probes that were expected to anneal to repetitive DNA to have sufficient fluorescent signal for detection (Rokka et al. 1998c, 1998d). Wolters et al. (1994) applied GISH, otherwise known as chromosome painting, to visualize the genomic composition of somatic hybrids between potato and tomato; tomato genomic DNA, sonicated and labeled with fluorescein isothiocyanate (FITC), was used as a probe on mitotic preparations after potato DNA was used to block hybridization of the tomato probe with potato chromosomes. Euploid triploid somatic hybrids as well as aneuploids were readily distinguished after counterstaining the probed slides with propidium iodide and 4′,6-diamidino-2-phenylindole (DAPI). Hexaploid somatic hybrids between tomato and potato were readily observed to have a complement of four potato and two tomato genomes by GISH (Garriga-Calderé et al. 1999b). Transmission of alien chromosomes after backcrossing the hexaploid hybrids to tetraploid potato pollinators could be followed with GISH and FISH (Garriga-Calderé et al. 1997, 1998, 1999a).

The construction of a bacterial artificial chromosome (BAC) library for potato expanded the range of possibilities for molecular cytogenetics (Song et al. 2000). The library was constructed from DNA of the diploid species, *Solanum bulbocastanum*. It consists of 23,808 clones with an average insert size of 155 kb. Most BAC clones generated distinct signals on specific potato chromosomes. By screening the library with restriction fragment length polymorphic (RFLP) probes, mapped to specific arms of each linkage group on the potato genetic map, Dong et al. (2000) used a set of biotin labeled BACs to orient potato linkage groups with specific chromosome arms by FISH, revolutionizing karyotypic analysis (Figure 2.3A). By using FISH on extended DNA fibers in interphase nuclei of transgenic potato, then labeling T-DNA with biotin and vector DNA with FITC, Wolters et al. (1998) visualized the complex pattern of integration of T-DNA in *Agrobacterium*-transformed potato cv. Karnico (Figure 2.3B).

2.4.3 Current Genetic Map

The first high-density molecular linkage map of potato was constructed with more than 1,000 RFLP markers, mostly derived from tomato probes, with an average spacing of ca. 900 cM (Tanksley et al. 1992). The map provided a tool for chromosome walking by assuring that any gene of interest was tightly linked to a marker and also opened the door for QTL (quantitative trait loci) detection through the even distribution of molecular markers across the genome. The potato RFLP map can be viewed at the Solanaceae Genomics Network: http://www.sgn.cornell.edu/cgi-bin/mapviewer/mapTop.pl?map_id=3. The more closely linked a marker is to a QTL, the greater the detectable phenotypic effect (Gebhardt and Valkonen 2001). Interspecific diploid hybrids have been used to develop molecular maps to have sufficient allelic polymorphism, especially for RFLPs (Gebhardt et al. 1991; Jacobs et al. 1995; Tanksley et al. 1992). Jacobs et al. (1995) integrated isozyme and morphological markers into a common map using both male and female maps of a backcross hybrid. Chen et al. (2001) constructed a molecular function map of potato for genes involved in carbohydrate metabolism using CAPS (cleaved amplified polymorphic sequences), SCARs (sequence characterized amplified regions) and RFLPs in two mapping populations to try to associate QTLs for carbohydrate metabolism with allelic variation for candidate genes (CG). Comparative mapping between *Arabidopsis thaliana* and potato revealed the presence of several conserved syntenic blocks consisting of at least three markers, suggesting a common ancestral genome structure that has been duplicated to various degrees in each species (Gebhardt et al. 2003).

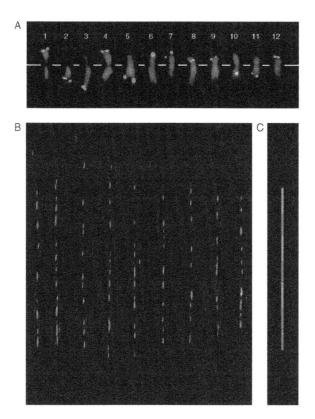

Figure 2.3 **(See color insert following page 304.)** (A) Molecular karyotype of potato showing twelve individual potato chromosomes with FISH signals derived from the chromosome-specific BAC clones. Biotin-labelled BAC probes were detected with 1% FITC-conjugated anti-biotin antibody (green). Propidium iodide (red) was used to counterstain the chromosomes. Distinctive FISH signals are specific to each pair of metaphase chromosomes. (From Dong, F. et al. *Theor. Appl. Genet.* 101, 1001, 2000. With permission.) (B) FISH analysis of extended DNA of transgenic potato cv. Karnico. Chromosomes in interphase leaf nuclei were simultaneously hybridized with the T-DNA probe (red fluorescence) and the vector DNA probe (green fluorescence). Fluorescence signals are shown on nine individual fibers. The interpretation is shown in 2C where the plasmid DNA integration consisted of four T-DNAs and three vector DNAs. (From Wolters, A. M. A., et al. *Plant J.*, 13, 837, 1998. With permission.)

Population size defines the total number of recombinations upon which a map can be based, a fact that leads to marker dense but low resolution maps, even when thousands of AFLP markers are used to construct a map (Isidore et al. 2003). Such marker dense AFLP maps require the posting of annotated gel images on websites to compare results among labs. The most recent map, a collaborative effort funded by the European Commission, can be found at: http://www.dpw. wageningen-ur.nl/uhd/index.html.

2.4.4 Marker-Assisted Selection and Association Genetics

The breeding structure of potato has made it a difficult crop for the application of marker-assisted selection (MAS) where selection for a marker closely linked to a trait obviates direct selection for the phenotype, thereby hastening the progress of plant breeding. Solomon-Blackburn and Barker (2001) predicted that the most likely use of MAS in potato would be for introgression of resistance genes from wild species. Hämäläinen et al. (1997) used bulk segregant analysis to link an RFLP marker on the proximal end of chromosome XI of potato with resistance to potato virus Y (PVY) to develop a molecular probe for PVY resistance. QTL analysis of resistance to black leg and soft rot

(*Erwinia* spp.) in potato revealed complex genetic control with significantly linked markers on all 12 chromosomes (Zimnoch-Guzowska et al. 2000) such that only the largest and most reproducible effects could be considered for MAS. Late blight resistance in both tubers and foliage of five hybrid families was associated with 21 PCR-based markers on ten chromosomes (Oberhagemann et al. 1999); a major QTL on linkage group V was associated with increased foliage but decreased tuber resistance. Marczewski et al. (2004) have proposed MAS for resistance to potato leafroll virus (PLRV) accumulation by using the PLRV.4 locus along with the closely linked molecular markers on chromosome XI. DNA markers tightly linked to genes for resistance to cyst nematodes, PVY, potato virus X (PVX) and potato wart are currently used to select resistant plants at the seedling stage (Gebhardt and Valkonen 2001). Environmental influence on detection of QTLs can hinder the efficacy of MAS, thus requiring QTL analysis of the same population over different environments to identify those markers that are significant across environments. Simko et al. (1999) however, found QTLs associated with early tuberization *in vitro* that, when used for MAS on greenhouse plants, resulted in a gain of nearly 12 days.

Although it can provide high resolution into genomic architecture and complement gene expression studies, QTL mapping traditionally explores only a limited range of the allelic diversity in a crop's gene pool. Using microsatellites to estimate diversity, Saghai Maroof et al. (1994), Maughan et al. (1995), and Ghislain et al. (2004) reported up to 18, 16, and 27 SSR alleles per locus in barley, soybean, and cultivated Andean potato, respectively; yet it is only feasible to capture a few of these in a single, biparental mapping progeny of an inbred diploid or even a heterozygous polyploid species. Association genetics examines historical recombinations that occur in present day gene-bank accessions to identify marker-trait associations through the analysis of linkage disequilibrium (Xiong and Guo 1997). Linkage disequilibrium mapping has only recently been applied to potato. Gebhardt et al. (2004) found highly significant association of late blight and plant maturity with PCR markers specific for *R1*, a major gene for resistance to late blight, in a collection of 600 potato cultivars released from 1850 to 1990 in various countries. Simko et al. (2004b) identified homologous sequences in potato to a CG for resistance to *Verticillium* in tomato where one allele explained 10–25% of the phenotypic variation in resistance in two potato subpopulations. Simko et al. (2004a) later characterized a set of 30 potato cultivars at the resistance locus and found three diverse haplotypes, indicating linkage disequilibrium. By developing haplotype specific primers, they characterized a segregating tetraploid population, finding a significant effect of haplotype on plant resistance. Expressed sequence tag (EST) libraries from which genes that are up- or down-regulated in specific developmental stages or in response to stress provide a means to identify CG with specific alleles associated with a desirable response (Ronning et al. 2003); information on gene expression studies in potato can be found at the website of The Institute for Genomic Research: http://www.tigr.org/tdb/potato/. Where MAS technology is available, a cost assessment should be carried out to determine its practicability in specific situations (Celebi-Toprak et al. 2005).

2.5 CONVENTIONAL BREEDING

Conventional breeding of potato and other crops is a critical component of the exploitation of genetic resources through modern technology (Knight 2003). The development of new and improved potato cultivars is a long-term, dynamic, and complex process. Several authors have reviewed conventional breeding (Bradshaw 2000; Caligari 1992; Douches and Jastrzebski 1993; Hoopes and Plaisted 1987; Tarn et al. 1992).

Because the development of a new potato cultivar takes approximately 12–15 years from the time that hybridization is carried out, breeding objectives must allow for changes in market demand. Processing is the fastest growing sector of the world potato economy (Bradshaw 2000; Guenthner 2001; Tarn et al. 1992). In the United States, the per capita consumption of frozen potato products (primarily french fries) had already surpassed that of fresh potatoes by the early 1990s (Lin et al. 2001). A similar trend is occurring in other developed countries. In addition to considering trends in potato consumption within his/her own country, the breeder must also consider trade with other countries when setting breeding objectives. The Netherlands, Canada, the United States and Belgium-Luxembourg are the world's major exporters of frozen potato products. Also, in the Netherlands, the export of seed potatoes far exceeds their domestic use (Guenthner 2001). Therefore the breeding objectives must include the demands of the export markets and evaluations of potential cultivars must include various trials in market countries (Struik and Wiersema 1999a). Douches et al. (1996) have assessed the potato breeding progress in the United States over the last century. Their data suggest that although a yield stasis has been found in potato cultivars in North America, the newer cultivars have been significantly improved for traits demanded by market utilization.

Many cultivars are available worldwide. The 2003 world catalogue of potato varieties lists about 3600 cultivars from about 100 countries (Hamester and Hils 2003). However, in most countries only a few cultivars are grown on a relatively large scale. This is illustrated for Canada in Figure 2.4 (CFIA 2000). In 1951 and 2000, five and eight cultivars, respectively, accounted for over 80% of the seed potatoes that passed inspection. In 2000, over 100 cultivars were registered in Canada (CFIA 2003). Similar situations occur in other countries. For example, in the Netherlands in 2002, the three top cultivars, Spunta, Bintje and Désirée accounted for 10%, 8%, and 7%, respectively, of the total area of seed potatoes of more than 300 cultivars grown (NAK, 2003). Nevertheless, there has been a major change in the relative importance of the major cultivars in

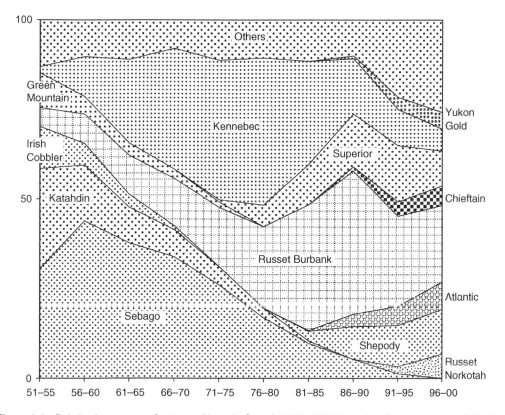

Figure 2.4 Relative importance of potato cultivars in Canada 1951–2000, based on five-year means. (Adapted from Canadian certified seed potatoes; area passing inspection 1951–2000. Canadian Food Inspection Agency. With permission.)

Canada. The rise and fall of several major cultivars can, to a large extent, be attributed to the change in market demand. Cultivars such as Sebago and Katahdin that in 1955 accounted for over 60% of all seed potatoes have been virtually phased out because of their unsuitability for processing. In 2000 cvs. Russet Burbank and Shepody, both of which excel for french fry production, comprised over 35% of the area passing inspection of Canadian seed potatoes. It is also interesting that cv. Russet Burbank, which was the most important cultivar in Canada in 2000, dates back to 1872. In the Netherlands, cv. Bintje, which was still the number two cultivar in 2002, was first commercialized in 1910.

The sequence of activities in a conventional breeding program usually involves: (1) establishment of objectives, (2) choice of parents in accordance with the objectives, (3) selection of seedlings, and (4) evaluation of clones that may have commercial potential. Although breeding programs may differ on some details, the basic principles are virtually the same (Bradshaw 2000; Bradshaw and Mackay 1994; Caligari 1992; Douches and Jastrzebski 1993; Hoopes and Plaisted 1987; Mackay 2005; Tarn et al. 1992). Many seedlings need to be grown to produce a new and improved cultivar. In North America in the 10-year period 1969–1978, approximately 200,000 seedlings were grown for each new cultivar released (Plaisted et al. 1984). In the Netherlands, the number of seedlings raised per newly registered cultivar rose from approximately 15,000 in the 1950s to approximately 250,000 in the 1980s (Zingstra 1983). Several authors have presented breeding schemes for the development of new cultivars (Ross 1986; Rousselle-Bourgeois and Rousselle 1996; Struik and Wiersema 1999a; Tarn et al. 1992). Programs for the identification of superior parents have been developed by Bradshaw and Mackay (1994), Brown and Dale (1998), Gopal (1998), and Tarn et al. (1992). Data processing programs for potato breeding programs have been developed by Tarn et al. (1992) and by Kozub et al. (2000).

The breeding objectives can generally be placed in the following three categories: (1) production, (2) utilization, and (3) protection. These traits, along with their respective screening and assessment methods were reviewed by Tarn et al. (1992). Simmonds (1969) has pointed out that although a new cultivar will likely not excel in all desirable traits, for acceptance in the industry it is important that a new cultivar does not have an extreme deficiency in any of the major economically important traits. In terms of industry acceptance, one major deficiency may cancel out many strong points in a new cultivar. Recently, breeding for disease resistance has been reviewed by Jansky (2000). For more detailed coverage of breeding potato for resistance to specific pests and pathogens, the reader is referred to the following reviews: *Phytophthora infestans* (Jones and Simko 2005; Swiezynski and Zimnoch-Guzowska 2001); *Verticillium* wilt and early blight (Jones and Simko 2005); *Meloidogyne* species and trichodorid-vectored virus (Brown and Mojtahedi 2005); viruses (Thieme and Thieme 2005); bacterial pathogens (Zimnoch-Guzowska et al. 2005). Conventional breeding methods have successfully been used to develop parents with combined resistances to several diseases (De Jong et al. 2001a; Murphy et al. 1999).

Breeding also involves an evaluation of promising clones over a wide range of environments. This involves the transportation of tubers to many different sites as well as synchronization of storage, transport and planting of the experiments. This can be expensive, especially when it is done on an international scale. A procedure has been developed in which environmental indices are obtained that measure the productivity of diverse potato-growing environments and then a prediction model for individual genotypes based on the environmental indices can be constructed. This should serve as an aid in the selection of cultivars that produce high yield over a wide range of environments (Tai and Young 1989; Tai et al. 1993).

Global warming may increase the growing season that in turn may lead to increased problems with insects and diseases (Hijmans 2003). Therefore, continued breeding for pest and disease resistance as well as an increased emphasis on breeding and selection for heat tolerance is warranted. Fortunately, there is considerable genetic diversity for heat tolerance among the germplasm resources (Veilleux et al. 1997).

2.6   GERMPLASM UTILIZATION AND ENHANCEMENT

Because of the close affinity between the cultivated potato and its wild relatives, it is relatively easy to incorporate related germplasm into cultivated forms (Peloquin et al. 1999). Many cultivars already contain one or more disease resistance genes that can be traced back to primitive cultivars or wild species (Ross 1986). If the utilization of germplasm is primarily as a donor of a specific trait then the need for a comprehensive improvement program is considerably less than if the germplasm source is expected to provide genetic variability for many traits. The latter approach does not have many parallels in other crops. It requires a relatively long-term commitment without immediate payoff in terms of new cultivars (Pavek and Corsini 2001; Plaisted and Hoopes 1989; Tarn et al. 1992). Spooner et al. (2004) provide an extensive list of potential uses of wild species in breeding programs.

2.6.1   Intraspecific and Interspecific Hybridization

To some extent, the taxonomic classification determines whether a certain type of hybridization is intraspecific or interspecific. In Dodds' (1962) classification system, *S. tuberosum* includes the primitive cultivated diploid Groups Stenotomum and Phureja as well as the tetraploid cultivated Groups Andigena and Tuberosum, so hybridization between these groups is intraspecific. This lack of species differentiation among cultivated types has been confirmed by molecular data, as mentioned earlier (Spooner et al. 2004; Sukhotu et al. 2004); however, the Group classification within *S. tuberosum* has not been broadly accepted, so that the separate species classification of cultivated types is used in much of the literature. Considerable use has been made of the Andigena Group in both broadening the genetic base of the Tuberosum Group as well as transferring specific traits. For example, in many cultivars with resistance to pathotype Ro1 of the potato cyst nematode, this resistance can be traced back to a single dominant gene from the Commonwealth Potato Collection CPC1673. This in turn was a collection of true seeds from the Bolivia Andigena cv. Polo (Plaisted and Hoopes 1989). By using several cycles of recurrent mass selection for adaptation to long days and other desirable traits, several breeders have recovered well-adapted populations of Andigena that have been used in cultivar development (Glendinning 1978; Munoz and Plaisted 1981; Plaisted et al. 1981, 2001; Simmonds 1969; Tarn and Tai 1983). The resemblance of these Andigena populations to cultivars of Tuberosum has been so great that they have been described as Neotuberosum.

Recurrent cyclical mass selection has also been applied to the primitive cultivated diploid Groups Phureja and Stenotomum (De Maine 1995; De Maine et al. 2000; Haynes 1980; Haynes and Lu 2005). The use of Tuberosum dihaploids* as crossing partners with the primitive cultivated groups as well as with the many diploid wild tuber-bearing species greatly accelerates the speed at which such material is enhanced and utilized. Both the primitive cultivated diploids and the wild species are usually poorly adapted to growing in the field under long-day conditions and are difficult to manage in cultivar development programs. Dihaploids can be used to capture, combine and retain the traits of interest at the diploid level (Watanabe et al. 1994). Diploid hybrid

---

* The term *haploid* is defined as an organism with the gametic chromosome number. This term is ambiguous in potato due to polyploidy; dihaploids ($2n=2x=24$) are defined as haploids extracted from a tetraploid whereas monoploids or monohaploids ($2n=1x=12$) refer to haploids with the basic chromosome number. However, potato breeders in North America have used the term haploid to refer to what Europeans generally describe as dihaploids by the above definition. To avoid confusion, we use dihaploid consistently in this review.

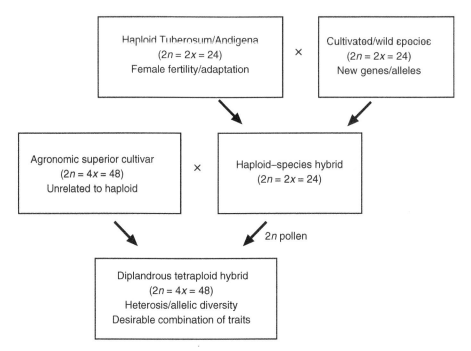

Figure 2.5   Breeding strategy to obtain 4x hybrids from 4x–2x crosses. (Adapted from Peloquin, S. J., *Genetics*, 153, 1493, 1999. With permission.)

populations (dihaploid Tuberosum × wild species) can be grown in the field and managed in similar ways as other breeding populations (Hermundstad and Peloquin 1985, 1986; Santini et al. 2000).

The key to the utilization of diploid hybrid populations in breeding lies in the use of 2n gametes. While selecting for adaptation and marketable traits, breeders can easily screen these populations for 2n gametes. Diploid parents with 2n gametes can be crossed with cultivars in efficient 4x–2x crossing combinations (Tai 1994). The use of 2n gametes has been more extensive in germplasm transfer for potato than for any other crop (Veilleux 1985). Sexual polyploidization as a breeding strategy for potato has recently been reviewed (Carputo et al. 2000; Ortiz 1998; Peloquin et al. 1999) and is illustrated in Figure 2.5. Many desirable traits from germplasm resources are quantitatively inherited. The 4x–2x breeding method lends itself to the transmission of such traits (Watanabe et al. 1999, 2005). This method also facilitates the transmission of transgenes (Johnson and Veilleux 2003). Ploidy manipulation can also be used in the study of gene action and mapping (Tai and Xiong 2005).

Although somatic hybridization is primarily used in accessing germplasm that is not sexually compatible with the cultivated potato (see wide hybridization below) such hybridization between sexually compatible diploids has been proposed as a method for new cultivar development (Gavrilenko et al. 1999; Mattheij and Puite 1992; Moellers and Wenzel 1992). Promising clones have been obtained through somatic hybridization; this approach, however, has not yet led to any new cultivars (Schwarzfischer et al. 1998).

In addition to their use in breeding, diploid potato hybrids represent a powerful tool for genetic analysis. The disomic inheritance pattern of the diploids results in much simpler segregation ratios compared to the tetrasomic inheritance of the cultivated potato (Ortiz and Peloquin 1994). Thus diploid potato has been used to determine the inheritance of economically important traits such as tuber shape (De Jong and Burns 1993; Van Eck et al. 1994b), tuber flesh and skin pigmentation (De Jong 1987; Van Eck et al. 1994a), and tuber skin texture (De Jong 1981). The genetic basis of some physiological mutants has also been analyzed with the use of diploids

(De Jong et al. 1998, 2001b). Diploids have been used in the development of an online catalogue of AFLP markers covering the potato genome (Rouppe van der Voort et al. 1998).

2.6.2 Wide Hybridization

There are several mechanisms in the reproductive biology of the various *Solanum* species that serve to isolate and preserve the identity of individual species. These mechanisms have implications for breeding and include differences in ploidy, as well as stylar and endosperm barriers (Hanneman 1999; Jackson and Hanneman 1999). Camadro et al. (2004) have suggested that several such barriers serve as substitutes for genome differentiation. The ploidy of the various *Solanum* species ranges from diploid to hexaploid. The stylar barriers include gametophytic incompatibility and, for some interspecific crosses, unilateral incompatibility (Jackson and Hanneman 1999). Several methods have been used successfully in wide hybridization. These include large numbers of pollinations, double pollination, embryo rescue, somatic hybridization, and bridging crosses (Hermsen 1994; Ramon and Hanneman 2002).

The endosperm balance hypothesis has greatly increased the understanding of endosperm related barriers. It is also used in predicting the success or failure of certain crosses and has been reviewed by Carputo et al. (1999). Under this hypothesis, a ratio of 2:1 of maternal:paternal genomes in the endosperm is necessary for the normal development of the seed (Figure 2.6). Apparently the genome of each species has an effective endosperm balance number that may not be a direct reflection of its ploidy. Through a series of testcrosses, the endosperm balance number (EBN) has been determined for over 80 species and subspecies of the tuber-bearing *Solanum* and their close non-tuber-bearing relatives (Hanneman 1994). Germplasm from sexually isolated relatives can be introgressed into the cultivated potato by manipulating the EBN (Adiwilaga and Brown 1991; Bamberg et al. 1994b; Barone et al. 2001; Brown and Adiwilaga 1990; Carputo et al. 1997a, 2003). There are some exceptions to the EBN theory where seed set occurs in crosses expected to fail (Jackson and Hanneman 1999).

Because diploid 1EBN and tetraploid 2EBN species are difficult to cross with either diploid 2EBN or tetraploid 4EBN cultivated potatoes, and at the same time contain several highly desirable traits, considerable efforts have been made to transfer such traits from these and other species to the

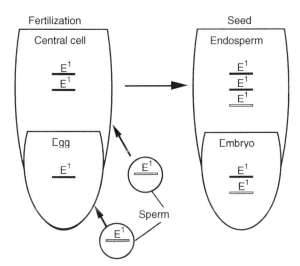

Figure 2.6 Endosperm balance number (EBN) hypothesis. (Adapted from Johnston, S. A. et al. *Theor. Appl. Genet.* 57, 5, 1980. With permission.)

cultivated potato via somatic hybridization. This area has recently been reviewed by Johnson and Veilleux (2001). Millam et al. (1995) have reviewed the progress and problems of integrating protoplast fusion-derived material into a potato breeding program. Many pest resistances are quantitatively inherited and do not lend themselves readily to gene transfer by bacteria-mediated transformation techniques (Watanabe and Watanabe 2000). An advantage of asymmetric somatic hybridization ($4x$ $tbr + 2x$ wild species) over the sexual method of hybridization is that the tbr genome remains intact while sexual hybridization results in a rearrangement of genetic information via meiosis and recombination. Potential bottlenecks of the use of somatic hybridization in a breeding program include the costs involved in developing appropriate skills, technologies and facilities, screening of somatic hybrid material, and identifying the products sought quickly and efficiently.

Sexual fertility in somatic hybrids is necessary for subsequent introgression of traits from the donor species into tbr. In the case of somatic hybrids involving S. *brevidens*, backcrosses of (hexaploid) asymmetric hybrids containing four genomes of potato and two genomes of S. *brevidens* with the cultivated potato as the male parent were successful whereas attempts to produce backcross progeny from (tetraploid) symmetric fusion products produced few, if any, seeds (Ehlenfeldt and Helgeson 1987). Similar results have been obtained with somatic hybrids between tomato and potato (Jacobsen et al. 1994). However, in some cases backcrosses of (tetraploid) symmetric somatic hybrids involving diploid 1EBN wild *Solanum* species with tbr were successful. This included somatic hybrids involving S. *circaeifolium* (Mattheij et al. 1992), S. *commersonii* (Carputo et al. 1998) and S. *etuberosum* (Thieme et al. 2000).

For several cross-combinations both somatic as well as sexual hybridization are possible. For example, in the case of S. *acaule* ($4x$, 2EBN), both hybridization methods have been applied successfully (Brown and Adiwilaga 1990; Rokka et al. 1998a, 1998b). This also applies to S. *commersonii* ($2x$, 1EBN) (Barone et al. 2001; Cardi 2001). Bamberg et al. (1994b) successfully used S. *commersonii* as a bridge species for sexual hybridization with several $4x$ 2EBN species; the hybrids in turn could then be crossed with tbr. Also, sexual hybrids have been reported between several nontuber-bearing and tuber-bearing *Solanum* species (Jackson and Hanneman 1999; Valkonen et al. 1995; Watanabe et al. 1995).

2.6.3 Genetic Transformation

Since genetic transformation protocols using disarmed *Agrobacterium tumefaciens* were first developed for potato (Ooms et al. 1987), innumerable alien genes have been integrated into the potato genome from diverse sources. Other methods of transformation, such as biolistic transformation (Romano et al. 2001), have been used but are of relatively minor importance, principally due to the complex integration patterns of transgenes (Romano et al. 2003). The objectives of genetic transformation have varied from practical to esoteric and the source organisms of transgenes as diverse as man and mouse (Figure 2.7). Commercial cultivars altered by a single trait, such as insect or pathogen resistance, have been released by Monsanto and cultivated extensively on more than 20,000 ha in the US in 1997 but subsequently withdrawn from the market in 2001 due to controversy over the safety of genetically altered food crops (Kaniewski and Thomas 2004). Research continues, however, at various institutions, on the methodology and possibilities of genetic transformation of potato. More sophisticated methods of transformation, such as minimizing the extraneous DNA segments within the T-DNA (Barrell et al. 2002), eliminating the selectable marker in the transgenic plants (de Vetten et al. 2003), chloroplast transformation (Sidorov et al. 1999), improved efficiency of *Agrobacterium*-mediated transformation for application to many cultivars (Heeres et al. 2002) and using a plant-derived DNA fragment (P-DNA) to replace *Agrobacterium*-derived T-DNA (Rommens et al. 2004) to obtain so-called *intragenic*

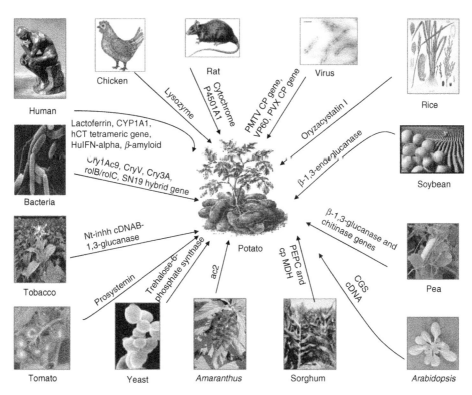

Figure 2.7 **(See color insert following page 304.)** Genes isolated from a wide range of organisms have been used to transform potato via *Agrobacterium*. Examples include (clockwise from upper left): human (*Homo sapiens*) *lactoferrin* (From Chong, D. K. X. and Langridge, W. H. R. *Transgenic Res.* 9, 71–78, 2000. With permission.), human *calcitonin tetrameric gene* (From Ofoghi, H., et al. *Biotechnol. Biotech. Equip.* 13, 20–24, 1999. With permission.), human *tumor necrosis factor-alpha* (From Osusky, M., et al. *Nature Biotechnol.* 18, 1162–1166, 2000. With permission.), human *beta-amyloid* peptide (From Kim, H.S., et al. *Plant Sci.* 165, 1445–1451, 2003. With permission.); chicken (*Gallus domesticus*) *lysozyme* gene (From Serrano, C., et al. *Am. J. Potato Res.* 77, 191–199, 2000. With permission.); rat (*Rattus Norvegicus*) *cytochrome P4501A1*(From Inui, H., et al. *Breeding Sci.* 48, 135–143, 1998. With permission.); potato mop-top virus coat protein gene (From Barker, H., et al. *Eur. J. Plant Path.* 104, 737 740, 1998; Barker, H., et al. *Phyton Ann. Rei Bot.* 39, 47–50, 1999; Germundsson, A., et al. *J. Gen. Virol.* 83, 1201–1209, 2002; Melander, M., et al. *Mol. Breed.* 8, 197–206, 2001. With permission.), potato virus X (PVX) coat protein gene (From Doreste, V., et al. *Phytoparasitica* 30, 177–185, 2002; Spillane, C., et al. *Irish J. Agric. Food Res.* 37, 173–182, 1998. With permission.), rabbit hemorrhagic disease virus capsid gene (From Castanon, S., et al. *Plant Sci.* 162, 87–95, 2002; Martin-Alonso, J. M., et al. *Transgenic Res.* 12, 127–130, 2003. With permission.); rice (*Oryza sativa*) cystatin I (From Lecardonnel, A. et al., *Plant Sci.*, 140, 71–79, 1999b. With permission.); soybean (*Glycine max*) beta-1,3-endoglucanase gene (From Borkowska et al. *Z. Naturforsch. C.* 53, 1012–1016, 1998. With permission.); pea (*Pisum sativum*) *beta-1,3-glucanase* and *chitinase* (From Chang, M. M., et al. *Plant Sci.* 163, 83–89, 2002. With permission.); *Arabidopsis thaliana* cystathionine gamma-synthase cDNA (From Di, R., et al. *J. Agric. Food Chem.* 51, 5695–5702, 2003. With permission.); sorghum (*Sorghum bicolor*) *C-4 phosphoenolpyruvate carboxylase* and *chloroplastic NADP(+)-malate dehydrogenase* (From Beaujean et al. *Plant Sci.* 160, 1199–1210, 2001. With permission.); *Amaranthus caudatus ac2* encoding the fungicidal peptide (defensin) (From Lyapkova, N. S., et al. *Appl. Biochem. Microbiol.* 37, 301–305, 2001. With permission.); yeast (*Saccharomyces cerevisiae*) *trehalose-6-phosphate synthase* (*TPS1*) gene (From Yeo, E.T. et al. *Mol. Cells* 10, 263–268, 2000. With permission.); tomato (*Lycopersicon esculentum*) *cathepsin D inhibitor* (From Brunelle et al. *Arch. Insect Biochem. Physiol.* 55, 103–113, 2004. With permission.), tomato *prosystemin* (From Narvaez-Vasquez, J. and Ryan, C. A. *Proc. Nat. Acad. Sci. U.S.A.* 99, 15818–15821, 2002. With permission.); tobacco (*Nicotiana tabacum*) *invertase inhibitor* (From Greiner, S., et al. *Nature Biotechnol.* 17, 708–711, 1999. With permission.); bacterial *cry1Ac9* gene (From Davidson, M.M., et al. *J. Am. Soc. Hort. Sci.* 127, 590–596, 2002. With permission.), *CryV* (From Douches, D. S., et al. *HortSci.* 33, 1053–1056, 1998; Li, W.B., et al. *J. Am. Soc. Hort. Sci.* 124, 218–223, 1999. With permission.), *Cry3A* (From Johnson, A. A. T., and Veilleux, R. E. *Euphytica* 133, 125–138, 2003; Lawson, E. C., et al. *Mol. Breed.* 7, 1–12, 2001. With permission.), *rolB* or *rolC* (From Aksenova, N.P., et al. *Russ. J. Plant Physiol.* 46, 513–519, 1999. With permission.), and SN19, a hybrid *Bacillus thuringiensis* delta-endotoxin (From Naimov, S., et al. *Plant Biotechnol. J.* 1, 51–57, 2003. With permission.)

plants (transformed plants that have only inserted plant-derived sequences) have been developed specifically using potato.

2.6.3.1 Insect Resistance

Codon optimization of *cry* genes to enhance their expression in plants (Adang et al. 1993) was the breakthrough that eventually led to commercialization of GMO potatoes with resistance to Colorado potato beetle. Monsanto released several cultivars under the trade name NewLeaf® followed by the original cultivar name. The NewLeaf products had been transformed with *cryIIIA* from *Bacillus thuringiensis* and produced a protein, delta-endotoxin, that interfered with insect feeding, presumably with no adverse effect on human consumption. Some cultivars had also been dually transformed with viral genes to convey resistance to PLRV or PVX in addition to the insect resistance [NewLeaf Plus®] (Lawson et al. 2001). The transgenic lines had been selected from many independently transformed plants for those with similar phenotypes to the original cultivars except for the resistances. They gained rapid popularity after their release in 1995, growing from an initial 600 to 20,000 ha over a few years but, because they were not accepted by fast food chains and chip makers, Monsanto withdrew them from the market in 2001. Another unfortunate finding was that expression of the innate golden nematode resistance of cv. Atlantic conveyed by the *H-1* gene had been inadvertently disrupted by the transformation process in Atlantic NewLeaf clone 6 (Brodie 2003). A highly entertaining novel written by Ruth Ozeki entitled *All Over Creation* features the thinly-disguised NewLeaf potato controversy (Ozeki 2003).

Despite the controversy, research has continued into insect resistance of potato transformed with *cry* genes, a topic that has been recently reviewed by Douches and Grafius (2005). Naimov et al. (2003) reported that a hybrid gene, *SN19*, comprised of *Cry1Ba/Cry1Ia* sequences to encode a hybrid toxin, increased the effectiveness of the delta-endotoxin such that resistance was conveyed to pests from two different orders, Coleoptera (potato tuber moth) and Lepidoptera (European corn borer larvae). Johnson and Veilleux (2003) documented adequate expression of *Cry3Aa* for CPB control in $4x–2x$ hybrids between wild type tetraploid cultivars and diploid single insert transgenic $2n$ pollen producers, thereby verifying that transgenic approaches could be integrated into true potato seed production schemes.

A different transgenic approach to control of Colorado potato beetle using oryzacystatin, a cysteine protease inhibitor that alters growth and development of herbivorous insects by inhibiting their proteolytic functions during digestion, was attempted by Lecardonnel et al. (1999b), who reported up to 53% mortality in larvae raised on transgenic leaves. Working with a construct free of sequences designed to inhibit insect feeding but containing a selectable marker and reporter gene, De Turck et al. (2002) found that Colorado potato beetle feeding on cv. Désirée transformed with an *nptII-gus* construct were more vigorous than controls feeding on untransformed Désirée. They concluded that *beta-glucuronidase* expression affected development and survival of the beetles. The *gus* protein was found to increase foliar consumption by Colorado potato beetle of similarly transformed Désirée (Lecardonnel et al. 1999a).

From 10 to 15 lines of two cultivars transformed with *cry1Ac9*, Davidson et al. (2002) selected one from each cultivar that either inhibited larval growth or prevented pupation of potato tuber moth (*Phthorimaea operculella* Zeller); the selected resistant lines were comparable phenotypically to the original cultivars. Douches et al. (1998) reported up to 96% control of potato tuber moth feeding on cv. Lemhi Russet that had been transformed with a codon-modified *CryV-Bt* gene. The resistance far exceeded that of controls with alternate forms of insect resistance (glandular trichomes or leptines). Ultimately multiple sources of resistance, including natural host plant resistance, must be explored to provide a sustainable potato production system that minimizes the environmental and human health costs (Douches and Grafius 2005).

2.6.3.2 Pathogen Resistance

Sanford and Johnston (1985) proposed the concept of pathogen-derived resistance (PDR), i.e., the expression of virus-derived sequences in plants to engineer resistance to viruses. Because of the numerous viral diseases to which potato succumbs and the potential of PDR, especially in regions of the world where isolation of seed certification plots is difficult or impossible, a flood of research into PDR for numerous potato viruses ensued (Martin 1994). Several of the virus genes including those for *replicase*, coat protein (CP) and viral movement, either used directly as they occur in the virus, codon modified or intentionally mutated, have been employed in this strategy. Barker et al. (1998, 1999) reported immunity to potato mop top virus (PMTV) in transgenic lines of two potato cultivars that had been transformed with the viral CP gene. Doreste et al. (2002) selected four clones with high resistance to PVX and high expression of the transgene among 100 clones transformed with the CP of PVX. A novel resistance that restricted the movement of PMTV from roots to leaves was found in transgenic plants expressing CP gene-mediated resistance (Germundsson et al. 2002). After transformation of potato with a mutated copy of one of the virus movement genes of PMTV, Melander et al. (2001) reported 79% virus reduction in the best of ten transformed clones of potato cv. Hulda in a field trial.

Maki-Valkama and Valkonen (1999) reviewed benefits and risks associated with PDR approaches to PVY. The discovery of RNA-mediated post-transcriptional gene silencing (PTGS) to provide natural protection against viral infections is among the benefits. Recombination between the transgene and the RNA genome of the infecting virus that can lead to altered viruses is among the risks. The CP gene of lettuce mosaic virus, another potyvirus related to PVY, conferred RNA-mediated resistance to PVYo transgenic potato (Hassairi et al. 1998). Li et al. (1999) demonstrated that it was possible to express multiple transgenes and resistances, one for CP-mediated resistance to PVYo and another for Bt-mediated resistance to CPB, into a single potato clone. A different isolate of PVY, that causing necrotic ringspot disease (NTN), was the target of transgenic research reported by Racman et al. (2001). Resistance varied among 34 transgenic clones of cv. Igor transformed with CP constructs in sense or in frame-shift orientations. Two of each were reported to be highly resistant whereas transformation of a different cultivar with the same constructs did not confer significant resistance.

Expression of a full-length unmodified replicase gene of potato leaf roll virus (PLRV) in cv. Russet Burbank resulted in a range of resistant clones from near immunity to full susceptibility (Thomas et al. 2000). The most resistant lines were considered suitable for commercial release. Rovere et al. (2001) subsequently demonstrated that the resistance was RNA-mediated as any of three versions of the gene (non-translatable sense, translatable sense with an engineered ATG and antisense) could result in resistant plants. A full-length copy of the cDNA of PLRV transformed into potato actually led to increased virus accumulation in transgenic plants compared to untransformed controls (Franco-Lara et al. 1999).

In a new transgenic approach to the study of gene function in potato, Faivre-Rampant et al. (2004) described a breakthrough technology where PVX was transformed with an antisense sequence of the potato *pds* (*phytoene desaturase*) gene. On infection of a susceptible potato cultivar with the transformed virus, the native *pds* gene was silenced, leading to distinct phenotypic changes associated with its lack of expression. This reverse-genetics approach to understanding gene function that exploits virus infectivity is expected to elucidate the role of expressed sequence tags in development.

An arsenal of alien genes has been raided in many attempts to transform potato for increased resistance against fungal and bacterial pathogens (Martin 1994). Borkowska et al. (1998) transformed potato with soybean *beta-1,3-endoglucanase* and reported increased resistance to late blight (*Phytophthora infestans*), still one of the most devastating pathogens of potato, in selected transformed plants. The same gene originating from *Nicotiana plumbaginifolia* was transformed into potato by Libantova et al. (1998); however, the resistance of transformed plants to fungal pathogens was not reported. Li and Fan (1999) transformed potato with the *harpin* gene from the apple fire blight pathogen (*Erwinia amylovora*) and reported that an engineered hypersensitive response to late blight had been obtained in two of 68 transformants. One of 23 transformants expressing the *glucose oxidase* gene of *Aspergillus niger* under a pathogen inducible promoter likewise exhibited the hypersensitive response to late blight infection of an originally susceptible potato cultivar (Zhen et al. 2000).

During et al. (1993) first reported reduced susceptibility of transgenic potato expressing bacteriophage *T4 lysozyme* against infection by *Erwinia carotovora atroseptica*. However the safety of agricultural utilization of these transgenic potato plants was subsequently questioned when the sensitivity of various soil-borne bacteria to T4 *lysozyme* was reported (de Vries et al. 1999) and the possible negative influence on soil microflora raised (Ahrenholtz et al. 2000). Chicken *lysozyme* has also been used to transform potato cv. Désirée with increased resistance to black leg and soft rot (*E. c. atroseptica* T7) reported in 13 of 63 and 20 of 69 clones, respectively (Serrano et al. 2000). The authors speculated that broad-spectrum disease resistance might be achieved through this strategy. Two insect genes, *attacin* and a synthetic *cecropin* (*sb-37*) both from silkmoth (*Hyalophora cecropia*) were transformed into potato cv. Désirée (Arce et al. 1999); four of 26 *att* clones and seven of 37 *sb-37* clones exhibited increased resistance to black leg) and seven of 35 *att* clones and seven of 31 *sb-37* clones exhibited increased resistance to soft rot infection. Osusky et al. (2000) confirmed the potential of broad-spectrum resistance against bacterial and fungal pathogens in transgenic potato expressing a synthetic *cecropin* gene (*CEMA*). Ray et al. (1998) hoped for broad-spectrum disease resistance in transgenic potato plants expressing the *anionic peroxidase* gene associated with induced resistance response in cucumber. However, there was no effect on disease symptoms induced by *Fusarium sambucinum, E. carotovora*, or *P. infestans*.

2.6.3.3 *Medicinal Use of Transgenic Potato*

Because potato produces considerable biomass in an easily harvested organ, the tuber, various studies have been undertaken to transform it with genes important to the pharmaceutical industry, either for the purpose of harvesting an expensive drug from tubers of an abundantly productive transformed cultivar or for using the nutritionally fortified edible tubers as a dietary supplement. Genes encoding human milk proteins (β-*casein* and *lactoferrin*) have been added in the hope of providing an inexpensive source of plant-synthesized hypoallergenic human milk for infant formula and baby foods. The enhanced digestibility and increased nutritional content could promote growth and protect neonates and young children against infectious diseases and the development of food allergies that may lead to autoimmune disease in later life. Chong and Langridge (2000) genetically engineered cv. Bintje with a gene encoding human *lactoferrin* (*hLF*) and observed that the transformed plants produced a full-length form of *hLF* that retained its biological activity based on immunoblot detection and bacteriostatic or bactericidal effects against a variety of human pathogenic bacterial strains. De Wilde et al. (2002) described the use of potato cv. Désirée transformed with either full size *immunoglobin G* (*IgG*) antibodies or its *Fab* antibody fragments, for molecular farming. *IgG* is a major effector molecule of the humoral immune response in man, accounting for about 75% of the total immunoglobulins in plasma of healthy individuals. The efficient transformation system available for potato and the amount and stability of the antibodies produced in tubers of the transformed plants led the authors to conclude that potato is a viable system for large-scale production of antibodies or other heterologous proteins.

β-Amyloid is a peptide that accumulates in plaques during the progression of Alzheimer's disease (AD) in humans. One strategy for developing an edible vaccine against AD is the use of potato plants expressing β-amyloid. Because the protein is toxic to animal cells, its production in *E. coli* or yeast is precluded. Kim et al. (2003) cloned human β-amyloid and transformed it into potato. Although expression levels of β-amyloid were low in the four transformed lines, there were no adverse phenotypic effects observed in the transgenic plants. Likewise, Ohya et al. (2002, 2001) expressed biologically active cytokines [human interferon (HuIFN-α2b and HuIFN-α8) and human tumor necrosis factor-α (HuTNF-α)] in potato in initial efforts to develop an inexpensive source of this antiviral pharmaceutical.

2.6.3.4 *Novel Use of Transgenics*

De Jong (2005) has reviewed different approaches to gene isolation. The candidate gene (CG) approach appears to have particular promise. Paal et al. (2004) identified candidate genes associated with root cyst nematode resistance in a resistant potato cultivar; transformation of a susceptible cultivar with one of three different CGs and testing the transgenics for resistance revealed the identity of the functional gene when 14 of 30 lines bearing one CG were resistant. None of the 30–37 lines bearing the other two CG was resistant. Hannapel et al. (2004) have overexpressed native potato transcription factors to elucidate their role in plant development.

2.6.4 Miscellaneous

2.6.4.1 *True Potato Seed*

The production of a potato crop from true potato seed (TPS) involves a technology that is different from the conventional method of producing potato from seed tubers (Almekinders et al. 1996; Struik and Wiersema 1999b; Upadhya et al. 1996). The primary focus of TPS technology is its use in developing countries, especially in Asia (Gaur et al. 2000; Upadhya et al. 1996). However, it can also be used in home gardens in developed countries (Renia et al. 2000). The applicability of TPS in developing countries depends upon the availability and quality of clonal seed potatoes. For example, the potential for TPS is greatest in situations where clonal seed costs exceed 20% of the value of production (Chilver et al. 2005). The production of potato from TPS in turn requires different breeding procedures and selection criteria (Clulow et al. 1995; Golmirzaie et al. 1994; Jackson 1987; Ortiz 1997; Ortiz and Golmirzaie 2004; Simmonds 1997). Several methods of producing commercial quantities of hybrid seed have been proposed to reduce production costs. This includes the use of $4x–2x$ crosses (Ortiz and Peloquin 1991) that can also be used to integrate transgenes into TPS (Johnson and Veilleux 2003). The tetrad type of male sterility is of interest in hybrid seed production because such flowers are still attractive to bumblebees that in turn are effective pollinators. Golmirzaie et al. (1994, 2003) have proposed the use of cybrid-derived cytoplasmic tetrad male sterility for the production of hybrid seed. Because the cultivated diploids are mostly self-incompatible, this system has also been proposed to produce TPS from $2x–2x$ crosses (De Maine 1996).

The availability of inbred lines in potato for hybrid seed production has been limited by severe inbreeding depression of cultivated types on self pollination (Simmonds 1997). The use of anther culture or prickle pollination has resulted in the extraction of monoploid potato clones from dihaploid or diploid germplasm (Lough et al. 2001). Homozygous diploids and tetraploids have been obtained by subsequent doubling or quadrupling the chromosome number of monoploids (De Maine and Simpson 1999; Fleming et al. 1992; Hulme et al. 1992). Although weaker than heterozygous $2x$ germplasm, doubled and quadrupled monoploids derived from *S. phureja* have set seed when cross-pollinated with diploid and tetraploid pollinators (M'Ribu and Veilleux 1992; Paz and Veilleux 1997), respectively, allowing the possibility of deriving more vigorous inbreds through recurrent cycles of monoploid extraction, diploidization, and cross-pollination (Foroughi-Wehr and Wenzel 1990). Such improved germplasm could form the basis of hybrid seed production in potato.

The culture of plant tissues via somatic embryos presents the possibility of the production of synthetically coated, clonal seed. Somatic embryogenesis has been attempted from various tissues of the potato plant (JayaSree et al. 2001; Seabrook and Douglass 2001). The regeneration of somatic embryos in potato is probably under nuclear control and the inheritance for regeneration may be relatively simple (Seabrook et al. 2001).

2.6.4.2   Apomixis

Hermsen (1980, 1983) has proposed gametophytic apomixis as an ideal means of producing true potato seed. Gametophytic apomixis is possible by diplospory (embryo sac formation from primary embryo sac mother cells by meiosis circumvention) and apospory (embryo sac formation from somatic cells in the ovule). The various mechanisms comprise a number of distinct and genetically controlled elements, most of which are known to occur in potato. Jongedijk (1987) and Jongedijk et al. (1991) have advocated the use of diplospory and Peloquin (1983) has proposed the use of apTSsporic apomixis in two different breeding schemes. Other suggested methods to produce aposporous apomictic seed in potato include the use of mutagenesis (Brown 1983) and of chemicals (Iwanaga 1983). The use of gametophytic apomixis in the production of TPS in potato has been reviewed by Jackson (1987) and by Upadhya (2000). Thus far this approach in potato remains an unfulfilled dream.

2.6.4.3   Mutation Breeding

General reviews on mutation breeding in potato have been done by Broertjes and van Harten (1988) and by Brown (1984). Thus far success in developing new potato cultivars by induced mutations has been limited. The FAO/IAEA mutant varieties database (Maluszynski et al. 2000) lists four cultivars (Mariline 2, Sarme, Desital, and Konkei No. 45) that have been obtained by mutation breeding. Only two of these (Sarme and Desital) are listed in the world catalogue of potato varieties (Hamester and Hils 2003). Several potato cultivars have originated from spontaneous mutation. These include cvs. Russet Burbank, Red Pontiac, and Red King Edward that arose from cvs. Burbank, Pontiac, and King Edward, respectively (Brown 1984). In addition, several cultivars have been derived from so-called "bolters" or "giant hills." These are spontaneous mutations in gene(s) responsible for the photoperiodic response. Such mutants generally have more vigorous vines, and a considerably later maturity. North American cultivars derived from such mutants include cv. Norgold strain M and several strains of cv. Russet Norkotah (Miller et al. 1995, 1999). The European cv. Jätte-Bintje is probably a similar mutant from Bintje (Bus 1990).

Because the potato germplasm resources contain extensive genetic diversity there is little need to induce mutations in potato to increase the genetic variability. Nevertheless, modifications of specific traits while leaving the genotype basically intact may have merit. Love et al. (1996a, 1996b) have attempted to improve cv. Russet Burbank and other clones by mutation breeding. Kowalski and Cassells (1999) focused on improving the late blight resistance of cv. Golden Wonder. A drawback of in vivo mutations is the relatively small number of genotypes that can be treated and evaluated. There have been few in vitro mutation breeding studies of potato (Ancora and Sonnino 1987; Sonnino et al. 1986; Upadhya et al. 1982). More recently this method is being used in the development of amylose-free starch potato cultivars (Hoogkamp et al. 2000).

2.6.4.4   Somaclonal Variation

Since Shepard et al. (1980) first documented the common occurrence of somaclonal variation in protoplast-derived regenerants (protoclones) of cv. Russet Burbank, many authors have reported the types and frequency of somaclonal variation in potato as well as attempted to understand its causes and the utility of molecular markers to expose it. The ability to change just a single trait of interest in a potato cultivar without disturbing the epistatic interactions underlying commercial traits popularized this approach to breeding; however, there is no control over the type of variation that can be generated nor over the ideal of changing just a single trait (Cassells et al. 1983). Therefore, subjecting potato cultivars to the range of tissue culture processes (Veilleux 2005) that provoke somaclonal variants is similar to mutation breeding in the randomness of effecting the desired changes. The belief that particular cultivars are especially prone to somaclonal variation has been supported in a few studies (Binsfeld et al. 1996; Rietveld et al. 1991, 1993). Whatever traits have been elicited as somaclonal variants must be examined through several vegetative cycles of propagation to verify their stability and distinguish them from epigenetic change. The frequent attribution of somaclonal variation to DNA methylation (Vazquez 2001) and its association with differing *in vitro* morphologies in potato (Joyce and Cassells 2002) increases the likelihood of instability of somaclonal variation and reversion of a somaclone to the parental type.

Protoclones

From more than 1,700 protoclones of cv. Russet Burbank Secor and Shepard (1981) selected 65 for replicated field studies based on favorable horticultural characters. Of 35 characters measured, statistically significant differences were found for 22 and each protoclone differed from the source cultivar for at least one trait. This study set the stage for exploiting somaclonal variation as a means to cultivar improvement in potato. Taylor et al. (1993) found that one of 37 protoclones of cv. Crystal had elevated resistance to soft rot (*Erwinia*), 20 were more resistant to bruising and five had enhanced chip color compared to the source cultivar. The authors were optimistic that somaclonal variation was a valid breeding approach, especially with cultivars that did not exhibit too many altered traits in a single somaclone. A protoclone of cv. Irish Cobbler (known as Danshakuimo in Japan) has been released as a new cultivar in Japan with the improvement that tubers do not brown after peeling compared to the source cultivar (Arihara et al. 1995).

Regeneration from Callus

Although the initial work on somaclonal variation had been conducted on protoclones, Wheeler et al. (1985) suggested that plant regeneration from explant cultures may be a better source of somaclones because of a reduced tendency towards ploidy change during regeneration and the relative ease of the procedure. Austin and Cassells (1983) demonstrated that the variation among plants derived from a single callus of cv. Golden Wonder was equivalent to that derived from multiple calluses from protoplasts, thus demonstrating that the variation was not due to preexisting somatic variation among cells in the explant but to variation arising in tissue culture. Evans et al. (1986) regenerated 346 clones from leaf, rachis, or stem explants of cv. Désirée and reported a range of variation among the regenerants from reduced yield to greater scab resistance and altered flesh color. They concluded that the changes were stable and that the method offered the possibility of effecting significant improvements in potato cultivars. Lentini et al. (1990) reported that two of 58 plants regenerated from callus of an interspecific hybrid between *S. tuberosum* and *S. berthaultii* surpassed the hybrid for insect resistance and marketable yield; altered segregation after meiosis was also reported in plants regenerated from callus. The ability of different clonal selections of a single cultivar to regenerate and yield somaclonal variants differed among plants obtained from tubers of cv. Record (Juned et al. 1991). Cassells et al. (1991) observed field resistance to late blight among clones regenerated from callus of cv. Bintje; however, most resistant clones also differed in maturity. Kowalski and Cassells (1999) screened 2,101 somaclones of cv. Golden Wonder, either mutagenized with X-rays or not, for late blight resistance and identified a single line with both improved yield and resistance after six years of field trials. Morphological variation of plants regenerated from suspension culture of cv. BP1 was obvious even *in vitro* (Lindeque et al. 1991).

Thieme and Griess (1996) conducted an extensive study of 13,000 somaclones regenerated from stem and leaf explants of 14 potato cultivars and found a "small, but relevant yield of positive variants in the range of 0.1–1%."

In Vitro *Selection*

Several authors have sought to direct somaclonal variation by imposing a selection criterion, such as an abiotic stress or culture filtrates derived from a pathogen to attempt to mimic infection, during the *in vitro* passage of potato explants or regenerating cultures. By selecting callus of an inter-specific hybrid (*S. tuberosum* × *S. chacoense*) for its ability to tolerate salt (NaCl) in the tissue culture medium, Burgutin et al. (1996) regenerated plants that exhibited salt tolerance evidenced by greater height, dry weight and root number compared to controls grown under salt stress. Increased frost tolerance and higher leaf proline content were reported for a regenerant from suspension culture of a diploid potato selected for resistance to hydroxyproline in the culture medium (van Swaaij et al. 1986); the regenerants were hypotetraploid ($2n = 44$–45; van Swaaij et al. 1987). Cerato et al. (1993) subjected suspension cultures of several cultivars of potato to various concentrations of culture filtrates of *Phytophthora infestans* and regenerated plants from the challenged cultures. More than one third of the regenerants were too abnormal to acclimate. Some of the regenerated somaclones exhibited improved resistance to late blight in greenhouse and laboratory tests; however, most grew poorly in the field due to other variation induced by the selection scheme. Sebastiani et al. (1994) selected nodal explants of callus-derived somaclones of cv. Désirée against culture filtrates of *Verticillium dahliae* and identified one resistant regenerant among 325 somaclones. Culture filtrate of *Alternaria solani* actually stimulated regeneration of five potato cultivars (Lynch et al. 1991).

Langille et al. (1993) selected leaf protoplasts that resisted the amino acid analog, 5-methyltryptophan, in culture to try to improve nutritional qualities of cv. Russet Burbank; however, there were no differences in free tryptophan in tubers of the protoclones compared to the source cultivar. Plants regenerated from salt tolerant callus of cv. Kennebec exhibited greater salt tolerance with higher fresh and dry weights and more tubers produced under salt stress compared to controls (Ochatt et al. 1998). Aluminum tolerant clones were obtained by Wersuhn et al. (1994) by alternating selective and nonselective media.

A recent example of *in vitro* selection for improved characters has been the release of a somaclone of cv. Lemhi Russet after selection of protoplasts in the presence of tyrosine (Gary Secor, North Dakota State University, personal communication). The resistant somaclone resulted in a processing cultivar that is resistant to black spot, a physiological disorder associated with potassium deficiency. However, evaluation of more than 10,000 somaclones of cv. Russet Burbank over several years did not result in a single selection that outperformed the parent cultivar. All somaclones that expressed one superior agronomic character were unfortunately deficient in agronomic performance. Russet Burbank may represent a cultivar that is too unstable in tissue culture such that somaclones with only a single altered trait cannot be obtained. Karp (1995) reviewed the potential of somaclonal variation for crop improvement and concluded that it represents a viable tool in crops with limited genetic systems or narrow genetic base.

Molecular Markers to Reveal Somaclones

Various attempts have been made to associate somaclonal variation with changes in molecular markers. Given that the two processes appear to be random, these attempts have met with limited success. Sometimes somaclones with obvious phenotypic abnormalities exhibit no marker poly-morphism compared to the source cultivar. Most authors have persisted until they find some minimal marker variation that they can attribute to somaclonal variation. Smith (1986) found novel electrophoretic banding patterns in protoclones of cv. Maris Piper but not in protoclones of cv.

Foxton. Allicchio et al. (1987) reported missing and additional isozyme bands in plants regenerated from callus of cv. Spunta. Sanford et al. (1984) could find no difference in isozyme patterns among cv. Russet Burbank, one of its protoclones and three bolters (genetic sports) similar in morphology to the protoclone. Using 25S rDNA as a probe, Landsmann and Uhrig (1985) found that 2 of 12 protoclones were variant, presumably deficient in rDNA genes.

Binsfeld et al. (1996) found that isozyme patterns differed among somaclones regenerated from internodes of three different cultivars of potato and that the frequency of individual somaclones with some difference varied from 19% to 86%, depending upon the cultivar, again verifying the varying susceptibility of cultivars to somaclonal variation. Albani and Wilkinson (1998) used ISSR (inter simple sequence repeat) polymorphism to study 40 regenerants of cv. Ukiraa from callus culture but found only two altered band patterns with 17 different primers. Mandolino et al. (1996) examined potato plants of two cultivars propagated by *in vitro* microtubers and 2-year-old suspension cultures with both RFLPs and RAPDs. With RFLPs, no variation was found among the microtuber-derived plants of either cultivar; however, the plants from suspension cultures of cv. Spunta exhibited an altered RFLP pattern. Deletion of an RFLP locus appeared to occur in 12 of 14 somatic hybrids between potato and *Solanum brevidens* (Williams et al. 1993). Evidence of presence/absence of the locus differed even between plants regenerated from different shoots originated from the same callus. The locus was described as highly mutable.

Using RFLPs, Potter and Jones (1991) found no variation in banding patterns of 128 micropropagated plants, nor in 18 and 21 plants, respectively, from low temperature growth or osmotic growth reduction; however, 6 of 46 plants regenerated from leaf callus exhibited variant banding patterns. Harding (1991) used RFLP polymorphism with rDNA probes to demonstrate that two of 16 DNA samples differed among plants recovered from slow growth whereas no changes were observed among control plants or those from cryopreservation.

Cytogenetic Variation in Somaclones

Cryogenic storage of potato germplasm has been used as a method to reduce maintenance of *in vitro* germplasm collections by eliminating the necessity of frequent subculture. However, in order for the process to be acceptable, genetic changes during storage must be held to a minimum. Several authors have studied the stability of potato germplasm released from cryogenic storage. A common effect of transition through tissue culture, especially a callus phase, has been a change in ploidy, with accompanying changes in phenotype. This has been exploited as a means to double the chromosome number of monoploid or diploid clones (M'Ribu and Veilleux 1990, 1992; Paz and Veilleux 1999). Cardi et al. (1992) found that chromosome doubling of diploid clones was more efficient if the explants were taken from *in vivo* compared to *in vitro* plants and a two-step regeneration protocol was used. Fleming et al. (1992), on the other hand, found that callus production was better from greenhouse grown plants but that regeneration did not differ between *in vitro* vs. *in vivo* leaf explants. Chauvin et al. (2003) used oryzalin applied to apical buds to enhance the frequency of doubled plants of diploid *Solanum* species but regenerated a high frequency of chimaeras. Benzine-Tizroutine et al. (1993) used floral bud morphology and stem dimensions to discern $2x$ from $4x$ regenerants of several potato clones. Regenerated tetraploids were found to be more frequently asyndetic than a control tetraploid. Imai et al. (1993) reported that *Agrobacterium* transformation increased the frequency of tetraploidization of diploid clones regenerated from tuber discs compared to untransformed controls that were similarly regenerated. Morphological instability in passage from initial somaclones of cv. Bintje to their vegetative progeny was attributed to chromosome mosaicism (Jelenic et al. 2001). Both euploids and aneuploids can be expected among regenerants (Tizroutine et al. 1994). An unexpected finding in this study was that some diploid regenerants from callus exhibited more regular meiotic pairing than the diploid source plant. Dihaploid plants that had been cryopreserved exhibited much greater ploidy stability than those regenerated from callus (Ward et al. 1993).

Callus and suspension cultures derived from potato are likely to exhibit even greater ploidy variation (polyploidy, aneuploidy, and chromosome rearrangements) than plants regenerated from them (Pijnacker et al. 1986; Pijnacker and Sree Ramulu 1990) and the frequency of these aberrations increased with the age of the callus (Hovenkamp-Hermelink et al. 1988; Ramulu et al. 1985), although most ploidy change occurred within a month of callus initiation and gradually leveled off (Hänisch ten Cate and Sree Ramulu 1987; Tizroutine et al. 1994). The lower the ploidy of the source plant, the more likely that polyploid callus would result and polyploid plants regenerate (Karp et al. 1984; Ramulu et al. 1989). After treating protoplasts with the chemical mutagen, ethyl methane sulphonate, Sadanandam (1991) identified a synaptic mutant among protoclones of a dihaploid potato.

Somaclonal Variation or Insertional Mutagenesis

Genetic changes that occur as a result of subjecting potato to transformation procedures using *Agrobacterium* can result either from insertion of the T-DNA into a region of the genome that disrupts normal gene function (insertional mutagenesis) or from somaclonal variation induced by the tissue culture process (usually leaf disc regeneration) required to regenerate the transformants. Separation of the two sources of variation is nearly impossible but the ambiguous variation has been observed in numerous studies. Dale and McPartlan (1992) compared populations of potato cv. Désirée that had been propagated from nodal cuttings with those regenerated from tuber discs, with or without *Agrobacterium*-mediated transformation with a construct that included *GUS* (*beta-glucuronidase*) and *NPTII* (*neomycin phosphotransferase*). Both of the populations of plants regenerated from tuber discs exhibited negative traits (reduced tuber yield, plant height); however, the transformed plants generally exhibited lower means than the untransformed somaclones, indicating that insertional mutagenesis had exaggerated somaclonal variation. The authors concluded that many transformations must be done to regenerate plants that have commercial characteristics equivalent to the mother cultivar. Conner et al. (1994) reported that 13 transgenic potato lines of three cultivars all expressing kanamycin resistance exhibited altered phenotypes or poor yield.

Presting et al. (1995) produced transgenic clones of cvs. Russet Burbank and Ranger Russet with the aim of increasing resistance to PLRV (potato leaf roll virus). Transgenic plants contained one of three constructs: (1) the native PLRV CP (coat protein) gene, (2) a modified PLRV CP gene that had been altered to decrease translation of the internal 17 kDa protein and concomitantly increase translation of the CP gene, and (3) a control construct with no CP gene whatsoever, only a *NPTII* gene. Surprisingly, many transgenics transformed with only the *NPTII* gene showed increased resistance to PLRV. The authors suggested that either transcription of any transgenic mRNA (PLRV CP or *NPTII*) regulated by a strong promoter may significantly reduce PLRV replication or that the increased resistance of transgenic plants carrying the control construct was due to somaclonal variation—perhaps the activation of quiescent retrotransposons during tissue culture. Kawchuk et al. (1997) likewise found resistance to PLRV in somaclones of cv. Russet Burbank that equaled that of transformants bearing coat protein cDNA. All 97 lines evaluated over several years exhibited abnormal phenotypes. The two most promising of 97 plants of cv. Russet Burbank that had been regenerated after *Agrobacterium*-mediated transformation with one of two constructs containing part of the PLRV CP gene were evaluated in field trials over several years in comparison with the single most promising somaclone from the same source. Although all three selections exhibited increased resistance after challenge with PLRV, the two transgenics had significantly lower yield than cv. Russet Burbank. By contrast, the somaclone was not significantly different in yield from cv. Russet Burbank. Again insertional mutagenesis appears to have amplified undesirable somaclonal variation.

Heeres et al. (2002) examined nearly 1,400 plants representing 16 cultivars transformed with an antisense gene coding for *granule-bound starch synthase* to obtain only 13 useful transformants with the desired amylose free starch phenotype, no kanamycin resistance gene or backbone vector sequences and field performance not significantly different from the original cultivars. Whether the discarded off-types were due to insertional mutagenesis or somaclonal variation was unknown but the conclusion was that many transformants must be produced to be able to select some with adequate transgene expression and minimal or undetectable somaclonal variation.

2.7 CONCLUSION

Progress in potato breeding has been slow in comparison to many crops with less complex breeding systems. The toolkits of genomic, molecular cytogenetic, and tissue culture technologies that have been applied to potato genetic research are numerous and increasing rapidly. Many new technologies have been readily applied to practical problems in propagation and genetic fingerprinting. There is hesitancy in accepting the more radical genomic technologies, not by scientists who generally revel in the application of new technologies, but by the industry and consumers who are more suspicious and perhaps more cautious. Whether we will see infant formula fortified with lactoferrin extracted from transgenic potatoes or similar products on the market has much less to do with our technical ability to make them happen than it does with the political and societal influences that allow them to come out of the laboratory and into the field.

ACKNOWLEDGMENTS

The authors would like to acknowledge the assistance of Richard Anderson, librarian at the Agriculture and Agri-Food Canada Potato Research Centre in Fredericton, and Rahul Gupta, former graduate student at Virginia Polytechnic Institute & State University.

REFERENCES

Adang, M. J. et al., The reconstruction and expression of a *Bacillus thuringiensis CryIIIa* gene in protoplasts and potato plants, *Plant Mol. Biol.*, 21, 1131–1145, 1993.

Adiwilaga, K. and Brown, C. R., Use of 2n pollen-producing triploid hybrids to introduce tetraploid Mexican wild species germplasm to cultivated tetraploid gene pool, *Theor. Appl. Genet.*, 81, 645–652, 1991.

Ahrenholtz, I. et al., Increased killing of *Bacillus subtilis* on the hair roots of transgenic T4 lysozyme-producing potatoes, *Appl. Environ. Microbiol.*, 66, 1862–1865, 2000.

Aksenova, N. P. et al., *In vitro* growth and tuber formation by transgenic potato plants harboring *rolC* or *rolB* genes under control of the patatin promoter, *Russ. J. Plant Physiol.*, 46, 513–519, 1999.

Albani, M. C. and Wilkinson, M. J., Inter simple sequence repeat polymerase chain reaction for the detection of somaclonal variation, *Plant Breed.*, 117, 573–575, 1998.

Allicchio, R. et al., Isozyme variation in leaf callus regenerated plants of *Solanum tuberosum, Plant Sci.*, 53, 81–86, 1987.

Almekinders, C. J. M. et al., Current status of the TPS technology in the world, *Potato Res.*, 39, 289–303, 1996.

Ancora, G. and Sonnino, A., *In vitro* induction of mutation in potato, In *Biotechnology in Agriculture and Forestry*, Bajaj, Y. P. S., Ed., Berlin, Heidelberg: Springer, chap. IV. 12, 1987.

Arce, P. et al., Enhanced resistance to bacterial infection by *Erwinia carotovora* subsp *atroseptica* in transgenic potato plants expressing the attacin or the cecropin SB-37 genes, *Am. J. Potato Res.*, 76, 169–177, 1999.

Arihara, A. et al., White Baron—a non-browning somaclonal variant of Danshakuimo (Irish Cobbler), *Am. Potato J.*, 72, 701–705, 1995.

Austin, S. and Cassells, A. C., Variation between plants regenerated from individual calli produced from separated potato stem callus cells, *Plant Sci. Letters*, 31, 107–114, 1983.

Bamberg, J. and del Rio, A., Conservation of potato genetic resources, In *Genetic Improvement of Solanaceous Crops. Volume 1: Potato*, Razdan, M. K. and Mattoo, A., Eds., Enfield, NH: Science Publishers, pp. 1–38, 2005.

Bamberg, J. B. et al., Elite selections of tuber-bearing *Solanum* species germplasm based on evaluations for disease, pest and stress resistance, In *Inter-Regional Potato Introduction Station, NRSP-6*, p. 56, 1994a.

Bamberg, J. B. et al., Using disomic 4*x* (2EBN) potato species' germplasm via bridge species *Solanum commersonii*, *Genome*, 37, 866–870, 1994b.

Bamberg, J. B. et al., International cooperation in potato germplasm, In *International Germplasm Transfer: Past and Present*, Duncan, R. R. et al., Eds., Madison, WI: Crop Science Society of America, chap. 13, 1995.

Barker, H. et al., High level of resistance in potato to potato mop-top virus induced by transformation with the coat protein gene, *Eur. J. Plant Path.*, 104, 737–740, 1998.

Barker, H. et al., A unique form of transgenic resistance to potato mop-top virus induced by transformation with the coat protein gene, *Phyton Ann. Rei Bot.*, 39, 47–50, 1999.

Barone, A. et al., Molecular marker-assisted introgression of the wild *Solanum commersonii* genome into the cultivated *S. tuberosum* gene pool, *Theor. Appl. Genet.*, 102, 900–907, 2001.

Barrell, P. J. et al., Alternative selectable markers for potato transformation using minimal T-DNA vectors, *Plant Cell Tiss. Org. Cult.*, 70, 61–68, 2002.

Beaujean, A. et al., Integration and expression of Sorghum C-4 phosphoenolpyruvate carboxylase and chloroplastic NADP(+)-malate dehydrogenase separately or together in C-3 potato plants, *Plant Sci.*, 160, 1199–1210, 2001.

Benzine-Tizroutine, S. et al., Somaclonal variation in potato—phenomenons correlated with flowering, *Acta Bot. Gallica*, 140, 5–16, 1993.

Binsfeld, P. C. et al., Isoenzymatic variation in potato somaclones (*Solanum tuberosum* L), *Brazilian J. Genet.*, 19, 117–121, 1996.

Borkowska, M. et al., Transgenic potato plants expressing soybean beta-1,3-endoglucanase gene exhibit an increased resistance to *Phytophthora infestans*, *Z. Naturforsch. C.*, 53, 1012–1016, 1998.

Bradshaw, J. E., Conventional breeding in potatoes: Global achievements, In *Potato, Global Research and Development*, Paul Khurana, S. M., Shekhawat, G. S., Singh, B. P., and Pandey, S. K., Eds., Shimla, India: Indian Potato Association, pp. 41–51, 2000.

Bradshaw, J. E. and Mackay, G. R., Breeding strategies for clonally propagated potatoes, In *Potato Genetics*, Bradshaw, J. E. and Mackay, G. R., Eds., Wallingford: CAB Int, chap. 21, 1994.

Brodie, B. B., The loss of expression of the *H-1* gene in *Bt* transgenic potatoes, *Am. J. Potato Res.*, 80, 135–139, 2003.

Broertjes, C. and van Harten, A. M., Potato, In *Applied Mutation Breeding for Vegetatively Propagated Crops*, Broertjes, C. and van Harten, A. M., Eds., Amsterdam: Elsevier, pp. 65–77, 1988.

Brown, C. R., Potato breeding through mutagenesis in aposporously generated true seed clones of established varieties, In *Research for the Potato in the Year 2000*, Hooker, W. J., Ed., Lima, Peru: International Potato Center, pp. 67–68, 1983.

Brown, C. R., Genetic modifications of established varieties of potato through mutagenesis, In *Utilization of Induced Mutations for Crop Improvement for Countries in Latin America*, Vienna: FAO/IAEA, pp. 307–321, 1984.

Brown, C. R., Origin and history of the potato, *Am. Potato J.*, 70, 363–374, 1993.

Brown, C. R. and Adiwilaga, K., Introgression of *Solanum acaule* germplasm from the endosperm balance number 2 gene pool into the cultivated endosperm balance number 4 potato gene pool via triplandroids, *Genome*, 33, 273–278, 1990.

Brown, C. R. and Mojtahedi, H., Breeding for resistance to Meloidogyne species and trichodorid-vectored virus, In *Genetic Improvement of Solanaceous Crops. Vol. 1: Potato*, Razdan, M. K. and Mattoo, A. K., Eds., Enfield, NH: Science Publishers, pp. 267–292, 2005.

Brown, J. and Dale, M. F. B., Identifying superior parents in a potato breeding program using cross prediction techniques, *Euphytica*, 104, 143–149, 1998.

Brunelle, F. et al., Colorado potato beetles compensate for tomato cathepsin D inhibitor expressed in transgenic potato, *Arch. Insect Biochem. Physiol.*, 55, 103–113, 2004.

Burgutin, A. B. et al., *In vitro* selection of potato for tolerance to sodium chloride, *Russ. J. Plant Physiol.*, 43, 524–531, 1996.

Bus, C. B., Jätte-Bintje: een Bintje type dat grotere knollen produceert (Transl.: Jätte-Bintje: a Bintje type which produces larger tubers), *Aardappelwereld*, 44, 16–17, 1990.

Caligari, P. D. S., Breeding new varieties, In *The Potato Crop. The Scientific Basis for Improvement*, Harris, P. M., Ed., London: Chapman and Hall, 1992, chap. 8.

Camadro, E. L. et al., Substitutes for genome differentiation in tuber-bearing *Solanum*: interspecific pollen-pistil incompatibility, nuclear-cytoplasmic male sterility, and endosperm, *Theor. Appl. Genet.*, 109, 1369–1376, 2004.

Cardi, T., Somatic hybridization between *Solanum commersonii* Dun. and *S. tuberosum* L (potato), In *Biotechnology in Agriculture and Forestry*, Nagata, T. and Bajaj, Y. P. S., Eds., Berlin, Heidelberg: Springer, pp. 245–263, 2001.

Cardi, T. et al., *In vitro* shoot regeneration and chromosome doubling in 2*x* potato and 3*x* potato clones, *Am. Potato J.*, 69, 1–12, 1992.

Carputo, D. et al., Endosperm balance number manipulation for direct *in vivo* germplasm introgression to potato from a sexually isolated relative (*Solanum commersonii* Dun.), *Proc. Natl. Acad. Sci. U.S.A.*, 94, 12013–12017, 1997a.

Carputo, D. et al., Resistance to blackleg and tuber soft rot in sexual and somatic interspecific hybrids with different genetic background, *Am. Potato J.*, 74, 161–172, 1997b.

Carputo, D. et al., Fertility of somatic hybrids *Solanum commersonii* (2*x*, 1EBN) (+) *S. tuberosum haploid* (2*x*, 2EBN) in intra- and inter-EBN crosses, *Genome*, 41, 776–781, 1998.

Carputo, D. et al., Uses and usefulness of endosperm balance number, *Theor. Appl. Genet.*, 98, 478–484, 1999.

Carputo, D. et al., 2*n* gametes in the potato: essential ingredients for breeding and germplasm transfer, *Theor. Appl. Genet.*, 101, 805–813, 2000.

Carputo, D. et al., The role of 2*n* gametes and endosperm balance number in the origin and evolution of polyploids in the tuber-bearing Solanums, *Genetics*, 163, 287–294, 2003.

Cassells, A. C. et al., Phenotypic variation in plants produced from lateral buds, stem explants and single-cell-derived callus of potato, *Potato Res.*, 26, 367–372, 1983.

Cassells, A. C. et al., Field resistance to late blight (*Phytophthora infestans* (Mont) Debary) in potato (*Solanum tuberosum* L) somaclones associated with instability and pleiotropic effects, *Euphytica*, 56, 75–80, 1991.

Castanon, S. et al., The effect of the promoter on expression of VP60 gene from rabbit hemorrhagic disease virus in potato plants, *Plant Sci.*, 162, 87–95, 2002.

Celebi-Toprak, F. et al., Molecular markers in identification of genotypic variation, In *Genetic Improvement of Solanaceous Crops. Vol. 1: Potato*, Razdan, M. K. and Mattoo, A. K., Eds., Enfield, NH: Science Publishers, pp. 115–141, 2005.

Cerato, C. et al., Resistance to late blight (*Phytophthora infestans* (Mont) Debary) of potato plants regenerated from *in vitro* selected calli, *Potato Res.*, 36, 341–351, 1993.

CFIA, Canadian certified seed potatoes; areas passing inspection 1951–2000, Compiled from annual reports, Canadian Food Inspection Agency, 2000.

CFIA. Potato. List of varieties which are registered in Canada, Canadian Food Inspection Agency, http://www.inspection.gc.ca/english/plaveg/variet/lovric.pdf (accessed 7/7/2006).

Chang, M. M. et al., *Agrobacterium*-mediated co-transformation of a pea *beta-1,3-glucanase* and *chitinase* genes in potato (*Solanum tuberosum* L. cv. Russet Burbank) using a single selectable marker, *Plant Sci.*, 163, 83–89, 2002.

Chauvin, J. E. et al., Chromosome doubling of 2*x* *Solanum* species by oryzalin: method development and comparison with spontaneous chromosome doubling *in vitro*, *Plant Cell Tiss. Org. Cult.*, 73, 65–73, 2003.

Chen, X. et al., A potato molecular-function map for carbohydrate metabolism and transport, *Theor. Appl. Genet.*, 102, 284–295, 2001.

Chilver, A. et al., On-farm profitability and prospects for true potato seed (TPS), In *Genetic Improvement of Solanaceous Crops. Vol. 1: Potato*, Razdan, M. K., Mattoo, A. K. et al., Eds., Enfield, NH: Science Publishers, 2005.

Chong, D. K. X. and Langridge, W. H. R., Expression of full-length bioactive antimicrobial human lactoferrin in potato plants, *Transgenic Res.*, 9, 71–78, 2000.

Clulow, S. A. et al., Producing commercially attractive, uniform true potato seed progenies: the influence of breeding scheme and parental genotype, *Theor. Appl. Genet.*, 90, 519–525, 1995.

Colon, L. T. et al., Partial resistance to late blight (*Phytophthora infestans*) in hybrid progenies of four South American *Solanum* species crossed with diploid *Solanum tuberosum*, *Theor. Appl. Genet.*, 90, 691–698, 1995.

Conner, A. J. et al., Field performance of transgenic potatoes, *N. Z. J. Crop Hort Sci.*, 22, 361–371, 1994.

Dale, P. J. and McPartlan, H. C., Field performance of transgenic potato plants compared with controls regenerated from tuber disks and shoot cuttings, *Theor. Appl. Genet.*, 84, 585–591, 1992.

Davidson, M. M. et al., Development and evaluation of potatoes transgenic for a *cry1Ac9* gene conferring resistance to potato tuber moth, *J. Am. Soc. Hort. Sci.*, 127, 590–596, 2002.

De Jong, H., Inheritance of russeting in cultivated diploid potatoes, *Potato Res.*, 24, 309–313, 1981.

De Jong, H., Inheritance of pigmented tuber flesh in cultivated diploid potatoes, *Am. Potato J.*, 64, 337–343, 1987.

De Jong, H. and Burns, V. J., Inheritance of tuber shape in cultivated diploid potatoes, *Am. Potato J.*, 70, 267–283, 1993.

De Jong, H. et al., Inheritance and mapping of a light green mutant in cultivated diploid potatoes, *Euphytica*, 103, 83–88, 1998.

De Jong, H. et al., The germplasm release of F87084, a fertile, adapted clone with multiple disease resistances, *Am. J. Potato Res.*, 78, 141–149, 2001a.

De Jong, H. et al., Development and characterization of an adapted form of *droopy*, a diploid potato mutant deficient in abscisic acid, *Am. J. Potato Res.*, 78, 279–290, 2001b.

De Jong, W., Approaches to gene isolation in potato, In *Genetic Improvement of Solanaceous Crops. Vol. 1: Potato*, Razdan, M. K. and Mattoo, A. K., Eds., Enfield, NH: Science Publishers, pp. 165–184, 2005.

De Maine, M. J., The adaptation and use of primitive cultivated diploid potato species, *Scottish Crop Res. Inst. Ann. Rept.*, 34–37, 1995.

De Maine, M. J., An assessment of true potato seed families of *Solanum phureja*, *Potato Res.*, 39, 323–332, 1996.

De Maine, M. J. and Simpson, G., Somatic chromosome number doubling of selected potato genotypes using callus culture or the colchicine treatment of shoot nodes *in vitro*, *Ann. Appl. Biol.*, 134, 125–130, 1999.

De Maine, M. J. et al., Long-day-adapted Phureja as a resource for potato breeding and genetic research, In *Potato, Global Research and Development*, Singh, B. P., Pandey, S. K. et al., Eds., Shimla, India: Indian Potato Association, pp. 134–137, 2000.

De Turck, S. et al., Transgenic potato plants expressing the *nptII-gus* marker genes affect survival and development of the Colorado potato beetle, *Plant Sci.*, 162, 373–380, 2002.

de Vetten, N. et al., A transformation method for obtaining marker-free plants of a cross-pollinating and vegetatively propagated crop, *Nature Biotechnol.*, 21, 439–442, 2003.

de Vries, J. et al., The bacteriolytic activity in transgenic potatoes expressing a chimeric *T4 lysozyme* gene and the effect of T4 lysozyme on soil- and phytopathogenic bacteria, *Syst. Appl. Microbiol.*, 22, 280–286, 1999.

De Wilde, C. et al., Expression of antibodies and *Fab* fragments in transgenic potato plants: a case study for bulk production in crop plants, *Mol. Breed.*, 9, 271–282, 2002.

del Rio, A. et al., Assessing changes in the genetic diversity of potato gene banks. 1. Effects of seed increase, *Theor. Appl. Genet.*, 95, 191–198, 1997a.

del Rio, A. et al., Assessing changes in the genetic diversity of potato gene banks. 2. In situ vs. ex situ, *Theor. Appl. Genet.*, 95, 199–204, 1997b.

Devine, K. J. and Jones, P. W., Potato cyst nematode hatching activity and hatching factors in inter-specific *Solanum* hybrids, *Nematology*, 3, 141–149, 2001.

Di, R. et al., Enhancement of the primary flavor compound methional in potato by increasing the level of soluble methionine, *J. Agric. Food Chem.*, 51, 5695–5702, 2003.

Dodds, K. S., Classification of cultivated potatoes, In *The Potato and its Wild Relatives*, Correll, D. S., Ed., Renner, TX: Texas Research Foundation, pp. 517–539, 1962.

Dong, F. et al., Development and applications of a set of chromosome-specific cytogenetic DNA markers in potato, *Theor. Appl. Genet.*, 101, 1001–1007, 2000.

Doreste, V. et al., Transgenic potato plants expressing the potato virus X (PVX) coat protein gene developed resistance to the viral infection, *Phytoparasitica*, 30, 177–185, 2002.

Douches, D. and Grafius, E. J., Transformation for insect resistance, In *Genetic Improvement of Solanaceous Crops. Vol. 1: Potato*, Razdan, M. K. and Mattoo, A. K., Eds., Enfield, NH: Science Publishers, pp. 235–266, 2005.

Douches, D. S. and Jastrzebski, K., Potato, In *Genetic Improvement of Vegetable Crops*, Kalloo, G. and Bergh, B. O., Eds., Oxford: Pergamon Press, 1993, chap. 44.

Douches, D. S. et al., Assessment of potato breeding progress in the USA over the last century, *Crop Sci.*, 36, 1544–1552, 1996.

Douches, D. S. et al., Potato transformation to combine natural and engineered resistance for controlling tuber moth, *HortSci.*, 33, 1053–1056, 1998.

During, K. et al., Transgenic potato plants resistant to the phytopathogenic bacterium *Erwinia carotovora*, *Plant J.*, 3, 587–598, 1993.

Ehlenfeldt, M. K. and Helgeson, J. P., Fertility of somatic hybrids from protoplast fusions of *Solanum brevidens* and *S. tuberosum*, *Theor. Appl. Genet.*, 73, 395–402, 1987.

Evans, N. E. et al., Somaclonal variation in explant-derived potato clones over three tuber generations, *Euphytica*, 35, 353–361, 1986.

Faivre-Rampant, O. et al., Potato virus X-induced gene silencing in leaves and tubers of potato, *Plant Physiol.*, 134, 1308–1316, 2004.

FAOSTAT, Food and Agriculture Organization of the United Nations Statistical Databases. Agricultural production, Primary crops. http://faostat.fao.org. (accessed Nov. 29, 2005).

Fleming, M. L. M. H. et al., Ploidy doubling by callus culture of potato dihaploid leaf explants and the variation in regenerated plants, *Ann. Appl. Biol.*, 121, 183–188, 1992.

Foroughi-Wehr, B. and Wenzel, G., Recurrent selection alternating with haploid steps—a rapid breeding procedure for combining agronomic traits in inbreeders, *Theor. Appl. Genet.*, 80, 564–568, 1990.

Franco-Lara, L. F. et al., Transformation of tobacco and potato with cDNA encoding the full-length genome of Potato leafroll virus: evidence for a novel virus distribution and host effects on virus multiplication, *J. Gen. Virol.*, 80, 2813–2822, 1999.

Garriga-Calderé, F. et al., Transmission of alien tomato chromosomes from BC1 to BC2 progenies derived from backcrossing potato (+) tomato fusion hybrids to potato: The selection of single additions for seven different tomato chromosomes, *Theor. Appl. Genet.*, 96, 155–163, 1998.

Garriga-Calderé, F. et al., Identification of alien chromosomes through GISH and RFLP analysis and the potential for establishing potato lines with monosomic additions of tomato chromosomes, *Genome*, 40, 666–673, 1997.

Garriga-Calderé, F. et al., Origin of an alien disomic addition with an aberrant homologue of chromosome-10 of tomato and its meiotic behaviour in a potato background revealed through GISH, *Theor. Appl. Genet.*, 98, 1263–1271, 1999a.

Garriga-Calderé, K. et al., Prospects for introgressing tomato chromosomes into the potato genome: An assessment through GISH analysis, *Genome*, 42, 282–288, 1999b.

Gaur, P. C. et al., True potato seed: Asian scenario, In *Potato, Global Research and Development*, Paul Khurana, S. M., Shekhawat, G. S., Singh, B. P., Pandey, S. K. et al., Eds., Shimla, India: Indian Potato Association, 2000.

Gavrilenko, T. et al., Assessment of genetic and phenotypic variation among intraspecific somatic hybrids of potato *Solanum tuberosum* L, *Plant Breed.*, 118, 205–213, 1999.

Gebhardt, C. et al., Assessing genetic potential in germplasm collections of crop plants by marker-trait association: a case study for potatoes with quantitative variation of resistance to late blight and maturity type, *Mol. Breed.*, 13, 93–102, 2004.

Gebhardt, C. et al., RFLP maps of potato and their alignment with the homoeologous tomato genome, *Theor. Appl. Genet.*, 83, 49–57, 1991.

Gebhardt, C. and Valkonen, J. P. T., Organization of genes controlling disease resistance in the potato genome, *Annu. Rev. Phytopathol.*, 39, 79–102, 2001.

Gebhardt, C. et al., Comparative mapping between potato (*Solanum tuberosum*) and *Arabidopsis thaliana* reveals structurally conserved domains and ancient duplications in the potato genome, *Plant J.*, 34, 529–541, 2003.

Germundsson, A. et al., Initial infection of roots and leaves reveals different resistance phenotypes associated with coat protein gene-mediated resistance to potato mop-top virus, *J. Gen. Virol.*, 83, 1201–1209, 2002.

Ghislain, M. et al., Selection of highly informative and user-friendly microsatellites (SSRs) for genotyping of cultivated potato, *Theor. Appl. Genet.*, 108, 881–890, 2004.

Glave, L. M., The conquest of highlands, In *The Potato Treasure of the Andes from Agriculture to Culture*, Graves, C., Ed., Lima: Peru: International Potato Center, 41, 42–51, 2001.

Glendinning, D. R., The potato gene-pool, and benefits deriving from its supplementation, In *Proc. Conf. Broadening Genet. Base Crops, 3–7 July*, van Harten, A. M. and Zeven, A. C., Eds., Wageningen: Pudoc, pp. 187–194, 1978.

Golmirzaie, A. M. et al., Breeding potatoes based on true seed propagation, In *Potato Genetics*, Bradshaw, J. E., Mackay, G. R. et al., Eds. Wallingford: CAB Int., chap. 22, 1994.

Golmirzaie, A. M. et al., Cybrids and tetrad sterility for developing true potato seed hybrids, *Ann. Appl. Biol.*, 143, 231–234, 2003.

Gopal, J., Identification of superior parents and crosses in potato breeding programmes, *Theor. Appl. Genet.*, 96, 287–293, 1998.

Greiner, S. et al., Ectopic expression of a tobacco invertase inhibitor homolog prevents cold-induced sweetening of potato tubers, *Nature Biotechnol.*, 17, 708–711, 1999.

Guenthner, J. F., *The International Potato Industry*. Cambridge: Woodhead, 2001.

Hämäläinen, J. H. et al., Mapping and marker-assisted selection for a gene for extreme resistance to potato virus Y, *Theor. Appl. Genet.*, 94, 192–197, 1997.

Hamester, W. and Hils., U., *World Catalogue of Potato Varieties*. Bergen, Dumme: Agri Media, 2003.

Hänisch ten Cate, C. H. and Sree Ramulu, K., Callus growth, tumour development and polyploidization in the tetraploid potato cultivar Bintje. *Plant Sci.* 49:209–216, 1987.

Hannapel, D. J. et al., Molecular controls of tuberization, *Am. J. Potato Res.*, 81, 263–274, 2004.

Hanneman, R. E., The potato germplasm resource, *Am. Potato J.*, 66, 655–668, 1989.

Hanneman, R. E., Assignment of endosperm balance numbers to the tuber-bearing Solanums and their close non-tuber-bearing relatives, *Euphytica*, 74, 19–25, 1994.

Hanneman, R. E., The reproductive biology of the potato and its implication for breeding, *Potato Res.*, 42, 283–312, 1999.

Harding, K., Molecular stability of the ribosomal RNA genes in *Solanum tuberosum* plants recovered from slow growth and cryopreservation, *Euphytica*, 55, 141–146, 1991.

Hassairi, A. et al., Transformation of two potato cultivars Spunta and Claustar (*Solanum tuberosum*) with lettuce mosaic virus coat protein gene and heterologous immunity to potato virus Y, *Plant Sci.*, 136, 31–42, 1998.

Hawkes, J. G., *The Potato Evolution, Biodiversity and Genetic Resources*. Washington, DC: Smithsonian Institute Press, 1990.

Hawkes, J. G., Origins of cultivated potatoes and species relationships, In *Potato Genetics*, Bradshaw, J. E. and Mackay, G. R., Eds., Wallingford: CAB Int, chap. 1, 1994.

Hawkes, J. G. and Hjerting, J. P., *The Potatoes of Bolivia their Breeding Value and Evolutionary Relationships*. Oxford: Clarendon Press, 1989.

Haynes, F. L., Progress and future plans for the use of Phureja-Stenotomum populations, In *Utilization of the Genetic Resources of the Potato. III*, Page, O. T., Ed., Lima, Peru: International Potato Center, pp. 80–88, 1980.

Haynes, K. G. and Lu, W., Improvement at the diploid species level, In *Genetic Improvement of Solanaceous Crops. Vol. 1. Potato*, Razdan, M. K. and Mattoo, A. K., Eds., Enfield, NH: Science Publishers, pp. 101–144, 2005.

Heeres, P. et al., Transformation of a large number of potato varieties: genotype-dependent variation in efficiency and somaclonal variability, *Euphytica*, 124, 13–22, 2002.

Hermsen, J. G. T., Breeding for apomixis in potato: pursuing a utopian scheme, *Euphytica*, 29, 595–607, 1980.

Hermsen, J. G. T., New approaches to breeding for the potato for the year 2000, In *Research for the Potato in the Year 2000*, Hooker, W. J., Ed., Lima, Peru: International Potato Center, 1983.

Hermsen, J. G. T., Introgression of genes from wild species, including molecular and cellular approaches, In *Potato Genetics*, Bradshaw, J. E. and Mackay, G. R., Eds., Wallingford: CAB Int, chap. 23, 1994.

Hermundstad, S. A. and Peloquin, S. J., Germplasm enhancement with potato haploids, *Heredity*, 76, 463–467, 1985.

Hermundstad, S. A. and Peloquin, S. J., Tuber yield and tuber traits of haploid-wild species F_1 hybrids, *Potato Res.*, 29, 289–297, 1986.

Hijmans, R. J., Global distribution of the potato crop, *Am. J. Potato Res.*, 78, 403–412, 2001.

Hijmans, R. J., The effect of climate change on global potato production, *Am. J. Potato Res.*, 80, 271–279, 2003.

Hoogkamp, T. J. H. et al., Development of amylose-free (amf) monoploid potatoes as new basic material for mutation breeding *in vitro*, *Potato Res.*, 43, 179–189, 2000.

Hoopes, R. W. and Plaisted, R. L., Potato, In *Principles of Cultivar Development*, Fehr, W., Ed., New York: Macmillan, pp. 385–436, 1987.

Hovenkamp-Hermelink, J. H. M. et al., Cytological studies on adventitious shoots and minitubers of a monoploid potato clone, *Euphytica*, 39, 213–219, 1988.

Huaman, Z. et al., The inter-genebank potato database and the dimensions of available wild potato germplasm, *Am. J. Potato Res.*, 77, 353–362, 2000.

Hulme, J. S. et al., An efficient genotype-independent method for regeneration of potato plants from leaf tissue, *Plant Cell Tiss. Org. Cult.*, 31, 161–167, 1992.

Imai, T. et al., High frequency of tetraploidy in *Agrobacterium*-mediated transformants regenerated from tuber disks of diploid potato lines, *Plant Cell Rep.*, 12, 299–302, 1993.

Inui, H. et al., Herbicide metabolism and resistance of transgenic potato plants expressing rat cytochrome *P4501A1*, *Breeding Sci.*, 48, 135–143, 1998.

Isidore, E. et al., Toward a marker-dense meiotic map of the potato genome: Lessons from linkage group I, *Genetics*, 165, 2107–2116, 2003.

Iwanaga, M., Chemical induction of aposporous apomictic seed production, In *Research for the Potato in the Year 2000*, Hooker, W. J., Ed., Lima, Peru: International Potato Center, pp. 104–105, 1983.

Jackson, M. T., Breeding strategies for true potato seed, In *The Production of New Potato Varieties: Technological Advances*, Jellis, G. J. and Richardson, D. E., Eds., Cambridge: Cambridge University Press, pp. 248–261, 1987.

Jackson, S. A. and Hanneman, R. E. J., Crossability between cultivated and wild tuber- and non-tuber-bearing Solanums, *Euphytica*, 109, 51–67, 1999.

Jacobs, J. M. E. et al., A genetic-map of potato (*Solanum tuberosum*) integrating molecular markers, including transposons, and classical markers, *Theor. Appl. Genet.*, 91, 289–300, 1995.

Jacobsen, E. et al., The first and second backcross progeny of the intergeneric fusion hybrids of potato and tomato after crossing with potato, *Theor. Appl. Genet.*, 88, 181–186, 1994.

Jansky, S., Breeding for disease resistance in potato, *Plant Breed. Rev.*, 19, 69–155, 2000.

JayaSree, T. et al., Somatic embryogenesis from leaf cultures of potato, *Plant Cell Tiss. Org. Cult.*, 64, 13–17, 2001.

Jelenic, S. et al., Mixoploidy and chimeric structures in somaclones of potato (*Solanum tuberosum* L.) cv. Bintje, *Food Technol. Biotechnol.*, 39, 13–17, 2001.

Johnson, A. A. T. and Veilleux, R. E., Somatic hybridization and applications in plant breeding, *Plant Breed. Rev.*, 20, 167–225, 2001.

Johnson, A. A. T. and Veilleux, R. E., Integration of transgenes into sexual polyploidization schemes for potato (*Solanum tuberosum* L.), *Euphytica*, 133, 125–138, 2003.

Johnston, S. A. et al., The significance of genic balance to endosperm development in interspecific crosses, *Theor. Appl. Genet.*, 57, 5–9, 1980.

Jones, R. W. and Simko, I., Resistance to late blight and other fungi, In *Genetic Improvement of Solanaceous Crops. Vol. 1: Potato*, Razdan, M. K. and Mattoo, A. K., Eds., Enfield, NH: Science Publishers, pp. 397–417, 2005.

Jongedijk, E., Desynapsis and FDR 2n-egg formation in potato: its significance to the experimental induction of diplosporic apomixis in potato, In *The Production of New Potato Varieties: Technological Advances*, Jellis, G. J. and Richardson, D. E., Eds., Cambridge: Cambridge University Press, pp. 225–228, 1987.

Jongedijk, E. et al., Formation of first division restitution (FDR) 2n-megaspores through pseudohomotypic division in *ds-1* (*desynapsis*) mutants of diploid potato: routine production of tetraploid progeny from 2x FDR×2x FDR crosses, *Theor. Appl. Genet.*, 82, 645–656, 1991.

Joyce, S. M. and Cassells, A. C., Variation in potato microplant morphology *in vitro* and DNA methylation, *Plant Cell Tiss. Org. Cult.*, 70, 125–137, 2002.

Juned, S. A. et al., Genetic variation in potato cv Record—evidence from *in vitro* regeneration ability, *Ann. Bot.*, 67, 199–203, 1991.

Kaniewski, W. K. and Thomas, P. E., The potato story, *AgBioForum*, 7, 41–46, 2004.

Karp, A., Somaclonal variation as a tool for crop improvement, *Euphytica*, 85, 295–302, 1995.

Karp, A. et al., Chromosome doubling in monohaploid and dihaploid potatoes by regeneration from cultured leaf explants, *Plant Cell Tiss. Org. Cult.*, 3, 363–373, 1984.

Kawchuk, L. M. et al., Field resistance to the potato leafroll luteovirus in transgenic and somaclone potato plants reduces tuber disease symptoms, *Can. J. Plant Pathol.*, 19, 260–266, 1997.

Kim, H. S. et al., Expression of human beta-amyloid peptide in transgenic potato, *Plant Sci.*, 165, 1445–1451, 2003.

Knight, J., A dying breed, *Nature*, 421, 568–570, 2003.

Kowalski, B. and Cassells, A. C., Mutation breeding for yield and *Phytophthora infestans* (Mont.) de Bary foliar resistance in potato (*Solanum tuberosum* L cv Golden Wonder) using computerized image analysis in selection, *Potato Res.*, 42, 121–130, 1999.

Kozub, J. G. et al., A relational database system for potato breeding programs, *Am. J. Potato Res.*, 77, 95–110, 2000.

Landsmann, J. and Uhrig, H., Somaclonal variation in *Solanum tuberosum* detected at the molecular level, *Theor. Appl. Genet.*, 71, 500–505, 1985.

Langille, A. R. et al., Effects of the amino acid analog, 5-methyltryptophan on protoplast survival, plating efficiency and free tryptophan levels in tubers of regenerated potato plants, *Am. Potato J.*, 70, 735–741, 1993.

Lawson, E. C. et al., NewLeaf Plus (R) Russet Burbank potatoes: Replicase-mediated resistance to potato leafroll virus, *Mol. Breed.*, 7, 1–12, 2001.

Lecardonnel, A. et al., Genetic transformation of potato with nptII-gus marker genes enhances foliage consumption by Colorado potato beetle larvae, *Mol. Breed.*, 5, 441–451, 1999a.

Lecardonnel, A. et al., Effects of rice cystatin I expression in transgenic potato on Colorado potato beetle larvae, *Plant Sci.*, 140, 71–79, 1999b.

Lee, H. K. and Hanneman, R. E., Identification of trisomics with Giemsa stain and comparison with pachytene chromosomes in potato, *Am. Potato J.*, 52, 281–282, 1975.

Lentini, Z. et al., Insect-resistant plants with improved horticultural traits from interspecific potato hybrids grown *in vitro*, *Theor. Appl. Genet.*, 80, 95–104, 1990.

Li, R. G. and Fan, Y. L., Reduction of lesion growth rate of late blight plant disease in transgenic potato expressing harpin protein, *Sci. China Ser. C: Life Sci.*, 42, 96–101, 1999.

Li, W. B. et al., Coexpression of potato PVYo coat protein and *cryV-Bt* genes in potato, *J. Am. Soc. Hort. Sci.*, 124, 218–223, 1999.

Libantova, J. et al., Transgenie tobacco and potato plants expressing basic vacuolar beta-1,3-glucanase from *Nicotiana plumbaginifolia*, *Biologia*, 53, 739–748, 1998.

Lin, B.-H. et al., Fast food growth boosts frozen potato consumption, *Food Rev.*, 24, 38–46, 2001.

Lindeque, J. M. et al., Variation in phenotype and proteins in plants regenerated from cell suspensions of potato cv BP1, *Euphytica*, 54, 41–44, 1991.

Lough, R. C. et al., Selection inherent in monoploid derivation mechanisns for potato, *Theor. Appl. Genet.*, 103, 178–184, 2001.

Love, S. L., Founding clones, major contributing ancestors, and exotic progenitors of prominent North American cultivars, *Am. J. Potato Res.*, 76, 263–280, 1999.

Love, S. L. et al., Induced mutations for reduced tuber glycoalkaloid content in potatoes, *Plant Breed.*, 115, 119–122, 1996b.

Love, S. L. et al., Mutation breeding for improved internal quality and appearance in Russet Burbank, *Am. Potato J.*, 73, 155–166, 1996a.

Lyapkova, N. S. et al., Transformed potato plants carrying the gene of the antifungal peptide of *Amaranthus caudatus*, *Appl. Biochem. Microbiol.*, 37, 301–305, 2001.

Lynch, D. R. et al., Effect of *Alternaria solani* culture filtrate on adventitious shoot regeneration in potato, *Plant Cell Rep.*, 9, 607–610, 1991.

M'Ribu, H. K. and Veilleux, R. E., Effect of genotype, explant, subculture interval and environmental conditions on regeneration of shoots from *in vitro* monoploids of a diploid potato species, *Solanum phureja* Juz & Buk.: Factors affecting shoot regeneration of monoploid potato, *Plant Cell Tiss. Org. Cult.*, 23, 171–179, 1990.

M'Ribu, H. K. and Veilleux, R. E., Fertility of doubled monoploids of *Solanum phureja*, *Am. Potato J.*, 69, 447–459, 1992.

Mackay, G. R., Propagation by traditional breeding methods, In *Genetic Improvement of Solanaceous Crops. Vol. 1: Potato*, Mattoo, M. K. and Mattoo, A. K., Eds., Enfield, NH: Science Publishers, pp. 65–81, 2005.

Maki-Valkama, T. and Valkonen, J. P. T., Pathogen derived resistance to Potato virus Y: Mechanisms and risks, *Agric. Food Sci. Finland*, 8, 493–513, 1999.

Maluszynski, M. et al., Officially released mutant varieties – the FAO/IAEA database, *Mutation Breed. Rev.*, 12, 1–84, 2000.

Mandolino, G. et al., Stability of fingerprints of *Solanum tuberosum* plants derived from conventional tubers and vitrotubers, *Plant Breed.*, 115, 439–444, 1996.

Marczewski, W. et al., Two allelic or tightly linked genetic factors at the PLRV.4 locus on potato chromosome XI control resistance to potato leafroll virus accumulation, *Theor. Appl. Genet.*, 109, 1604–1609, 2004.

Martin-Alonso, J. M. et al., Oral immunization using tuber extracts from transgenic potato plants expressing rabbit hemorrhagic disease virus capsid protein, *Transgenic Res.*, 12, 127–130, 2003.

Martin, R. R., Genetic engineering of potatoes, *Am. Potato J.*, 71, 347–358, 1994.

Matsubayashi, M., Species differentiation in *Solanum* Sect. *Petota*. XII. Intra- and inter-series genomic relationships in diploid *Commersoniana* and *Tuberosa* species, *Sci. Rep. Fac. Agr. Kobe Univ.*, 15, 203–216, 1983.

Mattheij, W. M. et al., Interspecific hybridization between the cultivated potato *Solanum tuberosum* subspecies *tuberosum* L. and the wild species *S. circaeifolium* subsp. *circaeifolium* Bitter exhibiting resistance to *Phytophthora infestans* (Mont.) and *Globodera pallida* (Stone) Behrens, *Theor. Appl. Genet.*, 83, 459–466, 1992.

Mattheij, W. M. and Puite, K. J., Tetraploid potato hybrids through protoplast fusions and analysis of their performance in the field, *Theor. Appl. Genet.*, 83, 807–812, 1992.

Maughan, P. J. et al., Microsatellite and amplified sequence length polymorphisms in cultivated and wild soybean, *Genome*, 38, 715–723, 1995.

Melander, M. et al., Reduction of potato mop-top virus accumulation and incidence in tubers of potato transformed with a modified triple gene block gene of PMTV, *Mol. Breed.*, 8, 197–206, 2001.

Millam, S. et al., The integration of protoplast fusion-derived material into a plant breeding programme— a review of progress and problems, *Euphytica*, 85, 451–455, 1995.

Miller, J. C. J. et al., Selection, evaluation, and identification of improved Russet Norkotah strains, *Am. J. Potato Res.*, 76, 161–167, 1999.

Miller, J. C. J. et al., Norgold Russet and Norgold Russet strain M—additional evidence for genetic dissimilarity, *Am. Potato J.*, 72, 273–286, 1995.

Moellers, C. and Wenzel, G., Somatic hybridization of dihaploid potato protoplasts as a tool for potato breeding, *Bot. Acta*, 105, 133–139, 1992.

Munoz, F. J. and Plaisted, R. L., Yield and combining abilities in Andigena potatoes after six cycles of recurrent phenotypic selection for adaptation to long day conditions, *Am. Potato J.*, 58, 469–480, 1981.

Murphy, A. M. et al., A multiple disease resistant potato clone developed with classical breeding technology, *Can. J. Plant Pathol.*, 21, 207–212, 1999.

Naimov, S. et al., A hybrid Bacillus thuringiensis delta-endotoxin gives resistance against a coleopteran and a lepidopteran pest in transigenic potato, *Plant Biotechnol. J.*, 1, 51–57, 2003.

NAK. 2003. Overzicht keuring van pootaardappelen van totaal Nederland over oogst 2002 (Overview of inspection of seed potatoes of total Netherlands over harvest 2002).

Narvaez Vasquez, J. and Ryan, C. A., The systemin precursor gene regulates both defensive and developmental genes in *Solanum tuberosum*, *Proc. Nat. Acad. Sci. U.S.A.*, 99, 15818–15821, 2002.

Novy, R. G. and Hanneman, R. E., Hybridization between Gp. Tuberosum haploids and 1EBN wild potato species., *Am. Potato J.*, 68, 151–169, 1991.

Oberhagemann, P. et al., A genetic analysis of quantitative resistance to late blight in potato: towards marker-assisted selection, *Mol. Breed.*, 5, 399–415, 1999.

Ochatt, S. J. et al., *In vitro* recurrent selection of potato: production and characterization of salt tolerant cell lines and plants, *Plant Cell Tiss. Org. Cult.*, 55, 1–8, 1998.

Ochoa, C. M., *The Potatoes of South America. Bolivia*, Cambridge: Cambridge University Press, 1990.

Ochoa, C. M., *Las papas de Sudamerica: Perú (parte 1)*. Lima, Peru: International Potato Center, 1999.

Ofoghi, H. et al., Human calcitonin tetrameric gene: Comparative expression in yeast and transgenic potato plants, *Biotechnol. Biotech. Equip.*, 13, 20–24, 1999.

Ohya, K. et al., Expression of two subtypes of human IFN-alpha in transgenic potato plants, *J. Interferon Cytokine Res.*, 21, 595–602, 2001.

Ohya, K. et al., Expression of biologically active human tumor necrosis factor-alpha in transgenic potato plant, *J. Interferon Cytokine Res.*, 22, 371–378, 2002.

Ooms, G. et al., Genetic transformation in two potato cultivars with T-DNA from disarmed *Agrobacterium*, *Theor. Appl. Genet.*, 73, 744–750, 1987.

Ortiz, R., Breeding for potato production from true seed, *Plant Breed. Abstr.*, 67, 1355–1360, 1997.

Ortiz, R., Potato breeding via ploidy manipulations, *Plant Breed. Rev.*, 16, 15–86, 1998.

Ortiz, R. and Golmirzaie, A. M., Combining ability analysis and correlation between breeding values in true potato seed, *Plant Breed.*, 123, 564–567, 2004.

Ortiz, R. and Peloquin, S. J., A new method of producing $4x$ hybrid true potato seed, *Euphytica*, 57, 103–107, 1991.

Ortiz, R. and Peloquin, S. J., Use of 24-chromosome potatoes (diploids and dihaploids) for genetic analysis, In *Potato Genetics*, Bradshaw, J. E. and Mackay, G. R., Eds. Wallingford: CAB Int, p. bib, chap. 6, 1994.

Osusky, M. et al., Transgenic plants expressing cationic peptide chimeras exhibit broad-spectrum resistance to phytopathogens, *Nature Biotechnol.*, 18, 1162–1166, 2000.

Ozeki, R. L., *All Over Creation*. New York: Viking, 2003.

Paal, J. et al., Molecular cloning of the potato Gro1-4 gene conferring resistance to pathotype Ro1 of the root cyst nematode *Globodera rostochiensis*, based on a candidate gene approach, *Plant J.*, 38, 285–297, 2004.

Pavek, J. J. and Corsini, D. L., Utilization of potato genetic resources in variety development, *Am. J. Potato Res.*, 78, 433–441, 2001.

Paz, M. M. and Veilleux, R. E., Genetic diversity based on randomly amplified polymorphic DNA (RAPD) and its relationship with the performance of diploid potato hybrids, *J. Am. Soc. Hort. Sci.*, 122, 740–747, 1997.

Paz, M. M. and Veilleux, R. E., Influence of culture medium and *in vitro* conditions on shoot regeneration in *Solanum phureja* monoploids and fertility of regenerated doubled monoploids, *Plant Breed.*, 118, 53–57, 1999.

Peloquin, S. J., New approaches to breeding for the potato in the year 2000, In *Research for the Potato in the Year 2000*, Hooker, W. J., Ed., Lima, Peru: International Potato Center, pp. 32–34, 1983.

Peloquin, S. J. et al., Meiotic mutants in potato: valuable variants, *Genetics*, 153, 1493–1499, 1999.

Pijnacker, L. P. and Ferwerda, M. A., Giemsa C-banding of potato chromosomes, *Can. J. Genet. Cytol.*, 26, 415–419, 1984.

Pijnacker, L. P. and Sree Ramulu, K., Somaclonal variation in potato—a karyotypic evaluation, *Acta Bot. Neerl.*, 39, 163–169, 1990.

Pijnacker, L. P. et al., Variability of DNA content and karyotype in cell cultures of an interdihaploid *Solanum tuberosum*, *Plant Cell Rep.*, 5, 43–46, 1986.

Plaisted, R. L. and Hoopes, R. W., The past record and future prospects for the use of exotic potato germplasm, *Am. Potato J.*, 66, 603–623, 1989.

Plaisted, R. L. et al., Rosa: a new golden nematode resistant variety for chipping and tablestock, *Am. Potato J.*, 58, 451–455, 1981.

Plaisted, R. L. et al., Selecting for resistance to diseases in early generations, *Am. Potato J.*, 61, 395–404, 1984.

Plaisted, R. L. et al., Eva: a midseason golden nematode- and virus-resistant variety for use as tablestock or chipstock, *Am. J. Potato Res.*, 78, 65–68, 2001.

Potter, R. and Jones, M. G. K., An assessment of genetic stability of potato *in vitro* by molecular and phenotypic analysis, *Plant Sci.*, 76, 239–248, 1991.

Presting, G. G. et al., Resistance to potato leafroll virus in potato plants transformed with the coat protein gene or with vector control constructs, *Phytopathology*, 85, 436–442, 1995.

Racman, D. S. et al., Strong resistance to potato tuber necrotic ringspot disease in potato induced by transformation with coat protein gene sequences from an NTN isolate of potato virus Y, *Ann. Appl. Biol.*, 139, 269–275, 2001.

Ramon, M. and Hanneman, R. E. J., Introgression of resistance to late blight (*Phytophthora infestans*) from *Solanum pinnatisectum* into *S. tuberosum* using embryo rescue and double pollination, *Euphytica*, 127, 421–435, 2002.

Ramulu, K. S. et al., Patterns of DNA and chromosome variation during *in vitro* growth in various genotypes of potato, *Plant Sci.*, 41, 69–78, 1985.

Ramulu, K. S. et al., Patterns of phenotypic and chromosome variation in plants derived from protoplast cultures of monohaploid, dihaploid and diploid genotypes and in somatic hybrids of potato, *Plant Sci.*, 60, 101–110, 1989.

Ray, H. et al., Transformation of potato with cucumber peroxidase: expression and disease response, *Physiol. Mol. Plant Pathol.*, 53, 93–103, 1998.

Renia, H. et al., Commercial acceptance of TPS by home gardeners in the USA, In *Potato, Global Research and Development*, Paul Khurana, S. M., Shekhawat, G. S., Singh, B. P., and Pandey, S. K., Eds., Shimla, India: Indian Potato Association, pp. 677–681, 2000.

Rietveld, R. C. et al., Somaclonal variation in tuber disk-derived populations of potato 1. Evidence of genetic stability across tuber generations and diverse locations, *Theor. Appl. Genet.*, 82, 430–440, 1991.

Rietveld, R. C. et al., Somaclonal variation in tuber disc-derived populations of potato 2. Differential effect of genotype, *Theor. Appl. Genet.*, 87, 305–313, 1993.

Rokka, V.-M. et al., Production of androgenic dihaploid lines of the disomic tetraploid potato species *Solanum acaule* ssp. *acaule*, *Plant Cell Rep.*, 18, 89–93, 1998a.

Rokka, V.-M. et al., Interspecific somatic hybrids between wild potato *Solanum acaule* Bitt. and anther-derived dihaploid potato (*Solanum tuberosum* L.), *Plant Cell Rep.*, 18, 82–88, 1998b.

Rokka, V. M. et al., Cytological and molecular characterization of repetitive DNA sequences of *Solanum brevidens* and *Solanum tuberosum*, *Genome*, 41, 487–494, 1998d.

Rokka, V. M. et al., Fluorescence in situ hybridization of potato somatohaploids and their somatic hybrid donors using two *Solanum brevidens* specific sequences, *Agric. Food Sci. Finland*, 7, 31–38, 1998c.

Romano, A. et al., Transformation of potato (*Solanum tuberosum*) using particle bombardment, *Plant Cell Rep.*, 20, 198–204, 2001.

Romano, A. et al., Transgene organisation in potato after particle bombardment-mediated (co-)transformation using plasmids and gene cassettes, *Transgenic Res.*, 12, 461–473, 2003.

Rommens, C. M. et al., Crop improvement through modification of the plant's own genome, *Plant Physiol.*, 135, 421–431, 2004.

Ronning, C. M. et al., Comparative analyses of potato expressed sequence tag libraries, *Plant Physiol.*, 131, 419–429, 2003.

Ross, H., *Potato Breeding—Problems and Perspectives*. Berlin: Parey, 1986.

Rouppe van der Voort, J. N. A. M. et al., An online catalogue of AFLP markers covering the potato genome, *Mol. Breed.*, 4, 73–77, 1998.

Rousselle-Bourgeois, F. and Rousselle, P., Amélioration génétique, In *La pomme de terre*, Rousselle, P., Robert, Y., and Crosnier, J. C., Eds., Paris: Institut National de la Recherche Agronomique (INRA), chap. 4, 1996.

Rovere, C. V. et al., Transgenic resistance in potato plants expressing potato leaf roll virus (PLRV) replicase gene sequences is RNA-mediated and suggests the involvement of post-transcriptional gene silencing, *Arch. Virol.*, 146, 1337–1353, 2001.

Sadanandam, A., Induced synaptic mutant from mesophyll cell protoclones of dihaploid *Solanum tuberosum*, *J. Plant Physiol.*, 138, 107–110, 1991.

Saghai Maroof, M. A. et al., Extraordinarily polymorphic microsatellite DNA in barley: species diversity, chromosomal locations, and population dynamics, *Proc. Nat. Acad. Sci. U.S.A.*, 91, 5466–5470, 1994.

Sanford, J. C. and Johnston, S. A., The concept of parasite-derived resistance—deriving resistance genes from the parasites own genome, *J. Theor. Biol.*, 113, 395–405, 1985.

Sanford, J. C. et al., Regarding the novelty and breeding value of protoplast-derived variants of Russet Burbank (*Solanum tuberosum* L.), *Euphytica*, 33, 709–715, 1984.

Santini, M. et al., Agronomic characterization of diploid hybrid families derived from crosses between haploids of the common potato and three wild Argentinian tuber-bearing species, *Amer. J. Potato Res.*, 77, 211–218, 2000.

Schwarzfischer, A. et al., Die erste Sorte aus fusionierten diploiden Kartoffeln? Erfolgreicher Einsatz der Protoplastenfusion in der praktischen Kartoffelzuechtung (Trans.: The first cultivar from fused diploid potatoes? Successful implementation of protoplast fusion in practical potato breeding), *Arbeitstagung der Vereinigung Öesterreichischer Pflanzenzüchter*, 49, 51–57, 1998.

Seabrook, J. E. A. and Douglass, L. K., Somatic embryogenesis on various potato tissues from a range of genotypes and ploidy levels, *Plant Cell Rep.*, 20, 175–182, 2001.

Seabrook, J. E. A. et al., Segregation for somatic embryogenesis on stem-internode explants from potato seedlings, *Plant Cell Tiss. Org. Cult.*, 65, 69–73, 2001.

Sebastiani, L. et al., Somaclonal variation for resistance to *Verticillium dahliae* in potato (*Solanum tuberosum* L.) plants regenerated from callus, *Euphytica*, 80, 5–11, 1994.

Secor, G. A. and Shepard, J. F., Variability of protoplast-derived potato clones, *Crop Sci.*, 21, 102–105, 1981.

Serrano, C. et al., Expression of the chicken lysozyme gene in potato enhances resistance to infection by *Erwinia carotovora* subsp. *atroseptica*, *Am. J. Potato Res.*, 77, 191–199, 2000.

Shepard, J. F. et al., Potato protoplasts in crop improvement, *Science*, 208, 17–24, 1980.

Sidorov, V. A. et al., Stable chloroplast transformation in potato: use of green fluorescent protein as a plastid marker, *Plant J.*, 19, 209–216, 1999.

Simko, I. et al., Similarity of QTLs detected for *in vitro* and greenhouse development of potato plants, *Mol. Breed.*, 5, 417–428, 1999.

Simko, I. et al., Mining data from potato pedigrees: tracking the origin of susceptibility and resistance to *Verticillium dahliae* in North American cultivars through molecular marker analysis, *Theor. Appl. Genet.*, 108, 225–230, 2004a.

Simko, I. et al., Linkage disequilibrium mapping of a *Verticillium dahliae* resistance quantitative trait locus in tetraploid potato (*Solanum tuberosum*) through a candidate gene approach, *Theor. Appl. Genet.*, 108, 217–224, 2004b.

Simmonds, N. W., Prospects of potato improvement, In *Scottish Society for Research in Plant Breeding 48th Annual Report*, Pentlandfield, Scotland: Scottish Plant Breeding Station, Pentlandfield, pp. 18–38, 1969.

Simmonds, N. W., Potatoes. *Solanum tuberosum* (Solanaceae), In *Evolution of Crop Plants*, Simmonds, N. W., Ed., London: Longman, pp. 279–283, 1976.

Simmonds, N. W., Potatoes. *Solanum tuberosum* (Solanaceae), In *Evolution of Crop Plants*, Smartt, J. and Simmonds, N. W., Eds., Singapore: Longman, chap. 93, 1995.

Simmonds, N. W., A review of potato propagation by means of seed, as distinct from clonal propagation by tubers, *Potato Res.*, 40, 191–214, 1997.

Smith, D. B., Variation in the electrophoretic band pattern of tuber proteins from somaclones of potato, *J. Agric. Sci.*, 106, 427–428, 1986.

Solomon-Blackburn, R. M. and Barker, H., Breeding virus resistant potatoes (*Solanum tuberosum*): a review of traditional and molecular approaches, *Heredity*, 86, 17–35, 2001.

Song, J. Q. et al., Construction of a bacterial artificial chromosome (BAC) library for potato molecular cytogenetics research, *Genome*, 43, 199–204, 2000.

Sonnino, A. et al., *In vitro* mutation breeding of potato, In *Nuclear Techniques and In Vitro* Culture for Plant Improvement, FAO/IAEA, Ed., Vienna: International Atomic Energy Agency, pp. 385–394, 1986.

Spillane, C. et al., Genetic engineering of the potato cultivar Glenroe for increased resistance to potato virus X (PVX), *Irish J. Agric. Food Res.*, 37, 173–182, 1998.

Spooner, D. M. and Bamberg, B. J., Potato genetic resources: sources of resistance and systematics, *Am. Potato J.*, 71, 325–338, 1994.

Spooner, D. M. and Hijmans, R. J., Potato systematics and germplasm collecting, 1989–2000, *Am. J. Potato Res.*, 78, 237–268, 2001.

Spooner, D. M. et al., *Wild potatoes (Solanum* section Petota; Solanaceae) of North and Central America. *Ann Arbor*, MI: American Society of Plant Taxonomists, 2004.

Struik, P. C. and Wiersema, S. G., Development of cultivars, In *Seed Potato Technology*, Struik, P. C. and Wiersema, S. G., Eds., Wageningen: Wageningen Pers, chap. 3, 1999a.

Struik, P. C. and Wiersema, S. G., True potato seed (TPS), In *Seed Potato Technology*, Struik, P. C. and Wiersema, S. G., Eds., Wageningen: Wageningen Pers, chap. 9, 1999b.

Sukhotu, T. et al., Nuclear and chloroplast DNA differentiation in Andean potatoes, *Genome*, 47, 46–56, 2004.

Swiezynski, K. M. and Zimnoch-Guzowska, E., Breeding potato cultivars with tubers resistant to *Phytophthora infestans*, *Potato Res.*, 44, 97–117, 2001.

Tai, G. C. C., Use of 2*n* gametes, In *Potato Genetics*, Bradshaw, J. E. and Mackay, G. R., Eds., Wallingford: CAB Int, chap. 5, 1994.

Tai, G. C. C. and Young, D. A., Performance and prediction of potato genotypes tested in international trials, *Euphytica*, 42, 275–284, 1989.

Tai, G. G. and Xiong, X., Ploidy manipulation—examination of gene action and method of gene mapping, In *Genetic Improvement of Solanaceous Crops. Vol. 1: Potato*, Razdan, M. K. and Mattoo, A. K., Eds., Enfield, NH: Science Publishers, pp. 143–164, 2005.

Tai, G. C. C. et al., Performance evaluations of varieties and selections in the northeastern region of North America, *Am. Potato J.*, 70, 685–698, 1993.

Tanksley, S. D. et al., High-density molecular linkage maps of the tomato and potato genomes, *Genetics*, 132, 1141–1160, 1992.

Tarn, T. R. and Tai, G. C. C., Tuberosum×Tuberosum and Tuberosum×Andigena potato hybrids: comparisons of families and parents, and breeding strategies for Andigena potatoes in long-day temperate climates, *Theor. Appl. Genet.*, 66, 87–91, 1983.

Tarn, T. R. et al., Breeding potatoes for long-day, temperate climates, *Plant Breed. Rev.*, 9, 217–332, 1992.

Taylor, R. J. et al., Tuber yield, soft rot resistance, bruising resistance and processing quality in a population of potato (cv. Crystal) somaclones, *Am. Potato J.*, 70, 117–130, 1993.

Thieme, R. and Griess, H., Somaclonal variation of haulm growth, earliness, and yield in potato, *Potato Res.*, 39, 355–365, 1996.

Thieme, T. and Thieme, R., Resistance to viruses, In *Genetic Improvement of Solanaceous Crops. Vol. 1: Potato*, Razdan, M. K. and Mattoo, A. K., Eds., Enfield, NH: Science Publishers, pp. 293–337, 2005.

Thieme, R. et al., Incorporation and testing of new genetic resources for virus resistance in potato, In *Potato, Global Research and Development*, Paul Khurana, S. M., Shekhawat, G. S., Singh, B. P., and Pandey, S. K., Eds., India: Indian Potato Association, pp. 271–278, 2000.

Thomas, P. E. et al., Extreme resistance to Potato leafroll virus in potato cv. Russet Burbank mediated by the viral replicase gene, *Virus Res.*, 71, 49–62, 2000.

Tizroutine, S. et al., Cytogenetic study of calli and regenerated clones in potato cv. Bf15 ($2x$) (H1), *Acta Bot. Gallica*, 141, 707–717, 1994.

Upadhya, M. D., Present and future research for true potato seed technology, In *Potato, Global Research and Development*, Paul Khurana, S. M., Shekhawat, G. S., Singh, B. P., and Pandey, S. K., Eds., Shimla, India: Indian Potato Association, Shimla, pp. 641–645, 2000.

Upadhya, M. D. et al., Mutation induction and isolation in potato through true seed and tuber mutagenesis and use of tissue culture, In *Induced Mutations in Vegetatively Propagated Plants II*, FAO/IAEA, Ed., Vienna: International Atomic Energy Agency, pp. 185–198, 1982.

Upadhya, M. D. et al., *Production and utilization of true potato seed in Asia*. Lima, Peru: International Potato Center, 1996.

Valkonen, J. P. T. et al., Resistance to viruses in F_1 hybrids produced by direct crossing between diploid *Solanum* series *Tuberosa* and diploid *S. brevidens* (series *Etuberosa*) using *S. phureja* for rescue pollination, *Plant Breed.*, 114, 421–426, 1995.

Van Eck, H. J. et al., The inheritance of anthocyanin pigmentation in potato (*Solanum tuberosum* L.) and mapping of tuber skin colour loci using RFLPs, *Heredity*, 73, 410–421, 1994a.

Van Eck, H. J. et al., Multiple alleles for tuber shape in diploid potato detected by qualitative and quantitative genetic analysis using RFLPs, *Genetics*, 137, 303–309, 1994b.

van Swaaij, A. C. et al., Selection, characterization and regeneration of hydroxyproline-resistant cell lines of *Solanum tuberosum*: tolerance to NaCl and freezing stress, *Physiol. Plant.*, 68, 359–366, 1986.

van Swaaij, A. C. et al., Increased frost tolerance and amino acid content in leaves, tubers and leaf callus of regenerated hydroxyproline resistant potato clones, *Euphytica*, 36, 369–380, 1987.

Vazquez, A. M., Insight into somaclonal variation, *Plant Biosyst.*, 135, 57–62, 2001.

Veilleux, R., Diploid and polyploid gametes in crop plants: mechanisms of formation and utilization in plant breeding, *Plant Breed. Rev.*, 3, 253–288, 1985.

Veilleux, R. E., Cell and tissue culture of potato, In *Genetic Improvement of Solanaceous Crops. Vol. 1: Potato*, Razdan, M. K. and Mattoo, A. K., Eds., Enfield, NH: Science Publishers, pp. 185–208, 2005.

Veilleux, R. et al., Potato germplasm development for warm climates: genetic enhancement of tolerance to heat stress, *Euphytica*, 98, 83–92, 1997.

Ward, A. C. W. et al., Flow cytometric assessments of ploidy stability in cryopreserved dihaploid *Solanum tuberosum* and wild *Solanum* species, *Cryo Letters*, 14, 145–152, 1993.

Watanabe, J. A. and Watanabe, K. N., Pest resistance traits controlled by quantitative loci and molecular breeding strategies in tuber-bearing *Solanum*, *Plant Biotechnol.*, 17, 1–16, 2000.

Watanabe, J. A. et al., Diploid potato germplasm derived from wild and land race genetic resources, *Am. Potato J.*, 71, 599–604, 1994.

Watanabe, K. N. et al., Overcoming crossing barriers between non tuber-bearing and tuber-bearing *Solanum* species: towards potato germplasm enhancement with a broad spectrum of solanaceous genetic resources, *Genome*, 38, 27–35, 1995.

Watanabe, J. A. et al., Frequency of potato genotypes with multiple quantitative pest resistance traits in $4x \times 2x$ crosses, *Breed. Sci.*, 49, 53–61, 1999.

Watanabe, K. N. et al., Breeding potential and transmission of traits in $4x$–$2x$ crosses, In *Genetic Improvement of Solanaceous Crops. Vol. 1: Potato*, Razdan, M. K. and Mattoo, A. K., Eds., Enfield, NH: Science Publishers, pp. 83–100, 2005.

Wersuhn, G. et al., Influence of regeneration and selection procedures on the production of aluminum tolerant potato regenerants, *Potato Res.*, 37, 423–428, 1994.

Wheeler, V. A. et al., Shoot formation from explant cultures of fourteen potato cultivars and studies of the cytology and morphology of regenerated plants, *Ann. Bot.*, 55, 309–320, 1985.

Williams, C. E. et al., RFLP analysis of chromosomal segregation in progeny from an interspecific hexaploid somatic hybrid between *Solanum brevidens* and *Solanum tuberosum*, *Genetics*, 135, 1167–1173, 1993.

Wolters, A. M. A. et al., Mitotic and meiotic irregularities in somatic hybrids of *Lycopersicon esculentum* and *Solanum tuberosum*, *Genome*, 37, 726–735, 1994.

Wolters, A. M. A. et al., Fluorescence in situ hybridization on extended DNA fibres as a tool to analyse complex T-DNA loci in potato, *Plant J.*, 13, 837–847, 1998.

Xiong, M. M. and Guo, S. W., Fine-scale genetic mapping based on linkage disequilibrium: Theory and applications, *Am. J. Hum.Genet.*, 60, 1513–1531, 1997.

Yeo, E. T. et al., Genetic engineering of drought resistant potato plants by introduction of the trehalose-6-phosphate synthase (TPS1) gene from *Saccharomyces cerevisiae*, *Mol. Cells*, 10, 263–268, 2000.

Zhen, W. et al., Enhanced late blight resistance of transgenic potato expressing glucose oxidase under the control of pathogen-inducible promoter, *Chin. Sci. Bull.*, 45, 1982–1986, 2000.

Zimnoch-Guzowska, E. et al., QTL analysis of new sources of resistance to *Erwinia carotovora ssp atroseptica* in potato done by AFLP, RFLP, and resistance-gene-like markers, *Crop Sci.*, 40, 1156–1167, 2000.

Zimnoch-Guzowska, E. et al., Resistance to bacterial pathogens, In *Genetic Improvement of Solanaceous Crops. Vol. 1: Potato*, Razdan, M. K. and Mattoo, A. K., Eds., Enfield, NH: Science Publishers, pp. 339–395, 2005.

Zingstra, H., *Vijftig jaar bevordering van het aardappelkweken en het onderzoek van aardappelrassen* (Trans,: Fifty years of promoting potato breeding and research of potato cultivars). Commission for the advancement of breeding and research of new potato cultivars (C.O.A.), Wageningen, 1983.

CHAPTER 3

Tomato

Yuanfu Ji and J. W. Scott

CONTENTS

3.1 INTRODUCTION

Tomato (*Lycopersicon esculentum* Mill) is an economically important vegetable that is widely used for both basic and applied research. Numerous advantageous features render tomato as a favorable model species for plant research in areas of classical genetics, cytogenetics, molecular genetics, and molecular biology. Its self-pollinating nature, easy crossability to the wild species, simple genome, and vast genetic variation render tomato an ideal species for classic genetic studies. The 12 chromosomes in the tomato genome can be readily identified through analyses of pachytene karyotype, synaptonemal complexes, and chromosome or chromosome arm-specific DNA sequences, making it an excellent species for cytogenetic research. Moreover, its small genome size (~ 950 Mb), various high-density molecular linkage maps, and numerous genomic databases and DNA libraries—such as expressed sequence tags and bacterial artificial chromosomes—make tomato a model system for molecular genetic and genomic studies. In addition, the ease of cell culture and genetic transformation by *Agrobacterium*-based vectors render tomato an excellent species for genetic engineering and molecular biology studies. Because of these outstanding features and its economic importance, tomato has been chosen by the recently initiated International Solanaceae Genome Project (SOL) as the Solanaceae model species for genome sequencing. The availability of whole genome sequences in the next decade will likely lead to a better understanding of plant adaptation and diversification. The utilization of potential natural variation discovered from the SOL project will accelerate tomato improvement, and thus improve the health and wellbeing of humans in a more environmentally friendly and sustainable manner.

3.2 TAXONOMY

3.2.1 Introduction

The tomato and its close wild relatives are believed to have originated in the mountainous regions of the Andes and the Galapagos Islands (Villand et al. 1998). The cultivated tomato became domesticated in the early civilizations of Mexico and then disseminated to the other parts of the world (Rick 1958). Alternative scientific names for tomato have appeared in the literature, such as *Solanum lycopersicum* L., *Lycopersicon lycopersicum* (L.) Karsten, or *Lycopersicon esculentum* (Broome et al. 1983). The last one was first proposed in the middle of the eighteenth century and has become the most frequently used (Taylor 1986).

3.2.2 Early Taxonomic Status of Tomato

The cultivated tomato, *Lycopersicon esculentum* Mill., belongs to the genus *Lycopersicon*, a relatively small genus within the extremely diverse and large Solanaceae family. This family consists of about 90 genera and is divided into two sub-families, the *Solanoideae* and the *Cestroideae*, based largely on different patterns of embryo development (D'Arcy 1979). All species in the genus *Lycopersicon* belong to the *Solanoideae* sub-family and have the same chromosome number ($2n = 2x = 24$).

Early studies divided the genus *Lycopersicon* into two sub-genera, *Eulycopersicon* (true *Lycopersicon*) and *Eriopersicon* (woolly *Persicon*; Muller 1940; Villand et al. 1998). The former includes the cultivated tomato (*L. esculentum*) and *L. pimpinellifolium* that bear glabrous and red- to orange-colored fruits; the latter includes four species (*L. peruvianum, L. cheesmaniae, L. hirsutum,* and *L. glandulosum*) that bear green fruits.

3.2.3 Crossing Relationships

The aforementioned early taxonomic treatments have become inadequate as the number of species and races collected from South America has increased (Rick 1971, 1991; Taylor 1986). Furthermore, this system of division is based on fruit color and fails to explore the fundamental differences among the species. A more meaningful classification is based on crossing relationships that divide the species into two complexes: the *esculentum* complex and the *peruvianum* complex (Rick et al. 1979b). The *peruvianum* group contains two species: *L. chilense* and *L. peruvianum*. The *esculentum* group is composed of the cultivated tomato and the other six *Lycopersicon* species (*L. cheesmannii, L. chmielewskii, L. hirsutum, L. parviflorum, L. pennellii,* and *L. pimpinellifolium*), that can be crossed easily to the cultivated tomato. Most sexual crosses within two groups can usually be achieved without difficulty, although they are often unilateral. However, crosses among members of different groups are difficult and the barrier can only be broken by embryo rescue (Rick et al. 1979b; Figure 3.1).

All the wild species of tomato in the genus *Lycopersicon* are diploid, have the same number of chromosomes as *L. esculentum* ($2n = 2x = 24$), and can be crossed to the latter—albeit with varying degrees of difficulty. Members of the *peruvianum* complex only hybridize to the cultivated tomato with great difficulty. *L. chilense* flowers have strongly exerted stigmas and are self-incompatible. Severe barriers prevent intercrossing and separate *L. chilense* from the cultivated tomato. The pistils of the *L. chilense* do not accept *L. esculentum* pollen and the flowers rapidly abscise. The reciprocal cross results in fruit development with some abnormal ovules with embryos. Embryos of sufficient size can be rescued with embryo culture, albeit with a low success rate (Rick and Lamm 1955; Scott et al. 1996). *L. peruvianum* generally shows even more severe barriers to intercrossing with the cultivated tomato than *L. chilense*; however, the barriers can be overcome

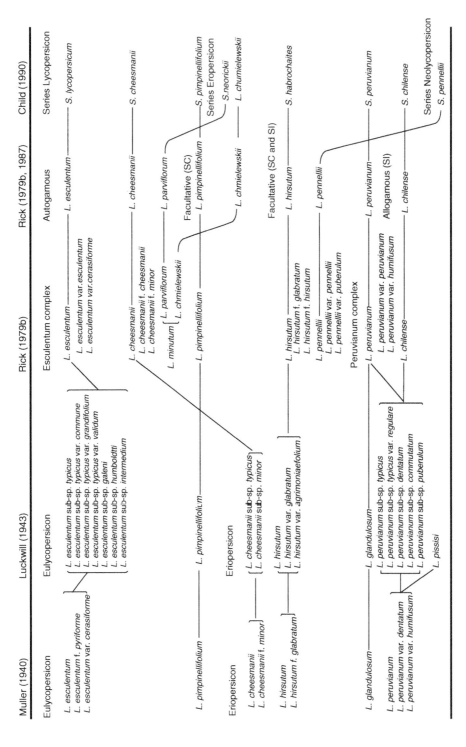

Figure 3.1   Comparison of various taxonomic treatments of genus *Lycopersicon*. The lines connect synonymous taxa. SC = self-compatible and SI = self-incompatible. *L. minutum* was classified into two species, *L. parviflorum* and *L. chmielewskii* by Rick, C. M., et al., *Theor. Appl. Genet.*, 47, 55–68, 1976b. (Adapted from Warnock, S. J., *HortScience*, 23, 669–673, 1988; Peralta, I. E., and Spooner, D. M., *Am. J. Bot.*, 88, 1888–1902, 2001.)

with several different approaches, such as using a series of bridge lines (Taylor and Al-Kummer 1982). Within the *peruvianum* complex, various barriers have been found to limit the intercrossing compatibility between the two member species (Rick and Lamm 1955; Rick et al. 1979b; Taylor and Al-Kummer 1982), though the barriers could be overcome via several bridging accessions. By contrast, taxa within the *esculentum* complex are more intercrossable and hybrids can be obtained in nearly all combinations (Rick et al. 1979b). Essentially, all the germplasm in the *Lycopersicon* genus is potentially transferable to the cultivated tomato. Introgression of many valuable genes of economic importance from the wild species, such as major disease resistance genes, has lead to the development of many superior tomato cultivars (Rick and Yoder 1988).

3.2.4 Mating Systems

Based on their mating systems, the nine *Lycopersicon* species can be classified into four categories (Rick et al. 1979b; Rick 1987): autogamous, including *L. esculentum*, *L. cheesmanii*, and *L. parviflorum*); facultative self-compatible, including *L. pimpinellifolium* and *L. chmielewskii*); facultative self-compatible and self-incompatible, including *L. hirsutum*, *L. pennellii*, and *L. peruvianum*; and allogamous self-incompatible, including *L. chilense*. The great diversity of mating systems in the *Lycopersicon* renders this genus a great system for studies on evolution and speciation.

3.2.5 A Recent Taxonomic Treatment

In a recent taxonomic treatment, Child (1990) placed tomato and its wild relatives under *Solanum* subgenus *Potatoe* (G. Don) D'Arcy, section *Lycopersicum*, subsection *Lycopersicon*, and segregated them into three series: *Lycopersicon*, including *S. lycopersicum*, *S. pimpinellifolium*, and *S. cheesmaniae*; *Neolycopersicon*, including *S. pennellii*; and *Eriopersicon*, including *S. chilense*, *S. peruvianum*, *S. chmielewskii*, *S. neorikii*, and *S. habrochaites*. In the new treatment, *S. neorickii* and *S. habrochaites* replace *L. chmielewskii* and *L. hirsutum*, respectively, as a change from the previous taxonomic treatment (Rick et al. 1979b). Phylogenetic studies on chloroplast DNA restriction sites, sequence data, and nuclear granule-bound starch synthase (GBSSI) sequence data (see discussion below for detail)—which place tomato and its wild relatives as a sister clade like potatoes—supports this taxonomic treatment. Although the majority of taxonomists are currently adopting *Solanum* as the generic name for tomatoes, the new taxonomic treatment has not been widely accepted among tomato breeders and researchers in applied fields due to various considerations. Thus, we maintain use of the genus name *Lycopersicon* in the present review.

3.3 GERMPLASM RESOURCES AND GENETIC VARIATION

3.3.1 Germplasm Resources

Wide-ranging germplasm resources are available in the cultivated tomato and related species, including genetic mutants, wild species, and other miscellaneous genetic or cytogenetic stocks. Germplasm resources in the United States are currently maintained at two major gene banks: the C.M. Rick Tomato Genetics Resources Center (TGRC), University of California at Davis, and the USDA's Plant Genetic Resources Unit (PGRU) in Geneva, New York. The TGRC maintains a collection of over 1100 wild species accessions representing nine *Lycopersicon* species and four closely related *Solanum* species (Chetelat and Peterson 2003; Chetelat 2004, 2005; Chetelat and Ji 2006). In addition, over 1000 monogenic mutants are preserved at the TGRC, including spontaneous and induced mutations affecting many aspects of plant development and morphology, disease resistance genes, and protein marker stocks. The remainder of the collection contains over 1400 miscellaneous genetic or cytogenetic stocks, including linkage testers, trisomics, tetraploids, translocations, landraces, and derivatives of wild species such as alien addition, substitution, and introgression lines. The PGRU contains ∼5800 accessions with ∼90% being *L. esculentum* and the rest wild species.* In addition to the two major gene banks in the United States, tomato germplasm are also preserved in other countries.** Germplasm collections in those gene banks have been extensively utilized as genetic resources for basic and applied research. The most comprehensive application is searching the germplasm collections for traits of economic interest, such as disease and insect resistances, abiotic stress tolerance, and improved horticultural and fruit characteristics (Rick and Chetelat 1995; Foolad and Chen 1998; Foolad et al. 2003; Kennedy 2003; Frelichowski and Juvik 2005; Ji and Scott 2005, 2006). Topic of a more fundamental nature include a wide range of studies in physiology and development, such as: leaf development, fruit ripening, and self-incompatibility (Chetelat and Deverna 1991; Azanza et al. 1995; Paran et al. 1997); genetics such as wide hybridization, linkage, and QTL mapping (Koorneef et al. 1993; Azanza et al. 1995; Imanishi et al. 1996); and genomics such as map-based cloning and comparative sequence analysis (Martin et al. 1993b; Pertuze et al. 2002).

3.3.2 Genetic Variation

Variation in the cultivated tomato can be classified into two general categories—genetic and chromosomal variations. Genetic variation is characterized by allelic differences, including spontaneous and induced mutations, whereas chromosome variation is characterized by changes in whole chromosomes or chromosome structures.

3.3.2.1 Mutation

Mutations constitute an important resource for tomato breeding and genetic research; they also affect many aspects of plant growth, such as morphology, disease resistance, and other traits of economic importance. The earliest mutants were mainly of spontaneous origin and this continues to be an important source of mutation. The mutant of greatest impact on the tomato industry in the early twentieth century is the *sp* mutant (determinate growth habit). Introduction of the recessive *sp* gene into tomato cultivars revolutionized tomato production because the determinate growth habit facilitates more concentrated fruit set and mechanical harvesting (Atherton and Harris 1986). A survey of cytogenetic causes of unfruitness in the tomato led to the identification and characterization of many genetic variants including male-sterile mutants and aneuploids (Rick 1945). With the introduction of mutagenic methods, such as treatment of seeds or pollen with ethylmethane sulfonate (EMS) or radiation (primarily x-rays and fast neutrons), and transposon tagging using the maize *Ac/Ds* elements, many more mutants have been generated (Meissner et al. 1997, 2000; Stubble 1972a, 1972b), resulting in a great expansion of the list of mutants. Currently, ∼1200 monogenic mutants at 1000 loci have been characterized, of which ∼400 have been mapped/ assigned to the chromosomes (Chetelat 2002; Chetelat and Ji 2006).

3.3.2.2 Chromosome Variation

Chromosome variation consists of either changes in whole chromosomes or changes in chromosome structure. The former group includes euploids (changes in whole sets of chromosomes) and aneuploids (changes within sets). The most important deviant euploids reported in tomato include haploids, triploids, and tetraploids (Rick 1945). Due to severe meiotic irregularity, haploids rarely produce

* (http://www.ars-grin.gov/ars/NoAtlantic/Geneva/lycopersiconmain.html)
** (see links at the TGRC web site)

viable gametes, and therefore do not set seeds. In contrast, both triploids and tetraploids are relatively fertile. Autotriploids are the most common unbalanced euploids that arise spontaneously. They can also be generated from crosses between a diploid and an autotetraploid at a low frequency (Mark 1966). Interspecific triploids or sesequidiploids between tomato and the wild species usually result in improved compatibility and seed yield (Rick et al. 1986). Although somewhat fertile, triploids produce unbalanced gametes due to meiotic irregularity; they are therefore genetically unstable and can only be maintained clonally. In contrast, tetraploids can be reproduced by seed, although subject to aneuploid variation due to abnormal meiosis. They have been the subject of many studies with respect to their morphology, physiology, sterility, and chromosome behavior (Rick and Butler 1956).

Aneuploid variation of the tomato at the diploid level consist of deletions (hapoploids) or additions (hyperploids) of whole chromosomes. Hapoploids, such as monosomics of tomato, have limited use in tomato research due to zero transmission of the monosomic condition (Gill 1983). Similarly, hyperploids have found very limited utilization in both basic and applied research, except for generating trisomics that have been the most useful aneuploids in tomato. A complete set of primary trisomics have been derived from several backgrounds; modified trisomics of secondary, tertiary, compensating, and telotrisomic types have also been generated (Khush and Rick 1968b). All types of trisomics are maintained via seed in the TGRC genebank and utilized for gene mapping and other cytogenetic studies. For example, the assignment of certain linkage groups of morphological markers to their respective chromosomes is based on distortion in segregation ratio in the F_2 populations of primary trisomics (Lesley 1932; Rick and Barton 1954). Other trisomic types can be used in finding information regarding arm location, position of centromeres, and orientation of linkage markers (Khush and Rick 1967a, 1967b; Khush and Rick 1968a). In addition, numerous trisomics have been used for chromosomal assignment of isozyme markers and RFLP markers by dosage analysis (Fobes 1980; Young et al. 1987), and for identification of individual chromosomes in synaptonemal complex spreads (Sherman and Stack 1992).

Another category of chromosome variation involves the addition of an alien chromosome to the tomato genome (monosomic addition) or the replacement of one tomato chromosome by its alien homeologue (heterozygous substitution). A complete set of monosomic additions, each containing one of the 12 S. lycopersicoides chromosomes in the background of tomato genome ($2n+1$), were identified from the progeny of the sesequidiploids (two genomes of L. esculentum and one genome of S. lycopersicoides) (Chetelat et al. 1998). Transmission rates and fertility are generally higher through the female than the male, and vary widely among the addition lines. For example, transmission for chromosome 10 was 24%, but 0% for chromosome 6 (Chetelat et al. 1998). Monosomic additions have been used for chromosomal assignment of genes and genetic markers, construction of chromosome specific libraries, investigation of gene expression of alien chromosomes in the genetic background of a wild relative, and studies on pairing and recombination of homeologous chromosomes (Ji and Chetelat 2003; Chang and de Jong 2005). Numerous heterozygous substitution lines, each containing a single S. lycopersicoides chromosome in replace of one of its tomato homeologues, were derived from the backcross progeny of L. esculentum × S. lycopersicoides (Chetelat et al. 2000; Ji and Chetelat 2003). In addition, a few monosomic additions and substitutions were also developed for the S. sitiens genome (Pertuze et al. 2002).

Changes in chromosome structures constitute another major type of chromosomal variation in tomatoes. Various kinds of structural alterations—including duplications, deficiencies, inversions, and translocations—have been identified. But the only transmissible and useful variants are reciprocal translocations. They have found numerous applications, such as assignment of gene loci to their respective chromosome arms and service as an important source of tertiary trisomics (Khush and Rick 1967a). A tester set of eight translocations involving all 12 tomato chromosomes have also been induced by chemical or x-ray treatment of seed (Gill 1983). Those translocation testers were used to identify other unknown translocations. A large set of induced tomato deficiencies were identified by the pollination of recessive marker stocks using the pseudo-dominant technique (Khush and Rick 1968b). These deficiencies permit more precise location of gene loci within chromosome arms, as well as mapping of the centromeres and establishment of the orientations of the linkage maps. However, deficiencies cannot be transmitted to the next generation because gametes with deficiencies are not viable in tomato. Therefore, they cannot be maintained in the genebanks and must be regenerated for each experiment.

Although genetic variation due to changes in chromosome structure within the tomato genome has found limited use in both basic and applied research, variation arising from the introgression of chromosome segments of the wild species into the tomato genome have found wide application. The vast amount of genetic variation in the wild species and their utilization in tomato research and improvement justify a separate section for discussion.

3.3.2.3 *Variation within the Wild Species*

The domestication of the tomato and the early history of improvement in Europe and North America greatly reduced genetic variability. The combination of natural self-pollination, reproduction in small populations, and the selection process further depleted variation within the cultivated tomato, as measured by allozymes (Rick and Fobes 1975) and RFLPs (Helentjaris et al. 1985). In contrast, the wild species possess much more genetic diversity which provides resources for improvement of many economically important traits, such as disease or insect resistance, stress or heat tolerance, and horticultural and fruit quality features (Rick 1987). Studies on evolution and speciation of *Lycopersicon* using morphological and allozyme markers indicated that cross-pollinating or allogamous species usually possess much higher variation than self-compatible species (Rick 1984; Rick and Fobes 1975; Rick and Tanksley 1983). This finding was later confirmed by many other studies on the phylogeny of *Lycopersicon* using a variety of DNA sequences (Palmer and Zamir 1982; McClean and Hanson 1986; Marshall et al. 2001; Peralta and Spooner 2001; Stadler et al. 2005), or molecular markers (Miller and Tanksley 1990; Williams and St. Clair 1993; Egashira et al. 2000; Alvarez et al. 2001; Tikunov et al. 2003; Nuez et al. 2004). In fact, more variation was found within a single accession of the self-incompatible species *L. peruvianum* than all accessions of the self-compatible species *L. esculentum* (Miller and Tanksley 1990; Stadler et al. 2005). The estimated level of DNA polymorphism detected with molecular markers within and among species was highly correlated with the respective mating system: more within-species variation in the cross-pollinating species than the self-compatible species (Alvarez et al. 2001; Stephan and Langley 1998). Similarly, higher levels of DNA polymorphism were found in molecular linkage maps based on interspecific crosses between *L. esculentum* and other cross-pollinating species than those based on crosses between two self-pollinating species (Chen and Foolad 1999). Similar findings were also obtained from gene sequence studies. The granule-bound starch synthase (GBSSI) gene sequences presented very low variation in the autogamous species, compared with the considerable diversity found in all allogamous species (Peralta and Spooner 2001). Due to the difficulty in finding polymorphisms among autogamous species, more variable molecular markers, such as microsatellites and single nucleotide polymorphisms (SNP), might be used to study their relationships (Alvarez et al. 2001; Tikunov et al. 2003; W. C. Yang et al. 2004).

3.3.2.4 *Introgression*

Introgression lines (ILs) that contain chromosome segments of wild species in the background of the related cultivated species increase genetic diversity of crop plants without losing the desirable elite phenotype and thus are great resources for crop improvement (Zamir 2001). ILs are identical for the entire genome except for a single introduced region, and thus variation in the ILs is due to the introgressed segment. The nearly isogenic nature of ILs offers great advantages in genetic studies such as QTL mapping and gene identification (Zamir and Eshed 1998; Zamir 2001; Gur and Zamir 2004). To enhance the progress of tomato breeding, Zamir and colleagues have produced a population of chromosomal introgression lines that comprised marker-defined regions of *L. pennellii* accession

LA716 in the genetic background of the cultivated tomato (Eshed and Zamir 1995). The entire genome of *L. pennellii* is covered with 76 ILs each carrying a single homozygous introgressed segment. Since their creation, those ILs have been extensively utilized in tomato research (Fridman et al. 2004; Liu et al. 2003; Paran and Zamir 2003). Similarly, a series of alien ILs has also been generated for *L. hirsutum* accession LA1777, which covers ~85% of the *L. hirsutum* genome (Monforte and Tanksley 2000). Introgression libraries are also being developed for other taxa, such as *L. chmielewskii** and *S. lycopersicoides* (Chetelat and Meglic 2000; Canady et al. 2005). Although ILs have proven useful for genome mapping and crop improvement, the mapping resolution is somewhat limited because each line contains a relatively large introgression. As an alternate approach, numerous types of mapping populations have been used in QTL mapping projects to improve mapping resolution. Recombination inbred lines (RILs) derived from intraspecific or interspecific crosses are useful for high-resolution QTL mapping purposes (Paran et al. 1995, 1997; Lecomte et al. 2004). A QTL analysis strategy proposed by Tanksley and Nelson (1996) integrates the processes of QTL mapping and variety development, and has found wide application in tomato studies using interspecific advanced backcross breeding populations in tomato involving *L. pimpinellifolium* (Tanksley and Nelson 1996), *L. peruvianum* (Fulton et al. 1997b), *L. hirsutum* (Bernacchi et al. 1998a, 1998b), *L. parvifolium* (Fulton et al. 2000), and *L. pennellii* (Frary et al. 2004). As another major objective, advanced backcross QTL analysis has been used to create near-isogenic lines (NILs) carrying a specific targeted region of a donor parent. NILs can then be used to fine-map QTLs, eliminate undesirable effects caused by linkage drag, and perform positional cloning of the QTLs (Alpert and Tanksley 1996; Bernacchi et al. 1998b; Monforte et al. 2001). Another type of unbalanced population that has been widely used in tomato studies is inbred backcross lines (IBLs), which combine desirable features of both backcross and single-seed descent breeding methods (Coaker et al. 2002; Doganlar et al. 2002; Kohler and St. Clair 2005).

3.4 CYTOGENETICS

3.4.1 Introduction

Tomato cytogenetics research made great progress around the middle of the twentieth century and is summarized by Rick and Butler (1956) and later by Gill (1983). During this period, a vast amount of genetic variation including mutants and various types of chromosome modifications were investigated and characterized. Tomato chromosomes and their direct application to tomato genetics and breeding is summarized by Quiros (1991). A more recent review by Chetelat and Ji (2006) summarized the most recent research achievements on tomato cytogenetics, including topics on classical and molecular linkage maps, centromere mapping, meiosis in wide hybridization, genome analysis using genomic *in situ* hybridization (GISH) and fluorescence *in situ* hybridization (FISH). The current review will focus on the study of genome relationships and chromosome mapping by cytogenetics with a brief discussion on the current situation for genetic linkage maps. The readers are advised to see the above-mentioned reviews for cytogenetic information not covered here.

3.4.2 Genome Relationship

All nine species in the genus *Lycopersicon* and four closely related *Solanum* species are diploid ($2n = 2x = 24$). Therefore, they share many common morphological and genetic features. Many studies have been carried out to elucidate the genomic relationships among species based on morphology and molecular investigations.

* (http://www.keygene.com/services/plants/services_plants_line.htm)

The phylogenetic studies of restriction-enzyme fragments on chloroplast DNA (cpDNA) of seven species of *Lycopersicon* and three closely related *Solanum* species demonstrated that the red-fruited self-compatible species (*L. esculentum*, *L. cheesmanii*, and *L. pimpinellifolium*) formed a monophyletic group (Palmer and Zamir 1982). This result was also supported by other studies using RFLP markers, microsatellites, and gene sequences (Alvarez et al. 2001; Marshall et al. 2001; Miller and Tanksley 1990; Peralta and Spooner 2001). Two green-fruited autogamous sibling species (*L. parviflorum* and *L. chmielewskii*) sharing many common genes (Breto et al. 1993) displayed close relationships to each other and to the red-fruited species. They usually form the same cluster as the red-fruited species (Egashira et al. 2000; Marshall et al. 2001; Peralta and Spooner 2001). As noted early by Chmielewski and Rick (1962), and Rick et al. (1976), the green-fruited *L. parviflorum* and *L. chmielewskii* are closer to the *esculentum* complex than the *peruvianum* complex. A close relationship was also found between *L. hirsutum* and *L. pennellii* (Palmer and Zamir 1982; Miller and Tanksley 1990; Marshall et al. 2001). These two self-incompatible species tend to form their own cluster, suggesting they have a different genetic background from other *Lycopersicon* species. In fact, *L. pennellii* was originally placed in the *Solanum* genus (Correll 1958), but transferred to the *Lycopersicon* genus later because of its morphological similarities and cross-compatibility with the latter (Rick 1960; D'Arcy 1979). Varied results have been reported regarding the relationships between *L. peruvianum* with other *Lycopersicon* species due to the vast amount of variation found within the accessions of *L. peruvianum* (Marshall et al. 2001). Similarly, large variation has also been observed within *L. chilense* accessions, albeit to a lesser degree (Rick et al. 1979a; Rick and Tanksley 1981; Rick 1984). Molecular systematic studies on the rDNA sequences suggested a close relationship between *L. peruvianum* and *L. chilense* (Marshall et al. 2001), supporting previous phylogenic studies on *Lycopersicon* species using RAPD analysis (Egashira et al. 2000). In an early study using chloroplast DNA, Palmer and Zamir (1982) observed that variation within accessions of *L. peruvianum* encompassed variation shown by *L. chilense* and suggested relegation of the latter to a subspecies within the *L. peruvianum* complex. In fact, *L. chilense* was treated as a synonym of *L. peruvianum* var. *denatum* in early taxonomic studies (Muller 1940) and was not regarded as a separate species until 1955 when Rick and Lamm (1955) showed that *L. chilense* was morphologically, reproductively, and geographically isolated from *L. peruvianum*.

Numerous studies have demonstrated that the genus *Solanum* is most closely related to *Lycopersicon* (Rick et al. 1979b; Olmstead and Palmer 1992, 1997; Tanksley et al. 1992). Especially, four species of *Solanum* (*S. juglandifolium* and *S. ochranthum*, *S. lycopersicoides* and *S. sitiens*) display a close resemblance to *Lycopersicon* species in many aspects including chromosome number and morphology (Rick et al. 1979b; Rick 1988). A higher pairing affinity was found for the *S. lycopersicoides* chromosomes to their *L. esculentum* homoeologues than those of the *S. sitiens*, suggesting the former might share a higher degree of sequence homology with *L. esculentum* (Ji et al. 2004). Evidence from crossing relationships and morphological affinities also suggests *S. lycopersicoides* is most closely related to *Lycopersicon* (Rick et al. 1979b, 1986). Supporting evidence can also be derived from phylogenic studies of the chloroplast DNA, rDNA, and other gene sequences (Palmer and Zamir 1982; Marshall et al. 2001; Peralta and Spooner 2001), in which the results demonstrate *S. juglandifolium* and *S. ochranthum* are in the closest outgroup to tomatoes, with *S. lycopersicoides* and *S. sitiens* as basal to these taxa.

3.4.3 Chromosome Mapping by Cytogenetics

3.4.3.1 *Cytological Identification of Tomato Chromosomes*

Individual chromosomes or chromosome arms can be identified by karyotype analysis of chromosome spreads prepared from pollen mother cells (PMC) or somatic root-tip cells. This analysis usually involves characterization of numerous cytogenetic landmarks such as centromere, heterochromatin regions, telomeres, and chromosome size. PMC spreads at the pachytene stage are ideal for tomato chromosome identification because pachytene chromosomes are usually 7–10 times longer than their mitotic metaphase counterparts and thus offer higher resolution of the cytological landmarks (Ramanna and Prakken 1967). The earliest cytological studies on tomato chromosomes presented a general picture of pachytene chromosome morphology (Lesley and Lesley 1935). It was not until the middle of the twentieth century that Brown (1949) performed a thorough investigation on tomato pachytene chromosomes and identified two chromosomes of the set. Later, Barton (1950) and Gottschalk (1951) were able to distinguish each of the 12 chromosomes of the tomato complement based on overall length, chromatic regions, centromere positions, and heterochromatic knobs. Barton (1950) identified chromosomes by number according to their lengths—number 1 being the longest, number 2 possessing the nucleolus organizer region (NOR), and chromosome 12 being the smallest. This system of numbering has been adopted as the standard for the tomato complement and for the ultimate designation of linkage groups (Barton et al. 1955). A handy key for identification of tomato pachytene chromosomes was created by Khush (1963), based on numerous criteria, including the presence of NOR, centromere position, cluster of heterochromatin regions on the short arms, proportion of chromatic and achromatic regions, and presence of specific knobs. More recently, karyotype analysis of tomato nuclei at the pachytene stage identified all 12 of its synaptonemal complexes (SCs) on the basis of relative lengths and arm ratios (Sherman and Stack 1992, 1995). The SC karyotype allows more accurate estimate of the size and chromatin composition of the tomato genome (Peterson et al. 1996, 1999). Additionally, description of recombination nodule (RN) distribution and frequency on SCs avoids many problems of chiasma maps (Sherman and Stack 1992, 1995).

A variety of computer software applications have been developed over the past several decades to facilitate and improve the accuracy of quantitative karyotype analysis. Gilbert (1966) was among the first to try to automate the identification of individual human chromosomes in a karyotype based on numerous parameters such as relative length and arm ratio. Computer programs developed subsequently allow various degrees of automation in chromosome measurement and identification (Green et al. 1980, 1984; McGurk and Rivlin 1983; Armstrong et al. 1987; Oud et al. 1987). However, these earlier programs generally require specific computer hardware configurations and are only applicable to limited types of applications. More fully automated computer-driven systems have also been developed, which have dedicated image processing systems for karyotype analysis (Fukui 1986, 2005; Fukui et al. 1998). Although chromosome imaging systems have been successfully used in plants with small and difficult-to-distinguish somatic chromosomes, they are not suitable for application requiring more complex analysis such as pachytene chromosomes or SCs (Peterson et al. 1999). A more recently developed image analysis software system MicroMeasures overcomes such limitations and offers several advantages in karyotype analysis (Reeves 2001). MicroMeasures incorporates routines for localizing loci and other point-like features and extends the ability to mark and measure multiple types of chromosome segments. Unlike similar applications, MicroMeasures may be adapted individually by the end users to suit their particular interests. Compatibility with Microsoft Excel provides users the ability to quantitatively and statistically analyze the collection of cytogenetic data. Lastly, MicroMeasure's compatibility with personal computers eliminates the need for more costly and complicated software and hardware.

3.4.3.2 Trisomic Analysis

Tomato primary trisomics were the main resource for assignment of linkage groups of morphological markers to their respective chromosomes in early research (Lesley 1932; Rick and Barton 1954). More recently, primary trisomics have been exploited for linkage maps of isozyme loci (Fobes 1980; Tanksley and Rick 1980) or chromosomal assignment of multiple genomic clones (Young et al. 1987). After assigning markers to individual chromosomes with primary trisomics, these markers can be further localized to their respective chromosome arms using other types of trisomics such as compensating trisomics, telotrisomics, secondary trisomics, and tertiary trisomics

(Khush and Rick 1967a, 1967b, 1968a, 1969). Besides arm assignment, these analyses are also able to map the centromeres and to ascertain the orientation of the linkage maps. Similarly, through dosage analysis of various trisomic stocks including complementary telo-, secondary, and tertiary trisomics, Frary et al. (1996) was able to localize the centromeres of chromosomes 7 and 9 to precise intervals on the molecular linkage map.

3.4.3.3 Chromosome Deletion Mapping

Chromosome deficiencies are useful for research such as gene identification and localization, and analysis of chromatin structure and function. Investigations on induced deficiencies of tomato chromosomes were first carried out during the early 1950s (Barton 1954; Sen 1952). The sizes of the deficiencies have significant impact on their transmission: the longer the deficiency, the less frequently it is transmitted through the gametophytic generation, with the male gametophyte usually being much less tolerant than the female (Sen 1952). Subsequently, Khush and coworkers applied the induced deficiency technique in cytogenetic analysis of several linkage groups (Rick and Khush 1961; Khush et al. 1964; Khush and Rick 1967c). In an attempt to integrate the genetic map of the whole tomato genome, Khush and Rick (1968b) mapped 35 gene loci to 18 of the 24 chromosome arms using 74 irradiation-induced segmental deficiencies in tomato. Their findings integrated data obtained from various types of trisomics and established centromere positions, orientation of linkage groups, and markers on all but three of the chromosome arms. More recently, induced chromosome deletions were used to construct a detailed locus-order map around the centromere of tomato chromosome 6 with molecular markers (Liharska et al. 1997). Mapping of the various chromosomal breakpoints in the isolated mutants permitted the resolution of a cluster of molecular markers from the centromeric heterochromatin that was hitherto irresolvable by genetic linkage analysis.

3.4.3.4 Chromosome Mapping by Molecular Cytogenetics

Tomato has a rather small genome with an estimate of $\sim$750 Mb (Anderson et al. 1985; Arumuganathan and Earle 1991). About 30% of the genome is composed of repetitive sequences, which are mainly located in the heterochromatin regions (Peterson et al. 1996; Van der Hoeven et al. 2002). These repeats can be broadly classified into two categories: (1) tandem repeats, including 45S and 5S rDNA, the Arabidopsis type (TTTAGGG) telomere heptamer, satellite, and other species-specific repeats that are generally confined to heterochromatin regions, and (2) dispersed repeats, such as transposable elements and retrotransposons found in both heterochromatin and euchromatin regions. Some of those repetitive sequences have been identified and characterized with respect to their copy number and chromosome location through various methods including the in situ hybridization (ISH) technique. This technique has been utilized in tomato research for genome analysis and physical mapping of DNA sequences. The first identified tomato repetitive DNA sequence is a 452-bp *Hind*-III repeat THG2, which is a member of a large complex repeat dispersed throughout the genome (Zabel et al. 1985). FISH to tomato metaphase and pachytene chromosomes showed that this repeat is located in the pericentromeric regions of all chromosomes with great variation in copy number on individual chromosomes (Zhong et al. 1996a).

Ribosomal DNAs, including 5S and 45S genes, represent the most abundant repetitive DNA family and comprise $\sim$3% of the tomato genome (Lapitan et al. 1991). Both 5S and 45S genes are tandem repeats with 1000 and 2300 copies, respectively. The 45S rDNA has been localized at the end of the short arm of chromosome 2 with additional minor loci on 2L, 6L, 9S, and 11S (Xu and Earle 1996b). The 5S rDNA was found on the short arm of chromosome 1, in a region next to the centromere (Lapitan et al. 1991; Xu and Earle 1996a).

A series of tomato genomic repeats (TGR) have also been identified and mapped to various chromosomal locations. TGR1, a tandemly repeated sequences of 162 bp with 77,000 copies in the genome, is clustered at the telomeres of most chromosomes, and is also found at centromeric and interstitial sites of a few chromosomes (Lapitan et al. 1989; Ganal et al. 1992). Two other tomato genomic repeats, TGR2 and TGR3, are less abundant with 4200 and 2100 copies, respectively (Ganal et al. 1988; Schweizer et al. 1988). TGR2 was randomly distributed on nearly all the chromosomes except chromosome 2 with an average of 133 kb between elements. In contrast, TGR3 is predominantly clustered in the centromeric regions of the chromosomes. Unlike TGR1 and TGR2, which are tomato specific, TGR3 hybridizes to all *Lycopersicon* and some *Solanum* species. More recently, Chang (2004) employed high-resolution FISH to map various repeats and confirmed the distribution of TGR2 and TGR3 on tomato chromosomes. TGR4, another tomato genomic repeat isolated from a bacterial artificial chromosome (BAC), is mainly clustered at all 12 centromeric regions and at a few pericentromeric heterochromatin blocks. In the same report, the author also investigated two other types of repeats: retrotransposon and microsatellites. The tomato *Ty1-copia*-like retrotransposon was dispersed within all heterochromatin blocks and also a few scattered at euchromatin, whereas the microsatellites are located entirely in the pericentromeric heterochromatin regions. Based on pachytene morphology and copy number of satellite repeats, Chang (2004) classified tomato chromatin into six categories:

1. Euchromatin—chromosomal regions with weak fluorescence and domains of most of the coding sequences.
2. Chromomeres and other minor heterochromatin islands in the euchromatin regions.
3. Distal heterochromatin and long arm interstitial heterochromatic blocks, including the *Arabidopsis* type telomere repeat and TGR1.
4. Pericentromeric heterochromatin—the largest chromatin class of the tomato chromosomes covering 75% of the total genomic DNA and encompassing domains of most of the repeat families such as TGR2, TGR3, some microsatellite repeats, and the *Ty1*-copia-like retrotransposon.
5. Functional centromeric heterochromatin, including domains of satellite repeat TGR4.
6. Nucleolar organizer region, including the 45S rDNA and microsatellite $(GACA)_4$.

In addition to mapping repetitive DNA sequences, FISH has also been used in localization of single and low copy sequences. This usually involves the use of yeast artificial chromosomes (YACs) or BACs carrying the target DNA inserts, the latter being more favorable and advantageous (Jiang et al. 1995; Fuchs et al. 1996). Using *Mi*-BAC clones and an *Aps-1* YAC clone as FISH probes to tomato pachytene chromosomes, Zhong et al. (1999) located the nematode resistance gene (*Mi*) and the acid phosphatase gene (*Asp-1*) near the junction of euchromatin and pericentromeric heterochromatin of chromosome arms 6S and 6L, respectively. Similarly, Tor et al. (2002) located the colorless non-ripening (*Cnr*) locus on chromosome 2 by FISH of individual BAC clones containing DNA sequences of molecular markers linked to the *Cnr* gene. Recently, the *jointless-2* gene was located within or near the centromeric region of tomato chromosome 12 based on high-resolution genetic and physical mapping of molecular markers linked to or co-segregating with the jointless gene (Budiman et al. 2004).

Other advances have prompted the application of FISH in physical mapping of single or low copy DNA sequences. For example, SC spreads possess some features more suitable for single-copy FISH than prefixed pachytene chromosomes, which was demonstrated by localization of several genomic clones of ∼14 kb containing RFLP DNA sequences on tomato SC 11 using FISH (Peterson et al. 1999). The resolution of mapping target DNA sequences on chromosomes using FISH can be further improved by hybridization to extended DNA fibers (fiber-FISH) (Fransz et al. 1996; Zhong et al. 1996b). The usefulness of fiber-FISH has been demonstrated in the localization of the *Sun* locus, which controls tomato fruit shape, to a 38- to 68-kb region on the short arm of chromosome 7 (van der Knaap et al. 2004), and in the accurate mapping of adjacent and overlapping BACs (Chang 2004). Besides its mapping purpose, fiber-FISH can also be used as a tool in measuring the size or copy number of DNA sequences (Jackson et al. 1998; Wolters et al. 1998; Li et al. 2002b), and in other studies such as the investigation on the molecular and chromosomal organization of individual telomere domains (Zhong et al. 1998).

3.4.3.5 Centromere Mapping

Numerous approaches have been employed in mapping the centromeres for each tomato chromosome. The centromeric positions of chromosomes 1 and 2 were determined by RFLP mapping and by in situ hybridization localization of 5S and 45S rDNAs (Tanksley et al. 1988; Lapitan et al. 1991). The centromeres of chromosomes 3 and 6 were placed on the integrated molecular-classical map by deletion mapping (van der Biezen et al. 1994; van Wordragen et al. 1994). Analysis of the potato/tomato inversions on chromosomes 5, 10, 11, and 12 located the centromeres of these chromosomes to the respective inversion breakpoints (Tanksley et al. 1992). The centromeric positions of chromosomes 4 and 8 were predicted based on the relationship between the cytological, genetic, and molecular maps. RFLP hybridization and dosage analysis of telo-, secondary, and tertiary trisomic stocks have more precisely localized the centromeres of chromosomes 7 and 9 (Frary et al. 1996).

3.4.3.6 Current Genetic Linkage Maps

For a comprehensive review on the current status of genetic linkage maps, readers are referred to Chetelat and Ji (2006). A brief summary of the linkage maps of various origins is presented in Table 3.1. The last comprehensive classical linkage map of morphological and isozyme markers was summarized ~20 years ago (Stevens and Rick 1986). Since then, a great number of new markers have been generated and identified, suggesting a need for updating. The classical maps have been revised for numerous chromosomes including the short arm of chromosome 1 (Balint Kurti et al. 1995), chromosome 3 (van der Biezen et al. 1994), chromosome 6 (Weide et al. 1993; van Wordragen et al. 1996), the long arm of chromosome 7 (Burbidge et al. 2001), chromosome 10 (van Tuinen et al. 1997), chromosome 11 (van Tuinen et al. 1998), and chromosome 12 (van Tuinen et al. 1997).

The first molecular linkage map, based on an F_2 population from the cross *L. esculentum* × *L. pennellii*, comprised more than 1000 RFLP markers with an average marker spacing of approximately 1.2 cm (Tanksley et al. 1992). The approximate positions of centromeres have been determined for each of the 12 linkage groups (Pillen et al. 1996). An integrated high density RFLP-AFLP map of tomato, based on two *L. esculentum* × *L. pennellii* F_2 populations, contains 67 RFLP markers and 1175 AFLP markers (Haanstra et al. 1999b). Recently, a new molecular linkage map has been developed for tomato using conserved ortholog set (COS) markers (Fulton et al. 2002). The COS markers are anchored to previous maps with a large number of RFLPs (~380) markers, and also include other types of markers such as SSR, CAPS, and ESTs, which add up to a total of over 1800 markers (relevant information available through Solanaceae Genomics Network at http://sgn.cornell.edu). Other genetic maps have also been developed from interspecific crosses of *L. esculentum* to all the wild *Lycopersicon* and two related *Solanum* Species (Table 3.1).

3.5 GERMPLASM UTILIZATION AND ENHANCEMENT

3.5.1 Conventional Breeding

Conventional breeding comprises two fundamental steps: (1) creation of a breeding population that is highly variable for traits of interest, and (2) selection among the segregating population for individuals that possess the trait of interest as well as other traits of agricultural importance.

Table 3.1 Summary of Genetic Maps of Tomato in *Lycopersicon* and Related *Solanum* Species

Species[a]	Mapping Population[b]	Type of Markers[c]	Number of Markers	Size (cM)[d]	Reference
Classical maps					
L. esculentum	Varied	Morphological, isozyme	319	1072	Rick and Yoder (1988) and Stevens and Rick (1986)
L. esculentum	F$_2$	Isozyme	36	1114	Tanksley and Rick (1980) and Tanksley and Bernatzky 1987)
Molecular maps ***intraspecific***					
L. peruvianum	BC$_1$	RFLP	73	1073	van Ooijen et al. (1994)
Interspecific					
L. esculentum × L. pennellii	BC$_1$, F$_2$	RFLP, isozyme	112	760	Bernatzky and Tanksley (1986) and Tanksley and Bernatzky 1987)
L. esculentum × L. pennellii	F$_2$	RFLP, isozyme, morphological	1030	1276	Tanksley et al. (1992)
L. esculentum × L. pennellii	F$_2$	RFLP	98	1141	deVicente and Tanksley (1993)
L. esculentum × L. pennellii	F$_2$	AFLP, RFLP	1145	1482	Haanstra et al. (1999b)
L. esculentum × L. pennellii	F$_2$	SSR, RFLP	20 (SSR)[e]	1276	Areshchenkova and Ganal 2002)
L. esculentum × L. pennellii[f]	F$_2$	COS, RFLP, SSR, CAPS, ESTs	1973	1340	Fulton et al. (2002)
L. esculentum × L. pennellii	BC$_2$	RFLP	110	703	Frary et al. (2004)
L. esculentum × L. hirsutum	BC$_1$	RFLP	135	1356	Bernacchi and Tanksley (1997)
L. esculentum × L. hirsutum	BC$_1$	RFLP, RGAs	171	1469	Zhang et al. (2002)
L. esculentum × L. hirsutum	BC$_1$	RFLP, RGAs	179	1298	Zhang et al. (2003)
L. esculentum × L. chmielewskii	BC$_1$	RFLP, isozyme, morphological	70	852	Paterson et al. (1990, 1988)
L. esculentum × L. peruvianum	F$_2$	RFLP, CAPS	11	45	Bonnema et al. (1997)
L. esculentum × L. peruvianum	BC$_3$	RFLP, SCAR	122	865	Fulton et al. (1997a)

continued

Table 3.1   Continued

Species[a]	Mapping Population[b]	Type of Markers[c]	Number of Markers	Size (cM)[d]	Reference
L. esculentum × L. pimpinellifolium	BC$_1$	RFLP, RAPD, morphological	120	1279	Grandillo and Tanksley (1996)
L. esculentum × L. pimpinellifolium	BC$_1$	RFLP	151	1192	Chen and Foolad (1999)
L. esculentum × L. pimpinellifolium	F$_2$	RFLP, CAPS	90	936	deVicente and Tanksley (1993)
L. esculentum × L. pimpinellifolium	IBLs	RFLP, morphological	127	1272	Doganlar et al. (2002)
L. esculentum × L. pimpinellifolium	F$_2$	SSR	112	808	Liu et al. (2005)
L. esculentum × L. pimpinellifolium	RILs	SSR, SCAR	132	457	Villalta et al. (2005)
L. esculentum × L. pimpinellifolium	F$_2$	RFLP	71	1023	Paterson et al. (1991b)
L. esculentum × L. cheesmannii	RILs	RFLP	132	1209	Paran et al. (1997), (1995)
L. esculentum × L. cheesmannii	RILs	SSR, SCAR	114	370	Villalta et al. (2005)
L. esculentum × L. cheesmannii	BC$_2$	RFLP, SCAR, morphological	133	940	Fulton et al. (2000)
L. esculentum × L. parviflorum	BC$_1$	RFLP, isozyme	135	1259	Stamova and Chetelat (2000)
L. esculentum × L. chilense	F$_2$	RAPD, morphological	14	44	Griffiths and Scott (2001)
L. esculentum × L. chilense	F$_2$	RAPD, morphological	26–29	69–118	Agrama and Scott (2006)
S. sitiens × S. lycopersicoides	F$_2$	RFLP, isozyme	101	1192	Pertuze et al. (2002)
Intergeneric					
L. esculentum × S. lycopersicoides	BC$_1$	RFLP, isozyme, morphological	93	867	Chetelat et al. (2000)

[a] Underlined maps cover only part of the genome.

[b] Population abbreviations are as follows: BC, backcross; RILs, recombinant inbred lines; IBLs, inbred backcross lines.

[c] Marker abbreviations are as follows: RFLP, restriction fragment length polymorphism; RAPD, random amplified polymorphic DNA; AFLP, amplified fragment length polymorphism; SSR, simple sequence repeat; CAPS, cleaved amplified polymorphic sequences; SCAR, sequence characterized amplified regions; COS, conserved ortholog set, RGA, resistance gene analogs; ESTs, expressed sequence tags.

[d] Figures are either copied from the references or calculated from the linkage maps in the references.

[e] SSR markers were placed in the RFLP map of Tanksley et al. (1992).

[f] For updated information, visit the Solanaceae Genomics Network http://sgn.cornell.edu.

Thus, conventional breeding is essentially the normal mating system that involves human manipulation and selection of parents and their progeny favoring human needs. Such a selection process has greatly advanced speciation and domestication of crop species.

Tomato improvement before the middle 1920s was largely achieved by simple selection of new genotypes within unimproved populations such as existing heterogeneous cultivars, or accidental variations resulted from spontaneous mutation, natural outcrossing, or recombination of pre-existing genetic variation (Boswell 1937). The following decade saw rapid progress in cultivar development as systematic mating and selection procedures were utilized. The predominant selection methods of handling successive generations in most crops were simple pedigree and bulk systems. The pedigree system, in which seeds of each single plant selection are maintained as separate lines, allows detailed monitoring of parentage and places maximum selection pressure in each segregating generation, and is the preferred breeding method for most crops. In contrast, the bulk system, in which seeds of single plant selection are bulked into one population for each original mating, permits manipulation of large numbers of plants and lines and thus is most appropriate for species grown in highly space-competitive conditions.

Early tomato breeding programs largely employed the pedigree system. Although this system offers numerous advantages, such as maximum selection pressure on each generation and the greatest control of populations, it has several drawbacks: (1) selection potentials are limited to the genes of the two parents; (2) selection efficiency is limited to the early generations because variation generated from the original hybridization are lost quickly due to intense selection pressure within small populations; and (3) selection progress is limited because a progeny test is required for each selected plant. Efforts have been made to improve the efficiency, the effectiveness, and the rate of progress in pedigree selection. One of the improved methods, backcross breeding, provides a rapid means for introducing single gene traits with the preservation of the qualities of the recurrent parent (Briggs 1930), and is frequently used in tomato breeding programs.

Bulk selection was rarely used in early tomato breeding projects because of the growth habit of tomato plant, the objectives of breeding programs, and the need for hand harvesting. Single seed descent (SSD), a method offering the ability to handle plants in a limited bulk system and to accelerate generation time (Brim 1966), was adapted in tomato breeding in the 1970s by Casali and Tigchelaar (1975). They found the pedigree selection through the F_4, followed by SSD, had the same effectiveness as the pedigree selection alone but reduced the development time by half. Due to its efficiencies and other advantages over the standard pedigree selection, SSD combined with pedigree selection has been utilized for tomato improvement of various traits such as fruit size, earliness, and yield (Peirce 1977). Recurrent reciprocal selection, an effective method of improving combining ability of lines in hybrid production (Hull 1945), was also utilized in tomato breeding programs to improve heterosis (Khotyleva and Kilchevskii 1984).

As with most crops, efforts have been made to breed tomato cultivars with improved traits of economic importance, including resistance to diseases, insects, stress tolerance, and fruit quality traits. The major achievements in various areas of tomato breeding have been summarized in several comprehensive reviews and book chapters (Stevens and Rick 1986; Tigchelaar 1986; Kalloo 1991a; Razdan and Mattoo 2006). In this chapter, we only briefly summarize the previous achievements and the current status.

3.5.1.1 Disease Resistance

Tomato is susceptible to all types of pathogens, including fungi, bacteria, and viruses. Chemical control is often too costly for growers and is ineffective with some pathogens; therefore, genetic control of pathogens with cultivars resistant to diseases is a very useful practice. Advances in tomato breeding for resistance are due mostly to the incorporation of major resistance genes identified from the wild *Lycopersicon* species (Razdan and Mattoo 2006; Scott 2006; Scott and

Gardner 2006). At present, more than 20 major genes for disease resistance have been reported, which are used in tomato improvement (Grube et al. 2000; Zhang et al. 2002) and cultivar development (Laterrot 1996; Lukyanenko 1991a; Table 3.2).

3.5.1.2 Insect Resistance

Resistance to most important insects in tomatoes has been reported in wild species (Kennedy 2006), particularly *L. hirsutum* that possesses resistance to at least 18 pest species (Gentile et al. 1969; Schalk and Stoner 1976; Farrar and Kennedy 1991; Momotaz et al. 2005), and *L. pennellii* (Muigai et al. 2002). Anthropod resistance is associated with several traits, including the physical and chemical properties of grandular trichomes, and chemical defenses associated with the leaf lamella (Fidantsef et al. 1999; Stout et al. 1999). High density of type IV glandular trichomes and the presence of high levels of toxic acylsugars in their exudate are correlated with high level of resistance to a number of arthropods, including aphids, whiteflies, tomato fruitworm, beet armyworm, and the agromyzid leafminer (Kennedy 2003). In spite of the rich source of natural resistance, only a few insect-resistant cultivars have been developed due to the complicated issues associated with the mobile organisms, which possess genetic diversity in preference for certain plants and plant parts and are influenced by a wide range of environmental conditions (Stevens and Rick 1986).

3.5.1.3 Environmental Stress Resistance

Environmental stress constitutes another major threat for tomato growth and production. Breeding cultivars tolerant to various adverse environmental conditions is of economic importance and has been a major practice in tomato breeding (Kalloo 1991b). Resistances to numerous stresses have been reported in the wild species of tomato and some have been utilized in tomato improvement, including cold (Wolf et al. 1986; Foolad et al. 1998), heat (Scott and George 1984; Abdulbaki 1991), drought conditions (Martin and Thorstenson 1988; Thakur 1990; Foolad et al. 2003), excessive moisture conditions (Kuo and Chen 1980), and soil salinity and alkalinity (Rush and Epstein 1981; Tal and Shannon 1983; Foolad et al. 2001; Foolad 2004).

3.5.1.4 Breeding Tomato for Fruit Quality

Although yield and adaptability are the primary concerns of most tomato breeding programs, efforts have also been taken seriously to develop cultivars with improved fruit quality (Berry and Uddin 1991), fruit color (Chalukova and Manuelyan 1991), and attributes suitable for processing and machine harvesting (Berry and Uddin 1991; Lukyanenko 1991b).

3.5.2 Wide Hybridization

3.5.2.1 Interspecific Hybridization

Due to limited variation in the cultivated tomato, breeders have been repeatedly forced to use exotic germplasm to find genes of interest for tomato improvement. Wild *Lycopersicon* species possess great genetic variation and many agriculturally important traits (see discussion above). Moreover, all these wild species can be crossed to tomato with varying degrees of difficulty; therefore, all of the germplasm in the genus is transferable to the cultivated tomato. In fact, many traits of economic

Table 3.2 Pathogens Controlled by Genetic Resistance in Tomato Cultivars or Breeding Lines Obtained through Conventional Breeding

Pathogens	Gene	Disease	Commercial Cultivars or Breeding Lines Examples[a]	Reference
Fungi				
Alternaria alternata f. sp. *lycopersici*	*Asc*	Stem Canker	'Reno', 'Propak'	Grogan et al. (1975) and Vakalounakis (1988)
Alternaria solani Sor.	Polygenic	Early Blight	'Manalee', 'Mountain Supreme'	Gardner (1984) and Gardner and Shoemaker (1999)
Fusarium oxysporum f. sp. *lycopersici*	*I, I-1, I-2, I-3*	Fusarium Wilt	'Walter', 'Escudeso'	Scott and Jones (1995), Scott et al. (2004), and Stall and Walter (1965)
Fusarium oxysporum f. sp. *radicis-lycopersici*	*Fr1*	Fusarium Crown and Root Rot	'Ohio CR-6', 'Sebring'	Berry and Oakes (1987) and Scott and Farley (1983)
Phytophthora infestans	*Ph1, Ph2, Ph3*	Late Blight	'Legand', 'Mini-Rose'	Gallegly and Marvel (1955) and Walter (1967)
Sclerotium rolfsii Sacc		Southern Blight	'Processor 278', '5635M'	Leeper et al. (1992) and Mohr and Watkins (1959)
Pyrenochaeta lycopersici	*Py-1*	Corky Root Rot	'Kyndia'	Hogenboom (1970) and Jones et al. (1989)
Verticillium dahliae	Ve	Verticillium Wilt	'Advance', 'All Star'	Baergen et al. (1993), and Schaible et al. (1951)
Cladosporium fulvum	*Cf* series	Leaf Mold	'Waltham', 'Pioneer'	Kerr et al. (1971) and Scott and Gardner (2006)
Stemphylium solani	*Sm*	Gray Leaf Spot	'Horizon', 'Legand'	Bashi et al. (1973) and Scott (1985)
Septoria lycopersici		Septoria Leaf Spot	'Advance'. 'Essary'	Barksdale and Stoner (1978) and Poysa and Tu (1993)
Bacteria				
Corynebacterium michiganense	Polygeneic (*Cm*) series	Bacterial Canker	'Saladmaster', 'H9144'	Anwar et al. (2004) and Hassan et al. (1968)
Ralstonia solanacearum	Polygenic (*Bw*) series	Bacterial Wilt	'Hawk', 'Heatmaster'	Acosta et al. (1964) and Thoquet et al. (1996a)
Pseudomonas syringae pv. *tomato*	*Pto*	Bacterial Speck	'Parentage', 'Ontario 7710'	Kozik (2002) and Yang and Francis (2005)
Xanthomonas campestris pv. *vesicatoria*	Polygeneic	Bacterial Spot	'Hawaii 7998', 'Puebla'	Scott et al. (2003), Whalen et al. (1993), and Yang et al. (2005)
Virus				
CTV	Polygenic	Curly Top Virus	'Owyhee', 'CVF4', 'C5'	Martin and Thomas (1986) and Martin and Thomas (1969)
ToLCV	*Tlc*	Tomato Leaf Curl Virus	'Sankranthi', 'Nandi', 'Vybhav'	Kalloo and Banerjee (1990) and Muniyappa et al. (2002)

continued

Table 3.2   Continued

Pathogens	Gene	Disease	Commercial Cultivars or Breeding Lines Examples[a]	Reference
ToMV	*Tm-1, Tm-2, Tm-2*[a]	Tomato Mosaic Virus	'Golden girl', 'Jetsetter'	Cirulli and Ciccarese (1975) and Ohmori et al. (1996)
TYLCV	*Ty-1*	Tomato Yellow Leaf Curl Virus	'Scala', 'Tygress', 'TY-52'	Friedmann et al. (1998) and Pilowsky and Cohen (1990)
TSWV	*Sw-5*	Tomato Spotted Wilt Virus	'Stevens', 'Pearl Harbor'	Clayberg et al. (1959) and Stevens et al. (1992)
Nematode				
Meloidogyne spp.	*Mi*	Root Knot Nematodes	'Styllus', 'Anahu'	Medina-Filho and Stevens (1980), and Williamson (1998)

[a] For a list of tomato cultivars in North America, see Scott (2002) or T.C. Wehner's webpages at http://cuke.hort.ncsu.edu/cucurbit/wehner/vegcult/vgclintro.html

importance have been bred from wild *Lycopersicon* species and many cultivars possess various combinations of such resistances (Rick 1986; Rick et al. 1987). Most notably, resistances to over 40 major diseases have been detected in wild accessions, at least 20 of which have been bred into horticultural tomatoes (Rick and Chetelat 1995). Improvements have also been made in fruit quality traits, tolerance of abiotic stresses (e.g., drought, temperature extremes), and resistance to insect pests. Examples of successful incorporation of genes from wild species into tomato cultivars include resistance to Fusarium wilt, root knot nematodes, and tomato mosaic virus. Recently, resistance to *Begomoviruses* have also been incorporated into tomato from *L. hirsutum* (Hassan et al. 1984), *L. pimpinellifolium* (Kasrawi 1989), *L. peruvianum* (Pilowsky and Cohen 1990), and *L. chilense* (Pilowsky and Cohen 1990; Zamir et al. 1994; Griffiths and Scott 2001; Ji and Scott 2005).

3.5.2.2 Intergeneric Hybridization

In addition to wild *Lycopersicon* species, potential genetic resources for tomato improvement can also be found in the huge ancestral genus *Solanum*, which possesses vast amounts of genetic variation. Of special interest to tomato breeders is the array of traits in *Solanum* species not found in *Lycopersicon*, including resistances to certain diseases and tolerance of extreme aridity or chilling stress (Rick 1988). Four nightshade species in this genus—*S. juglandifolium*, *S. ochranthum, S. lycopersicoides* and *S. sitiens*, all members of series *Juglandofolia* within subsection *Potatoe* of section *Petota*—display close resemblance to members of *Lycopersicon* in respect to chromosome number and morphology (Rick et al. 1979b; Rick 1988). Although they have many features in common, these four species are extremely diverse in their growth habit, fruit characteristics, and autoecology. Among those species, S. *juglandifolium* and S. *ochranthum* are closely related to each other in many respects, but they are distinct from the other two species, S. *lycopersicoides* and S. *sitiens*. The latter two are closely-related sister taxa, as evidenced from classical and molecular systematics, ecology, and crossing relationships (Rick 1988; Peralta and Spooner 2001). In fact, the only successful cross between the four species is between S. *lycopersicoides* and S. *sitiens*. The resulting interspecific F_1 hybrids are easily synthesized and displayed normal meiotic behavior and high fertility (Rick 1979; Pertuze et al. 2002; Ji et al. 2004). A comparative linkage map of the S. *lycopersicoides* and S. *sitiens* genomes revealed that all chromosomes were collinear with the tomato map, except for chromosome 10, where a paracentric inversion on the long arm was detected (Pertuze et al. 2002). The breakpoint of this inversion was most likely identical to the one reported for potato chromosome 10L (Tanksley et al. 1992), one of the five such rearrangements that differentiate it from tomato. Due to the presence of this inversion, recombination events were absent in this region (Chetelat et al. 2000; Ji and Chetelat 2003).

Hybridization of *S. sitiens* and *S. lycopersicoides* with cultivated tomato is more problematic. In the case of *S. lycopersicoides*, F_1 hybrids are readily obtained by embryo culture, but are generally infertile (Chetelat et al. 1997), owing largely to meiotic abnormalities including reduced chiasma formation and frequent univalents (Rick 1951; Menzel 1962). However, there is complete synapsis in most pachytene nuclei of the $2x$ hybrid and no evidence of structural differentiation among the parents (Menzel and Price 1966). Pairing in synthetic allotetraploid and allotriploid hybrids is mostly between homologous chromosomes, indicating the two genomes are homeologous (Menzel 1964; Rick et al. 1986). Besides hybridization to the tomato, S. *lycopersicoides* has also been hybridized unilaterally with other *Lycopersicon* species, including *L. pimpinellifolium, L. cheesmanii,* and *L. chilense*, to produce small but viable seeds. In contrast, *S. sitiens* does not cross directly to tomato, but the equivalent F_1 has been obtained indirectly using polyploid and bridging line methods (Deverna et al. 1990; Pertuze et al. 2003). Chromosomes or chromosome segments from *S. lycopersicoides* and *S. sitiens* have been successfully introgressed to varying degrees into the tomato genome (Chetelat and Meglic 2000; Ji and Chetelat 2003). Transfer of the S. *lycopersicoides* genome to tomato has been accomplished to a large extent through a set of introgression lines (Canady et al. 2005; Chetelat and Meglic 2000). Progress with S. *sitiens* has been hampered by more severe barriers (Deverna et al. 1990). Nonetheless, recent experiments have resulted in the transfer of specific chromosomal regions into tomato (Pertuze et al. 2003). These nightshades therefore represent largely unexplored but promising resources for tomato improvement.

3.5.3 Hybrid Tomato Breeding

3.5.3.1 Introduction

Heterosis refers to the phenomenon in which a hybrid outperforms over its parents for a particular trait (Shull 1908; East 1936). Heterosis manifests itself most strongly in F_1 and decreases progressively in the following generations. Also known as hybrid vigor or outbreeding enhancement, heterosis is caused by heterozygosity, as opposed to inbreeding depression, associated with the uncovering of deleterious recessive lethals in homozygous genotypes (Shull 1908; East 1909).

3.5.3.2 Breeding of Tomato Hybrids

The advantage of F_1 hybrids in tomato was recognized as early as the beginning of the last century. Practical application was first seen in the early 1930s with the release of the first tomato hybrid in Bulgaria, and later in England, Holland, France, the United States, and other countries (Georgiev 1991; Atanassova and Georgiev 2002). Tomato F_1 hybrids display advantages in many valuable traits including high yields and fruit quality, earliness, ripeness, and uniformity of plants and fruits (Burdick 1954; Dhillon et al. 1974; Balibrea et al. 1997; Bhatt et al. 2001). More importantly, in some cases F_1 hybrids possess multiple resistance to diseases and demonstrate strong ability to adapt different environments (Griffing and Zsiros 1971; Christakis and Fasoulas 2002; Frelichowski and Juvik 2005). Because of these advantages, hybrid tomato breeding has become a major practice in vegetable seed industry and will continue to predominate in high input agricultural systems (Scott and Angell 1998).

3.5.3.3 Hybrid Seed Production

One problem in tomato hybrid seed production is the pre-anthesis manipulation of the female parent to prevent self-pollination. A standard procedure used in current hybrid seed production is to emasculate the flower buds of normal fertile parents, which inevitably increases the price of the hybrid seeds. An alternative way is to use male-sterile or other sterile mutants as female parents.

Tomato possesses over 40 male-sterile (ms) mutants in a wide variety of types and genetic backgrounds (Stevens and Rick 1986). Among this series, only a few have styles accessible for cross-pollination without emasculation and can be applied in hybrid seed production (Stevens and Rick 1986). The stamenless (sl) series develop only vestiges or reduced stamens. Owing to the lack of stamens, they are the most suitable for cross-pollination, but could not be used in practice due to low receptivity of the style and poor seed production (Nash et al. 1985). The functional sterility type *ps* is characterized by unmodified production of functional pollen but inability to release it (Larson and Paur 1948). Frequently, the emasculation provokes self-pollination due to the presence of free-dispersing viable pollen. The potential self-pollination, together with the lower receptivity of the exserted style, render the *ps* sterility not very suitable for hybrid seed production. In contrast, the functional sterility type *ps-2* develops morphologically normal flowers, but cannot be self-pollinated due to the non-opening anthers. Thus artificial selfing can usually produce 100% *ps-2* progeny. The easy maintenance of the sterile lines under homozygous state by artificial selfing and high yield of hybrid seeds at low cost constitute the major advantages in utilizing this type of sterility in tomato hybrid seed production (Atanassova 1999). Other functional types of sterility, including exserted style (*ex*) and short style (*sh st*), also find limited application in tomato hybrid seed production (Georgiev and Atanassova 1980; Tikoo and Anand 1980).

3.5.4 Marker Assisted Selection

Although conventional breeding has contributed greatly to tomato improvement, particularly to disease resistance, it has several limitations: (1) the process of making crosses and backcrosses and the selection of desired resistant progeny are usually time-consuming and expensive; (2) there are always susceptible plants that escape attack and get selected, thus deteriorating the efficiency of selection; (3) screening different pathogens and their pathotypes simultaneously or sequentially is usually difficult or impossible because the pathogens have to be maintained either on hosts or alternate hosts if they are obligate parasites; (4) some resistances are difficult to utilize because the diagnostic tests are hard to develop due to the challenges posed by inoculum production and maintenance; (5) in cases where symptoms are only detectible on adult plants or fruits the diagnostic tests could be particularly expensive and hard to perform (Mohan et al. 1997; Barone 2004).

The limitations with conventional breeding procedures could be overcome with the application of molecular markers. The development and utilization of molecular markers in plant genomes since the early 1980s has greatly improved some breeding approaches, since these markers directly reveal genetic variability through DNA analysis and their detection is not affected by environmental effects (Staub and Serquen 1996). The use of molecular markers in plant breeding is called marker-assisted selection (MAS) (Paterson et al. 1991a). Since their emergence, molecular markers have been widely used as a principal tool for breeding of many crops, including tomato. Particularly, great efforts have been made to identify and develop molecular markers linked to genes resistant to various diseases. Thus far, more than 50 genes that confer resistance to all major classes of pathogens in tomato have been mapped and/or cloned from Solanaceous species (Grube et al. 2000; Chunwongse et al. 2002; Parrella et al. 2002; Bai et al. 2003, 2004, 2005; Table 3.3). A linkage map of tomato based on resistance gene analogues (RGAs) was also constructed, which

Table 3.3 Resistance Genes Single Gene and Quantitative Resistance Loci Mapped in *Lycopersicon*

Gene[a]	Pathogen	Disease	Source	Chromosomal Location	Reference
Asc	*Alternaria alternata* f. sp. *Lycopersici*	Alternaria Stem Canker	*L. pennellii*	3	van der Biezen et al. (1995)
Am	AMV	Alfalfa Mosaic Virus	*L. hirsutum*	6	Parrella et al. (2004)
Bm-2a, 2c, 3, 9, 12 (Q)	*Alternaria alternate*	Blackmold	*L. cheesmanii*	2, 3, 9, 12	Robert et al. (2001)
Bs4	*Xanthomonas campestris* pv. *vesicatoria*	Bacterial Spot	*L. pennellii*	5	Ballvora et al. (2001a, 2001b)
Bw-1, 3 (Q)	*Ralstonia solanacearum*	Bacterial Wilt	*L. pimpinellifolium*	6, 10	Danesh et al. (1994) and Mangin et al. (1999)
Bw-4, 5 (Q)	*R. solanacearum*	Bacterial Wilt	*L. pimpinellifolium*	4, 6	Mangin et al. (1999) and Thoquet et al. (1996b)
Cf-1	*Cladosporium fulvum*	Leaf Mold	*L. esculentum*	1	Rick and Butler (1956)
Cf-2	*C. fulvum*	Leaf Mold	*L. pimpinellifolium*	6	Dickinson et al. (1993) and Jones et al. (1993)
Cf-4	*C. fulvum*	Leaf Mold	*L. hirsutum*	1	Balint-Kurti et al. (1994) and Jones et al. (1993)
Cf-5	*C. fulvum*	Leaf Mold	*L. esculentum* var. *cerasiforme*	6	Jones et al. (1993)
Cf-9	*C. fulvum*	Leaf Mold	*L. pimpinellifolium*	1	Balint-Kurti et al. (1994) and Jones et al. (1993)
Cf-ECP2	*C. fulvum*	Leaf Mold	*L. pimpinellifolium*	1	Haanstra et al. (1999a)
Cf-ECP3	*C. fulvum*	Leaf Mold	*L. esculentum*	1	Yuan et al. (2002)
Cm1.1-10.1 (Q)	*Clavibacter michiganensis*	Bacterial Canker	*L. peruvianum*	1, 6, 7, 8, 9, 10	Sandbrink et al. (1995)
Cmr	CMV	Cucumber Mosaic Virus	*L. chilense*	12	Stamova and Chetelat (2000)
EBR3.1-11.1 (Q)	*Alternaria solani*	Early Blight	*L. hirsutum*	3, 4, 5, 6, 8, 10, 11	Foolad et al. (2002) and Zhang et al. (2003)
Fen	HF to Fenthion	Bacterial Speck	*L. pimpinellifolium*	5	Martin et al. (1994)
Fr1	*Fusarium oxysporum* f. sp. *radidics-lycopersici*	Fusarium Crown and Root Rot	*L. peruvianum*	9	Vakalounakis et al. (1997)
Hero	*Globodera rostochiensis*	Potato Cyst Nematode	*L. pimpinellifolium*	4	Ganal et al. (1995)
I	*F. oxysporum* f. sp. *lycopersici*	Fusarium Wilt	*L. pimpinellifolium*	11	Bohn and Tucker (1939) and Paddock (1950)

continued

Table 3.3 Continued

Gene[a]	Pathogen	Disease	Source	Chromosomal Location	Reference
I1	F. oxysporum	Fusarium Wilt	L. pimpinellifolium	7	Sarfatti et al. (1991) and Scott et al. (2004)
I-2	F. oxysporum	Fusarium Wilt	L. pimpinellifolium	11	Sarfatti et al. (1989) and Simons et al. (1998)
I2C	F. oxysporum	Fusarium Wilt	L. pimpinellifolium	11	Ori et al. (1997)
I-3	F. oxysporum	Fusarium Wilt	L. pennellii	7	Hemming et al. (2004)
LB-1, 2 (Q)	Phytophthora infestans	Late Blight	L. pimpinellifolium	6	Frary et al. (1998)
lb1-1b12 (Q)	P. infestans	Late Blight	L. hirsutum	All 12 chrom.	Brouwer et al. (2004)
Lv	Leveillula taurica	Powdery Mildew	Lycopersicon	12	Chunwongse et al. 1994, 1997)
Mi	Meloidogyne spp., Macrosiphum euphorbiae	Root Knot Nematode	L. peruvianum	6	Messeguer et al. (1991) and Williamson (1998)
Mi-1	Meloidogyne spp.	Root Knot Nematode	L. peruvianum	6	Vos et al. (1998)
Mi-3	M. incognita, M. javanica	Root Knot Nematode	L. peruvianum	12	Yaghoobi et al. (1995)
Mi-9	M. incognita, M. javanica, M. arenaria	Root Knot Nematode	L. peruvianum	6	Ammiraju et al. (2003)
Ol-1	Oidium lycopersici	Powdery Mildew	L. hirsutum	6	Huang et al. (2000) and van der Beek et al. (1994)
ol-2	O. lycopersici	Powdery Mildew	L. esculentum var. cerasiforme	4	Ciccarese et al. (1998) and De Giovanni et al. (2004)
Ol-3	O. lycopersici	Powdery Mildew	L. hirsutum	6	Huang et al. (2000)
Ol-4	O. lycopersicon	Powdery Mildew	L. peruvianum	6	Bai et al. (2004)
Ol-5	O. lycopersicon	Powdery Mildew	L. hirsutum	6	Bai (2004)
Ol-6	O. lycopersicon	Powdery Mildew	Unknown	6	Bai (2004)
Ol-qtl1, - 2, - 3 (Q)	O. lycopersici	Powdery Mildew	L. parviflorum	6, 12	Bai et al. (2003)
Ph-1	Phytophthora infestans	Late Blight	L. pimpinellifolium	7	Pierce (1971)
Ph-2	P. infestans	Late Blight	L. pimpinellifolium	10	Moreau et al. (1998)
Ph-3	P. infestans	Late Blight	L. pimpinellifolium	9	Chunwongse et al. (2002)
Pot-1	Potyvirus	Potyvirus	L. hirsutum	3	Parrella et al. (2002)
Prf	Required for Pto/Fen		L. pimpinellifolium	5	Salmeron et al. (1996)

Gene	Pathogen	Common disease name	Source species	Chromosome	Reference
Pto	*Pseudomonas syringae* pv. *tomato*	Bacterial Speck	*L. pimpinellifolium*	5	Martin et al. (1993a)
Py-1	*Pyrenochaeta lycopersici*	Corky Root Rot	*L. peruvianum*	3	Doganlar et al. (1998)
rx-1, 2, 3	*Xanthomonas campestris* pv. *vesicatoria*	Bacterial Spot	*L. esculentum*	1, 5	Yu et al. (1995)
Rcm2.0, 5.1 (Q)	*C. michiganensis* subsp. *michiganensis*	Bacterial Canker	*L. hirsutum*	2, 5	Coaker and Francis (2004) and Kabelka et al. (2002)
Rcm-5, 7, 9 (Q)	*C. michiganensis* subsp. *michiganensis*	Bacterial Canker	*L. peruvianum*	5, 7, 9	van Heusden et al. (1999)
Sm	*Stemphylium spp.*	Gray Leaf Spot	*L. pimpinellifolium*	11	Behare et al. (1991)
Sw-5	TSWV	Tomato Spot Wilt Virus	*L. peruvianum*	9	Stevens et al. (1995)
Tm-1	TMV	Tobacco/Tomato Mosaic Virus	*L. hirsutum*	2	Levesque et al. (1990)
Tm-2[a] (*Tm-2²*)	TMV	Tobacco/Tomato Mosaic Virus	*L. peruvianum*	9	Young and Tanksley (1989) and Young et al. (1988)
To-qtl1 (Q)	ToMoV	Tomato Mottle Virus	*L. chilense*	6	Griffiths and Scott (2001) and Ji and Scott (2005)
Ty-1 (Q)	TYLCV	Tomato Yellow Leaf Curl Virus	*L. chilense, L. hirsutum, L. pimpinellifolium, L. cheesmanii*	6	Chagué et al. (1997) and Zamir et al. (1994)
Ty-2	TYLCV	Tomato Yellow Leaf Curl Virus	*L. hirsutum*	11	Hanson et al. (2000)
Ty-qtl1 (Q)	TYLCV	Tomato Yellow Leaf Curl Virus	*L. chilense*	6	Ji and Scott (2005)
Ve	*Verticillium dahliae*	Verticillium Wilt	*L. esculentum*	9	Diwan et al. (1999)

[a] Underlined gene names given by Grube et al. (2000) or present authors for simplicity; Q in parentheses indicates quantitative trait loci.

Source: Adapted from Grube, R. C., et al., *Genetics*, 155, 873–887, 2000.

provides a basis for further identification and mapping of genes and quantitative trait loci for disease resistance (Zhang et al. 2002). Some of the markers linked to these resistance gene loci have been or will be useful for marker-assisted selection.

3.5.5 Tissue Culture as a Tool for Tomato Improvement

Tissue culture techniques have been used extensively in plant research and improvement since their emergence in the 1930s. Among many different plant species, tomato is one of the most favorable experimental materials for tissue culture. Numerous studies on plant regeneration from a wide range of tissues and organs of wild and cultivated tomato germplasm have been conducted (Norton and Boll 1954; Padmanabhan et al. 1974; Cassells 1979; Novak and Maskova 1979; Koblitz 1991a; Bhatia et al. 2004). *In vitro* plant regeneration techniques have been used in tomato improvement for selection of stress tolerance and disease resistance (Rahman and Kaul 1989; Steinitz et al. 1993; Toyoda et al. 1989), development of haploids (Shtereva et al. 1998; Zagorska et al. 1998, 2004), production of somatic and sexual hybrids (Wijbrandi et al. 1990; Wolters et al. 1994), and genetic transformation (Hille et al. 1991).

3.5.5.1 Selection for Abiotic Stresses Tolerance

In vitro culture of tomato has been used for selection of cell lines tolerant to various abiotic stresses (Koblitz 1991a). Tolerance to sodium chloride in the callus and suspended cells of *Lycopersicon* has been reported from numerous sources (Tal et al. 1978; Rahman and Kaul 1989). Response *in vitro* and *in vivo* to salt in *Lycopersicon* was positively correlated, and thus *in vitro* screening for salt tolerance can be used for improvement of this trait (Li et al. 1988). Selection of cells tolerant to water stress was achieved by gradual exposure of cells to increased stress levels in the growth medium (Handa et al. 1983a,1983b; 1986; Srivastava et al. 1995), as increased dry weight/fresh weight ratio as well as increased intracellular concentrations of proline and other compounds were found under water stress conditions (Handa et al. 1983a, 1986). Other applications of the tissue culture technique include studies of toxic effects (Zilkah and Gressel 1977, 1978; Handa et al. 1982), herbicide tolerance (Thomas and Pratt 1982), heavy metal tolerance (Steffens and Williams 1987), and heat stress adaptation (Staraci et al. 1987; Smith et al. 1989).

3.5.5.2 Selection for Disease Resistance

In vitro selection of mutants among populations of cultured tomato cells has been conducted in the field of resistance to numerous fungal or bacterial pathogens (Warren and Routley 1970; Toyoda et al. 1984, 1989; Shahin and Spivey 1986; Witsenboer et al. 1988). In a study of a single gene resistance of tomato to *Phytophthora infestans* (*Ph1*), Warren and Routley (1970) observed that callus from plants homozygous for *Ph1* maintained at least partial resistance to *P. infestans*, while callus from susceptible plants were apparently infected and destroyed by the pathogen, suggesting that at least some aspects of *Ph1* resistance can operate in undifferentiated tissue. Chromatographic comparisons of phenolics from resistant and susceptible tissue culture lines showed a similar level of phenolic content before and up to six days after inoculation, but nearly total loss of phenolics was observed for the susceptible plants ten days after inoculation. Exposure of cultured cells to toxins produced from *F. oxysporum* or *R. solanacearum* allowed selection of tomato cell lines resistant to these pathogens (Toyoda et al. 1984, 1989). A single dominant gene for resistance to Fusarium wilt was recovered from cotyledonary protoplasts of tomato cultivars (Shahin and Spivey 1986, 1987). Investigations on the effects of host-specific toxins of *Alternaria alternata f.* sp. *lycopersici* on various tissues of tolerant and susceptible tomato genotypes indicated that the toxins are active at the cellular level and that the tolerance to the toxins is expressed only in higher organized morphological structures (Witsenboer et al. 1988). To facilitate testing large numbers of tomato somaclones for resistance to bacterial canker, a fast screening procedure and a criterion for selection of resistant plants, based on rating scores of the severity of wilting symptoms, were developed (van den Bulk et al. 1991). Somaclones from tissue explants of the susceptible tomato cultivar Moneymaker were evaluated for resistance. Although somaclonal variation was found for morphological characters in these populations, no differences were detected between the somaclone population and controls in either mean susceptibility or distribution of plants over disease severity categories, suggesting that the potential of somaclonal variation as a source of resistance to bacterial canker may be limited. Most recently, the possibility of obtaining tomato plants resistant to *Clavibacter michiganense* subsp. *michiganense* (*Cmm*) through anther and tissue culture was investigated (Sotirova et al. 1999). Regenerants from anther and tissue culture of some tomato genotypes displayed varied resistance to *Cmm*, suggesting the gametoclonal and somaclonal variation may be efficiently applied to obtain tomato plants resistant to *Cmm*.

Tissue culture has also been utilized in the selection for resistance to other pathogens, including viruses and root-knot nematode. Selection for cell lines resistant to tobacco mosaic virus (TMV) were explored by physically injecting viruses into cells (Toyoda et al. 1985). Somaclonal variants were also exploited for the development of TMV-resistant lines (Barden et al. 1986). Tomato leaf discs from a TMV-susceptible isogenic line were used in regenerating somaclones. Of the somaclones inoculated with TMV-Flavum, ~1.5% showed no visible symptoms and TMV could not be detected by various test methods. Tomato plants generated adventitiously from cotyledon explants retained their resistance to the root-knot nematode, which reaffirms the application of tissue culture in screening and isolating nematode-resistant variants (Ammati et al. 1984).

3.5.5.3 Development of Haploids

In vitro culture of anthers and isolated microspores has been an important way to generate haploids (Koblitz 1991a; Bhatia et al. 2004). Gresshof and Doy (1972) were the fisrt to successfully produce tomato haploid plants through anther culture. A differentiation of haploid callus cultures occured in darkness as well as under illumination, and was controlled by hormones added to the growth media. The plantlets developed only under the light and possessed a haploid set of chromosomes in the root tip cells. This procedure was successfully reproduced for isolated uninuclear microspores, which was later converted into haploid embryoids with or without suspensor (Varghese and Yadav 1986). In an effort to produce tomato haploid plants, Thomas and Mythili (1995) found up to 38% of the anthers formed callus in a growth media supplemented with hormones. The frequency of haploid cells in anther-derived callus cultures could be as high as 27.5%, but decreases considerably in subsequent subcultures (Levenko et al. 1977). Following the first success in the generation of tomato haploids using anther culture, numerous groups have reported some success in obtaining anther-derived haploid plants from various tomato cultivars (Zamir et al. 1980, 1981; Ziv et al. 1984).

Other factors affecting anther callus production and regenerant development have been identified in recent research. First, light illumination, as mentioned before, is a key factor in anther callus induction. Callus number and diameter increased as the dark period duration increased, but callus quality and appearance decreased noticeably after eight weeks of dark incubation (Jaramillo and Summers 1991). Second, microspore developmental stage is another factor influencing the induction of tomato anther callus (Summers et al. 1992; Ma et al. 1999). Although calli were induced at all stages of anther development, anthers containing prophase I microspores produced the highest frequency of calli, and fewer calli were produced as microspores approached the uninucleate and binucleate pollen stage (Jaramillo and Summers 1991). Callus diameter also decreased as anther development progressed, with significantly larger calli produced from prophase I than later-stage anthers. Third, anther callus induction is also affected by the anther length

(Summers et al. 1992). Anther and flower bud length both were significantly correlated with anther developmental stage, the number of anthers producing calli, and mean calli diameter. In each case, anther length exhibited a significantly better correlation than bud length. Fourth, tomato genotypes play an important role in anther callus induction and subsequent regeneration. Callus induction capacity was significantly improved by the presence of male-sterile genes that disrupt microsporogenesis at meiosis (Zagorska et al. 1998; Ma et al. 1999; Shtereva and Atanassova 2001). In addition, the frequency of callus induction in the anthers of two other mutants characterized by anther abnormalities, solanifolia (*sf*) and trifoliate (*tf*), was significantly higher than in the wild type ones (Shtereva and Atanassova 2001). These results suggested that in tomato, processes leading to or resulting from anther abnormalities might be favorable for enhancing anther callus induction and subsequent regeneration. Other factors influencing anther callus induction include donor plant growth conditions, medium composition and anther treatment before being subject to culture (Shtereva et al. 1998; Shtereva and Atanassova 2001).

3.5.5.4 Development of Somatic Hybrids through Protoplast Fusion

Wild relatives of *Lycopersicon* and *Solanum* possess valuable traits of economic importance for tomato improvement. However, some of the wild species of *Lycopericon* and *Solanum* are not sexually compatible with the cultivated tomato. In such cases, somatic hybrids generated through protoplast fusion would be one of the best choices to utilize genes of interest in the wild species and to obtain novel cytoplasm–nucleus combinations (Koblitz 1991b). The first somatic hybrids produced with tomato were the tomato (+) potato hybrids (Melchers et al. 1978). Since then, many somatic hybrids between tomato and other *Lycopersicon* species, or between tomato and *Solanum* or *Nicotiana* species have been reported (for a recent review, see Wolters et al. 1994).

More recently, a hexaploid potato (+) tomato fusion hybrid was generated and successfully backcrossed to a tetraploid tomato, which yielded a BC_1 plant with only nine tomato chromosomes instead of the expected haploid set of 12 (Jacobsen et al. 1994). An additional backcross yielded BC_2 progeny with varied numbers of extra tomato chromosomes (Jacobsen et al. 1995; Garriga-Caldere et al. 1997). Following further backcrosses, a complete series of monosomic tomato chromosome addition lines in the cultivated potato were established (Ali et al. 2001a; Garriga-Caldere et al. 1998). In another report, a tetraploid potato (+) tomato fusion hybrid was successfully backcrossed with a diploid *L. pennellii* to generate a trigenomic hybrid possessing three different genomes of potato, tomato, and *L. pennellii*. During metaphase I, bivalents were formed predominantly between tomato and *L. pennellii* chromosomes, the univalents of potato chromosomes were most common. Trivalents formed in rare cases, including homeologous chromosomes of potato, tomato, and *L. pennellii*. Homeologous pairing and recombination between potato and tomato chromosomes occurred at low rates, suggesting the potential for gene transfer between the two crops (Garriga-Caldere et al. 1999; Ali et al. 2001b).

3.5.5.5 Development of Sexual Hybrids through Embryo Rescue

Among the various barriers in the hybridization involving the distant species of *Lycopersicon*, the post fertilization mechanisms—including zygote degeneration, hybrid sterility and abnormality in post syngamy restriction—are a decisive factor for hybridization failure (Kalloo 1991c). The failure of hybridization is primarily due to the abnormal development of embryos through lack of nutrition caused by disturbances in the development of the endosperm and its subsequent degeneration, and mitotic abnormities of the hybrid endosperm and the zygote (Kosova and Kiku 1979). This hybridization barrier could be overcome by the embryo rescue technique, which has been used to hybridize tomato with numerous wild species, including *L. peruvianum*, *L. chilense*, and *S. lycopersicoides*. Smith (1944) made the first successful attempt to obtain a hybrid of

L. esculentum×*L. peruvianum* through embryo culture. Since then, numerous scientists have attempted embryo culture to produce a *L. esculentum*×*L. peruvianum* hybrid (Choudhury 1955; Alexander 1956; Guan et al. 1988; Segeren et al. 1993; Chen and Adachi 1996; Pico et al. 2002). Pollen mixture crosses, as well as embryo rescue, allowed the recovery of interspecific hybrids between tomato and various *L. chilense* accessions (Scott et al. 1996; Pico et al. 2002). Intergeneric hybrids of *L. esculentum*×*S. lycopersicoides* were first reported by Rick (1951) through embryo culture, but the resulting F_1 could not be backcrossed directly to tomato due to its pollen sterility and unilateral incompatibility.

3.5.6 Genetic Transformation of Tomato

Transfer of genes from heterologous species by means of genetic transformation offers direct access to a vast pool of genes that are not accessible to traditional plant breeders due to genetic barriers, or which do not exist in the crops of interest. With the recent advances in genetic engineering techniques, simultaneous manipulation of numerous desirable genes in a single event is feasible, thus allowing the concurrent introduction of novel genes or traits into an elite background. Genetic engineering also offers the advantage of introducing a desirable trait without deleterious effects due to linkage drag, as is often the case with the classical breeding system.

Various methods are available in the production of transgenic plants, including the most widely used *Agrobaterium*-mediated transformation, microprojectile bombardment with DNA or biolistics, microinjection of DNA, and direct DNA transfer into isolated protoplasts (Raybould and Gray 1993; Sharma et al. 2005). In the past decades, important breakthroughs have been achieved with the various procedures for the development of transgenic plants, including efficient methods of gene cloning, reliable tissue culture regeneration systems, improved gene constructs and transformation techniques, appropriate organ-specific promoters for gene expression, and a series of selectable marker genes (Horsch et al. 1985; Cheng et al. 1998; Cortina and Culianez-Macia 2004; Lima et al. 2004; Sharma et al. 2005). Those significant advances have led to the production of a large number of transgenic plants for various major crop plants including tomato, and a series of genes governing traits of economic importance have thus been transferred through various transformation techniques (Babu et al. 2003; Raybould and Gray 1993). For tomato, genetic engineering has successfully led to the transfer of numerous genes resistant to insects, herbicides, and various pathogens.

3.5.6.1 Herbicide Tolerance

The use of herbicides has become an essential practice in modern agriculture for weed control. Continuous efforts have been carried out in search for new herbicides with increasing efficiency and utilization in crop production, though it is very difficult to discover better and more economical weed control compounds than those that are already available. Moreover, the cost of regulatory approval has increased dramatically. Most important, most of the new herbicides do not distinguish between weeds and crops, thus destroy the crop along with the target weeds.

In contrast, herbicide-tolerant genes offer a more efficient way for crops to resist herbicides. Many major crops have been genetically modified to be resistant to certain herbicides, thus expanding the potential market and application of the established herbicides (Paoletti and Pimentel 2000). The most successful method to introduce resistance genes is through genetic engineering. Numerous independent strategies have been employed to obtain transgenic plants tolerant to the most commonly used herbicides (Quinn 1990). The first approach involves overproduction of the sensitive target enzyme. It has been demonstrated that over-expression of the plant enzyme 5-enol-pyruvylshikimate-3 phosphate synthase conferred glyphosate tolerance in transgenic petunia plants (Shah et al. 1986). In a second approach, a mutant form of the sensitive target enzyme less sensitive to the herbicide is produced in the transgenic plants. An example in tomato involves the transfer of a glyphosate tolerance gene using a binary *Agrobacterium tumefaciens* vector (Fillatti et al. 1987). The target protein of glyphosate is modified in the engineered tomato plants, thus rendering its activity insensitive to the herbicide. The third strategy for engineering herbicide resistance in plants involves the expression of an enzyme that detoxifies the herbicide. As an example in tomato, the bialaphos resistance gene (*bar*) from *Streptomyces hygroscopicus* has been introduced into tomato plants through an *Agrobacterium* vector (Deblock et al. 1987). The *bar* gene encodes a phosphino-thricin acetyltransferase (PAT) that converts phosphinothricin into the nontoxic acetylated form. Therefore, the resulting transgenic plants showed complete resistance towards high doses of the commercial formulations of phosphinothricin and bialaphos. More recently, tomato plants were engineered to confer resistance to thiazopyr, a member of the pyridine herbicide family, via an esterase deactivation mechanism (Feng et al. 1998). Transgenic tomato seedlings demonstrated both *in vitro* and *in vivo* deactivation of thizopyr to the monoacid, resulting in the loss of herbicidal activity (Feng et al. 1995a, 1995b).

3.5.6.2 Insect Resistance

A wide variety of insecticides have been used to control insect damage in modern agriculture. Chemical control is expensive, and large application of insecticides could result in toxic residues in food and food products and have adverse effects on non-target organisms and environment. Thus, efforts have been focused on improving plant defense against insect attack though conventional plant breeding and biotechnological approaches. Among those, transgenic resistance to insects has been demonstrated in plants expressing insecticidal genes such as δ-endotoxins from *Bacillus thuringiensis* (*Bt*), protease inhibitors, enzymes, secondary plant metabolites, and plant lectins (Sharma et al. 2004).

Transgenic tomatoes carrying *Bt* genes were one of the first examples of genetically modified plants with resistance to insects (Fischhoff et al. 1987). The field tests of the transgenic tomato plants revealed a significant reduction in damage to tomato fruit by the lepidopteran insects including tobacco hornworm (*Manduca sexta*), tomato pinworm (*Keiferia lycopersicella*), and tomato fruitworm (*Helicoverpa zea*) (Delannay et al. 1989; van der Salm et al. 1994). However, expression of the protein was too low for commercial use. This problem was overcome through the use of a redesigned synthetic *Bt* gene, which resulted in a 100-fold increase in expression of the protein as compared with the wild-type gene (Perlak et al. 1991). Transgenic tomato plants with a chimeric δ-endotoxin gene of *Bt* subsp. *tenebrionis* (*B.t.t.*) displayed a high level of resistance to Colorado potato beetle larvae (Rhim et al. 1995). More recently, expression in tomato of a synthetic *cry1Ac* gene coding for an insecticidal crystal protein (ICP) of *Bt thuringiensis* was highly effective against the larvae of tomato fruit borer (*Helicoverpa armigera*) (Mandaokar et al. 2000).

The second category of transgenes carrying insecticidal activity is plant-derived genes, such as proteinase inhibitors and lectins. Those insecticidal proteins of plant origin are of particular interest because they are part of the natural defense systems in plant against insect attack (Casaretto and Corcuera 1995). In tomato plants, proteinase inhibitors have been found to accumulate in response to infection by pathogenic microorganisms (Peng and Black 1976; Li et al. 2002a). Expression of a synthetic prosystemin gene in the transgenic tomato plants resulted in the constitutive expression and accumulation of proteinase inhibitor proteins, which, in turn, constitutively activates the octadecanoid pathway. The activation of the octadecanoid signaling pathway promotes the trans-genic tomato plants resistant to a broad spectrum of herbivores including various chewing insects and some pathogenic fungi (McGurl et al. 1994; Li et al. 2002a; Madureira et al. 2006).

The proteinase inhibitor genes have several limitations. Within a given insect species, some inhibitors are effective anti-metabolites, but others are not (Sharma and Ortiz 2000). Besides, inhibitors can display varied degrees of effectiveness as an anti-metabolite against different insects (Murdock and Shade 2002). More importantly, the proteinase inhibitor genes do not provide the acute mortality afforded by *Bt* proteins (Babu et al. 2003). An appropriate way to optimize the interactions between a proteinase inhibitor and its target proteinase is by utilizing two or more inhibitors in the transgenic plants to affect different digestive proteinase in the pest (Sharma and Ortiz 2000). Recently, Abdeen et al. (2005) demonstrated that leaf-specific over-expression of two distinct proteinase inhibitors (the potato Pi-II and varboxypeptidase inhibitor) that had been engineered into the tomato plants resulted in increased resistance to *Heliothis obsoleta* and *Liriomyza trifolii* larvae in homozygous tomato lines expressing high levels of the transgenes.

Plant lectins are a heterogenous group of carbohydrate-binding proteins that have a protective function against a range of insects. Wu et al. (2000) introduced the snowdrop lectin gene (*Galanthus nivalis agglutinin*) into tomato. The transgenic tomato plants displayed a high level of resistance to aphid (*Myzus persicae Sulzer*) larvae.

3.5.6.3 *Virus Resistance*

Cross-protection, in which plants are infected with a mild virus strain to protect against infection with virulent strains, has been used in agriculture to increase plant yield (Tumer et al. 1987; Nelson et al. 1988; Praveen et al. 2005). Following the introduction of genetic engineering procedures, new genes derived from viruses themselves have been exploited for engineering virus resistance into plants, a concept referred to as pathogen-derived resistance (PDR) (Sanford and Johnston 1985). The resistance is usually mediated either by the transgene proteins (protein mediated) or transgenic RNAs (RNA mediated). Expression of a coat protein (CP) virus gene in transgenic tobacco plants resulted in the introduction of resistance to tobacco mosaic virus (TMV) (Abel et al. 1986). This coat-protein-mediated resistance (CP-MR) strategy was subsequently employed in tomato to engineer resistances to alfalfa mosaic virus, tobacco mosaic virus, and Physallis mottle tymovirus (Nelson et al. 1988; Tumer et al. 1987; Vidya et al. 2000). The field trail of tomato plants exhibited CP-MR against TMV, demonstrating that the genetically modified plants were highly resistant to this disease (Nelson et al. 1988).

The RNA-mediated resistance approach operates via a post-transcriptional gene silencing (PTGS) mechanism (Nelson et al. 1988). Research has demonstrated that the expression of the *N* gene from tomato spotted wilt virus (TSWV) was effective in promoting resistance to TSWV in transgenic tomato plants and tomato hybrids (Kim et al. 1994; Ultzen et al. 1995; Fedorowicz et al. 2005). Field trails using these transgenic tomato hybrids demonstrated functional resistance to TSWV under high inoculum pressure (Gonsalves et al. 1996; Haan et al. 1996). RNA-mediated resistance to TSWV can usually compete or even over-perform the natural resistance conferred by the *Sw-5* gene (Haan et al. 1996; Gubba et al. 2002). An evaluation of a non-transgenic tomato hybrid harboring the naturally occurring resistance gene *Sw-5*, and a transgenic tomato hybrid expressing the *N* gene demonstrated that no transgenic hybrid plants became infected, but TSWV replication was found in the leaf discs of the commercial hybrid containing the *Sw-5* gene (Accotto et al. 2005). Besides the *N* gene, the *NSm* gene of TSWV and the *NSs* gene of Potato virus X (PVX) have also been engineered in plants, and the resulting transgenic plants conferred resistance to TSWV albeit at lower levels (Accotto et al. 2005).

Resistance to *Begamovirus* has also been engineered in tomato via RNA-mediated strategy using partial, entire, or mutated *Begomovirus* replication-associated protein (*Rep*) genes. Transgenic expression of a truncated form of the tomato yellow leaf curl Sardinia virus (TYLCSV) *C1* gene encoding *Rep*'s first 210 amino acids conferred resistance to the homologous virus in tomato plants (Brunetti et al. 2001). Polston and Heibert (2001) demonstrated a single copy of the tomato mottle virus (ToMoV) *Rep* gene engineered into tomato conferred stable and high levels of resistance to ToMoV under field conditions. Recently, resistance to tomato leaf curl virus (TLCV) and tomato yellow leaf curl virus (TYLCV) were also engineered in tomato using respective viral *Rep* gene sequences (Bendahmane and Gronenborn 1997; Y. Yang et al. 2004b; Praveen et al. 2005).

In general, RNA-mediated resistance is effective only against viral strains that have a high degree of sequence homology to the transgene. For example, transgenic plants expressing the *N* gene showed a high level of resistance to TSWV, but not to the groundnut ringspot virus (GRSV), which is distantly related to TSWV at the nucleotide level (Pang et al. 1994). This problem may be overcome by introduction of multiple virus resistances through transforming plants with a construct containing gene segments from several different viruses (Jan et al. 2000). A single chimeric gene construct containing gene segments of two distinct viruses, TLCV and cucumber mosaic virus (CMV), was transformed into tomato, and the resulting transgenic tomato plants conferred resistance to both viruses (Antony et al. 2005). Furthermore, the strategy of multiple virus resistance can be combined with the natural resistance in tomato conferred by the *Sw-5* gene, leading to a durable, oligogenic form of virus resistance (Prins and Goldbach 1998). This approach was recently demonstrated in transgenic tomato, which combined transgenic and natural resistance to tospovirus in a single plant resulting in broad resistance to various tospoviruses including a lettuce isolate of TSWV, a tomato isolate from Hawaii (TSWV-H), and a Brazillian isolate of GRSV (Gubba et al. 2002).

3.5.6.4 *Other Disease Resistance*

Progression in the development of bacterial or fungal pathogen resistant crops has lagged behind development of insect and virus resistance crops. However, advances in the cloning of several bacterial resistance genes, such as the tomato *Pto* gene, may provide insights into plant-bacterial interactions at the molecular level. The *Pto* gene specifies race-specific resistance to the bacterial pathogen *Pseudomonas syringae* pv *tomato* carrying the *avrPto* gene. Tang et al. (1999) demonstrated that overexpression of the *Pto* gene in tomato activates defense responses and general resistance in the absence of the *Pto-AvrPto* interaction. Transformation of *Nicotiana benthamiana* with *Pto* results in specific resistance to *Pseudomonas syringae* pv *tabaci* strains carrying *avrPto*, suggesting that the functionality of host-specific resistance genes can be extended by intergeneric transfer (Rommens et al. 1995). Besides *Pto*, *Pfr* is also required for resistance to the bacterial pathogen. Overexpression of *Prf* leads to enhanced resistance to a number of normally virulent bacterial and viral pathogens and to increased sensitivity to the organophosphate insecticide fenthion (Oldroyd and Staskawicz 1998). Based on the results from an assay of bacterial and fungal plant pathogens using three cationic lytic peptides—including MSI-99, magainin II (MII), and cecropin B (CB)—Alan and Earle (2002) suggested that a cationic lytic peptides (MSI-99) can be used as a transgene to generate tomato lines with enhanced resistance to both bacterial and fungal diseases.

Enzymes that degrade the major constituents of the fungal cell wall (chitin and β-1, 3- glutin) play a key role in plant protection against fungal disease. Transgenic tomato plants expressing a *L. chilense* chitinase gene demonstrated improved resistance to Verticillium wilt (Tabaeizadeh et al. 1999). Increased resistance to Fusarium wilt was found in transgenic tomato plants that expressed tobacco β-1, 3-glucanase, and bean chitinase (Ouyang et al. 2003). Co-expression of chitinase and glucanase genes in transgenic tomato plants showed significantly enhanced resistance to Fusarium wilt compared to those that only expressed a single transgene (Zhu et al. 1994; Jongedijk et al. 1995). Similarly, co-expression of a tobacco osmotin gene and a bean chitinase gene in transgenic tomato plants conferred improved resistance to Fusarium wilt (Ouyang et al. 2005). Disease assays of transgenic tomato plants expressing an *Arabidopsis thionin* (*Thi2.1*) gene driven by fruit-inactive promoter revealed significant levels of enhanced resistance to both bacterial wilt and Fusarium wilt (Chan et al. 2005).

The tomato *Hero A* gene is the only member of a multigene family that confers a high level of resistance to all the economically important pathotypes of potato cyst nematode (PCN) species

Globodera rostochiensis and *G. pallida*. Transgenic tomato plants expressing the *Hero* gene showed a similar level of resistance to PCN as tomato line LA1792 that contains the introgressed *Hero* multigene family (Sobczak et al. 2005). The tomato root-knot nematode resistance gene, *Mi-1.2*, was transformed into two tomato lines (Goggin et al. 2004). In both lines a reduction in resistance was noted in the T_2 generation, and this was more pronounced in the T_3 generation. In contrast, the transgenic plants did not show reduced transcript levels, and *Mi-1.2* mRNA levels in the transgenic plants were comparable to levels observed in resistant control plants. Reductions in nematode resistance could be due to silencing of other endogenous gene(s) involved in resistance. Alternatively, these reductions could be caused by previously uncharacterized mechanisms of transgene inactivation that target the *Mi-1.2* protein or block its production.

3.5.6.5 *Abiotic Stress Tolerance*

A plant's response to abiotic stress—particularly cold, drought, and salt stress—involves the functionality of many genes that lead to a series of biochemical and physiological changes. Numerous genes have been engineered into various plant species, resulting in stress-tolerant plants (Foolad 2004). In tomato, limited efforts have been carried out to develop plants with enhanced salt tolerance using the genetic engineering approach. Expression of the yeast *HAL1* gene in transgenic tomato plants minimizes the reduction in fruit production caused by salt stress (Gisbert et al. 2000; Rus et al. 2001). Enhanced water and K^+ contents have been observed in the transgenic lines under salt stress conditions, which results in enhanced growth and maintenance of the fruit production. Similarly, transgenic tomato plants engineered with the yeast *HAL2* gene also showed increased tolerance to salt stress (Arrillaga et al. 1998). However, these transgenic plants, when grown under normal conditions (i.e., without salt stress), exhibited reduced growth in the short-term (Gisbert et al. 2000) as well as in the long term as reflected by lower fruit yield (Rus et al. 2001). Improving salt tolerance of tomato plants has also been achieved by metabolic engineering of the betaine aldehyde dehydrogenase (*BADH*) gene from *Atriplex hortensis* (Jia et al. 2002) and from sorghum (Moghaieb et al. 2000). The transgenic plants exhibited tolerance to salt stress, and grew normally at salt concentrations up to 120 mM. Another recent significant achievement was the development of transgenic tomato plants overexpressing *AtNHX1*, a single gene controlling vacuolar Na^+/H^+ antiport protein from *Arabidopsis thaliana* (Zhang and Blumwald 2001). Transgenic plants overexpressing this gene were able to grow, produce flowers, and set fruits in the presence of 200 mM sodium chloride, whereas the nontransgenic control plants did not survive. The high concentration of salt was only found in leaves and not in the fruit, demonstrating the potential use of these transgenic tomato plants for agricultural use in saline soils.

Chilling injury, another limiting factor in crop production worldwide, is controlled by many genes (McKersie and Leshem 1994). Improving chilling tolerance by using traditional breeding approaches has been limited. With recent advances in plant transformation technology, researchers have explored the genetic engineering approach for enhancing cold tolerance in various plant species including tomato (Park et al. 2004; Xu et al. 2004). Park et al. (2004) demonstrated that genetic engineering of a metabolic pathway in tomato protects seeds, plants, and flowers from chilling damage. Tomato was transformed by a chloroplast-targeted *codA* gene of *Arthrobacter globiformis*, which encodes choline oxidase to catalyze the conversion of choline to glycinebetaine (GB), a metabolite playing a key role in enhancing cold stress (Kishitani et al. 1994). The transgenic plants expressing *codA* gene accumulate GB and exhibit enhanced chilling tolerance at several development stages.

Expression or overexpression of transgenes conferring tolerance to multiple stresses has also been demonstrated in engineered tomato plants. Hsieh et al. (2002a, 2002b) reported that the use of a strong constitutive 35S cauliflower mosaic virus (CaMV) promoter to drive the expression of *Arabidopsis CBF1* in tomato improved tolerance to chilling, drought, and salt stress, but the tolerance was achieved at the expense of plant growth and yield. In a later study, Lee et al. (2003) used three copies of a stress-inducible ABRC1 (ABA-responsive complex) promoter from the barley *HAV22* gene to drive the expression of *Arabidopsis CBF1*, in order to improve the agronomic performance of the transgenic tomato plants. Constitutive expression of ABRC1-*CBF1* in transgenic tomato plants exhibited enhanced tolerance to chilling, drought, and salt stress in comparison with untransformed plants. Furthermore, plants with ABRC1-*CBF1* also maintained normal growth and yield that was equivalent to the untransformed plants under normal growth conditions. In another report, Xu et al. (2004) transformed tomato with animal antiapoptotic genes *bcl-xL* and *ced-9* through *Agrobacterium*-mediated transformation. The transgenic tomato plants expressing *bcl-xL* and *ced-9* demonstrated enhanced tolerance to both biotic and abiotic stress including virus infection and chilling, but plant development was also affected. More recently, the *yeast trehalose-6-phosphate synthase* (*TPS1*) gene, under the control of the cauliflower mosaic virus regulatory sequences (CaMV35S), was engineered in tomato (Cortina and Culianez-Macia 2005). Transgenic tomato plants overexpressing the *TPS1* gene exhibited an increase of tolerance to salt, drought, and oxidative stress, as compared to the nontransformed plants. Due to increased drought tolerance, transgenic plants usually display reduced dehydration and wilting symptoms, permitting floral and vegetative development after rewatering, and thus obtaining a tomato production similar to standard growth conditions. These results demonstrate the potential use of a transgenic approach in developing new tomato cultivars tolerant to abiotic stress without decreased tomato productivity.

3.5.6.6 Genetic Engineering of Fruit Quality and Other Attributes

Great advances have been achieved in manipulating components of tomato fruit ripening using the genetic engineering approach. Numerous biochemical changes take place during tomato fruit ripening, many of which affect the intrinsic quality of the fruit. The most important processes include synthesis and action of ethylene, induction of ripening by hormones, inhibition of carotene biosynthesis, metabolism of sugars, organic acids and lipids involved in the generation of flavor, and modifications of the structure and composition of the cell walls affecting fruit firmness and solids content (Oke et al. 2003). These changes are primarily the result of the expression of genes controlling quality attributes. Modification of the expression of these genes through genetic engineering can therefore lead to improvement of fruit quality.

Fruit firmness and texture are major quality attributes of fresh tomato. Genetic engineering has been used to improve the processing attributes by reducing the activity of cell-wall degrading enzymes such as cellulose, polygalacturonase (PG), and pectinesterase (PE) (Sheehy et al. 1988; Smith et al. 1988; Hall et al. 1993), or by inhibiting the activity of phospholipase D (PLD), a key enzyme that initials membrane deterioration and leads to the loss of compartmentalization and homeostasis during fruit ripening and senescence (Paliyath and Droillard 1992; Oke et al. 2003; Pinhero et al. 2003).

Flavor is another important quality attribute of tomato and its processed products, but its genetic basis is complex, making it difficult to alter by conventional plant breeding. Genetic engineering provides an alternative to modify flavor characteristics. Transgenic tomato plants expressing a thaumatin gene from *Thaumatococcus daniellii Benth* were sweeter than the controls and possessed a specific aftertaste as determined by sensory evaluation (Bartoszewski et al. 2003). Reduced activity of PLD in the transgenic tomato plants can tremendously improve the flavor quality (Oke et al. 2003).

Another area of great interest is the ability to manipulate ethylene production. Ethylene is involved in many plant processes ranging from seed germination to leaf and flower senescence and fruit ripening. Genetic and molecular analyses of fruit development, especially fruit ripening, have resulted in significant gains in knowledge in the areas of ethylene biosynthesis and response and are summersized in numerous reviews (Jiang and Fu 2000; Giovannoni 2001, 2004; Kepczynski and Kepczynska 2005).

3.6 CONCLUDING REMARKS

Extensive germplasm resources and a vast amount of genetic variation in tomato and its wild relatives are the primary assets for tomato improvement. The wild species in *Lycopersicon* genus and two *Solanum* species (*S. lycopersicoides* and *S. sitiens*) possess many traits of economic importance, including resistance to disease, insects, and nematodes, tolerance to abiotic stress, and superior fruit quality. All these favorable attributes are potentially transferable to the cultivated tomato via conventional breeding although various approaches have to be taken to break the crossing barriers between some of these species. Somatic hybridization using protoplast fusion offers the opportunity to transfer germplasm from more distant *Solanum* species into tomato. Marker-assisted selection (MAS) has the potential to greatly speed up the breeding process, and has been used in tomato breeding programs for selection of numerous traits. Transfer of genes through genetic engineering offers several advantages such as direct access to a vast pool of genes and concurrent introduction of novel genes without deleterious effects due to linkage drag. In conclusion, both conventional breeding and molecular breeding approaches utilizing natural and generated genetic variation have played a key role in tomato improvement, and will continue to do so in the future.

REFERENCES

Abdeen, A. et al., Multiple insect resistance in transgenic tomato plants over-expressing two families of plant proteinase inhibitors, *Plant Mol. Biol.*, 57, 189–202, 2005.

Abdulbaki, A. A., Tolerance of tomato cultivars and selected germplasm to heat-stress, *J. Am. Soc. Hort. Sci.*, 116, 1113–1116, 1991.

Abel, P. P. et al., Delay of disease development in transgenic plants that express the tobacco mosaic virus coat protein gene, *Science*, 232, 738–743, 1986.

Accotto, G. P. et al., Field evaluation of tomato hybrids engineered with tomato spotted wilt virus sequences for virus resistance, agronomic performance, and pollen-mediated transgene flow, *Phytopathology*, 95, 800–807, 2005.

Acosta, C. J. et al., Heredity of bacterial wilt resistance in tomato, *Proc. Am. Soc. Hort. Sci.*, 84, 455–462, 1964.

Agrama, H. A., Scott, J. W., Quantitative trait loci for tomato yellow leaf curl virus and tomato mottle virus resistance in tomato, *J. Amer. Sco. Hort. Sci.*, 131, 267–273, 2006.

Alan, A. R. and Earle, E. D., Sensitivity of bacterial and fungal plant pathogens to the lytic peptides, MSI-99, magainin II, and cecropin B, *Mol. Plant Microbe Interact.*, 15, 701–708, 2002.

Alexander, L. J., Embryo culture of tomato interspecific hybrids (Abstr), *Phytopathology*, 46, 6, 1956.

Ali, S. N. H. et al., Establishment of a complete series of a monosomic tomato chromosome addition lines in the cultivated potato using RFLP and GISH analyses, *Theor. Appl. Genet.*, 103, 687–695, 2001a.

Ali, S. N. H. et al., Genomic in situ hybridization analysis of a trigenomic hybrid involving *Solanum* and *Lycopersicon* species, *Genome*, 44, 299–304, 2001b.

Alpert, K. B. and Tanksley, S. D., High-resolution mapping and isolation of a yeast artificial chromosome contig containing *fw2.2*: a major fruit weight quantitative trait locus in tomato, *Proc. Natl. Acad. Sci. U.S.A.*, 93, 15503–15507, 1996.

Alvarez, A. E. et al., Use of microsatellites to evaluate genetic diversity and species relationships in the genus *Lycopersicon*, *Theor. Appl. Genet.*, 103, 1283–1292, 2001.

Ammati, M. et al., Retention of resistance to the root-knot nematode, *Meloidogyne incognita*, by *Lycopersicon* plants reproduced through tissue culture, *Plant Sci. Lett.*, 35, 247–250, 1984.

Ammiraju, J. S. S. et al., The heat-stable root-knot nematode resistance gene *Mi-9* from *Lycopersicon peruvianum* is localized on the short arm of chromosome 6, *Theor. Appl. Genet.*, 106, 478–484, 2003.

Anderson, L. K. et al., The relationship between genome size and synaptonemal complex length in higher plants, *Exp. Cell Res.*, 156, 367–378, 1985.

Antony, G. et al., A single chimeric transgene derived from two distinct viruses for multiple virus resistance, *J. Plant Biochem. Bio.*, 14, 101–105, 2005.

Anwar, A. et al., Bacterial canker (*Clavibacter michiganensis* subsp. *michiganensis*) of tomato in commercial seed produced in Indonesia, *Plant Dis.*, 88, 680, 2004.

Areshchenkova, T. and Ganal, M. W., Comparative analysis of polymorphism and chromosomal location of tomato microsatellite markers isolated from different sources, *Theor. Appl. Genet.*, 104, 229–235, 2002.

Armstrong, K. C. et al., *Hordeum chilense* ($2n = 14$) computer-assisted Giemsa karyotypes, *Genome*, 29, 683–688, 1987.

Arrillaga, I. et al., Expression of the yeast *HAL2* gene in tomato increases the *in vitro* salt tolerance of transgenic progenies, *Plant Sci.*, 136, 219–226, 1998.

Arumuganathan, K. and Earle, E. D., Nuclear DNA content of some important plant species, *Plant Mol. Biol. Rep.*, 9, 208–218, 1991.

Atanassova, B., Functional male sterility (*ps-2*) in tomato (*Lycopesicon esculentum* Mill.) and its application in breeding and hybrid seed production, *Euphytica*, 107, 13–21, 1999.

Atanassova, B. and Georgiev, H., Using genic male sterility in improving hybrid seed production in tomato (*Lycopersicon esculentum* Mill.), *Acta Hort.*, 579, 185–188, 2002.

Atherton, J. G. and Harris, G. P., Flowering, In *The Tomato Crop*, Atherton, J. G. and Rudich, J., Eds., New York: Chapman and Hall, p. 167, 1986.

Azanza, F. et al., Genes from *Lycopersicon chmielewskii* affecting tomato quality during fruit ripening, *Theor. Appl. Genet.*, 91, 495–504, 1995.

Babu, R. M. et al., Advances in genetically engineered (transgenic) plants in pest management—an over view, *Crop Prot.*, 22, 1071–1086, 2003.

Baergen, K. D. et al., Resistance of tomato genotypes to 4 isolates of *Verticillium dahliae* race 2, *HortScience*, 28, 833–836, 1993.

Bai, Y., The genetics and mechanisms of resistance to tomato powdery mildew (*Oidium neolycopersici*) in *Lycopersicon* species, PhD diss., Wageningen University, The Netherlands, 2004.

Bai, Y. et al., QTLs for tomato powdery mildew resistance (*Oidium lycopersici*) in *Lycopersicon parviflorum* G1.1601 co-localize with two qualitative powdery mildew resistance genes, *Mol. Plant Microbe Interact.*, 16, 169–176, 2003.

Bai, Y. et al., Mapping *Ol-4*, a gene conferring resistance to *Oidium neolycopersici* and originating from *Lycopersicon peruvianum* LA2172, requires multi-allelic, single-locus markers, *Theor. Appl. Genet.*, 109, 1215–1223, 2004.

Bai, Y. et al., Tomato defense to *Oidium neolycopersici*: dominant *Ol* genes confer isolate-dependent resistance via a different mechanism than recessive *ol-2*, *Mol. Plant Microbe Interact.*, 18, 354–362, 2005.

Dalibrea, M. E. et al., Salinity effects on some postharvest quality factors in a commercial tomato hybrid, *J. Hort. Sci.*, 72, 885–892, 1997.

Balint-Kurti, P. J. et al., RFLP linkage analysis of the *Cf-4* and *Cf-9* genes for resistance to *Cladosporium fulvum* in tomato, *Theor. Appl. Genet.*, 88, 691–700, 1994.

Balint-Kurti, P. J. et al., Integration of the classical and RFLP linkage maps of the short arm of tomato chromosome 1, *Theor. Appl. Genet.*, 90, 17–26, 1995.

Ballvora, A. et al., Chromosome landing at the tomato *Bs4* locus, *Mol. Genet. Genomics*, 266, 639–645, 2001a.

Ballvora, A. et al., Genetic mapping and functional analysis of the tomato *Bs4* locus governing recognition of the *Xanthomonas campestris* pv. *vesicatoria* AvrBs4 protein, *Mol. Plant Microbe Interact.*, 14, 629–638, 2001b.

Barden, K. A. et al., Regeneration and screening of tomato somaclones for resistance to tobacco mosaic virus, *Plant Sci.*, 45, 209–213, 1986.

Barksdale, T. H. and Stoner, A. K., Resistance in tomato to *Septoria lycopersici*, *Plant Dis Rep*, 62, 844–847, 1978.

Barone, A., Molecular marker-assisted selection for potato breeding, *Am. J. Potato Res.*, 81, 111–117, 2004.

Barton, D. W., Pachytene morphology of the tomato chromosome complement, *Am. J. Bot.*, 37, 639–643, 1950.

Barton, D. W., Comparative effects of x-ray and ultraviolet radiation on the differentiated chromosomes of the tomato, *Cytologia*, 19, 157–175, 1954.

Barton, D. W. et al., Rules for nomenclature in tomato genetics, *J. Hered.*, 46, 22–26, 1955.

Bartoszewski, G. et al., Modification of tomato taste in transgenic plants carrying a thaumatin gene from Thaumatococcus daniellii Benth, *Plant Breed.*, 122, 347–351, 2003.

Bashi, E. et al., Resistance in tomatoes to *Stemphylium floridanum* and *S. botryosum* f. sp. *lycopersici.*, *Phytopathology*, 63, 1542–1544, 1973.

Behare, J. et al., Restriction fragment length polymorphism mapping of the *Stemphylium* resistance gene in tomato, *Mol. Plant Microbe Interact.*, 4, 489–492, 1991.

Bendahmane, M. and Gronenborn, B., Engineering resistance against tomato yellow leaf curl virus (TYLCV) using antisense RNA, *Plant Mol. Biol.*, 33, 351–357, 1997.

Bernacchi, D. and Tanksley, S. D., An interspecific backcross of *Lycopersicon esculentum* × *L. hirsutum*: linkage analysis and a QTL study of sexual compatibility factors and floral traits, *Genetics*, 147, 861–877, 1997.

Bernacchi, D. et al., Advanced backcross QTL analysis in tomato. I. Identification of QTLs for traits of agronomic importance from *Lycopersicon hirsutum.*, *Theor. Appl. Genet.*, 97, 381–397, 1998a.

Bernacchi, D. et al., Advanced backcross QTL analysis of tomato. II. Evaluation of near-isogenic lines carrying single-donor introgressions for desirable wild QTL-alleles derived from *Lycopersicon hirsutum* and *L. pimpinellifolium*, *Theor. Appl. Genet.*, 97, 170–180, 1998b.

Bernatzky, R. and Tanksley, S. D., Toward a saturated linkage map in tomato based on isozymes and random cDNA sequences, *Genetics*, 112, 887–898, 1986.

Berry, S. Z. and Oakes, G. L., Inheritance of resistance to fusarium crown and root rot in tomato, *HortScience*, 22, 110–111, 1987.

Berry, S. Z. and Uddin, M. R., Breeding tomato for quality and processing attributes, In *Genetic Improvement of Tomato Monographs Theoretical and Applied Genetics*, Kalloo, G., Ed., Vol. 14, Berlin: Springer-Verlag, p. 197, 1991.

Bhatia, P. et al., Tissue culture studies of tomato (*Lycopersicon esculentum*), *Plant Cell Tiss. Org.*, 78, 1–21, 2004.

Bhatt, R. P. et al., Heterosis, combining ability and genetics for vitamin C, total soluble solids and yield in tomato (*Lycopersicon esculentum*) at 1700 m altitude, *J. Agric. Sci.*, 137, 71–75, 2001.

Bohn, G. W. and Tucker, C. M., Immunity to Fusarium wilt in the tomato, *Science*, 89, 603–604, 1939.

Bonnema, G. et al., Tomato chromosome 1: high resolution genetic and physical mapping of the short arm in an interspecific *Lycopersicon esculentum* × *L. peruvianum* cross, *Mol. Gen. Genet.*, 253, 455–462, 1997.

Boswell, V. R., Improvement and genetics of tomatoes, peppers, and eggplant, In *Yearbook of Agriculture*, Washington, DC: United States Government Printing Office, pp. 176–206, 1937.

Breto, M. P. et al., Genetic variability in *Lycopersicon* species and their genetic relationships, *Theor. Appl. Genet.*, 86, 113–120, 1993.

Briggs, F. N., Breeding wheat resistant to bunt by backcross method, *J. Am. Soc. Agro.*, 22, 239–244, 1930.

Brim, C. A., A modified pedigree method of selection in soybeans, *Crop Sci.*, 6, 220, 1966.

Broome, C. R. et al., Proposal to conserve *Lycopersicon esculentum* Miller as the scientific name of the tomato, *Tomato Genet. Coop. Rep.*, 33, 55, 1983.

Brouwer, D. J. et al., QTL analysis of quantitative resistance to *Phytophthora infestans* (late blight) in tomato and comparisons with potato, *Genome*, 47, 475–492, 2004.

Brown, S. W., The structure and meiotic behavior of the differentiated chromosomes of tomato, *Genetics*, 34, 437–461, 1949.

Brunetti, A. et al., Transgenically expressed T-Rep of tomato yellow leaf curl Sardinia virus acts as a trans-dominant-negative mutant, inhibiting viral transcription and replication, *J. Virol.*, 75, 10573–10581, 2001.

Budiman, M. A. et al., Localization of *jointless-2* gene in the centromeric region of tomato chromosome 12 based on high resolution genetic and physical mapping, *Theor. Appl. Genet.*, 108, 190–196, 2004.

Burbidge, A. et al., Re-orientation and integration of the classical and interspecific linkage maps of the long arm of tomato chromosome 7, *Theor. Appl. Genet.*, 103, 443–454, 2001.

Burdick, A. B., Genetics of heterosis for earliness in the tomato, *Genetics*, 39, 488–505, 1954.

Canady, M. A. et al., A library of *Solanum lycopersicoides* introgression lines in cultivated tomato, *Genome*, 48, 685–697, 2005.

Casali, V. W. D. and Tigchelaar, E. C., Breeding progress in tomato with pedigree selection and single seed descent, *J. Am. Soc. Hort. Sci.*, 100, 362–364, 1975.

Casaretto, J. A. and Corcuera, L. J., Plant proteinase inhibitors: a defensive response against insects, *Biol. Res.*, 28, 239–249, 1995.

Cassells, A. C., Effect of 2,3,5-triiodobenzoic acid on caulogenesis in callus-cultures of tomato and pelargonium, *Physiol. Plantarum*, 46, 159–164, 1979.

Chagué, V. et al., Identification of RAPD markers linked to a locus involved in quantitative resistance to TYLCV in tomato by bulked segregant analysis, *Theor. Appl. Genet.*, 95, 671–677, 1997.

Chalukova, M. and Manuelyan, H., Breeding for carotenoid pigments in tomato, In *Genetic Improvement of Tomato Monographs Theoretical and Applied Genetics*, Kalloo, G., Ed., Vol. 14, Berlin: Springer-Verlag, p. 179, 1991.

Chan, Y. L. et al., Transgenic tomato plants expressing an *Arabidopsis thionin (Thi2.1)* driven by fruit-inactive promoter battle against phytopathogenic attack, *Planta*, 221, 386–393, 2005.

Chang, S. B., Cytogenetic and molecular studies on tomato chromosomes using diploid tomato and tomato monosomic additions in tetraploid potato, PhD diss., Wageningen University, The Netherlands, 2004.

Chang, S. B. and de Jong, H., Production of alien chromosome additions and their utility in plant genetics, *Cytogenet. Genome Res.*, 109, 335–343, 2005.

Chen, G. Q. and Foolad, M. R., A molecular linkage map of tomato based on a cross between *Lycopersicon esculentum* and *L. pimpinellifolium* and its comparison with other molecular maps of tomato, *Genome*, 42, 94–103, 1999.

Chen, L. Z. and Adachi, T., Efficient hybridization between *Lycopersicon esculentum* and *L. peruvianum* via "embryo rescue" and *in vitro* propagation, *Plant Breed.*, 115, 251–256, 1996.

Cheng, X. Y. et al., Agrobacterium-transformed rice plants expressing synthetic *cryIA(b)* and *cryIA(c)* genes are highly toxic to striped stem borer and yellow stem borer, *Proc. Natl. Acad. Sci. U.S.A.*, 95, 2767–2772, 1998.

Chetelat, R. T., Revised list of monogenic stocks, *Tomato Genet. Coop. Rep.*, 52, 41–62, 2002.

Chetelat, R. T., Revised list of wild species stocks, *Tomato Genet. Coop. Rep.*, 54, 52–76, 2004.

Chetelat, R. T., Revised list of monogenic stocks, *Tomato Genet. Coop. Rep.*, 55, 48–69, 2005.

Chetelat, R. T. and Deverna, J. W., Expression of unilateral incompatibility in pollen of Lycopersicon pennellii is determined by major loci on chromosomes 1, 6 and 10, *Theor. Appl. Genet.*, 82, 704–712, 1991.

Chetelat, R. T. and Ji, Y., Cytogenetics and evolution, In *Genetic Improvement of Solanaceous Crops. Vol. 2: Tomato*, Razdan, M. K. and Mattoo, A. K., Eds., New Delhi, India: Science Publisher, 2006.

Chetelat, R. T. and Meglic, V., Molecular mapping of chromosome segments introgressed from *Solanum lycopersicoides* into cultivated tomato (*Lycopersicon esculentum*), *Theor. Appl. Genet.*, 100, 232–241, 2000.

Chetelat, R. T. and Peterson, J., Revised list of miscellaneous stocks, *Tomato Genet. Coop. Rep.*, 53, 44–61, 2003.

Chetelat, R. T. et al., A genetic map of tomato based on BC$_1$ *Lycopersicon esculentum* × *Solanum lycopersicoides* reveals overall synteny but suppressed recombination between these homeologous genomes, *Genetics*, 154, 857–867, 2000.

Chetelat, R. T. et al., A male-fertile *Lycopersicon esculentum* × *Solanum lycopersicoides* hybrid enables direct backcrossing to tomato at the diploid level, *Euphytica*, 95, 99–108, 1997.

Chetelat, R. T. et al., Identification, transmission, and cytological behavior of *Solanum lycopersicoides* Dun. monosomic alien addition lines in tomato (*Lycopersicon esculentum* Mill.), *Genome*, 41, 40–50, 1998.

Child, A., A synopsis of *Solanum* subgenus *Potatoe* (G. DON) D'Arcy (*Tuberarium* (Dun.) Bitter (s.l.), *Feddes Repert.*, 101, 209–235, 1990.

Chmielewski, T. M. and Rick, C. M., *Lycopersicon* minutum, *Tomato Genet. Coop. Rep.*, 12, 21–22, 1962.

Choudhury, B., Embryo culture technique. III. Growth of hybrid embryos (*Lycopersicon esculentum* × *Lycopersicon peruvianum*) in culture medium, *Indian J. Hort.*, 12, 155–156, 1955.

Christakis, P. A. and Fasoulas, A. C., The effects of the genotype by environmental interaction on the fixation of heterosis in tomato, *J. Agric. Sci.*, 139, 55–60, 2002.

Chunwongse, J. et al., Chromosomal localization and molecular marker tagging of the powdery mildew resistance gene (*Lv*) in tomato, *Theor. Appl. Genet.*, 89, 76–79, 1994.

Chunwongse, J. et al., High-resolution genetic map of the *Lv* resistance locus in tomato, *Theor. Appl. Genet.*, 95, 220–223, 1997.

Chunwongse, J. et al., Molecular mapping of the *Ph-3* gene for late blight resistance in tomato, *J. Hort. Sci. Biotech.*, 77, 281–286, 2002.

Ciccarese, F. et al., Occurrence and inheritance of resistance to powdery mildew (*Oidium lycopersici*) in *Lycopersicon* species, *Plant Pathol.*, 47, 417–419, 1998.

Cirulli, M. and Ciccarese, F., Interaction between TMV isolate, temperature, allelic condition and combination of the *Tm* resistance genes in tomato, *Phytopathol. Medit.*, 14, 100–105, 1975.

Clayberg, C. D. et al., List of genes as of January 1959, *Tomato Genet. Coop. Rep.*, 9, 6–18, 1959.

Coaker, G. L. and Francis, D. M., Mapping, genetic effects, and epistatic interaction of two bacterial canker resistance QTLs from *Lycopersicon hirsutum*, *Theor. Appl. Genet.*, 108, 1047–1055, 2004.

Coaker, G. L. et al., A QTL controlling stem morphology and vascular development in *Lycopersicon esculentum*×*Lycopersicon hirsutum* (*Solanaceae*) crosses is located on chromosome 2, *Am. J. Bot.*, 89, 1859–1866, 2002.

Correll, D. S., A new species and some nomenclatural changes in *Solanum*, section *Tuberarium*, *Madrono*, 14, 232–236, 1958.

Cortina, C. and Culianez-Macia, F. A., Tomato transformation and transgenic plant production, *Plant Cell Tiss. Org.*, 76, 269–275, 2004.

Cortina, C. and Culianez-Macia, F. A., Tomato abiotic stress enhanced tolerance by trehalose biosynthesis, *Plant Sci.*, 169, 75–82, 2005.

Danesh, D. et al., Genetic dissection of oligogenic resistance to bacterial wilt in tomato, *Mol. Plant Microbe Interact.*, 7, 464–471, 1994.

D'Arcy, W. G., The classification of *Solanaceae*, In *The Biology and Taxanomy of the Solanaceae*, Hawkes, J. G. et al., Eds., London: Academic Press, p. 3, 1979.

Deblock, M. et al., Engineering herbicide resistance in plants by expression of a detoxifying enzyme, *Embo J.*, 6, 2513–2518, 1987.

De Giovanni, C. et al., Identification of PCR-based markers (RAPD, AFLP) linked to a novel powdery mildew resistance gene (*ol-2*) in tomato, *Plant Sci.*, 166, 41–48, 2004.

Delannay, X. et al., Field performance of transgenic tomato plants expressing the *Bacillus thuringiensis Var Kurstaki* insect control protein, *Bio-Technol.*, 7, 1265–1269, 1989.

Deverna, J. W. et al., Sexual hybridization of *Lycopersicon esculentum* and *Solanum rickii* by means of a sesquidiploid bridging hybrid, *Proc. Natl. Acad. Sci. U.S.A.*, 87, 9486–9490, 1990.

deVicente, M. C. and Tanksley, S. D., QTL analysis of transgressive segregation in an interspecific tomato cross, *Genetics*, 134, 585–596, 1993.

Dhillon, G. S. et al., Heterosis in tomato, *J. Res.*, 12, 1974.

Dickinson, M. J. et al., Close linkage between the *Cf-2/Cf-5* and *Mi* resistance loci in tomato, *Mol. Plant Microbe Interact.*, 6, 341–347, 1993.

Diwan, N. et al., Mapping of *Ve* in tomato: a gene conferring resistance to the broad-spectrum pathogen, *Verticillium dahliae* race 1, *Theor. Appl. Genet.*, 98, 315–319, 1999.

Doganlar, S. et al., Molecular mapping of the *Py-1* gene for resistance to corky root rot (*Pyrenochaeta lycopersici*) in tomato, *Theor. Appl. Genet.*, 97, 784–788, 1998.

Doganlar, S. et al., Mapping quantitative trait loci in inbred backcross lines of *Lycopersicon pimpinellifolium* (LA1589), *Genome*, 45, 1189–1202, 2002.

East, E. M., The distinction between development and heredity in inbreeding, *Am. Nat.*, 43, 173–181, 1909.

East, E. M., Heterosis, *Genetics*, 21, 375–397, 1936.

Egashira, H. et al., Genetic diversity of the "peruvianum-complex" (*Lycopersicon peruvianum* (L.) Mill. and *L. chilense* Dun.) revealed by RAPD analysis, *Euphytica*, 116, 23–31, 2000.

Eshed, Y. and Zamir, D., An introgression line population of *Lycopersicon pennellii* in the cultivated tomato enables the identification and fine mapping of yield-associated QTL, *Genetics*, 141, 1147–1162, 1995.

Farrar, F. R. and Kennedy, G. G., Insect and mite resistance in tomato, In *Genetic Improvement of Tomato Monographs Theoretical and Applied Genetics*, Kalloo, G., Ed., Vol. 14, Berlin: Springer-Verlag, p. 121, 1991.

Fedorowicz, O. et al., Pathogen-derived resistance to tomato spotted wilt virus in transgenic tomato and tobacco plants, *J. Am. Soc. Hort. Sci.*, 130, 218–224, 2005.

Feng, P. C. C. et al., Inhibition of thiazopyr metabolism in plant seedlings by inhibitors of monooxygenases, *Pesticide Sci.*, 45, 203–207, 1995a.

Feng, P. C. C. et al., Metabolic deactivation of the herbicide thiazopyr by animal liver esterases, *Xenobiotica*, 25, 27–35, 1995b.

Feng, P. C. C. et al., Engineering plant resistance to thiazopyr herbicide via expression of a novel esterase deactivation enzyme, *Pestic. Biochem. Physiol.*, 59, 89–103, 1998.

Fidantsef, A. L. et al., Signal interactions in pathogen and insect attack: expression of lipoxygenase, proteinase inhibitor II, and pathogenesis-related protein P4 in the tomato, *Lycopersicon esculentum*, *Physiol. Mol. Plant Pathol.*, 54, 97–114, 1999.

Fillatti, J. J. et al., Efficient transfer of a glyphosate tolerance gene into tomato using a binary agrobacterium tumefaciens vector, *Bio-Technol.*, 5, 726–730, 1987.

Fischhoff, D. A. et al., Insect tolerant transgenic tomato plants, *Bio-Technol.*, 5, 807–813, 1987.

Fobes, J. F., Trisomic analysis of isozymic loci in tomato species—segregation and dosage effects, *Biochem. Genet.*, 18, 401–421, 1980.

Foolad, M. R., Recent advances in genetics of salt tolerance in tomato, *Plant Cell Tiss. Org.*, 76, 101–119, 2004.

Foolad, M. R. and Chen, F. Q., RAPD markers associated with salt tolerance in an interspecific cross of tomato (*Lycopersicon esculentum×L. pennellii*), *Plant Cell Rep.*, 17, 306–312, 1998.

Foolad, M. R. et al., RFLP mapping of QTLs conferring cold tolerance during seed germination in an interspecific cross of tomato, *Mol. Breed.*, 4, 519–529, 1998.

Foolad, M. R. et al., Identification and validation of QTLs for salt tolerance during vegetative growth in tomato by selective genotyping, *Genome*, 44, 444–454, 2001.

Foolad, M. R. et al., Identification of QTLs for early blight (*Alternaria solani*) resistance in tomato using backcross populations of a *Lycopersicon esculentum×L. hirsutum* cross, *Theor. Appl. Genet.*, 104, 945–958, 2002.

Foolad, M. R. et al., Genetics of drought tolerance during seed germination in tomato: inheritance and QTL mapping, *Genome*, 46, 536–545, 2003.

Fransz, P. F. et al., High-resolution physical mapping in Arabidopsis thaliana and tomato by fluorescence in situ hybridization to extended DNA fibres, *Plant J.*, 9, 421–430, 1996.

Frary, A. et al., Molecular mapping of the centromeres of tomato chromosomes 7 and 9, *Mol. Gen. Genet.*, 250, 295–304, 1996.

Frary, A. et al., Identification of QTL for late blight resistance from *pimpinellifolium* L3708, *Tomato Genet. Coop. Rep.*, 48, 19–21, 1998.

Frary, A. et al., Advanced backcross QTL analysis of a *Lycopersicon esculentum×L. pennellii* cross and identification of possible orthologs in the Solanaceae, *Theor. Appl. Genet.*, 108, 485–496, 2004.

Frelichowski, J. E. and Juvik, J. A., Inheritance of sesquiterpene carboxylic acid synthesis in crosses of *Lycopersicon hirsutum* with insect-susceptible tomatoes, *Plant Breed.*, 124, 277–281, 2005.

Fridman, E. et al., Zooming in on a quantitative trait for tomato yield using interspecific introgressions, *Science*, 305, 1786–1789, 2004.

Friedmann, M. et al., Novel source of resistance to tomato yellow leaf curl virus exhibiting a symptomless reaction to viral infection, *J. Am. Soc. Hort. Sci.*, 123, 1004–1007, 1998.

Fuchs, J. et al., In situ localization of yeast artificial chromosome sequences on tomato and potato metaphase chromosomes, *Chromosome Res.*, 4, 277–281, 1996.

Fukui, K., Standardization of karyotyping plant chromosomes by a newly developed chromosome image analyzing system (CHIAS), *Theor. Appl. Genet.*, 72, 27–32, 1986.

Fukui, K., Recent development of image analysis methods in plant chromosome research, *Cytogenet. Genome Res.*, 109, 83–89, 2005.

Fukui, K. et al., Quantitative karyotyping of three diploid Brassica species by imaging methods and localization of 45S rDNA loci on the identified chromosomes, *Theor. Appl. Genet.*, 96, 325–330, 1998.

Fulton, T. M. et al., Introgression and DNA marker analysis of *Lycopersicon peruvianum*, a wild relative of the cultivated tomato, into *Lycopersicon esculentum*, followed through three successive backcross generations, *Theor. Appl. Genet.*, 95, 895–902, 1997a.

Fulton, T. M. et al., QTL analysis of an advanced backcross of *Lycopersicon peruvianum* to the cultivated tomato and comparisons with QTLs found in other wild species, *Theor. Appl. Genet.*, 95, 881–894, 1997b.

Fulton, T. M. et al., Advanced backcross QTL analysis of a *Lycopersicon esculentum×Lycopersicon parviflorum* cross, *Theor. Appl. Genet.*, 100, 1025–1042, 2000.

Fulton, T. M. et al., Identification, analysis, and utilization of conserved ortholog set markers for comparative genomics in higher plants, *Plant Cell*, 14, 1457–1467, 2002.

Gallegly, M. E. and Marvel, M. E., Inheritance of resistance to tomato race-O of *Phytophthora infestans*, *Phytopathology*, 45, 103–109, 1955.

Ganal, M. W. et al., A molecular and cytogenetic survey of major repeated dna-sequences in tomato (*Lycopersicon esculentum*), *Mol. Gen. Genet.*, 213, 262–268, 1988.

Ganal, M. W. et al., Genetic mapping of tandemly repeated telomeric DNA sequences in tomato (*Lycopersicon esculentum*), *Genomics*, 14, 444–448, 1992.

Ganal, M. W. et al., Genetic mapping of a wide spectrum nematode resistance gene (*Hero*) against *Globodera rostochiensis* in tomato, *Mol. Plant Microbe Interact.*, 8, 886–891, 1995.

Gardner, R. G., Use of *Lycopersicon hirsutum* PI 126445 in breeding early blight resistant tomatoes (Abstr), *HortScience*, 19, 208, 1984.

Gardner, R. G. and Shoemaker, P. B., "Mountain supreme" early blight-resistant hybrid tomato and its parents, NC EBR-3 and NC EBR-4, *HortScience*, 34, 745–746, 1999.

Garriga-Caldere, F. et al., Identification of alien chromosomes through GISH and RFLP analysis and the potential for establishing potato lines with monosomic additions of tomato chromosomes, *Genome*, 40, 666–673, 1997.

Garriga-Caldere, F. et al., Transmission of alien tomato chromosomes from BC_1 to BC_2 progenies derived from backcrossing potato (+) tomato fusion hybrids to potato: the selection of single additions for seven different tomato chromosomes, *Theor. Appl. Genet.*, 96, 155–163, 1998.

Garriga-Caldere, K. et al., Prospects for introgressing tomato chromosomes into the potato genome: an assessment through GISH analysis, *Genome*, 42, 282–288, 1999.

Gentile, A. G. et al., *Lycopersicon* and *Solanum* spp. resistance to carmine and two spotted spider mite, *J. Econ. Entomol.*, 62, 834–836, 1969.

Georgiev, H., Heterosis in tomato breeding, In *Genetic Improvement of Tomato Monographs Theoretical and Applied Genetics*, Kalloo, G., Ed., Vol. 14, Berlin: Springer-Verlag, p. 83, 1991.

Georgiev, H. and Atanassova, B., Genetical analysis of the characteristics determing a long style in the tomato flowers, *Genet. Plant Breed., Sofia*, 13, 126–133, 1980.

Gilbert, C. W., A computer programme for analysis of human chromosomes, *Nature*, 212, 1437, 1966.

Gill, B. S., Tomato cytogenetics-a search for new frontiers, In *Cytogenetics of Crop Plants*, Swaminathan, M. S. et al., Eds., New Delhi, India: MacMillan India, p. 456, 1983.

Giovannoni, J., Molecular biology of fruit maturation and ripening, *Annu. Rev. Plant Phys.*, 52, 725–749, 2001.

Giovannoni, J. J., Genetic regulation of fruit development and ripening, *Plant Cell*, 16, S170–S180, 2004.

Gisbert, C. et al., The yeast *HAL1* gene improves salt tolerance of transgenic tomato, *Plant Physiol.*, 123, 393–402, 2000.

Goggin, F. L. et al., Instability of *Mi*-mediated nematode resistance in transgenic tomato plants, *Mol. Breed.*, 13, 357–364, 2004.

Gonsalves, C. et al., Breeding transgenic tomatoes for resistance to tomato spotted wilt virus and cucumber mosaic virus, *Acta Hort.*, 431, 442–448, 1996.

Gottschalk, W., Undersuchungen am pachytan normaler und rontgenbestrahlter pollenmuttersellen von *Solanum lycopersicum*, *Chromosoma*, 4, 298–341, 1951.

Grandillo, S. and Tanksley, S. D., Genetic analysis of RFLPs, GATA microsatellites and RAPDs in a cross between *L. esculentum* and *L. pimpinellifolium*, *Theor. Appl. Genet.*, 92, 957–965, 1996.

Green, D. M. et al., An interactive, microcomputer-based karyotype analysis system for phylogenetic cyto-taxonomy, *Comput. Biol. Med.*, 10, 219–227, 1980.

Green, D. M. et al., Chrompac-III—an improved package for microcomputer-assisted analysis of karyotypes, *J. Hered.*, 75, 143, 1984.

Gresshof, P. M. and Doy, C. H., Development and differentiation of haploid *Lycopersicon esculentum* (tomato), *Planta*, 107, 161–170, 1972.

Griffing, B. and Zsiros, E., Heterosis associated with genotype-environment interactions, *Genetics*, 68, 443–455, 1971.

Griffiths, P. D. and Scott, J. W., Inheritance and linkage of tomato mottle virus resistance genes derived from *Lycopersicon chilense* accession LA 1932, *J. Am. Soc. Hort. Sci.*, 126, 462–467, 2001.

Grogan, R. G. et al., Stem canker disease of tomato caused by *Alternaria Alternata f. sp lycopersici*, *Phytopathology*, 65, 880–886, 1975.

Grube, R. C. et al., Comparative genetics of disease resistance within the *Solanaceae*, *Genetics*, 155, 873–887, 2000.

Guan, G. R. et al., A study of backcrossing the interspecific F$_1$ hybrid of *Lycopersicon esculentum* Mill×
   Lycopersicon peruvianum Mill to *Lycopersicon esculentum* Mill, *Acta Hort. Sin.*, 15, 39–44, 1988.

Gubba, A. et al., Combining transgenic and natural resistance to obtain broad resistance to tospovirus infection
   in tomato (*Lycopersicon esculentum* Mill), *Mol. Breed.*, 9, 13–23, 2002.

Gur, A. and Zamir, D., Unused natural variation can lift yield barriers in plant breeding, *Plos Biol.*, 2,
   1610–1615, 2004.

Haan, P. et al., Transgenic tomato hybrids resistant to tomato spotted wilt virus infection, *Acta Hort.*, 431,
   417–426, 1996.

Haanstra, J. P. et al., The *Cf-ECP2* gene is linked to, but not part of, the *Cf-4/Cf-9* cluster on the short arm of
   chromosome 1 in tomato, *Mol. Gen. Genet.*, 262, 839–845, 1999a.

Haanstra, J. P. W. et al., An integrated high density RFLP-AFLP map of tomato based on two *Lycopersicon
   esculentum*×*L. pennellii* F$_2$ populations, *Theor. Appl. Genet.*, 99, 254–271, 1999b.

Hall, L. N. et al., Antisense inhibition of pectin esterase gene expression in transgenic tomatoes, *Plant J.*, 3,
   121–129, 1993.

Handa, A. K. et al., Use of plant-cell cultures to study production and phytotoxicity of Alternaria dolani
   toxin(s), *Physiol. Plant Pathol.*, 21, 195–309, 1982.

Handa, A. K. et al., Clonal variation for tolerance to polyethylene glycol-induced water-stress in cultured
   tomato cells, *Plant Physiol.*, 72, 645–653, 1983a.

Handa, S. et al., Solutes contributing to osmotic adjustment in cultured plant-cells adapted to water-stress,
   Plant Physiol., 73, 834–843, 1983b.

Handa, S. et al., Proline accumulation and the adaptation of cultured plant cells to water stress, *Plant Physiol.*,
   80, 938–945, 1986.

Hanson, P. M. et al., Mapping a wild tomato introgression associated with tomato yellow leaf curl virus
   resistance in a cultivated tomato line, *J. Am. Soc. Hort. Sci.*, 15, 15–20, 2000.

Hassan, A. A. et al., Application of cotyledonary symptoms in screening for resistance to tomato bacterial
   canker and in host range studies, *Phytopathology*, 58, 233–239, 1968.

Hassan, A. A. et al., Inheritance of resistance to tomato yellow leaf curl virus derived from *Lycopersicon
   cheesmanii* and *Lycopersicon hirsutum*, *HortScience*, 19, 574–575, 1984.

Helentjaris, T. et al., Restriction fragment polymorphisms as probes for plant diversity and their development
   as tools for applied plant breeding, *Plant Mol. Biol.*, 5, 109–118, 1985.

Hemming, M. N. et al., Fine mapping of the tomato *I-3* gene for *Fusarium* wilt resistance and elimination of a
   co-segregating resistance gene analogue as a candidate for *I-3*, *Theor. Appl. Genet.*, 109, 409–418, 2004.

Hille, J. et al., Genetic transformation of tomato and prospects for gene transfer, In *Genetic Improvement of
   Tomato Monographs Theoretical and Applied Genetics*, Kalloo, G., Ed., Vol. 14, Berlin: Springer-Verlag,
   p. 283, 1991.

Hogenboom, N. G., Inheritance of resistance to corky root in tomato (*Lycopersicon esculentum* Mill), *Euphytica*, 19, 413–425, 1970.

Horsch, R. B. et al., A simple and general method for transferring genes into plants, *Science*, 227, 1229–1231,
   1985.

Hsieh, T. H. et al., Tomato plants ectopically expressing Arabidopsis *CBF1* show enhanced resistance to water
   deficit stress, *Plant Physiol.*, 130, 618–626, 2002a.

Hsieh, T. H. et al., Heterology expression of the Arabidopsis C-repeat/dehydration response element binding
   factor 1 gene confers elevated tolerance to chilling and oxidative stresses in transgenic tomato, *Plant
   Physiol.*, 129, 1086–1094, 2002b.

Huang, C. C. et al., Characterization and mapping of resistance to *Oidium lycopersicum* in two Lycopersicon
   hirsutum accessions: evidence for close linkage of two *Ol*-genes on chromosome 6 of tomato, *Heredity*,
   85, 511–520, 2000.

Hull, F. H., Recurrent selection for specific combining ability in corn, *J. Am. Soc. Agro.*, 37, 134–145, 1945.

Imanishi, S. et al., Development of interspecific hybrids between *Lycopersicon esculentum* and *L. peruvianum
   var humifusum* and introgression of *L. peruvianum* invertase gene into *L. esculentum*, *Breeding Sci.*, 46,
   355–359, 1996.

Jackson, S. A. et al., Application of fiber-FISH in physical mapping of Arabidopsis thaliana, *Genome*, 41,
   566–572, 1998.

Jacobsen, E. et al., The first and 2nd backcross progeny of the intergeneric fusion hybrids of potato and tomato
   after crossing with potato, *Theor. Appl. Genet.*, 88, 181–186, 1994.

Jacobsen, E. et al., Genomic in situ hybridization (GISH) and RFLP analysis for the identification of alien chromosomes in the backcross progeny of potato(+)tomato fusion hybrids, *Heredity*, 74, 250–257, 1995.

Jan, F. J. et al., A single chimeric transgene derived from two distinct viruses confers multi-virus resistance in transgenic plants through homology-dependent gene silencing, *J. Gen. Virol.*, 81, 2103–2109, 2000.

Jaramillo, J. and Summers, W. L., Dark-light treatments influence induction of tomato anther callus, *HortScience*, 26, 915–916, 1991.

Ji, Y. and Chetelat, R. T., Homoeologous pairing and recombination in *Solanum lycopersicoides* monosomic addition and substitution lines of tomato, *Theor. Appl. Genet.*, 106, 979–989, 2003.

Ji, Y. and Scott, J. W., Development of SCAR and CAPS markers linked to tomato begomovirus resistance genes introgressed from *Lycopersicon chilense*, *HortScience*, 40, 1090, 2005.

Ji, Y. and Scott, J. W., Identification of RAPD markers linked to *Lycopersicon chilense* derived geminivirus resistance genes on chromosome 6 of tomato, *Acta Horticulturae*, 695, 407–416, 2006.

Ji, Y. et al., Genome differentiation by GISH in interspecific and intergeneric hybrids of tomato and related nightshades, *Chromosome Res.*, 12, 107–116, 2004.

Jia, G. X. et al., Transformation of tomato with the *BADH* gene from Atriplex improves salt tolerance, *Plant Cell Rep.*, 21, 141–146, 2002.

Jiang, Y. M. and Fu, J. R., Ethylene regulation of fruit ripening: molecular aspects, *Plant Growth Regul.*, 30, 193–200, 2000.

Jiang, J. M. et al., Metaphase and interphase fluorescence in-situ hybridization mapping of the rice genome with bacterial artificial chromosomes, *Proc. Natl. Acad. Sci. U.S.A.*, 92, 4487–4491, 1995.

Jones, R. A. et al., A greenhouse screening method for *Pyrenochaeta lycopersici* resistance in tomatoes, *Euphytica*, 40, 187–191, 1989.

Jones, D. A. et al., Two complex resistance loci revealed in tomato by classical and RFLP mapping of the *Cf-2,Cf-4, Cf-5*, and *Cf-9* genes for resistance to *Cladosporium fulvum*, *Mol. Plant Microbe Interact.*, 6, 348–357, 1993.

Jongedijk, E. et al., Synergistic activity of chitinases and beta-1,3-glucanases enhances fungal resistance in transgenic tomato plants, *Euphytica*, 85, 173–180, 1995.

Kabelka, E. et al., Two loci from *Lycopersicon hirsutum* LA407 confer resistance to strains of *Clavibacter michiganensis* subsp. *michiganensis*, *Phytopathology*, 92, 504–510, 2002.

Kalloo, G., In *Genetic Improvement of Tomato Monographs Theoretical and Applied Genetics*, Vol. 14, Berlin: Springer-Verlag, p. 1, 1991a.

Kalloo, G., Breeding for environmental stress resistance in tomato, In *Genetic Improvement of Tomato Monographs Theoretical and Applied Genetics*, Kalloo, G., Ed., Vol. 14, Berlin: Springer-Verlag, p. 153, 1991b.

Kalloo, G., Interspecific and intergeneric hybridization in tomato, In *Genetic Improvement of Tomato Monographs Theoretical and Applied Genetics*, Kalloo, G., Ed., Vol. 14, Berlin: Springer-Verlag, p. 73, 1991c.

Kalloo, G. and Banerjee, M. K., Transfer of tomato leaf curl virus resistance from *Lycopersicon hirsutum* f, *glabratum* to *L. esculentum*, *Plant Breed.*, 105, 156–159, 1990.

Kasrawi, M. A., Inheritance of resistance to tomato yellow leaf curl virus (TYLCV) in *Lycopersicon pimpinellifolium*, *Plant Dis.*, 73, 435–437, 1989.

Kennedy, G. G., Tomato, pests, parasitoids, and predators: Tritrophic interactions involving the genus *Lycopersicon*, *Annu. Rev. Entomol.*, 48, 51–72, 2003.

Kennedy, G. G., Resistance in tomato and other *Lycopersicon* species to insect and mite pests, In *Genetic Improvement of Solanaceous Crops. Vol. 2: Tomato*, Razdan, M.K. and Mattoo, A.K., Eds., New Delhi, India: Science Publisher, Inc., 2006.

Kepczynski, J. and Kepczynska, E., Manipulation of ethylene biosynthesis, *Acta Physiol. Plant.*, 27, 213–220, 2005.

Kerr, E. A. et al., Resistance in tomato species to new races of leaf mold (*Cladosporium fulvum cke*), *Hort. Res.*, 11, 84–92, 1971.

Khotyleva, L. V. and Kilchevskii, A. V., Efficiency of the first cycle of recurrent reciprocal selection on the basis of intervarietal tomato hybrid, *Genetika*, 20, 1511–1518, 1984.

Khush, G. S., Identification key for pachytene chromosomes of *L. esculentum*, *Tomato Genet. Coop. Rep.*, 13, 12–13, 1963.

Khush, G. S. and Rick, C. M., Novel compensating trisomics of the tomato: cytogenetics, monosomic analysis, and other applications, *Genetics*, 56, 297–307, 1967a.

Khush, G. S. and Rick, C. M., Tomato tertiary trisomics—origin identification morphology and use in determining poiition of centromeres and arm location of markers, *Can. J. Genet. Cytol.*, 9, 610–631, 1967b.

Khush, G. S. and Rick, C. M., Studies on linkage map of chromosome 4 of tomato and on transmission of induced deficiencies, *Genetica*, 38, 74–94, 1967c.

Khush, G. S. and Rick, C. M., Tomato telotrisomics: origin, identification, and use in linkage mapping, *Cytologia*, 33, 137–148, 1968a.

Khush, G. S. and Rick, C. M., Cytogenetic analysis of tomato genome by means of induced deficiencies, *Chromosoma*, 23, 452–484, 1968b.

Khush, G. S. and Rick, C. M., Tomato secondary trisomics—origin identification morphology and use in cytogenetic analysis of genome, *Heredity*, 24, 129–146, 1969.

Khush, G. S. et al., Genetic activity in heterochromatic chromosome segment of tomato, *Science*, 145, 1432–1434, 1964.

Kim, J. W. et al., Disease resistance in tobacco and tomato plants transformed with the tomato spotted wilt virus nucleocapsid gene, *Plant Dis.*, 78, 615–621, 1994.

Kishitani, S. et al., Accumulation of glycinebetaine during cold-acclimation and freezing tolerance in leaves of winter and spring barley plants, *Plant Cell Environ.*, 17, 89–95, 1994.

Koblitz, H., Cell, tissue and organ culture of *Lycopersicon*, In *Genetic Improvement of Tomato Monographs Theoretical and Applied Genetics*, Kalloo, G., Ed., Vol. 14, Berlin: Springer-Verlag, p. 121, 1991a.

Koblitz, H., Protoplast culture and somatic hybridization in *Lycopersicon*, In *Genetic Improvement of Tomato Monographs Theoretical and Applied Genetics*, Kalloo, G., Ed., Vol. 14, Berlin: Springer-Verlag, p. 247, 1991b.

Kohler, G. R. and St. Clair, D. A., Variation for resistance to aphids (*Homoptera: Aphididae*) among tomato inbred backcross lines derived from wild *Lycopersicon* species, *J. Econ. Entomol.*, 98, 988–995, 2005.

Koorneef, M. et al., Characterization and mapping of a gene controlling shoot regeneration in tomato, *Plant J.*, 3, 131–141, 1993.

Kosova, A. I. and Kiku, V. N., Features of fertilization and seed development in tomato following interspecific hybridization, *Bull Akad Stiince RSS Mold Ser Biol Khim*, 1, 10–15, 1979.

Kozik, E. U., Studies on resistance to bacterial speck (*Pseudomonas syringae* pv *tomato*) in tomato cv. Ontario 7710, *Plant Breed.*, 121, 526–530, 2002.

Kuo, C. G. and Chen, B. W., Physiological response of tomato cultivars to flooding, *J. Am. Soc. Hort. Sci.*, 105, 751–755, 1980.

Lapitan, N. L. V. et al., Somatic chromosome karyotype of tomato based on in situ hybridization of the TGR1 satellite repeat, *Genome*, 32, 992–998, 1989.

Lapitan, N. L. V. et al., Organization of the 5S ribosomal RNA genes in the genome of tomato, *Genome*, 34, 509–514, 1991.

Larson, R. E. and Paur, S., The description and inheritance of a functionally sterile mutant in tomato and its probable value in hybrid tomato seed production, *Proc. Am. Soc. Hort. Sci.*, 52, 355–364, 1948.

Laterrot, H., Breeding strategies for disease resistance in tomatoes with emphasis on the topics: current status and research challenges, In *Proc 1st International Conference on the Processing Tomato/1st International Symposium on Tropical Tomato Diseases*, Recife, Brazil: ASHS Press, p.126, 1996.

Lecomte, L. et al., Marker-assisted introgression of five QTLs controlling fruit quality traits into three tomato lines revealed interactions between QTLs and genetic backgrounds, *Theor. Appl. Genet.*, 109, 658–668, 2004.

Lee, J. T. et al., Expression of Arabidopsis *CBF1* regulated by an ABA/stress inducible promoter in transgenic tomato confers stress tolerance without affecting yield, *Plant Cell Environ.*, 26, 1181–1190, 2003.

Leeper, P. W. et al., Southern blight resistant tomato breeding lines—5635M, 5707M, 5719M, 5737M, 5876M, and 5913M, *HortScience*, 27, 475–478, 1992.

Lesley, J. W., Trisomic types of the tomato and their relation to the genes, *Genetics*, 17, 545–559, 1932.

Lesley, M. M. and Lesley, J. W., Heteromorphic a chromosomes of the tomato differing in satellite size, *Genetics*, 20, 568–580, 1935.

Levenko, B. A. et al., Studies on callus-tissue from anthers. 1. Tomato, *Phytomorphology*, 27, 377–383, 1977.

Levesque, H. et al., Ientification of a short rdna spacer sequence highly specific of a tomato line containing *Tm-1* gene introgressed from *Lycopersicon hirsutum*, *Theor. Appl. Genet.*, 80, 602–608, 1990.

Li, N. J. et al., *In vitro* and *in vivo* evaluation of tolerance to salt in tomato (Abstr), *Genet. Agr.*, 42, 79, 1988.

Li, C. et al., Resistance of cultivated tomato to cell content-feeding herbivores is regulated by the octadeca-noid-signaling pathway, *Plant Physiol.*, 130, 494–503, 2002a.

Li, Z. Y. et al., Determination of copy number for 5S rDNA and centromeric sequence RCS2 in rice by Fiber-FISH, *Chinese Sci. Bull.*, 47, 214–217, 2002b.

Liharska, T. B. et al., Molecular mapping around the centromere of tomato chromosome 6 using irradiation-induced deletions, *Theor. Appl. Genet.*, 95, 969–974, 1997.

Lima, J. E. et al., Micro-MsK: a tomato genotype with miniature size, short life cycle, and improved *in vitro* shoot regeneration, *Plant Sci.*, 167, 753–757, 2004.

Liu, Y. S. et al., There is more to tomato fruit colour than candidate carotenoid genes, *Plant Biotechnol. J.*, 1, 195–207, 2003.

Liu, Y. et al., Construction of a genetic map and localization of QTLs for yield traits in tomato by SSR markers, *Prog. Nat. Sci.*, 15, 793–797, 2005.

Lukyanenko, A. N., Diseae resistance in tomato, In *Genetic Improvement of Tomato Monographs Theoretical and Applied Genetics*, Kalloo, G., Ed., Vol. 14, Berlin, Heidelberg: Springer-Verlag, p. 99, 1991a.

Lukyanenko, A. N., Breeding tomato for mechanized harvesting, In *Genetic Improvement of Tomato Mono-graphs Theoretical and Applied Genetics*, Kalloo, G., Ed., Vol. 14, Berlin: Springer-Verlag, p. 213, 1991b.

Ma, Y. H. et al., Efficient callus induction and shoot regeneration by anther culture in male sterile mutants of tomato (*Lycopersicon esculentum* Mill. Cv. First), *J. Jpn. Soc. Hort. Sci.*, 68, 768–773, 1999.

Madureira, H. C. et al., Immunolocalization of a defense-related 87 kDa cystatin in leaf blade of tomato plants, *Environ. Exp. Bot.*, 55, 201–208, 2006.

Mandaokar, A. D. et al., Transgenic tomato plants resistant to fruit borer (*Helicoverpa armigera Hubner*), *Crop Prot.*, 19, 307–312, 2000.

Mangin, B. et al., Temporal and multiple quantitative trait loci analyses of resistance to bacterial wilt in tomato permit the resolution of linked loci, *Genetics*, 151, 1165–1172, 1999.

Mark, G. E., The enigma of triploid potatoes, *Euphytica*, 15, 285–290, 1966.

Marshall, J. A. et al., Molecular systematics of *Solanum* section *Lycopersicum* (*Lycopersicon*) using the nuclear ITS rDNA region, *Theor. Appl. Genet.*, 103, 1216–1222, 2001.

Martin, M. W. and Thomas, P. E., C5, a new tomato breeding line resistant to curly top virus, *Phytopathology*, 59, 1754, 1969.

Martin, M. W. and Thomas, P. E., Levels, dependability, and usefulness of resistance to tomato curly top disease, *Plant Dis.*, 70, 136–141, 1986.

Martin, B. and Thorstenson, Y. R., Stable carbon isotope composition (delta-c-13), water-use efficiency, and biomass productivity of *Lycopersicon esculentum*, *Lycopersicon pennellii*, and the F_1 hybrid, *Plant Physiol.*, 88, 213–217, 1988.

Martin, G. B. et al., High-resolution linkage analysis and physical characterization of the *Pto* bacterial resistance locus in tomato, *Mol. Plant Microbe Interact.*, 6, 26–34, 1993a.

Martin, G. B. et al., Map-based cloning of a protein-kinase gene conferring disease resistance in tomato, *Science*, 262, 1432–1436, 1993b.

Martin, G. B. et al., A member of the tomato *Pto* gene family confers sensitivity to fenthion resulting in rapid cell death, *Plant Cell*, 6, 1543–1552, 1994.

McClean, P. E. and Hanson, M. R., Mitochondrial-DNA sequence divergence among *Lycopersicon* and related *Solanum* species, *Genetics*, 112, 649–667, 1986.

McGurk, J. and Rivlin, K., A basic computer-program for chromosome measurement and analysis, *J. Hered.*, 74, 304, 1983.

McGurl, B. et al., Overexpression of the prosystemin gene in transgenic tomato plants generates a systemic signal that constitutively induces proteinase-inhibitor synthesis, *Proc. Natl. Acad. Sci. U.S.A.*, 91, 9799–9802, 1994.

McKersie, B. D. and Leshem, Y. Y., *Stress and Stress Coping in Cultivated Plants*, Dordrecht, The Nether-lands: Kluwer Academic Publishers, p. 77, 1994.

Medina-Filho, H. P. and Stevens, M. A., Tomato breeding for namatode resistance: survey of resistant varieties for horticultural characteristics and genotype of acid phosphates, *Acta Hort.*, 100, 383–393, 1980.

Meissner, R. et al., A new model system for tomato genetics, *Plant J.*, 12, 1465–1472, 1997.

Meissner, R. et al., A high throughput system for transposon tagging and promoter trapping in tomato, *Plant J.*, 22, 265–274, 2000.

Melchers, G. et al., Somatic hybrid plants of potato and tomato regenerated from fused protoplasts, *Carlsberg Res. Commun.*, 43, 203–218, 1978.

Menzel, M. Y., Pachytene chromosomes of intergeneric hybrid *Lycopersicon esculentum × Solanum lycopersicoides*, *Am. J. Bot.*, 49, 605–615, 1962.

Menzel, M. Y., Preferential chromosome pairing in allotetraploid *Lycopersicon esculentum-Solanum lycopersicoides*, *Genetics*, 50, 855–862, 1964.

Menzel, M. Y. and Price, J. M., Fine structure of synapsed chomosomes in F_1 *Lycopersicon esculentum-Solanum lycopersicoides* and its parents, *Am. J. Bot.*, 53, 1079–1086, 1966.

Messeguer, R. et al., High-resolution RFLP map around the root-knot nematode resistance gene (*Mi*) in tomato, *Theor. Appl. Genet.*, 82, 529–536, 1991.

Miller, J. C. and Tanksley, S. D., RFLP analysis of phylogenetic-relationships and genetic-variation in the genus *Lycopersicon*, *Theor. Appl. Genet.*, 80, 437–448, 1990.

Moghaieb, R. E. A. et al., Expression of betaine aldehyde dehydrogenase gene in transgenic tomato hairy roots leads to the accumulation of glycine betaine and contributes to the maintenance of the osmotic potential under salt stress, *Soil Sci. Plant Nutr.*, 46, 873–883, 2000.

Mohan, M. et al., Genome mapping, molecular markers and marker-assisted selection in crop plants, *Mol. Breed.*, 3, 87–103, 1997.

Mohr, H. C. and Watkins, G. M., The nature of resistance to southern blight in tomato and the influence of nutrition on its expression, *Proc. Am. Soc. Hort. Sci.*, 74, 484–493, 1959.

Momotaz, A. et al., Searching for silverleaf whitefly and geminivirus resistance genes from *Lycopersicon hirsutum* accession LA 1777, *Acta Hort.*, 695, 417–422, 2005.

Monforte, A. J. and Tanksley, S. D., Development of a set of near isogenic and backcross recombinant inbred lines containing most of the *Lycopersicon hirsutum* genome in a *L. esculentum* genetic background: a tool for gene mapping and gene discovery, *Genome*, 43, 803–813, 2000.

Monforte, A. J. et al., Comparison of a set of allelic QTL-NILs for chromosome 4 of tomato: deductions about natural variation and implications for germplasm utilization, *Theor. Appl. Genet.*, 102, 572–590, 2001.

Moreau, P. et al., Genetic mapping of *Ph-2*, a single locus controlling partial resistance to *Phytophthora infestans* in tomato, *Mol. Plant Microbe Interact.*, 11, 259–269, 1998.

Muigai, S. G. et al., Mechanisms of resistance in *Lycopersicon* germplasm to the whitefly *Bemisia argentifolii*, *Phytoparasitica*, 30, 347–360, 2002.

Muller, C. H., A revision of the genus *Lycopersicon*, *USDA Misc. Publ.*, 382, 1–28, 1940.

Muniyappa, V. et al., Tomato leaf curl virus resistant tomato lines TLB111, TLB130, and TLB182, *HortScience*, 37, 603–606, 2002.

Murdock, L. L. and Shade, R. E., Lectins and protease inhibitors as plant defenses against insects, *J. Agric. Food Chem.*, 50, 6605–6611, 2002.

Nash, A. F. et al., Evaluation of allelism and seed set of 8 stamenless tomato mutants, *HortScience*, 20, 440–442, 1985.

Nelson, R. S. et al., Virus tolerance, plant-growth, and field performance of transgenic tomato plants expressing coat protein from tobacco mosaic virus, *Bio-Technology*, 6, 403–409, 1988.

Norton, J. P. and Boll, W. G., Callus and shoot formation from tomato roots *in vitro*, *Science*, 119, 220–221, 1954.

Novak, F. J. and Maskova, I., Apical shoot tip culture of tomato, *Sci. Hort.-Amsterdam*, 10, 337–344, 1979.

Nuez, F. et al., Relationships origin, and diversity of Galapagos tomatoes: implications for the conservation of natural populations, *Am. J. Bot.*, 91, 86–99, 2004.

Ohmori, T. et al., Molecular characterization of RAPD and SCAR markers linked to the *Tm-1* locus in tomato, *Theor. Appl. Genet.*, 92, 151–156, 1996.

Oke, M. et al., The effects of genetic transformation of tomato with antisense phospholipase D cDNA on the quality characteristics of fruits and their processed products, *Food Biotechnol.*, 17, 163–182, 2003.

Oldroyd, G. E. D. and Staskawicz, B. J., Genetically engineered broad-spectrum disease resistance in tomato, *Proc. Natl. Acad. Sci. U.S.A.*, 95, 10300–10305, 1998.

Olmstead, R. G. and Palmer, J. D., A chloroplast DNA phylogeny of the *Solanaceae*—subfamilial relationships and character evolution, *Ann. Mo. Bot. Gard.*, 79, 346–360, 1992.

Olmstead, R. G. and Palmer, J. D., Implications for the phylogeny, classification, and biogeography of *Solanum* from cpDNA restriction site variation, *Syst. Bot.*, 22, 19–29, 1997.

Ori, N. et al., The *I2C* family from the wilt disease resistance locus *I2* belongs to the nucleotide binding, leucine-rich repeat superfamily of plant resistance genes, *Plant Cell*, 9, 521–532, 1997.

Oud, J. L. et al., Computerized analysis of chromosomal parameters in karyotype studies, *Theor. Appl. Genet.*, 73, 630–634, 1987.

Ouyang, B. et al., Increased resistance to Fusarium wilt in transgenic tomato expressing bivalent hydrolytic enzymes, *J. Plant Physiol. Mol. Biol.*, 29, 179–184, 2003.

Ouyang, B. et al., Transformation of tomatoes with osmotin and chitinase genes and their resistance to *Fusarium* wilt, *J. Hort. Sci. Biotech.*, 80, 517–522, 2005.

Paddock, E. F., A tentative assignment of the Fusarium-immunity locus to linkage group-5 in tomato, *Genetics*, 35, 683–684, 1950.

Padmanabhan, V. et al., Plantlet formation from *Lycopersicon esculentum* leaf callus, *Can. J. Bot.*, 52, 1429–1432, 1974.

Paliyath, G. and Droillard, M. J., The mechanisms of membrane deterioration and disassembly during senescence, *Plant Physiol. Bioch.*, 30, 789–812, 1992.

Palmer, J. D. and Zamir, D., Chloroplast DNA evolution and phylogenetic relationships in *Lycopersicon*, *Proc. Natl. Acad. Sci.-Biol.*, 79, 5006–5010, 1982.

Pang, S. Z. et al., Resistance of transgenic *Nicotiana-benthamiana* plants to tomato spotted wilt and impatiens necrotic spot tospoviruses—evidence of involvement of the N-protein and N-gene RNA in resistance, *Phytopathology*, 84, 243–249, 1994.

Paoletti, M. G. and Pimentel, D., Environmental risks of pesticides versus genetic engineering for agricultural pest control, *J. Agr. Environ. Ethic.*, 12, 279–303, 2000.

Paran, I. and Zamir, D., Quantitative traits in plants: beyond the QTL, *Trends Genet.*, 19, 303–306, 2003.

Paran, I. et al., Recombinant inbred lines for genetic mapping in tomato, *Theor. Appl. Genet.*, 90, 542–548, 1995.

Paran, I. et al., QTL analysis of morphological traits in a tomato recombinant inbred line population, *Genome*, 40, 242–248, 1997.

Park, E. J. et al., Genetic engineering of glycinebetaine synthesis in tomato protects seeds, plants, and flowers from chilling damage, *Plant J.*, 40, 474–487, 2004.

Parrella, G. et al., Recessive resistance genes against potyviruses are localized in colinear genomic regions of the tomato (*Lycopersicon* spp.) and pepper (*Capsicum* spp.) genomes, *Theor. Appl. Genet.*, 105, 855–861, 2002.

Parrella, G. et al., The Am gene controlling resistance to Alfalfa mosaic virus in tomato is located in the cluster of dominant resistance genes on chromosome 6, *Phytopathology*, 94, 345–350, 2004.

Paterson, A. H. et al., Resolution of quantitative traits into Mendelian factors by using a complete linkage map of restriction fragment length polymorphisms, *Nature*, 335, 721–726, 1988.

Paterson, A. H. et al., Fine mapping of quantitative trait loci using selected overlapping recombinant chromosomes, in an interspecies cross of tomato, *Genetics*, 124, 735–742, 1990.

Paterson, A. H. et al., DNA markers in plant improvement, *Adv. Agron.*, 46, 39–90, 1991a.

Paterson, A. H. et al., Mendelian factors underlying quantitative traits in tomato: comparison across species, generations, and environments, *Genetics*, 127, 181–197, 1991b.

Peirce, L. C., Impact of single seed descent in selecting for fruit size, earliness and total yield in tomato, *J. Am. Soc. Hort. Sci.*, 102, 520–522, 1977.

Peng, J. H. and Black, L. L., Increased proteinase-Inhibitor activity in response to infection of resistant tomato plants by *Phytophthora infestans*, *Phytopathology*, 66, 958–963, 1976.

Peralta, I. E. and Spooner, D. M., Granule-bound starch synthase (GBSSI) gene phylogeny of wild tomatoes (*Solanum* L. section *Lycopersicon* (Mill.) Wettst. subsection *Lycopersicon*), *Am. J. Bot.*, 88, 1888–1902, 2001.

Perlak, F. J. et al., Modification of the coding sequence enhances plant expression of insect control protein genes, *Proc. Natl. Acad. Sci. U.S.A.*, 88, 3324–3328, 1991.

Pertuze, R. A. et al., Comparative linkage map of the *Solanum lycopersicoides* and *S. sitiens* genomes and their differentiation from tomato, *Genome*, 45, 1003–1012, 2002.

Pertuze, R. A. et al., Transmission and recombination of homeologous *Solanum sitiens* chromosomes in tomato, *Theor. Appl. Genet.*, 107, 1391–1401, 2003.

Peterson, D. G. et al., DNA content of heterochromatin and euchromatin in tomato (*Lycopersicon esculentum*) pachytene chromosomes, *Genome*, 39, 77–82, 1996.

Peterson, D. G. et al., Localization of single- and low-copy sequences on tomato synaptonemal complex spreads using fluorescence in situ hybridization (FISH), *Genetics*, 152, 427–439, 1999.

Pico, B. et al., Widening the genetic basis of virus resistance in tomato, *Sci. Hort.-Amsterdam*, 94, 73–89, 2002.

Pierce, L. C., Linkage tests with Ph conditioning resistance to race O, *Phytophthora infestans.*, *Tomato Genet. Coop. Rep.*, 21, 30, 1971.

Pillen, K. et al., Status of genome mapping tools in the taxon *Solonaceae*, In *Genome Mapping in Plants*, Paterson, A. H., Ed., Austin, TX: RG Landes, p. 282, 1996.

Pilowsky, M. and Cohen, S., Tolerance to tomato yellow leaf curl virus derived from *Lycopersicon peruvianum*, *Plant Dis.*, 74, 248–250, 1990.

Pinhero, R. G. et al., Developmental regulation of phospholipase D in tomato fruits, *Plant Physiol. Bioch.*, 41, 223–240, 2003.

Polston, J. E. and Hiebert, E., Engineered resistance to tomato germini viruses. In *Proc. Fla. Tomato Inst.*, Gilreath, P., Ed., University of Florida, Gainsville, Naples, FL, pp.19–22, 2001.

Poysa, V. and Tu, J. C., Response of cultivars and breeding lines of *Lycopersicon* spp. to *Septoria lycopersici*, *Can. Plant. Dis. Surv.*, 73, 9–13, 1993.

Praveen, S. et al., Engineering tomato for resistance to tomato leaf curl disease using viral rep gene sequences, *Plant Cell Tiss. Org.*, 83, 311–318, 2005.

Prins, M. and Goldbach, R., The emerging problem of tospovirus infection and nonconventional methods of control, *Trends Microbiol.*, 6, 31–35, 1998.

Quinn, J. P., Evolving strategies for the genetic-engineering of herbicide resistance in plants, *Biotechnol. Adv.*, 8, 321–333, 1990.

Quiros, C. F., *Lycopersicon* cytogenetics, In *Chromosome Engineering in Plants: Genetics, Breeding, Evolution, Part B*, Tsuchiya, T. and Gupta, P. K., Eds., Amsterdam: Elsevier, p. 119, 1991.

Rahman, M. M. and Kaul, K., Differentiation of sodium-chloride tolerant cell-lines of tomato (*Lycopersicon esculentum* Mill) cv jet star, *J. Plant Physiol.*, 133, 710–712, 1989.

Ramanna, M. S. and Prakken, R., Structure of and homology between pachytene and somatic metaphase chromosomses of the tomato, *Genetica*, 38, 115–133, 1967.

Raybould, A. F. and Gray, A. J., Genetically-modified crops and hybridization with wild relatives—a UK perspective, *J. Appl. Ecol.*, 30, 199–219, 1993.

Razdan, M. K. and Mattoo, A. K., *Genetic Improvement of Solanaceous Crops. Vol. 2: Tomato*, Enfield, NH: Science Publishers, Inc., 2006.

Reeves, A., MicroMeasure: a new computer program for the collection and analysis of cytogenetic data, *Genome*, 44, 439–443, 2001.

Rhim, S. L. et al., Development of insect resistance in tomato plants expressing the delta-endotoxin gene of *Bacillus-thuringiensis* subsp. *tenebrionis*, *Mol. Breed.*, 1, 229–236, 1995.

Rick, C. M., A survey of cytogenetic causes of unfruitfulness in the tomato, *Genetics*, 30, 347–362, 1945.

Rick, C. M., Hybrids between *Lycopersicon esculentum* Mill and *Solanum lycopersicoides* Dun, *Proc. Natl. Acad. Sci. U.S.A.*, 37, 741–744, 1951.

Rick, C. M., The role of natural hybridization in the derivation of cultivated tomatoes of western South America, *Econ. Bot.*, 12, 346–367, 1958.

Rick, C. M., Hybridization between *Lycopersicon esculentum* and *Solanum pennellii*—phylogenetic and cytogenetic significance, *Proc. Natl. Acad. Sci. U.S.A.*, 46, 78–82, 1960.

Rick, C. M., *Lycopersicon*, In *Flora of the Galapsgos Islands*, Wiggins, I. L. and Porter, D. M., Eds., Palo Alto, CA: Stanford University Press, p. 468, 1971.

Rick, C. M., Evolution of interspecific barriers in *Lycopersicon*, In *Proceedings of Conference Broadening the Genetic Base of Crops*, Wageningen, The Netherlands: Pudoc, p. 283, 1979.

Rick, C. M., Evolution of mating systems: evidence from allozyme variation, In *Proceedings of XV International Congress on Geneties*, Genetics: New Delhi, India: New Frontiers, p. 216, 1984.

Rick, C. M., Germplasm resources in the wild tomato species, *Acta Hort.*, 190, 39–47, 1986.

Rick, C. M., Genetic resources in *Lycopersicon*, In *Tomato biology. Plant biology*, Nevins, D. J. and Jones, R. A., Eds., Vol. 4, New York: Alan R. Liss, Inc., p. 17, 1987.

Rick, C. M., Tomato-like nightshades—affinities, autoecology, and breeders opportunities, *Econ. Bot.*, 42, 145–154, 1988.

Rick, C. M., Tomato genetic resources of South America reveal many genetic treasures, *Diversity*, 7, 54–56, 1991.

Rick, C. M. and Barton, D. W., Cytological and genetical identification of the primary trisomics of the tomato, *Genetics*, 39, 640–666, 1954.

Rick, C. M. and Butler, L., Cytogenetics of the tomato, *Adv. Genet.*, 8, 267–382, 1956.

Rick, C. M. and Chetelat, R. T., Utilization of related wild species for tomato improvement, *Acta Hort.*, 412, 21–38, 1995.

Rick, C. M. and Fobes, J. F., Allozyme variation in the cultivated tomato and closely related species, *Bull. Torrey Bot. Club*, 102, 376–384, 1975.

Rick, C. M. and Khush, G. S., X-ray-induced deficiencies of chromosome 11 in the tomato, *Genetics*, 46, 1389–1393, 1961.

Rick, C. M. and Lamm, R., Biosystematic studies on the status of *Lycopersicon chilense*, *Am. J. Bot.*, 42, 663–675, 1955.

Rick, C. M. and Tanksley, S. D., Genetic variation in *Solanum pennellii*—comparisons with two other sympatric tomato species, *Plant Syst. Evol.*, 139, 11–45, 1981.

Rick, C. M. and Tanksley, S. D., Isozyme monitoring of gnetic variation in *Lycopersicon, Isozymes. Curr. Top. Biol. Med. Res.*, 11, 269–284, 1983.

Rick, C. M. and Yoder, J. I., Classical and molecular-genetics of tomato—highlights and perspectives, *Annu. Rev. Genet.*, 22, 281–300, 1988.

Rick, C. M. et al., Genetic and biosystematic studies on two new sibling species of *Lycopersicon* from Interandean Peru, *Theor. Appl. Genet.*, 47, 55–68, 1976.

Rick, C. M. et al., Evolution of mating systems in *Lycopersicon hirsutum* as deduced from genetic variation in electrophoretic and morphological characters, *Plant Syst. Evol.*, 132, 279–298, 1979a.

Rick, C. M. et al., Biosystematic studies in *Lycopersicon* and closely related species of *Solanum*, In *The Biology and Taxonomy of the Solanaceae*, Hawkes, J. G. et al., Eds., New York: Academic Press, p. 667, 1979b.

Rick, C. M. et al., Meiosis in sesquidiploid hybrids of *Lycopersicon esculentum* and *Solanum lycopersicoides*, *Proc. Natl. Acad. Sci. U.S.A.*, 83, 3580–3583, 1986.

Rick, C. M. et al., Potential contributions of wide crosses to improvement of processing tomatoes, *Acta Hort.*, 200, 45–55, 1987.

Robert, V. J. M. et al., Marker-assisted introgression of blackmold resistance QTL alleles from wild *Lycopersicon cheesmanii* to cultivated tomato (*L. esculentum*) and evaluation of QTL phenotypic effects, *Mol. Breed.*, 8, 217–233, 2001.

Rommens, C. M. T. et al., Intergeneric transfer and functional expression of the tomato disease resistance gene *Pto*, *Plant Cell*, 7, 1537–1544, 1995.

Rus, A. M. et al., Expressing the yeast *HAL1* gene in tomato increases fruit yield and enhances K+/Na+ selectivity under salt stress, *Plant Cell Environ.*, 24, 875–880, 2001.

Rush, D. W. and Epstein, E., Breeding and selection for salt tolerance by the incorporation of wild germplasm into a domestic tomato, *J. Am. Soc. Hort. Sci.*, 106, 699–704, 1981.

Salmeron, J. M. et al., Tomato *Prf* is a member of the leucine-rich repeat class of plant disease resistance genes and lies embedded within the *Pto* kinase gene cluster, *Cell*, 86, 123–133, 1996.

Sandbrink, J. M. et al., Localization of genes for bacterial canker resistance in *Lycopersicon peruvianum* using RFLPs, *Theor. Appl. Genet.*, 90, 444–450, 1995.

Sanford, J. C. and Johnston, S. A., The concept of parasite-derived resistance—deriving resistance genes from the parasites own genome, *J. Theor. Biol.*, 113, 395–405, 1985.

Sarfatti, M. et al., An RFLP marker in tomato linked to the *Fusarium oxysporum* resistance gene *I2*, *Theor. Appl. Genet.*, 78, 755–759, 1989.

Sarfatti, M. et al., RFLP mapping of *I1*, a new locus in tomato conferring resistance against *Fusarium oxysporum* f. sp. *lycopersici* race 1, *Theor. Appl. Genet.*, 82, 22–26, 1991.

Schaible, L. et al., Inheritance of resistance to *Verticillium* wilt in a tomato cross, *Phytopathology*, 41, 986–990, 1951.

Schalk, J. M. and Stoner, A. K., A bioassay differentiates resistance to the Colorado potato beetle on tomatoes, *J. Am. Soc. Hort. Sci.*, 101, 74–76, 1976.

Schweizer, G. et al., Species-specific DNA-sequences for identification of somatic hybrids between *Lycopersicon esculentum* and *Solanum acaule*, *Theor. Appl. Genet.*, 75, 679–684, 1988.

Scott, J. W., Horizon a fresh market tomato with concentrated fruit set, *Tomato Genet. Coop. Rep.*, 35, 61, 1985.

Scott, J. W., Vegetable varieties from North America—Tomato, *HortScience*, 37, 70–75, 2002.

Scott, J. W., Breeding for resistance to viral pathogens, In *Genetic Improvement of Solanaceous Crops. Vol. 2: Tomato*, Razdan, M. K. and Mattoo, A. K., Eds., New Delhi, India: Science Publisher, Inc., 2006.

Scott, J. W. and Angell, F. F., Tomato, In *Hybrid Cultivar Development*, Banga, S. S. and Banga, S. K., Eds., New Delhi, India: Narosa Publishing House, pp. 451–475, 1998.

Scott, J. W. and Farley, J. D., Ohio CR-6 tomato, *HortScience*, 18, 114–115, 1983.

Scott, J. W. and Gardner, R. G., Breeding for resistance to fungal pathogens. In *Genetic Improvement of Solanaceous Crops. Vol. 2: Tomato*, Razdan, M. K. and Mattoo, A. K., Eds., New Delhi, India: Science Publisher, 2006.

Scott, J. W. and George, W. L., Enhancement of tomato fruit-set by combining heat tolerance and parthenocarpic genes, *HortScience*, 19, 546, 1984.

Scott, J. W. and Jones, J. P., Fla. 7547 and Fla. 7481 tomato breeding lines resistant to *Fusarium oxysporum* f. sp. *lycopersici* races 1, 2, and 3, *HortScience*, 30, 645–646, 1995.

Scott, J. W. et al., Introgression of resistance to whitefly-transmitted geminiviruses from *Lycopersicon chilense* to tomato, In *Bemisia: Taxonomy, Biology, Damage Control, and Management*, Gerling, D. and Mayer, R. T., Eds., Andover, Hants, UK: Intercept, pp. 357–367, 1996.

Scott, J. W. et al., Tomato bacterial spot resistance derived from PI 114490; Inheritance of resistance to race T2 and relationship across three pathogen races, *J. Am. Soc. Hort. Sci.*, 128, 698–703, 2003.

Scott, J. W. et al., RFLP-based analysis of recombination among resistance genes to Fusarium wilt races 1, 2, and 3 in tomato, *J. Am. Soc. Hort. Sci.*, 129, 394–400, 2004.

Segeren, M. I. et al., Tomato breeding .1. Embryo rescue of interspecific hybrids between *Lycopersicon esculentum* Mill and *Lycopersicon peruvianum* (l) Mill, *Rev. Bras. Genet.*, 16, 367–380, 1993.

Sen, N. K., Isochromosomes in tomato, *Genetics*, 37, 227–241, 1952.

Shah, D. M. et al., Engineering herbicide tolerance in transgenic plants, *Science*, 233, 478–481, 1986.

Shahin, E. A. and Spivey, R., A single dominant gene for *Fusarium wilt* resistance in protoplast-derived tomato plants, *Theor. Appl. Genet.*, 73, 164–169, 1986.

Shahin, E. A. and Spivey, R., *In vitro* breeding for disease resistance in tomato, In *Tomato Biotechnology. Plant Biology*, Nevins, D. J. and Jones, R. A., Eds., Vol. 4, New York: Alan R. Liss., p. 89, 1987.

Sharma, H. C. and Ortiz, R., Transgenics, pest management, and the environment, *Curr. Sci.*, 79, 421–437, 2000.

Sharma, H. C. et al., Genetic transformation of crops for insect resistance: potential and limitations, *Crit. Rev. Plant Sci.*, 23, 47–72, 2004.

Sharma, K. K. et al., Genetic transformation technology: status and problems, *In Vitro Cell. Dev. Biol. Plant*, 41, 102–112, 2005.

Sheehy, R. E. et al., Reduction of polygalacturonase activity in tomato fruit by antisense RNA, *Proc. Natl. Acad. Sci. U.S.A.*, 85, 8805–8809, 1988.

Sherman, J. D. and Stack, S. M., Two-dimensional spreads of synaptonemal complexes from solanaceous plants.V. tomato (*lycopersicon-esculentum*) karyotype and idiogram, *Genome*, 35, 354–359, 1992.

Sherman, J. D. and Stack, S. M., Two-dimensional spreads of synaptonemal complexes from solanaceous plants.VI. high-resolution recombination nodule map for tomato (*Lycopersicon esculentum*), *Genetics*, 141, 683–708, 1995.

Shtereva, L. and Atanassova, B., Callus induction and plant regeneration via anther culture in mutant tomato (*Lycopersicon esculentum* Mill.) lines with anther abnormalities, *Israel J. Plant Sci.*, 49, 203–208, 2001.

Shtereva, L. A. et al., Induced androgenesis in tomato (*Lycopersicon esculentum* Mill). II. Factors affecting induction of androgenesis, *Plant Cell Rep.*, 18, 312–317, 1998.

Shull, G. H., The composition of a field of maize, *Am. Breed. Assoc. Rep.*, 4, 296–301, 1908.

Simons, G. et al., Dissection of the Fusarium *I2* gene cluster in tomato reveals six homologs and one active gene copy, *Plant Cell*, 10, 1055–1068, 1998.

Smith, P. G., Embryo culture of a tomato species hybrid, *Proc. Am. Soc. Hort. Sci.*, 44, 413–416, 1944.

Smith, C. J. S. et al., Antisense RNA inhibition of polygalacturonase gene-expression in transgenic tomatoes, *Nature*, 334, 724–726, 1988.

Smith, M. A. L. et al., Cell osmolarity adjustment in *Lycopersicon* in response to stress pretreatments, *J. Plant Nutr.*, 12, 233–244, 1989.

Sobczak, M. et al., Characterization of susceptibility and resistance responses to potato cyst nematode (*Globodera* spp.) infection of tomato lines in the absence and presence of the broad-spectrum nematode resistance *Hero* gene, *Mol. Plant Microbe Interact.*, 18, 158–168, 2005.

Sotirova, V. et al., Resistance responses of plants regenerated from tomato anther and somatic tissue cultures to *Clavibacter michiganense* subsp *michiganense*, *Israel J. Plant Sci.*, 47, 237–243, 1999.

Srivastava, D. K. et al., *In vitro* selection and characterization of water stress tolerant callus cultures of tomato (*Lycopersicon esculentum* L.), *Indian J. Plant Physiol.*, 38, 99–104, 1995.

Stadler, T. et al., Genealogical footprints of speciation processes in wild tomatoes: demography and evidence for historical gene flow, *Evolution*, 59, 1268–1279, 2005.

Stall, R. E. and Walter, J. M., Selection and inheritance of resistance in tomato to isolates of races 1 and 2 of Fusarium wilt organism, *Phytopathology*, 55, 1213–1215, 1965.

Stamova, B. S. and Chetelat, R. T., Inheritance and genetic mapping of cucumber mosaic virus resistance introgressed from *Lycopersicon chilense* into tomato, *Theor. Appl. Genet.*, 101, 527–537, 2000.

Staraci, L. C. et al., Stress adaptation in cultured tomato cells, In *Tomato Biotechnology. Plant biology*, Nevins, D. J. and Jones, R. A., Eds., Vol. 4, New York: Alan R. Liss, p. 99, 1987.

Staub, J. E. and Serquen, F. C., Genetic markers, map construction, and their application in plant breeding, *HortScience*, 31, 729–741, 1996.

Steffens, J. C. and Williams, B. G., Heavy metal tolerance in tomato cells, In *Tomato Biotechnology. Plant Biology*, Nevins, D. J. and Jones, R. A., Eds., Vol. 4, New York: Alan R. Liss, Inc., p. 109, 1987.

Steinitz, B. et al., Expression of insect resistance in *in-vitro*-derived callus-tissue infested with lepidopteran larvae, *J. Plant Physiol.*, 142, 480–484, 1993.

Stephan, W. and Langley, C. H., DNA polymorphism in *Lycopersicon* and crossing-over per physical length, *Genetics*, 150, 1585–1593, 1998.

Stevens, M. A. and Rick, C. M., Genetics and breeding, In *The Tomato Crop: a Scientific Basis for Improvement*, Atherton, J. G. and Rudich, J., Eds., London: Chapman and Hall, p. 35, 1986.

Stevens, M. R. et al., Inheritance of a gene for resistance to tomato spotted wilt virus (TSWV) from *Lycopersicon peruvianun* Mill, *Euphytica*, 59, 9–17, 1992.

Stevens, M. R. et al., Mapping the *Sw-5* locus for tomato spotted wilt virus resistance in tomatoes using RAPD and RFLP analyses, *Acta Hort.*, 431, 385–392, 1995.

Stout, M. J. et al., Signal interactions in pathogen and insect attack: systemic plant-mediated interactions between pathogens and herbivores of the tomato, *Lycopersicon esculentum*, *Physiol. Mol. Plant Pathol.*, 54, 115–130, 1999.

Stubble, H., Mutanten der Kulturtomate *Lycopersicon esculentum* Miller VI, *Kulturpflanze*, 19, 185–230, 1972a.

Stubble, H., Mutanten der Wildtomate *Lycopersicon pimpinellifolium* (Jusl.) Mill IV, *Kulturpflanze*, 19, 231–263, 1972b.

Summers, W. L. et al., Microspore developmental stage and anther length influence the induction of tomato anther callus, *HortScience*, 27, 838–840, 1992.

Tabaeizadeh, Z. et al., Transgenic tomato plants expressing a *Lycopersicon chilense* chitinase gene demonstrate improved resistance to *Verticillium dahliae* race 2, *Plant Cell Rep.*, 19, 197–202, 1999.

Tal, M. and Shannon, M. C., Salt tolerance in the wild relatives of the cultivated tomato—responses of *Lycopersicon esculentum*, *Lycopersicon cheesmanii*, *Lycopersicon peruvianum*, *Solanum pennellii* and F$_1$ hybrids to high salinity, *Aust. J. Plant Physiol.*, 10, 109–117, 1983.

Tal, M. et al., Salt tolerance in the wild relatives of the cultivated tomato: response of callus tissue of *Lycopersicon esculentum*, *L. peruvianum*, *Solanum pennellii* to high salinity, *Z. Pflanzenphysiol.*, 86, 231–240, 1978.

Tang, X. Y. et al., Overexpression of *Pto* activates defense responses and confers broad resistance, *Plant Cell*, 11, 15–29, 1999.

Tanksley, S. D. and Bernatzky, R., Molecular markers for the nuclear genome of tomato, In *Tomato Biology*, Nevins, D. J. and Jones, R. A., Eds., New York: Alan R. Liss, 1987.

Tanksley, S. D. and Nelson, J. C., Advanced backcross QTL analysis: a method for the simultaneous discovery and transfer of valuable QTLs from unadapted germplasm into elite breeding lines, *Theor. Appl. Genet.*, 92, 191–203, 1996.

Tanksley, S. D. and Rick, C. M., Isozymic gene linkage map of the tomato—applications in genetics and breeding, *Theor. Appl. Genet.*, 57, 161–170, 1980.

Tanksley, S. D. et al., Conservation of gene repertoire but not gene order in pepper and tomato, *Proc. Natl. Acad. Sci. U.S.A.*, 85, 6419–6423, 1988.

Tanksley, S. D. et al., High density molecular linkage maps of the tomato and potato genomes, *Genetics*, 132, 1141–1160, 1992.

Taylor, I. B., Biosystematics of the tomato, In *The Tomato Crop: a Scientific Basis for Improvement*, Atherton, J. G. and Rudich, J., Eds., London: Chapman and Hall, p. 1, 1986.

Taylor, I. B. and Al-Kummer, M. K., The formation of complex hybrids between *Lycopersicon esculentum* and *L. peruvianum*, and their potential use in promoting interspecific gene transfer, *Theor. Appl. Genet.*, 61, 59–63, 1982.

Thakur, P. S., Different physiological responses of tomato (*Lycopersicon esculentum* Mill) cultivars to drought, *Acta Physiol. Plant.*, 12, 175–182, 1990.

Thomas, P. and Mythili, J. B., Development of cultured tomato anther to a fruit-like structure accompanied by *in-vitro* ripening, *Curr. Sci.*, 69, 94–95, 1995.

Thomas, B. R. and Pratt, D., Isolation of paraquat-tolerant mutants from tomato cell-cultures, *Theor. Appl. Genet.*, 63, 169–176, 1982.

Thoquet, P. et al., Quantitative trait loci determining resistance to bacterial wilt in tomato cultivar Hawaii7996, *Mol. Plant Microbe Interact.*, 9, 826–836, 1996a.

Thoquet, P. et al., Polygenic resistance of tomato plants to bacterial wilt in the French West Indies, *Mol. Plant Microbe Interact.*, 9, 837–842, 1996b.

Tigchelaar, E. C., Tomato breeding, In *Breeding Vegetable Crops*, Bassett, M. J., Ed., Westpoint, CT: AVI Publishing Company, p. 135, 1986.

Tikoo, S. K. and Anand, N., Development of tomato genotypes with exserted stigma and a seedling marker for use as female parents to exploit heterosis, *Curr. Sci.*, 49, 326–327, 1980.

Tikunov, Y. M. et al., Application of ISSR markers in the genus *Lycopersicon*, *Euphytica*, 131, 71–80, 2003.

Tor, M. et al., Genetic analysis and FISH mapping of the colourless non-ripening locus of tomato, *Theor. Appl. Genet.*, 104, 165–170, 2002.

Toyoda, H. et al., Effects of the culture filtrate of *Fusarium oxysporum f.sp. lycopersici* on tomato callus growth and the selection of resistant callus cells to the filtrate, *Ann. Phytopathol. Soc. Jpn.*, 50, 53–62, 1984.

Toyoda, H. et al., Resistance mechanism of cultured plant cells to tobacco mosaic virus (III) Efficient micro-injection of tobacco mosaic virus into tomato callus cells, *Ann. Phytopathol. Soc. Jpn.*, 51, 32–38, 1985.

Toyoda, H. et al., Selection of bacterial wilt-resistant tomato through tissue culture, *Plant Cell Rep.*, 8, 317–320, 1989.

Tumer, N. E. et al., Expression of alfalfa mosaic virus coat protein gene confers cross protection in transgenic tobacco and tomato plants, *Embo J.*, 6, 1181–1188, 1987.

Ultzen, T. et al., Resistance to tomato spotted wilt virus in transgenic tomato hybrids, *Euphytica*, 85, 159–168, 1995.

Vakalounakis, D. J., Cultivar reactions and the genetic basis of resistance to *Alternaria* stem canker (*Alternaria-alternata f sp lycopersici*) in tomato, *Plant Pathol.*, 37, 373–376, 1988.

Vakalounakis, D. J. et al., Linkage between *Frl* (*Fusarium oxysporum f sp radicis-lycopersici resistance*) and *Tm-2* (tobacco mosaic virus resistance-2) loci in tomato (*Lycopersicon esculentum*), *Ann. Appl. Biol.*, 130, 319–323, 1997.

van den Bulk, R. W. et al., Screening of tomato somaclones for resistance to bacterial canker (*Clavibacter-michiganensis subsp michiganensis*), *Plant Breed.*, 107, 190–196, 1991.

van der Beek, J. G. et al., Resistance to powdery mildew (*Oidium lycopersicon*) in *Lycopersicon hirsutum* is controlled by an incompletely dominant gene *Ol-1* on chromosome 6, *Theor. Appl. Genet.*, 89, 467–473, 1994.

van der Biezen, E. A. et al., Integrated genetic map of tomato chromosome 3, *Tomato Genet. Coop. Rep.*, 44, 8–10, 1994.

van der Biezen, E. A. et al., Inheritance and genetic mapping of resistance to *Alternaria alternata f sp lycopersici in Lycopersicon pennellii*, *Mol. Gen. Genet.*, 247, 453–461, 1995.

van der Hoeven, R. et al., Deductions about the number, organization, and evolution of genes in the tomato genome based on analysis of a large expressed sequence tag collection and selective genomic sequencing, *Plant Cell*, 14, 1441–1456, 2002.

van der Knaap, E. et al., High-resolution fine mapping and fluorescence in situ hybridization analysis of *sun*, a locus controlling tomato fruit shape, reveals a region of the tomato genome prone to DNA rearrangements, *Genetics*, 168, 2127–2140, 2004.

van der Salm, T. et al., Insect resistance of transgenic plants that express modified *Bacillus thuringiensis cryIa(b)* and *cryIc* genes—a resistance management strategy, *Plant Mol. Biol.*, 26, 51–59, 1994.

van Heusden, A. W. et al., Three QTLs from *Lycopersicon peruvianum* confer a high level of resistance to *Clavibacter michiganensis* ssp. *michiganensis*, *Theor. Appl. Genet.*, 99, 1068–1074, 1999.

van Ooijen, J. W. et al., An RFLP linkage map of *Lycopersicon peruvianum*, *Theor. Appl. Genet.*, 89, 1007–1013, 1994.

van Tuinen, A. et al., The mapping of phytochrome genes and photomorphogenic mutants of tomato, *Theor. Appl. Genet.*, 94, 115–122, 1997.

van Tuinen, A. et al., Mapping of the *Pro* gene and revision of the classical map of chromosome 11, *Tomato Genet. Coop. Rep.*, 48, 62–70, 1998.

van Wordragen, M. F. et al., Genetic and molecular organization of the short arm and pericentromeric region of tomato chromosome 6, *Euphytica*, 79, 169–174, 1994.

van Wordragen, M. F. et al., Tomato chromosome 6: a high resolution map of the long arm and construction of a composite integrated marker order map, *Theor. Appl. Genet.*, 92, 1065–1072, 1996.

Varghese, T. M. and Yadav, G., Production of embryoids and calli from isolated microspores of tomato (*Lycopersicon esculentum* Mill) in liquid-media, *Biol. Plantarum*, 28, 126–129, 1986.

Vidya, C. S. S. et al., Agrobacterium-mediated transformation of tomato (*Lycopersicon esculentum* var. Pusa ruby) with coat-protein gene of Physalis mottle tymovirus, *J. Plant Physiol.*, 156, 106–110, 2000.

Villalta, I. et al., Comparative microsatellite linkage analysis and genetic structure of two populations of F6 lines derived from *Lycopersicon pimpinellifolium* and *L. cheesmanii*, *Theor. Appl. Genet.*, 110, 881–894, 2005.

Villand, J. et al., Genetic variation among tomato accessions from primary and secondary centers of diversity, *Crop Sci.*, 38, 1339–1347, 1998.

Vos, P. et al., The tomato *Mi-1* gene confers resistance to both root-knot nematodes and potato aphids, *Nat. Biotechnol.*, 16, 1365–1369, 1998.

Walter, J. M., Hereditary resistance to diseases in tomato, *Annu. Rev. Phytopathol.*, 51, 131–162, 1967.

Warnock, S. J., A review of taxonomy and phylogeny of the genus *Lycopersicon*, *HortScience*, 23, 669–673, 1988.

Warren, R. S. and Routley, D. G., Use of tissue culture in study of single gene resistance of tomato to *Phytophthora infestans*, *J. Am. Soc. Hort. Sci.*, 95, 266–269, 1970.

Weide, R. et al., Integration of the classical and molecular linkage maps of tomato chromosome 6, *Genetics*, 135, 1175–1186, 1993.

Whalen, M. C. et al., Avirulence gene *Avrrxv* from *Xanthomonas campestris* pv. *vesicatoria* specifies resistance on tomato line Hawaii 7998, *Mol. Plant Microbe Interact.*, 6, 616–627, 1993.

Wijbrandi, J. et al., Selection and characterization of somatic hybrids between *Lycopersicon esculentum* and *Lycopersicon peruvianum*, *Plant Sci.*, 70, 197–208, 1990.

Williams, C. E. and St. Clair, D. A., Phenetic relationships and levels of variability detected by restriction fragment length polymorphism and random amplified polymorphic DNA analysis of cultivated and wild accessions of *Lycopersicon esculentum*, *Genome*, 36, 619–630, 1993.

Williamson, V. M., Root-knot nematode resistance genes in tomato and their potential for future use, *Annu. Rev. Phytopathol.*, 36, 277–293, 1998.

Witsenboer, H. M. A. et al., Effects of *Alternaria alternata* f. sp. *lycopersici* toxins at different levels of tomato plant cell development, *Plant Sci.*, 56, 253–260, 1988.

Wolf, S. et al., Cold temperature tolerance of wild tomato species, *J. Am. Soc. Hort. Sci.*, 111, 960–964, 1986.

Wolters, A. M. et al., Somatic hybridization as a tool for tomato breeding, *Euphytica*, 79, 265–277, 1994.

Wolters, A. M. A. et al., Fluorescence in situ hybridization on extended DNA fibres as a tool to analyse complex T-DNA loci in potato, *Plant J.*, 13, 837–847, 1998.

Wu, C. Y. et al., Genetic transformation of tomato with snowdrop lectin gene (GNA), *Acta Bot. Sin.*, 42, 719–723, 2000.

Xu, J. and Earle, E. D., Direct FISH of 5S rDNA on tomato pachytene chromosomes places the gene at the heterochromatic knob immediately adjacent to the centromere of chromosome 1, *Genome*, 39, 216–221, 1996a.

Xu, J. and Earle, E. D., High resolution physical mapping of 45S (5.8S, 18S and 25S) rDNA gene loci in the tomato genome using a combination of karyotyping and FISH of pachytene chromosomes, *Chromosoma*, 104, 545–550, 1996b.

Xu, P. et al., Expression of antiapoptotic genes *bcl-xL* and *ced-9* in tomato enhances tolerance to viral-induced necrosis and abiotic stress, *Proc. Natl. Acad. Sci. U.S.A.*, 101, 15805–15810, 2004.

Yaghoobi, J. et al., Mapping a new nematode resistance locus in Lycopersicon peruvianum, *Theor. Appl. Genet.*, 91, 457–464, 1995.

Yang, W. C. and Francis, D. M., Marker-assisted selection for combining resistance to bacterial spot and bacterial speck in tomato, *J. Am. Soc. Hort. Sci.*, 130, 716–721, 2005.

Yang, W. C. et al., Discovery of single nucleotide polymorphisms in *Lycopersicon esculentum* by computer aided analysis of expressed sequence tags, *Mol. Breed.*, 14, 21–34, 2004.

Yang, Y. et al., Use of tomato yellow leaf curl virus (TYLCV) *Rep* gene sequences to engineer TYLCV resistance in tomato, *Phytopathology*, 94, 490–496, 2004.

Yang, W. C. et al., Resistance in *Lycopersicon esculentum* intraspecific crosses to race T1 strains of *Xanthomonas campestris* pv. *vesicatoria* causing bacterial spot of tomato, *Phytopathology*, 95, 519–527, 2005.

Young, N. D. and Tanksley, S. D., RFLP analysis of the size of chromosomal segments retained around the *Tm-2* locus of tomato during backcross breeding, *Theor. Appl. Genet.*, 77, 353–359, 1989.

Young, N. D. et al., Rapid chromosomal assignment of multiple genomic clones in tomato using primary trisomics, *Nucleic Acids Res.*, 15, 9339–9348, 1987.

Young, N. D. et al., Use of isogenic lines and simultaneous probing to identify DNA markers tightly linked to the *Tm-2-alpha* gene in tomato, *Genetics*, 120, 579–585, 1988.

Yu, Z. H. et al., Genomic localization of tomato genes that control a hypersensitive reaction to *Xnthomonas campestris pv. vesicatoria* (doidge) dye, *Genetics*, 141, 675–682, 1995.

Yuan, Y. N. et al., The Cladosporium fulvum resistance gene *Cf-ECP3* is part of the Orion cluster on the short arm of tomato Chromosome 1, *Mol. Breed.*, 10, 45–50, 2002.

Zabel, P. et al., Towards the construction of artificial chromosomes for tomato, In *Molecular Form and Function of the Plant Genome. Nato Asi Series, Life Sciences*, van Vloten-Doting, L. et al., Eds., Vol. 83, New York: Plenum Press, p. 609, 1985.

Zagorska, N. A. et al., Induced androgenesis in tomato (*Lycopersicon esculentum* Mill.). I. Influence of genotype on androgenetic ability, *Plant Cell Rep.*, 17, 968–973, 1998.

Zagorska, N. A. et al., Induced androgenesis in tomato (*Lycopersicon esculentum* Mill.). III. Characterization of the regenerants, *Plant Cell Rep.*, 22, 449–456, 2004.

Zamir, D., Improving plant breeding with exotic genetic libraries, *Nat. Rev. Genet.*, 2, 983–989, 2001.

Zamir, D. and Eshed, Y., Case history in germplasm introgression: tomato genetics and breeding using nearly isogenic introgression lines derived from wild species, In *Molecular Dissection of Complex Traits*, Paterson, A., Ed., Boca Raton, FL.: CRC Press, p. 207, 1998.

Zamir, D. et al., Anther culture of male-sterile tomato (*Lycopersicon esculentum* Mill) mutants, *Plant Sci. Lett.*, 17, 353–361, 1980.

Zamir, D. et al., Genetic analysis of the origin of plants regenerated from anther tissues of *Lycopersicon esculentum* Mill, *Plant Sci. Lett.*, 21, 223–227, 1981.

Zamir, D. et al., Mapping and introgression of a tomato yellow leaf curl virus tolerance gene *Ty-1*, *Theor. Appl. Genet.*, 88, 141–146, 1994.

Zhang, H. X. and Blumwald, E., Transgenic salt-tolerant tomato plants accumulate salt in foliage but not in fruit, *Nat. Biotechnol.*, 19, 765–768, 2001.

Zhang, L. P. et al., A molecular linkage map of tomato displaying chromosomal locations of resistance gene analogs based on a *Lycopersicon esculentum* × *Lycopersicon hirsutum* cross, *Genome*, 45, 133–146, 2002.

Zhang, L. P. et al., Mapping QTLs conferring early blight (*Alternaria solani*) resistance in a *Lycopersicon esculentum* × *L. hirsutum* cross by selective genotyping, *Mol. Breed.*, 12, 3–19, 2003.

Zhong, X. B. et al., Preparation of tomato meiotic pachytene and mitotic metaphase chromosomes suitable for fluorescence in situ hybridization (FISH), *Chromosome Res.*, 4, 24–28, 1996a.

Zhong, X. B. et al., High-resolution mapping on pachytene chromosomes and extended DNA fibres by fluorescence in-situ hybridisation, *Plant Mol. Biol. Rep.*, 14, 232–242, 1996b.

Zhong, X. B. et al., FISH studies reveal the molecular and chromosomal organization of individual telomere domains in tomato, *Plant J.*, 13, 507–517, 1998.

Zhong, X. B. et al., FISH to meiotic pachytene chromosomes of tomato locates the root knot nematode resistance gene *Mi-1* and the acid phosphatase gene *Aps-1* near the junction of euchromatin and pericentromeric heterochromatin of chromosome arms 6S and 6L, respectively, *Theor. Appl. Genet.*, 98, 365–370, 1999.

Zhu, Q. et al., Enhanced protection against fungal attack by constitutive coexpression of chitinase and glucanase genes in transgenic tobacco, *Bio-Technology.*, 12, 807–812, 1994.

Zilkah, S. and Gressel, J., Cell cultures vs whole plants for measuring phytotoxicity .1. establishment and growth of callus and suspension cultures—definition of factors affecting toxicity on calli, *Plant Cell Physiol.*, 18, 641–655, 1977.

Zilkah, S. and Gressel, J., Estimation of cell-death in suspension cultures evoked by phytotoxic compounds—differences among techniques, *Plant Sci. Lett.*, 12, 305–313, 1978.

Ziv, M. et al., *Lycopersicon esculentum*—trifoliate plants recovered from anther cultures of heterozygous *Tftf* plants, *Plant Cell Rep.*, 3, 10–13, 1984.

CHAPTER 4

Brassica Vegetable Crops

N. Inomata

CONTENTS

4.1 INTRODUCTION

The tribe Brassiceae consists of four subtribes: Brassicinae, Raphaniae, Vellinae, and Moricandiinea. These subtribes contain 18 genera. Of these genera, genus *Brassica* in Brassicinae is economically the most important. *Brassica* crops consist of species with many different chromosome numbers. The tribe Brassiceae constitutes over 200 wild and allied species. These species provide a diverse gene pool (Gómez-Campo 1980). *Brassica* crops are under cultivation from tropical zones to temperate zones. Chinese cabbage (*Brassica rapa* ssp. *pekinensis*) and cabbage

(*B. oleracea* var. *capitata*) are sold in the market of Papua New Guinea and the Solomon Islands (Ogo, Inomata, and Yamamoto 1987).

Maintenance of this primary gene pool is essential for crop improvement. Selective introgression of alien genes broadens the genetic base of cultigens. Hybridization is one method used to incorporate economically important genetic information from secondary (GP-2) and tertiary (GP-3) gene pools. It is difficult to produce hybrids between the GP-1, GP-2, or GP-3 due to compatibility barriers. These barriers may result from failure to bypass sexual incompatibility or from hybrid breakdown. Production of interspecific and intergeneric hybrids has a great potential for improvement of brassicas. Successful crosses result in hybrids that can be used to introduce exogenous genes in *brassica* crops. Phylogenetic relationship between *Raphanus sativus* and its allied genera, *Brassica*, *Moricandia*, and *Sinapis*, has revealed that *R. sativus* is closer to its allied genera than formerly established by conventional taxonomy (Bang et al. 1997).

This paper describes the origin, distribution, and genomic and molecular relationships among *brassica* crops. Sexual hybridization, embryo rescue, and protoplast fusion are used to overcome crossing incompatibility. The scope of this review discusses the introgression of genes among *brassica* crops and from wild relatives to *brassica* crops, chromosome associations in F_1 hybrids, and the F_1 hybrid as a bridge plants. The ecological impact of transgenic plants and introgression of genes and their cytogenetic stability are discussed.

4.2 ORIGIN AND DISTRIBUTION

Brassica crops consist of three fundamental genomes: *B. nigra* ($2n = 16$; BB), *B. oleracea* ($2n = 18$; CC), *B. rapa* ($2n = 20$; AA), and, additionally, three of their amphidiploids: *B. carinata* ($2n = 34$; BBCC), *B. juncea* ($2n = 36$; AABB), and *B. napus* ($2n = 38$; AACC). The cytogenetic relationship among the *Brassica* species was presented by U (Figure 4.1). Genome analysis in *brassica* crops and allied genera were conducted by Mizushima (1980).

Brassica nigra probably originated in the Asia Minor–Iran areas and it was commonly used as a commercial spice. It is widespread in Europe, Africa, India and Far East (Smartt and Simonds 1995). *Brassica nigra* grows wild in the cultivated area of the Mediterranean region, in semi-cultivated condition in Morocco, Rhodes, Crete, Sicily and Turkey and is cultivated for seed oil,

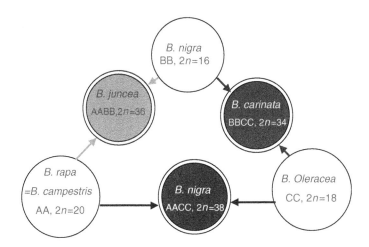

Figure 4.1 Genomic relationships of *brassica* crops. (Adapted from U.N., *Jpn. J. Bot.*, 7, 389–542, 1935.)

spice, as a vegetable and for medicinal uses in Ethiopia (Mizushima and Tsunoda 1967; Tsunoda 1980; Öztürk and Dogan 1997).

The *Brassica oleracea* group originated in the coastal rocky cliffs of the Mediterranean region and of northern Spain, western France, and southern and southwestern Britain (Mizushima and Tsunoda 1967; Tsunoda 1980; IBPGR 1981; Song, Osborn, and Williams 1990; Smartt and Simonds 1995). *Brassica* contains many varieties such as var. *acephala*, *botrys*, *bullata*, *capitata*, *gongylodes* and *gemmifera* (Hosoda 1961).

The wild *B. oleracea* is distributed throughout the Mediterranean region (Snogerup 1980; Gustafsson 1982; Gustafsson and Snogerup 1983; Gustafsson, Gómez-Campo, and Zamenis 1983; Perrino and Hammer 1985, Hammer, Cifarelli, and Perrino 1986, Snogerup, Gustafsson, and von Bothmer 1990; Perrino, Pignone, and Hammer 1992; Xhuveli 1995; Branca and Lapichino 1997). The relationship between wild relatives of *B. oleracea* ($2n = 18$) and cultivated crops, and among wild relatives of *B. oleracea*, has been established by morphological and cytogenetic investigations (Gustafsson et al. 1976; Snogerup and Persson 1983; von Bothmer, Gustafsson, and Snogerup 1995).

The origin of *B. rapa* (syn. *campestris*) is the Mediterranean region, which is thought to be a primary center of origins of European forms. Eastern Afghanistan and the adjoining portion of Pakistan are considered to be other primary centers of origin. Asia Minor, Transcaucasus and Iran are secondary centers of origin (Sinskaia 1928). A great colony of wild *B. rapa* was also observed near Ankara (Mizushima and Tsunoda 1967; Tsunoda 1980). The history of *B. rapa* in the Far East is obscure, but there are many cultivars of subspecies. Subspecies *dichotoma* and *trilocularis* are cultivated in India and Pakistan. Brown Sarson, an ecotype of ssp. *trilocular*, is the prevalent crop, while the ecotype Toria of ssp. *dichotoma* is confined to the foothills of the Himalayas (Prakash 1980).

The subspecies of *B. rapa* (*oleifera*, *rapifera*, *chinensis*, *pekinensis*, *narinosa*, and *nipposinica*) are present in Japan. The weedy plant, *B. rapa* ssp. *sylvestris*, is found in cultivated lands of Europe (Jørgensen and Andersen 1994; Landbo, Andersen, and Jørgensen 1996), Turkey (Öztürk and Dogan 1997) and New Zealand (Tsunoda 1980).

Brassica carinata is an amphidiploid between *B. nigra* and *B. oleracea* (Figure 4.1). Its cultivation is limited to the Ethiopian plateau, with extensions into part of Kenya (Tsunoda 1980).

Brassica juncea is an amphidiploid between *B. nigra* and *B. rapa*. This mustard species is a pungent species with a variety of cultivated forms. Many variations are found in Far East Asia and Japan. Several classified varieties are *juncea*, *napiformis*, *foliosa*, *japonica*, *sabellica*, *interglifolia*, and *rugosa* (Kumazawa and Abe 1955). They are cultivated in the north of the Indian subcontinent as an oilseed and spice plant with many recorded types (Nishi 1980; Prakash 1980; Tsunoda 1980; Smartt and Simonds 1995).

Brassica napus is an amphidiploid between *B. rapa* and *B. oleracea* and evolved on the coast of northern Europe where the distribution of *B. oleracea* and *B. rapa* overlap. The growth rate of *B. napus* is intermediate between the rapid growth rate of *B. rapa* and the very slow growth rate of *B. oleracea*. This may be the reason why *B. napus* cannot compete with common crop plants (Tsunoda 1980). There are two subspecies: ssp. *oleifera* (oil rape and forage rape) and *rapifera* (rutabaga) (Hosoda 1961).

4.3 GENOMIC AND MOLECULAR RELATIONSHIPS AMONG *BRASSICA* CROPS

Phylogenetic analysis based on molecular taxonomy (nuclear and chloroplast DNA markers) indicated a clear division of the subtribe Brassicinae into two ancient evolutionary lineages: the Nigra and the Rapa/Oleracea. The Nigra lineage contains *B. nigra*, *B. fruticulosa*, *B. tournefortii*, *Sinapis pubescens*, *S. alba*, *S. flexuosa*, *S. arvensis*, *Coincya cheiranthos*, *Erucastrum canariense*, and *Hirschfeldia incana*. The Rapa/Oleracea lineage contains *B. rapa, B. oleracea* ssp. *oleracea* and *B. oleracea* ssp. *alboglabra, B. rupestris, B. drepanensis, B. macrocarpa, B. villosa, B. barrelieri, B. deflexa, B. oxyrrhina, B. gravinae, Diplotaxis erucides, D. tenuifolia, Eruca sativa, Raphanus raphanistrum, R. sativus,* and *Sinapis aucheri* (Song, Osborn, and Williams 1988a; Warwick and Black 1991). Members of the *Brassica, Diplotaxis, Erucastrum,* and *Sinapis* genera are represented in both lineages. Molecular phylogenetic relationship is established among the wild relatives (Yanagino, Takahata, and Hinata 1987; Pradhan et al. 1992; Warwick, Black, and Aguinagalde 1992; Warwick and Black 1993). The comparative analysis of phenolic compounds helped determine the phylogenetic relationship of some wild species of the genus *Brassica* (Sánchez-Yélamo 2002).

Molecular cytogenetics clustered species into the broccoli, cauliflower, and cabbage groups (Nienhuis et al. 1993). Based on nuclear restriction fragment length polymorphisms (RFLP) analysis, the cultivated *B. oleracea* group can be divided into three district groups, represented by thousand-headed kale, broccoli, and cabbage (Song, Osborn, and Williams 1988b). Genetic markers of random amplified polymorphic DNA (RAPD) exhibited similarities between *B. oleracea* and *B. incana,* whereas *B. oleracea* and *B. cretica* or *B. hilarionis* are distinct (Lazaro and Aguinagalde 1996).

A wild *B. rapa* from India was positioned intermediatly between European and East Asian types based on RFLPs (Song, Osborn, and Williams 1988a, 1990). In contrast, RAPD markers failed to separate *B. rapa* into European or Asian groups (Warwick and McDonald 2001). Wild *B. rapa* dominates in the high plateaus of the Irano-Turanian region.

Brassica carinata is characterized as an amphidiploid of recent multiple origins (Quiros et al. 1985; Song, Osborn, and Williams 1988a) based on ribosomal genes and RFLP analysis. There are many cultivars in Ethiopia (Astley, Gyorgis, and Toll 1982; Seegeler 1983) but no major subclustering is detected by RFLP markers (Warwick and Soleimani 2001). In Ethiopia, recent studies have identified and registered mustard cultivars with zero erucic acid levels as genetic stock (Fernández-Escobar et al. 1988; Fernández-Martínez et al. 2001).

Nuclear RFLP results clearly revealed that *B. napus* is of multiple origins. Plants from this group can be clustered into three subgroups: oilseed cultivars from Canada (group 1), from Europe (group 2) and rutabaga (group 3) (Song and Osborn 1992). Rutabaga (forage rape) is an important fodder crop in Europe, Russia, and New Zealand. The acreage of oilseed rape has increased significantly in Europe since the mid-1960s. *Brassica rapa* is of particular economic importance in Canada and China. Ma et al. (2000) identified genetic diversities among Chinese, Japanese, and European varieties by RAPD markers. The recently published reviews on taxonomy, origin and domestication of *brassica* crops and wild allies are extremely informative (Gómez-Campo 1999; Gómez-Campo and Prakash 1999).

4.4 OVERCOMING CROSSING-INCOMPATIBILITY THROUGH SEXUAL HYBRIDIZATION, EMBRYO RESCUE, AND PROTOPLAST FUSION

Interspecific and intergeneric hybrids in *brassica* crops produced by sexual and asexual hybridization, embryo rescue, and genetic manipulation were described in several reviews (Namai 1987; Inomata 1997a). Only recent and relevant studies are presented in this chapter.

4.4.1 Conventional Crossing

The cytogenetic relationships in the *brassica* crops and allied species were based on sexual pollination (Mizushima 1950). Production of interspecific and intergeneric hybrids among *brassica* crops and the wild relatives is difficult, with the exception of the cross between *B. napus* and *B. rapa*.

Table 4.1   Recent Study of Interspecific and Intergeneric Hybrid Production by Sexual Pollination in *Brassica* Crops

Cross Combination	Author
Diplotaxis muraris × *B. rapa*	Hinata and Konno (1979)
B. napus × *Sinapis arvensis*	Inomata (1988)
B. fruticulosa × *B. rapa*	Nanda Kumar et al. (1988)
B. napus × *Erucastrum gallium*	Lefol et al. (1997)
B. rapa × *E. gallium*	
B. napus × *Raphanus raphanistrum*	
B. oxyrrhina × *Raphanus sativus*	Bang et al. (1997)
B. juncea × *Orychophragmus violaceus*	Li et al. (1998)
B. carinata × *O. violaceus*	
B. napus × *Diplotaxis catholica*	Bijral and Sharma (1998)
B. juncea × *Eruca sativa*	Bijral and Sharma (1999a)
B. juncea × *B. oxyrrhina*	Bijral and Sharma (1999b)
Sinapis alba × *B. napus*	Brown et al. (1999)
B. rapa × *Orychophragmus violaceus*	Li and Heneen (1999)
B. nigra × *O. violaceus*	
B. napus × *Orychophragmus violaceus*	Cheng and Seguin-Swartz (2000)
B. napus × *Orychophragmus violaceus*	Cheng et al. (2002)
Sinapis alba × *B. nigra*	Choudhary and Joshi (2000)
B. tournefortii × *Raphanus caudatus*	Choudhary, Joshi, and Singh (2000a)
B. carinata × *B. rapa*	Choudhary et al. (2000b)
B. tournefortii × *B. rapa*	Choudhary and Joshi (2001)
B. juncea × *Erucastrum virgatum*	Inomata (2001)
B. napus × *Raphanus sativus*	Huang et al. (2002)

In the cross of *B. rapa* × *B. oleracea*, only four and eight hybrids were obtained in 732 and 913 flowers pollinated, respectively (U 1935; Sarashima 1964). Table 4.1 contains recently studied interspecific and intergeneric hybrids reported by sexual pollination.

4.4.2   Embryo Culture

Sexual hybridization is useful for introgression of genes among varieties of the same species. But gene introgression is difficult in interspecific or intergeneric crosses. To overcome barriers, embryo culture was employed to produce many interspecific and intergeneric hybrids from combinations with low compatibility (Inomata 1996b). In the genus *Brassica*, Nishi, Kawata, and Toda (1959) were the first to use embryo culture for producing hybrids between *B. rapa* and *B. oleracea*. Hybrid plants were also obtained from a cross of *B. oleracea* × *B. rapa*. In reciprocal crosses, hybrid embryos degenerated soon after fertilization, and no viable embryos were recovered. Hybrids were more easily obtained when *B. oleracea* was used as the female parent than when used as the male parent (Takeshita, Kato, and Tokumasu 1980). Triazine resistance and resistance to race 2 of *Plasmodiophora brassicae* were transferred from *B. napus* to *B. oleracea* (Ayotte, Harney, and Machado 1989; Chiang, Chiang, and Grant 1997;). Cabbage aphid (*Brevicoryne brassicae*) resistance was transferred from *B. oleracea* to *B. napus* (Quazi 1988). Embryo culture was used to produce intergeneric hybrids in diploid *B. oleracea* × *R. sativus*, autotetraploid *B. oleracea* × *R. sativus* and *B. napus* × *R. sativus* (Inomata 1993a), *B. napus* × *Orychophragmus violaceus* (Li, Liu, and Luo 1995), and *R. sativus* × *B. fruticulosa* (Bang et al. 1997).

Successful culture of the embryo becomes more difficult when attempted at stages earlier than the heart-shaped stage, which is about two weeks after pollination. Earlier embryos had difficulty in differentiating into plantlets. However, recently, improved culture media and methods were formulated and pro-embryos were developed into normal mature embryos with high frequency (Liu, Xu, and Chua 1993; Monnier 1995). These procedures have revolutionized the production of interspecific and intergeneric hybrids.

4.4.3 Embryo Rescue

Improved embryo rescue protocols have been used to produce interspecific and intergeneric hybrids. The fertilized ovules were harvested from the ovary and cultured. Hybrids were obtained in reciprocal crosses between *B. rapa* and *B. oleracea* at the developmental stage when the embryos were globular to early torpedo-shaped (Takeshita, Kato, and Tokumasu 1980). Hybridization was also successful between the following: *B. fruticulosa*×*B. rapa* (Nanda Kumar, Shivanna, and Prakash 1988), *Sinapis alba*×*B. napus* (Ripley and Arnison 1990), *Eruca sativa*×*B. rapa*, *B. napus*×*Raphnobrassica* (Agnihotri et al. 1990a, 1990b), and in the reciprocal cross between *B. oleracea*×*R. sativus* (Hossain, Inden, and Asahira 1990), *B. rapa*×*Sinapis turgida* or *S. arvensis, B. oleracea*×*S. alba or S. turgida, B. carinata*×*S. alba, S. arvensis or S. turgida* (Momotaz, Kato, and Kakihara 1998), and reciprocal crosses between *B. rapa*×*B. oleracea* (Lu et al. 2001; Zhang et al. 2001) and *B. napus*×*B. oleracea* (Ripley and Beversdorf 2003).

Pre-fertilization barriers in many crossing-combinations have prevented successful hybridization between cultivated species and wild relatives. However, *in vitro* fertilization, followed by culturing the fertilized ovules seems promising. The technique of ovule culture was also used to study test tube fertilization in *Brassica* species (Kameya, Hinata, and Mizushima 1966; Kameya and Hinata 1970; Zenkteler 1990).

In many cases, degeneration of the hybrid embryo occurred in early embryogenesis, preventing ovule dissection and embryo selection. Under these circumstances, it was much easier to culture ovaries excised from pollinated flowers. Hybrids may also be obtained from ovary culture *in vitro*. Inomata (1993a) listed early stages of ovary culture. The media conditions for ovary culture were standardized for *Brassica* species (Inomata 1977, 1978a, 1978b, 1979, 1985a; Matsuzawa 1978). A large number of interspecific and intergeneric hybrids were successfully produced (Inomata 1985a, 1990, 1993a). When hybrid seeds and/or well-developed embryos were obtained after cultivation of fertilized ovaries, they were further cultured in the medium leading to development of plantlets from embryos and, eventually, hybrid seeds. Table 4.2 summarizes recent studies on embryo rescue through ovary culture and production of interspecific and intergeneric hybrids.

The development of embryo and endosperm are visible in the ovule three days after pollination in sib crosses of *B. rapa* (Nishiyama and Inomata 1966). When ovaries of the interspecific cross between *B. rapa* and *B. oleracea* were cultured at two days after pollination, seeds and well-developed embryos were obtained and were capable of maturing in the field (Inomata 1993a).

4.4.4 Somatic Hybrids by Protoplast Fusion

Gleba and Hoffmann's (1978, 1979) attempted to produce intertribal hybrids (*Arabidobrassica*) between *Arabidopsis thaliana* and *B. rapa* by protoplast fusion but were unsuccessful. Somatic hybrid plants were abnormal compared with the parents. More recently, fertile somatic hybrids between *B. napus* and *A. thaliana* (Forsberg, Landgren, and Glimelius 1994) and between *B. carinata* and *B. tournefortii* (Mukhopadhyay 1994) were produced. The male sterility and male fertility-restorer genes were introduced from *R. sativus* to *B. napus* (Sakai and Imamura 1990; Sakai 1996). Regenerated plants from intertribal hybridization between *Capsella-bursa pastoris* and *B. oleracea* were intermediate for morphological traits. *Capsella-bursa pastoris* was highly resistant to *Alternaria brassicae, A. brassicola*, flea beetles (*Phylotreta cruciferae* and *P. striolata*) as well as cold tolerant (Sigareva and Earle 1997a). Regeneration of plants

Table 4.2 Production of Interspecific and Intergeneric Hybrids by Embryo Rescue in *Brassica* Crops

Cross Combination	Author
B. rapa× *B. cretica*	Inomata (1985a)
B. rapa× *B. bourgeaui*	Inomata (1986)
B. rapa× *B. Montana*	Inomata (1987)
Diplotaxis erucoides× *B. napus*	Delourme et al. (1989)
Eruca sativa× *B. rapa*	Matsuzawa and Sarashima (1986)
E. sativa× *B. oleracea*	
E. sativa× *Raphanus sativus*	
B. napus× *Sinapis arvensis*	Inomata (1988)
B. fruticulosa× *B. rapa*	Nanda Kumar et al. (1988)
Enarthrocarpus lyratus× *B. oleracea*	Gundimeda et al. (1992)
E. lyratus× *B. rapa*	
E. lyratus× *B. carinata*	
E. lyratus× *B. napus*	
B. juncea× *E. lyratus*	
Diplotaxis siettiana× *B. rapa*	Nanda Kumar and Shivanna (1993)
B. napus× *Sinapis pubescens*	Inomata (1994a)
B. juncea× *Diplotaxis virgata*	Inomata (1994b)
Erucastrum abyssunicum× *B. juncea*	Sarmah and Sarla (1994)
Raphanus sativus× *Sinapis turgida*	Bang et al. (1996a)
R. sativus× *S. alba*	
R. sativus× *S. arvensis*	
R. sativus× *S. pubescens*	
S. arvensis× *R. sativus*	
S. turgida× *R. sativus*	
Moricandia arvensis× *Raphanus sativus*	Bang et al. (1996b)
B. rapa× *Sinapis turgida*	Inomata (1996a)
B. tournefortii× *B. rapa*	Nagpal et al. (1996)
B. tournefortii× *B. nigra*	
Raphanus sativus× *B. fruticulosa*	Bang et al. (1997)
B. fruticulosa× *R. sativus*	
B. maurorum× *R. sativus*	
B. oxyrrhina× *R. sativus*	
Sinapis alba× *B. napus*	Brown et al. (1997)
B. oxyrrhina× *Raphanus sativus*	Matsuzawa et al. (1997)
B. juncea× *Diplotaxis erucoides*	Inomata (1998)
B. juncea× *Crambe abyssinica*	Youping and Peng (1998)
B. maurorum× *Raphanus sativus*	Bang et al. (1998)
B. rapa× *Moricandia arvensis*	Matsuzawa et al. (1998)
Diplotaxis tenuisiliqua× *B. rapa*	Sarma and Sarla (1998)
B. oleracea× *Orychophragmus violaceus*	Li and Heneen (1999)
B. rapa× *Diplotaxisis virgata*	Inomata (1999)
B. rapa× *Eruca sativa*	Robers et al. (1999)
B. rapa× *B. spinescens*	Inomata (2000)
B. napus× *Raphanus sativus*	Luo et al. (2000)
B. juncea× *Erucastrum virgatum*	Inomata (2001)
Erucastrum canariensis× *B. rapa*	Prakash et al. (2001)
Erucastrum canariensis× *B. rapa*	Bhaskar et al. (2002)
B. napus× *B. bourgeaui*	Inomata (2002a)
B. napus× *B. Montana*	
B. napus× *B. oleracea*	
B. napus× *B. cretica*	
B. napus× *Diplotaxis virgata*	Inomata (2002b)
Diplotaxis siifolia× *B. rapa*	Ahuja et al. (2003)
D. siifolia× *B. juncea*	
Diplotaxis catholica× *B. rapa*	Banga et al. (2003)
D. catholica× *B. juncea*	

continued

Table 4.2   Continued

Cross Combination	Author
B. juncea×Diplotaxis virgata	Inomata (2003)
Erucastrum cardaminoides×B. rapa	Chandra et al. (2004a)
E. cardaminoides×B. nigra	
B. fruticulosa×B. rapa	Chandra et al. (2004b)
B. juncea×B. oleracea	Tonguç, and Griffiths (2004)
B. napus×Diplotaxis harra	Inomata (2005)

from intertribal somatic hybridization between *Camelina sativa* and *B. oleracea* was successful and many hybrids were obtained. Somatic hybrids were intermediate for morphological traits. *Camelina sativa* was resistant to *Alternaria brassicae* and *Leptosphaeria maculans* as well as tolerant to drought and cold (Hansen 1997; Sigareva and Earle 1997b).

Somatic hybrids with $2n=46$ and $2n=92$ chromosomes were produced between *B. oleracea* ($2n=18$) and *Moricandia nitens* ($2n=28$). In most cases, anthers were stunted and produced low pollen fertility. However, high pollen fertility rates (over 70%) were obtained in the hybrids with $2n=92$ chromosomes (Yan et al. 1999). Many hybrids with sesquitetraploid chromosomes were produced when *B. napus* was crossed with the fertile somatic hybrid ($2n=46$) between *B. oleracea* ($2n=18$) and *Moricandia nitens* ($2n=28$; Zhang 2004).

Somatic hybrids, resistant to bacterial soft rot (*Erwinia carotovora* subsp. *catotovora*), were obtained using protoplast fusion between *B. rapa* and *B. oleracea*, followed by backcrossing (Ren, Dickson, and Earle 1999). Sigareva, Ren, and Earle (1999) transferred resistance to *Alternaria barssicicola* from *Sinapis alba* to *B. oleracea* by somatic hybridization and backcrossing.

Male-sterile lines, corrected for chlorosis, were developed by fusing protoplasts of cytoplasmic male sterile (CMS) *B. juncea* (AABB) with *B. oxyrrhina* cytoplasm and normal *B. oleracea* (CC). Male-sterile AABBCC somatic hybrids were produced (Arumugam 2000). Heat-tolerant *B. carinata* was obtained from the cross between heat-tolerant *R. sativus* (RR) and *B. oleracea* (CC). The RC intergeneric hybrid was fused with the protoplasts of *B. nigra* (BB) to produce a RCBB somatic hybrid. The somatic hybrid was backcrossed to *B. carinata* (Arumugam 2002). The somatic hybrid between *B. napus* and *O. violaceus* could be selfed or backcrossed to *B. napus* (Hu et al. 2002).

Amphidiploid plants, with $2n=46$ chromosomes, were obtained by somatic fusion between *B. oleracea* ($2n=18$) and *Moricandia arvensis* ($2n=28$). The somatic hybrids showed 23 bivalents at metaphase I, with about 50% fertile pollen (Ishikawa et al. 2003).

Somatic hybrids between *B. napus* ($2n=38$) and *Crambe abyssinica* ($2n=90$) were obtained by UV-irradiated protoplast of *C. abyssinica*. Nine hybrids contained $2n=38$ chromosomes and the others contained $2n=40-78$ chromosomes. An analysis of fatty acid composition in the seeds, showed significantly greater amount of erucic acid than *B. napus* (Wang, Sonntag, and Rudloff 2003). Recent reviews of the protoplast culture and somatic hybridization in brassicas are very informative (Glimelius 1999; Cardoza and Steward 2004; Christey 2004).

4.5 CONCEQUENCE OF ESCAPED GENES FROM TRANSGENIC PLANTS

Advances in molecular biotechnology have facilitated the transfer of new traits of economic importance from foreign organisms into crops. The transference of herbicide resistance to *Brassica napus* is likely having the greatest economic impact among the Brassica crops.

Misra and Gedamu (1989) produced heavy metal tolerance into a transgenic line of *B. napus* by use of Ti-plasmid of *Agrobacterium tumefaciens*. A transgenic line of *B. napus*, with increased levels of methionine in the seed was transferred from methionine-rich seed of Brazil nut (Altenbach 1992).

Studies on transgenic plants are new and more transgenic plants will continue to appear. Production of transgenic brassicas has been reviewed by Christey and Braun (2004) and Pua and Lim (2004).

The escape of traits from genetically modified transgenic lines of *B. napus* to weedy plants through artificial crosses, spontaneous crosses, and backcrossing, possesses a potential risk (Table 4.3). Care must be taken to prevent escape of transgenic traits from crop plants, which may have negative impact on non-crop species. In a field of transgenic Indian mustard (*B. juncea*), gene escape was limited to 35 m (GhoshDastidar and Varma 1990, GhoshDastidar, Varma, and Kapui 2000). Pollen movement caused an unacceptable level of gene flow from genetically modified *B. napus* (cv. Canola) to non-genetically modified *B. napus* crops or to related weedy species. Pollen movement occurred at a low frequency but at a considerable distance (Rieger et al. 2002). Insect pollinators did not discriminate between non-transgenic and transgenic oilseed rape (Pierre 2003).

Many interspecific and intergeneric hybrids have been obtained between transgenic *B. napus* and the wild relatives of the tribe Brassiceae (Table 4.3). Genes can be dispersed from commercial crops to the wild weedy population. In the interspecific F_1 hybrid between *B. napus* × *B. adpressa* (syn. *Hirschfelia incana*), maximum bivalent associations at metaphase I was from 7 to 10 (Harberd and McArthur 1980; Kerlan, Chèvre, and Eber 1993) and from 4 to 18 (Eber 1994). Maximum bivalent association in F_1 hybrids of *B. napus* × *R. raphanistrum*, *B. napus* × *S. arvensis* and *B. napus* × *B. nigra* was 10 (Kerlan, Chèvre, and Eber 1993). The range was from 8 to 18 in the F_1 hybrid between *B. napus* and *R. raphanistrum* (Eber 1994). Meiotic association in interspecific F_1 hybrids of *B. napus* × *B. rapa* ssp. *sylvestris* (weed) and *B. napus* × *B. kaber* has not been reported. However, high levels of spontaneous hybridization between *B. napus* and weedy *B. rapa* ssp. *rapa* or ssp.

Table 4.3   Hybrids Obtained between Transgenic *Brassica napus* and *Brassica* Crops or the Wild Relatives

Transgenic Plant	*Brassica* Crop or Wild Relative	Remark
B. napus	*B. oleracea* var. *acephala* and var. *capitata* *B. adpressa* (syn. *Hirschfelia incana*) *B. nigra* (black mustard) *Sinapis arvensis* (wild mustard) *Raphanus raphanistrum*	Kerlan et al. (1992)
B. napus	*B. adpressa* (syn. *Hirschfelia incana*) *Raphanus raphanistrum*	Eber et al. (1994)
B. napus	*B. adpressa* (syn. *Hirschfelia incana*)	Lefol et al. (1995)
B. napus	*Raphanus raphanistrum*	Darmency et al. (1995)
B. napus	*B. rapa* (field mustard)	Mikkelsen et al. (1996a, 1996b)
B. napus	*Sinapis arvensis* (wild mustard)	Lefol et al. (1996a)
B. napus	*Hirschfelia incana* (wild hoary mustard)	Lefol et al. (1996b)
B. napus	*B. rapa* (field mustard)	Brown et al. (1996a)
B. napus	*B. rapa* (field mustard) *B. kaber* (wild mustard) *B. nigra* (black mustard)	Brown et al. (1996b)
B. napus	*B. kaber* (wild mustard) *B. nigra* (black mustard) *B. rapa* (field mustard)	Nair et al. (1996)
B. napus	*Raphanus raphanistrum*	Chevre et al. (1997)
B. napus	*B. rapa* (weedy species) *B. juncea*	Jorgensen et al. (1998)
B. napus	*Hischfelia incana* (syn. *B. adpressa*)	Darmency and Fleury (2000)
B. napus	*B. rapa* (field mustard)	Halfhill et al. (2001)
B. napus	*B. rapa* (wild population)	Warwick et al. (2003)

sylvestris may occur in the field (Jørgensen and Andersen 1994; Hauser, Jørgensen, and Østergård 1998). F_1 hybrids between *B. napus* and *B. rapa* ssp. *pekinensis* showed $10_{II} + 9_I$ chromosome configuration (Mizushima 1950). The introgression of genes between *B. napus* and *B. rapa* occurred under natural conditions (Hansen, Siegismund, and Jørgensen 2001). Hybridization between *B. rapa* and *B. napus* occurred in commercial fields, however, gene flow between *Raphanus raphanistrum* and *B. napus* was very rare. Hybrids between *Sinapis arvensis* or *Erucastrum gallicum* × *B. napus* hybrids were not detected (Warwick 2003). F_1 hybrids between *B. napus* and other *brassica* crops or the wild relatives describe above and in Table 4.3 are partially fertile and may become useful as bridge plants.

4.6 INTROGRESSION OF GENES FROM WILD RELATIVES TO *BRASSICA* CROPS

Brassica crops are attacked by a number of biotic and abiotic stresses. The yield loss varies, depending upon the type and nature of pathogen association, time and severity of the attack, resistance level of the cultivars and the prevailing environmental conditions (Saharan 1997). Table 4.4 shows the introgression of genes among the *brassica* crops and from wild relatives to *brassica* crops by sexual hybridization, embryo rescue, protoplast fusion, and genetic transformation.

Many genes have been transferred into the economically important *B. napus*. Breeding for resistance to pathogens and for improvement of meal and oil quality is also important. In recent studies, interspecific crosses were used to produce yellow-seeded *B. napus* to improved feeding value of the meal. Meal from this seed contained low erucic acid and low glucosinolate content (Rahman, Joersbo, and Poulsen 2001). Black rot-resistant characteristics were transferred from *B. juncea* to *B. oleracea* (Tonguç and Griffiths 2004a).

Satisfactory genetic control of pathogens and virus diseases has not been achieved. This is primarily due to the absence of adequate sources of resistance. Thus, there is an urgent need to develop methods for identifying resistance genes in the wild species (Saharan 1993). Informative *brassica* crop reviews have been published for disease resistance (Saharan 1997; Tewari and Mithen 1999; Dixelius, Bohman, and Wretblad 2004), insect resistance (Earle, Cao, and Shelton 2004), and herbicide resistance (Warwick and Miki 2004).

4.7 CHROMOSOME ASSOCIATION IN THE F_1 HYBRIDS AND THE F_1 HYBRIDS AS A BRIDGE PLANT

Production of interspecific and intergeneric hybrids is important for two reasons: to establish the phylogenic relationships in the tribe Brassiceae, and to determine the feasibility of introgression of useful genes from the wild relatives into *brassica* crops.

Many interspecific and intergeneric hybrids have been produced by sexual hybridization, embryo rescue culture, and protoplast fusion. Meiotic chromosome association of the F_1 hybrids has been described in five reviews (Harberd and McArthur 1980; Prakash and Hinata 1980; Namai 1987; Inomata 1997a; Prakash et al. 1999). Interspecific and intergeneric F_1 hybrids showed more frequent bivalents and univalents configurations but few trivalents and tetravalents (quadrivalents) have been recorded. Multivalent association is attributed in part to allosyndesis. This suggests that recombination may occur among these species. Figure 4.2 shows the morphology of the interspecific and intergeneric F_1 hybrid plants. Chromosome configuration at metaphase I of the intergeneric F_1 hybrids is shown in Figure 4.3.

Somatic chromosome numbers of A-, B-, and C-genome (fundamental genome constitution) *brassica* crops are $2n = 20$, 16, and 18, respectively. Maximum number of bivalents at metaphase I in dihaploid of AB and BC genome species was 5, in both hybrids (Olsson 1960; Sarla and Raut

Table 4.4   Introgression of Characteristics and Genes among *Brassica* Crops and from Wild Relatives to *Brassica* Crops

From	To	Trait	Reference
B. rapa	*B. napus*	Early maturing character, resistance to high moisture	Shiga (1970); Namai et al. (1980)
B. rapa	*B. napus*	*Plasmodiophora brassicae* resistance (Clubroot disease resistance)	Lammerink (1970)
B. rapa	*B. napus*	*Plasmodiophora brassicae* resistance	Johnston (1974)
B. napus	*B. oleracea*	*Plasmodiophora brassicae* resistance	Chiang et al. (1997)
B. napus	*B. oleracea*	*Plasmodiophora brassicae* resistance	Chiang and Crete (1983)
B. juncea	*B. napus*	*Leptoshaeria maculans* resistance (Black leg disease resistance)	Roy (1984)
B. juncea	*B. napus*	*Phoma lingam* resistance (Black leg disease resistance)	Sacristán and Gerdemann (1986)
Raphanus sativus	*B. napus*	Fertility restoring gene	Paulmann and Robbelen (1988)
Sinapis alba	*B. napus*	*Alternaria brassicae* resistance (Alternaria black spot)	Primard et al. (1988)
B. napus	*B. carinata*	Erucic acid content	Fernandez-Escobar et al. (1988)
B. juncea	*B. napus*	Pod shattering	Prakash and Chopra (1990)
Raphanus sativus	*B. napus*	Cytoplasmic male sterility	Sakai and Imamura (1990)
R. sativus	*B. oleracea*	*Plasmodiophora brassicae* resistance	Hagimori et al. (1992)
R. sativus	*B. napus*	*Heterodera schachtii* resistance (Beat cyst nematode resistance)	Lelivelt and Krens (1992)
Sinapis alba	*B. napus*	*Heterodera schachtii* resistance	Lelivelt et al. (1993)
B. nigra	*B. napus*	*Leptoshaeria maculans* resistance	Gerdemann-Knorck et al. (1994)
B. juncea	*B. carinata*	Erucic acid content	Getinet et al. (1994)
B. nigra	*B. napus*	*Phoma lingam* resistance	Gerdemann-Knorck et al. (1995)
B. nigra	*B. napus*	*Plasmodiophora brassicae* resistance	
B. nigra	*B. napus*	*Leptoshaeria maculans* resistance	Chevre et al. (1996)
B. juncea	*B. napus*	*Leptoshaeria maculans* resistance	Pang and Halloran (1996)
Raphanus sativus	*B. napus*	Fertility restoring gene	Sakai et al. (1996)
B. nigra	*B. napus*	*Phoma lingam* resistance	Struss et al. (1996)
B. rapa	*B. napus*	*Plasmodiophora brassicae* resistance	Bradshaw, Gemmell, and Wilson (1997)
Trachystoma ballii	*B. juncea*	Fertility restoring gene	Kirti et al. (1997)
Moricandia arvensis	*B. juncea*	Fertility restoring gene	Prakash et al. (1998)
Eruca sativa	*B. rapa*	Fertility restoring gene	Matsuzawa et al. (1999)
Trichoderma harzianum	*B. oleracea* var. *italica*	Fungal pathogen resistaces	Mora amd Earle (1999)
Sinapis turgida	*B. carinata*	Fatty acid composition	Momotaz et al. (2000)
S. arvensis	*B. carinata*	Fatty acid composition	
B. napus	*B. rapa*	Long pod genotype	Lewis, Woods, and Cheng (2001)
Synthesized *B. napus*	*B. napus*	Fatty acid composition	Lu et al. 2001
Erucastrum canariens	*B. juncea*	Cytoplasmic male sterility and fertility restoring gene	Prakash et al. (2001)
B. alboglbra× *B. rapa*	*B. napus*	Yellow-seeded color	Rahman (2001)
B. carinata× *B. rapa*	*B. napus*	Brown-seeded color	
B. napus (yellow seed)	*B. napus*	Low erucic acid and glucocinolate content	Rahman et al. (2001)

continued

Table 4.4　Continued

From	To	Trait	Reference
Arabidopsis thaliana	*B. napus*	*Leptosphaeria maculans* resistance (Black leg disease resistance)	Bohman et al. (2002)
Wheat germ agglutinin	*B. juncea*	*Lipahis erysimi* resistance (Mustard aphid resistance)	Kanrar et al. (2002)
B. juncea	*B. oleracea*	*Xanthomonas campestris* pv. *campestris* (Black rot disease resistance)	Tonguç and Griffiths (2004a)
B. carinata	*B. oleracea*	*Erysiphe polygoni* (*Powdery mildew*)	Tonguç and Griffiths (2004b)
B. fruticulosa	*B. rapa*	Aphid resistance	Chandra et al. (2004b)

1991; Inomata 1997a), and the maximum number of bivalents was 9 in AC genome (Inomata 1980, 1997a; Attia and Röbbelen 1986). Recombination occurred frequently among genomes of A, B, and C. If the useful gene(s) from the wild relatives could be introduced into each of A, B, and C genome of *brassica* crops, the majority of genes could be transferred to other genomes or amphidiploid crops. Mizushima (1980) reported genome analysis in *brassica* crops and allied genera. Chromosome association at metaphase I and pollen fertility in interspecific and intergeneric F_1 hybrids between *brassica* crops and wild relatives are shown in Table 4.5 and Table 4.6.

The intergeneric F_1 hybrid between *B. napus* ($2n = 38$) and *Orychophragmus violaceus* ($2n = 24$) was obtained from embryo culture. Chromosome instability was observed in somatic tissues (Li, Liu, and Luo 1995), and the maximum bivalent association was 12. *Orychophragmus violaceus* is an autotetraploid species with $x = 6$ basic chromosome number, suggesting that allosyndesis may have occurred between the AC genome and *O. violaceus* genome (Li, Liu, and Heneen 1996).

The intergeneric F_1 hybrids, *B. juncea* ($2n = 36$) × *O. violaceus* ($2n = 24$) and *B. carinata* ($2n = 34$) × *O. violaceus* ($2n = 24$), consisted of mixoploid-ranging chromosomes from 12–24 to 12–34, respectively. Pollen stainability was high in both hybrids (Li 1998). The somatic chromosome number varied for intergeneric F_1 hybrids between three cultivated *Brassica* diploids (*B. rapa*, *B. oleracea*, *B. nigra*) and *O. violaceus*. Variation of the mitotic cells in F_1 hybrids between *B. rapa* and *O. violaceus* was larger than that of F_1 hybrids between *B. oleracea* and *O. violaceus* or between *B. nigra* and *O. violaceus*. The A-genome was more stable than C- and B-genomes (Li and Heneen 1999). Chromosome elimination and duplication in the intergeneric F_1 hybrids of *B. napus* × *O. violaceus* also occurred spontaneously. *Orychophragmus violaceus* contained high levels of palmitic, oleic and linoleic acids, and low levels of linolenic acid and erucic acid (Hu and Hansen 1998; Cheng, Séguin-Swartz, and Somers 2002).

Bivalents of intergeneric F_1 hybrids ($2n = 18$) between *R. sativus* ($2n = 18$) and three species of genus *Sinapis* (*S. arvensis*, *S. pubescens*, *S. turgida*; $2n = 18$), ranged from 0 to 7. Bivalents ranged from 6 to 18 in the amphidiploids with $2n = 36$ chromosomes and pollen fertility rate exceeded 50%. Bivalents of intergeneric F_1 hybrids ($2n = 21$) between *R. sativus* and *S. alba* ($2n = 24$), ranged from 2 to 4. The occurrence of bivalents may be attributed, in part, to allosyndesis (Bang, Kaneko, and Matsuzawa 1996a).

Three intergeneric F_1 hybrids of *B. tournefortii* ($2n = 20$) × *B. rapa* ($2n = 20$), *B. tournefortii* × *B. nigra* ($2n = 16$) and *B. tournefortii* × *B. oleracea* ($2n = 18$) were male sterile and did not set seed. However, pollen viability was high in these amphidiploids. Some seed set was obtained by back-crossing and open pollination, when these hybrids were crossed with *B. napus* and *B. juncea* (Nagpal et al. 1996).

The intergeneric F_1 hybrids ($2n = 23$) between *M. arvensis* ($2n = 28$) and *R. sativus* ($2n = 18$) showed 1.9% fertile pollen. When the F_1 hybrid was backcrossed to *R. sativus*, no BC_1 plants were obtained. On the other hand, colchicine-treated amphidiploid F_1 hybrids had 66.6% fertile pollen. The BC_1 plants obtained by backcrossing to *R. sativus* were sesquidiploid, with $2n = 32$, and

Figure 4.2 Intergeneric F₁ hybrid plants and their parents. (A) *B. rapa* ssp. *pekinensis* (2n=20), (B) *Diplotaxis virgata* (2n=18) and (C) Its hybrid plant (2n=19) (Inomata 1999). (D) *B. carinata* (2n=34), (E) *Sinapis turgida* (2n=18) and (F) Its hybrid plant (2n=26) (Inomata 1992). (G) *B. juncea* (2n=36), (H) *Diplotaxis erucoides* (2n=14) and (I) Its hybrid plant (2n=25) (Inomata 1998). (J) *B. napus* (2n=38), (K) *Sinapis pubescens* (2n=18) and (L) Its hybrid plant (2n=28) (Adapted from Inomata, N., *Theor. Appl. Genet.*, 89, 540–544, 1994. With permission.)

hyperploid, with $2n=55$ chromosomes. Many seeds were obtained in BC₂ plants (Bang, Kaneko, and Matsuzawa 1996b).

The intergeneric F₁ hybrid ($2n=18$) of *B. oxyrrhina* ($2n=18$)×*R. sativus* ($2n=18$) was completely pollen and seed sterile. Amphidiploid plants were intermediate for morphological

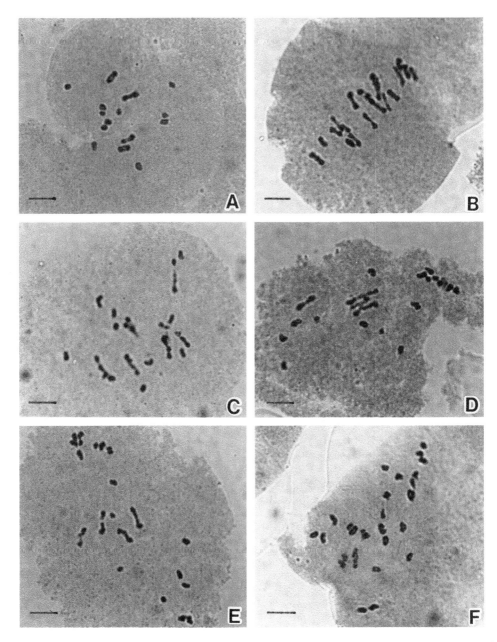

Figure 4.3   Meiotic chromosomes pairing in the intergeneric F_1 hybrid. (A) $4_{II} + 11_I$ of the F_1 hybrid ($2n = 19$) of *B. rapa* ($2n = 20$) $\times$ *Diplotaxis virgata* ($2n = 18$) (From Inomata, N., *Cruciferae Newsl.*, 21, 39–40, 1999. With permission.) (B) 19_{II} of the amphidiploid of F_1 hybrid ($2n = 38$) of *B. rapa* $\times$ *D. virgata* (From Inomata, N., *Cruciferae Newsl.*, 21, 39–40, 1999. With permission.) (C) $1_{III} + 6_{II} + 11_I$ of the F_1 hybrid ($2n = 26$) of *B. carinata* ($2n = 34$) $\times$ *Sinapis turgida* ($2n = 18$) (From Inomata, N., *Gent. (Life Sci. Adv.)*, 11, 129–140, 1992. With permission.) (D) $4_{II} + 17_I$ of the F_1 hybrid ($2n = 25$) of *B. juncea* ($2n = 36$) $\times$ *Erucastrum virgatum* ($2n = 14$) (From Inomata, N., *Cruciferae Newsl.*, 23, 17–18, 2001. With permission.) (E) $4_{II} + 17_I$ of the F_1 hybrid ($2n = 25$) of *B. juncea* ($2n = 36$) $\times$ *Diplotaxis erucoides* ($2n = 14$) (From Inomata, N., *Cruciferae Newsl.*, 20, 17–18, 1998. With permission.) (F) $7_{II} + 18_I$ of the F_1 hybrid ($2n = 32$) of *B. napus* ($2n = 38$) $\times$ *Diplotasis harra* ($2n = 26$) (From Inomata, N., *Euphytica*, 145, 87–93, 2005. With permission.) Bar = 5 μm.

Table 4.5   Chromosome Pairing at Meiotic Metaphase I and Pollen Fertility in Interspecific F_1 Hybrids in *Brassica* Crops

Hybrid Combination	Hybrid Chromosome Number (2n)	Bivalents		Pollen Fertility (%)	Author
		Range	Mean		
B. rapa × *B. cretica*	19	—	9^a	—	Inomata (1985a)
B. rapa × *B. bourgeaui*	19	—	9^a	0.1	Inomata (1986)
B. rapa × *B. bourgeaui*	19	—	9^a	0.8	Inomata (1987)
B. fruticulosa × *B. rapa* Amphidiploid	18	0–5	0.8	0	Nanda Kumar et al. (1988)
B. fruticulosa × *B. rapa*	36	Almost 18_{II}		ca. 50	
B. juncea × *B. hirta* (syn. *Sinapis alba*)	30	10–14	12.31	>90	Bijral et al. (1991)
B. napus × *B. adpressa*	26	1–7	3.86	—	Kerlan et al. (1993)
B. adpressa × *B. napus*	26	1–7	3.77	—	
B. napus × *B. oleracea*	28	6–12	9.00	—	
B. oleracea × *B. napus* Amphidiploid	28	8–10	9.00	—	
B. napus × *B. oleracea*	56	24–28	26.73	—	
B. cossoneana × *B. carinata*	33	—	11.08	—	Bijral et al. (1994)
B. tournefortii × *B. rapa*	20	0	0	0	Nagpal et al. (1996)
B. tournefortii × *B. nigra* Amphidiploid	18	0	0	0	
B. tournefortii × *B. rapa*	40	Almost 20_{II}		76	
B. tournefortii × *B. nigra*	36	Almost 18_{II}		93	
B. tournefortii × *B. oleracea*	38	Almost 19_{II}		ca. 10	
B. juncea × *B. oxyrrhina*	27	—	4.52	40	Bijral and Sharma (1999b)
B. rapa × *B. spinescens*	18	0–4	0.9	0	Inomata (2000)
B. tournefortii × *B. rapa*	20	0–7	1.97	6.5	Choudhary and Joshi (2001)
B. fruticulosa × *B. rapa*	18	1–3	2.1	0	Chandra et al. (2004b)

Note: —, not examined.

[a] The number shows the mode.

traits. The relatively high frequency of meiocytes resulted in 18 bivalents. Plants also showed high pollen and seed fertility (Bang et al. 1997; Matsuzawa et al. 1997).

Meiotic bivalents in the intergeneric F_1 hybrids ($2n=28$) of *B. napus* ($2n=38$) × *D. catholica* ($2n=18$) ranged from 3 to 6. The presence of few ring bivalents indicated the partial homology between parental genomes (Bijral and Sharma 1998). Morphology of most intergeneric F_1 hybrids ($2n=29$) of *B. juncea* ($2n=36$) × *E. sativa* ($2n=22$) resembled the female parent. A few traits, such as the shape of silique and beak articulation, were those of the male parent. The number of ring bivalents ranged from 2 to 11. F_2 and BC_1 seeds were obtained (Bijral and Sharma 1999a) Interspecific F_1 hybrids ($2n=27$) of *B. juncea* ($2n=36$) × *B. oxyrrhina* ($2n=18$), exhibited 40% pollen stainability and hybrid seeds were obtained (Bijral and Sharma 1999b).

Pollen fertility in the intergeneric F_1 hybrids between *R. sativus* and three species of the genus *Brassica* (*fruticulosa* ssp. *cossoneana* ($2n=32$), ssp. *mauritanica* ($2n=32$), *B. maurorum* ($2n=16$) and *B. oxyrrhina* ($2n=18$) ranged from 0 to 70.2%. The induced amphidiploids of these F_1 hybrids showed relatively stable chromosome association as shown in Table 4.5 (Bang et al. 1997). The amphidiploids of *B. maurorum* ($2n=16$) × *R. sativus* ($2n=18$) and *B. fruticulosa* ssp. *mauritanica* ($2n=32$) × *R. sativus* ($2n=18$) showed high seed fertility (Bang, Kaneko, and Matsuzawa 1998, 2000).

Table 4.6   Chromosome Pairing at Meiotic Metaphase I and Pollen Fertility in Intergeneric F_1 Hybrids in *Brassica* Crops

Hybrid Combination	Hybrid Chromosome Number (2n)	Bivalents		Pollen Fertility (%)	Author
		Range	Mean		
B. napus × *Sinapis arvensis*	37[a]	17–18	17.94	95.2	Inomata (1988)
Diplotaxis erucoides × *B. napus*	26	3–10	6.13	—	Delourme et al. (1989)
D. erucoides × *B. napus*	33[b]	5–13	10.30	—	
Eruca sativa × *B. rapa*	21	Majority 21$_{II}$		—	Agnihotri et al. (1990a)
B. napus × *Raphanobrassica*	37	Very low 9$_{II}$		0	Agnihotri et al. (1990b)
Sinapis alba × *B. napus*	31	—	6.00	0	Ripley and Arnison (1990)
S. alba × *B. napus*	43[b]	—	16.00	—	
Moricandia arvensis × *B. oleracea*	23	0–6	2.17	0 and 56.2	Takahata (1990)
B. rapa × *Moricandia arvensis*	24	0–5	2.60	0	Takahata and Takeda (1990)
M. arvensis × *B. rapa*	24	0–6	2.67	0	
M. arvensisi × *B. nigra*	22	0–5	2.52	0	
B. nigra × *Sinapis allioni*	17	—	4.25	—	Banga and Labana (1991)
B. juncea × *Sinapis pubescens*	27	0–6	—	0	Inomata (1991b)
B. napus × *S. pubescens*	28	2–11	—	0	
B. rapa × *Raphanus sativus*	19	0–5	1.58	—	Sarashima (1991)
R. sativus × *B. nigra*	19	0–4	1.42	—	
B. oleracea × *R. sativus*	18	0–4	1.52	—	
R. sativus × *B. oleracea*	18	0–5	1.53	—	
B. carinata × *Sinapis turgida*	26	4–10	—	0–19.0	Inomata (1992)
Enarthrocarpus lyratus × *B. oleracea*	19	0–6	3.14	0	Gundimeda et al. (1992)
E. lyratus × *B. rapa*	20	0–5	2.20	0	
E. lyratus × *B. carinata*	37[b]	10–16	12.89	2.0	
E. lyratus × *B. napus*	29	3–12	6.10	0	
B. juncea × *E. lyratus* Amphidiploid	28	1–10	6.04	0	
E. lyratus × *B. oleracea*	38	7–19	15.13	82.0	
B. napus × *Sinapis arvensis*	28	3–10	5.42	—	Kerlan et al. (1993)
B. napus × *Raphanus raphanistrum*	28	4–10	6.40	—	
R. raphanistrum × *B. napus* Amphidiploid	28	4–10	6.88	—	
B. napus × *Sinapis arvensis*	56	23–28	25.70	—	
Diplotaxis siettiana × *B. rapa*	18	0–5	3.50	0	Nanda Kumar and Shivanna (1993)
Diplotaxis siettiana × *B. rapa* Amphidiploid	36	8–17	14.1	50.0	
B. napus × *Sinapis pubescens*	28	2–11	10[c]	0	Inomata (1994a)
B. juncea × *Diplotaxis virgata*	36	7–12	10.3	15.3	Inomata (1994b)
Erucastrum abyssinicum × *B. juncea*	34	—	12[c]	10.94	Sarmah and Sarla (1994)
E. abyssinicum × *B. oleracea*	25	—	7[c]	6.5	
B. napus × *Orychophragmus violaceus*	31	0–12	ca. 8.22	—	Li et al. (1996)
Sinapis arvensis × *Raphanus sativus*	18	0–6	1.99	0–47.5	Bang et al. (1996a)

continued

Table 4.6   Continued

Hybrid Combination	Hybrid Chromosome Number (2*n*)	Bivalents Range	Bivalents Mean	Pollen Fertility (%)	Author
R. sativus × *S. arvensis*	18	0–5	1.83	0	
R. sativus × *S.pubescens*	18	0–4	1.19	14.3	
S. turgida × *R. sativus*	18	0–7	1.85	0–74.3	
R. sativus × *S. turgida*	18	0–7	1.84	0–74.3	
R. sativus × *S. alba* Amphidiploid	21	2–4	2.7	0	
S. turgida × *H. sativus*	36	—	—	50.0	
R. sativa × *S. arvensis*	36	17–18	17.9	58.0	
R. sativa × *S. pubescens*	36	16–18	17.74	69.9	
Moricandia arvensis × *Raphanus stivus* Amphidiploid	23	0–5	1.76	1.9	Bang et al. (1996b)
M. arvensis × *R. stivus*	46	22–23	22.72	66.6	
B. rapa × *Sinapis turgida*	19	0–5	ca. 2.64	0	Inomata (1996a)
B. fruticulosa × *Raphanus sativus*	25	2–9	7.34	22.8–70.2	Bang et al. (1997)
B. maurorum × *R. sativus*	17	0–6	2.44	0–43.3	
B. oxyrrhina × *R. sativus* Amphidiploid	18	0–6	2.61	0–0.7	
B. fruticulosa × *R. sativus*	50	19–25	22.48	55.1	
B. maurorum × *R. sativus*	34	14–17	16.67	75.5–87.5	
B. oxyrrhina × *R. sativus*	36	15–18	17.59	70.0–82.0	
B. napus × *Diplotaxis catholica*	28	3–7	ca. 5.0	—	Bijral and Sharma (1998)
B. juncea × *Orychohphragmus violaceus*	30	2–12	6.27	39.3	Li et al. (1998)
B. carinata × *O. violaceus*	29	—	17.0	97.3	
B. rapa × *Moricandia arvensis*	24	0–5	—	0–32	Matsuzawa, Bang, and Kaneko (1998)
Amphidiploid					
B. rapa × *M. arvensis*	48	22–24	—	83.0	
B. juncea × *Diplotaxis erucoides*	25	0–6	3.17	0	Inomata (1998)
B. juncea × *Eruca sativa*	29	2–11	—	98.0	Bijral and Sharma (1999a)
B. rapa × *Diplotaxis virgata* Amphidiploid	19	0–7	2.8	0	Inomata (1999)
B. rapa × *D. virgata*	38	16–19	17.1	48.9	
B. oleracea × *Orychophragmus violaceus*	21	0–8	3.93	—	Li and Heneen (1999)
B. rapa × *O. violaceus*	23–42	—	—	51.0	
B. rapa × *O. violaceus*	16–26	0–11	—	0–99	
B. rapa × *Eruca sativa*	—	—	—	0	Robers et al. (1999)
B. oleracea + *Moricandia nitens*[d]	46	—	—	few	Yan et al. (1999)
B. oleracea + *M. nitens*[d]	92	—	—	>70	
Sinapis alba × *B. nigra*	20	0–5	—	9.3	Choudhary and Joshi (2000)
B. tournefortii × *Raphanus caudatus* Amphidiploid	18	0–6	—	2.7	Choudhary et al. (2000a)
B. tournefortii × *R. caudatus*	36–42	13–19	—	77.3	
B. juncea × *Erucastrum virgatum*	25	1–8	4.3	0	Inomata (2001)

continued

Table 4.6 Continued

Hybrid Combination	Hybrid Chromosome Number (2n)	Bivalents		Pollen Fertility (%)	Author
		Range	Mean		
Erucastrum canariensis× B. rapa Amphidiploid	19	1–8	1.89	0	Bhaskar et al. (2002)
E. canariensis×B. rapa	38	Predomin- ance 19ᵢᵢ		65–80	
B. napus×Orychophragmus violaceus	29–39	9–19	—	—	Cheng et al. (2002)
B. napus×Raphanus sativus	28	—	6.29	—	Huang et al. (2002)
B. napus×Diplotaxis virgata	28	4–11	7.52	0	Inomata (2002b)
Dplotaxis siifoloa×B. rapa	20	1–6	3.64	26.7	Ahuja et al. (2003)
D. siifoloa×B. juncea	28	3–10	4.81	28.0	
Diplotaxis catholica×B. rapa Amphidiploid	19	2–5	3.26	34	Banga et al. (2003)
D. catholica×B. rapa	38	16–19	18.03	73	
D. catholica×B. juncea	27	1–6	4.41	32	
B. juncea×Diplotaxis virgata	36[a]	7–12	10.12	15.3	Inomata (2003)
B. oleracea+Moricandia arvensis[d]	46	20–23	>70%, 23ᵢᵢ	ca. 50	Ishikawa et al. (2003)
Erucastrum cardaminoides× B. rapa	19	1–5	1.87	38.0	Chandra et al. (2004a)
E. cardaminoides×B. nigra Amphidiploid	17	1–5	1.78	32.0	
E. cardaminoides×B. rapa	38	16–19	18.7	75.0	
E. cardaminoides×B. nigra	34	15–17	16.8	67	
B. napus×Diplotaxis harra	32	2–10	7.22	7.8	Inomata (2005)

Note: —, not examined.
[a] Spontaneous chromosome doubling in male nucleus.
[b] Spontaneous chromosome doubling in female nucleus.
[c] The number shows the mode.
[d] Protoplast fusion.

Pollen fertility in the intergeneric F_1 hybrids ($2n=24$) of *B. rapa* ($2n=20$)×*M. arvensis* ($2n=28$) ranged from 0 to 32%. Their amphidiploids ($2n=48$) had high seed fertility (Matsuzawa, Bang, and Kaneko 1998). The intergeneric F_1 hybrid ($2n=27$) of *B. juncea* ($2n=36$)×*B. oxyrrhina* ($2n=18$) was successfully backcrossed to *B. juncea* (Bijral and Sharma 1999b).

The intergeneric F_1 hybrid ($2n=20$) of *S. alba* ($2n=24$)×*B. nigra* ($2n=16$) was vigorous and intermediate for morphological traits. Some seeds were obtained from the hybrids (Choudhary and Joshi 2000). The intergeneric F_1 hybrid ($2n=19$) of *B. tournefortii* ($2n=20$)×*R. caudatus* ($2n=18$) showed very low pollen fertility. On the other hand, most of the amphidiploid plants exhibited fairly high pollen and seed fertility (Choudhary, Joshi, and Singh 2000a). F_1 plants of *B. tournefortii* ($2n=20$)×*B. rapa* ($2n=20$) were vigorous, but seed set was not achieved after self-pollination and backcrossing (Choudhary and Joshi 2001). The intergeneric F_1 hybrids ($2n=28$) of *Ogura* CMS *B. napus* ($2n=38$)×*R. sativus* ($2n=18$) showed a variable seed set in different cross combinations, and some progenies were produced in backcross with *B. napus* (Huang et al. 2002).

All the interspecific and intergeneric F_1 hybrids are not necessarily bridge plants. Pollen fertility was not observed in many interspecific and intergeneric F_1 hybrids, although low frequency of bivalent association at metaphase I was obtained (Table 4.5 and Table 4.6). If amphidiploidy of these F_1 hybrids is produced, they may show more regular meiotic chromosome association and higher pollen and seed fertility than that of F_1 as described for the cross of *R. sativus*×*S. arvensis* and *R. sativus*×*S. pubescens* (Bang, Kaneko, and Matsuzawa 1996a) and *B. rapa*×*R. sativus* (Matsuzawa 2000). These F_1 hybrids may be useful as bridge plants.

Interspecific conventional pollination crosses of *B. rapa*×*B. oleracea* and *B. carinata*× *B. oleracea*, resulted in hybrids with $2n = 28$ and $2n = 35$ chromosomes, respectively (U.N. 1935). Embryo rescue techniques were used to produce interspecific and intergeneric F_1 hybrids ($2n = 37$) with the following results: *B. napus* ($2n = 38$)×*S. arvensis* ($2n = 18$) (Inomata 1988), *B. napus*×*B. oleracea* var. *acephala* ($2n = 18$), *B. napus*×*B. oleracea* var. *alboglabra* ($2n = 18$) and *B. napus*×*B. cretica* ($2n = 18$) (Inomata 2002a). Apparently, the normal female gamete was fertilized with an unreduced male gamete producing triploid hybrids.

In other crosses, when the interspecific F_1 hybrid ($2n = 19$) between *B. rapa* and *B. oleracea* was crossed with *B. rapa* and *B. napus*, BC_1 plants were produced with $2n = 29$ and $2n = 38$ chromosomes, respectively (Inomata 1983). An intergeneric F_1 hybrid from the cross of *R. sativus* ($2n = 18$)×*S. turgida* ($2n = 18$) contained $2n = 27$ chromosomes; produced from an unreduced female gamete (Bang, Kaneko, and Matsuzawa 1996a). When the intergeneric F_1 hybrid ($2n = 21$) of *E. sativa* ($2n = 22$)×*B. rapa* ($2n = 20$) was backcrossed to *B. rapa*, a sesquidiploid ($2n = 31$) chromosomes was obtained (Matsuzawa 1999). It is evident that unreduced normal female gametes may be fertilized by a normal male spore producing the triploid hybrids. The intergeneric F_1 hybrids, described above, were partially fertile and may be useful as bridge plants. Heyn (1977) described the occurrence of a polyploid resulting from unreduced gametes in the interspecific and intergeneric crosses among the tribe Brassiceae.

4.8 INTROGRESSION OF GENES AND THEIR CYTOGENETIC STABILITY

With the development of embryo rescue and somatic hybridization techniques, many interspecific and intergeneric hybrids have been produced in Brassiceae (see Bajaj 1990; Inomata 1990, 1997a; Pelletier 1990; Vambing and Glimelius 1990; Glimelius 1999). The meiotic chromosome association of these F_1 hybrids was reported (Inomata 1997a; Prakash et al. 1999). However, little information is available on the successive generation of the F_1 hybrids, because F_1 hybrids have very low pollen and seed fertility.

The pollen fertility of the F_1 hybrids produced by ovary culture, and having $2n = 19$ chromosomes from the cross between *B. rapa*×*B. oleracea*, ranged from 0 to 26.8% with a mean of 2.6%. Spontaneous amphidiploid had 91.7% pollen fertility. The pollen fertility rate of the F_1 hybrid in the cross of *B. rapa*×autotetraploid *B. oleracea* var. *acephala* ranged from 0 to 38.0% with a mean of 14.9%. When the F_1 ($2n = 19$) with total pollen sterility was backcrossed to *B. rapa*, many BC_1 plants with $2n = 29$ chromosomes were produced, but only a few seeds were obtained after backcrossing with *B. oleracea*. Self-pollination in spontaneous amphidiploids with $2n = 38$ chromosomes resulted in progenies with $2n = 29$ and 38 chromosomes. Many seeds were obtained when F_1 hybrids with $2n = 28$ chromosomes were crossed with *B. rapa*, autotetraploids *B. oleracea* and *B. napus*. In these hybrids, a *B. rapa*-type plant with $2n = 20$ chromosomes, a *B. oleracea*-type plant with $2n = 18$ chromosomes, and a *B. napus*-type plant with $2n = 38$ chromosomes, occurred. The genes from *B. rapa* and *B. oleracea* can be exchanged reciprocally, and the genes of both species *B. rapa* and *B. oleracea* can be introgressed into *B. napus* (Inomata 1983; Inomata 1991a).

The intergeneric F_1 hybrid ($2n = 26$) between *B. carinata* ($2n = 34$) and *S. turgida* ($2n = 18$) showed a 6.3% pollen fertility rate. Two types of morphological traits were expressed in the F_1. One trait resembled a trait of the pistillate parent and the other trait resembled a trait of the staminate parent. When F_1 plants were backcrossed to *B. carinata*, a BC_1 plant with $2n = 34$ chromosomes was identified (Inomata 1992).

The interspecific F_1 hybrid ($2n = 19$) between *B. rapa* and wild relatives of *B. oleracea* (*B. bourgeaui, B. cretica, B. montana*; $2n = 18$) showed almost no pollen fertility (Inomata 1985b, 1986, 1987). However, many seeds were obtained when the F_1 hybrid was backcrossed to *B. rapa, B. napus* and when open pollinated. Many *B. rapa*-type and *B. napus*-type progeny were harvested along with plants of different chromosome numbers (Inomata 1993b).

The intergeneric F_1 ($2n = 28$) produced from *B. napus* ($2n = 38$) and *Sinapis pubescens* ($2n = 18$) cross. Seed set was not recorded from self pollinated plants or plants derived after backcrossing to *B. napus*. Seed was harvested from open pollinated plants. Hybrids with $2n = 38$ and 44 chromosomes were isolated. Many *B. napus*-type plants were obtained from self-pollination and open pollination plants and from backcrosses with *B. napus* (Inomata 1994a).

The intergeneric amphidiploid ($2n = 34$) of *B. maurorum* ($2n = 16$) $\times R.$ *sativus* ($2n = 18$) had high pollen fertility rates (Bang et al. 1997). While this amphidiploid exhibited higher pollen fertility rates, seed set decreased after self-pollination with subsequent generations. When F_1 hybrids were backcrossed to *B. maurorum*, BC_1 plants with $2n = 25$ chromosomes were obtained. Amphidiploid and BC_1 plants were fertile. These plants may be useful as bridge plants for the breeding of *brassica* crops and *R. sativus* (Bang, Kaneko, and Matsuzawa 1998).

An amphidiploid plant from an intergeneric F_1 hybrid ($2n = 24$) of *B. rapa* ($2n = 20$) $\times$ *M. arvensis* ($2n = 28$) was treated with colchicine. Bivalents ranged from 22 to 24 and the pollen fertility rate was 85%. Sesquidiploids with $2n = 34$ chromosomes occurred in F_3 progenies, and bivalents ranged from 8 to 12. The sesquidiploids produced fertile seeds in sib crosses, open pollinations, and in backcrosses with *B. rapa* (Matsuzawa, Bang, and Kaneko 1998).

The intergeneric F_1 hybrid of *E. sativa* ($2n = 22$) $\times B.$ *rapa* ($2n = 20$) was obtained for producing a cytoplasmic male sterile of *B. rapa* (Matsuzawa and Sarashima 1986). The F_1 and F_2 plants obtained were amphihaploid ($2n = 21$) and sesquidiploid ($2n = 31$), respectively. Male-sterile plants with $2n = 20$ were selected in each generation, and many types of male sterility of *B. rapa* ($2n = 20$) were isolated in F_7 and F_8 generations (Matsuzawa 1999).

Allohexaploid plants between *B. carinata* $\times S.$ *arvensis* and *B. carinata* $\times S.$ *turgida* were produced by colchicine treatment and tested for stable seed fertility and fatty acid composition during F_2–F_4 generations. Progeny of both crosses contained higher levels of linoleic acid than that of either parent (Momotaz, Kato, and Kakihara 2000).

Brassica juncea and *Erucastrum canariense* ($2n = 18$) were pollinated with *B. rapa* to produce alloplasmic male sterility. The synthetic allotetraploid was crossed with *B. juncea*. Cytoplasmic male sterile (CMS) plants were selected in the BC_4 generation (Prakash 2001).

To obtain brown-seeded *B. napus, B. alboglabra* was crossed with yellow-seeded *B. rapa* and the F_1 hybrid was diplodized using a colchicine treatment. The resynthesized *B. napus* was crossed with a trigenomic diploid, which was obtained by crossing it with *B. carinata* (yellow seed) $\times$ *B. rapa* (yellow seed). To obtain the yellow-seeded *B. napus, B. carinata* (yellow seed) was crossed with *B. rapa* (yellow seed) and the F_1 was crossed with *B. napus* (black seed). The gene for the yellow seed was transferred from A-genome to C-genome and subsequently the yellow-seeded *B. napus* was produced (Rahman, Joersbo, and Poulsen 2001).

A hybrid between *B. napus* and *S. arvensis* ($2n = 18$) produced two types of plants with $2n = 28$ and $2n = 37$ chromosomes. Many *B. napus*-type plants with $2n = 38$ chromosomes were obtained from the progenies of the plants with $2n = 37$ chromosomes (Inomata 1997b). In the F_1 hybrids between *B. napus* and *B. oleracea, B. bourgeaui, B. cretica* or *B. montana*, two types of plants with $2n = 28$ and $2n = 37$ chromosomes were also identified. Many *B. napus*-type plants with $2n = 38$ chromosomes were obtained from the progenies of plants with $2n = 37$ chromosomes (Inomata 2002a).

An intergeneric F_1 hybrid between *B. juncea* ($2n = 36$) and *D. virgata* ($2n = 18$) produced plants with $2n = 36$ chromosomes. Chromosome configuration of these plants may consist of one genome of *B. juncea* and two genomes of *D. virgata*. The morphological characteristics resembled those of *B. juncea*. Chromosome association at metaphase I ranged from 8 to 12, and the pollen fertility rate was 15.3%. The F_2 and BC_1 (backcrossed with *B. juncea*) seedlings showed different chromosome constitutions and a high frequency of bivalent association. Many F_3 and BC_2 seeds were obtained after self-pollination, open pollination, and backcross of BC_1 to *B. juncea*. Many *B. juncea*-type plants were obtained with $2n = 36$, $2n = 37$, and $2n = 38$ chromosomes from F_3 and BC_2 plants. An F_1 hybrid between *B. juncea* and *D. virgata* with $2n = 36$ chromosomes may be useful as a bridge plant for breeding of other *brassica* crops (Inomata 2003).

An interspecific F_1 hybrid ($2n = 18$) between *B. fruticulosa* ($2n = 16$) and *B. rapa* ($2n = 20$) was totally pollen sterile. But the amphidiploid showed 18 bivalents at metaphase I and normal pollen fertility (Chandra et al. 2004b).

Morphological characteristics of intergeneric F_1 hybrid ($2n = 32$) between *B. napus* ($2n = 38$) and *D. harra* ($2n = 26$) resembled that of *B. napus*. Many F_2 seeds and BC_1 seeds, backcrossed with *B. napus*, were harvested from conventional pollination. The meiotic bivalent chromosome association of the F_2 and BC_1 plants with $2n = 38$ chromosomes ranged from 13 to 18, and pollen fertility rate ranged from 0 to 66.1%, with a mean of 28.1%. Many F_3 and BC_2 plants with $2n = 38$ chromosomes were obtained (Inomata 2005).

4.9 SUMMARY

1. The *brassica* crops among the tribe Brassiceae originated in the Middle East, Near East Asia, and the Mediterranean Sea areas.
2. Brassiceae is divided into two ancient evolutionary lineages: the Nigra and the Rapa/Oleracea; the *Rahanus sativus* and the Rapa/Oleracea lineage.
3. Conventional crossing, embryo rescue, and protoplast fusion were employed for crossing-incompatibility. Many interspecific and intergeneric hybrids were produced.
4. Transgenic brassicas have been produced. They have the potential to impact the ecosystem. However, studies on the transgenic plants should be promoted for the introgression of economically useful genes.
5. Interspecific and intergeneric F_1 hybrids showed many bivalents and sometimes, trivalents and tetravalents. Multivalent association may be attributed to allosyndesis. F_1 intergeneric hybrids showed low pollen fertility; however, many F_2 and BC_1 seeds were obtained. Progeny showed different morphological and somatic chromosome arrangements. Cultivated type plants were isolated in the hybrid progenies.
6. Production of fertile F_1 hybrids was important.
7. It is suggested to preserve the ancient species, wild forms, land races and primitive cultivars of *brassica* crops and *Raphanus sativus* found throughout the world.
8. Preservation and investigation of economically useful traits of the tribe Brassiceae in the wild forms are essential.

Alien additional lines, genome organization, and recent molecular work have been described in *Brassica* Oilseeds (Chapter 7) in *Oilseed Crops*, volume 4 of this series by Rod Snowdon, Wilfried Lühs, and Wolfgang Friedt.

ACKNOWLEDGMENTS

The author would like to thank Dr. Ram J. Singh, the editor, and Joe Nicholas for valuable suggestion and critical reading of the manuscript.

REFERENCES

Agnihotri, A. et al., Production of *Eruca-Brassica* hybrid by embryo rescue, *Plant Breed.*, 104, 281–289, 1990a.

Agnihotri, A. et al., Production of *Brassica napus* × *Raphanobrassica* hybrids by embryo rescue: an attempt to introduce shattering resistance into *B. napus*, *Plant Breed.*, 105, 292–299, 1990b.

Ahuja, I., Bhaskar, P. B., Banga, S. K., and Banga, S. S., Synthesis and cytogenetic characterization of intergeneric hybrids of *Diplotaxis siifolia* with *Brassica rapa* and *B. juncea*, *Plant Breed.*, 122, 447–449, 2003.

Altenbach, S. B. et al., Accumulation of a Brazil albumin in seeds of transgenic canola results in enhanced levels of seed protein methionine, *Plant Mol. Biol.*, 18, 235–245, 1992.

Arumugam, N. et al., Somatic cell hybridization of '*oxy*' CMS *Brassica juncea* (AABB) with *B. oleracea* (CC) for correction of chlorosis and transfer of novel organelle combinations to allotetraploid brassicas, *Theor. Appl. Genet.*, 100, 1043–1049, 2000.

Arumugam, N. et al., Synthesis of somatic hybrids (RCBB) by fusion heat-tolerant *Raphanus sativus* (RR) and *Brassica oleracea* (CC) with *Brassica nigra* (BB), *Plant Breed.*, 121, 168–170, 2002.

Astley, D., Gyorgis, A. M. H., and Toll, J., Collecting brassicas in Ethiopia, *Plant Genet. Resour. Newsl.*, 51, 15–20, 1982.

Attia, T. and Röbbelen, G., Meiotic pairing in haploids and amphidiploids of spontaneous versus synthetic origin in rape *Brassica napus* L, *Can. J. Genet. Cytol.*, 28, 330–334, 1986.

Ayotte, R., Harney, P. M., and Machado, V. S., The transfer of triazine resistance from *Brassica napus* L. to *B. oleracea* L. IV. Second and third backcrosses to *B. oleracea* and recovery of an 18-chromosome, triazine resistant BC_3, *Euphytica*, 40, 15–19, 1989.

Bajaj, Y. P. S., Wide hybridization in legumes and oilseed crops through embryo, ovule, and ovary culture, *Biotechnology in Agriculture and Forestry: Legumes and Oilseed Crops I*, Bajaj, Y. P. S., Ed., Vol. 10, Berlin: Springer, pp. 3–37, 1990, chap. 1.

Bang, S. W., Iida, D., Kaneko, Y., and Matsuzawa, Y., Production of new intergeneric hybrids between *Rapahnus sativus* and *Brassica* wild species, *Breed. Sci.*, 47, 223–228, 1997.

Bang, S. W., Kaneko, Y., and Matsuzawa, Y., Production of intergeneric hybrids between *Raphanus* and *Sinapis* and the cytogenetics of their progenies, *Breed. Sci.*, 46, 45–51, 1996a.

Bang, S. W., Kaneko, Y., and Matsuzawa, Y., Production of intergeneric hybrids between *Raphanus* and *Moricandia*, *Plant Breed.*, 115, 385–390, 1996b.

Bang, S. W., Kaneko, Y., and Matsuzawa, Y., Cytogenetical stability and fertility of an intergeneric amphidiploid line synthesized from *Brassica maurorum* Durieu. and *Raphanus sativus* L, *Bull. Coll. Agri. Utsunomiya Univ.*, 17, 23–29, 1998.

Bang, S. W., Kaneko, Y., and Matsuzawa, Y., Cytogenetical stability and fertility in an intergeneric amphidiploid line synthesized from *Brassica fruticulosa* Cyr. ssp. *auritanica* (Coss.) Maire. and *Raphanus sativus* L, *Bull. Coll. Agri. Utsunomiya Univ.*, 17, 67–73, 2000.

Banga, S. S., Bhaskar, P. B., and Ahuja, I., Synthesis of intergeneric hybrids and establishment of genomic affinity between *Diplotaxis catholica* and crop *Brassica* species, *Theor. Appl. Genet.*, 106, 1244–1247, 2003.

Banga, S. S. and Labana, K. S., Cytoplasmic–genetic relationship between *Brassica nigra* and *Sinapis allioni*, *Cruciferae Newsl.*, 14/15, 12–13, 1991.

Bhaskar, P. B., Ahuja, I., Janeja, H. S., and Banga, S. S., Intergeneric hybridization between *Erucastrum canariense* and *Brassica rapa*: genetic relatedness between E^c and A genomes, *Theor. Appl. Genet.*, 105, 754–758, 2002.

Bijral, J. S., Gupta, B. B., Singh, K., and Sharma, T. R., Interspecific hybridization between *Brassica juncea* (L.) Czern & Coss and *Brassica hirta* Monch, *Indian J. Genet.*, 1, 476–478, 1991.

Bijral, J. S., Kanwal, K. S., and Sharma, T. R., *Brassica cossoneana* × *Brassica carinata* hybrids, *Cruciferae Newsl.*, 16, 22, 1994.

Bijral, J. S. and Sharma, T. R., Production and cytology of intergeneric hybrids between *Brassica napus* and *Diplotaxis catholica*, *Cruciferae Newsl.*, 20, 15, 1998.

Bijral, J. S. and Sharma, T. R., *Brassica juncea*—*Eruca sativa* sexual hybrids, *Cruciferae Newsl.*, 21, 33–34, 1999a.

Bijral, J. S. and Sharma, T. R., Morpho-cytogenetics of *Brassica juncea* × *Brassica oxyrrhina* hybrids, *Cruciferae Newsl.*, 21, 35–36, 1999b.

Bohman, S., Wang, M., and Dixelius, C., *Arabidopsis thaliana*-derived resistance against *Leptosphaeria maculans* in a *Brassica napus* genomic background, *Theor. Appl. Genet.*, 105, 498–504, 2002.

Bradshaw, J. E., Gemmell, D. J., and Wilson, R. N., Transfer of resistance to clubroot (*Plasmodiophora brassicae*) to swedes (*Brassica napus* L. var. *napobrassica* Petrm) from *B. rapa*, *Ann. Appl. Biol.*, 130, 337–348, 1997.

Branca, F. and Lapichino, G., Some wild and cultivated Brassicaceae exploited in Sicily as vegetables, *Plant Genet. Resour. Newsl.*, 110, 22–28, 1997.

Brown, J., Brown, A. P., Erickson, D. A., Davis, J. B., and Seip, L., Competitive and reproductive fitness of transgenic Canola × weed species hybrids, *Cruciferae Newsl.*, 18, 34–35, 1996b.

Brown, J., Brown, A. P., and Davis, J. B., Developing alternative *Brassica* crops from interspecific hybrids, *Cruciferae Newsl.*, 21, 37–38, 1999.

Brown, J., Brown, A. P., Thill, D. C., and Brammer, T. A., Gene transfer between canola (*Brassica napus*) and related weed species, *Cruciferae Newsl.*, 18, 36–37, 1996a.

Cardoza, V. and Stewart, C. N., *Brassica* biotechnology: progress in cellular and molecular biology, *In Vitro Cell. Dev. Biol. Plant*, 40, 542–551, 2004.

Chandra, A., Gupta, M. L., Ahuja, I., and Kaur, G., Intergeneric hybridization between *Erucastrum cardaminoides* and two diploid crop *Brassica* species, *Theor. Appl. Genet.*, 108, 1620–1626, 2004a.

Chandra, A., Gupta, M. L., Banga, S. S., and Banga, S. K., Production of an interspecific hybrid between *Brassica fruticulosa* and *B. rapa*, *Plant Breed.*, 123, 497–498, 2004b.

Cheng, B. F. and Séguin-Swartz, G., Meiotic studies on intergeneric hybrids between *Brassica napus* and *Orychophragmus violaceus*, *Cruciferae Newsl.*, 22, 11–12, 2000.

Cheng, B. F., Séguin-Swartz, G., and Somers, D. J., Cytogenetic and molecular characterization of intergeneric hybrids between *Brassica napus* and *Orychophragmus violaceus*, *Genome*, 45, 110–115, 2002.

Chèvre, A. M. et al., Characterization of *Brassica nigra* chromosomes and of blackleg resistance in *B. napus*–*B. nigra* addition lines, *Plant Breed.*, 115, 113–118, 1996.

Chèvre, A. M., Eber, F., Baranger, A., and Renard, M., Gene flow from transgenic crops, *Nature*, 389, 924, 1997.

Chiang, M. S., Chiang, B., and Grant, W. F., Transfer of resistance to race 2 of *Plasmodiophora brassicae* from *Brassica napus* to cabbage (*B. oleracea* var. *capitata*). I. Interspecific hybridization between *B. napus* and *B. oleracea* var. *capitata*, *Euphytica*, 26, 319–336, 1997.

Chiang, M. S. and Crête, R., Transfer of resistance to race 2 of *Plasmodiophora brassicae* from *Brassica napus* to cabbage (*B. oleracea* ssp. *capitata*). V. The inheritance of resistance, *Euphytica*, 32, 479–483, 1983.

Choudhary, B. R. and Joshi, P., Cytomorphology of intergeneric hybrid *Sinapis alba* × *Brassica nigra*, *J. Genet. Breed.*, 54, 157–160, 2000.

Choudhary, B. R. and Joshi, P., Crossability of *Brassica tournefortii* and *B. rapa*, and morphology and cytology of their F₁ hybrids, *Theor. Appl. Genet.*, 102, 1123–1128, 2001.

Choudhary, B. R., Joshi, P., and Ramarao, S., Interspecific hybridization between *Brassica carinata* and *Brassica rapa*, *Plant Breed.*, 119, 417–420, 2000b.

Choudhary, B. R., Joshi, P., and Singh, K., Synthesis, morphology and cytogenetics of *Raphanofortii* (TTRR, $2n=38$): a new amphidiploid of hybrid *Brassica tournefortii* (TT, $2n=20$) × *Raphanus caudatus* (RR, $2n=18$), *Theor. Appl. Genet.*, 101, 990–999, 2000a.

Christey, M. C., Brassica protoplast culture and somatic hybridization, *Biotechnology in Agriculture and Forestry*, Pua, E. C. and Douglas, C. J., Eds., Vol. 54, Berlin: Springer, p. 1, 2004, chap. 3

Christey, M.C. and Braun, R., Production of transgenic vegetable brassicas, In *Biotechnology in Agriculture and Forestry*, vol. 54, Pua, E. C. and Douglas, C. J., Eds., Berlin: Springer, 5:169–194, 2004.

Darmency, H. and Fleury, A., Mating system in *Hirschfeldia incana* and hybridization to oilseed rape, *Weed Res.*, 40, 231–238, 2000.

Darmency, H., Fleury, A., and Lefol, E., Effect of transgenic release on weed biodiversity: oilseed rape and wild radish, *Brighton Crop Protection Conf. Weeds*, 433–438, 1995.

Delourme, R., Eber, F., and Chevre, A. M., Intergeneric hybridization of *Diplotaxis erucoides* with *Brassica napus* I. Cytogenetic analysis of F₁ and BC₁ progeny, *Euphytica*, 41, 123–128, 1989.

Dixelius, C., Bohman, S., and Wretblad, S., Disease resistance, In *Biotechnology in Agriculture and Forestry*, vol. 54, Pua, E. C. and Douglas, C. J., Eds., Berlin: Springer, 2:253–271, chap. 4, 2004.

Earle, E.D., Cao, J., and Shelton, A.M., Insect-resistant transgenic brassicas, In *Biotechnology in Agriculture and Forestry*, vol. 54, Pua, E. C. and Douglas, C. J., Eds., Berlin: Springer, 1: 227–251, chap. 4, 2004.

Eber, F. et al., Spontaneous hybridization between a male-sterile oilseed rape and two weeds, *Theor. Appl. Genet.*, 88, 362–368, 1994.

Fernández-Martínez, J. M., Del Río, M., Velasco, L., Domínguez, J., and De Haro, A., Registration of zero erucic acid Ethiopian mustard genetic stock 25X-1, *Crop Sci.*, 41(282), 2001, 2001.

Fernández-Escobar, J., Dominguez, J., Martin, A., and Fernández-Martinez, J. M., Genetics of the erucic acid content in interspecific hybrids of Ethiopian mustard (*Brassica carinata* Braun) and rapeseed (*B. napus* L.), *Plant Breed.*, 100, 310–315, 1988.

Forsberg, J., Landgren, M., and Glimelius, K., Fertile somatic hybrids between *Brassica napus* and *Arabidopsis thaliana*, *Plant Sci.*, 95, 213–223, 1994.

Gerdemann-Knörck, M., Sacristán, M. D., Braatz, C., and Schieder, O., Utilization of asymmetric somatic hybridization for the transfer of disease resistance from *Brassica nigra* to *Brassica napus*, *Plant Breed.*, 113, 106–113, 1994.

Gerdemann-Knörck, M. et al., Transfer of disease resistance within the genus *Brassica* through asymmetric somatic hybridization, *Euphytica*, 85, 247–253, 1995.

Getinet, A., Rakow, G., Raney, J. P., and Downey, R. K., Development of zero erucic acid Ethiopian mustard through an interspecific cross with zero erucic acid Oriental mustard, *Can. J. Plant Sci.*, 74, 793–795, 1994.

Gleba, Y. Y. and Hoffmann, F., Hybrid cell lines *Arabidopsis thaliana* + *Brassica campestris*: no evidence for specific chromosome elimination, *Mol. Gen. Genet.*, 165, 257–264, 1978.

Gleba, Y. Y. and Hoffmann, F., 'Arabidobrassica': plant-genome engineering by protoplast fusion, *Naturwissenschaften*, 66, 547–554, 1979.

GhoshDastidar, N. G. and Varma, N. S., A study on extent of cross pollination in field trials of transgenic Indian mustard, *Cruciferae Newsl.*, 20, 49–50, 1998.

GhoshDastidar, N. G., Varma, N. S., and Kapur, A., Gene escape studies on transgenic Indian mustard (*Brassica juncea*), *Cruciferae Newsl.*, 22, 27–28, 2000.

Glimelius, K., Somatic hybridization, In *Biology of Brassica Coenospecies*, Gómez-Campo, C., Ed., Amsterdam, The Netherlands: Elsevier, pp. 107–148, 1999, chap. 4.

Gómez-Campo, C., Morphology and morpho-taxonomy of the tribe *Brassiceae*, *Brassica Crops and Wild Allies*, Tsunoda, S., Hinata, K., and Gómez-Campo, C., Eds., Tokyo: Japan Science Societies Press, pp. 3–31, 1980, chap. 1.

Gómez-Campo, C., Taxonomy, In *Biology of Brassica Coenospecies*, Gómez-Campo, C., Ed., Amsterdam: Elsevier, pp. 3–32, 1999, chap. 1.

Gómez-Campo, C. and Prakash, S., Origin and domestication, In *Biology of Brassica Coenospecies*, Gómez-Campo, C., Ed., Amsterdam: Elsevier, pp. 33–58, 1999, chap. 2.

Gundimeda, H. R., Prakash, S., and Shivanna, K. R., Intergeneric hybrids between *Enarthrocarpus lyratus*, a wild species, and crop brassicas, *Theor. Appl. Genet.*, 83, 655–662, 1992.

Gustafsson, M., Germplasm conservation of wild (*n* = 9) Mediterranen *Brassica* species, *Sveriges Utsadesforenings Tidskrift*, 92, 133–142, 1982.

Gustafsson, M., Bentzer, B., von Bothmer, R., and Snogerup, S., Meiosis in Greek *Brassica* of the *oleracea* group, *Bot. Notiser*, 129, 73–84, 1976.

Gustafsson, M., Gómez-Campo, C., and Zamenis, A., Report from the first *Brassica* germplasm exploration in Greece 1982, *Sveriges Utsadesforeinings Tidskrift*, 93, 151–160, 1983.

Gustafsson, M. and Snogerup, S., A new subspecies of *Brassica cretica* from Peloponnisos, Greece, *Bot. Chron.*, 3, 7–11, 1983.

Hagimori, M., Nagaoka, M., Kato, N., and Yoshikawa, H., Production and characterization of somatic hybrids between the Japanese radish and cauliflower, *Theor. Appl. Genet.*, 84, 819–824, 1992.

Halfhill, M. D., Richards, H. A., Mabon, S. A., and Stewart, C. N., Expression of GFP and Bt transgenes in *Brassica napus* and hybridization with *Brassica Rapa*, *Theor. Appl. Genet.*, 103, 659–667, 2001.

Hammer, K., Cifarelli, S., and Perrino, P., Collection of land-races of cultivated plants in South Italy, 1985, *Kulturpflanze*, 34, 261–273, 1986.

Hansen, L. N., Intertribal somatic hybridization between *Brassica oleracea* L. and *Camelina sativa* (L.) Crantz, *Cruciferae Newsl.*, 19, 55–56, 1997.

Hansen, L. B., Siegismund, H. R., and Jørgensen, R. B., Introgression between oilseed rape (*Brassica napus* L.) and its weedy relative *B. rapa* L. in a natural population, *Genet. Resour. Crop Evol.*, 48, 621–627, 2001.

Harberd, D. J. and McArthur, E. D., Meiotic analysis of some species and genus hybrids in the *Brassiceae*, *Brassica Crops and Wild Allies*, Tsunoda, S., Hinata, K., and Gómez-Campo, C., Eds., Tokyo, Japan: Science Societies Press, pp. 65–87, 1980, chap. 4.

Hauser, T. P., Jørgensen, R. B., and Østergård, H., Fitness of backcross and F_2 hybrids between weedy *Brassica rapa* and oilseed rape (*B. napus*), *Heredity*, 81, 436–443, 1998.

Heyn, F. W., Analysis of unreduced gametes in the *Brassiceae* by crosses between species and ploidy levels, *Z. Pflanzenzüchtg.*, 78, 13–30, 1977.

Hinata, K. and Konno, N., Studies on a male sterile strain having the *Brassica campestris* nucleus and the *Diplotaxis muralis* cytoplasm. I. On the breeding procedure and some characteristics of the male sterile strain, *Jpn. J. Breed.*, 29, 305–311, 1979.

Hosoda, T., Studies on the breeding of new types of *Napus* crops by means of artificial synthesis in genomes of genus *Brassica*, *Mem. Fac. Agri. Tokyo Kyoiku Univ.*, 7, 1–94, 1961.

Hossain, M. M., Inden, H., and Asahira, T., *In vitro* ovule culture of intergeneric hybrids between *Brassica oleracea* and *Raphanus sativus*, *Scientia Horticulturae*, 41, 181–188, 1990.

Hu, Q. and Hansen, L. N., Protoplast fusion between *Brassica napus* and related species for transfer of new genes for low erucic acid, *Cruciferae Newsl.*, 20, 37–38, 1998.

Hu, Q., Hansen, L. M., Laursen, J., and Dixelius, C., Intergeneric hybrids between *Brassica napus* and *Orychophragmus violaceus* containing traits of agronomic importance for oilseed rape breeding, *Theor. Appl. Genet.*, 105, 834–840, 2002.

Huang, B., Liu, Y., Wu, W., and Xue, X., Production and cytogenetics of intergeneric hybrids between *Ogura* CMS *Brassica napus* and *Raphanus sativus*, *Cruciferae Newsl.*, 24, 25–27, 2002.

IBPGR, *Genetic Resources of Cruciferous Crops*, Rome, Italy, pp. 1–48, 1981.

Inomata, N., Production of interspecific hybrids between *Brassica campestris* and *Brassica oleracea* by culture *in vitro* of excised ovaries. I. Effects of yeast extract and casein hydrolysate on the development of excised ovaries, *Jpn. J. Breed*, 27, 295–304, 1977.

Inomata, N., Production of interspecific hybrids between *Brassica campestris* and *Brassica oleracea* by culture *in vitro* of excised ovaries. II. Effects of coconut milk and casein hydrolysate on the development of excised ovaries, *Jpn. J. Genet.*, 53, 1–11, 1978a.

Inomata, N., Production of interspecific hybrids in *Brassica campestris* × *B. oleracea* by culture *in vitro* of excised ovaries. I. Development of excised ovaries in the crosses of various cultivars, *Jpn. J. Genet.*, 53, 161–173, 1978b.

Inomata, N., Production of interspecific hybrids in *Brassica campestris* × *B. oleracea* by culture *in vitro* of excised ovaries. II. Development of excised ovaries on various culture media, *Jpn. J. Breed.*, 29, 115–120, 1979.

Inomata, N., Hybrid progenies of the cross, *Brassica campestris* × *B. oleracea*. I. Cytogenetical studies on F_1 hybrids, *Jpn. J. Genet.*, 55, 189–202, 1980.

Inomata, N., Hybrid progenies of the cross, *Brassica campestris* × *B. oleracea*. II. Crossing ability of F_1 hybrids and their progenies, *Jpn. J. Genet.*, 58, 433–449, 1983.

Inomata, N., A revised medium for *in vitro* culture of *Brassica* ovaries, *The Experimental Manipulation of Ovule Tissue*, Chapman, G. P., Mantell, S. H., and Daniels, R. W., Eds., New York: Longman, pp. 164–176, 1985a, chap. 13.

Inomata, N., Interspecific hybrids between *Brassica campestris* and *B. cretica* by ovary culture *in vitro*, *Cruciferae Newsl.*, 10, 92–93, 1985b.

Inomata, N., Interspecific hybrids between *Brassica campestris* and *B. bourgeaui* by ovary culture *in vitro*, *Cruciferae Newsl.*, 11, 14–15, 1986.

Inomata, N., Interspecific hybrid between *Brassica campestris* and *B. montana* by ovary culture *in vitro*, *Cruciferae Newsl.*, 12, 8–9, 1987.

Inomata, N., Intergeneric hybridization between *Brassica napus* and *Sinapis arvensis* and their crossability, *Cruciferae Newsl.*, 13, 22–23, 1988.

Inomata, N., Interspecific hybridization in *Brassica* through ovary culture, In *Biotechnology in Agriculture and Forestry: Legumes and Oilseed Crops I*, vol. 10, Bajaj, Y. P. S., Ed., Berlin: Springer, pp. 367–384, 1990, chap. IV.

Inomata, N., Hybrid progenies of the cross, *Brassica campestris* × *B. oleracea*. IV. Crossability of F_2, B_1 and hybrid plants, and their progenies, *Jpn. J. Genet.*, 66, 449–460, 1991a.

Inomata, N., Intergeneric hybridization in *Brassica juncea* × *Sinapis pubescens* and *B. napus* × *S. pubescens*, and their cytological studies, *Cruciferae Newsl.*, 14/15, 10–11b, 1991b.

Inomata, N., Intergeneric hybridization between *Brassica carinata* and *Sinapis turgida* through ovary culture, and cytology and crossability of their progenies, *Gent. (Life Sci. Adv.)*, 11, 129–140, 1992.

Inomata, N., Embryo rescue techniques for wide hybridization, *Breeding Oilseed Brassicas*, Labana, K. S., Banga, S. S., and Banga, S. K., Eds., Berlin: Springer, pp. 94–107, 1993a, chap. 7.

Inomata, N., Crossability and cytology of hybrid progenies in the cross between *Brassica campestris* and three wild relatives of *B. oleracea*. *B. bourgeaui, B. cretica* and *B. Montana, Euphytica*, 69, 7–17, 1993b.

Inomata, N., Intergeneric hybridization between *Brassica napus* and *Sinapis pubescens*, and the cytology and crossability of their progenies, *Theor. Appl. Genet.*, 89, 540–544, 1994a.

Inomata, N., Intergeneric hybridization between *Brassica juncea* and *Diplotaxis virgata*, and their cytology, *Cruciferae Newsl.*, 16, 30–31, 1994b.

Inomata, N., Cytogenetical studies and crossability on the F_1 hybrid between *Brassica campestris* and *Sinapis turgida, Cruciferae Newsl.*, 18, 14–15, 1996a.

Inomata, N., Overcoming the cross-incompatibility through embryo rescue and the transfer of characters within the genus *Brassica* and between wild relatives and *Brassica* crops, *Sci. Report Fac. Agri. Okayama Univ.*, 85, 79–88, 1996b.

Inomata, N., Wide hybridization and meiotic pairing, *Recent Advances in Oilseed Brassicas*, Kalia, H. R. and Gupta, S. K., Eds., Ludhiana, India: Kalyani, pp. 53–76, 1997a, chap. 4.

Inomata, N., Hybrid progenies of the cross in *Brassica napus* × *Sinapis arvensis, Cruciferae Newsl.*, 19, 23–24, 1997b.

Inomata, N., Production of the hybrids and progenies in the intergeneric cross between *Brassica juncea* and *Diplotaxis erucoides, Cruciferae Newsl.*, 20, 17–18, 1998.

Inomata, N., Production of intergeneric hybrids between *Brassica campestris* and *Diplotaxis virgata*, and cytology and crossability of their hybrids, *Cruciferae Newsl.*, 21, 39–40, 1999.

Inomata, N., Interspecific hybridization between *Brassica campestris* and *B. spinescens* and the cytogenetic analysis of their progenies, *Cruciferae Newsl.*, 22, 13–14, 2000.

Inomata, N., Intergeneric hybridization between *Brassica juncea* and *Erucastrum virgatum* and the meiotic behavior of F_1 hybrids, *Cruciferae Newsl.*, 23, 17–18, 2001.

Inomata, N., A cytogenetic study of the progenies of hybrids between *B. napus* and *B. oleracea, B. bourgeaui, B. cretica* or *B. Montana, Plant Breed.*, 121, 174–176, 2002a.

Inomata, N., Production of intergeneric hybrids between *Brassica napus* and *Diplotaxis virgata* and meiotic chromosome association of the F_1 hybrids, *Cruciferae Newsl.*, 24, 29–30, 2002b.

Inomata, N., Production of intergeneric hybrids between *Brassica juncea* and *Diplotaxis virgata* through ovary culture, and the cytology and crossability of their progenies, *Euphytica*, 133, 57–64, 2003.

Inomata, N., Intergeneric hybrid between *Brassica napus* and *Diplotaxis harra* through ovary culture and the cytogenetic analysis of their progenies, *Euphytica*, 145, 87–93, 2005.

Ishikawa, S., Bang, S. W., Kaneko, Y., and Matsuzawa, Y., Production and characterization of intergeneric somatic hybrids between *Moricandia arvensis* and *Brassica oleracea, Plant Breed.*, 122, 233–238, 2003.

Johnston, T. D., Transfer of disease resistance form *Brassica campestris* L. to rape (*B. napus* L.), *Euphytica*, 23, 681–683, 1974.

Jørgensen, R. B. and Andersen, B., Spontaneous hybridization between oilseed rape (*Brassica napus*) and weedy *B. campestris* (Brassicaceae): a risk of growing genetically modified oilseed rape, *Amer. J. Bot.*, 81, 1620–1626, 1994.

Jørgensen, R. B. et al., Introgression of crop genes from oilseed rape (*Brassica napus*) to related wild species— an avenue for the escape of engineered genes, *Acta Hort.*, 459, 211–217, 1998.

Kameya, T. and Hinata, K., Test-tube fertilization of excised ovules in *Brassica, Jpn. J. Breed.*, 20, 253–260, 1970.

Kameya, T., Hinata, K., and Mizushima, U., Fertilization *in vitro* of excised ovules treated with calcium chloride in *Brassica oleracea* L, *Proc. Jpn. Acad.*, 42, 165–167, 1966.

Kanrar, S., Venkateswari, J., Kirti, P. B., and Chopra, V. L., Transgenic Indian mustard (*Brassica juncea*) with resistance to the mustard aphid (*Lipaphis erysimi*) Kalt, *Plant Cell Rep.*, 20, 976–981, 2002.

Kerlan, M. C. et al., Risk assessment of outcrossing of transgenic rapeseed to related species: I. Interspecific hybrid production under optimal conditions with emphasis on pollination and fertilization, *Euphytica*, 62, 145–153, 1992.

Kerlan, M. C., Chèvre, A. M., and Eber, F., Interspecific hybrids between a transgenic rapeseed (*Brassica napus*) and related species: cytogenetical characterization and detection of the transgene, *Genome*, 36, 1099–1106, 1993.

Kirti, P. B. et al., Introgression of a gene restoring fertility to the CMS (*Trachystoma*) *Brassica juncea* and the genetics of restoration, *Plant Breed.*, 116, 259–262, 1997.

Kumazawa, S. and Abe, S., Studies on varieties of mustard, *J. Jpn. Soc. Hort. Sci.*, 24, 69–84, 1955.

Lammerink, J., Inter-specific transfer of clubroot resistance from *Brassica campestris* L. to *B. napus*, *N.Z.J. Agri. Res.*, 13, 103–110, 1970.

Landbo, L., Andersen, B., and Jørgensen, R. B., Natural hybridization between oilseed rape and a wild relative: hybrids among seeds from weedy *B. campestris*, *Hereditas*, 25, 89–91, 1996.

Lazaro, A. and Aguinagalde, I., Molecular characterization of *Brassica oleracea* L. and wild relatives ($n = 9$) using RAPD's, *Cruciferae Newsl.*, 18, 24–25, 1996.

Lefol, E. et al., Gene dispersal from transgenic crops. I. Growth of interspecific hybrids between oilseed rape and the wild hoary mustard, *J. Appl. Ecol.*, 32, 803–808, 1995.

Lefol, E., Danielou, V., and Darmency, H., Predicting hybridization between transgenic oilseed rape and wild mustard, *Field Crop Res.*, 45, 153–161, 1996a.

Lefol, E., Fleury, A., and Darmency, H., Gene dispersal from transgenic crops. II. Hybridization between oilseed rape and the wild hoary mustard, *Sex. Plant Reprod.*, 9, 189–196, 1996b.

Lefol, E., Séguin-Swartz, G., and Downey, R. K., Sexual hybridization in crosses of cultivated *Brassica* species with the crucifers *Erucastrum gallicum* and *Raphanus raphanistrum*: potential for gene introgression, *Euphytica*, 95, 127–139, 1997.

Lelivelt, C. L. C. and Krens, F. A., Transfer of resistance to the beet cyst nematode (*Heterodera schachtii* Schm.) into the *Brassica napus* L. gene pool through intergeneric somatic hybridization with *Raphanus sativus* L, *Theor. Appl. Genet.*, 83, 887–894, 1992.

Lelivelt, C. L. C. et al., Transfer of resistance to the beet cyst nematode (*Heterodera schachtii* Schm.) from *Sinapis alba* L. (white mustard) to the *Brassica napus* L. gene pool by means of sexual and somatic hybridization, *Theor. Appl. Genet.*, 85, 688–696, 1993.

Lewis, L. J., Woods, D. L., and Cheng, B. F., Introgression of long pod genotype from spring rape (*Brassica napus* L.) into summer turnip rape (*Brassica rapa* L.), *Can. J. Plant Sci.*, 81, 59–60, 2001.

Li, Z. Y., Liu, H. L., and Heneen, W. K., Meiotic behavior in intergeneric hybrids between *Brassica napus* and *Orychophragmus violaceus*, *Hereditas*, 125, 69–75, 1996.

Li, Z., Liu, H. L., and Luo, P., Production and cytogenetics of intergeneric hybrids between *Brassica napus* and *Orychophragmus violaceus*, *Theor. Appl. Genet.*, 91, 131–136, 1995.

Li, Z. et al., Production and cytogenetics of the intergeneric hybrids *Brassica juncea* × *Orychophragmus violaceus* and *B. carinata* × *O. violaceus*, *Theor. Appl. Genet.*, 96, 251–265, 1998.

Li, Z. and Heneen, W. K., Production and cytogenetics of intergeneric hybrids between the three cultivated *Brassica* diploids and *Orychophragmus violaceus*, *Theor. Appl. Genet.*, 99, 694–704, 1999.

Liu, C. M., Xu, Z. H., and Chua, N. H., Proembryo culture: *in vitro* development of early globular-stage zygotic embryos from *Brassica juncea*, *Plant J.*, 3, 291–300, 1993.

Lu, C. M., Zhang, B., Kakihara, F., and Koto, M., Introgression of genes into cultivated *Brassica napus* through resynthesis of *B. napus* via ovule culture and the accompanying change in fatty acid composition, *Plant Breed.*, 120, 405–410, 2001.

Ma, C. et al., Genetic diversity of Chinese and Japanese rapeseed (*Brassica napus* L.) varieties detected by RAPD markers, *Breed. Sci.*, 50, 257–265, 2000.

Matsuzawa, Y., Studies on the interspecific hybridization in genus *Brassica*. I. Effects of temperature on the development of hybrid embryos and the improvement of crossability by ovary culture in interspecific cross, *B. campestris* × *B. oleracea*, *Jpn. J. Breed.*, 28, 186–196, 1978.

Matsuzawa, Y., Bang, S. W., and Kaneko, Y., Production of hybrid progenies between *Brassica campestris* and *Moricandia arvensis*, *Cruciferae Newsl.*, 20, 21–22, 1998.

Matsuzawa, Y., Minami, T., Bang, S. W., and Kaneko, Y., A new *Brassicoraphanus* ($2n = 36$); the true-breeding amphidiploid line of *Brassica oxyrrhina* Coss. ($2n = 18$) × *Raphanus sativus* L. ($2n = 18$), *Bull. Coll. Agri. Utsunomiya Univ.*, 16, 1–7, 1997.

Matsuzawa, Y. and Sarashima, M., Intergeneric hybrids of *Eruca*, *Brassica* and *Raphanus*, *Cruciferae Newsl.*, 11, 17, 1986.

Matsuzawa, Y. et al., Male sterility in allopasmic *Brassica rapa* L. carrying *Eruca sativa* cytoplasm, *Plant Breed.*, 118, 82–84, 1999.

Matsuzawa, Y. et al., Synthetic *Brassica rapa-Raphanus sativus* amphidiploid lines developed by reciprocal hybridization, *Plant Breed.*, 119, 357–359, 2000.

Mikkelsen, T. R., Angersen, B., and Jørgensen, R. B., The risk of crop transgene spread, *Nature*, 380, 31, 1996a.

Mikkelsen, T. R., Jensen, J., and Jørgensen, R. B., Inheritance of oilseed rape (*Brassica napus*) RAPD markers in a backcross progeny with *Brassica campestris*, *Theor. Appl. Genet.*, 92, 492–497, 1996b.

Misra, S. and Gedamu, L., Heavy metal tolerant transgenic *Brassica napus* L. and *Nicotiana tabacum* L. plants, *Theor. Appl. Genet.*, 78, 161–168, 1989.

Mizushima, U., Karyogenetic studies of species and genus hybrids in the tribe *Brassiceae* of *Cruciferae*, *Tohoku Agri. Res.*, 1, 1–14, 1950.

Mizushima, U., Genome analysis in *Brassica* and allied genera, *Brassica Crops and Wild Allies*, Tsunoda, S., Hinata, K., and Gómez-Campo, C., Eds., Tokyo: Science Societies Press, pp. 89–106, 1980, chap. 5

Mizushima, U. and Tsunoda, S., A plant exploration in *Brassica* and allied genera, *Tohoku Jour. Agri. Res.*, 17, 249–277, 1967.

Momotaz, A., Kato, M., and Kakihara, F., Production of intergeneric hybrids between *Brassica* and *Sinapis* species by mean of embryo rescue techniques, *Euphytica*, 103, 123–130, 1998.

Momotaz, A., Kato, M., and Kakihara, F., Variation in seed fertility and fatty acid composition in the allohexaploids between *Brassica carinata* and *Sinapis* species with the advance of generations, *Breed. Sci.*, 50, 91–99, 2000.

Monnier, M., Culture of zygotic embryos, *In vitro* embryogenesis in plants, Thorpe, T. A., Ed., Dordrecht, Netherlands: Kluwer Academic, pp. 117–153, 1995, chap. 4.

Mora, A. and Earle, E. D., Transformation of broccoli with a *Trichoderma harzianum* endochitinase gene, *Cruciferae Newsl.*, 21, 57–58, 1999.

Mukhopadhyay, A. et al., Somatic hybrids with substitution type genomic configuration TCBB for the transfer of nuclear and organelle genes from *Brassica tournefortii* TT to allotetraploid oilseed crop *B. carinata* BBCC, *Thor. Appl. Genet.*, 89, 19–25, 1994.

Nagpal, S. N. et al., Transfer of *Brassica tournefortii* (TT) genes to allopteraploid oilseed *Brassica* species (*B. juncea* AABB, *B. napus* AACC, *B. carianta* BBCC): homoeologous pairing is more pronounced in the three-genome hybrids (TACC, TBAA, TCBB) as compared to allodiploids (TA, TB, TC), *Theor. Appl. Genet.*, 92, 566–571, 1996.

Nair, H. S., Brown, J., and Brown, A. P., Interspecific and intergeneric hybridization between genetically engineered canola (*Brassica napus* L.) and related weed species, *Cruciferae Newl.*, 18, 38–39, 1996.

Namai, H., Inducing cytogenetical alterations by means of interspecific and intergeneric hybridization in brassica crops, *Gamma Field Symp.*, 26, 41–89, 1987.

Namai, H., Sarashima, M., and Hosoda, T., Interspecific and intergeneric hybridization breeding in Japan, *Brassica Crops and Wild Allies, Biology and Breeding*, Tsunoda, S., Hinata, K., and Gómez-Campo, C., Eds., Tokyo: Japan Science Societies Press, pp. 191–203, 1980, chap. 11.

Nanda Kumar, P. B. A. and Shivanna, K. R., Intergenetic hybridization between *Diplotaxis siettiana* and crop brassicas for the production of alloplasmic lines, *Theor. Appl. Genet.*, 85, 770–776, 1993.

Nanda Kumar, P. B. A., Shivanna, K. R., and Prakash, S., Wide hybridization in *Brassica*. Crossability barriers and studies on the F_1 hybrid and synthetic amphidiploid of *B. fruticulosa*×*B. campestris*, *Sex. Plant Reprod.*, 1, 234–239, 1988.

Nienhuis, J., Slocum, M. K., DeVos, D. V., and Muren, R., Genetic similarity among *Brassica oleracea* L. genotypes as measured by restriction fragment length polymorphisms, *J. Am. Soc. Hort. Sci.*, 118, 298–303, 1993.

Nishi, S., Differentiation of *Brassica* crops in Asia and the breeding of 'Hakuran' a newly synthesized leafy vegetable, *Brassica Crops and Wild Allies*, Tsunoda, S., Hinata, K., and Gómez-Campo, C., Eds., Tokyo: Japan Science Societies Press, pp. 133–150, 1980, chap. 8.

Nishi, S., Kawata, J., and Toda, M., On the breeding of interspecific hybrids between two genomes, 'c' and 'a' of *Brassica* through the application of embryo culture techniques, *Jpn. J. Breed.*, 8, 215–222, 1959.

Nishiyama, I. and Inomata, N., Embryological studies on cross-incompatibility between 2x and 4x in *Brassica*, *Jpn. J. Genet.*, 41, 27–42, 1966.

Ogo, T., Inomata, N., and Yamamoto, Y., Plant species used in Papua New Guinea and Solomon Islands, *Jpn. J. Trop. Agri.*, 31, 16–23, 1987.

Olsson, G., Species crosses within the genus *Brassica*. I. Artificial *B. juncea Coss*, *Hereditas*, 40, 171–223, 1960.

Öztürk, M. and Dogan, Y., Preliminary studies on some endangered *Brassica* species in Turkey, *Cruciferae Newsl.*, 19, 9–10, 1997.

Pang, E. C. K. and Halloran, G. M., The genetics of adult-plant blackleg (*Leptosphaeria maculans*) resistance from *Brassica juncea* in *B. napus*, *Theor. Appl. Genet.*, 92, 382–387, 1996.

Paulmann, W. and Röbbelen, G., Effective transfer of cytoplasmic male fertility from radish (*Raphanus sativus* L.) to rape (*Brassica napus* L.), *Plant Breed.*, 100, 299–309, 1988.

Pelletier, G., Cybrids in Oilseed *Brassica* crops through protoplast fusion, In *Biotechnology in Agriculture and Forestry: Legumes and Oilseed Crops I*, Vol. 10, Bajaj, Y. P. S., Ed., Berlin: Springer, pp. 418–433, 1990, chap. IV.

Perrino, P. and Hammer, K., Collection of land-races of cultivated plants in South Italy, 1984, *Kulturpflanze*, 33, 225–236, 1985.

Perrino, P., Pignone, D., and Hammer, K., The occurrence of a wild *Brassica* of the *oleracea* group ($2n = 18$) in Calabria (Italy), *Euphytica*, 59, 99–101, 1992.

Pierre, J. et al., Effects of herbicide-tolerant transgenic oilseed rape genotypes on honey bees and other pollinating insects under field conditions, *Entomologia Experimentalis et Applicata*, 108, 159–168, 2003.

Pradhan, A. K., Prakash, S., Mukhopadhyay, A., and Pental, D., Phylogeny of *Brassica* and allied genera based on variation in chloroplast and mitochondrial DNA patterns: molecular and taxonomic classifications are incongruous, *Theor. Appl. Genet.*, 5, 331–340, 1992.

Prakash, S., Cruciferous oilseed in India, *Brassica Crops and Wild Allies*, Tsunoda, S., Hinata, K., and Gometz-Campo, C., Eds., Tokyo: Japan Science Societies Press, pp. 151–163, 1980, chap. 9.

Prakash, S. and Chopra, V. L., Reconstruction of alloploid brassicas through non- homologous recombination: introgression of resistance to pod shatter in *Brassica napus*, *Genet. Res. Camb.*, 56, 1–2, 1990.

Prakash, S. and Hinata, K., Taxonomy, cytogenetics and origin of crops Brassicas, a review, *Opera Botanica*, 55, 1–57, 1980.

Prakash, S., Takahata, Y., Kirti, P. B., and Chopra, V. L., Cytogenetics, *Biology of Brassica Coenospecies*, Gómez-Campo, C., Ed., Amsterdam: Elsevier, pp. 59–106, 1999, chap. 3.

Prakash, S. et al., *Moricandia arvensis* based cytoplasmic male sterility and fertility restoration system in *Brassica juncea*, *Theor. Appl. Genet.*, 97, 488–492, 1998.

Prakash, S. et al., Expression of male sterility in alloplasmic *Brassica juncea* with *Erucastrum canariense* cytoplasm and the development of a fertility restoration system, *Plant Breed.*, 120, 479–482, 2001.

Primard, C. et al., Interspecific somatic hybridization between *Brassica napus* and *Brassica hirta* (*Sinapis alba* L.), *Theor. Appl. Genet.*, 75, 546–552, 1988.

Pua, E. C. and Lim, T. S., Transgenic oilseed brassicas, *Biotechnology in Agriculture and Forestry*, Pua, E. C. and Douglas, C. J., Eds., vol. 54, Berlin: Springer, pp. 195–224, 2004, chap. 3.

Quazi, M. H., Interspecific hybrids between *Brassica napus* L. and *B. oleracea* L. developed by embryo culture, *Theor. Appl. Genet.*, 75, 309–318, 1988.

Quiros, C. F., Kianian, S. F., Ochoa, O., and Douches, D., Genome evolution in *Brassica*: use of molecular markers and cytogenetic stocks, *Cruciferae Newsl.*, 10, 21–23, 1985.

Rahman, M. H., Production of yellow-seeded *Brassica napus* through interspecific crosses, *Plant Breed.*, 120, 463–472, 2001.

Rahman, M. H., Joersbo, M., and Poulsen, M. H., Development of yellow-seeded *Brassica napus* of double low quality, *Plant Breed.*, 120, 473–478, 2001.

Ren, J., Dickson, M. H., and Earle, E. D., Re-synthesis of *B. napus* by protoplast fusion to improve resistance to bacterial soft rot, *Cruciferae Newsl.*, 21, 133–134, 1999.

Rieger, M. A. et al., Pollen- mediated movement of herbicide resistance between commercial Canola fields, *Science*, 296, 2386–2388, 2002.

Ripley, V. L. and Arnison, P. G., Hybridization of *Sinapis alba* L. and *Brassica napus* L. *via* embryo rescue, *Plant Breed.*, 104, 26–33, 1990.

Ripley, V. L. and Beversdorf, W. D., Development of self-incompatible *Brassica napus*: (I) introgression of S-alleles from *Brassica oleracea* through interspecific hybridization, *Plant Breed.*, 122, 1–5, 2003.

Robers, M. B., Williams, P. H., and Osborn, T. C., Cytogenetics of *Eruca sativa/Brassica rapa* hybrids produced by embryo rescue, *Cruciferae Newsl.*, 21, 41–42, 1999.

Roy, N. N., Interspecific transfer of *Brassica juncea*-type high blackleg resistance to *Brassica napus*, *Euphytica*, 33, 295–303, 1984.

Sacristán, M. D. and Gerdemann, M., Different behavior of *Brassica juncea* and *B. carinata* as sources of *Phoma lingam* resistance in experiments of interspecific transfer to *B. napus*, *Plant Breed.*, 97, 304–314, 1986.

Saharan, G. S., Disease resistance, *Breeding OilseedbBrassicas*, Labana, K. S., Banga, S. S., and Banga, S. K., Eds., Berlin: Springer, pp. 181–205, 1993, chap. 12.

Saharan, G. S., Disease resistance, *Recent Advances in OilseedbBrassicas*, Kalia, H. R. and Gupta, S. K., Eds., Ludhiana, India: Kalyani Publishers, pp. 233–259, 1997, chap. 1.

Sakai, T. and Imamura, J., Intergeneric transfer of cytoplasmic male sterility between *Raphanus sativus* (cms line) and *Brassica napus* through cytoplast–protoplast fusion, *Theor. Appl. Genet.*, 80, 421–427, 1990.

Sakai, T. et al., Introduction of a gene from fertility restored radish (*Raphanus sativus*) into *Brassica napus* by fusion of X- irradiated protoplasts from a radish restorer line and iodacetomide-treated protoplasts from a cytoplasmic male-sterile hybrid of *B. napus*, *Theor. Appl. Genet.*, 93, 373–379, 1996.

Sánchez-Yélamo, M. D., Comparative analysis of phenolic compounds among some species of the genus *Brassica* from Sect. *Sinapistrum* and Sect. *Micropodium*, *Cruciferae Newsl.*, 24, 19–20, 2002.

Sarashima, M., Studies on the breeding of artificially synthesized rape (*Brassica napus*). I. F_1 hybrids between *B. campestris* group and *B. oleracea* group and the derived F_2 plants, *Jpn. J. Breed.*, 14, 226–237, 1964.

Sarashima, M., *Plant Breeding of Cruciferae II- A Basis of Interspecific and Intergeneric Hybridization*, Ochiai Shoten, Utsunomiya, Japan, 1991.

Sarla, N. and Raut, R. N., Cytogenetical studies on *Brassica nigra* × *B. oleracea* hybrids, *Indian J. Genet.*, 51, 408–413, 1991.

Sarmah, B. K. and Sarla, N., Hybridization of *Diplotaxis* and *Erucastrum* with crop *Brassica*, *Cruciferae Newsl.*, 16, 34–35, 1994.

Sarmah, B. K. and Sarla, N., *Diplotaxis tenuisiliqua* × *Brassica campestris* hybrids obtained by embryo rescue, *Indian J. Genet.*, 58, 35–39, 1994.

Seegeler, C. P. J., *Oil Plants in Ethiopia, Their Taxonomy and Agricultural Significance*, Wageningen: Center Agri. Pub. & Docum, 1983 pp. 1–368.

Shiga, T., Rape breeding by interspecific crossing between *Brassica napus* and *B. campestris* in Japan, *Jpn. Agri. Res. Q.*, 5, 5–10, 1970.

Sigareva, M. and Earle, E. D., *Capsella bursa-pastoris*: regeneration of plants from protoplasts and somatic hybridization with rapid cycling *Brassica oleracea*, *Cruciferae Newsl.*, 19, 57–58, 1997a.

Sigareva, M. and Earle, E. D., Intertribal somatic hybrids between *Camelina sativa* and rapid cycling *Brassica oleracea*, *Cruciferae Newsl.*, 19, 49–50, 1997b.

Sigareva, M., Ren, J., and Earle, E. D., Introgression of resistance to *Alternaria brassicicola* from *Sinapis alba* to *Brassica oleracea* via somatic hybridization and back-crosses, *Cruciferae Newsl.*, 21, 135–136, 1999.

Sinskaia, E. N., The oleiferous plants and root crops of the family Cruciferae, *Bull. Appl. Bot. Genet. Plant Breed.*, 19, 1–648, 1928.

Smartt, J. and Simonds, N. W., *Evolution of Crop Plants* 2nd ed., United Kingdom: Longman Scientific Technical, p. 531, 1995,.

Snogerup, S., The wild forms of the *Brassica oleracea* group ($2n = 18$) and their possible relations to the cultivated ones, *Brassica Crops and Wild Allies*, Tsunoda, S., Hinata, K., and Gómez-Campo, C., Eds., Tokyo: Japan Science Societies Press, pp. 121–132, 1980, chap. 7.

Snogerup, S., Gustafsson, M., and von Bothmer, R., *Brassica* sect. *Brassica* (*Brassicaseae*) I. Taxonomy and variation, *Willdenowia*, 19, 271–365, 1990.

Snogerup, S. and Persson, D., Hybridization between *Brassica insularis* Moris and *B. balearica* Pers, *Hereditas*, 99, 187–190, 1983.

Song, K. and Osborn, T. C., Polyphyletic origins of *Brassica napus*: new evidence based on organelle and nuclear RFLP analyses, *Genome*, 35, 992–1001, 1992.

Song, K. M., Osborn, T. C., and Williams, P. H., *Brassica* taxonomy based on nuclear restriction fragment length polymorphisms (RFLPs).1. Genome evolution of diploid and amphidiploid species, *Theor. Appl. Genet.*, 75, 784–794, 1988a.

Song, K. M., Osborn, T. C., and Williams, P. H., *Brassica* taxonomy based on nuclear restriction fragment length polymorphisms (RFLPs). 2. Preliminary analysis of subspecies within *B. rapa* (syn. *campestris*) and *B. oleracea*, *Theor. Appl. Genet.*, 76, 593–600, 1988b.

Song, K. M., Osborn, T. C., and Williams, P. H., *Brassica* taxonomy based on nuclear restriction fragment length polymorphisms (RFLPs). 3. Genome relationships in *Brassica* and related genera and the origin of *B. oleracea* and *B. rapa* (syn. *campestris*), *Theor. Appl. Genet.*, 79, 497–506, 1990.

Struss, D., Quiros, C. F., Plieske, J., and Röbbelen, G., Construction of *Brassica* B genome synteny groups based on chromosomes extracted from three different sources by phenotypic, isozyme and molecular markers, *Theor. Appl. Genet.*, 93, 1026–1032, 1996.

Takahata, Y., Production of intergeneric hybrids between a C_3–C_4 intermediate species *Moricandia arvensis* and a C_3 species *Brassica oleracea* through ovary culture, *Euphytica*, 46, 259–264, 1990.

Takahata, Y. and Takeda, T., Intergeneric (intersubtribe) hybridization between *Moricandia arvensis* and *Brassica* A and B genome species by ovary culture, *Theor. Appl. Genet.*, 80, 38–42, 1990.

Takeshita, M., Kato, M., and Tokumasu, S., Application of ovule culture to the production of intergeneric or interspecific hybrids in *Brassica* and *Raphanus*, *Jpn. J. Genet.*, 55, 373–387, 1980.

Tewari, J. P. and Mithen, R. F., Disease, *Biology of Brassica Coenospecies*, Gómez-Campo, C., Ed., Amsterdam: Elsevier, pp. 305–411, 1999, chap. 12

Tonguç, M. and Griffiths, P. D., Development of black rot resistant interspecific hybrids between *Brassica oleracea* L. cultivars and *Brassica accession* A using embryo rescue, *Euphytica*, 136, 313–318, 2004a.

Tonguç, M. and Griffiths, P. D., Transfer of powdery mildew resistance from *Brassica Carinata* to *Brassica Obracea* through embryo rescue, *Plant Breed*, 123, 587–589, 2004b.

Tsunoda, S., Eco-physiology of wild and cultivated forms in *Brassica* and allied genera, *Brassica Crops and Wild Allies*, Tsunoda, S., Hinata, K., and Gómez-Campo, C., Eds., Tokyo: Japan Science Societies Press, pp. 109–120, 1980, chap. 6.

U.N., Genome-analysis in *Brassica* with special reference to the experimental formation of *B. napus* and peculiar mode of fertilization, *Jpn. J. Bot*, 7, 389–542, 1935.

Vambing, K. and Glimelius, K., Regeneration of plants from protoplasts of oilseed *Brassica* crops, In *Biotechnology in Agriculture and Forestry: Legumes and Oilseed Crops I*, Vol. 10, Bajaj, Y. P. S., Ed., Springer, Berlin, 2: 385–417, 1990, chap. IV.

von Bothmer, R., Gustafsson, M., and Snogerup, S., *Brassica* sect. *Brassica* (*Brassicaceae*). II. Inter- and intraspecific crosses with cultivars of *B. oleracea*, *Genetic Resour. Crop Evol.*, 42, 165–178, 1995.

Wang, Y. P., Sonntag, K., and Rudloff, E., Development of rapeseed with high erucic acid content by asymmetric somatic hybridization between *Brassica napus* and *Crambe abyssinica*, *Theor. Appl. Genet.*, 106, 1147–1155, 2003.

Warwick, S. I. and Black, L. D., Molecular systematics of *Brassica* and allied genera (subtribe Brassicinae, Brassiceae)- chloroplast genome and cytodeme congruence, *Theor. Appl. Genet.*, 82, 81–92, 1991.

Warwick, S. I. and Black, L. D., Molecular relationships in subtribe Brassicinae Cruciferae, tribe Brassiceae), *Can. J. Bot.*, 71, 906–918, 1993.

Warwick, S. I., Black, L. D., and Aguinagalde, I., Molecular systematics of *Brassica* and allied genera (Subtribe Brassicinae, Brassiceae)—chloroplast DNA variation in the genus *Diplotaxis*, *Theor. Appl. Genet.*, 83, 839–850, 1992.

Warwick, S. I. and McDonald, T., Molecular characterization of genetic relationships in *Brassica rapa* based on RAPD markers, *Cruciferae Newsl.*, 23, 13–14, 2001.

Warwick, S. L. and Soleimani, V., Genetic diversity in *Brassica carinata*, *B. juncea* and *B. nigra* based on molecular AFLP markers, *Cruciferae Newsl.*, 23, 15–16, 2001.

Warwick, S. I. et al., Hybridization between transgenic *Brassica napus* L. and its wild relatives: *Brassica rapa* L., *Raphanus raphanistrum* L., *Sinapis arvensis* L., and *Erucastrum gallicum* (Willd.) O.E. Schulz, *Theor. Appl. Genet.*, 107, 528–539, 2003.

Warwick, S. and Miki, S., Herbicide resistance, In *Biotechnology in Agriculture and Forestry*, Vol. 54, Pua, E. C. and Douglas, C. J., Eds., Springer, Berlin, 3: 273–295, 2004, chap. IV.

Xhuveli, L., Preliminary study about an endangered wild *Brassica* population in Albania, *Cruciferae Newsl.*, 17, 8–9, 1995.

Yan, Z. et al., Production of somatic hybrids between *Brassica oleracea* and the C_3–C_4 intermediate species *Moricandia nitens*, *Theor. Appl. Genet.*, 99, 1281–1286, 1999.

Yanagino, T., Takahata, Y., and Hinata, K., Chloroplast DNA variation among diploid species in *Brassica* and allied genera, *Jpn. J. Genet.*, 62, 119–125, 1987.

Zenkteler, M., *In-vitro* fertilization of ovules of some species of *Brassicaceae*, *Plant Breed.*, 105, 221–228, 1990.

Zhang, B., Hondo, K., Kakihara, F., and Koto, M., Production of amphidiploids between A and C genomic species in *Brassica*, *Breed. Res.*, 3, 31–41, 2001.

Zhang, C. et al., A dominant *gdcP*-specific marker derived from *Moricandia nitens* used for introducing the C_3–C_4 character from *M. nitens* into *Brassica* crops, *Plant Breed.*, 123, 438–443, 2004.

CHAPTER **5**

Okra (*Abelmoschus* spp.)

I. S. Bisht and K. V. Bhat

CONTENTS

5.1   INTRODUCTION

The cultivated okra [*Abelmoschus esculentus* L. (Moench)] is an important vegetable crop throughout tropical and subtropical low altitude regions of Asia and Africa. It is valued very highly for its mature, tender, and green fruits. Okra fruits are rich in calcium (90 mg per 100 g fresh weight) and provide a valuable supplementary item in the tropical diet. The young fruits are consumed as cooked vegetable, mostly fresh but sometimes sun dried or frozen. The fruit is fairly rich in protein and minerals. Its seed contains 13–22% edible oil and 20–24% edible protein, and is viewed as alternative source for edible oil. Stem fiber of some wild *Abelmoschus* species is used for making rope. Edible roots are used as food and for medicinal purpose. In Papua New Guinea and the Solomon and Fiji Islands, some forms of *A. manihot* are used as a leafy vegetable. The roasted and powdered seed of okra is substituted for coffee in parts of China and Africa, while in Japan, okra pods are pickled.

5.2   DESCRIPTION AND CROP PRODUCTION

5.2.1   Botany and Floral Biology

Bailey (1969) described the morphology of *A. esculentus* in detail. Okra is an erect, herbaceous annual, 1–2 m tall with green stems or with a reddish tinge. Leaves alternate, broadly cordate, palmately 3–7 lobed, hirsute, serrate. Flower solitary, axillary with about 2 cm long peduncle; epicalyx up to 10, narrow hairy bracteoles that fall before the fruit reaches maturity; calyx split longitudinally as flower opens; petals 5, yellow with crimson spot on claw, 5–7 cm long; staminal column united to the base of petals with numerous stamens; ovary superior, stigma 5–9 deep red. Fruit is a capsule, light green, or sometimes red in color, pyramidal-oblong, beaked, longitudinally furrowed, 10–30 cm long, dehisces longitudinally when ripe. Seeds are green to dark brown, rounded, or sub-reniform. The greatest increase in fruit weight, length, and diameter occurs during 4–6 days after pollination. Generally, the fiber formation in the fruit starts from the fifth to sixth day of fruit formation, and a sudden increase in fiber content from the ninth day is observed.

Okra flowers occur singly in the leaf axils and the flowers are hermaphrodite. Flowers take about 22 days to fully develop. Anthesis takes place in the morning, and most flowers open between 9:00 and 10:00 a.m. During August high temperature and low humidity hasten anthesis. Anther dehiscence begins when the flowers are opening and is completed before flowers are finished. Dehiscence occurs from 6:00 to 11:00 a.m. and is at a maximum between 8:00 and 9:00 a.m. Pollen grains are round in shape. Pollen fertility is optimum between 6:30 and 8:30 a.m. Stigma receptivity varies in duration and is maximally recepetive on the day of anthesis.

5.2.2   World Production Area and Utilization

The cultivation of okra extends throughout the tropics and warmer parts of temperate Asia. It is grown commercially in India, Turkey, Iran, western Africa, Yugoslavia, Bangladesh, Afghanistan, Pakistan, Burma, Japan, Malaysia, Brazil, Ghana, Ethiopia, Cyprus, and the southern United States. India ranks first in the world with 3.5 million tones (70% of the total world production) of okra produced from over 0.36 million ha land (FAOSTAT 2004). It accounted for 70% of the earnings from fresh vegetable exports (excluding onion) during 1998–1999 (APEDA 2000). The exports are limited to Middle East, Europe, UK, and Singapore, etc. Other major okra producing countries in the world are Nigeria (0.73 million tones), Pakistan (0.12 million tones), Ghana (0.10 million tones), and Egypt (0.08 million tones). The USA is the fifteenth largest producer of okra in the world. The area, production, and yield of some major okra producing countries are presented in Table 5.1.

5.3   ORIGIN, DOMESTICATION, AND DISPERSION

The genus *Abelmoschus* is of Asiatic origin, but the exact ancestral home of the cultigen *A. esculentus* is disputed. It is thought to have been domesticated in the Ethiopian region (Vavilov 1936) or in western Africa (Murdock 1959; Joshi, Gadwal, and Hardas 1974). A late nineteenth century record shows its occurrence in the wild state on the White Nile in Sudan (Singh, Swarup, and Singh 1975). Joshi and Hardas (1976) suggested that the cultigen, *A. esculentus*, might have originated in Asia, or it might originally have been present in Africa and Asia as a polyphyletic species. van Borssum Waalkes (1966) considered southeast Asia as the center of diversity for the genus *Abelmoschus*. An Indian origin of *A. esculentus* has also been advocated by Masters (1875), Zeven and Zhukovasky (1975) and Zeven and de Wet (1982). Vredebregt (1990) observed that rich diversity in *A. ficulneus* in India combined with that of *A. tuberculatus* may be an argument for an Asian origin of *A. esculentus. Abelmoschus ficulneus* is thought to have contributed the second genome of cultivated okra, one genome being from *A. tuberculatus*. This view is supported by the existence of Sanskrit words *tindisha* and *gandhamulla*, which mean "okra."

Nine *Abelmoschus* species, namely *A. angulosus, A. tuberculatus, A. manihot, A. moschatus, A. ficulneus, A. esculentus, A. tetraphyllus* var. *tetraphyllus, A. tetraphyllus* var. *pungens*, and

Table 5.1   Area, Production, and Yield of Some Major Okra Producing Countries in 2003

Country	Area Harvested (ha)	Production (mt)	Yield (hg/ha)
Barbados	131	1,466	111,908
Belize	27	240	88,889
Benin	24,000	72,000	30,000
Burkina Faso	2,600	26,000	100,000
Cameroon	18,000	32,000	17,778
Cyprus	73	1,300	178,082
Egypt	6,000	85,000	141,667
Fiji Islands	200	1,200	60,000
Ghana	18,000	100,000	55,556
Guatemala	800	6,000	75,000
India	360,000	3,500,000	97,222
Jordan	900	16,000	177,778
Kuwait	350	2,400	68,571
Lebanon	1,000	3,000	30,000
Mexico	6,600	35,000	53,030
Nigeria	276,000	730,000	26,449
Pakistan	12,500	110,000	88,000
Palestine, Occupied Tr.	1,000	4,200	42,000
Qatar	250	950	38,000
Saudi Arabia	6,600	52,000	78,788
Syrian Arab Republic	3,800	12,000	31,579
Turkey	6,100	30,000	49,180
United States of America	1,200	8,100	67,500
Yemen	3,800	22,000	57,895
World	768,931	4,940,071	64,246

Source: From FAOSTAT 2004. (www.fao.org)

A. crinitus occur in India. Of these only *A. esculentus* is known to be cultivated. *A. moschatus* occurs wild though it is also cultivated (for its aromatic seeds) while the rest are truly wild taxa.

High variation has been observed in cultivated species, *A. esculentus*. The countries with the largest collection or diversity are Turkey and India, and to a lesser extent Afghanistan, Iran, Pakistan, Bangladesh, and Yugoslavia. In cultivated okra, genetic diversity has been observed for absence or presence of number of ridges on fruits, extent of hairiness and pigmentation on fruit and other plant parts, length:width (diameter) ratio of fruits, fruit attachment, lobing of leaves, branching, and plant height. Some of these characteristics are highly influenced by environment. In hilly areas (warmer parts of temperate region) smooth fruited variety develops hard hairs, and sometimes non-pigmented ones develop partial pigmentation.

The other cultivated group, *A. caillei* from western Africa, is clearly distinguished by several morphological characteristics. It is considered to be a natural amphidiploid of *A. esculentus* and *A. manihot*. In the western Ghat Mountains of India, such forms are also available.

In India, maximum variability in plant type and fruit type has been collected from and is available in eastern (Orissa, West Bengal, and Bihar) and northeastern regions (Assam, Meghalaya, Sikkim) where landraces and primitive types can still be found with substistence farmers and in certain tribal areas, though they have poor yield and varying degrees of tolerance to stresses.

Abelmoschus esculentus forms are adapted to different agro-climates and consumer preferences. In western Africa, both Sudanian (*A. esculentus*) and Guinnean (*A. caillei*) forms are under cultivation. The west African okra (*A. caillei*) is an unconventional okra type that grows naturally in the region. It is bigger than the conventional okra, *A. esculentus*, and it is essentially photoperiod sensitive. It is an annual but can grow vegetatively as long as the day length remains long. It is relatively tolerant to diseases and insect pests and commonly found around kitchen gardens and along roads, but rarely in undisturbed forests. As a result of its high yield and hardiness, it has become a major source of okra pods in Nigeria and other parts of western Africa, and its cultivation is progressively replacing the conventional type. The production of west African okra is, however, still limited to the four months of September to December because of its photoperiodism. In the Philippines, Sounde (extra-early) and Gbogboligbo (both *A. esculentus*) are the two major types. In Papua New Guinnea, Fiji, and Solomon Islands, flowering types as well as nonflowering types (Aidiba and Aibika) of *A. manihot* are used as leafy vegetables. These rejuvenate after multiple cuttings and are propagated through rooted stem cuttings. The presence of some forms having round, nonlobed leaves and short fruits (probably with $2n = 72$) in Gujarat (Saurashtra and Dwarka region), in India, and some parts of Japan (Teshima 1933; Ugale, Patil, and Khupse 1976) indicate considerable genetic variability within *A. manihot*.

5.4 TAXONOMY

Cultivated okra and related wild species were originally included in the genus *Hibiscus*, section *Abelmoschus*, of the family Malvaceae, by Linnaeus (1753). Medikus (1787) proposed to raise this section to the rank of a distinct genus, but the reference to the genus *Hibiscus* remained until the middle of twentieth century. Not until the rehabilitation of the genus *Abelmoschus* by Hochreutiner (1924) was its use accepted in the taxonomic and contemporary literature. This genus is distinguished from the genus *Hibiscus* by the characteristics of the calyx; spathulate, with five short teeth, connate to the corolla and caducous after flowering (Kundu and Biswas 1973; Terrell and Winters 1974). About 50 species have been described by taxonomists. The taxonomical revision undertaken by van Borssum Waalkes (1966) and its continuation by Bates (1968) constitutes the most fully documented studies of the genus *Abelmoschus*. Taking the classification of van Borssum Waalkes as the starting point, the classification adopted by the International Okra Workshop held at NBPGR in 1990 is as follows (IBPGR 1991):

Classification Developed by van Borssum Waalkes (1966)	Classification Adopted by International Okra Workshop IBPGR (1991)
A. moschatus Medikus	*A. moschatus* Medikus
—subsp. *moschatus*	—subsp. *moschatus*
var. *moschatus*	var. *moschatus*
—subsp. *Moschatus*	—subsp. *Moschatus*
var. *betulifolius* (Mast.) Hochr.	var. *betulifolius* (Mast.) Hochr.
—subsp. *biakensis* (Hochr.) Borss.	—subsp. *biakensis* (Hochr.) Borss.
—subsp. *tuberosus* (Span.) Borss.	—subsp. *tuberosus* (Span.) Borss.
A. manihot (L.) Medikus	*A. manihot* (L.) Medikus
—subsp. Manihot	
subsp. *tertraphyllus* (Roxb.ex Hornem.) Borss.	—subsp. *tetraphyllus* (Roxb.ex Hornem.) Borss.
var. *tetraphyllus*	var. *tetraphyllus*
—subsp. *Tetraphyllus*	—var. *pungens*
var. *pungens* (Roxb.) Hochr.	
A. esculentus (L.) Moench.	*A. esculentus* (L.) Moench.
(including *A. tuberculatus* Pal & Singh)	*A. tuberculatus* Pal & Singh
A. ficulneus (L.) W. & A. ex Wight	*A. ficulneus* (L.) W. & A. ex Wight
A. crinitus Wall.	*A. crinitus* Wall.
A. angulosus Wall. ex. W. & A.	*A. angulosus* Wall. ex. W. & A.
	A. caillei (A. Chev.) Stevels

It was noted that the previous denominations corresponds to the following species:

- "Guinean" type of okra = *A. caillei*
- "Soudanian" type of okra = *A. esculentus*

- *A. manihot* var. *caillei* A. Chev. = *A. caillei*
- *A. caillei* was incorrectly identified earlier as *A. manihot* subsp. *manihot* (Flora of Tropical West Africa, Kew Botanic Gardens)

The adoption of this new classification requires the amendment of the determination key of *Abelmoschus* to accommodate the distinction between *A. esculentus* and *A. tuberculatus* as well as the distinction between *A. manihot/A. tetraphyllus* and *A. callei*. The existing botanical descriptors (*A. tuberculatus*, *A. manihot*, and *A. tetraphyllus*) need to be compared with the variation in the accessions of the global base collection and other existing collections. The intraspecific classification in *A. moschatus*, *A. tetraphyllus*, *A. esculentus*, and *A. angulosus* should, however, receive further attention (IBPGR 1991). The diagnostic characteristics in *Abelmoschus* are presented in Table 5.2.

Table 5.3 summarizes observations on chromosome numbers and ploidy levels of different species in the genus. The lowest number reported is $2n = 56$ for *A. angulosus* (Ford 1938). Charrier (1984) and Hamon (1987) reported $2n = 38$ as the lowest number, based on observations by Skovsted (1935, 1941) for *Hibiscus coccineus* Walter and *Hibiscus grandiflorus* Michx. However, *H. coccineus*, supposedly a synonym of *A. moschatus* ssp. *tuberosus*, and *H. grandiflorus*, supposedly a synonym of *Abelmoschus angulosus*, are North American species and do not belong to the genus *Abelmoschus*.

The highest chromosome numbers reported are close to 200 for *A. manihot* var. *caillei* (Singh and Bhatnagar 1975; Siemonsma 1982a, 1982b). The numbers reported for *A. esculentus* vary greatly. $2n = 58$ is reported for a form described by Pal, Singh, and Swarup (1952) as *A. tuberculatus*. van Borssum Waalkes (1966) includes it in *A. esculentus*, although he notes that it may be one of the ancestors. There are three reports on a race with $2n = 72$. The most frequently observed somatic chromosome number, however, is $2n = 130$, although Datta and Naug (1968) suggest that the numbers $2n = 72$, 108, 120, 132, and 144 are an indication of a regular series of polyploids with $x = 12$.

The enormous morphological variation in *A. manihot* seems to have a genetic basis, as evidenced by chromosome counts ($2n = 130$, 138) on *A. manihot* ssp. *tetraphyllus*, which are of a higher ploidy level than the more frequently observed $2n = 60$, 68.

5.5   DISTRIBUTION OF *ABELMOSCHUS* SPECIES

5.5.1   *Abelmoschus angulosus* Wall ex W. and A.

It is distributed in wet and temperate regions from eastern Kerala (India) down to Bali (Indonesia) through Sri Lanka, and through northern Myanmar and Sumatra (Indonesia), between 750 and 2,000 masl. Among the intraspecific taxa, two entities are currently recognized: ssp. *purpureus* and ssp. *grandiflorus*. In addition, two varieties are proposed within the ssp. *purpureus* as var. *purpureus* and var. *setinervis*. *A. angulosus* is the only wild *Abelmoschus* species with a pronounced tolerance to low temperatures and light night frosts that occasionally occur above the altitude of 2,000 masl.

5.5.2   *Abelmoschus crinitus* Wall

It is distributed from northern Pakistan down to Java (Indonesia) through India, Myanmar, and Thailand. It occurs in open forest and in areas subject to a pronounced dry season up to 1,400 masl. Although van Borssum Waalkes (1966) reported it as "a common species of Indo-Chinese peninsula," *A. crinitus* is also distributed on the southern slopes of Himalayas and in western China (Yunnan). In the Hainan province of China, it produces fragrant flowers.

Table 5.2 Diagnostic Characteristics in *Abelmoschus*

Taxon	Epicalyx Segments				Capsule		Pedicel	Indumentum
	Number	Length (mm)	Width (mm)	Persistence	Shape	Length (cm)	Length (cm)	
1. *A. moschatus*								
a. ssp. *moschatus* var. *moschatus*	7–10	8–15	1–2	Through fruit	Ovoid-oblong	5–8	3–8	Stems devoid of prickly hairs
b. ssp. *moschatus* var. *betulifolius*	6–8	17–25	2.5–5			5–8	3–8	Stems with prickly hairs; epicalys soft-margined
c. ssp. *biakensis*	8	15–20	0.35–4			5–8	17–19	
d. ssp. *tuberosus*	9–10	10–25	?			2–5	?	
2. *A. manihot*								
a. ssp. *manihot*	4–8	10–30	5–10	Through fruit	Ovoid-prismatical	3.5–6	?	Stems with prickly hairs; epicalyx hispid-margined
b. ssp. *tetraphyllus* var. *tetraphyllus*								
c. ssp. *tetraphyllus* var. *pungens*								
3. *A. esculentus*	7–15	5.25	0.5–3	Through flowering	Fusiform-deltoid	5–35	0.5–1.5	
4. *A. ficulneus*	5–6	4–12	0.5–1.5	Through early flower	Ovoid	3–3.5	?	
5. *A. crinitus*	10–16	25–40	0.5–1	Through fruit	Ovoid-globose	3.5–5	?	
6. *A. angulosus*	4–5	20–35	10–20	Through fruit	Ovoid-prismatical	3–5	?	
A. caillei	5–10	10–35	4–13	Through early fruit	Ovoid	5–35	1–13	

Source: From van Borssum Waalkes, J., *Blumea* 14, 1, 1966; Siemonsma, J. S., *La culture du gombo (Abelmoschus spp.), legume-fruit tropical (avec reference speciale a la Cote d' Ivoire).* Thesis, Wageningen Agricultural University, The Netherlands, 1982.

Table 5.3 Chromosome Numbers (2n) in *Abelmoschus*

Species	Chromosome Numbers (2n)	Authors	Ploidy Level	Genepool (GP)
A. esculentus	±66	Ford (1938)		GP1
	72	Teshima (1933), Ugale et al. (1976) and Kamalova (1977)		
	108	Datta and Naug (1968)	2	
	118	Krenke In: Tischler (1931)	2	
	120	Krenke In: Tischler (1931), Purewal and Randhawa (1947), and Datta and Naug (1968)	2	
	122	Krenke In: Tischler (1931)	2	
	124	Kuwada (1957a, 1966)	2	
	126–134	Chizaki (1934)	2	
	130	Skovsted (1935) and Joshi and Hardas (1953); Gadwal In: Joshi and Hardas (1976), Gadwal et al. (1968), Joshi et al. (1974), and Singh and Bhatnagar (1975)	2	
	131–143	Siemonsma (1982a,1982b)	2	
	132	Medwedewa (1936) and Roy and Jha (1958)	2	
	±132	Breslavetz et al. (1934) and Ford (1938)	2	
	144	Datta and Naug (1968)	2	
A. manihot				
—ssp. *manihot*	60	Teshima (1933) and Chizaki (1934)	1	GP3
	66	Skovsted (1935) and Kamalova (1977)	1	
	68	Kuwada (1957a, 1974)	1	
—ssp. *tetraphyllus* var. *tetraphyllus*	130	Ugale et al. (1976)	2	GP3
	138	Gadwal In: Joshi and Hardas (1976)	2	GP3
—ssp. *tetraphyllus* var. *pungens*	138	Gadwal In: Joshi and Hardas (1976)	2	GP3
A. moschatus	72	Skovsted (1935), Gadwal et al. (1968); Joshi et al. (1974)	1	GP3
A. ficulneus	72	Hardas and Joshi (1954), Kuwada (1966, 1974), Gadwal et al. (1968) and Joshi et al. (1974)	1	GP2
	78	Skovsted (1935)	1	
A. angulosus	56	Ford (1938)	1	GP3
A. tuberculatus	58	Joshi and Hardas (1953), Kuwada (1966, 1974), Gadwal, Joshi, and Iyer (1968) and Joshi, Gadwal, and Hardas (1974)	1	GP2
A. caillei	194	Singh and Bhatnagar (1975)	3	GP3
(A. manihot var. caillei)	185–199	Siemonsma (1982a, 1982b)	3	

Ploidy level 1: 2n=56–72; Ploidy level 2: 2n=108–144; Ploidy level 3: 2n=185–199.
Source: From Charrier, A., Genetic resources of the genus *Abelmoschus* Med. (Okra), IBPGR, Rome, 1984; Siemonsma, J. S., *International Crop Network Series*. Report of an international workshop on okra genetic resources, IBPGR, Rome, 5:52–68 1991.

5.5.3 *Abelmoschus ficulneus* (L.) W. and A. ex Wight

It is the wild, *Abelmoschus* species have the widest distribution from southern Chad down to central Australia through Sudan, Ethiopia, Madagaskar, India, and Timor (Indonesia). It occurs in low lands up to 600 masl with pronounced dry season, in open vegetations, and on waste lands.

5.5.4 *Abelmoschus manihot* (L.) Medikus

Separated from the wild forms (ssp. *tetraphyllus*), *A. manihot* ($2n=60-68$) appears to be a more uniform entity. It is a species with uncertain distribution. *Flora of West Tropical Africa*

(Hutchinson and Dalziel 1958) makes mention of the collection of *A. manihot* ssp. *manihot*, but in spite of many collecting missions, it has not been observed recently.

It remains to be decided where the important leaf vegetable "Aibika" (*A. manihot*) from Papua New Guinea belongs. Its appearance is rather different from the typical forms of *A. manihot* in Indonesia. No chromosome counts are available for this collection.

5.5.5 *Abelmoschus manihot* ssp. *tetraphyllus* var. *tetraphyllus* (Roxb.ex Hornem.) Borss.

This taxon shows a disjunction between its distributional area in southern Asia (India, Myanmar, and Thailand) and southwest Pacific (Philippines, Moluccas, and Papua New Guinea), requiring annual dry season, from sea levels up to 400 m.

5.5.6 *Abelmoschus manihot* ssp. *tetraphyllus* var. *pungens* (Roxb.) Hochr.

It is distributed in Indonesia, Philippines, Christmas Islands, Japan, China, India, Nepal, the Indo-Chinese Peninsula, and Myanmar, in forest glades and in hilly slopes, between 400 and 3,000 masl. It is tolerant to low temperature.

5.5.7 *Abelmoschus manihot* ssp. *tetraphyllus* var. *megaspermus* Hemadri

It is reported only from Indian states of Maharashtra, Gujarat, and Madhya Pradesh on shady hill slopes and foot hills.

5.5.8 *Abelmoschus moschatus* Medikus

Polymorphic species, cultivated and wild spp. *moschatus* seems to have pantropical distribution (several South American floras mention its occurrence); the other forms are confined to Asia and northern Australia. *A. moschatus* ssp. *tuberosus* (Span) Borss. has very restricted distribution in India and is confined to midlands of Kerala. In Western Ghats, it is confined up to 800 masl elevations, whereas *A. moschatus* ssp. *moschatus* is widely distributed and occurs both in cultivated and wild form. Bates (1968) proposes to elevate *A. moschatus* spp. *tuberosus* to specific ranking (correct name: *A. rugosus* Wall. Ex W. & A.). Some forms of *A. moschatus* were also described as *Hibiscus coccineus* Walter ($2n = 38$). However, *H. coccineus* cannot be maintained as a synonym of *A. moschatus* ssp. *tuberosus* (Hamon 1987), and therefore the hypothesis that *Abelmoschus* spp. *moschatus* ($2n = 72$) might be a polyploidy derived from *A. moschatus* ssp. *tuberosus* is highly speculative. Hamon (1987) proposed to elevate *A. moschatus* ssp. *moschatus* var. *betulifolius* to specific ranking (*A. betulifolius*) on the hypothesis that its morphological characteristics suggest it to be an amphidiploid between *A. moschatus* and *A. manihot*. Limited chromosomal studies conducted so far indicate somatic chromosome number of this species to be $2n = 72$.

5.5.9 *Abelmoschus tuberculatus* Pal and Singh

Abelmoschus tuberculatus is endemic to India and is the close relative of cultivated okra, *A. esculentus*. It is widely distributed in semi-arid regions of north and northwestern parts of India including Uttar Pradesh, Rajasthan, Madhya Pradesh, Maharashtra, Gujarat, and parts of Andhra Pradesh. The somatic chromosome number of this species has been reported to be $2n = 58$ (Pal, Singh, and Swarup 1952).

5.5.10 *Abelmoschus esculentus* (L.) Moench.

Cultigen (2*n* = 72 − 108 − 144) is of uncertain origin, but with pan (sub)tropical distribution. Most probably an amphiploid of *A. tuberculatus* (2*n* = 58) and *A. ficulneus* (2*n* = 72). Involvement of *A. esculentus* (2*n* = 72) race in the cultivar evolution cannot be completely discarded.

5.5.11 *Abelmoschus caillei* (A. Chev.) Stevels

This second edible okra species has a distribution limited to West and central Africa. Based on chromosome numbers and morphological characteristics, Siemonsma (1982a, 1982b) put forward the hypothesis that *A. caillei* might be an amphiploid of *A. esculentus* (2*n* = 130) and *A. manihot* (2*n* = 60–68). Production of fertile amphiploid "Nori-Asa" between these two species, according to Kuwada (1957b,1961), resembles *A. caillei* in morphological characteristics and crossing behavior supporting the hypothesis. However, the apparent absence of *A. manihot* in West Africa nowadays is not fully in favor of this hypothesis. *Abelmoschus caillei* cultivars in Guinea show characteristics (indumentum of seed and fruit) that are derived from *A. manihot* (Hamon 1987).

The wild species (Figure 5.1) occupy diverse habitats in India. *Abelmoschus ficulneus* and *A. tuberculatus* in semi-arid areas in north and northwestern India; *A. crinitus* and *A. manihot* (*tetraphyllus* and *pungens* types) in *tarai* range and lower Himalayas; *A. manihot* (*tetraphyllus* types), *A. angulosus*, and *A. moschatus* in western and eastern Ghats; and *A. crinitus* and *A. manihot* (mostly *pungens* types) in the northeastern region depicts their broad range of distribution in different phytogeographical regions of the country. Intra- as well as interspecific variations do exist in different phyto-geographic areas.

Recently, Mishra and Rawat (2000a, 2000b) reported certain *Abelmoschus* species from Western Himalayas and Uttaranchal (tarai belt), Haryana and Rajasthan, that were not previously documented. Different *Abelmoschus* species with variants were collected even from beyond the established areas of their occurrence. Existence of these species in different areas in India observed in a recent survey is presented in Table 5.4.

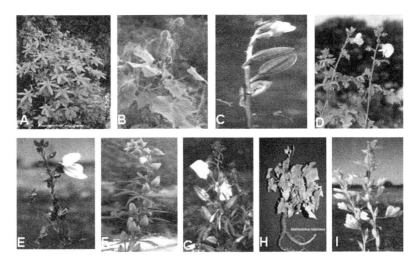

Figure 5.1   (A–I) Wild Abelmoschus species occurring in India. (A) *A. angulosus*, (B) *A. crinitus*, (C) *A. ficulneus*, (D) *A. manihot* ssp. *manihot*, (E) A. *manihot* ssp. *tetraphyllus* var. *tetraphyllus*, (F) *A. manihot* ssp. *tetraphyllus* var. *pungens*, (G) *A. moschatus* ssp. *moschatus*, (H) A. *moschatus* ssp. *tuberosus*, (I) *A. tuberculatus*.

Table 5.4   Distribution of Wild *Abelmoschus* Species in Different Phytogeographical Regions of India Based on a Recent Survey

S. No.	Species	Distribution
1.	A. angulosus	Tamil Nadu, Kerala
2.	A. cancellatus?	Uttaranchal, Himachal Pradesh, Uttar Pradesh, Orissa
3.	A. criniturs	Uttaranchal. Madhya Pradesh, Orissa
4.	A. ficulneus	Jammu & Kashmir, Rajasthan, Madhya Pradesh, Chhatisgarh, Maharashtra, Tamil Nadu, Andhra Pradesh, Uttar Pradesh
5.	A. manihot ssp. tetraphyllus var. tetraphyllus	Uttar Pradesh, Rajasthan, Madhya Pradesh, Maharashtra, Orissa, Chhatisgarh
6.	A. manihot ssp. tetraphyllus var. pungens	Uttaranchal, Himachal Pradesh, Jammu & Kashmir, Assam, Andaman & Nicobar islands
7.	A. moschatus ssp. moschatus	Uttaranchal, Orissa, Kerala, Karnataka, Andaman & Nicobar islands
8.	A. moschatus ssp. tuberosus	Kerala and parts of Western Ghats in Tamil Nadu
9.	A. tuberculatus	Uttar Pradesh, Rajasthan, Madhya Pradesh, Maharashtra

5.6   CYTOGENETICS

5.6.1   Cytogenetic Relationship among Taxa

On the basis of cytogenetic studies, affinities between cultivated okra and related wild taxa have been determined. Joshi and Hardas (1956) and Joshi, Gadwal, and Hardas (1974) studied meiosis in hybrids obtained by crossing *A. esculentus* ($n=65$) and *A. tuberculatus* ($n=29$). They observed that 29 of the 65 chromosomes of *A. esculentus* had complete homology with 29 chromosomes of *A. tuberculatus*. The remaining set of 36 chromosomes (genome Y) of *A. esculentus* showed greater (but still incomplete) homology with 36 chromosomes of *A. ficulneus* as compared to those of *A. moschatus*. It was concluded that one of the parents of *A. esculentus* ($n=65$) should have been *A. tuberculatus* ($n=29$). However, the other genome of 36 chromosomes that was possibly unlike either of the two Indian species, namely *A. ficulneus* and *A. moschatus*, could not be established. These studies established that cultivated okra was an amphidiploid (29 T + 36 Y). Another group of polyploidy species showing genetic affinity includes *A. esculentus*, *A. tetraphyllus*, and *A. pungens*. The latter two were also found to behave like amphidiploids. Joshi, Gadwal, and Hardas (1974) have summarized the observed chromosomal homologies among different species.

Another species of interest to India is the "Ghana" okra (*A. manihot* ssp. *manihot*), introduced by the NBPGR. Chromosomal status of this West African okra was studied by Nerkar and his group at Marathwada Agricultural University (MAU); Parbhani and crosses were also made between *A. manihot* ssp. *manihot* and *A. tetraphyllus*. Two experimentally sysnthesized amphiploids, namely *A. esculentus-manihot* and *A. esculentus-tetraphyllus*, were also studied for their meiotic behavior.

Amphidiploids using wild forms were also developed while studying the interrelationship between different *Abelmoschus* species (*A. esculentus*×*A. manihot*, *A. esculentus*×*A. tuberculatus*, *A. manihot*×*A. tuberculatus*, *A. esculentus*×*A. tetraphyllus*, *A. ficulneus*×*A. tuberculatus*, and *A. esculentus*×*A. manihot* ssp. *manihot*), but these were not stable (Jambhale and Nerkar 1981b). The F_1s were largely sterile but fertility could be restored to varying degrees through polyploidisation or backcrossing. The existence of natural amphiploids of *A. esculentuts* and *A. manihot* was also been reported (Siemonsma 1982a). However, Kuwada (1966) was able to produce fertile artificial amphidiploid ($2n=192$) of *A. esculentus* ($2n=124$)×*A. manihot* ($2n=68$).

Siemonsma (1991) presented in greater detail the taxonomical and cytogenetical overview of *Abelmoschus* spp.

5.6.2   Interspecific Hybridization

5.6.2.1   Between Ploidy Level 1 Species

Four of the six ploidy levels in one species have been subjected to interspecific crosses. The results are summarized in Table 5.5. Viable seed was only obtained in the crosses between *A. manihot* $2n=68$ and *A. tuberculatus* $2n=58$, and between *A. ficulneus* $2n=72$ and *A. tuberculatus* $2n=58$. Resulting plants were sterile. Study of the meiosis of the hybrids showed very little affinity between the genomes of the parent species (Joshi and Hardas 1956; Joshi, Gadwal, and Hardas 1974; Kuwada 1974).

The artificial amphiploid $2n=130$ between *A. ficulneus* and *A. tuberculatus* (reconstruction of *A. esculentus* $2n=130$), realized by Joshi, Gadwal, and Hardas (1974) was genetically unbalanced and completely sterile.

5.6.2.2   Between Ploidy Level 1 and Ploidy Level 2 Species

In the search for the parental species of *A. esculentus*, a large number of interspecific crosses have been performed (Table 5.6). Study of the meiosis of hybrids between *A. esculentus* $2n=124$, 130 and *A. tuberculatus* $2n=58$ revealed the almost perfect pairing of the genome of *A. tuberculatus* with 29 chromosomes of *A. esculentus* $n=62$, 65. *A. tuberculatus* is generally accepted as one of the ancestors of *A. esculentus*, apparently an amphiploid. Concerning the complementary genome, considerable but incomplete pairing was observed with *A. ficulneus*. The possibility of an *A. esculentus* (?) race $2n=72$, as reported by Teshima (1933), Ugale, Patil, and Khipse (1976), and Kamalova (1977) should not be completely discarded. It would constitute the most logical source of the missing genome $n=36$. Kuwada (1957b, 1961) obtained a fertile artificial amphiploid $2n=192$ between *A. esculentus* $2n=124$ and *A. manihot* $2n=68$, called "Nori-Asa."

Crosses between ploidy level 1 species and *A. tetraphyllus* are less documented. Pal, Singh, and Swarup 1952 obtained sterile hybrids in crosses between different forms of *A. manihot*. No viable hybrid seed was obtained by Hamon and Yapo (1986) in crosses between *A. tetraphyllus* and *A. manihot* and *A. moschatus*. Ugale, Patil, and Khupse (1976) reported on the hybridization of *A. esculentus* ($2n=72$) and *A. tetraphyllus* ($2n=130$). Almost perfect pairing of the genome of *A. esculentus* with 36 chromosomes of the other species was observed.

Table 5.5   Results of Crosses between Ploidy-Level 1 Species (Positive = Viable Hybrid Seed)

	A. manihot	*A. moschatus*	*A. ficulneus*	*A. tuberculatus*
A. manihot		Negative (Skovsted 1935; Hamon and Yapo 1986)	Negative (Pal et al. 1952)	Positive (Pal et al. 1952), Negative (Kuwada 1974)
A. moschatus	Negative (Skovsted 1935; Hamon and Yapo 1986)		Negative (Gadwal et al. 1968)	
A. ficulneus	Negative (Pal et al. 1952)	Negative (Gadwal et al. 1968)		Positive (Joshi and Hardas 1956; Joshi et al. 1974)
A. tuberculatus	Positive (Pal et al. 1952; Kuwada 1974)	Negative (Gadwal et al. 1968)	Positive (Joshi and Hardas 1956; Joshi et al. 1974)	

Source: From Charrier, A., Genetic resources of the genus *Abelmoschus* Med. (Okra), IBPGR, Rome, 1984; Siemonsma, J. S. *International Crop Network Series*. Report of an international workshop on okra genetic resources, IBPGR, Rome, 5, 52–68 1991.

Table 5.6 Results of Crosses between *Abelmoschus esculentus* and Ploidy Level 1 Species (Positive= Viable Seed)

Cross A. esculentus X	Chromosome Numbers	Authors	Indicated Cross	Reciprocal Cross	Bivalents in Meiosis
A. tuberculatus		Pal, et al.	Positive	Positive	
	130×58	Joshi and Hardas (1956) and Joshi et al. (1974)	Positive	Positive	28.8 (2.29)
	124×58	Kuwada (1966)	Positive	Positive	27–29
A. manihot	72×60	Teshima (1933)	Positive	Negative	0
	(126–134)×60	Chizaki (1934)	Positive		0–7
	100)(00	Okovsted (1935)	Positive	Positive	
		Ustinova (1937)	Positive	Negative	
		Singh et al. (1938)[a]	Positive		
		Ustinova (1949)	Positive	Negative	
		Pal et al. (1952)	Positive	Positive	
	124×68	Kuwada (1957a)	Positive	Positive	7
		Hamon and Yapo (1986)	Negative	Negative	
A. ficulneus		Pal, Singh, and Swarup (1952)	Negative	Negative	
	130×72	Gadwal et al. (1968) and Joshi et al. (1974)	Negative		27.5 (26–28)[b]
A. moschatus	130×72	Skovsted (1935)	Positive	Negative	
	130×72	Gadwal et al. (1968) and Joshi et al. (1974)	Negative		8.3 (3–16)[b]
		Hamon and Yapo (1986)	Positive	Negative	

[a] The article deals with a cross between *A. esculentus* and *A. ficulneus*, but the description of the latter species corresponds to *A. manihot*.
[b] Hybrids obtained by embryo- and/or ovule-culture.
Source: From Charrier, A., Genetic resources of the genus *Abelmoschus* Med. (Okra), IBPGR, Rome, 1984; Siemonsma, J. S. *International Crop Network Series*. Report of an international workshop on okra genetic resources, IBPGR, Rome, 5, 52–68 1991.

5.6.2.3 Between Ploidy Level 2 Species

Viable seed, but sterile hybrids, in the cross *A. esculentus*×*A. tetraphyllus*, were reported by Gadwal (cf. Joshi and Hardas 1976) and Hamon and Yapo (1986). No data on genome affinity were presented. Artificial and spontaneous amphiploids between these two species have been realized in India in attempts to transfer YVMV resistance to cultivated okra (Jambhale and Nerkar 1981a, 1981b).

5.6.2.4 With the Ploidy Level 3 Species

Hybridization between *A. caillei* and *A. manihot* (Asian origin) did yield hybrid seed (Siemonsma 1982a, 1982b), although it germinated poorly and showed abnormal growth. Similar results were reported by Jambhale and Nerkar (1981a) and Hamon and Yapo (1986).

Crosses between *A. esculentus* and *A. caillei* produced viable hybrids with strongly reduced fertility (Singh and Bhatnagar 1975; Joshi and Hardas 1976; Siemonsma 1982a, 1982b; Hamon and Yapo 1986; Hamon 1987). Hamon and Yapo (1986) reported viable and sterile hybrids in the cross *A. caillei*×*A. tetraphyllus*. No data on genome affinity were presented. It is interesting to note that the fertile artificial amphiploid "Nori-Asa" [between *A. esculentus* ($2n=124$) and *A. manihot* ($2n=68$)] showed morphological characteristics and crossing behavior similar to *A. caillei* (Kuwada 1957b, 1961; Siemonsma, 1982b). Cytogenetical relations in the genus *Abelmoschus* as described by Charrier (1984) are summarized in Figure 5.2.

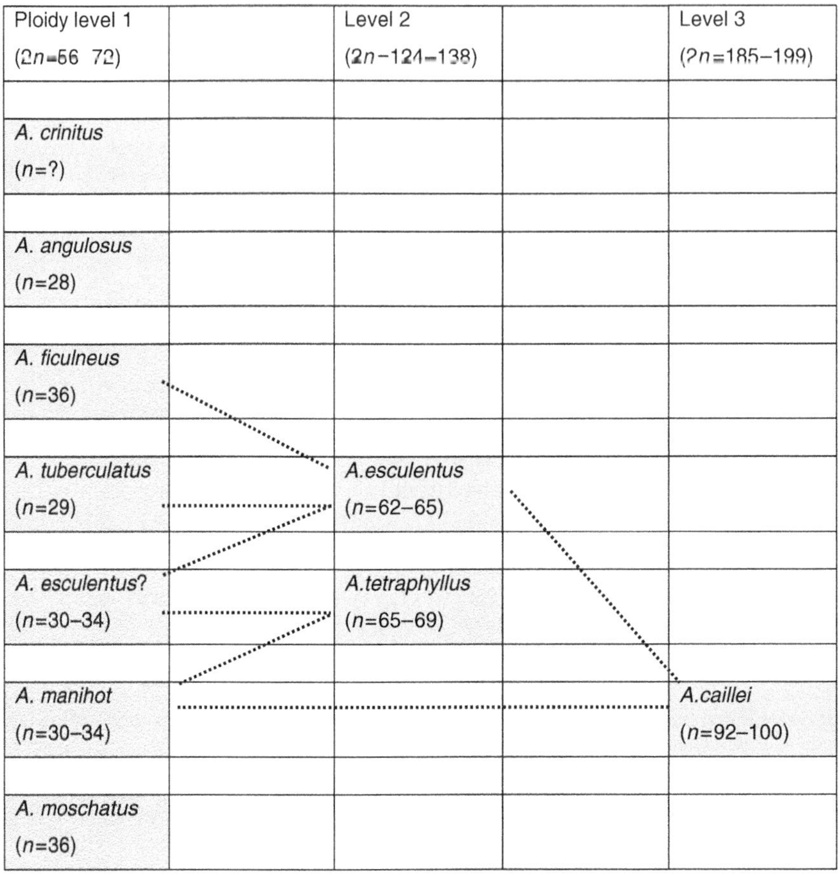

Ploidy level 1 ($2n=66$ 72)		Level 2 ($2n-124-138$)		Level 3 ($?n=185-199$)
A. crinitus ($n=?$)				
A. angulosus ($n=28$)				
A. ficulneus ($n=36$)				
A. tuberculatus ($n=29$)		A.esculentus ($n=62-65$)		
A. esculentus? ($n=30-34$)		A.tetraphyllus ($n=65-69$)		
A. manihot ($n=30-34$)				A.caillei ($n=92-100$)
A. moschatus ($n=36$)				

Figure 5.2 Cytogenetical relations in *Abelmoschus*. (From Charrier, A., Genetic resources of the genus *Abelmoschus* Med. (Okra), IBPGR, Rome, 1984; Siemonsma, J. S. *International Crop Network Series*. Report of an international workshop on okra genetic resources, IBPGR, Rome, 5, 52–68 1991.)

5.7 GENETICS

Studies concerning genetics of different traits in okra are limited. Ustinova (1949) reported that yellow of the corolla, a characteristic of *A. manihot*, predominated over the cream of *A. esculentus*. According to Venkitaramani (1952), dark-green plants were dominant over light green, while green was recessive to greenish-red. Characteristics such as petal blotch, intense stem pigmentation; and faint petiole pigmentation showed simple digenic segregation (Erickson and Couto 1963). Monogenic control of pigmentation of calyx, corolla, and fruit was inferred by Kolhe and D'Cruz (1966). The dominance of cut leaves over lobed leaves and white fruit over green was reported by Jasim (1967). Pod spininess was governed by a single gene while pod shape (angular vs. round) was digenic with epistatic effects. Jasim and Fontenot (1967) observed that leaf shape, pod; and spiny pods were governed by single genes. Fruit pubescense and leaf lobing were reported to be controlled by a single recessive gene, hirsute seed by four dominant genes, hirsute hilum by one major gene, and a second gene controlling hilum trichome number if the major allele was absent; smooth testa was found in one major gene, and in a second gene controlling testa trichome number if the major allele was absent. There was a pleiotropic or modifying relationship between the two pairs of genes because certain phenotypes were not found in the F_2.

Malik (1968) reported high heritability and genetic advance for fruit diameter, fruit length, and contents of crude fiber, total sugars, and vitamin C. High heritability and genetic advance were also observed for plant height and number of days to flowering by Rao (1972). Rao (1977) reported additive genetic effects for days to flower, dominance for plant height, and additive and dominance both for number of fruits per plant. Additive genetic effects for days to flower, number of pods and yield, and nonadditive effects for plant height and number of seeds per pod were also reported by Rao and Ramu (1977). Number of fruits had a predominance of additive genetic effects, whereas days to flower and plant height were controlled by nonadditive gene effects (Kulkarni, Rao, and Virupakshappa 1976, 1978). Both additive and nonadditive components of variance were reported to be high for plant height, number of fruits, and yield per plant (Ramu 1976). Additive gene effects for number of pods and nonadditive for days to flowering and plant height have also been reported. Dominance of additive gene action was observed for number of pods by Kulkarni and Thimmappaiah (1977). Both additive and dominant gene effects were observed for days to anthesis, fruit number, plant height, days to first flowering and marketable maturity, number of fruits per plant, fruit length, weight and diameter, number of ridges per fruit, plant height, and yield (Sharma and Mahajan 1978). Additive gene effects were indicated by Partap and Dhankhar (1980) for fruit diameter, length, weight, and yield. Similarly, Arora (1980) also reported the presence of additive gene effects for days to first flowering, length and weight of fruit, seed weight, number of ridges, fruit per plant, nodes per plant, days to marketable maturity, total yield, and protein content. Predominance of additive gene effects for fruit length, diameter and weight, number of ridges, and internodal length, and also dominance for days to flowering and marketable maturity, were observed by Singh (1979). Additive gene effects were observed for days to flower, fruit weight and length, number of ridges, internodal length, and plant height (Singh 1983). According to Randhawa and Sharma (1988), heritability and genetic advance were high for number of fruits and branches per plant in F_3, indicating the possibility of following selection in the early generation with developing plants that contain a higher number of fruits.

In an intraspecific cross involving *A. esculentus* and Guinean strains from Africa and Japan, Arumugam and Muthukrishnan (1979) inferred the presence of additive and dominant gene effects for plant height, number of branches, days to flowering, and leaf index. According to Randhawa (1989), yield displayed overdominance, while other traits showed partial dominance. Additive gene effects were predominant for all the horticultural traits. It was suggested that selection for high yield should be made in early generations.

While studying interrelationships among different economic characters, Mahajan and Sharma (1979) found that fruit yield was positively correlated with plant height, number of fruits per plant, and fruit length, in both parents and hybrids. Thus, selection based on number of fruits per plant, fruit length, and diameter would be effective for increasing yield. According to Singh and Sharma (1983), a shift of association from one generation to another and heterogeneity of correlations of a few traits indicate the possibility of obtaining desirable recombinants in biparentals rather than in F_2. They also observed realization of higher vigor in F_1 than in the F_2 hybrids for early maturity, number of pods, pod weight, and yield, possibly due to breakage of both coupling and repulsion-phase linkages as indicated by the changes in the phenotypic correlation matrix (Singh and Sharma 1983). Correlation coefficients were found to vary between seasons by Ariyo, Akenova, and Fatokun (1987). It was observed that edible pod weight had the greatest positive effect on pod yield in both early and late season. Although pod weight was the best index of yield because pod weight cannot be assessed visually in the field, edible pod width may be a better criterion for selection in view of its close association with pod weight.

According to Sharma, Kumar, and Bajaj (1981), yellow vein mosaic resistant parent "Ghana" and its F_1 with Pusa Sawani contained higher levels of moisture, orthodihydroxyphenols, and total chlorophyll content than susceptible cultivars. The overdominance of biochemical constituents was indicated for all the traits except total cholorophyll. Heterosis over the better parent was manifested for phenolic compounds, orthodihydroxyphenols, and moisture content. Phenolic compounds were also positively associated with total chlorophyll in the resistant parent and F_1. However, no significant differences between susceptible and resistant cultivars were observed for the levels of moisture, chlorophyll content, total sugars, and nonreducing sugars (Ahmed, Thakur, and Bajaj 1987).

5.8 GENETIC RESOURCES

5.8.1 Germplasm Exploration and Collection

An extensive collection of *Abelmoschus* was made by the NBPGR in New Delhi, India. This included the introduction of a number of wild species from Africa. Explorations were conducted in parts of Brazil and Africa (Nigeria, Ghana) for diverse germplasm. From the Ivory Coast, an intensive collection was made that has been discussed by Siemonsma (1982b). The germplasm collection in the United States is generally introduced from Asia and Africa. In 1946, work on okra germplasm exploration and collection in India was initiated at the Indian (then Imperial) Agricultural Research Institute, New Delhi. From 1970 onwards, germplasm acquisition through indigenous collections and exotic sources was strengthened. About 1,000 accessions were acquired and maintained at the NBPGR until 1989. An International Board for Plant Genetic Resources [(IBPGR), now the International Plant Genetic Resources Institute (IPGRI)], funded project on exploration and collection of okra and its wild relatives, and it was operated at the NBPGR from 1989 to 1992. Explorations were undertaken in all parts of the country. In all, 31 explorations were conducted including three major ones in neighbouring countries, namely Bangladesh, Nepal, and Sri Lanka. A total of 1,184 accessions of cultivated okra (*A. esculentus*) and 613 accessions of wild *Abelmoschus* species were collected. The cultivated types from southern India were predominantly thick, short, and multi-edged. The collections from the northwestern plains, especially from Rajasthan and adjoining areas, were the predominantly hairy and multi-edged type with tolerance to insect-pests and diseases. Accessions from eastern regions included some primitive landraces having field resistance to insect-pests and diseases. Accessions from the northeastern region possessed both five-edged and multi-edged types. More than 500 accessions from northern and central India, mainly from tribal-dominated belts, were collected, and these exhibited wide variability for multiple traits. Wild *Abelmoschus* species were also collected from across the country. These included mainly by *A. tuberculatus*, *A. manihot* ssp. *tetraphyllus*, and *A. ficulneus*. Other species with more restricted distributions included *A. angulosus* (mainly from southern region and higher hills of Western Ghats), *A. crinitus* (Orissa, West Bengal and eastern Ghats), and *A. moschatus* (northern plains). In recent exploration missions, different *Abelmoschus* species with variants were collected even outside the established areas of their occurrence (Mishra and Rawat 2000a).

In India, exotic okra germplasm have been introduced from various countries since 1949. Initially, introductions were mainly commercial varieties that were promoted for cultivation after evaluating their response to local growing conditions. Such varieties were Clemson's spineless, Perkins's long green, smooth long, velvet green, and white velvet. These introductions were made through United States Department of Agriculture (USDA). Later, one collection of *A. manihot* ssp. *manihot* from Ghana received through the USDA was observed to have a high degree of resistance to yellow vein mosaic virus (YVMV). This species had broad leaves, perennial growth habit, thick stocky fruits, and a long fruiting season. It has been used by various breeders in developing YVMV-resistant okra. Some of the distinct types (Guinnean) of West African okra have also been introduced. Over the years introductions have also been made from countries such as Brazil, Turkey, Mexico, earstwhile USSR, U.K., Belgium, France, Japan, Singapore, Philippines, Nepal, Bangladesh, Pakistan, Sri Lanka, Indonesia, Nigeria, Ghana, Senegal, Sudan, Zambia, Malawi, Ivory Coast, etc. A total of 148 exotic germplasm accessions from Nigeria, 150 from Brazil, and 177 of the World Okra Core Collection were obtained by NBPGR through IPGRI.

5.8.2   Germplasm Evaluation

5.8.2.1   *Morphological Characterization*

The primary objectives of okra germplasm characterization have typically been to identify high-yielding genotypes with resistance to YVMV, fruit borer, jassid, and higher vitamin C content in the wild species that can be utilized for the improvement of *A. esculentus*. Martin et al. (1981) evaluated 585 varieties and Siemonsma (1982a) assessed 314 varieties of okra for yield-related characteristics. Okra germplasm consisting of 94 lines (Sandhu et al. 1974) and 74 genotypes (Sharma and Sharma 1984) was screened against field resistance to yellow vein mosaic virus. Great variability with respect to fruit shape was observed in 718 samples of *A. esculentus*, and except for *A. moschatus*, the risk of genetic erosion seemed slight (Hamon and Charrier 1983).

At NBPGR in New Delhi, the evaluation of germplasm collected from different locations continues. An on-going goal in the evaluation of the okra germplasm is to find sources of resistance to YVMV, which is the most severe problem of okra. The wild species *A. manihot* showed complete resistance to YVMV and the variety "Ghana Red" also showed complete immunity to this disease. The primitive spiny types "Best-1" and "IC-7174" were promising sources of insect and disease resistance. *A. tetraphyllus* var. *pungens* and *A. crinitus* were found resistant to yellow vein mosaic disease. The description of 224 samples of *A. esculentus* for 11 characteristics has been listed in the USDA catalogue. The main characteristics are phenology, plant type, fruits, and resistance to pests and diseases. The critical assessment of 314 samples has been reported by Siemonsma (1982a) for earliness of flowering, plant height, dimension, weight of fruit, yield of fresh fruits and seeds, and susceptibility to pests and diseases.

Over 1,800 accessions of *A. esculentus* maintained at the NBPGR have been systematically characterized for more than 40 morphological descriptors, both qualitative and quantitative. Frequency distributions of qualitative traits, and mean and range of expression of some important quantitative traits were recorded (Thomas et al. 1990, 1991; Bisht et al. 1993). The pigmentation and pubescence of stem, leaf, fruit, and seed were important components of variability in the germplasm. Some of the qualitative descriptors such as red leaf vein, epicalyx number, shape, size and persistence, red coloration at the petal base, position of fruit on main stem, and immature/mature fruit color expressed low variability. The accessions were predominantly erect in growth habit with traits such as low-branching potential, green leaves, epicalyx segments more than 8, linear in shape, nonpersistent, and fruits green, 5-edged, erect on stem. These traits were, however, of great significance in characterizing the West African taxon, *A. caillei* (Hamon and van Sloten 1989). Accessions with robust plant type, thick stem, short internode, considerable amount of anthocyanin in stems and leaves, day length sensitivity, reduced number of epicalyx segments, extremely wide and short fruits, fruits mounted at an angle to the stem (rather than parallel to the stem as in *A. esculentus*), large and showy flowers, and large pollen similar to the West African taxon (*A. callei*) naturalized in parts of Western Ghats in southern India. Their occurrence may be ascribed to the probable outcrossing between *A. esculentus* and *A. caillei*, as the latter is naturalized in these areas (Rana et al. 1994). The range of variation for quantitative traits is presented in Table 5.7. The observed pattern of variability revealed that more than 60% of the accessions in the collection resembled with Pusa Sawani in plant habit, inflorescence, and fruit characteristics. This was attributed to frequent outcrossing between landraces and improved types (mainly Pusa Sawani). Pusa Sawani has been in commercial cultivation in India for the last three to four decades and is still popular in India and other countries, leading to significant introgression with and genetic erosion of traditional landraces.

A detailed study was performed on a representative set of 260 okra germplasm accessions at the NBPGR (Bisht, Mahajan, and Rana 1995b), representing all agro-ecological areas of the country. Table 5.8 lists the range of variation for various morphological traits of the above representative set (Figure 5.3A and Figure 5.3B).

Table 5.7   Range of Variation for Quantitative and Qualitative Characters of 1628 Okra Accessions
Studied at the NBPGR

Characters	Min.	Max.	Mean	CV (%)
Quantitative traits				
Plant height (cm)	5.7	226.0	58.7	47.0
No. of internodes	2.0	33.5	12.1	32.7
Leaf length	2.3	33.0	14.0	3.6
Leaf width	1.3	41.6	12.0	30.0
Days to flowering	42.0	128.0	64.2	10.6
Days to maturity	81.0	154.0	106.4	7.7
First fruiting node	4.0	24.0	7.4	48.4
Fruits on main stem	1.0	22.0	3.4	52.4
Fruit length (cm)	1.7	24.3	12.1	36.7
Fruit width (cm)	1.0	7.8	2.2	97.0
Fruits/plant	1.0	134.5	6.7	15.9
Seeds/fruit	2.0	102.0	37.1	6.1
Qualitative traits				
Branching habit	Profuse 19.2%; low: 80.8%			
Stem pubescence	Glabrous: 46.5%; slight: 51.2%; conspicuous: 2.9%			
Stem color	Green: 77.7%; green with red patches: 21.6%; purple:2.9%			
Leaf color	Green: 85.7%; green with red veins: 12.0%; red: 2.3%			
Fruit pubescence	Downy: 56.5%; slightly rough: 41.2%; prickly: 2.3%			
Seed surface	Downy:48.1%; glabrous: 51.9%			

Min., minimum, Max., maximum, *CV*, coefficient of variation.
Sources: From Thomas, T. A. et al., Catalogue on okra (*Abelmoschus esculentus* (L.) Moench) germplasm, Part I. NBPGR, New Delhi, 1990; Thomas, T. A., Catalogue on okra (*Abelmoschus esculentus* (L.) Moench) germplasm, Part II. NBPGR, New Delhi, 1991; Bisht, I. S. et al., Catalogue of okra (*Abelmoschus esculentus* (L.) Moench) germplasm. Part III. NBPGR, New Delhi, 1993.

Correlations between pairs of quantitative variables were also recorded (Table 5.9). Plant height was correlated with number of internodes and various fruit characteristics such as number of ridges/fruit, fruits on main stem, fruit length, and number of fruits/plant. The strongest and most persistent correlations were between flowering and fruiting parameters, such as size and weight and seeds/fruit. Close correlations between various fruiting parameters were also determined.

A multivariate analysis (Bisht, Mahajan, and Rana 1995b) was also performed on this representative set of 260 accessions. Principal component analysis (PCA) revealed that days to flowering, plant height, and various fruit characteristics were important components of variability in these accessions. Cluster analysis revealed the relative contribution of the various quantitative parameters to the total variability. A moderately high correlation between fruit length and number of fruits on the main stem indicated the possibility of selecting highly prolific types with longer fruits. A negative correlation between fruit length and days to flowering suggested the preponderance of a favorable, negative selection pressure for duration of the crop, while simultaneously selecting for prolific and long-podded types. A correlation between other quantitative parameters also indicated the possibility of effective selection for alternate phenotypes for either of the linked pair of characters. Earliness, plant height, fruit length, first fruiting node, and other fruit characteristics have also been reported to be important components of heterosis for yield (Singh, Srivastava, and Singh 1975; Singh and Singh 1979; Elangovan, Muthukrishnan, and Irulappan 1981). Mahajan and Sharma (1979) also found a positive and significant association of fruit yield with plant height, number of fruits/plant, and fruit length, in both parents and hybrids.

Table 5.8   Mean, Range, and *CV* for Various Quantitative Characters of 260 Representative Accessions (core set) of Okra at the NBPGR

Characters	Min.	Max.	Mean	*CV* (%)
Plant height (cm)	5.7	202.5	96.0	33.4
No. of internodes	5.0	33.5	16.5	23.6
Days to flowering	40.0	98.0	54.0	24.1
First fruiting node	2.0	24.0	8.6	46.5
Fruits on main stem	1.0	22.0	6.5	63.1
Ridges/fruit	5.0	8.0	5.6	14.2
Fruit length (cm)	3.2	23.3	12.3	33.6
Fruit width (cm)	1.2	4.7	3.5	17.1
Fruits/plant	1.0	134.5	14.3	93.6
Seeds/fruit	10.0	102.0	45.5	41.1
Branching habit	Profuse 23.2%; low: 76.8%			
Stem pubescence	Glabrous: 45.0%; slight: 48.2%; conspicuous: 6.8%			
Stem color	Green: 69.0%; green with red patches: 25.7%; purple:5.3%			
Leaf color	Green: 79.0%; green with red veins: 16.5.0%; red: 4.5%			
Fruit pubescence	Downy: 51.3%; slightly rough: 43.2%; prickly: 5.5%			
Seed surface	Downy:45.1%; glabrous: 54.9%			

Min., minimum, Max., maximum, *CV*, coefficient of variation.
Source: From Bisht, I. S., Mahajan, R. K. and Rana, R. S. *Ann. Appl. Biol.*, 126, 539–550, 1995.

Germplasm characterization provides useful information for germplasm collectors in their efforts to designing future collection strategies.

Days to flowering has been described to be important in discriminating varieties of *A. esculentus* (Ariyo and Odulaja 1991). Sneath and Sokal (1973) also emphasized flowering behavior in numerical taxonomy. The accessions of okra collected from India that flowered within 40–90 days were mostly photoperiod insensitive. Flower initiation and flowering were hardly affected by day length. A few photo-sensitive accessions, however, also existed, probably due to the natural out-crossing of *A. esculentus* with *A. callei* (Rana et al. 1994). Further, pigmentation of various plant aprts of the accessions and fruit characteristics contributed significantly to the total variation explained in a set of accessions of West African Okra, *A. caillei*, studied by Ariyo (1993).

Detailed characterization of 241 accessions comprised of *A. ficulneus* (68 accessions), *A. manihot* ssp. *tetraphyllus* (81 accessions), *A. moschatus* (7 accessions), and *A. tuberculatus* (85 accessions), was also undertaken at the NBPGR. Diversity in various qualitative and quantitative traits was studied (Bisht, Mahajan, and Rana 1995a, 1997). Diversity for a range of morphological characters, namely epicalyx segments, shape, size and persistence, petal color and petal blotch, stem and fruit pubescence, pigmentation of various plant parts, days to flowering, plant height, and the fruit yield/plant were important components of variation and contributed significantly to the total variation in germplasm accessions of the above four species studied. In *A. ficulneus* much of the within species variation was recorded for traits such as epicalyx segments (3–12), epicalyx length (5–15 mm), fruit color, position of fruit on main stem, fruit pubescence, plant height (12–75 cm), days to maturity (85–173), and fruits/plant (22–131). Much of the variation in *A. manihot* ssp. *tetraphyllus* accessions were recorded for plant height (38–109 cm), days to maturity (103–193), and number of fruits/plant (19–121). Within species variations for *A. moschatus*, they also recorded for traits such as plant height (81–184 cm), days to maturity (130–175), number of epicalyx segments (5–10), fruits/plant (15–86), and seeds/fruit (31–70). Wide variation was recorded in *A. tuberculatus* accessions with regard to plant height (31–135 cm), stem pubescence, stem color, epicalyx shape, epicalyx persistence, epicalyx

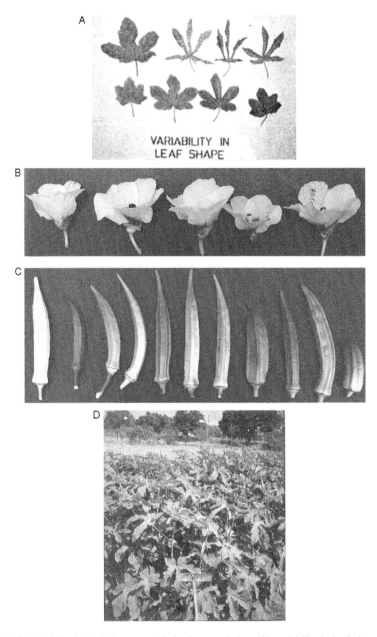

Figure 5.3   (A–D) Variability in leaf, flower and fruit characteristics. (A) variability in leaf shape and size, (B) variability in flower shape and size, (C) variability in fruit characteristics, (D) field view of a promising genotype TAT-1.

segments (4–12), epicalyx length (4–15 mm), number of internodes (6–32), days to maturity (86–151), fruit color, fruit pubescence, fruits/plant (10–98), and seeds/fruit (16–41). A comparative range of variation for the above wild *Abelmoschus* species is presented in Table 5.10.

Based on characterization data, three catalogues on cultivated okra containing information on a total of 1,560 accessions and one catalogue on 241 accessions of wild *Abelmoschus* species for 40 descriptors have been published by the NBPGR (Thomas et al. 1990, 1991; Bisht, Mahajan, and Rana 1993, 1995b). In these catalogues, IBPGR descriptors (Charrier 1984) with minor modifications were used.

Table 5.9 Correlation Matrix of Various Quantitative Variables

Variable	Ridges/Fruit	Plant Height	Internodes	Days to Flowering	First Fruiting Node	Fruits on Main Stem	Fruit Length	Fruit Width	No. of Fruits/Plant	No. of Seeds/Fruit
Ridges/fruit	1.00									
Plant height	−0.31[a]	1.00								
Internodes	−0.19[a]	0.36[a]	1.00							
Days to flowering	0.28[a]	−0.28[a]	−0.19[a]	1.00						
First fruiting node	0.11	−0.20[a]	−0.02	0.63[a]	1.00					
Fruits on main stem	−0.33[a]	0.64[a]	0.38[a]	−0.50[a]	−0.39[a]	1.00				
Fruit length	−0.35[a]	0.58[a]	0.37[a]	−0.57[a]	−0.39[a]	0.62[a]	1.00			
Fruit width	−0.08	0.15[b]	0.12	−0.52[a]	−0.31[a]	0.23[a]	0.41[a]	1.00		
No. of fruits/plant	−0.23[a]	0.47[a]	0.15[b]	−0.23[a]	−0.11	0.47[a]	0.39[a]	0.19[a]	1.00	
No. of seeds/fruit	−0.09	0.19[a]	0.07	−0.39[a]	−0.32[a]	0.28[a]	0.38[a]	0.31[a]	0.17	1.00

[a] $P < 0.01$.
[b] $P < 0.05$.

Source: From Bisht, I. S., Mahajan, R. K. and Rana, R. S. *Ann. Appl. Biol.*, 126, 539–550, 1995.

Table 5.10   Range of Variation for Morphological Traits of Wild *Abelmoschus* Species

Characters		*A. ficulneus*	*A. manihot* ssp. *tetraphyllus*	*A. moschatus*	*A. tuberculatus*
Quantitative traits					
Days to flowering	Range	40–143	75–149	101–135	49–113
	Mean	110.7	99.9	111.7	81.5
	CV	27.8	16.4	10.2	19.1
Days to maturity	Range	85–173	103–193	130–175	86–151
	Mean	131.7	136.1	144.0	117.4
	CV	26.9	14.6	10.58	13.2
Plant height	Range	12–75	38–109	81–184	31–135
	Mean	46.5	64.3	133.0	66.4
	CV	33.9	25.9	30.1	39.3
No. of epicalyx segments	Range	3–12	4–9	5–10	4.-12
	Mean	7.1	4.4	8.0	7.9
	CV	16.7	23.6	20.4	31.6
Fruiting nodes on main stem	Range	2–41	2–13	9–15	3–18
	Mean	27.4	5.1	12.4	8.5
	CV	11.3	34.2	15.9	38.8
Fruits/plant	Range	22–131	19–121	15–86	10–98
	Mean	78.1	56.9	63.1	33.9
	CV	45.2	42.6	41.6	76.1
Seeds/fruit	Range	19–154	17–41	31–70	16–41
	Mean	45.3	19.6	53.7	25.0
	CV	16.7	52.0	31.5	21.2
Qualitative traits (predominant descriptor states)					
Grwoth habit		Medium to procumbent	Medium to procumbent	Erect	Erect
Branching		Profuse	Profuse	Profuse	Low
Stem pubescence		Slight	Slight	Conspicuous	Conspicuous
Leaf lobing		5-lobed	5-lobed	7-lobed	5-lobed
Epicalyx shape		Linear	Triangular	Linear	Linear
Epicalyx persistence		Nonpersistent	Nonpersistent	Persistent	Partially persistent
Red coloration at petal base		Inside only	Both side	Both side	Inside only
Fruit pubescence		Slightly rough	Slightly rough	Slightly rough	Prickly
Seed shape		Round	Round	Round	Reniform

CV, coefficient of variation.
Source: From Bisht, I. S. et al., *Catalogue of Wild Abelmoschus Species Germplasm*, NBPGR, New Delhi, 1995.

5.8.2.2   Molecular Characterization

Genetic diversity in *A. esculentus* and four related species (*A. ficulneus*, *A. manihot* ssp. *tetraphyllus*, *A. moschatus*, and *A. tuberculatus*) represented by 71 accessions assembled from different parts of southern Asia and Africa was studied at the NBPGR using isozyme electrophoresis and RAPD techniques (Bhat et al. 1996). The study covered allelic variations at 13 isozyme loci and 189 amplification products obtained by random amplification of genomic DNA using 22 random primers of 10 nucleotides' length. Genetic diversity within *A. ficulneus* and *A. moschatus* as revealed by both isozyme and RAPD analysis was moderate. Diversity within *A. manihot* ssp. *tetraphyllus*, *A. tuberculatus*, and *A. esculentus* was low. The *A. moschatus* genome was observed to be quite distinct from that of the other four species. High similarities among *A. esculentus*, *A. ficulneus*, *A. tuberculatus*, and *A. manihot* ssp. *tetraphyllus* accessions were indicated by the parsimony analyses. Gene duplication was common in all the species studied.

5.8.2.3 Sources/Donors for Various Traits

Variation for most of the agronomic and horticultural traits is present in both the cultivated and semi-cultivated forms. These include the desired types of fruit, earliness, adaptability, tolerance to abiotic stresses, and qualitative traits of fruits. A number of accessions were evaluated for reaction to selected biotic stresses in many studies. The accessions identified to possess resistance/tolerance are listed in Table 5.11 (Dhankhar et al. 2005). Resistance/tolerance to major biotic stresses is also available in related wild species of okra. Though it is not always easy to transfer such characters to cultivated okra through conventional breeding methods, modern biotechnological methods have been successful to a considerable extent.

Table 5.11   Germplasm Accessions Identified to Be Promising for Various Traits in Okra

Trait	Accessions
Biotic stress	
Diseases	
Fusarium wilt (*Fusarium oxysporum* f. *vasinfectum*)	*A. manihot*, P.I. 379584
Powdery mildew (*Erysiphe cichoracearum*)	*A. tetraphyllus*, *A. manihot*, *A. manihot* ssp. *manihot*, *A. moschatus* (immune), *A. esculentus* cv. Nigeria, *A. angulosus*
Cercospora and Alternaria blight	*A. crinitus*, *A. angulosus*, *A. moschatus*, cv. 7-1, Round selection, cv. Nigeria, EC 32598, IC 10238, IC 8248, IC 1542
Yellow Vein Mosaic Virus	EC 305616 (tolerant), *A. angulosus*, *A. manihot*, *A. manihot* ssp. *manihot*, *A. manihot* ssp. *manihot* var. ghana, *A. tetraphyllus* (some forms), *A. crinitus* (immune), *A. pungens* (immune), cvs. Vaishali Vadhu & Red Wonder (symptoms suppressed), NIC 9303A, 6308, 3322, 3325, 9408, EC 329375
Insect-pests	
Jassids (*Amrasca bigutulla bigutulla*)	IC 7194, IC 8889 (*A. esculentus*); *A. manihot* ssp. *manihot* var. Ghana, *A. moschatus*, *A. crinitus*, EC 305656, 305694, 305695, 305714, 306731
Shoot and fruit Borer (*Earias* spp.)	*A. tuberculatus*, *A. caillei*, cv. Narnaul Special
Mites (*Tetranychus* spp.)	EC 305656, EC 305664, EC 305696
Abiotic stress	
Low temperature and frost	*A. angulosus*
Salinity	Pusa Sawani (moderate tolerance)
Photo-sensitivity	Pusa Sawani (moderate tolerance)
Horticultural characters	
Fruits/plant	*A. tetraphyllus*, *A. manihot* ssp. *manihot*, *A. caillei*
Branches/plant	*A. tetraphyllus*
Dark green fruit and extended fruiting period	*A. manihot* ssp. *manihot*
Attractive smooth green fruits	EC 187251 stock 510,171 (from the Philippines)

Source: From Dhankhar, B. S., Mishra, J. P. and Bisht, I. S. *in Plant Genetic Resources: Horticultural Crops*, edited by B. S. Dhillon, R. K. Tyagi, S. Saxena and G. J. Randhawa, Narosa Publishing House, New Delhi, 59–74, 2005.

5.8.2.4 Systematic Collection and Evaluation Efforts from Other Countries

In Cote d'Ivorie, okra is grown in all climatic zones, with more production in central regions of the country due to dietary preferences. Young pods are eaten fresh, dried, or in powder for meat or fish sauces. Fruits that are already somewhat lignified are cut into slices and consumed in the form of powder after drying. In some regions, fresh leaves are also eaten. Many varieties can be found in Cote d'Ivore that differ from each other in earliness, vegetative growth, and fruit characters. "Sounde" (very early) and "Gbogboligbo" are the most widespread varieties. A total of 226 acessions are maintained in addition to 2,149 accessions collated by ORSTOM with IPGRI financial support. Characterization and evaluation of joint ORSTOM/IBPGR okra collection (world collection) based in Cote d'Ivoire and comprising 2,283 accessions was undertaken (Hamon and van Sloten 1989). The African continent (with 2,029 accessions) and West Africa in particular (with 1,769) is far more heavily represented than other continents and countries. Wild and cultivated species from Asia, other than *A. esculentus*, are absent in the collection. A core collection of 189 accessions was established on the basis of representative variability as described by passport, characterization, and evaluation data, but also including rare types. This core collection has already been distributed to several countries for further evaluation.

In Nigeria, 300 genotypes from 13 countries including 81 Nigerian accessions are maintained at University of Ibadan (Chheda and Fatokun 1991). The diversity studies revealed that Nigerian indigenous collection have high variation. All the cultivars with origin other than West Africa belong to *A. esculentus* group (including "Soudanian" type), whereas the most of the local material belong to "Guinean" okra or *A. caillei*. The Guinean type okra generally flowers very late and many are short-day plants and less susceptible to YVMV. Hybrids between Soudanian and Guinean types produce vigorous plants but seed set is very poor (15–20 seeds per pod). The partial sterility of F_1 hybrids coupled with their production of few viable seeds suggests reproductive isolation of these two okra types. Gene transfer, however, can be affected between them, and few cultivars with some of the characters of Guinean type have been developed through a series of backcrosses. In Senegal, local and introduced germplasm from IIRSDA (ORSTOM, Cote d'Ivore) was evaluated for cool weather adaptation, biotic stresses particularly against root knot nematode (*Meloidogyne* spp.) and Fusarium wilt, and morphological traits. Germlasm comprises four species: *A. esculentus*, *A. caillei*, *A. moschatus*, and *A. manihot*. A wide range of varieties including Clemson spineless and local cultivars are grown in the Niayes, an area from Dakar to St. Louis, along the Atlantic Ocean, with a variable width and a special climate and in other parts of the country. In Sudan, 132 accessions of three *Abelmoschus* species, *A. esculentus*, *A. manihot*, and *A. ficulneus* are maintained. These accessions were characterized for various vegetative and fruit characters. The collections made from rainlands around Sennar and central Blue Nile resulted in development of three varieties: Rahida, Higairat and, Sennar, with good export potential. Yet there is a great potential of landrace diversity existing in various parts of Sudan e.g., Blue Nile Province, parts of Kordofan, Darfur, southern region, Khartoum area, and White Nile Province.

In 1986, systematic collection of indigenous germplasm started in Sri Lanka. A total of 130 accessions from different parts of the country were collected. Three wild species *A. ficulneus*, *A. moschatus*, and *A. angulosus* also reported to occur, but the diversity is yet to be collected. Yellow vein mosaic is a serious disease affecting okra. Of the germplasm screened against YVMV, 13 accessions were found to be free from disease symptoms. The breeding program is aimed at developing varieties with high yield, less pubescence, improved shelf life, less mucilage content and fiber, medium height (100–150 cm), and resistance to powdery mildew, YVMV, and pod borer. Several improved varieties such as MI5, MI7, and MI18 have been released. Pusa Sawani, an introduction from India, is also being grown and used in breeding program (Hindagala et al. 1991). In the Philippines, the Institute of Plant Breeding maintains over 700 accessions of okra, majority of them originate from Brazil, the Philippines, and Turkey. The germplasm accessions were evaluated for specific traits as resistance to biotic stresses, shade tolerance, and suitability as ratoon crop, in addition to routine evaluation for horticultural and quality traits. Characterization data on 12 plant characters and 9 fruit characters have been computerized and accessions stored in genebank at −20°C and 6% relative humidity. In China, okra is grown as a minor vegetable. The edible okra, *A. esculentus*, also called as coffee okra, is grown for its tender fruits as vegetables. Several wild species occur in China. Much variability occurs in *A. manihot*. Other species are *A. crinitus* and *A. moschatus*. Two taxonomically different types locally described as *A. sagittifolius* and *A. muliensis* are also reported. In Brazil, several institutions hold about 200 accessions of germplasm collections. Okra cultivar Santa Cruz 47 is widely grown in the country. The other cultivars are Amarelinho, Campinas 2, Piranema, and Chifre de Veado. These cultivars are mainly used as a fresh market vegetable. Recently work has been done involving food processing technology for canning fruits for export.

5.8.2.5 Genetic Erosion

Pioneering work done at the NBPGR followed by active breeding programmes undertaken by IIHR (Bangalore), MAU (Parbhani), PAU (Ludhiana), and more private seed companies led to development of several improved okra varieties. These varied in maturity from nearly 60 days to around 120 days and were well suited to prevalent agronomic/seasonal patterns. Farmer's seed requirements are generally met by the National Seeds Corporation (NSC), State Seed Corporations, and many leading private seed companies. The annual seed production by NSC alone is estimated to be about 7,000 tonnes comprising 5,000 tonnes of Pusa Sawani, 1,500 tonnes of Parbhani Kranti, and 500 tonnes of other varieties. Since the 1950s, there has been gradual replacement of okra landraces because of the rapid spread of high yielding and the commercially more acceptable varieties, particularly "Pusa Sawani." "Parbhani Kranti" is now well accepted in Maharashtra and in adjoining areas of central India. The NBPGR began collecting native variability in okra so as to overcome the growing threat of genetic erosion.

5.8.2.6 Germplasm Conservation

More than 2,500 accessions of cultivated and wild species are maintained both as base collection in the National Genebank at NBPGR (long-term storage at −20°C) and as active collection under medium term storage (4°C) at NBPGR Regional Station in Akola, Maharashtra. In addition, working collections are maintained at Indian Institute of Horticultural Research (IIHR), Bangalore; Marathwada Agricultural University (MAU), Parbhani; Orissa University of Agriculture & Technology (OUAT), Bhubaneshwar; Chaudhary Charan Singh Haryana Agricultural University (CCSHAU), Hissar; Punjab Agricultural University (HAU), Ludhiana; Chandra Shekhar Azad University of Agriculture & Technology (CSAUAT) Kanpur; Tamil Nadu Agricultural University (TNAU), Coimbatore; Gujarat Agricultural University (GAU), Anand; and Indian Agricultural Research Institute (IARI), New Delhi.

As per recent IPGRI germplasm database, more than 46 institutions in different countries worldwide possess about 11,000 accessions of cultivated okra and wild related species. Major institutions, holding more than 100 accessions, are listed in Table 5.12.

5.8.2.7 Germplasm Utilization

Until 1950, there were no improved varieties of okra in India, and local cultivars, both 5-edged and multi-edged types, were in cultivation. During 1950s under the leadership of late Dr. Harbhajan Singh, research on okra was initiated for germplasm collection and varietal

Table 5.12   Worldwide Holdings of Okra Germplasm at Major Centers (with More Than 100 Accessions)

Institute	Accessions
Lab. de Recursos Geneticos (CCTA), Universidade Estadual do Norte Fluminense, Av. Alberto Lamego No.2000, Bairro Horto, 28015–620 Campos dos Goytacazes, Brazil	159
Departamento de Fitotecnia - Universidad Federal de Vicosa, 36571-000 Vicosa, Minas Gerais, Brazil	192
Institut Int. de la Recherche Scientifique pour Dev en Afrique, BP V 51, Abidjan, Cote d'Ivoire	1,430
Lab. Ress. Genetiques et Amelior. des Plantes Tropicales, ORSTOM F-34032 Montpellier Cedex, France	965
Crop Research Institute Plant Genetic Resources Unit, P. O. Box 7, Bunso- East Akim, Ghana	198
National Bureau of Plant Genetic Resources, New Delhi-110 012, India	1,650
Miglior. Genetico e Prod. Sementi DIVRAPA, Universita di Torino Via Pietro Giuria 15, 10126 Torino, Italy	131
National Centre for Genetic Resources and Biotechnology, Ibadan, Nigeria	374
Jericho Reservation Area, National Horticultural Research Inst., Ibadan, Nigeria	258
Agril. Rehabilitation Programme, Dept. of Agric. & Livestock, Konedobu, Papua New Guinea	112
Institute of Plant Breeding, College of Agriculture UPLB, 4031, Laguna, Philippines	477
National Plant Genetic Resources Laboratory, IPB/UPLB College 4031 Laguna, Philippines	905
Plant Genet. Resources Centre, Gannoruwa, Peradeniya, Sri Lanka	193
Horticultural Research Section Agricultural Research Corporation Wad Medani, Sudan	173
Institut Togolais de Recherches Agronomique, Lome, Togo	170
Plant Genetic Resources Dept. Aegean Agricultural Research Inst., Menemen, 35661 Izmir, Turkey	104
Southern Regional Plant Introduction Station, USDA-ARS-SAA, Griffin, USA	3,379
Mount Makulu Agric. Research Sta., Chilanga, Zambia	106

Source: From IPGRI database (http://www.ipgri.org).

improvement. As a result, Pusa Makhmali was developed from a collection from West Bengal in 1955 and released for cultivation. During 1960, Pusa Sawani was developed from an intervarietal cross between IC-1542 (symptomless carrier for YVMV from West Bengal) and Pusa Makhmali. Pusa Sawani had field resistance to YVMV and had excellent agronomic performance. It became very popular throughout the country because of its wide adaptability, particularly due to low photoperiod sensitivity and soil salinity tolerance. It replaced most of the landraces, causing extensive "gene erosion." The search for resistance to YVMV in the Plant Introduction Division of the IARI (now the NBPGR) led to the identification of an introduction from Ghana to be highly resistant to YVMV. The introduction belonged to *A. mahihot* ssp. *mainhot*. Utilizing it, many resistant lines were developed by different breeding programs in the country during 1980s and a number of cultivars were released, namely G-2 and G-2-4 (NBPGR), Punjab Padmini, Punjab-7 (PAU), Parbhani Kranti (MAU), and IIHR Sel-4 and Sel-10 (IIHR), etc. Apart from these cultivars were Sel-2, Varsha Uphar and Hissar Unnat (CCSHAU), and Pusa A-4 (IARI). These were also developed as YVMV resistant derivative selections of interspecific hybrids between *A. manihot* ssp. *manihot* and *A. esculentus*. Okra is one classical example in which wild species have been used as donor for resistance to YVMV, and as a result commercial YVMV resistant varieties have been developed.

At present, at the national level, some of the important varieties are Pusa Sawani, Pusa Makhmali, Punjab Padmini, Parbhani Kranti, Co-1, and introductions such as Perkins long green, white velvet and Clemson spineless. Red Bhindi (Selection AE 106) is a choice variety for kitchen gardens and nonseasonal. Pusa Makhmali and Perkins long green are early maturing types. Co-1 is a selection from Red Wonder from Hyderabad. Other varieties include T1, T2, T3, and T4 from Uttar Pradesh; Vaishali Vadhu from Sabour, Bihar; Lam Hybrid Lam, Andhra Pradesh; Shankerpatti and Somaltotte also from southern India; No. 13 from Punjab; besides Satpani, Bandanwar, Silari, Patna, Jhabua, Panchdhar, Satshari, Dasdhari, etc.

Interspecific hybridization has led to a large-scale variation in okra. Use of related wild species in transferring disease and insect pest resistance genes to cultivated okra is well illustrated by the work done at NBPGR, IIHR, MAU, and HAU (Rana, Thomas, and Arora 1991; Dutta 1991; Nerkar 1991; Dhankhar, Saharan, and Pandita 1997, 1998).

Clemson spineless is the standard open pollinated variety grown for over 40 years in the United States and other countries. It is still used because of low seed cost and wide adaptation. Clemson spineless 80 is an open-pollinated selection from Clemson spineless with shorter plants and greater uniformity. Other commercially grown culivars are Emerald, Lee, Prelude, Annie Oakley (hybrid) etc.

The constraints in germplasm utilization in okra are as follows:

- Lack of detailed evaluation of available germplasm accessions, as such potential of the germplasm held in genebank not fully known
- Erosion of genes in primitive types and landraces due to spread of high yielding varieties
- Lack of transferable resistance/tolerance against biotic and abiotic stresses, to cultivated types
- Difficulty in maintaining wild related species due to differences in their adaptability and seed dormancy
- Non-crossability between cultivated okra and wild related species with desired genes
- High degree of sterility in F_1s involving certain partially crossable forms
- Difficulty in removal of undesirable genes from wild parent during generation advancement in resistance breeding
- Non-availability of potential sources of resistance against insect-pests, particularly borers and jassids

5.9 BREEDING AND CROP IMPROVEMENT

5.9.1 Breeding Objectives

The genetic improvement of the following traits should result in increased productivity in terms of time and area of cultivation. Breeders' objectives should be as follows (Sharma 1993):

1. To develop high-yielding varieties capable of an inceased marketable yield of dark-green, tender, long, smooth pods. High yield of seed would be an added advantage.
2. To breed early-maturing and late-senescing varieties.
3. To develop varieties resistant to virus diseases such as okra mosaic virus, YVMV, and leaf curl; fungal diseases such as vascular wilt, Cercospora blight, powdery mildew, fruit rot, and damping off; root-knot nematodes; insect pests such as shoot and fruit borer, leafhopper, aphids, red spider mite, flower beetles, white fly, etc.
4. To develop varieties with multiple resistance to diseases and insect pests, with special emphasis on combining yellow vein mosaic virus resistance with resistance to okra mosaic virus, leaf curl, fruit and shoot borer, jassids and leafhopper, and root-knot nematodes.
5. To develop the most suitable ideotype possessing characters such as short plant with more nodes and short internodal length, which is more productive than tall plant with long internodal distance. Plants and fruits should be devoid of conspicuous hairs. Fruits should snap easily from the stalk, facilitating easy and economic harvest. Photoinsensitive type would be more desirable.
6. To breed varieties with optimum seed-setting ability for rapid multiplication.
7. To evolve varieties tolerant to abiotic stresses, especially tolerance to low temperature, drought, excessive rain, saline and alkaline soils, and damage by fungicides, insecticides, and other environmental pollutants.
8. To develop varieties suitable for export market.

9. To evolve varieties with better nutritive attributes and also varieties suitable for dehydration, canning, and freezing.
10. To develop the most economic hybrids.
11. To evolve varieties and hybrids for wider adaptibility.

5.9.2 Breeding Methods

Because the predominant breeding system prevalent in okra is autogamous accompanied by allogamy and breeding, methods common to self-pollinated crops can be employed for improvement. The methods commonly employed are plant introduction, pure line selection, intraspecific and interspecific hybridization using backcross techniques, mutation, and polyploidy breeding.

5.9.3 Plant Introduction

The cultivar "Pusa Sawani" evolved in India and has been introduced to the different okra-growing regions of the world; it is widely cultivated. This variety was bred as a symptomless carrier of yellow vein mosaic virus. Similarly the variety "Clemson Spineless Louisiana," bred in the United States, has been introduced to a large number of okra-growing countries both for fresh market and canning purposes. Indigenous introductions have played a significant role in improving the okra wealth of many countries. A cultivar from Ghana identified as *A. manihot* ssp. *mainhot* introduced into India has served as a source of resistance to YVMV.

5.9.4 Pure Line Selection

In India, the first improved variety of okra, "Pusa Makhmal," was developed through pure line selection using material from West Bengal (Singh and Sikka 1955). The cultivar Co. 1 is also a single plant selection from a heterozygous population of "Red Wonder." "Gujarat Bhinda 1" is a pure line selection of unknown origin bulked seed.

5.9.5 Hybridization

In India, intervarietal hybridization followed by pedigree selection produced the widely cultivated, high yielding, yellow vein mosaic virus-tolerant cultivar "Pusa Sawani" (Singh et al. 1962). Similarly, Selection 2 is a derivative of multiple intervarietal crossing (Thomas and Prashad 1985). Interspecific hybridization has been followed in the development of "Punjab Padmini" (Sharma 1982), "Winter Bush" (Martin 1982), "Parbhani Kranti" (Jambhale and Nerkar 1986), "Punjab-7" (Thakur and Arora 1988), "Arka Anamika," and "Arka Abhay" (Dutta 1991). In all the above cases except "Punjab Padmini" backcrossing has been followed. Using West African species, Martin (1982) inferred that backcrossing is useful for transfer of the traits of vigor, woodiness, perennialism, branched inflorescence, and high number of seeds per pod, as well as disease resistance.

5.9.6 Distant Hybridization in *Abelmoschus* Species

Abelmoschus manihot is being extensively involved in the breeding programm of okra for yellow vein mosaic virus disease resistance. Quite a few studies were carried out on interspecific hybridization in genus *Abelmoschus* with varying degree of success (Pal, Singh, and Swarup 1952; Joshi and Hardas 1956; Kuwada 1957a, 1957b; Gadwal, Joshi, and Iyer 1968; Jambhale and Nerkar 1982; Martin 1982; Siemonsma 1982a). The results of hybridization between *A. esculentus* and allied species are summarized in Table 5.5, Table 5.6, and Figure 5.2. *A. esculentus* ($2n = 124$) is crossable with *A. manihot* ($2n=68$) and produces amphiploids ($2n=192$) with the help of colchicine. *A. esculentus* is crossable with *A. moschatus* and *A. ficulneus* with special aids such as embryo and ovule cultures. *A. esculentus* ($2n=130$) is crossable in both direction with *A. tuberculatus* ($2n=58$).

5.9.7 Mutation Breeding

Efforts have been made to generate more variability through mutations (Kuwada 1970; Nandpuri, Sandhu, and Randhawa 1971; Nirmala Devi 1982; Jambhale and Nerkar 1982, 1985; Abraham and Bhatla 1984; Abraham 1985; Dalia 1986; Fatokun 1987; Sharma and Arora 1991). Kuwada (1970) isolated a few promising lines in X_{10} generation. According to Fatokun, Aken Ora, and Chhedha (1979), supernumerary inflorescences were found in several plants of north Nigerian ecotype. The mutant was controlled by a single dominant gene. It was believed that the mutant could be grown commercially to produce a large number of fruits of acceptable size on each plant. Of all the above studies, the report of Sharma and Arora (1991) seems to be of most significance, as a mutant EMS 8 carrying resistance to YVMV and tolerance to fruit borer has been developed through the use of ethyl methansulfonate.

5.9.8 Polyploid Breeding

Failure of many species crosses due to hybrid sterility prompted researchers to try amphidiploids. Kuwada (1961) obtained a fertile amphidiploid by duplicaton of the F_1 between *A. esculentus* × *A. manihot* called "Nori-Asa." Kuwada (1966) later synthesized an amphidiploid *A. tubercular esculentus*. Joshi, Gadwal, and Hardas (1974) reported an amphidiploid between *A. tuberculatus* and *A. ficulneus*. An amphidiploid between *A. esculentus* and *A. manihot* ssp. *manihot* was synthesized by Jambhale and Nerkar (1982). It had 81.8% seed fertility, against 7.07% in the F_1. The amphidiploid was resistant to yellow vein mosaic virus. In 1983 these workers also made a successful attempt at inducing polyploidy in three interspecific hybrids. It has been inferred by Siemonsma (1982a) that the West African okra Guinean is a natural amphidiploid of *A. esculentus* and *A. manihot*. Suresh Babu (1987) also induced amphidiploids in a cross between *A. esculentus* and *A. manihot* ssp. *tetraphyllus*. The amphidiploid was highly fertile and gave 94.14% pollen stainability.

5.9.9 Heterosis

Okra has been intensively studied for heterosis. However, practical utilization of heterosis is highly restricted. Heterosis has been very well-documented for yield (Singh, Swarup, and Singh 1975; Singh et al. 1977; Singh and Singh 1979; Elangovan, Muthukrishnan, and Irulappan 1981). Number of fruits was the most important component of yield heterosis. The other components attributing to yield heterosis were plant height, fruit width, fruit length, earliness, and first fruiting node. Twenty four hybrids were studied by Singh, Swarup, and Singh (1975) and ranges of heterosis obtained were 6.62 to 32.27% for yield, 0.03 to 39.03% for fruit number, 4.19 to 8.09% for fruit width, 0.31 to 14.27% for fruit length, and 14.29 to 32.11% for plant height over better parent. Kulkarni and Virupakshappa (1977) studied heterosis for number of fruits, plant height, and days to flowering, and found heterosis up to 12.4, 19.9 and 11.17%, respectively, over best parent.

5.9.10 Breeding for Disease/Pest Resistance

Yellow vein mosaic (YVM) is the most serious disease of okra in the tropics and sub-tropics. *Abelmoschus manihot* was found resistant to this disease. Inheritance studies involving

"Pusa Sawani" as a susceptible and *A. manihot* as a resistant parent have indicated that resistance is controlled by two complementary dominant genes, Yv 1 and Yv 2 (Thakur 1976). The inheritance of field resistance to yellow vein mosaic virus indicated that susceptibility was controlled by two complementary dominant genes (Chand and Walker 1964).

In *A. esculentus* Jassids and spotted boll worm (*Earias* spp.) are the serious pests. "Crimson smooth long" IC-7194, IC-8899, and *A. manihot* var. *Ghana* were resistant to Jassid (Sandhu et al. 1974). Regarding resistance to spotted bollworm, although varietal differences have been observed the degree of resistance was low against this pest. "Narnaul Special" "Perkins long green," "Clemson spineless," "white snow," and "red round" had less than 10% infestation against shoot and fruit borers (Kashyap and Verma 1983). Resistance to leaf hopper (*E. devastans*) was classified on the basis of insect number and rate of damage in the field. Resistance was observed in AE 11 (Uthamasamy 1980).

5.9.11   Breeding for Quality and Processing Traits

For dehydration of okra (*A. esculentus*), less fiber, less mucilaginous substances, high protein, high dry matter, and mineral are important attributes. For canning and freezing, however, high chlorophyll, less crude fiber and mucilage substance, low dry matter and high protein, vitamins, and minerals are required. Small tender pods should be taken for canning. "Pusa Sawani," "Dwarf Green Smooth," and "Vaishali Vadhu" (Kalra et al. 1983) were found to have the best quality for canning. The variety "Pusa Sawani" was found to be the best for dehydration followed by "Sel. 2" for color, texture, taste, flavor, and dehydration and rehydration ratio. Distant hybridization can be used for the transfer of quality traits like green color from *A. manihot* to *A. esculentus*. "Sel. 2" is suitable for freezing.

5.9.12   Biotechnology and Its Implication

5.9.12.1   Tissue Culture

The family Malvaceae, to which cultivated okra belongs, is a recalcitrant species for plant regeneration through tissue culture. Reynolds, Blackmoa, and Postek (1981) reported the first instance of establishment of an embryogenic callus from cotyledonary explants, cultured on Nitsch medium containing 40 g/l glucose instead of sucrose, 8 g/l agar, 0.1 mg/l 2,4-dichlorophenoxyacetic acid (2,4-D), and 1 mg/l dimethylallyl aminopurine (2 ip). The callus subsequently differentiated into somatic embryos under a 16-hr photoperiod. Mangat and Roy (1986) used explants such as hypocotyls, cotyledon, cotyledonary nodes, and leaf segments from aseptically grown seedlings of *A.esculentus* cv. "Perkins Mammoth Longpod." The explants were cultured on Murashige and Skoog (MS) basal nutrient medium supplemented with naphthalene acetic acid (NAA) 0.1 mg/l, indolacetic acid (IAA) 1.0 mg/l, benzyladenine (BA) 1 mg/l and kinetin 1 mg/l. Callus formation and root differentiation occurred on the media containing NAA or IAA. The addition of 2,4-D suppressed root formation. Kinetin and zeatin proved ineffective in inducing shoot buds, but the combination of NAA and BA was effective in inducing shoot buds in the callus established from cotyledons and cotyledonary nodes. The shoot buds grew further and roots were formed in the same medium, and the plantlets on transfer to soil grew normally. The method can be used for rapid vegetative propagation of okra.

5.9.12.2   Genetic Engineering

Genetic engineering would form a suitable system for the incorporation of specific genes such as *CP* (coat protein) gene and antisense RNA gene for viruses for elevated viral resistance,

Bt (*Bacillus thuringiensis*) δ endotoxin gene for resistance against fruit borer, and other genes of interest.

5.10   FUTURE PERSPECTIVES

Useful information has been generated on economic uses of wild species. There has been an emphasis to collect and introduce the existing variability in *Abelmoschus* species available, particularly in southern/southeast Asia. Chromosomal variability within *A. esculentus* needs to be explored, collected, and conserved. Landraces, wild and weedy relatives including different forms in polytypic *Abelmoschus* species, need to be augmented. For efficient and effective management of genetic diversity in okra including utilization, emphasis needs to be laid on the following aspects:

- The spread of high yielding varieties caused genetic erosion in most of the accessible areas. Still variability has been found to exist in temperate, subtropical as well as tropical agro-climates for different *Abelmoschus* species. In India, the primitive locally adapted types in cultivated and semi-cultivated forms are available only in areas difficult to reach particularly in eastern and northeastern region. These possess desirable characteristics associated with sustainable agriculture, such as adaptation to stresses. These materials need to be collected.
- Wild relatives of okra are important for their inherent resistance to one or more biotic and abiotic stresses. A vast diversity present in wild okra in India, other parts in Asia, and in Africa is yet to be collected. Further, intermediate forms resulting from natural hybridization do occur and such materials are easy to utilize; these should also be collected. Some important traits for which germplasm should also be introduced along with source country are shade tolerance, short day types, resistance/tolerance against leaf hopper, aphid and cotton bug from the Philippines; *Fusarium*, wilt and nematode resistance, short day types adapted to cool weather (up to 22°C) from Senegal; *A. caillei*, drought tolerant types with YVMV resistance and long duration from Sudan and other drier parts of Africa; multiple/prolonged fruiting and tolerance to high humidity from West Africa and adaptation to high altitudes in Nepal. With the change in the concept of plant type from nonbranching habit to upright branching, accessions with these traits need to be introduced. Germplasm resources for intercropping under partial shade, arid conditions and for cooler climate will help in introducing cultivation of okra in nontraditional areas.
- Studies on seed dormancy and regeneration of wild relatives, and strengthening germplasm enhancement activities to promote the use of wild relatives through conventional as well as biotechnological methods.
- The variability created through interspecific hybridization utilizing *A. manihot* and *A. tetraphyllus* needs to be fully exploited.
- The existing germplasm accessions need detailed evaluation for various traits by multidisciplinary team of scientists. In the breeding programs the characters that need to be given emphasis include medium plant height, upright branching, high number of fruiting nodes, low position of first fruiting node, short internode and deeply lobed medium sized leaf for enhanced productivity; smooth as well as 5-ridged fruits of medium to short size, dark green color, early maturity, low fiber content, and low mucilage content for enhanced fruit quality and appearance, tolerance to abiotic stresses (water logging, drought, and high and low temperatures, salinity/alkalinity), and resistance/tolerance to biotic stresses (YVMV, wilt, powdery mildew, rhizoctonia, leaf hopper, borers, and red spider mite) for stable and sustainable production.
- Further, developing suitable cultivars for canning, freezing, and dehydration for home consumption and export and cultivars having high seed oil and protein for their use as an

alternative source of edible oil and protein are also important breeding objectives. Germplasm needs to be evaluated for these traits to identify accessions to feed the breeding programs.

• Development of core collection and its systematic evaluation in multilocation testing.

REFERENCES

Abraham, V., Inheritance of ten induced mutants in okra, *Curr. Sci.*, 54, 931–935, 1985.

Abraham, V. and Bhatia, C. R., Induced mutation in okra (*A. esculentus*), *Mutation Breed. Newsl.*, 24, 1–3, 1984.

Ahmed, N., Thakur, M. R., and Bajaj, K. L., Effect of yellow vein mosaic virus on moisture, chlorophyll, cholopophyllase and sugar contents of okra [*Abelmoschus esculentus* (L. Moench)], *J. Pl. Sci. Res.*, 3, 1–5, 1987.

APEDA (Agricultural and Processed Food Products Export Development Authority). *Agro-Export Statistics.* New Delhi: Agriculture and Processed Food Export Development Agency, 2000.

Ariyo, O. J., Genetic diversity in West African okra (*Abelmoschus caillei* (AS.Chev.) Stevels): multivariate analysis of morphological and agronomic characteristics, *Genet Resour. Crop Evol.*, 40, 25–32, 1993.

Ariyo, O. J. and Odulaja, A., Numerical analysis of variation among accessions of okra [*Abelmoschus esculentus* (L. Moench)], Malvaceae, *An. Bot.*, 67, 527–531, 1991.

Ariyo, O. J., Akenova, M. E., and Fatokun, C. A., Plant character correlations and path analysis of pod yield in okra, *Euphytica*, 36, 677–686, 1987.

Arora, S. K., "Genetic analysis for some qualitative and quantitative characters in okra." Ph.D. diss., Himachal Pradesh Krishi Vishvavidyalaya, Solan, 1980.

Arumugam, R. and Muthukrishnan, C. R., Gene effects of some quantitative characters in okra, *Indian J. Agric. Sci.*, 49, 602–604, 1979.

Bates, D. M., Notes on the cultivated Malvaceae. 2, *Abelmoschus. Baileya*, 16, 99–112, 1968.

Bailey, J. H., *Manual of Cultivated Plants*, Canada: Colier McMillan, 1969.

Bhat, K. V. et al., Analysis of the genetic relationship among *Abelmoschus* species using isozyme and RAPD markers, Abstract of paper presented in 2nd International Crop Science Congress, New Delhi, India, Nov. 17–24, 1996.

Bisht, I. S. et al., *Catalogue of Okra (Abelmoschus Esculentus* (L.) Moench) Germplasm. Part III, New Delhi: NBPGR, 1993.

Bisht, I. S., Mahajan, R. K., and Rana, R. S., Genetic diversity in South Asian okra (*Abelmoschus esculentus*) germplasm collection, *Ann. Appl. Biol.*, 126, 539–550, 1995a.

Bisht, I. S. et al., *Catalogue of Wild Abelmoschus Species Germplasm*, New Delhi: NBPGR, 1995b.

Bisht, I. S., Patel, D. P., and Mahajan, R. K., Classification of genetic diversity in *Abelmoschus tuberculatus* germplasm collection using morphometric data, *An. Appl. Biol.*, 130, 325–335, 1997.

Breslavetz, L., Medvedeva, G., and Magitt, M., Zytologische untersuchungen der bastpflanzen, *Z. Züchtung*, 19, 229–234, 1934.

Chand, J. N. and Walker, W. C., Inheritance of resistance to angular leaf spot of cucumber, *Phytopathology*, 54, 51–53, 1964.

Charrier, A., *Genetic Resources of the Genus Abelmoschus* Med. (Okra), Rome: IBPGR, 1984.

Chheda, H. R. and Fatokun, C. A., Studies on olkra germplasm in Nigeria, In *International Crop Network Series*, Report of an international workshop on okra genetic resources, Rome: IBPGR, 5 (21–23), 1991.

Chizaki, Y., Breeding of a new interspecific hybrid between *Hibiscus esculentus* L. and *H. manihot* L., *Proc. Crop Sci. Soc. (Japan)*, (in Japanese), 6, 164–172, 1934.

Dalia, C., Radiation induced variability in interspecific hybrids involving *Abelmoschus esculentus* (L.) Moench) and *A. manihot* (L.) Medik. M.Sc. (Hort.) (thesis submitted to the Kerala Agric. Univ., Vellanikkara), 1986.

Dhankhar, B. S., Mishra, J. P., and Bisht, I. S., Okra, In *Plant Genetic Resources: Horticultural Crops*, Dhillon, B. S., Tyagi, R. K., Saxena, S., and Randhawa, G. J., Eds., New Delhi: Narosa Publishing House, pp. 59–74, 2005.

Dhankhar, B. S., Saharan, B. S., and Pandita, M. L., Okra, In *varsha uphar is resistant to YVMV*, *Indian Hort.*, 41, 50–51, 1997.

Dhankhar, B. S., Saharan, B. S., and Sharma, N. K., Hissar unnat: new YVMV resistant okra, *Indian Hort.*, 44, 6, 1998.

Datta, P. C. and Naug, A., A few strains of *Abelmoschus esculentus* (L.) Moench. Their karyological study in relation to phylogeny and organ development, *Beitr. Biol. Pflanzen*, 45, 113–126, 1968.

Dutta, O. P., Okra germplasm utilization at IIHR, Bangalore, In *International Crop Network Series*, Report of an International Workshop on Okra Genetic Resources. Rome: International Board for Plant Genetic Resources, 5: 114–116, 1991.

Elangovan, M., Muthukrishnan, C. R., and Irulappan, I., Hybrid vigour in bhindi (*Abelmoschus esculentus* (L.) Moench) for some economic characters, *S. Indian Hortic.*, 29, 4–14, 1981.

Erickson, H. T. and Couto, F. A. A., Inheritance of four plant and floral characters in okra, *Proc. Am. Soc. Hortic. Sci.*, 83, 605–608, 1963.

FAOSTAT., 2004. http://www.fao.org

Ford, C. E., A contribution to a cytogenetical survey of the *Malvaceae*, *Genetica*, 20, 431–452, 1938.

Fatokun, C. A., Wide hybridization in okra, *Theor. Appl. Genet.*, 74, 483–485, 1987.

Fatokun, C. A., Aken Ora, M. E., and Chhedha, H. R., Supernumerary inflorescence—a mutation of agronomic significance in okra, *J. Hered.*, 70, 270–271, 1979.

Gadwal, V. R., Joshi, A. B., and Iyer, R. D., Interspecific hybrids in *Abelmoschus* through ovule and embryo culture, *Indian J. Genet. Plant Breed.*, 28, 269–274, 1968.

Hamon, S., Organization genetique du genre *Abelmoschus* (gombo): co-evolution de deux especes cultivees de gombo en Afrique de l'Ouest (*A.esculentus et A.caillei*), Paris: These Universite de Paris-Sud, Centre d'Orsay, 1987.

Hamon, S. and Charrier, A., Large variation of okra in Benin & Togo (Ivory Coast), plant genet, *Resour. Newsl.*, 56, 52–58, 1983.

Hamon, S. and van Sloten, D. H., Characterisation and evaluation of okra, In *The Use of Plant Genetic Resources*, Brown, A. H. D., Frankel, O. H., Marshal, D. R., and Williams, J. T., Eds., London: Cambridge University Press, pp. 173–196, 1989.

Hamon, S., Yapo, A., Peturbation induced within the genus *Abelmoschus* by the discovery of a second edible okra species in West Africa, In *Maesen, L.J.G. Van Der, First International Symposium on Taxonomy of Cultivated Plants. Acta Hort.* 182, 133–144, 1986.

Hardas, M. W. and Joshi, A. B., A note on the chromosome numbers of some plants, *Indian J. Genet. Plant Breed*, 14, 47–49, 1954.

Hindagala, C. B., Jayawardena, S. D. G., Siriwardena, K. P. D., and Liyanage, A. S. V., et al., Conservation and utilization of okra genetic resources in Sri Lanka. In *International Crop Network Series*, Report of an international workshop on okra genetic resources. IBPGR: Rome, 5, 15–20, 1991.

Hochreutiner, B. P. G., Genres nouveaux et genres discutes de la famille des Malvacees, *Candollea*, 2, 79–90, 1924.

Hutchinson, J. and Dalziel, J. M., *Flora of West Tropical Africa*, 2nd ed., 1, pp. 343–348, 1958.

IBPGR., *International Crop Network Series*, Report of an International Workshop on Okra Genetic Resources, Rome: International Board for Plant Genetic Resources, 5, 1991.

Jambhale, N. D. and Nerkar, Y. S., Inheritance of resistance to okra yellow vein mosaic disease in interspecific crosses of *Abelmoschus*, *Theor. Appl. Genet.*, 60, 313–316, 1981a.

Jambhale, N. D. and Nerkar, Y. S., Occurrences of spontaneous amphiploidy in an interspecific cross between *Abelmoschus esculentus* and *A. tetraphyllus*, *J. MAU.*, 6, 167, 1981b.

Jambhale, N. D. and Nerkar, Y. S., Induced amphidiploidy in the cross *Abelmoschus esculentus* (L.) Moench × *A. manihot* (L.) Medik spp. manihot, *Genet. Agr.*, 36, 19–22, 1982.

Jambhale, N. D. and Nerkar, Y. S., An unstable gene controlling developmental variegation in okra, *Theor. Appl. Genet.*, 71, 122–125, 1985.

Jambhale, N. D. and Nerkar, Y. S., Parbhani Kranti, a yellow vein mosaic resistant okra, *Hort. Sci.*, 21, 1470–1471, 1986.

Jasim, A. J., Inheritance of certain characters in okra (*H. esculentus* L.), *Diss. Abstr. Sect.*, 28, 3, 1967.

Jasim, A. J. and Fontenot, J. F., Inheritance of certain characters in okra (*H. esculentus*), *Diss. Abstr. Sect.*, 28, 211, 1967.

Joshi, A. B. and Hardas, M. W., Chromosome number in *Abelmoschus tuberculatus* Pal & Singh—a species related to cultivated bhindi, *Curr. Sci.*, Bangalore, 22, 384–385, 1953.

Joshi, A. B. and Hardas, M. W., Alloploid nature of Okra, *Abelmoschus esculentus* (L.) Moench, *Nature*, 178, 1190, 1956.

Joshi, A. B. and Hardas, M. W., *Okra Simmonds, N.W. Evolution of Crop Plants*, London: Longman, pp. 194–195, 1976.

Joshi, A. B., Gadwal, V. R., and Hardas, M. W., Evolutionary studies in world crops, In *Diversity and Change in the Indian Sub-Continent*, Hutchinson, J. B., Ed., London: Cambridge University Press, pp. 99–105, 1974.

Kalra, C. L. et al., The influence of varieties on the quality of dehydrated okra (*Abelmoschus esculentus* (L.) Moench), *Indian Food Packer*, 37, 47–55, 1983.

Kamalova, G. V., Cytological studies of some species of the Malvaceae, *Uzbekistan Biologija Zurnali*, 3, 66–69, 1977.

Kashyap, R. K. and Verma, A. N., Relative susceptibility of okra to shoot and fruit borer, *Indian J. Ecol.*, 10, 303–309, 1983.

Kolhe, A. K. and D'Cruz, R., Inheritance of pigmentation in okra, *Indian J. genet. Plant Breed*, 26, 112–117, 1966.

Kulkarni, R. S. and Thimmappaiah, Inheritance of number of pods in *Bhindi, Agric. Res. J. Kerala*, 15, 174–176, 1977.

Kulkarni, R. S. and Virupakshappa, K., Heterosis and inbreeding depression in okra, *Indian J. Agric. Sci.*, 47, 552–555, 1977.

Kulkarni, R. S., Rao, T. S., and Virupakshappa, K., Gene action in *Bhindi, Agric. Res. J. Kerala*, 14, 13–20, 1976.

Kulkarni, R. S., Rao, T. S., and Virupakshappa, K., Genetics of important yield components in Bhindi, *Indian J. genet. Plant Breed.*, 38, 160–162, 1978.

Kundu, B. C. and Biswas, C., Anatomical characters for distinguishing *Abelmoschus* spp. and *Hibiscus* spp., *Proc. Indian Sci. Cong.*, 60, 295–298, 1973.

Kuwada, H., *Crosscompatibility in the reciprocal crosses between Abelmoschus esculentus and A. manihot, and the characters and meiosis in F1 hybrids, Jap. J. Breed.*, 7, 93–102, 1957a.

Kuwada, H., Cross compatibility in the reciprocal crosses between amphidiploid and its parents (*Abelmoschus esculentus and A. manihot*) and the characters and meiotic division in the hybrids obtained among them, *Jap. J. Breed*, 7, 103–111, 1957b.

Kuwada, H., Studies on the interspecific crossing between *Abelmoschus esculentus* (L.) *Moench* and *A. manihot* (L.) Medikus, and the various hybrids and polyploids derived from the above two species, *Fac. Agric. Kagawa Univ. Mem.*,8, 91, 1961 (in Japanese).

Kuwada, H., The new amphidiploid plant named "*Abelmoschus tubercular-esculentus*," obtained from the progeny of the reciprocal crossing between *A. tuberculatus* and *A. esculentus, Jap. J. Breed*, 16, 21–30, 1966.

Kuwada, H., X-ray induced mutation in okra, *Tech. Bull. Fac. Agric. Kagawa*, 21, 2–8, 1970.

Kuwada, H., F1 hybrids of *Abelmoschus tuberculatus x A. manihot* with reference to the genome relationship, *Jap. J. Breed*, 24, 207–210, 1974.

Linnaeus, C., *Species Plantarum*, Vol. I & II, Stockholm, 1753.

Mahajan, Y. P. and Sharma, B. R., Parent-offspring correlations and heritability of some characters in okra, *Scientia Hort.*, 10, 135–139, 1979.

Malik, Y. S., Genetic variability and correlation studies in okra (*Abelmoschus esculentus* (L.) Moench). M.Sc. Thesis, Haryana Agric. Univ., Hissar, 1968.

Mangat, B. S. and Roy, M. K., Tissue culture and plant regeneration of okra (*Abelmoschus esculentus*), *Plant Sci.*, 47, 57–61, 1986.

Martin, F. W., A second edible okra species and its hybrids with common okra, *Ann. Bot.*, 50, 277–283, 1982.

Martin, F. W., Ortiz, M., Diaz, F. et al., Variation in okra, *Euphytica*, 30, 697–705, 1981.

Masters, M. T., *Flora of British India Ashford Kent*, J. D. Hooker, Ed., 1, 320–348, 1875.

Medikus, F. K., Ueber einige kunstliche Geschlechter aus der Malvenfamilie, denn der Klasse der, *Monadelphien.*, 45–46, 1787.

Medwedewa, G. B., Karyological review of 15 species of the genus *Hibiscus, J. Bot. Urss*, 21, 533–550, 1936. (in Russian)

Mishra, J. P., Rawat, G. S., et al., Conservation of biodiversity in wild germplasm of okra, International Conference on Managing Natural Resources for Sustainable Agricultural Production in the 21st Century, 4, Feb, 14–18, 2000a.

Mishra, J. P., Rawat, G. S., et al. Yellow vein mosaic virus resistance in okra-Utilization of wild relatives for better stability, International Conference on Managing Natural Resources for Sustainable Agricultural Production in the 21st century, 4, 14–18, 2000b.

Murdock, G. P., *Africa, Its People and Their Cultural History*, New York: McGraw-Hill, 1959.

Nandpuri, K. S., Sandhu, K. S., and Randhawa, K. S., Effect of irradiation on variability in okra, *J. Res. Punjab Agric. Univ.*, 8(2), 183–188, 1971.

Nerkar, Y.S. The use of related species in transferring disease and pest resistance genes to okra, *International Crop Network Series*, Report of an International Workshop on Okra Genetic Resources, Rome: International Board for Plant Genetic Resources, 5: 110–113, 1991.

Nirmala Devi, S. Induction of variability in *Abelmoschus manihot* var, Ghana by irradiation. M.Sc. (Hort.) Thesis, Kerala Agric. Univ., Vellanikkara, 1982.

Pal, B. P., Singh, H. B., and Swarup, V., Taxonomic relationships and breeding possibilities of species of *Abelmoschus* related to okra (*A. esculentus*), *Bot. Gaz.*, 113, 455–464, 1952.

Partap, P. S. and Dhankhar, B. S., Heterosis studies in okra (*Abelmoschus esculentus* (L.) Moench), *Haryana Agric. Univ. J. Res.*, 10, 336–341, 1980.

Purewal, S. S. and Randhawa, G. S., Studies in *Hibiscus esculentus* (Lady's finger), Chromosome and pollination studies, *Indian J. Agric. Sci.*, 17, 129–136, 1947.

Ramu, P. M., Breeding investigation in bhindi (*Abelmoschus esculentus* (L.) Moench), *Mysore J. Agric. Sci.*, 10, 146–149, 1976.

Rana, R. S., Thomas, T. A., Arora, R. K. Plant genetic resources activities in okra- an Indian perspective. In *International Crop Network Series*, Report of an International Workshop on Okra Genetic Resources. Rome: International Board for Plant genetic Resources, 5: 38–47, 1991.

Rana, R. S, et al. Germplasm collection of okra and eggplant and their wild relatives from South Asia. Project Report, New Delhi: NBPGR, 1994.

Randhawa, J. S., Genetics of economic characters in an intervarietal cross of okra, *Indian J. Agric. Sci.*, 59, 120–122, 1989.

Randhawa, J. S. and Sharma, B. R., Correlation, heritability and genetic advance studies in an intervarietal cross of okra (*Abelmoschus esculetus* (L.) Moench), *Punjab Agric. Univ. J. Res.*, 25, 389–392, 1988.

Rao, T. S., Note on the natural variability for some qualitative and quantitative characters in okra, *Indian J. Agric. Sci.*, 42, 437–438, 1972.

Rao, T. S., Line X tester analysis of heterosis and combining ability in bhindi, *Agric. Res. J. Kerala*, 15, 112–118, 1977.

Rao, T. S. and Ramu, P. M., Combining ability in bhindi, *Prog. Hort.*, 9, 5–11, 1977.

Reynolds, B. D., Blackmoa, W. J., and Postek, C. E., Production of somatic embryos from cellused okra hypocotyls explants, *Hort. Sci.*, 16, 87–90, 1981.

Roy, R. P. and Jha, R. P., A semi-asynaptic plant of *Abelmoschus esculentus* (L.) Moench (=*Hibiscus esculentus* L.), *Cytologia*, 23, 356–361, 1958.

Sandhu, G. S. et al., Sources of resistance to jassid and whitefly in okra germplasm, *Crop Improvement*, 1, 77–81, 1974.

Sharma, B. R., "Punjab Padmini"-a new variety of okra, *Prog. Farming*, 18, 15–18, 1982.

Sharma, D. R., Okra, *Abelmoschus* spp., In *Genetic Improvement of Vegetable Crops*, Kalloo, G. and Bergh, B. O., Eds., Oxford: Pergamon Press, UK, pp. 751–769, 1993.

Sharma, B. R., Arora, S. K., Mutation breeding in okra (Abst.), Golden Jubilee Symp, on Genetic Reseasrch and Education—Current Trends and the Next 50 Years, New Delhi, Feb. 12–15, 1991.

Sharma, B. R. and Mahajan, Y. P., Line x tester analysis of combining ability and heterosis for some economic characters in okra, *Sci. Hort.*, 9, 111–118, 1978.

Sharma, B. R., Kumar, V., and Bajaj, K. L., Biochemical basis of resistance to yellow vein mosaic virus in okra, *Genet. Agr.*, 35, 121–130, 1981.

Sharma, B. R. and Sharma, O. P., Field evaluation of okra germplasm against yellow vein mosaic virus, *Punjab Hort. J.*, 24, 131–133, 1984.

Siemonsma, J. S., West—African okra—morphological and cytogenetical indications for the existence of a natural amphidiploid of *Abelmoschus esculentus* (L.) Moench and *A. manihot* (L.) Medikus, *Euphytica*, 31, 241–252, 1982a.

Siemonsma, J. S. La culture du gombo (*Abelmoschus* spp.), legume-fruit tropical (avec reference speciale a la Cote d' lvoire). Thesis Wageningen Agricultural University, The Netherlands, 1982b.

Siemonsma, J. S. *Abelmoschus*: a taxonomical and cytogenetical overview, In *International Crop Network Series*, Report of an international workshop on okra genetic resources, Rome: IBPGR, 5, 52–68, 1991.

Singh, S. B. Genetical studies in okra (*Abelmoschus esculentus* (L.) Moench). Ph.D. thesis, Punjab Agricultural University, Ludhiana, 1979.

Singh, D. Biometrical and genetical studies in okra, *Abelmoschus esculentus* (L.) Moench. Ph.D. diss., Punjab Agricultural University, Ludhiana, 1983.

Singh, H. B. and Bhatnagar, A., Chromosome number in an okra from Ghana, *Indian J. Genet. Plant Breed*, 36, 26–27, 1975.

Singh, B. N., Chakravarthi, S. C., and Kapoor, G. O., An interspecific hybrid between *Hibiscus ficulneus* and *H. esculentus*, *J. Hered.*, 29, 37–41, 1938.

Singh, S. B. and Sharma, B. R., Relative efficiency of different mating systems for improvement of okra, *SABRAO*, 15, 125–131, 1983.

Singh, H. B. and Sikka, S. M., "Pusa Makhmali"- a new find in lady's finger, *Indian Farming*, 4, 27, 1955.

Singh, S. P. and Singh, H. N., Hybrid vigour for yield and its components in okra, *Indian J. Agric. Sci.*, 49, 596–601, 1979.

Singh, S. P., Srivastava, J. P., and Singh, H. N., Heterosis in bhindi (*Abelmoschus esculentus* (L.) Moench), *Progr. Hort.*, 7, 5–15, 1975.

Singh, H. B., Joshi, B. S., Khanna, P. O., Gupta, P. S. et al., Breeding for field resistance to YVM in bhindi, *Indian J. Genet. Plant Breed*, 22, 137–139, 1962.

Singh, S. P. et al., Genetic divergence and nature of heterosis in okra, *Indian J. Agric. Sci.*, 47, 546–551, 1977.

Singh, H.B., Swarup, V., and Singh, B., Three decades of vegetable research in India. ICAR Technical Bulletin, New Delhi, 1975.

Skovsted, A., Chromosome numbers in the family *Malvaceae* 1, *J. Genet*, 31, 263–296, 1935.

Skovsted, A., Chromosome numbers in the family *Malvaceae*, *C.R. Labs. Carls. S. Physiol*, 23, 195–242, 1941.

Sneath, P. H. A. and Sokal, R., *Numerical Taxonomy*, San Francisco: W.H. Freeman, 1973.

Suresh Babu, K. V. Cytogenetical studies in okra (*Abelmoschus esculentus* (L.) Moench). Ph.D. diss., University of Agricultural Sciences, Bangalore, 1987.

Terrell, E. E. and Winters, H. F., Change in scientific names for certain crop plants, *Hort. Sci.*, 9, 324–325, 1974.

Teshima, T., Genetical and cytological studies in an interspecific hybrid of *Hibiscus esculentus* and *H. manihot*, *J.Fac. Agric. Hokkaido Univ.*, 34, 1–155, 1933.

Thakur, M. R., Inheritance of resistance to yellow vein mosaic in a cross of okra species, *Abelmoschus esculentus* X *A. manihot* subsp. manihot, *SABRAO J.*, 8, 69–73, 1976.

Thakur, M. R. and Arora, S. K., "Punjab-7" a virus resistant variety of okra, *Progr. Farm.*, 24, 13, 1988.

Thomas, T. A. and Prashad, R., New okra selections, *Indian Hort.*, 29, 19–21, 1985.

Thomas, T. A., Bisht, I. S., Patel, D. P., Sapra, R. L., and Rana, R. S., Catalogue on okra (*Abelmoschus esculentus* (L.) Moench) germplasm, Part I, New Delhi: NBPGR, 1990.

Thomas, T. A. et al., *Catalogue on Okra (Abelmoschus esculentus (L.) Moench) germplasm*, Part II, New Delhi: NBPGR, 1991.

Tischler, G., Pflanzliche chromosomen-Zahlen (Nachtrag no.1), *Tab. Biol.*, 7, 109–226, 1931.

Ugale, S. D., Patil, R. C., and Khupse, S. S., Cytogenetic studies in the cross between *Abelmoschus esculentus* and *A. tetraphyllus*, *J. Maharashtra Agric. Univ.*, 1(2–6), 106–110, 1976.

Ustinova, E. I., Interspecific hybridization in the genus *Hibiscus*, *Genetica*, 19, 356–366, 1937.

Ustinova, E. I., A description of the interspecific hybrid of *Hibiscus esculentus* and *H. manihot*, *Priroda (Nature)*, 6, 58–60, 1949.

Uthamasamy, S., Studies in host resistance in certain okra (*Abelmoschus esculentus* L Moench) varieties to the laef hopper, *Amrasca devastans* (Dist.), *Pesticides*, 14(8), 39–42, 1980.

van Borssum Waalkes, J., Malesian Malvaceae revised, *Blumea*, 14, 1, 1966.

Vavilov, N. I., Studies on the origin of cultivated plants, *Bull. Appl. Bot.*, 26, 1–248, 1936.

Venkitaramani, K. S., A preliminary study on some intervarietal crosses and hybrid vigour in *Hibiscus esculenta* (L.), *J. Madras Uni.*, 22, 183–200, 1952.

Vredebregt, J. H. Taxonomic and ecological observations on species of *Abelmoschus* Medik. Report of an International workshop on okra Genetic Resources (NBPGR), 1990.

Zeven, A. C. and de Wet, J. M. J., *Dictionary of Cultivated Plants and Their Regions of Diversity*, Wageningen: PUDOC, 1982.

Zeven, A. C. and Zhukovasky, P. M., *Dictionary of Cultivated Plants and Their Centres of Diversity*, The Netherlands: Centre of Agricultural Publishing and Documentation, 1975.

CHAPTER 6

Capsicum

Caroline Djian-Caporalino, Véronique Lefebvre, Anne-Marie Sage-Daubèze,
and Alain Palloix

CONTENTS

6.1 INTRODUCTION

Pepper (*Capsicum* L.) is a member of the Solanaceae family and originates from South and Central America where American Indians domesticated it around 7000 BC. It was introduced in Europe at the end of fifteenth century, after the first voyage of Christopher Columbus, and its use spread rapidly over the Old World continents providing a new source of fresh vegetable and spice. Pepper is one of the most important Solanaceous crops, largely cultivated through the tropics, subtropics, and temperate countries. Its high vitamin content helps improve the human diet and it has the ability to grow in diverse agroclimatic conditions.

Capsicum terminology is complex. The plant and its fruit have several names. For example, in Mexico, it is called *chile* (Nahuatl language from Aztec) and in American English it goes by *chilli*. In the Taino language of the Carib and West Indies, it is called *aji*. It is also known as *pimiento*

(Spanish), *red pepper* and *Pepper* (English), *peperone* (Italian), *piment* (French), *paprika* (German and Northern European), and *Capsicum*.

Fuchs first proposed the botanical name *Capsicum* in 1543, and Linné (1753) further adopted it. The name is a neo-Latin derivation of Greek *kapsa* [κάψα] "box, capsule" and refers to the shape of the fruit. Currently, red pepper or chile pepper designates the small-fruited, pungent varieties. Bell pepper or sweet pepper generally refer to large fruited non-pungent varieties, as do the names *poivron* (French) and *peperone* (Italian) that are augmentative forms of the terms *poivre* and *pepe* (black pepper), meaning "large pepper." Distinction between pungent and sweet fruits did not always lead to confusing terminology: the German *paprika*, Spanish *pimiento* or Turkish *biber* designate all the variety types in the *Capsicum* genus.

Various authors ascribe around 25 species to the genus. This estimate will likely increase with the anticipated discovery of new species. Several authors classified this enormous diverse group of peppers by species, "pod-type," and cultivars. Probably no other crop plant has such a diversity of fruits originating from initial gene pools. This diversity has further increased through domestication and selection by numerous nations that have adopted it for different uses.

6.2 CROP DESCRIPTION

6.2.1 World Production and Utilization

6.2.1.1 Economics of Pepper Production

Peppers have been domesticated for 9000 years. This places *Capsicum* among the oldest cultivated crops of the Americas. Pepper dominates the world hot spice trade. The cultivation of pepper extends over five continents and grows primarily in tropical and sub-tropical countries. *Capsicum* species are cultivated everywhere the climate conditions allow. The growing range extends from 55° south to 52° north latitude. Of the five domesticated species, *C. annuum* is the most cultivated worldwide. It is also the most important species from an economic and nutritional viewpoint.

The production of pepper for the use as spice and vegetable has consistently increased during the last 30 years. World production was estimated at 5.5 billion tons in 1970. Production increased to 11 billion tons in 1990, and more than 24 billion tons in 2004 (FAOSTAT 2004, F.A.O., http://faostat.fao.org) (Table 6.1). World production is distributed unequally on different continents (Figure 6.1, Table 6.1). China, Mexico, and Turkey are the largest producers. In recent years, China has led the world production of peppers (43% of world production), with 639,000 hectares cultivated annually when considering fresh and dried peppers together. Annually, world pepper production takes place on more than 3.6 million hectares (about half of which are in Asia). Land demand for pepper cultivation has doubled in the last 20 years. One may attribute this increased to the intensification and extension of the cultivation areas in Asia. Yield variation is a function of the country, the cultivar type, and the cultivation methods. Yield of large fruited F_1 hybrids is always much higher than that of small fruited and pungent cultivars. Most harvests of fresh pepper take place over four to 12 months. However, in The Netherlands, pepper harvests extend over 10 months through use of heated glasshouses. The Netherlands achieves the world's highest harvest yield of 24 tons/hectare. This yield exceeds that of industrial harvests of determinate cultivars that are harvested in a single manual or mechanical harvest.

In developing countries, pepper production challenges that of the tomato, as the leading vegetable species. The majority of fresh and dried products are sold in local markets. Quantification of the volume of commercial exchanges attached to pepper is difficult in the subtropical countries. Western Europe and North America are the major importing regions of the world. In Europe, fresh pepper importations and exportations reached 680,000 tons in 2000, with a value of EUR 600 million. Spain, Turkey, Italy, Morocco, and the Netherlands are the main European importers and account for 98% of trade volume. Within the last 10 years, sweet pepper trade increased by 60%

Table 6.1   World Production of Peppers and Main Fresh Fruit or Dried Powder Producers in the Different Continents

Total: Pimento, Allspice, Chillies and Peppers, Green	Pepper, Fresh Market		Dried Pepper	
	Production (10^6 t)	Harvested Area (10^3 ha)	Production (10^6 t)	Harvested Area (10^3 ha)
World	24,027	1,656	2,461	1,975
Africa	2,104	277	395	470
Nigeria	720	91	47	31
Egypt	390	26	46	15
Ghana	270	75	22	17
Tunisia	255	20	7	3
Ethiopia	0	0	116	290
Asia	15,578	1,002	1,843	1,406
China	12,028	603	230	36
Turkey	1,790	88	20	9
Indonesia	629	155	0	0
Korea	340	65	0	0
India	0	0	1,100	940
America	3,412	231	130	44
Mexico	1,854	141	55	34
United States of America	978	34	0	0
Europe	2,887	143	138	54
Spain	1,006	22	0	0
Italy	362	14	0	0
The Netherlands	318	1	0	0
Hungary	0	0	57	6
Romania	237	18	30	30

Source: Adapted from FAOSTAT data, Food and Agriculture Organization (FAO), Publishing Management Service, Information Division, Food and Agricultural Organization of the United Nations (FAO), Viale delle Terme di Caracalla, 00100 Rome, Italy, http://faostat.fao.org.

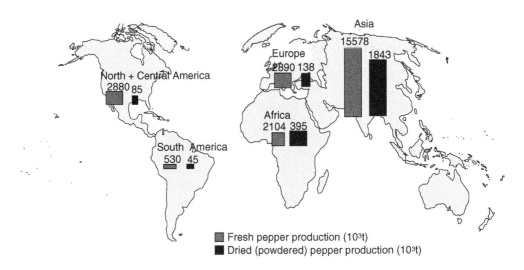

Figure 6.1   Distribution of pepper production (fresh and dried products) over the word cultivation areas. (Adapted from FAOSTAT data, Food and Agriculture Organization (FAO), Publishing Management Service, Information Division, Food and Agricultural Organization of the United Nations (FAO), Viale delle Terme di Caracalla, 00100 Rome, Italy, Last updated 20 December 2004, http://faostat.fao.org.)

(Erard 2002). In the United States, pepper is the fourth leading vegetable crop, following lettuce, tomato, and onions. The estimated United States profit from sale of peppers is US$506 million (http://usda.mannlib.cornell.edu/reports/nassr/fruit/pvg-bban/Vgan0100.txt). Pakistan, Mexico, India, China, and Chile are the main pepper exporters to North America. This does not include the dehydrated spice, extract markets (paprika, cayenne), or production of pungent peppers for fresh market and processing. Salsa and hot sauce also represent a large market in the United States (more than 500 million dollars, Morse 2000). In Hungary, during the nineties, income from dried pepper and oleoresin production for food dye exceeded the income for all other plants.

There are 749 pepper cultivars presently registered in the official European Union catalogue (G.N.I.S. 2006). This ranks pepper fourth in the number of registered cultivars, among vegetable crops in Europe, behind tomatoes, lettuce, and beans for breeding activity. The primary outlets for seed markets are Spain, Italy, Turkey, and the Netherlands. The United States, Mexico, Asia, and Australia comprise the newer seed markets. More than 50% of seeds sold in the Mediterranean countries are from hybrid cultivars from France and the Netherlands. The total value of sweet pepper seed sales by the French companies is around EUR 50 million. Seventy percent of the sales are domestic and 30% are international.

6.2.1.2 Diversity of Use from Spirit Nourishment to Diet Enrichment

Consumers used red peppers in pre-Columbian times in tropical regions of the Americas, possibly as a spice or vegetable to complement taste and augment the poor vitamin content of starchy food. The pepper was also a symbol and an important component of culture and religion. In 1609, Garcilao de la Vega published several myths and tales from the ancient Peruvians of the Cuzco region. He reported that four brothers were the origin of the Inca people. One of them named *Ajar Uchu* was the embodiment of joy and beauty. *Uchu* means "pepper" in Quechuan and still designates the wild *Capsicum* berries (*C. cardenasii*) harvested in the forests of the Amazonian slopes of the Andes. The Aztec priests rubbed their body incisions with hot peppers, taking advantage of the congestional effect of capsaicin, to increase the blood offertory to the gods during times of unfavorable astral conjunctions. Similarly, in ritual preparations of young Caribbean warriors, pepper was rubbed on their wounds. In both cases, they took advantage of the antiseptic and healing power of the high levels of vitamin C found in pepper. A pre-Columbian manuscript (Codice Mendocino) depicts a boy held above an asphyxiating smoke emanating from a blazing pepper inferno. This was probably for medicinal purposes or part of an initiation practice.

A secondary diversification of fruit types, plant adaptation, and modes of consumption resulted from the spread of domesticated peppers throughout the world during the last 500 years. People have used three of the domesticated species, *C. annuum*, *C. frutescens* and *C. chinense*, on a global scale. Both *C. baccatum* var. *pendulum* and *C. pubescens* are extensively used in South America but remain largely confined to that market.

Cooks may use peppers as a fresh or cooked vegetable, a condiment, or a spice. A large variety of pepper recipes and dishes are dependent on the local variety types and on regional cooking traditions. On the world level, pepper provides a large variety of tastes and colors. In North America and Northern Europe, consumption is increasing due, in part, to the increased popularity of Mexican and Mediterranean diet. The food industry also uses peppers more extensively as spice or a coloring commodity worldwide.

At least three chemicals are present in large amounts (vitamin C), or are unique to pepper (the capsaicinoids and the red pigments capsanthin and capsorubin). They increase the nutritive quality and market value of pepper. Several books have reported the chemical compositions of *Capsicum* fruits. Govindarajan (1985–1991) published a complete review.

Pepper fruit, when consumed as a vegetable or condiment, is a concentrated source of vitamins from the A, B, C, E, and K groups. It is an important vitamin source for the world population, particularly in the countries where starch is predominant in the diet. Vitamin C (ascorbic acid) concentrations range from 0.5 to 3 g per kg of fresh weight, depending upon the variety. This exceeds the vitamin C content all other fruits. In 1929, Hungarian scientist, Saent-Gyorgyi, first identified ascorbic acid after it crystallized from sweet pepper fruits. The antioxidant vitamins A, C, and E are also present in high concentrations in various pepper types. Considerable research has focused on antioxidants in *Capsicum* because antioxidants reportedly offer protection against cancer (Hartwell 1971).

Capsanthine and capsorubine are unique carotenoids from the pepper fruits (Davies, Matthews, and Kirk 1970). Those red pigments are widely used in the food industry as an alternative to artificial red food dyes known for their toxicity. The natural red pigments are concentrated in the oleoresin, a complex mixture of compounds extracted from the dried powdered fruits using solvents containing the compounds responsible for the color. The pepper's red pigments have an advantage of their natural origin and thermostability. They are an ingredient in many manufactured food products including beverages, milk-based products, processed meats, candies and sweets, pastries, dressings, and sometimes ketchup. Specific cultivars are genetically improved for their capsanthine content, providing red pigment concentrations up to 100 times that of nonimproved cultivars (Levy et al. 1995; Nuez, Gil Ortega, and Costa 1996). Those cultivars grow in open fields, in high density, using grouped and mechanical harvest, in India, Spain, Hungary, the Southern United States.

Pungency is the trait that made the red pepper famous to consumers and contributed to its spectacular spread around the world. The nature of the pungency has been established as a mixture of several homologous branded-chain alkyl vanillylamides (Hoffman, Lego, and Galetto 1983) named capsaicinoids that are unique to the *Capsicum* genus. The evolutionary role of these compounds was recently established thanks to molecular and physiological studies. The role may serve as a defense against herbivorous mammals and aid in seed dispersal by birds (Tewksbury and Nabhan 2001). Capsaicinoids develop in glands of the fruit placenta. Capsaicinoid concentrations vary greatly between cultivars, from traces (below 10 ppm) in sweet peppers, to a few percentage of dry weight in the hottest cultivars that belong to *C. chinense*. The Scoville test measures organoleptic pungency (ASTA 1985). However, content of the different capsaicinoid compounds must be measured by high pressure liquid chromatography (HPLC) (Collins, Wasmund, and Bosland 1995). Besides its use as a spice, capsaicinoids also possess medicinal properties. Its congestive effect (increase of blood flux in the contact zone) is used for local metabolism stimulation, for example, ointments for sportsmen. It also increases salivation, gastric secretion and peristaltic contractions resulting in activated digestion. Its fibrinolytic activity is also used against blood embolism.

Capsaicinoids were part of traditional medicine in pre-Columbian civilizations. They are also used by the modern pharmaceutical industry (Bosland and Votava 2000). More recently, capsaicin was shown to stimulate the sensory receptor of the anandamide, which is the molecular signal for pain and heat in mammals. Capsaicin is used for medical research on antalgic molecules and pain relief. Conversely, capsaicin is a powerful repellent when sprayed at high concentration. The "bear spray" is already known by backcountry hikers who adventure in bear country without protection of firearms.

Gardeners have used peppers as garden ornamentals, because of their unique and extremely diversified fruit shapes and colors. Most ornamental cultivars are also pungent. Although not poisonous, pungency remains a potential hazard for uninformed users and sweet ornamental cultivars remain to be selected (Shifriss et al. 1998).

6.2.2 Plant Traits and Agroclimatic Requirements

6.2.2.1 *Growth Habit and Plant Structure*

Official descriptors for pepper plants and fruits have been published, in an attempt to classify the tremendous variability and to help in genetic resource characterization (IPGRI, AVRDC, and

CATIE 1995). Pepper plants have a straight, woody primary stem with two to more than 20 internodes, resulting from the growth of its primary apex, which terminates in a flower. Plant growth and development follow a pseudo-dichotomy scheme with one fork (two secondary stems) developing at each internode, whereas the initial shoot turns into a flower (Figure 6.2). The plant may present an erected, prostrate, or bushy habit, depending on the development of auxiliary shoots on the main stem, on the internode length, and the plagiotropy of secondary stems. Plant growth can be determinate or indeterminate. Peppers grown in temperate regions are herbaceous annuals, but are herbaceous perennials in the tropics, where temperatures do not drop below freezing (Yamaguchi 1983).

The fruit is a berry with colors ranging from dark green to ivory-white when immature, and from ivory-yellow to red when mature. There are many variants of yellow, orange, purple, and brown color (Figure 6.3). Fruit shape is highly variable; from round to ovate berries weighing 0.1 g in wild plants and up to 500 g for large-rectangular fruits of traditional Mediterranean cultivars. Seeds develop inside the fruits, attach to the placental veins. Mature seeds are yellow and kidney shaped, with 120–150 seeds per gram. Despite some tendency to partial parthenocarpy in some

Figure 6.2 Polymorphism of *Capsicum* for plant architecture. (1) Main axis and first branching of young plant; (2) indeterminate growth and erected habit with successive fruiting; (3) determinate growth with erected fruits and simultaneous maturing; (4) bushy habit of wild plants (*C. chacoense*); (5) prostrate habit with plagiotropic growth; (6) indeterminate growth and erected habit submitted to two-branch pruning for progressive fruiting under glasshouse.

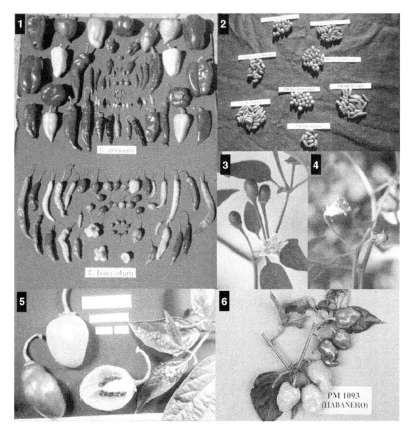

Figure 6.3 Phenotypic variability among *Capsicum* species, samples from the germplasm collection of INRA-Avignon, France. (1) Diversity in fruit shape and color in *C. annuum* (upper part) and *C. baccatum* (lower part of the picture), wild fruit types in the center and domesticated types at the periphery of each group; (2) typical fruits from wild accessions of *C. annuum* var *glabriusculum* with erected and deciduous fruits, fruit weight from 0.1 to 0.2 g; (3) flower and fruit from *C. microcarpum* with yellow spots in the petals characteristic of the *C. baccatum* group; (4) campanulate flower and fruit of *C. cardenasii*; (5) purple flower and fruit with black seeds from *C. pubescens*; (6) fruits from *C. chinense* (var. "Habanero") with multiple fruits (flowers) per node.

cultivars, fruit development depends on fertilization. Fully seeded fruits display a larger, more regular shape and higher capsaicinoid content.

6.2.2.2 Reproductive System

Most cultivated peppers are autogamous. Pollen grains are delivered with the characteristic longitudinal deciduous of the anthers and transported passively by wind or gravity to the stigma at anthesis of the flower. Significant inbreeding effects have not resulted from successive selfing. However, pepper flowers are rich in nectar and attractive for many insects. Rates of out-crossing in open fields depend on the presence of natural insect pollinators. Wild honeybees, domesticated honeybees, flies, thrips, and other insects may act as pollinators. Reproductive behavior of pepper is controversial. A tendency to allogamy increases in some domesticated *Capsicum*, due to protogyny. The style and stigmata extrude from the flower bud one or two days before anthesis. Inversely, other cultivars display a cleistogamous-like behavior with the dehiscence of the anthers preceding flower opening. In open field, out-crossing commonly ranges from 7 to 90%. This argues that *Capsicum* should be considered facultative cross-pollinating species (Odland and Porter 1941; Franceschetti

1971; Tanksley 1984). The amount of cross-pollination has an effect not only on the precautions needed for seed production, but also on the breeding methodologies used by the plant breeder (Bosland 1992). The use of honeybees for the pollination of sweet pepper has become common practice in the Netherlands because it results in larger and heavier fruits with more seeds, and fewer malformed fruits. A few wild species (*C. eximium, C. cardenasii*) are self-incompatible with a gametophytic system as are other wild solanaceae species.

The multiplication rate of pepper plants is high, due to the large number of flowers and fruits produced by a single plant and the numerous seeds in each fruit. One single plant can produce from one to several thousands seeds, depending on the cultivar and growing conditions.

6.2.2.3 Agroclimatic Requirements

Writers have reviewed agroclimatic requirements and cultural in various books, in several languages (Siviero and Gallerani 1992; De Witt and Bosland 1993; Nuez, Gil Ortega, and Costa 1996; Erard 2002). Although peppers, especially hot pepper, are well adapted to hot climates, not all fruit fair well when the night temperature is too high. The optimum temperature for seed germination is 25–30°C. For growth and fruit quality, *C. annuum* should grow in areas with a temperature ranging from 21 to 29°C (Nonnecke 1989). The accepted temperature threshold below, which no vegetative growth occurs is 14°C. When temperature falls below 15°C or exceeds 32°C, growth is usually retarded, blossoms drop, fruit-set ceases, and yield decreases (Knott and Deanon 1967). However, flowering and fruit-set return to normal as temperatures return to the optimum (Greenleaf 1986). The exception is *C. chinense*, which originated from the Amazonian lowlands and is well adapted to growth and fruit setting under very hot and humid climates.

Well-drained, friable sandy loam soil with a pH between 6.5 and 7.5 is optimum for production. Soils below pH 6.0 should be limed to raise the pH before planting. The salt content should be low in the soil. Added organic matter will increase water holding capacity and supply nutrients and minerals for physical characteristics of the plant. Peppers require high soil fertility early in the growing cycle and frequent side dressings to supplement nitrogen. They are particularly responsive to nitrogen and do not have any serious element problems (Greenleaf 1986). The cultivation requires weed control, irrigation, and insect and disease management for maximum fruit production. The plant must be irrigated during root development. Constant irrigation is also important during flowering and fruit set. An excess or deficit of water during this period induces flower abortion or further blossom end rot of the fruits. Steady irrigation practices while the fruit is growing and maturing will enhance fruit quality. Researchers have that drip irrigation is the most effective for crop production. Peppers are considered photoperiod-insensitive (Knott and Deanon 1967).

6.2.3 Diseases and Pests

The geographical dispersion of the cultivations exposes pepper to several parasites. In the tropics and subtropics, humidity and heat favor conditions for multiplication of insect pests and diseases. Intensive and monovarietal cultivations in Mediterranean and temperate regions also favor development of epidemics. Several authors have performed surveys of pepper diseases (Sherf and Macnab 1986; Yoon et al. 1989; Green and Kim 1991) and Pernezny et al. published a recent review (2003). Those reviews highlight the main problems challenging pepper growing. They focus on pathogens, diseases, and cultivations.

6.2.3.1 Viral Diseases

Viral diseases constitute the major limiting factor in successful pepper cultivation throughout the world (Figure 6.4) (Martelli and Quacquarelli 1983; Florini and Zitter 1987; Yoon et al. 1989; Green

Figure 6.4   Symptoms of virus infection in pepper. (1) TMV symptoms; (2) and (3): CMV symptoms with filiformism of leaves and discolored ringspot on fruits; (4) and (5) PVY symptoms with leaf mosaic or vein necrosis; (6) symptoms of CVMV; (7) symptoms of PVMV; (8) discoloration of fruits due to TSWV; (9) leaf mosaic due to AlMV.

and Kim 1991). More than 20 viruses, belonging to 15 different taxonomic groups reportedly cause damage in pepper. Mechanically transmitted viruses such as the tobamoviruses prevail in protected crops, whereas insect-transmitted viruses akin to members of the poty-, cucumo- and tospovirus groups are more frequent and severe in open fields. Virus complexes are frequent in infected plants, increasing the symptoms and making diagnosis very uncertain without serological characterization.

Several tobamoviruses infect pepper. The ubiquitous tobacco mosaic virus (TMV), tomato mosaic virus (ToMV), and pepper mild-mosaic virus (PMMV) are transmitted mechanically by humans when manipulating plants and harvesting fruits, and by seeds on the external tegument. The virus is very stable and sanitation including seed and material disinfection is required for the control of all these tobamoviruses. Several seed disinfection methods provide efficient control. These methods include the use of ethanol, sodium hypochlorite, and sodium triphosphate solutions (Duffé et al. 1989). Infection symptoms may include strong or weak mosaic, necrosis of vegetative parts, sterility, and malformation and discoloration of fruits. A single locus L confers a dominant hypersensitive resistance to these tobamoviruses with several pathotype-specific alleles (Holmes 1937; Cook 1963; Boukema 1980; Csillery and Rusko 1980; Boukema 1982; Kovacs et al. 2004). The allele L^1 from *C. annuum* confers resistance to the common strains of TMV and ToMV (pathotype 0) and the allele L^2 from *C. frutescens* confers resistance to virulent strains of these viruses. However, the latter was not used commercially. The alleles L^3 from *C. chinense* and L^4

from *C. chacoense* confer resistance against the PMMV strains (pathotypes 1–2 and 1–2–3). This overcomes the preceding resistance alleles and has recently spread in protected and intensive cultivations of the Netherlands and Mediterranean regions. PMMV strains express weaker symptoms and are generally restricted to glass- and plastic-protected crops. Many commercial cultivars adapted to these cultivations, possess resistance. In field cultivation, the L^1 allele remains efficient and most of the present cultivars possess this allele.

The cucumber mosaic virus (CMV), cucumovirus, causes severe damage to peppers worldwide. Symptoms include punctiform mosaic dull leaves, filiformism of young leaves, necrotic "oak-leaf" syndrome in older leaves, misshaped fruits with annular discolorations, and sterility, when infection occurred at plantlet stage. CMV may be transmitted from perennial weeds by aphids in a nonpersistent manner. That makes the prevention of virus spread difficult. In nurseries where aphicides or nets can prevent primary contamination, control may be established. No major genes for CMV resistance have been identified in pepper germplasm, but several sources of partial resistance and field tolerance, all with polygenic control and quantitative expression, are documented (Dufour et al. 1989; Pochard and Daubèze 1989; Nono-Womdim et al. 1993a; Caranta et al. 1997b, 2002; Ben Chaim et al. 2001b). These resistant sources only partially restricted the virus spread, but proved to confer a good level of resistance in the field, particularly when different sources combined in a cultivar and a few resistant cultivars were released (Nono-Womdim et al. 1993b; Lapidot et al. 1997; Palloix et al. 1997).

Many different members from the potyvirus genus infect pepper. The potato virus Y (PVY) is common wherever pepper is cultivated, with strains presenting different pathotypes (ability to overcome resistance genes). The pepper veinal mottle virus (PVMV) is the major potyvirus in sub-Saharan Africa. The chili veinal mottle virus (CVMV) is prevalent in Asia. The tobacco etch virus (TEV) and pepper mottle virus (PepMoV) are mostly present in the North, Central, and South Americas.

The pepper severe mosaic (PSMV) and pepper yellow mosaic (PYMV) are common in South-Central America (Green et al. 1991). Aphids transmit these viruses in a nonpersistent manner from solanaceous weeds and crops often in mixture with CMV. Symptoms are highly variable, from vein mosaic to mottling, leaf, fruit and stem necrosis, and plant collapse, depending on the host genotype and environment conditions. Many diversified *Capsicum* cultivars and local populations carry resistance against potyviruses, attesting from the coevolution between this virus group and *Capsicum* species. Resistance to common PVY strains is present in 37% of 782 tested pepper accessions sampled in the INRA germplasm (A. Palloix unpublished data) Researchers have identified several major recessive or dominant genes (Kyle and Palloix 1997) and polygenic resistance with quantitative expression for these viruses (Kuhn, Nutter, and Padgett 1989; Caranta, Lefebvre, and Palloix 1997a,b). Some resistance genes and QTLs confer resistance to different potyviruses such as PVY and TEV for *pvr2*, PVY, PepMoV and PSMV for *Pvr4*, PVMV, and CVMV for the combination *pvr2–pvr6*. The recessive resistance genes of the host and the corresponding avirulence gene in the virus have been identified (Ruffel et al. 2002; Moury et al. 2004). Pepper-potyvirus is an interesting model system to progress in durable management of resistance (Lecoq et al. 2004). The recessive resistance (at *pvr1* or *pvr2*) against PVY and TEV is present in many local populations and hybrid cultivars. Recently, scientists introduced the dominant resistance (*Pvr4*) against PVY, PepMoV, and PSMV. The large choice of available resistance factors is favorable to the construction of genotypes with distinct combination and deployment in the different cultivation regions (Palloix et al. 1998).

Several tospoviruses also cause severe damages to the pepper crops worldwide. The tomato spotted wilt virus (TSWV) is the most ubiquitous. Other tospoviruses, including the impatiens necrotic spot virus (INSV), the groundnut ringspot virus (GRSV) in South Africa and America, and the groundnut bud-necrosis virus (GBNV) in India, display restricted areas of distribution (Moury et al. 1998a, 1998b). Different thrips species transmit these tospoviruses. *Frankliniella occidentalis* caused severe epidemics in Europe during the 1990s. Many weeds species and vegetable or ornamental crops form the virus as well as vector reservoir. Symptoms vary from leaf mosaic to leaf, stem necrosis, and display of annular discoloration, ringspot of fruits. Sanitation in and around greenhouses may achieve disease control. It is essential to discard all infected plants and potential secondary hosts. Biological control of the vector is the only effective way, because *F. occidentalis* develops very rapidly insecticide resistance. In pepper, a single dominant gene, *Tsw*, was shown to confer a dominant hypersensitive resistance to TSWV, but not to the other tospoviruses. This gene has been introduced from *C. chinense* in *C. annuum* (Black, Hobbs, and Gatti 1991; Moury et al. 1997). Like TMV hypersensitive resistance, this gene is not effective at high temperature, inducing the systemy of the virus and collapse of the plant when temperature drops back (Moury et al. 1998). Several recent hybrids possessing the *Tsw* gene were commercialized but virulent TSWV strains rapidly occurred and became prevalent in the fields (Roggero, Masenga, and Tavella 2002; Margaria et al. 2004). The germplasm identified no other major resistance genes. Therefore, looking for alternative sources of quantitative resistance and observing strict sanitation rules, are the priority.

The begomovirus group has to be noticed as a potential threat for infecting pepper crops worldwide. Several begomoviruses were isolated from pepper, particularly in Latin America and India and more rarely in Africa and Southern Europe (Lotrakul et al. 2000; Faria, Bezerra, and Zerbini 2000; Chakraborty, Pandey, and Banerjee 2003; Hussain, Mansoor, and Amin 2003; Hussain, Mansoor, and Iram 2004; Faten et al. 2004).

6.2.3.2 *Fungal Diseases*

Pepper is particularly susceptible to soilborne fungi and fungi like (Oomycetes) complexes causing wilt and collar or root rot in field as well as in protected cropping systems. Some fungi and Oomycetes, like *Phytophthora capsici*, the major cause of pepper root rot, are distributed worldwide. Primary infection occurs mainly on collar and roots at any development stage. Necrotic lesions on lower stems, branches, leaves, and fruits may also be observed (Figure 6.5). Plants wilt suddenly and die. *Phytophthora parasitica*, *Fusarium oxysporum*, and *Rhizoctonia solani* are secondary parasites also causing plant wilt. In temperate climate, *Verticillium* wilt caused by *Verticillium dahliae* induces a progressive plant wilt with an initial dryness of the margin and tip of older leaves. In intertropical regions, the southern blight caused by *Sclerotium rolfsii* is common during the hot season, causing sudden wilt and collar rot similar symptoms to *P. capsici* but producing brownish sclerots at the collar. Water management and crop rotation are of primary importance to control these soilborne diseases because these parasites can survive several years in the soil, due to their necrotrophic ability. Flooding and furrow irrigation are very favorable to all these wilt diseases, as well as any system that maintain a moist soil at the contact of the plant collar. Drip irrigation, well-drained soils, and transplanting in elevated soil layers greatly decrease the disease incidence. Soil treatments with chemicals are recommended before transplanting, but fungicides are not effective if peppers are planted in poorly drained fields with a history of the disease. Several sources of partial and polygenic resistance *against P. capsici and V. dahliae* were characterized in many local hot pepper cultivars and were introduced in a few large fruited cultivars (Palloix et al. 1990a, 1990b; Poulos 1994; Thabuis et al. 2004a, 2004b). The complex genetic control of resistance to these numerous soilborne pathogens makes it difficult to conciliate high resistance level and horticultural traits. Breeding for multiresistant rootstocks is now considered with the additional objective to improve the vigour of the rooting system.

Airborne fungi also damage pepper crops (Figure 6.5). Powdery mildew (*Leveillula taurica*) causes plant defoliation in Mediterranean climates, dry climates, and under glasshouses in northern Europe. *Cercospora capsici* causes foliar necrotic ringspots and plant defoliation in most intertropical regions, and *anthracnose* caused by *Colletotrichum* spp. is the main cause of mature fruit rot and harvest losses in hot and wet climates. Several fungicides may be used against those pathogens, however; their efficiency is reduced because the primary infection occurs several weeks before symptoms appearance. Yellow sulphur can be preventively applied against *L. taurica*. Only systemic fungicides may be efficient against these intercellular fungi. However,

Figure 6.5 Symptoms of fungal and bacterial infections. (1) and (2) Wilt symptoms along irrigation furrows and collar necrosis due to *Phytophthora capsici*; (3) and (4) defoliation and white sporulations on leaves due to *Leveillula taurica*; (5) and (6) leaf necrosis and spots on fruits due to *Xanthomonas vesicatoria*; (7) and (8) wilt symptoms and necrosis of vascular tissue of the primary stem due to *Ralstonia solanacearum*.

their use is not recommended during the harvest periods spread over several months. Several sources of oligogenic resistance were recently characterized against powdery mildew, *Cercospora capsici* and anthracnose in local hot pepper landraces of *C. annuum* and *C. chinense* (Daubèze, Palloix, and Hennart 1995; Lefebvre et al. 2003; Lim and Kim 2003; Pakdeevaraporn et al. 2004; Voorrips et al. 2004). Introgression in large fruited cultivars is underway.

6.2.3.3 Bacterial Diseases

Bacterial leaf spot, caused by *Xanthomonas campestris* pv. *Vesicatoria* and bacterial wilt, caused by *Ralstonia solanacearum*, are the two main bacterial diseases that cause severe damage to pepper crops within the intertropical belt, under hot and humid climates (Figure 6.5).

Xanthomonas campestris causes necrotic or water-soaked lesions on pepper leaves, fruits and stems. It can survive on leaf surface of many weeds and plant debris and it is transmitted by contact, rain splashes, and seeds, as the bacteria remain at the surface of seed tegument from infected fruits. Crop rotation and seed disinfection can help in the control of the disease. Copper spray and bactericide are also used but resistant strains of the bacteria were generated by these practices (Ritchie and Dittapongpitch 1991). Several major genes (*Bs1, Bs2, Bs3*) controlling a dominant race specific resistance were introgressed in pepper cultivars and confer differential crop protection levels, depending on the prevalent races of the bacteria. A high ability to generate multi-virulent races is noted (Hibberd, Stall, and Bassett 1987; Minsavage et al. 1990; Kousik and Ritchie 1998). Alternative sources of resistance were also identified that do not display race specificity and may substitute the former ones in the future cultivars (Poulos, Reifschneider, and Coffman 1991; Szarka et al. 2002).

Ralstonia solanacearum is a soilborne bacterium that infects plant roots, spreads in the plant stems and causes plant wilt at any development stage. *R. solanacearum* also infects many other hosts including solanaceous crops. Very few preventive measures are efficient, except complex organic soil amendments (Hayward 1991; Fegan and Prior 2005). Partial resistance to bacterial wilt was identified in several pepper germplasm collections with oligogenic control and is mostly available in local cultivars (Singh and Sood 2003; Kim and Kim 2004; Lafortune et al. 2005).

6.2.3.4 Nematodes

Root-knot nematodes (RKN) (*Meloidogyne* spp.) have a wide host range and are pathogenic to several solanaceous crops, especially peppers, potatoes, and tomatoes (Khan and Haider 1991; Sasser 1977). Affected plants have a poor appearance, often showing symptoms of stunting, wilting or chlorosis (yellowing) (Figure 6.6) resulting in yield reduction. RKN causes lumps or galls, ranging in size from 1 to 10 mm in diameter. Soil moisture conditions that are optimum for plant growth are also ideal for the development of RKN (Stirling 1991). They are harmful pests particularly under hot climates and under plastic tunnels and greenhouses, where their multiplication occurs continuously (Babaleye 1987). Four species of RKN, *Meloidogyne incognita* (Kofoid and White) Chitwood, *M. hapla* Chitwood, *M. javanica* (Treub) Chitwood, *M. arenaria* (Neal) Chitwood, cause severe attacks, particularly in the Mediterranean area and in central Africa and America (Lindsey and Clayshulte 1982; Fery and Dukes 1984; Thies et al. 1997). The most frequently used nematicid, methyl bromide, has been included in the Montreal Protocol as substances that deplete the ozone layer (UNEP 2004) and this fumigant has been prohibited in industrialized countries since 2005, and in developing countries since 2015. Pepper and tomato were the main crops concerned with methyl bromide soil disinfection. Among recommendations, a mid and long term objective was proposed to develop new alternatives. Soil solarization was recommended where the climatic conditions permit this technique (soil temperature of 50–55°C should be maintained for at least thirty days), alone, or even better, combined with other chemicals like metham sodium, 1,3-dichloropropene, or dazomet.

Biological methods, particularly releasing resistant new cultivars or rootstocks are the main alternative. Researchers have identified many resistance genes from diverse origins in pepper germplasm. The first dominant gene for resistance to *M. incognita*: the *N* gene was identified in the *C. frutescens* L. "Santanka XS" line. This gene had variable efficiency limited to certain *Meloidogyne* species when transferred into susceptible cultivars (Hare 1956). More recently, Di Vito and Saccardo (1979) and Di Vito et al. (1992) have discovered high levels of resistance to RKN in some lines of *C. chacoense, C. chinense*, and *C. frutescens*. Hendy et al. (1983) also found that some *C. annuum* accessions were resistant to RKN populations. Resistance is associated with several dominant genes (Djian-Caporalino et al. 2001; Submitted for publication). Some are highly specific, whereas others are effective against a wide range of species and RKN populations.

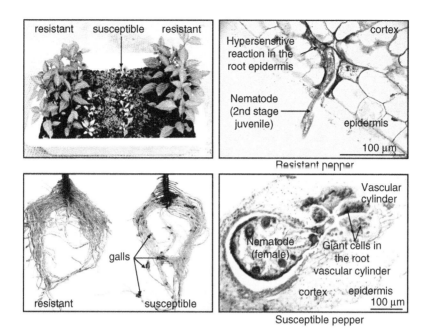

Figure 6.6 **(See color insert following page 304.)** Symptoms caused by root-knot nematodes on susceptible and resistant pepper *Capsicum annuum* L. Galls develop all over the roots of susceptible peppers, plant growth is reduced, and juveniles induce specialized feeding structures (giant cells) which are essential to the nutrition and development of the nematode. In resistant peppers, nematode penetration is limited and post-penetration biochemical responses, including the hypersensitive reaction, block nematode migration and thereby prevent juvenile development and reproduction.

They also present different resistance expression from initial limitation of nematode penetration, post-penetration biochemical responses, or hypersensitive response (Djian-Caporalino et al. 1999; Pegard et al. 2005).

6.2.3.5 Insects and Mites

More than 35 species of insects and mites attack pepper plants, and fruit at pre- and post-harvest stages (Chaney et al. 2003). Young pepper plants are commonly infested with aphids, white flies, thrips, and mites, in the greenhouse, cold frame, and field crops, causing yield decrease up to 75% (Figure 6.7) (Ahamad, Gouse-Mohamed, and Murthy 1987).

There are many species of thrips (*Tripidae*) affecting flowers, fruits, and leaves of pepper plants. Besides TSWV transmission, insect symptoms are heavy, with distorted and curled leaves, silvering or corky fruits. Two main mite species also proliferate under hot and dry conditions: the red arachnids (*Tetranychidae*) that causes leave or fruit bronzing, and the white mite *Polyphagotarsonemus latus* that causes CMV-like symptom with dull appearance of leaves and filiformism. Heavy infections induce necrosis or abortion of vegetative shoots and flower. Several white flies (*Aleyrodidae*) species also damage pepper and can transmit begomoviruses. *Bemisia tabaci* is able to develop and colonize pepper plants, producing strong mottling and vein chlorosis in the absence of any virus. This is probably due to toxic effect of salivary compounds. Many additional insects commonly infect peppers, particularly in hot and humid climates. These include the leaf miner *Liriomyza trifolii*, the flesh-colored caterpillar *Ostrinia nubilalis* (*Pyralidae*), the green caterpillar of *Heliotis* spp. that feeds on the fruit, the European corn borer that commonly feed on pepper stems and fruits, the white larvae of the Pepper maggot *Zonosemata electa* (*Tephritidae*), and the larvae of the fruit fly *Ceratitis* spp. that feeds within the young pods and cause fruit rot (Figure 6.7).

Figure 6.7 Symptoms of insects and pests in pepper (*Capsicum* spp). (1) leaf filiformism and shoot abortion due to mite (*Polyphagotarsonemus latus*) infestation; (2) extremely corky and woody fruits in a bell pepper cultivar (Yolo type) due to thirps (*Frankliniella occidentalis*) infestation; (3) fruit and seed rot with characteristic exit holes of the caterpillar of *Heliotis* spp.; (4) fruit rot due to the fruit fly (*Ceratitis* spp.); (5) vein chlorosis on leaves due to heavy infestation of *Bemisia tabacci*.

The use of insecticides may control most of these pests. One may use synthetic pyrethrinoids or avermectins with short-term efficiency. However, most insecticide and acaricide treatments were shown to destroy auxiliary insects (predators) and other natural control. Increasing pest populations due to over use of chemicals has been demonstrated for thrips, mites, and white flies, which easily develop chemical resistance. Against these pests, biological controls with added or natural beneficial insects or *Bacillus thuringiensis* toxins (against beetles and borers) are the most effective. A few sources of low-level genetic resistance were reported against thrips and white mites (Fery and Schalk 1991; Escher et al. 2002; Maris et al. 2003). However, natural sources of resistance to pests (except nematodes) are lacking in available germplasm (Talekar and Berke 1998).

6.3 ORIGIN, DOMESTICATION, AND DISPERSION

6.3.1 Centers of Origin and Domestication

Several species of pepper were cultivated in America long before the arrival of the Europeans. Consequently, their native countries cannot be unambiguously determined. A prehistoric site in

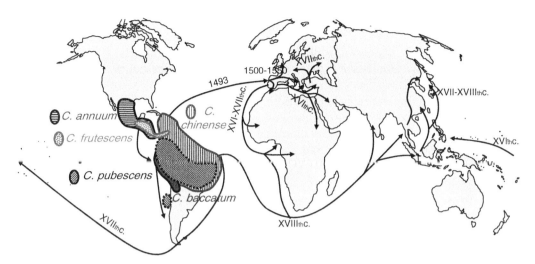

Figure 6.8 Areas of origin and of presumed domestication of the five cultivated *Capsicum* species (colored/ hatched areas in Central and South America) and main routes of spread of the pepper in the world from the fifteenth to the eighteenth centuries. (Adapted from Andrews, J., *Pepper, the Domesticated Capsicum* Austin, University of Texas Press, 163, 1984; IBPGR, AGPG/IBPGR/82/12, Rome, Italy, 49, 1983; Eshbaugh, W. H. et al., *Ethnobiology*, 3, 49, 1983; Somos, A., Akademiai Kiado, Budapest, 302; Pickersgill, B. In *Precolumbian Plant Migration*, Stone, D., Ed., Haward University Press, p. 109, 1984. With permission.)

Peru (dated 8500 to 5500 BC), gave evidence of human utilization of wild plants that were subsequently domesticated. *Capsicum* traces were also found at the site of Guitarrero and the cave of Pachamamay in Peru. In Mexico, the valley of Tehuacan also provided evidence of human use of plants including pepper, amaranth, gourd, and millet, as soon as 7000 to 5000 BC. Archaeological data and collections that are more recent showed that different pepper species were domesticated independently in different places from North to South America (Figure 6.8) and that pepper is one of the oldest cultivated crops of the Americas. This also argues in favor of the hypothesis that agriculture arose at different times and in distinct areas independently. Five domesticated species have been characterized, and both cultivated varieties and wild progenitors are known for four of the varieties (McNeish 1964; Smith 1967; Pickersgill 1969, 1991; Pickersgill, Heiser, and McNeill 1979; Eshbaugh et al. 1983). *Capsicum annuum* can still be collected in North and Central America, and *C. frutescens* in Central America *C. chinense*, whose wild ancestors originated in the Amazon basin may still be collected in the Caribbean and South America. *Capsicum baccatum* (var. *pendulum*) is mostly cultivated in the Andean countries and its wild progenitors (*C. baccatum* var. *baccatum*, *C. praetermissum*, and *C. microcarpum*) are distributed in the Amazonian slopes of the Andes and in the South Amazon basin. The fifth cultivated species *C. pubescens* is mainly cultivated in the Andes and highlands of Central America. Contrary to the other domesticated species, no closely related wild progenitors are known. It was probably domesticated in the Central Andes of Peru or Bolivia, where some wild species (*C. cardenasii, C. eximium*) may be candidates for progenitors of *C. pubescens*, because they maintained a partial sexual compatibility with this cultivated species.

6.3.2 Further Migration and Secondary Centers of Diversification

By the time the Europeans arrived in the Carib and America, natives had already developed a large number of cultivars as attested by the drawings published by the botanists Besler in *Hortus Eystettensis* (1613) and Parkinson in *Theatrum Botanicum* (1640). The main routes of dissemination of pepper in the Old World are summarized in Figure 6.8 and have been described in Somos (1984) and Palloix et al. (2004). The early pepper introductions from Portugal and Spain further spread to Europe thanks to the medical officers attached to the Royal Courts. In the early sixteenth century, pepper was cultivated in the Mediterranean, Central Europe, the Middle East, and West Africa, where religious missions imported it. In the seventeenth century, the Portuguese sailors developed exportations from Brazil to their trading posts in East Africa and the Guinea Gulf. The introductions expanded to include the Indian posts of Goa and Calicut (Berke 1997). Local cultivations have further allowed the spread of American pepper cultivars to China and Japan. Because of the Treaty of Tordesillas (1494), the Spanish sailors had to sail the opposite route from Acapulco and Lima, through the Pacific Ocean, to Indonesia and South China. Local traders from Arabia, Persia, and Malaysia also promoted the spread of pepper to East and West, so that by the eighteenth century, pepper were extensively cultivated in Asia and Africa which became secondary centres of diversification. In Europe, complex exchanges continued, including new introductions from Americas and from Asia and Middle East. European botanists also described some species after secondary importation from Asia. That explains some nomenclature mistakes like *C. chinense*. Several Portuguese trading posts from the Middle East, occupied by the Ottomans, promoted the spread of pepper in their Empire. Today, the long or bell fruits with light green or ivory color are still very popular in both Eastern Europe and Turkey.

6.3.3 Present Distribution of Cultivated Species and of Genetic Diversity

Different pepper species display unequal pattern of distribution. *Capsicum annuum* is grown all over the world with thousands of cultivar types because of secondary diversification. The large phenotypic polymorphism is partially related to the genetic variability (McLeod et al. 1983; Loaiza-Figueroa et al. 1989; Prince et al. 1995; Lefebvre et al. 2001). Genetic polymorphism between cultivars from distant origins (Europe, Africa, Mexico, and India) was measured with biochemical or molecular markers. It represents 20–40% of the polymorphism found between wild forms of *C. annuum* or closely related species. However, modern hybrid cultivars, with large fruits, look much more closely related. This suggests a very narrow bottleneck due to modern selection, compared to the relatively large bottleneck resulting from the successive introductions of *C. annuum* into the old world.

Two other species are also largely spread. *Capsicum chinense* is largely cultivated in West and Central Africa, probably due to its adaptation to hot and humid conditions of tropical lowlands. It results from successive importations from Brazil, by the Portuguese, but also from the Carib during the slave trade period. *C. frutescens* is present in most continents, although cultivated in smaller quantities.

The two Andean species, *C. baccatum*, and *C. pubescens* remained mainly confined to South America, although a few *C. pubescens* populations are cultivated in Northwest China and Tibetan mountains. Agroclimatic conditions in those regions were similar to their geographical origin in the Andean highlands.

No wild species were found outside the centre of origin, except the "Bird pepper" (*C. annuum* var *glabriusculum*) that is cultivated in small gardens. All the accessions described display domesticated traits although some primitive traits (deciduous fruits, smooth flesh) can be observed in some locally cultivated populations.

6.4 TAXONOMY AND GERMPLASM RESOURCES

6.4.1 Taxonomy, Relationships between Domesticated and Wild Species

Joseph Pitton de Tournefort made the first botanical description of *Capsicum* in 1700. He preserved the initial record and name **Capsicum** from Fuchs (1542). Linné (1753) preserved the "*Capsicum*" denomination for the genus and made public three of the first species (*C. annuum*, *C. chinense*, and *C. baccatum*) in 1753 (Species Plantarum) and 1767 (Mantisa Plantarum). Due to the high variation of plant types, floral and particularly fruit morphology that resulted from selection under domestication, the list of *Capsicum* species exceeded one hundred at the end of the nineteenth century. However, taking into account the taxonomic and nomenclatural synonyms, twenty-five distinct species are now currently accepted under the *Capsicum* genus (Baral and Bosland 2002b). All species are diploid and possess $2n = 24$ chromosomes. Only two species (*C. ciliatum* H., B., K. and *C. lanceolatum* Green) were reported to possess $2n = 26$ chromosomes and curiously represent the only sweet (nonpungent) fruited wild species. Nevertheless, their assignment to the *Capsicum* genus is strongly contested. *Capsicum ciliatum* should probably be assigned to the *Witheringia* genus and *C. lanceolatum* to the *Brachistus* genus (Baral and Bosland 2002b). A few cultivars were also reported to be tetraploids resulting from spontaneous or induced chromosome doubling of cultivated *C. annuum*. Differentiation is mainly based on selected morphological characters. Morphology characteristic use for cultivar definition may include flower color, seed color, shape of the calyx and corolla, and the number of flowers per node and their orientation. A key for determination of the main species was published by International Board for Plant Genetic Resources (IBPGR) in (1983). Numerical analysis of morphological characters has been used for taxonomic identification and determination of relationships within the genus *Capsicum* (Pickersgill, Heiser, and McNeill 1979; Eshbaugh 1993). The recognition of distinct species in some taxa with a common ancestral gene pool (e.g., *C. annuum*, *C. frutescens*, and *C. chinense*) is complicated because of overlapping morphology and partial sexual compatibility (Pickersgill, Heiser, and McNeill 1979; Pickersgill 1991). Currently, molecular markers are increasingly being used in phylogenetic analysis. They are reliable in measuring genetic diversity and may better reflect actual genetic differences than morphological information. Combining the results from geographic origin, morphological traits, reproductive behavior, karyotype analysis, biochemical and molecular markers brought most authors to classify the *Capsicum* species into three main complexes (the "true Capsicums") including five cultivated species and their wild relatives. These three groups are organized in two main phylogenetic branches which were named the "white flowered" group including *C. annuum* and *C. baccatum* complexes and the "purple flowered" group including only the *C. pubescens* complex (Eshbaugh 1977, 1980; Pickersgill, Heiser, and McNeill 1979; McLeod, Eshbaugh, and Guttman 1979; Loaiza-Figueroa et al. 1989). Flower color is not a strict distinction trait because several accessions from *C. annuum* possessing the *A* allele for anthocyanin present purple corolla, and some *C. eximium* accessions present white corolla with bluish margins. Among the 20 additional species that were known only as wild, some are tentatively related to the previous complexes (summarized in Table 6.2), but structuration of the *Capsicum* genus complex requires more study.

The *C. annuum* group includes three species: *C. annuum*, *C. frutescens*, and *C. chinense* that can hybridize with each other and are probably derived from a recent common ancestor (Pickersgill 1971). They form a morphological continuum especially at a primitive level (McLeod, Eshbaugh, and Guttman 1979). Genetic evidence from isoenzymes (McLeod et al. 1983; Jensen et al. 1979) and numerical taxonomy (Pickersgill, Heiser, and McNeill 1979) also confirm the close relationship of these three taxa. The wild Bird Pepper *C. annuum* var. *glabriusculum* (Dun.), (also named var. *aviculare* Dierb), grows from Northern South America (Colombia) to Southern United States and the Carib. It is the most probable progenitor of the domesticated *C. annuum*. Crossability studies indicate a very close relationship between these two subspecies (Smith and Heiser 1957; Emboden 1961; Pickersgill 1971). *Capsicum frutescens*, in its primitive form, may be the ancestor of *C. chinense* (Eshbaugh et al. 1983) or *C. frutescens* may be a weedy offshoot of *C. chinense* or *C. annuum*. Isozyme studies and DNA polymorphism, together with hybridization/pollen viability, confirmed the clustering of the three distinct species and their relationships between wild and domesticated *C. annuum* (Loaiza-Figueroa et al. 1989; Prince et al. 1995; Rodriguez et al. 1999; Baral and Bosland 2004). The quantitative data from multiloci

204 GENETIC RESOURCES, CHROMOSOME ENGINEERING, AND CROP IMPROVEMENT

Table 6.2 Described Species of *Capsicum*

Sections	Species Name	Area of Distribution (or Collection)
Capsicum annuum complex	*C. annuum* L. var. *glabriusculum* (wild) var. *annuum* (domesticated)	From North Colombia to Southern United States
	C. chinense Jacq.	Caribbean, Central and South America
	C. frutescens L.	Central America
C. baccatum complex	*C. baccatum* L. var. *baccatum* (wild) (syn: *C. microcarpum*) var. *pendulum* (domesticated)	Argentina, Bolivia, Brazil, Paraguay, Peru
	C. praetermissum Heiser & Smith	South Brazil
C. pubescens complex	*C. pubescens* Ruiz & Pav.	Highlands of South America (Bolivia, Peru)
	C. eximium Hunz.	Argentina, Bolivia
	C. cardenasii Heiser & Smith	Bolivia
Additional *Capsicum* wild species	*C. chacoense* Hunz.	Argentina, Bolivia, Paraguay
	C. galapagoense Hunz.	Galapagos Islands, Ecuador
	C. buforum Hunz.	Brazil
	C. campylopodium Sendt.	South Brazil
	C. coccineum (Rusby) Hunz.	Bolivia, Peru
	C. cornutum (Hiern) Hunz.	South Brazil
	C. dimorphum (Miers)	Colombia
	C. dusenii Bitter	Southeast Brazil
	C. hookerianum (Miers)	Ecuador
	C. leptopodum (Dunal)	Brazil
	C. minutiflorum (Rusby) Hunz.	Argentina, Bolivia, Paraguay
	C. mirabile Mart ex. Sendt	South Brazil
	C. parvifolium Sendt.	Colombia, Northeast Brazil, Venezuela
	C. scolnikianum Hunz.	Peru
	C. schottianum Sendt.	Argentina, South Brazil, Southeast Paraguay
	C. tovarii (Eshbaugh)	Peru
	C. villosum Sendt.	South Brazil

Sources: Baral, J. B. and Bosland, P. W., An updated synthesis of the *Capsicum* genus, *Capsicum and Eggplant Newsl.*, 21, 11, 2002b. Eshbaugh, W. H. The taxonomy of the genus *Capsicum–Solanaceae*, Pochard, E., Ed., *Capsicum 77, Presented at Comptes Rendus 3me Congres Eucarpia Piment*, Avignon-Montfavet, France, 13,1977. Eshbaugh, W. H., The taxonomy of the genus *Capsicum* (*Solanaceae*), *Phytologia*, 47, 153,1980. IBPGR. Genetic resources of *Capsicum*, *International Board for Plant Genetic Resources*, AGPG/IBPGR/82/12, Rome, Italy, 49,1983. Pickersgill, B., Migration of chile peppers, *Capsicum* spp., in the Americas, *PreColumbian Plant Migration*, Stone, D., Ed., Haward University Press, 109, 1984. With permission.

analyses also indicate that intermediate genotypes between semi-domesticated and wild forms of *C. annuum* var. *glabriusculum* and *C. frutescens* were collected in Mexico. That may result from recent genetic exchanges or from the very recent (concomitant to domestication) speciation of these groups.

The *C. baccatum* group is clearly distinguished from the *C. annuum* group and the first includes the domesticated forms (var. *pendulum*), its close wild relative (var. *baccatum* previously known as *C. microcarpum*) and a related wild species *C. praetermissum* (Pickersgill, Heiser, and McNeill 1979; Pickersgill 1991). One particular trait of this group is the presence of yellow spots at the base of the petals. This complex of species shows a restriction of its distribution area to the lowlands of South America, from Peru to Brazil.

The *C. pubescens* group includes *C. pubescens* and the wild species *C. eximium* and *C. cardenasii*. *C. pubescens* has black seeds, a trait unique in the *Capsicum* genus and is known only under its domesticated form. These taxa grow sympatrically in the Andean highlands (from Equator to Bolivia) and can produce fertile hybrids. Nevertheless, their fertility is reduced. Moreover, when *C. cardenasii* and *C. eximium* appear closely related, from the morphological, biological, and genetic points of view, the molecular distance between *C. pubescens* and the wild species is not close. These three species are considered to derive from a common ancestral pool (Eshbaugh 1979, 1982; Jensen et al. 1979; McLeod et al. 1983; Walsh and Hoot 2001).

Among the wild species that were not initially related to the previous group (Table 6.2), *C. chacoense* and *C. galapagoense* should be related to the white flowered group. Chloroplast DNA indicated that *C. galapagoense* probably belongs to the *C. annuum* complex (Walsh and Hoot 2001). *Capsicum chacoense* can produce natural hybrids with *C. annuum* and presented similar genetic distances with the *C. baccatum* and *C. annuum* complexes when considering isozymic and molecular markers (McLeod, Eshbaugh, and Guttman 1979, 1983; Panda, Kumar, and Rao 1986; Paran, Aftergoot, and Shifriss 1998; Rodriguez et al. 1999). *Capsicum tovarii* initially attached to the *C. pubescens* complex, based on allozymic analysis (Jensen et al. 1979; McLeod et al. 1983) but crossability data related this taxon to the *C. baccatum* complex (Tong and Bosland 1999).

6.4.2 Germplasm Resources

Research and technical groups in more than 140 countries are involved in pepper improvement and maintain germplasm collections of various sizes and representativeness. An exhaustive listing of these collections is not available. In all these countries, working collections are maintained by botanical gardens, local or national institutes, and universities and are regularly updated with popular local cultivars. Such local collections are a precious source of local landraces derived from early pepper introductions from the primary and secondary diversification centers, particularly in Africa, Asia, and Europe. Larger collections, including local and international accessions are also available in several countries (Table 6.3). The two main challenges in pepper genetic resources are the phenotypic and molecular characterization of the accessions, and the availability of information to the international community.

IBPGR, in collaboration with the FAO, played a major role in the collection of pepper accessions in the 1980's. This is particularly so in primary diversification centers (Latin America) and in funding genetic resource programs in different countries. The IBPGR gathered information on the content of existing germplasm collections throughout the world and published this in the form of directories of collections (Toll and van Sloten 1982; IBPGR 1983). As the legal successor of IBPGR, the International Plant Genetic Resource Institute (IPGRI) continued the efforts toward the coordination of germplasm characterization by the different national actors and the diffusion of information. IPGRI encourages the collection of data for descriptors on the first four categories of this list: passport, management, environment and site and characterization. Descriptors for *Capsicum* accessions were published in Spanish and English, facilitating the phenotypic characterization and classification by researchers and encouraging the interconnection between the different collections (IPGRI, AVRDC and CATIE 1995).

Major collections for *Capsicum* are listed in Table 6.3. All these collections maintain accessions from the different cultivated species and several wild species. Collections are preserved through seed conservation with middle term storage at 5–6°C with controlled relative humidity (40% rH) that allow seed conservation for more than 20 years. Long-term storage and duplicates are still rarely available. These collections are also heterogeneous in the set of data describing morphological and resistance characters. Among these collections, the European germplasm centers for pepper recently joined their efforts under the coordination of IPGRI to share and centralize passport and descriptor data in a single database (Daunay, Maggioni, and Lipma 2003). The European database is centralized by the Aegean Agricultural Research Institute (AARI) in Izmir (Turkey) with the objective to gather data from the 22,000 accessions maintained in 18 pepper collections from 14 different countries.

Table 6.3   Major Collections for *Capsicum*

Country	Institute	Nb Accessions
Azerbaijan	Genetic Resources Institue, Baku[a]	300
Bulgaria	Institute for Plant Genetic Resources, Sadovo[a]	643
	Maritza Vegetable Crops Research Institute, Plodiv[a]	437
Czech Republic	Research Institute of Crop Production Genebank, Olomouc[a]	514
France	INRA, Montfavet[a]	1,150
Germany	IPK Genebank, Gatersleben[a]	1,433
Hungary	A Institute for Agrobotany, Tapioszele[a]	734
	Vegetable Research Institute, Budapest[a]	1,400
Italy	Istituto del Germoplasma, Bari[a]	167
	Universita di Torino, facolta di Agraria, Torino[a]	400
The Netherlands	Center for Genetic resources, Wageningen[a]	1,048
	Botanical and Experimental garden, University of Nijmegen[a]	90
Poland	Research Institute of Vegetable Crops, Skierniewice[a]	189
Russian Federation	Vavilov Institute Research, St Petersburg[a]	2,313
Spain	UPV, Valencia[a]	907
Turkey	Aegean Agricultural Research Institute (AARI), Izmir[a]	9,980
Ukrain	IVMP, Seletsijne[a]	625
Serbia-Montenegro	Center forVegetable Crops, Smederevska Palanka[a]	352
USA	United States Department of Agricultural (USDA) Plant Genetic Resources Conservation Unit,1109 Experiment Street, Griffin, Georgia 30223-1797	4,710
	The Chile Pepper Institute: New Mexico State University Box 30003 MSC 3Q: Las Cruces, NM 88003	N/A
People's Republic of China	Institute of Crop Germplasm Resources, CAAS12 Zhongguancun nan dajie, Beijing 100081	1,932
Taiwan	Genetic Resources and Seed Unit AVRDC—the World Vegetable Center P.O. Box 42, Shanhua, Tainan 741	5,117
India	National Research Center on DNA Fingerprinting NBPGR, New Delhi-110012	N/A
Costa Rica	Centro Agronomico Tropical de Investigations y Ensenanza (CATIE) 7170, Turrialba	N/A

[a] Collections included in the European Database for pepper.
Source: Adapted from http://www.bgard.science.ru.nl/WWW-IPGRI-Capsicum/Pepperdb.htm.

6.5   CYTOGENETICS

6.5.1   Chromosome Structure of Capsicum

6.5.1.1   *Basic Capsicum Karyotypes*

With a basic chromosome number of 12, the *Capsicum* genome joins most of the other Solanaceous crops (tomato, potato, eggplant, and petunia). The physical size of its genome, with 5.6–7.5 pg DNA/cell (approximately 2,700–3,400 Mbp per haploid genome), averages three to four times that of tomato (750 Mbp) or diploid potato (1,000 Mbp) (Arumuganathan and Earle 1991). Comparing different species for DNA content, a 25% variation was estimated between the smallest (*C. annuum*) and the largest (*C. pubescens*) genomes (Belletti, Marzachi, and Lanteri 1998). All the *Capsicum* species present a very homogeneous karyotype with 1 pair of acrocentric chromosome and 11 pairs of metacentric or submetacentric chromosomes (Figure 6.9). Pickersgill (1971, 1988, 1991) documented one major variation to this karyotype.

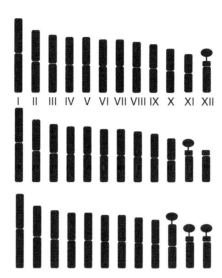

Figure 6.9 Basic Karyotype of *Capsicum*. Top, karyotype of most *Capsicum* species including most wild *C. annuum* and all *C. chinense* and *C. frutescens*; middle, Karyotype specific to all domesticated and a few wild *C. annuum*; bottom, possible variations in the location and number of chromosome satellites (1–3) in the genus (Adapted from Pickersgill, B., Tsuchiya, T., and Gupta, P. K., Eds., *Chromosome Engineering in Plants, Genetics, Breeding and Evolution*, Elsevier, Amsterdam, p. 139, 1991; and Pochard, E., *Ann. Amélior. Plantes*, 20, 233, 1970. With permission.

All domesticated *C. annuum* collected in the Americas and in Europe presented two pairs of acrocentric chromosomes. This karyotype was restricted to a few wild accessions from Central Mexico, suggesting that the domesticated *C. annuum* originated from a few wild forms in this diversification center. Other variations between karyotypes, concerning the number of chromosome satellites (bearing the nuclear organizer and ribosomal DNA) that vary from one to three and can be positioned on the acrocentric chromosome(s) or other chromosome pairs (Figure 6.9). This trait does not correlate with the botanical taxa, but seems more variable within wild accessions than within cultivated ones. This was also true of variation in chromosome size. Again, this argues in favor of a significant genetic bottleneck due to the domestication of *C. annuum* and the other species (*C. chinense, C. baccatum*). Variation in size and morphology of mitotic chromosomes were minor. Vein-banding techniques were not successful in pepper. The 11 (or 10) metacentric chromosomes are hardly recognizable with cytological techniques (Pochard 1970).

6.5.1.2 *Chromosome Interchanges between Capsicum Species*

Meiotic chromosomes were also analyzed in interspecific hybrids. Frequencies of hexavalents and tetravalents in meiotic figures of interspecific hybrids indicated exchanges of chromosome ends affecting seven of the 12 chromosome pairs when considering the five domesticated species and related wild forms. The detailed information on the exchanges between chromosome pairs (translocations) was related to speciation. The number and position of translocation correlates with distances between taxa and fertility of hybrids (Gonzalez de Leon 1986; Pickersgill 1991). Isozyme markers were linked with chromosomal interchanges between *C. annuum, C. chinense*, and *C. baccatum* (Tanksley 1984; Gonzalez de Leon 1986). Localization of chromosomal interchanges in the present genetic maps remains to be established. This information needs to be integrated into the genome mapping data using interspecific hybrids because chromosome translocations generate skewed segregation in such mapping progenies. *Capsicum annuum* (domesticated) and *C. chinense* were shown to differ by one or two translocations plus one chromosome satellite, affecting chromosome segregation (Lanteri 1991). That may explain the discrepancies between the chromosomes P1 and P8 in the interspecific map of Livingstone et al. (1999) and the *C. annuum* intraspecific maps of Lefebvre et al. (1995, 2002).

6.5.2   Chromosome Mapping by Cytogenetics and Molecular Methods

6.5.2.1   *Pepper Genetic Linkage Maps*

Pochard (1970) selected a complete set of the 12 primary trisomics in the self-progeny of a *C. annuum* haploid plant, with the aim to further develop a linkage map and assign phenotypical and biochemical markers to physical chromosomes. These 12 primary trisomics received French coding names representing an array of colors (*violet, indigo, bleu, vert, jaune, orange, rouge, pourpre, noir, brun, bistre, gris*). A few isozyme loci and 10 monogenic traits were mapped in seven of the 12 primary trisomics (Pochard 1977a; Pochard and Dumas de Vaulx 1982; Tanksley 1984). That allowed assigning seven of the further molecular genetic linkage groups to seven physical chromosomes (Tanksley et al. 1988; Lefebvre et al. 2002; Chaim et al. 2003). The first embryonic linkage map of pepper consisted in nine isozyme markers arranged in four linkage groups (Tanksley et al. 1988). With the advent of DNA-based markers in the 1990s, several detailed genetic linkage maps for pepper were constructed. These took advantage of the different tools to detect natural polymorphism, from hybridization-based analysis of restriction fragment length polymorphism (RFLP), to polymerase chain reaction (PCR) based markers (RAPD, AFLP, SSR, and SSAP...). The latest tool is the detection of single nucleotide polymorphism (SNP) by comparative sequencing of alleles. Different molecular genetic markers were developed on *Capsicum* for diverse applications going from diversity studies to marker-assisted selection (reviewed in Lefebvre et al. 2005). The first molecular pepper genetic linkage maps were constructed with RFLP markers (Tanksley et al. 1988; Prince, Pochard, and Tanksley 1993). That enabled the detection of more polymorphism among accessions than isozyme markers (Lefebvre, Palloix, and Rives 1993). Later, RAPD markers, then AFLP markers were exploited because they were easier to reveal in mass than RFLPs. SSR markers publicly available recently increased (Nagy, Polley, and Ganal 1998, 2004; Huang et al. 2001; Lee et al. 2004). More than 450 SSRs were located on pepper maps and should be available in the near future (SGN database: http://www.sgn.cornell.edu/).

Detailed molecular genetic linkage maps have been constructed for the 12 pepper chromosomes in several mapping populations as summarized in Table 6.4. The parents crossed for generating the first mapping populations included *C. annuum* and another sexually compatible species such as *C. chinense* in most of the cases and *C. frutescens*, in a few cases. Interspecific crosses were often preferred to intraspecific crosses because they revealed a high level of polymorphism that is close to 80%. *Capsicum annuum* × *C. chinense* crosses were frequently used because of their relative good compatibility and fertility. Intraspecific linkage maps based on *C. annuum* crosses between bell pepper lines and small-fruited exotic lines were also constructed. These presented high fertility, high recombination rates, and undistorted segregations. Indeed as mentioned above, interspecific hybrids exhibited end-chromosome exchanges between chromosome pairs (Pickersgill 1991). These reciprocal chromosomal translocations may cause errors in mapping process, and link by artifact different linkage groups, as it was supposed in the map of Livingstone et al. (1999). Indeed the Livingstone's linkage group P1 contained the isozyme marker *Idh-1* that was assigned to the *pourpre* trisomics carrying three doses of the P8 chromosome (Tanksley 1984). Most of the individual maps are unsaturated because they displayed more linkage groups than the basic chromosome number. However, alignment of individual intraspecific maps permitted

Table 6.4 Molecular Maps Constructed for Pepper Genome

Parental Species	Population Name	Size and Type of Progeny	Marker Type	No. of Marker Loci	References
Interspecific maps					
[C.a×C.c]×C.a	?	46 BC	Isozyme+RFLP	80	Tanksley et al. (1988)
C.a×C.c	Pepper-AC99 (NuMex-PI159)	46–100 F2	Isozyme, RFLP, RAPD, AFLP, SSR	424	Prince et al. (1993), Livingstone et al. (1999), available on SGN*
C.a×C.c	SNU/SNU2	86–107 F2	RFLP, AFLP, SSR	333	Kim et al. (1997), Kang et al. (2001), Lee et al. (2004), available on SGN*
[[C.a×C.f]×C.f]×C.f	?	248 BC2	RFLP	92	Rao et al. (2003)
C.a×C.f	Pepper-FA03	100 F2	RFLP, RAPD, AFLP, SSR	713	Available on SGN*
Intraspecific maps					
C.a×C.a	DH591	44 DH	Morphological, RFLP	61	Lefebvre et al. (1995)
C.a×C.a	DH702	31 DH	Morphological, RFLP	57	Lefebvre et al. (1995)
C.a×C.a	HV	101 DH	Morphological, RFLP, RAPD, AFLP	543	Lefebvre et al. (1997, 2002)
C.a×C.a	PY (=DH200)	114 DH	Morphological, RFLP, RAPD, AFLP	630	Lefebvre et al. (1995, 1997, 2002)
C.a×C.a	YC	151 F2	RFLP, AFLP	208	Lefebvre et al. (2002)
C.a×C.a	Maor-Perennial	180 F2	Morphological, RFLP, RAPD, AFLP	177	Ben Chaim et al. (2001a)
C.a×C.a	F5YC	297 RIL	RFLP, AFLP, SSR, SSAP	>800	Unpublished data from INRA Montfavet, France
Integrated map					
4 [C.a×C.a]+1 [C.a×C.c]+1 [C.a×C.c]×C.a	HV+PY+YC+Maor-Perennial+NuMex-PI159	101+114 DH +151 F2+180 F2+75 F2+83 BC1	Morphological, RFLP, RAPD, AFLP	2262	Paran et al. (2004) database available on http://www.keygene.com/pdf/in_map_pepper_complete.pdf

SGN*, SGN: Solanaceae Genomics Network database, http://url:sgn.cornell.edu/, Compendium first written in January 2004 by Christine Bunth of (PhD), researcher at Wageningen University, Laboratory for Bioinformatics, PO Box 8128, 6700 ET Wageningen, The Netherlands. The last Major update of the Compendium text was on March 2, 2004.

the arrangement of 12 consensus major linkage groups (Lefebvre et al. 2002) presumed to correspond to the 12-pepper chromosomes, named P1 to P12. The RFLP-based anchoring of pepper maps allowed demonstrating that co-migrating RAPD and AFLP markers correspond to homologous loci across *Capsicum* crosses (Lefebvre et al. 2002; Paran et al. 2004). Integration of the four intraspecific and two interspecific maps developed by the French, Israeli, and American laboratories allowed the gathering of more than 2,200 markers that were ordered and arranged in 13 distinct linkage groups (Paran et al. 2004). The average marker density reached up one marker per 0.8 cM on this map. Three hundred twenty common markers between at least two individual maps facilitated the integration. As this integrated map included the data from interspecific crosses, the chromosome P1 and P8 are not congruent with the intraspecific map of Lefebvre et al. (2002). Once again, we hypothesized that the P1 linkage group of the integrated map resulted from a pseudo-linkage between markers. These markers were close to the interchange breaks between the chromosomes P1 and P8 of the intraspecific maps.

Seven linkage groups were assigned to seven pepper chromosomes. Two linkage groups were assigned to the trisomics *pourpre* and *noir* by allele dosage of the *Idh-1* and *6Pgdh-1* isozyme markers in trisomic F$_1$ hybrids (Tanksley 1984). Six linkage groups were assigned through mapping the six phenotypic markers *C*, *pvr2*, *y*, *L*, *up* (Lefebvre et al. 2002) and *A* (Ben Chaim et al. 2003b) previously assigned to the trisomics *jaune*, *Orange*, *Indigo*, *Brun*, *Noir* and *Rouge*, respectively, by segregation analysis of the primary trisomics of Pochard (1970). Assignment of linkage groups to primary trisomics exhibited the correspondence between chromosome nomenclatures: P2-*Jaune*, P4-*Orange*, P6-*Indigo*, P8-*Pourpre*, P10-*Rouge*, P11-*Brun*, and P12-*Noir* (reviewed in Lefebvre et al. 2004).

6.5.2.2 *Synteny of Capsicum with Other Solanaceae Species*

The RFLP assay is based on nucleic acid hybridization between a labeled nucleic probe and a target DNA sequence. Depending on the experimental conditions, cross-hybridization is detected between similar DNA sequences. Thus, RFLP probes originating from one species can be used in more or less related species, upon condition that both DNA sequences share more than ~70% of similarity. RFLP markers made it possible to compare the genome structure between sexually incompatible species. Comparative genome analysis is a well-developed area of study in the Solanaceae where synteny relationships were thoroughly explored by using mainly tomato-derived probes in the four most important Solanaceae species, pepper, tomato, potato, and eggplant (Tanksley et al. 1992; Prince, Pochard, and Tanksley 1993; Doganlar et al. 2002). Comparative mapping revealed that the tomato and potato genomes differ by only five paracentric inversions on five distinct chromosomes (Tanksley et al. 1992). Most of the tomato and pepper clones reciprocally hybridized to each other, demonstrating that the gene repertoire is also conserved between these two species and that no major losses have occurred during the divergence of these genomes. However, the linear order of the common markers was not conserved along the chromosomes of pepper and tomato. Comparative mapping showed that 18 homologous blocks cover 98% of the tomato genome and 95% of the pepper genome and, that a minimum of 22 chromosome breaks were responsible of the chromosomal translocations and inversions between the both genomes (Livingstone et al. 1999). Comparative mapping revealed that the chromosomes P2, P6, P7, and P10 were roughly colinear to the entire tomato chromosomes T2, T6, T7, and T10, respectively, with some paracentric inversions inside. For the eight other pepper chromosomes, the two chromosome arms were colinear to two distinct tomato chromosome segments. It is supposed that the chromosome breakage preferentially occurred at centromeres, resulting in centric fusion. However, at a micro-synteny level, certain pepper chromosomes would result from more complex rearrangements of the tomato chromosomes (Lefebvre et al. 2002). By comparing the tomato and the pepper maps,

Tanksley et al. (1988) determined a minimum of 32 breakages of tomato chromosomes to account for the position of orthologous genes in pepper.

6.5.3 Mapping Traits of Interest

Pepper molecular maps are used to genetically map loci-determining traits of interest. Many major genes and quantitative trait loci governing qualitative and quantitative resistances to various pests and pathogens have been mapped. These include resistance to the oomycete *Phytophthora capsici*, which is responsible for downy mildew (root rot and stem blight), resistance to the fungus *Leveillula taurica*, which is responsible for powdery mildew, and resistance to the *tobacco mosaic virus* (TMV). In addition, genes for resistance to diverse Potyviruses (PVY, TEV, PVMV, and PepMoV), the *cucumber mosaic virus* (CMV), the *tomato spotted wilt virus* (TSWV), the root nematodes of the genus *Meloigogyne*, and to the bacterium *Xanthomonas campestris* have been identified. Several horticultural traits of primary importance for fertility and fruit characteristics were recently mapped.

6.5.3.1 Genetic Mapping of Mendelian Traits

Currently, a set of useful markers narrowly linked to major genes determining agronomic traits are yet available for marker-assisted selection (Table 6.5). They concern disease resistance traits as well as horticultural traits (fruit color, pungency…). Markers for 19 major genes mapped on nine chromosomes (except P5, P7, and P8) have been developed in different progenies. The gene *cl* (chlorophyll retainer), mapped on the chromosome P1 of an intraspecific cross, is colinear to the tomato chromosome T8 (Efrati et al. 2005) and could correspond to the chromosome P8 of the intraspecific maps. These major gene markers are of interest in plant breeding although the sequence polymorphism used in mapping population may not always be conserved in the breeding populations. SSRs, SNPs, or Indels (insertion/deletion polymorphisms) markers are more promising for this goal. Already six pepper genes determining agronomic traits were isolated, and their nucleotide sequence determined. Cloning of such genes delivers allele-specific markers that undoubtedly provide a valuable resource for marker-assisted selection. Scientists also used the gene-linked markers assigned to chromosomes (listed on Table 6.5) to assign linkage groups of the further linkage maps to the nine concerned chromosomes.

6.5.3.2 Quantitative Trait Loci Identification

Maps permitted a real advancement in the understanding of complex traits (Table 6.6). Numerous resistance reactions to pathogens are under polygenic control in pepper. QTL analyses were performed for resistance against *Phytophthora capsici*, *Leveillula taurica*, Potyviruses (PVY and potyvirus E), and *cucumber mosaic virus* (CMV). In all these cases, a major effect QTL was detected. QTLs acting in epistasis were often identified in the determination of disease resistance in pepper. Results of QTL mapping for disease resistance were reviewed in Lefebvre (2004).

Anchor markers such as common RFLPs make it possible to compare the positions of resistance factors across mapping experiments and to integrate this information into a pepper-integrated map (Figure 6.10). A number of R genes (major effect resistance genes) and QTLs (quantitative trait loci) for resistance to different types of pathogens maps to similar positions, so-called hot spots for resistance. There is such a hot spot on chromosome P10, which includes major dominant R genes *Pvr4* and *Pvr7* both for resistance to *potyviruses*, the *Tsw* gene for resistance to TSWV, the QTL *Pc_10.1* for resistance to *Phytopththora capsici*, and the QTL *Lt_10.1* for resistance to *Leveillula taurica*. At least five further resistance hot spots against different types of pathogens are located on chromosomes P2, P5, P9, P11 and P12. Resistance hot spots may result from allelic series at single

Table 6.5   Major Genes Mapped on the Pepper Genome (Classified by Chromosome Number)

Carrier Chromosome	Locus Name	Effect (Gene)	Useful Marker	References
P1	cl	Chlorophyll retainer mutation	RFLP (CT28) at 3.8 cM	Efrati et al. (2005)
P2	Bs3	Resistance to Xanthomonas campestris race 2	SCAR at 2.1 cM	Pierre et al. (2000)
Unknown (P2?)	Bs2[a]	Resistance to Xanthomonas campestris race 1 (gene encoding a NBS-LRR)	AFLP at 0 cM Cloned gene + SCAR at 4.9 and 5.3 cM	Tai et al. (1999a, 1999b), Kim et al. (2001); Chromosome assignment: unpublished data from INRA-Montfavet
P2	C[a]	Presence/absence of Capsaicinoids (gene encoding acyltransferase 3	AFLP at 5 cM RFLP (TG205) at 0 cM CAPS at 0.4 cM Cloned gene: 5 SCARs in the gene	Lefebvre et al. (1995) Lefebvre et al. (2002), Blum et al. (2002) Lee et al. (2005) Stewart et al. (2005)
P3	Pvr6	Resistance to PVMV when associated to pvr2² gene	RFLP marker (tg057)	Caranta et al. (1996)
P4	C2[a]	Orange fruit color (gene encoding phytoene synthase)	Cloned gene: RFLP	Thorup et al. (2000), Huh et al. (2001)
P4	Pvr2[a]	Resistance to PVY(0), PVY(1) (gene encoding eIF4E)	Cloned gene: SCAR	Caranta et al. (1997a), Ruffel et al. (2002)
P6	y[a]	Red versus yellow fruit color (gene encoding capsanthin-capsorubin synthase)	Cloned gene: SCARs	Lefebvre et al. (1998), Popovsky and Paran (2000)
P6	Rf	Fertility restorer	RAPD at 0.37 cM	Zhang et al. (2000)
P9	Me3, Me4	Heat-stable resistance to root-knot nematodes (Meloidogyne spp)	AFLP at 0.5 cM and at 10.0 cM	Djian-Caporalino et al. (2001)
P10	Pvr4	Resistance to PVY(0), PVY(1), PVY(1–2)	CAPS at 2.1 cM + SCAR	Caranta et al. (1999), Arnedo-Andres et al. (2002)
P10	Pvr7	Resistance to PVY(0), PVY(1), PVY(1–2)	CAPS linked to Pvr4	Grube et al. (2000a)
P10	Tsw	Resistance to TSWV	CAPS at 0.9 cM	Moury et al. (2000)
P10	A	Anthocyanin pigments in the tissues	RFLP (TG63)	Chaim et al. (2003)
P10	S[a]	Soft flesh and deciduous fruit (gene encoding a polygalacturonase)	RFLP at 0 cM (PG)	Rao and Paran (2003)
P11	L	Resistance to TMV	RFLP (tg036) at 6 cM	Lefebvre et al. (1995)
P12	up	Erected fruit	AFLP at 5 cM	Lefebvre et al. (1995), Lefebvre et al. (2002)
Unknown	(Without name)	Stunted growth when associated to the cytoplasm of C. chinense	RAPD at 6 cM	Inai et al. (1993)

[a] Cloned gene.

Table 6.6 Number, Effect, and Map Position of QTLs Associated with Agronomic Interest Traits in Pepper

Trait	Number of QTLs Detected	Effect of the QTLs	Map Position of the Major Effect QTL	References
Number of flowers per node	2 QTLs	28.8 + 39.9%	P2	Prince et al. (1993)
Fruit-related traits	55 QTLs	6–67% according to the QTL	(—)	Ben Chaim et al. (2001a)
Yield and fruit-related traits	58 QTLs	1–25% according to the QTL	(—)	Rao et al. (2003)
Fruit shape	1 QTL /s3.7	2.9–66.7 % according to the trait and to the cross	P3	Ben Chaim et al. (2003a)
Capsaicinoid content	1 QTL	34–38% according to the experimental year	P7	Blum et al. (2003)
Restoration of cytoplasmic male sterility	5 QTLs	8–69% according to the QTL and trait	P6	Wang et al. (2004)
Resistance to *Phytophthora capsici*	7 to 9 QTLs + digenic interactions (depending on the crosses)	43 to 81% according to the resistance components	P5	Lefebvre and Palloix (1996), Thabuis et al. (2003)
Resistance to potyviruses	11 QTLs + 1 digenic interaction	66 to 76% according to the potyvirus strain	P4	Caranta et al. (1997a)
Restriction of cucumber mosaic virus installation in host cells	2 QTLs + 1 digenic interaction	together explaining 57% of the phenotypic variation	P12 (upper arm)	Caranta et al. (1997b)
Resistance to cucumber mosaic virus	4 QTLs + 2 digenic interactions	7–33% according to the QTL	P11	Ben Chaim et al. (2001b)
Restriction of cucumber mosaic virus long-distance movement	4 QTLs + 2 digenic interactions	4.0–63.6% according to the QTL	P12 (lower arm)	Caranta et al. (2002)
Resistance to *Leveillula taurica*	5 QTLs + 2 digenic interactions	Together explaining more than 50% of the phenotypic variation	P6 and P10	Lefebvre et al. (2003)

locus. This is the case for the *pvr1/pvr2/pvr5* (now renamed *pvr2³*), which proved to be alleles at the *pvr2* locus on the chromosome P4 (Ruffel et al. 2002; Ruffel 2004; Kang et al. 2005). Related genes located next to each other in the same narrow genome segment, were similar. Such clustered gene families are supposed to evolve from common ancestors by local gene duplications followed by structural and functional diversification (Michelmore and Meyers 1998). Meanwhile, a single gene with a pleiotropic effect against different pathogens or closely unrelated genes are not to banish too.

Most of the horticultural traits show continuous phenotypic variation because they are controlled by many genes and influenced by the environment. However, fruit pungency (pungent versus sweet), fruit color (orange or yellow versus red), fruit habit (erected versus pendant), fruit texture (soft versus firm), deciduous fruit, anthocyanin pigments in the tissues (tissues reddish or not), and abnormal

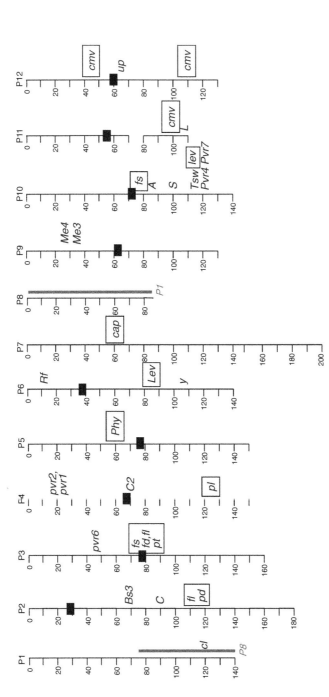

Figure 6.10   Genetic map of pepper with localization of major genes and major QTLs for disease resistance and horticultural traits. Frame map from Paran, I. et al., *Mol. Breed.*, 13, 251, 2004, and Lefebvre, V. et al., *Genome*, 45, 839, 2002. The chromosome P1 is that of intraspecific maps, its lower part corresponds to the chromosome P8 in interspecific maps (in italics). Reciprocally, the chromosome P8 is that of intraspecific map and was designated as chromosome P1 in interspecific maps (in italics). Those two chromosomes probably correspond to the reciprocal translocation that was detected between the two species. Only the major genes and the major QTLs (those with the highest R^2 value for each trait) were positionned from litterature references in Table 6.5 and Table 6.6. Major QTLs are indicated in rectangular boxes. Abbreviations for QTLs: *cap*= capsaicin content, *fd*= fruit diameter, *fl*= fruit length, *fs*= fruit shape (ration length/width), *pd*= pedicel diameter, *pl*= pedicel length, *cmv*= resistance to CMV, *Phy*= resistance to *P. capsici*, *Lev*= resistance to *L. taurica*.

growth (stunted or not) segregated as Mendelian traits were mapped (Table 6.5). The soft flesh and deciduous fruit of pepper is controlled by the same dominant *S* locus. DNA-based markers also made it possible to dissect the genetic components of the quantitative horticultural traits (Table 6.6). QTL analyses have been done for the number of flowers per node, the restoration of cytoplasmic male sterility, fruit-related traits (yield, shape, color intensity), and the capsaicinoid content. A major effect QTL influencing the fruit shape located on the chromosome P3 is conserved in intraspecific crosses and in crosses of *C. annuum* with *C. chinence* and *C. frutescens* (Ben Chaim et al. 2003).

6.5.4 Comparative Genetic Mapping

6.5.4.1 Pest and Pathogen Resistance

The genomic positions of mapped loci in pepper were compared to those of tomatoes and potatoes. Colocalization of resistance loci is frequent within the Solanaceae when rough mapping data are confronted (Grube et al. 2000b) but only few were further confirmed with fine mapping or candidate gene mapping. The genes L in pepper, and Tm-1 and Tm-2 in tomato control resistance to TMV. These three genes do not map to syntenic regions in the two species (Young et al. 1988; Levesque et al. 1990; Lefebvre et al. 1995). The pepper *Tsw* locus and the tomato *Sw-5* locus, controlling both the resistance to TSWV, do not map to corresponding genomic regions too (Jahn et al. 2000). Despite these two examples, several cases of functional colinearities were described that is to say that a colinear position in at least two genera of Solanaceae controls a similar trait. The tomato *pot-1* gene, conferring resistance to PVY and TEV, was identified in a colinear genomic region to the *pvr2* pepper locus (Parrella et al. 2002) and was demonstrated to correspond to the same coding gene (Ruffel 2004; Ruffel et al. 2005a). The *Me3–Me4* locus cluster located on the pepper chromosome P9 was supposed to be in a colinear region of the tomato *Mi-3* locus mapped on the chromosome T12 and the potato *Gpa2* locus on the chromosome XII (Djian-Caporalino et al. 2001). *Me3*, *Me4*, *Mi-3*, and *Gpa2* are involved in the resistance to different genus of nematodes. Colinearities were also suspected for QTLs. The major effect resistance QTL to *P. capsici* on the chromosome P5 in pepper is at a syntenic position to resistance QTLs to *P. infestans* on the potato chromosome IV (Pflieger et al. 2001b; Thabuis et al. 2003) and on the tomato chromosome T4 (Brouwer and St. Clair 2004; Brouwer et al. 2004). The two resistance QTLs to powdery mildew due to *L. taurica* mapped on the pepper chromosomes P6 and P9 show colinear positions with R genes and QTLs involved in the resistance to powdery mildew on the tomato chromosomes T6 and T12 (Lefebvre et al. 2003).

6.5.4.2 Horticultural Traits

Several pepper fruit-related QTLs were located in genome regions corresponding to loci controlling the same traits in tomato, suggesting that these genes may be orthologous in the two species (Ben Chaim et al. 2001a). More recently, comparison of the genomic locations of the eggplant fruit weight, shape and color QTLs with the positions of similar loci in tomato, potato, and pepper revealed that 40% of these loci have putative orthologous counterparts in at least one of these other crop species (Doganlar et al. 2002). Thorup et al. (2000) observed two cases of putative orthologous loci involved in the organ pigmentation. The *CrtZ-2* marker locus, encoding the β-carotene hydroxylases, cosegregated in pepper with QTLs affecting the red chroma and the red lightness in fruits located on chromosome P3, and in potato with the *Y* locus responsible of the white to orange or yellow-turning color of the tuber flesh located on chromosome III. The *CCS* marker locus, encoding the capsanthin capsorubin synthase, mapped to pepper chromosome P6 and perfectly co-segregated with the *y* locus responsible for the white or green to red turning of the fruit color (Lefebvre et al. 1998). The *CCS* gene reportedly possesses lycopene β-cyclase activity (Hugueney et al. 1995).

In tomato, the colinear position of the pepper *CCS* gene corresponds to the *B* locus located on chromosome T6 and defined morphologically by the hyperaccumulation of beta-carotene in fruits. The *B* gene has been cloned and shown to encode a lycopene β-cyclase too, like the CCS gene. These results suggest that the colinear genes originated from common ancestral genes. These observations support the hypothesis that orthologous genetic networks would control related complex phenotypes.

6.5.5 Loci Characterization: Toward Genes and Functions

Three resistance genes were cloned in pepper and proved to belong to two distinct gene families. The *Bs2* gene encodes a putative nucleotide-binding site and leucine rich repeat domain (NBS–LRR; Tai et al. 1999b), that were already described for several dominant resistance genes against various types of pathogens in numerous species. The *pvr2* and *pvr6* genes for resistance to potyviruses were cloned utilizing the candidate gene strategy and validated using transient expression with the PVX virus vector. These loci were shown to encode, respectively, the eukaryotic initiation factor 4E (eIF4E) and an isoform of this gene (*eIFiso4E*) (Ruffel 2004; Ruffel et al. 2002, submitted). These proteins are involved in the formation of translation initiation complexes of eukaryotic mRNAs, and are recruited by viruses for their replication/translation cycle. They were further demonstrated to correspond to recessive resistance genes in diverse plant-potyvirus interactions and draw the attention to a new class of natural resistance genes (Caranta et al. 2003; Whitham and Wang 2004). Candidate gene approach (Pflieger et al. 2001a) based on mapping of DNA fragments with similarity to these two types of genes revealed close linkages of RGAs (resistance gene analogues, with homology to NBS–LRR domains) and different members of the translation initiation complexes to resistance loci (Pflieger et al. 1999; Ruffel 2004, sumbitted). This suggests that NBS–LRR type genes or genes of the translation initiation complexes are candidates for being the molecular basis of a considerable proportion of qualitative and quantitative resistance factors in pepper. Candidates for controlling quantitative resistance may be, in addition, genes functional in pathogenesis or in the signal transduction. A number of such genes have also been mapped that allowed highlighting colocalizations with QTLs (Pflieger et al. 2001b).

Similar to resistance factors, candidate genes for horticultural and fruit traits were mapped, providing perfect colocalizations between candidate genes and loci for phenotypic traits. The locus y/y^+ controlling yellow versus red fruit color, and one of the loci controlling orange fruit ($C2/C2^+$) were, respectively, identified as capsanthin–capsorubin (*CCs*) synthase and Phytoene synthase (*Psy*) genes, both involved in the carotenoid biosynthetic pathway. (Lefebvre et al. 1998; Huh et al. 2001) Similarly, cosegregation and expression analyses permitted to characterize a polygalacturonase gene as candidate genes for the *S* locus controlling soft flesh and deciduous fruits, and a *MYB* transcription factor for the *A* locus controlling anthocyanin in pepper plant (Rao and Paran 2003; Borovsky et al. 2004). Recently, the *C* locus (determining the fruit pungency) has been characterized thanks to a candidate gene approach performed by a Korean group. Kim et al. (2001) isolated 39 cDNAs specifically expressed in pungent fruit placenta by suppression substractive hybridization (SSH). They identified three candidate genes coding for enzymes of the capsaicinoid biosynthesis pathway. One of them, the capsaicinoid synthetase gene (*CS*), shared a homology with an acyltransferase gene and cosegregated perfectly with the *C* locus in the mapping progeny. The comparison between a large range of sweet fruited and pungent fruited cultivars demonstrated that sweetness was associated with a large deletion in the *CS* gene resulting in the absence of mRNA accumulation in all the sweet cultivars (Lee et al. 2005). Stewart et al. (2005) further confirmed that the same candidate *CS* gene (named in their manuscript *AT3*) cosegregated with the *C* locus (named *pun1* in their manuscript) and obtained a decrease in capsaicinoid accumulation using virus-induced gene silencing (VIGS) methodology.

6.6 GENE POOLS

6.6.1 Primary Gene Pools

Gene exchanges within the *Capsicum* genus has mainly focused toward introgression into cultivated species, and particularly *C. annuum*, although many interspecific crosses were attempted
between wild species by taxonomists. Within each species, wild, semi-domesticated and domesticated accessions form the primary genetic pool with reciprocal crossability and nearly fully fertile
hybrids. Considering *C. annuum*, thousands of accessions from the American centers of origin and
from the secondary diversification centers in Asia, Europe and Africa provide usable genetic
resources. This genetic pool presents an extremely high phenotypic diversity but a rather restricted
genetic variability. The very large phenotypic diversity of *C. annuum* results from its spread over
the world since the sixteenth century and the selection for various uses in contrasting environments.
When considering genetic variability, isozyme as well as DNA markers reveal a significant and
progressive restriction of genetic diversity in relation with domestication and selection. The wild
accessions of *C. annuum* (var. *glabriusculum*) and the semi-domesticated accessions are phenotypically homogeneous, but genetically most diversified. The cultivated local populations from
different continents that present the largest phenotypic heterogeneity display a poor genetic polymorphism that is restricted to 20–40% of the wild accessions. Finally, the modern large and sweet-
fruited cultivars only retained approximately 5 to 10% of the genetic polymorphism from the wild
forms (Loaiza-Figueroa et al. 1989; Prince et al. 1992; Lefebvre, Palloix, and Rives 1993, Posh et
al. 1994; Paran et al. 1998; Rodriguez et al. 1999; Lefebvre et al. 2001; Hernandez-Verdugo et al.
2001; Baral and Bosland 2002a). Using phenotypic and molecular markers, the large fruited
cultivars were further divided into two main genetic pools: the Mediterranean group with very
large fruits and vigorous vegetative growth and the American-Dutch group with blocky fruits and
compact plants. Most of the modern F_1 are cultivars derived from hybridization between those
groups (Lefebvre et al. 2001). The pool of cultivated varieties remains a valuable source of
variability for horticultural and adaptation traits and short to mid-term genetic gain. However,
wild relatives of *C. annuum* (var. *glabriusculum*) still represent a larger and under-exploited
primary gene pool.

6.6.2 Secondary Gene Pool

For every cultivated species, the secondary gene pool includes the wild and domesticated species
within the same complex. Most interspecific crosses can be achieved within the three
species complexes (Figure 6.11) producing partially fertile hybrids without biotechnology tools
(i.e., embryo rescue). However, the fertility of hybrids is heterogeneous (pollen stainability rarely
overpass a few percent), and crossability between species is not always reciprocal. For example,
reciprocal crosses between *C. annuum* and *C. frutescens* and between *C. chinense* and *C. frutescens*
delivered fertile hybrids. However, genetic exchanges between *C. chinense* and *C. annuum* were
less successful, leading to abnormal hybrids (virus-like syndrome) and complete sterility when
C. chinense is used as female. This was due to unfavorable interactions between *C. chinense*
cytoplasm and *C. annuum* nuclear genes (Lanteri and Pickersgill 1993; Inai et al. 1993). Within
the same species complex, hybrid seeds were also obtained from crosses between *C. annuum*
(female) and the wild *C. chacoense* or *C. galapagoense*. Male sterility of the hybrids was observed
when *C. chacoense* was used as female, and most hybrids obtained with *C. galapagoense* displayed
flower shedding. Considering the *C. baccatum* complex, all reciprocal crosses between the three
species delivered hybrid seeds (Tong and Bosland 1999), but most hybrids obtained by Pickersgill
(1991) between *C. tovari* and *C. baccatum* proved lethal at the plantlet stage. Considering the
C. pubescens complex, genetic exchanges can be successful between the three species, although

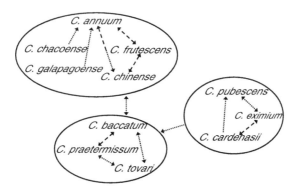

Figure 6.11 Reported exchanges between and within gene pools in the *Capsicum* genus. Arrows point in direction of the female parent. Dashed lines indicate that fertile hybrids are relatively easily obtained using different accessions. Dotted lines indicate that very few hybrids with very low fertility were obtained or depended on the accessions used in the parental species.

the fertility of hybrids between *C. pubescens* and the two wild species is much lower. *Capsicum pubescens* must be used as female when crossed with *C. cardenasii* (Eshbaugh 1975, 1979).

6.6.3 Tertiary Gene Pool

Genetic exchanges between species complexes (tertiary gene pools; Figure 6.11) are much more restricted with frequent expression of unilateral or bilateral incompatibility such as pollen germination or when tube growth is inhibited. Between the *C. annuum* and *C. baccatum* complexes, hybrids with low fertility were obtained only between *C. baccatum* and *C. chinense*, *C. chacoense*, and a few wild accessions of *C. annuum* (var. *glabriusculum*) (Pochard 1977b; Pickersgill 1991; Zijlstra, Purimahua, and Lindhout 1991). A few fertile hybrids were also obtained with cultivated *C. annuum* using doubled pollinations or chemical treatments (Dumas de Vaulx and Pitrat 1977). However, using the wild genetic pool of the cultivated species was shown to facilitate genetic exchanges between distant cultivated species. The *C. pubescens* complex is genetically isolated and breeders have not succeeded with any exchange with the *C. annuum* complex, despite cold and altitude adaptation and for disease resistance. However, hybrids with low fertility were reported between *C. baccatum* or *C. tovari* and *C. eximium* and between *C. praetermissum* and *C. cardenasii* (Esbaugh 1975, 1979; Pickersgill 1988). This suggests that *C. baccatum* complex may be used as a bridge toward *C. annuum*.

6.7 GERMPLASM ENHANCEMENT—BREEDING OBJECTIVES AND METHODS

Open-pollinated varieties have traditionally dominated domestic and commercial pepper cultivation. In the last 25 years, commercial F_1 hybrids of sweet and hot peppers have progressively dominated in temperate climatic regions, but most of the world production still relies on landraces or commercial open-pollinated inbred lines, particularly in Africa, Asia, and Latin America. Considering modern cultivars (F_1 hybrids), breeding programs, and cultivar release are mainly operated by private seed companies with a large investment of several multinational and national companies, originating from the Netherlands, France, and the United States. Competition is sharp between those companies to capture the biggest parts of the seed market in developed countries where the seed price can reach of EUR 0.25 per seed (US$0.33) for the best-performing, large-fruited cultivars.

Companies now compete to dominate new seed markets in developing countries. Market share motivates breeders to develop hybrids in local cultivar types. However, hybrid seed costs remain high and many farmers will benefit more from growing improved open-pollinated types until they can afford the investment in hybrid varieties. Several universities and national institutes in developing countries also maintain breeding programs to improve homogeneity and productivity of local cultivar types and to enhance the value of their genetic resources. The Asian Vegetable Research and Development Center (AVRDC) supports pepper breeding initiatives in developing countries throughout tropical regions. In addition, it promotes information and genetic exchanges between local programs.

6.7.1 Breeding Objectives

One may grow peppers under many different environmental conditions and cultivation practices throughout the world with various end-use products including fresh vegetable, dried spice, food coloring, and pharmaceuticals. Diverse breeding objectives, growing conditions, and end-use products and are referenced in different books (Somos 1984; Greenleaf 1986; Nuez, Gil Ortega, and Costa 1996; Bosland and Votava 2000) and in the proceedings of the Eucarpia Meetings on Genetics and Breeding of *Capsicum* and Eggplant. Palloix (1992) and Poulos (1994) also published syntheses of pepper breeding achievements and perspectives. The main breeding objectives in pepper concern the adaptation to abiotic constraints, the plant ideotypes and cultivation practices, horticultural types and fruit quality, and the resistance to pests and diseases.

6.7.1.1 Adaptation to Abiotic Constraints

Several abiotic stresses affect pepper growth and fruit production. These include the abiotic stresses of flooding, drought, soil or water salinity, low light intensity, and supra- or suboptimal temperatures. These stresses can affect directly the plant growth and fruit yield or can decrease the fruit quality by inducing fruit malformation, blossom end rot or discoloration (stip). The objective shared by many breeders is to extend the pepper cultivation area and period toward "low energy" conditions (northern temperate climates or winter cultivation in Mediterranean climates) or toward the very hot and humid season in intertropical areas when fresh vegetables are lacking. Breeding for tolerance to the abiotic stresses is the most challenging objective for breeders who have to take into consideration complex criteria with late expression such as pollen production and fertility, fruit set, flowering earliness, vegetative growth, canopy, and rooting system. Those traits are under quantitative and polygenic inheritance. Moreover, most damages are due to interactions between different stresses. For example, blossom end rot is due to an interaction between water stresses and mineral nutrition at fruit setting. Stip symptom results from an interaction between low light intensity, low temperature, and mineral nutrition during fruit development. These different environmental conditions are hard to control. Therefore, researchers may not easily reproduce testing conditions. This makes multilocal and repeated trials necessary to evaluate breeding populations. *Capsicum annuum* is adaptable to large abiotic variances. Higher levels of stress tolerance may occur in local populations of *C. chinense* and *C. baccatum*. These abilities to adapt, should allow breeders to surmount the present limitations to *C. annuum* cultivation.

6.7.1.2 Plant Ideotypes and Adaptation to Cultivation Practices

Distinct plant ideotypes exist in sweet as well as hot peppers. These vary with the cropping system and cultivation practices. Peppers grown for dried powder or oleoresine production thrive at various densities in open fields (250,000–300,000 plants per hectare). Harvesting methods for this type of production requires a short, determinate growth habit with early and simultaneous fruit maturation. Conversely, pepper cultivars raised for the fresh consumption market are grown at lower densities (25,000–50,000 plants per hectare) and harvested over a long period. Harvesting may extend from several months in open fields to a full year under glasshouse. These cultivars require breeding for vigorous indeterminate or semi determinate plants with continuous fruit set at successive plant nodes. Breeding may be for plant architecture traits including branching, internode length, and canopy density. Breeding characteristics need to be further adjusted for particular cultivation methods. Higher vegetative vigor is required for open field than for cultivars grown in protected cultivation. Branching ability may depend on the trellising and pruning practices. These different practices require breeding for distinct morphological and specialized plant ideotypes that are not interchangeable.

6.7.1.3 Horticultural Types and Fruit Quality

Considering the wide range in plant genetic diversity, in ethnic preferences and in end-use products, it is not realistic to numerate exhaustively the horticultural types. A grouping in 7–8 main horticultural types has been proposed for genetic resource characterization by IPGRI (IPGRI, AVRDC, and CATIE 1995) that was completed in the 2001 report (Daunay et al. 2001). In breeding programs, specific selection criteria are considered, with distinct levels of requirements, particularly in fruit morphology and chemical composition (color and pigment content, pungency and capsaicinoid content).

Every horticultural type is stringently selected for a specific fruit shape and size that have to meet standard requirements in various criteria. The main criteria are fruit width and length, number locules, pointed, spherical, or quadrangular shape, pericarp thickness, external appearance (smooth, dented, corkiness). For example, fruits for fresh market in Mediterranean countries are bred for their quadrangular (long or half-long) and smooth shape, four locules, maximum fresh fruit weight (from 300 to 500 g per fruit) that requires a thick pericarp. Conversely, fruits for dried powder are selected for various shapes (conical, spherical), all with a moderate size (below 50 g), but high dry matter content and thin pericarp for easy drying.

Breeding objectives for chemical composition also depend on the end-use product. Pepper fruit color mainly depends on carotenoids composition, content (more than 20 distinct carotenoids were identified in pepper fruits) and on chlorophyll and phenolics (anthocyanins). These traits are controlled by major genes y, $c1$, $c2$ (Hurtado-Hernandez and Smith 1985) that confer presence/absence of some pigments, and quantitative genetic factors (content in each pigment). Presence of pungency is also governed by a major gene C and by quantitative factors that control the composition and the content in the different capsaicinoids. This permits breeding for different color classes and color intensity in each class as well as for sweet or pungent fruits and quantified degrees of pungency. Ornamental and fresh market peppers are bred for a very wide range of fruit colors at the immature as well as mature stages to diversify the market and attract the consumer. Peppers for dried powder and oleoresin (food dye) production are bred for red carotenoid content, capsorubin, and particularly capsanthin, and for capsaicinoid content in the final product. As chemical composition of the fruit changes during ripening and post harvest processing, sampling fruits for screening and selection must be rigorous. Color evaluation may be performed with rapid and simplified procedures (eye evaluation, reflectance measurements) but quantitative measures of red pigments in the extracts uses standard procedures with spectrophotometric process (ASTA method) or HPLC. Similarly, pungency is measured with the standard organoleptic test of Scoville but precise capsaicinoid content requires standardized HPLC procedures (ASTA 1985; Collins, Wasmund, and Bosland 1995; Levy et al. 1995).

6.7.1.4 Resistance to Pests and Diseases

Breeding for genetic control of pepper pests and disease must consider the diversity of pepper pathogens (see Section 6.2.3). Only major challenges in terms of incidence and cultivated areas will be commented upon in this chapter. Breeding for disease resistance in pepper has been detailed in previous reviews (Greenleaf 1986; Palloix 1992; Poulos 1994). Review updates should include recent releases of resistant cultivars, new pathogens, or virulent strains adapted to previously resistant cultivars, and new policies in chemical treatments that enhance the interest for genetic resistance to soilborn pests and disease.

The release of resistant cultivars in the past decade attests to the efforts and success of breeders in meeting this objective. Considering the range of pepper pathogens in Section 6.2.3 of this chapter, genetic resistance to the tomamoviruses (TMV, ToMV, PMMV) controlled by the dominant alleles at the L locus has been introduced in most of the present cultivars, polygenic resistance to CMV is available in many local cultivars and has recently been available in F_1 hybrid large fruited types. A large variety of major genes or polygenic resistances against potyviruses has been identified and made available in several horticultural types. Resistance to the major fungal diseases powdery mildew (due to *L. taurica*), anthracnose (due to *Colletotrichum spp*), and *Phytophthora* root rot, was identified in several accessions. For all these resistance traits, objectives are mainly turned toward the introgression of the resistance into the different horticultural types in relation to the pathogenic status of the production regions. Most efforts have been developed in the large fruited and sweet pepper types adapted to the temperate or Mediterranean cultivation regions. However, few multi resistant cultivars were bred for adaptation to intertropical regions where either ubiquitous pathogens (TMV, CMV, PVY, *L. taurica*, *P. capsici*) or region-specific pathogens (*Colletotrichum*, potyviruses other than PVY, bacteria) cause severe yield damages. For this purpose, concerted or network programs with multiple partners were initiated (Poulos 1994; Palloix et al. 1998; Ahmed et al. 2001) and need to be sustained.

Relatively few novel pepper diseases or newly virulent pathogen strains have recently emerged or become prevalent in the pepper fields. Several begomoviruses were isolated from pepper, particularly in Latin America and India, and less frequently, in Africa and Southern Europe (Faria, Bezerra, and Zerbini 2000; Lotrakul et al. 2000; Chakraborty, Pandey, and Banerjee 2003; Hussain, Mansoor, and Amin 2003, 2004; Faten et al. 2004). Accessions with partial resistance were also characterized. Prevalence and virus spread remained much lower than in tomatoes but pathologist and breeders have to be aware because tomato and pepper are commonly co-cultivated in many cropping systems. Several tospoviruses including tomato spotted wilt virus (TSWV) spread over pepper cultivation regions in the past 10 years as a result from the spread of its favorite vector *Frankiniella occidentalis* (Moury et al. 1998b). Rapid and wide dispersion of the TSWV, required breeders to deploy quickly the single dominant allele *Tsw* previously identified in the germplasms. However, virulent strains emerged rapidly within these cultivars and are now prevalent in several cultivation regions. Researchers predicted this from laboratory experiments (Moury et al. 1997; Roggero et al. 1999; Roggero, Masenga, and Tavella 2002; Margaria et al. 2004). Similarly, virulent strains on the tobamovirus PMMV and of the bacteria *Xanthomonas campestris* emerged after intensive cultivation of resistant cultivars in disease favorable environments. The virulent strains of PMMV did not become prevalent and remains confined to intensive glasshouse cultivation. The virulent races of *X. campestris* can be controlled by polygenic field tolerance (Poulos, Reifschneider, and Coffman 1991) or by a new major gene (the GDR gene of Szarka et al. 2002) whose durability remains to be tested. However, for these diseases and for TSWV, no or very few alternative major resistance genes or alleles have been identified and prophylactic practices have to be considered while breeders are looking for alternative sources of resistance including polygenic and quantitative resistance, or field tolerance.

Breeding for resistance or field tolerance to soilborne pathogens has beena relevant challenge for years. The challenges have increased with the enforcement of new policies in European

Community concerning chemicals and the eradication of the major soil fumigants. Pepper is particularly susceptible to wilt diseases caused by ubiquitous soilborne microorganisms that cause severe damages worldwide. In most of the cases, wilting results from a pathogenic complex including one or more primary parasites. In hot and tropical climates, these may include *Phytophthora capsici, Verticillium dahliae, Meloidogyne* spp. *Sclerotium rolsfii, and Ralstonia solanacearum*. Secondary or opportunist parasites infect pepper previously infected by a primary parasite or when submitted to abiotic stresses. These may include *Fusarium* spp., *Pythium* spp. *Phytophthora parasitica, Rhizoctonia solani*, and *Sclerotinia sclerotiurum*. Breeding for resistance to *P. capsici* and/or *V. dahliae* has been achieved by several partners with moderate success due to the polygenic control of these partial resistances and difficulties in recovering a high gain in horticultural traits. More recently, breeders introduced major genes controlling a wide spectrum of resistance to the root knot nematodes in different horticultural types and should be successful. Genetics of partial and oligogenic resistance to bacterial wilt (*R. solanacearum*) was recently analyzed and should be soon introduced into large fruited varieties. However, breeding for resistance to the secondary parasites do not seem realistic because only low tolerance levels were identified in germplasm and artificial inoculations are poorly repeatable. Breeding objectives in this matter have to consider resistance to primary soilborne pathogens, tolerance to abiotic stresses and particularly structure of the rooting system to increase the vigor of the weak rooting system of large fruited peppers. In this objective, introgressions from related species (*C. chinense, C. baccatum*) are also considered. For the short-term selection, breeders recently considered the release of multi resistant rootstocks, avoiding the problem due to genetic drag and horticultural performances.

6.7.2 Breeding Methods

Pepper is an autogamous genus (with exception of a few wild species) and traditional landraces or open pollinated populations are regarded as inbred cultivars. Until the middle of the twentieth century, farmers selected and maintained the inbred landraces through massal selection. This lead to the constitution of several secondary diversification centers in Mediterranean, Sub-Saharan, and Asian production regions, with adaptation of the cultivars to the local biotic environments and consumption customs. Landraces or open pollinated populations are presently the main cultivars in the pepper species *C. chinense, C. frutescens, C. baccatum*, and *C. pubescens*. In *C. annuum*, they are still largely cultivated in the developing countries. Moreover, locally adapted open pollinated populations were derived from recent inbred or F_1 cultivars. Breeding on a scientific basis started in the extension services of universities from the United States, with the introgression of monogenic resistances to viruses (Greenleaf 1956; Cook and Anderson 1959; Cook 1960) and the selection of pure inbred lines with an improved homogeneity and yield potential that are still cultivated in many regions like "Yolo Wonder," "Early Calwonder." A new step was reached in the pepper breeding activity with the release of the F_1 hybrids. INRA (France) released the first hybrid, "Lamuyo," in 1973 and it is still used commercially. Since that time, 90% of officially registered cultivars in Europe are F_1 hybrids. Private seed companies are the leaders in pepper breeding. Heterosis has been demonstrated for a few traits in pepper (Lippert 1975) including some yield components. The main one is the combination of earliness and sustained production of fruits with a constant quality (fruit weight and shape) over the whole cultivation period. Competition for allocation of resources toward reproductive organs (fruits) or vegetative growth is very strong in large fruited pepper types. The hybrid genetic structure permitted to obtain the best compromise from crosses between two main horticultural pools, the early cultivars with compact plant habit and the vigorous cultivars with late but sustained production. However, parental lines are primarily selected for their inbred line performance including horticultural and resistance traits and further for their specific crossing ability measured in the commercial yield of hybrid combinations. Some parental lines display high crossability and are used in several hybrid combinations.

6.7.2.1 Conventional (Sexual) Breeding, Exploitation of Intraspecific and Interspecific Variability, Impact of Haploidization and Marker-Assisted Selection

Pure line and pedigree selection remain the most common methods used by pepper breeders and development of doubled haploids has proven to be of particular interest in reducing the time for homozygous line fixation. Haploids have been obtained in various ways in pepper, including polyembryony, and *in vitro* androgenesis proved to be the most reliable method. The technique was initially set up by Sibi et al. (1979), Dumas de Vaulx and Chambonnet (1980), and Dumas de Vaulx et al. (1981) and was further improved by several authors (Mityko et al. 1995) to increase the rate of haploid regenerations from anthers and to reduce the genotype-dependency of the response. Doubled haploid production is now commonly used to generate new genitors from heterozygous plants in the early phases of pedigree selection (Pochard, Palloix, and Daubeze 1986). It is also used to estimate genetic gain across selection cycles (Thabuis et al. 2004a) and to accelerate and secure the fixation of homozygous lines.

To exploit the variability of primary and secondary gene pools, both backcrossing and genotype construction through recurrent selection of populations has been widely used in pepper. Backcrossing remains an essential method for introgressing genes into improved cultivars: most of the original traits, particularly disease resistance have to be introduced into large fruited cultivars from small fruited and pungent accessions of intraspecific or interspecific origin. Recovering the favorable genetic background requires many backcross cycles and has to be repeated in every horticultural type. Most of the monogenic traits for disease resistance or fruit quality were successfully introduced from distant but intraspecific crosses due to the high variability of the *C. annuum* species. The traits introduced from interspecific crosses essentially concern monogenic disease resistance (Table 6.7). Recurrent selection of populations has been performed for polygenic and quantitative disease resistance where backcross selection proved to be ineffective. In pepper, polygenic resistance with quantitative expression was used against several major pathogens. Formerly, there were no major resistance genes in the available germplasm (particularly for *P. capsici*, *V. dahliae*, CMV). Polygenic resistance was also useful when virulent strains challenged the resistance of released cultivars (*X. campestris*, potyviruses) (Padgett et al. 1990; Palloix et al. 1990a, 1990b; Poulos, Reifschneider, and Coffman 1991; Palloix et al. 1997). Looking for quantitative resistance in pepper germplasm generally reveals a large variability of resistance sources carrying partial resistance components, each one restricting one or several steps of the infection cycle and finally reducing the disease rate in the field. In pepper, such partial resistance components

Table 6.7 Traits Introduced into *C. annuum* Cultivars through Interspecific Srosses

Donor Species (Accessions)	Trait (Gene)	References
C. frutescens (cv Tabasco)	Hypersensitivity to TMV (L^2)	Cook (1960)
C. chinense (PI 152225, PI 159230)	Hypersensitivity to PMMV (L^3)	Boukema (1980)
	Hypersensitivity to TSWV (*Tsw*)	Black et al. (1991), Moury et al. (1997)
	Resistance to PVY and TEV (*pvr1*)	Greenleaf (1956), Kyle and Palloix (1997)
PRI 95030	Resistance to *Colletotrichum* spp.	Voorrips et al. (2004)
C. chacoense (PI 260429)	Hypersensitivity to PMMV (L^4)	Boukema (1982)
	Hypersensitivity to *Xanthomonas vesicatoria* (*Bs2*)	Cook and Guevara (1984)
C. baccatum (Pen 3–4)	Partial resistance to CMV (polygenic)	Pochard (1977b), Dufour et al. (1989), Caranta et al. (2002)

were characterized and combined in breeding populations with large genetic basis under recurrent selection. After several intermating cycles alternating with increasing selection pressure, transgressive genotypes were delivered that displayed high resistance levels (Palloix et al. 1990a, 1990b; Palloix 1992). The demonstration was very conclusive for CMV resistance were the combination of genes conferring a delay in virus vascular movement, a reduction or virus multiplication and a lower probability of virus installation in host cells lead to inbred lines close to virus immunity whereas those individual components only resulted in a slight delay of plant infection (Figure 6.12). At the end of the recurrent selection process for resistance to *P. capsici*, molecular markers were used to analyze the changes in resistance allele composition in the population (Thabuis et al. 2004a). This experiment clearly showed how the frequency of resistant alleles and their progressive accumulation in genotypes increased during selection. The genetic gain also increased when interactions between plant genotypes and environment or pathogen genotypes were taken into account. Similar recurrent selection of multiparental populations was also used to deliver multi-resistant genotypes (Ahmed et al. 2001).

Marker assisted selection (MAS) in pepper has taken advantage from the DNA markers delivered for numerous major genes governing horticultural as well as resistance traits (Table 6.5). Most of these markers were designed for use in selection (SCAR or CAPS markers) and are commonly used by breeders in their pedigree or backcrossing selection. A few genes have been cloned and PCR-based markers specific to the targeted allele are now available. These markers were possible because of the single nucleotide polymorphisms (SNPs) between the different alleles as for the allelic series at the *pvr1/pvr2* locus for potyvirus resistance. The deletions differentiated the alleles at the *pvr6* (resistance to PVMV), *y* (red versus yellow fruit color) and *C* (pungent versus sweet fruits) loci. Such allele-specific markers allow direct screening for the favorable alleles. This subsequently allows selection in any genetic background, and escapes risks due to recombination between the targeted locus and the marker. Many quantitative trait loci were also flanked with markers (Table 6.6). This promoted marker assisted selection for the quantitative horticultural and resistance traits. The use of MAS for polygenic traits is new in pepper for resistance to *P. capsici* (Thabuis 2002; Thabuis et al. 2004b). In this experiment, four and three QTLs were introgressed from two small-fruited peppers into large fruited cultivars using markers bordering and inside the QTL confidence intervals and markers of the recipient genetic background. MAS proved to be effective because the targeted QTLs and expected resistant phenotypes (foreground selection) were recovered after three backcross cycles, validating the additive and epistatic effects of the QTLs. This experiment also demonstrated that MAS accelerated the recovery of the genetic background (background selection) compared to phenotypic selection. This was particularly true for chromosomes that did not carry the resistant QTLs (Figure 6.13). However, recombination in the chromosomes carrying the resistance QTLs was poor and indicated both the necessity to reduce the confidence intervals and to increase the selection effort for recombinant chromosomes. Moving from one genetic background to another was also limited because several markers flanking the QTL confidence interval were no longer polymorphic in the new genetic background that differed from that of the mapping experiments.

6.7.2.2 Mutation Breeding and Male Sterility

Mutation breeding was attempted in pepper using both chemicals and x-rays or gamma irradiation (Saccardo and Vitale 1982; Csillery 1983) with the objective to knock out genes and generate new traits including disease resistance, horticultural traits and male sterility. Many other mutants were obtained with altered phenotypes, but their expected use as phenotypic markers for genetic mapping was restricted because of the development of molecular markers. Among mutants of horticultural interest, only a few male sterile mutants presented a stable sterile phenotype. These were commercially exploited for hybrid seed production (Milkova and Daskalov 1984; Shifriss 1997).

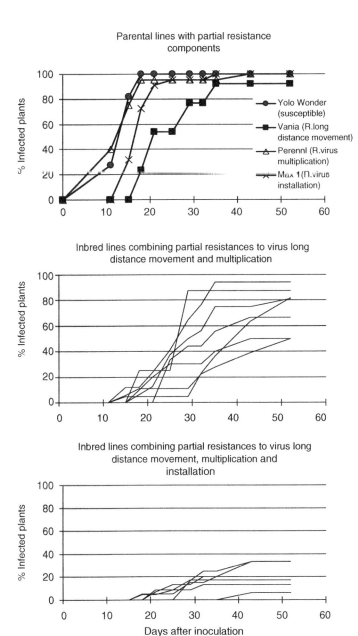

Figure 6.12 Breeding for CMV resistance in pepper. Partial resistance components affecting distinct steps of the virus infection cycle were characterized in different pepper accessions: partial resistance to virus installation in host cells, restriction of virus multiplication, restriction of long distance movement of the virus through the vascular tissue. These individual resistance components only delay the infection process of 10–15 days after artificial inoculation of young plantlets (top graphic), although they confer field tolerance at the adult plant stage. When partial resistances to virus movement and to virus multiplication are combined in a single genotype, plantlet infection is largely delayed affect 50–90% instead of 100% (middle graphic). When the three components are combined in a single inbred line, plantlet infections never surpass 0–30%. Those genotypes are close to immunity in field epidemics, attesting of the transgression obtained by favorable allele combination in a population under recurrent selection. (From Palloix, A., et al., *C. R. Acad. Agric. Fr.*, 83(7), 87, 1997. With permission.)

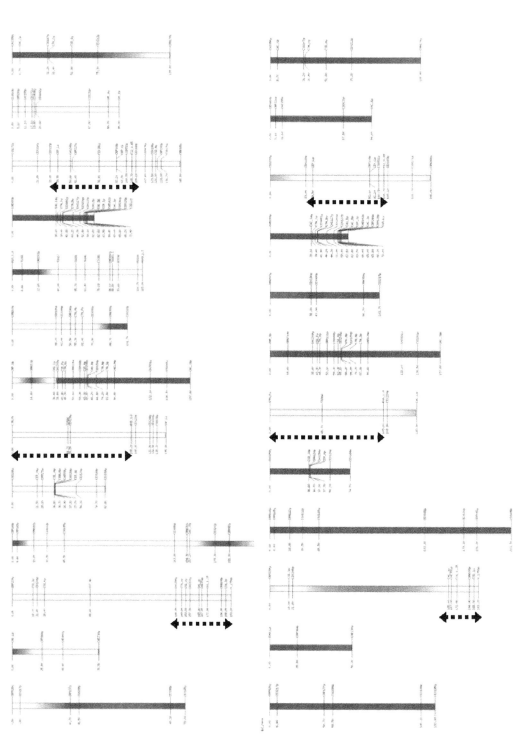

Figure 6.13 Graphical genotype of the pepper inbred line obtained after three cycles of marker-assisted backcrosses. (From Thabuis, A. et al., *Mol. Breed.*, 14, 9, 2004. With permission.) Top figure: donor parent for *P. capsici* resistance (genome from the original source of resistance in white). Lower figure: genotype of the BC₃M 3 plant (recipient genome in black). Hatched arrows: genome intervals containing the resistance QTLs and submitted to foreground selection with markers.

Cytoplasmic sterility of maternal inheritance ensures 100% sterility in the female parent and is the preferred system for hybrid seed production, when sterility is stable and restorer genes are available. In pepper, CMS lines were obtained from interspecific crosses between *C. annuum* and *C. chinense* or *C. baccatum* or *C. chacoense*, but the search for restorer genes remained unsuccessful (Shifriss 1997). Only the CMS source from Peterson (1958), originating from an Indian *Capsicum annuum* accession (USDA P.I. 164835), was successfully restored. Dominant restorer alleles were found in several hot and small-fruited pepper genotypes. Many sweet and large fruited genotypes were shown to possess recessive maintainer alleles (Peterson 1958; Zhang et al. 2000). Fertility restoration was shown to be quantitative and controlled by one major QTL (*Rf*) and several additional QTLs whose expression were environment dependant (Wang et al. 2004). Sterility instability necessitates tolerance of a very low frequency of inbred seeds (maternal parent) in the commercial hybrids that was not acceptable for private seed companies. However, genetic as well as cytoplasmic male sterility is already used in India, China, and Bulgaria for low cost production of hybrid cultivars. This is particularly true of cultivars used in the food processing industry.

6.7.2.3 Gene Transfer through Biotechnology

Stable genetic transformation and regeneration of pepper remains a difficult task and has resisted the efforts of many laboratories for many years. Transient gene expression has been successful using bacterial or virus vectors, thus allowing functional validation of cloned genes (see Section 6.5). First transformation events were directed toward establishing virus resistance. The next transformation challenge will be developing varieties with increased pest and insect resistance. This kind of resistance was not found in pepper germplasms.

Few instances of successful transformation have been reported (Zhu et al. 1996; Kim et al. 1997; Manoharan et al. 1998; Mihálka et al. 1998, 2000) since reports of transformation were first published (Dong et al. 1995). Stable transformation events were reported in the bibliography, although the transformed plants probably never get out from their original laboratory and were never confirmed by external laboratories.

Indeed, except for anther-culture, pepper is recalcitrant to *in vitro* plant regeneration from somatic explants. Fertile transgenic *C. annuum* plants were first successfully regenerated from explants that were co-cultivated with *Agrobacterium tumefaciens* harboring a plasmid that contains the cucumber mosaic cucumovirus (CMV) satellite RNA (Dong et al. 1995; Kim et al. 1997) or CMV coat protein (CP) gene (Zhu et al. 1996). The authors showed that the rate of plant regeneration depended on the type of explant cultivated and medium used. Cotyledons and young leaves were most effective for bud induction and subsequent plant elongation hypocotyls, while hypocotyls were the most ineffective. Kang et al. (1998) also demonstrated that plant regeneration was genotype-dependent. *Capsicum baccatum* accessions were found to be the most responsive, while *C. annuum* showed clear-cut differences among genotypes. Wolf et al. (2001) showed that the scarce plant regeneration seems to be a consequence of sporadic differentiation of normal apical shoot meristems, but not due to a defect in shoot development from available meristems. They also observed that no spatial overlap exists between regeneration-competent cells and those cells that undergo stable *A. tumefaciens*-mediated transformation. The regeneration of entire and fertile plants *in vitro* was also successfully induced from the hypocotyls of zygotic embryos in Tunisia (Arous et al. 1998) and in Hungary (Borychowski et al. 2002).

From the successfully regenerated transgenic plants that were self-fertilized, Kim et al. (1997) demonstrated that CMV genes could be stably transmitted and expressed in the progeny, which showed lower disease index and slow CMV development. Nevertheless, the authors of these publications reported about a low transformation efficiency and the published regeneration and transformation protocols were rarely reproduced in other laboratories.

Highly efficient regeneration systems for *Agrobacterium* mediated genetic transformation have been developed since 1998 in Hungary (Mihálka et al. 1998), in Spain (Pozueta-Romero et al. 2001), and in India (Shivegowda et al. 2002), using the basal part of young cotyledons for *in vitro* shoot induction. Mihálka et al. (1998) showed that alternative drugs (geneticin, hygromycin, methotrexate, phosphinotricin) proved to be better selection agents than kanamycin used before for producing transgenic pepper plants. They also investigated the transformation capacity of vectors and *Agrobacterium* strains, length and type of the co-cultivation and pre-incubation of the explants. Dabauza and Pena (2003) also gave the suitability of various *A. tumefaciens* strains for inducing tumors in different *C. annuum* varieties. The application of the protocol of Mihálka et al. (1998) resulted in whole plant regeneration in many different genotypes, but only a line was selected and bred for high *in vitro* response.

Currently, transformation seems relatively frequent but occurs at low frequencies. Mihálka et al. (2000, 2003) have published optimized protocols for efficient plant regeneration and gene transfer in pepper (*C. annuum*) that used a binary transformation systems based on "shooter" mutants of *A. tumefaciens*, and permits markers gene elimination. Using this technology, they introduced traits into pepper genome including CMV and potyvirus resistance. Shin et al. (2002) described the potential use of viral coat protein (CP) genes as transgene screening markers. They developed a simple and effective screening procedure, differentiating positive transformants from the non-transformed plants, using hypersensitive response upon a virus challenge inoculation. They also showed that the transgenic pepper plants expressing the CP genes of CMV and ToMV were also resistant to other viruses. Cai et al. (2003) also transferred both CMV and TMV CP genes to a *C. annuum* cultivar by a modified procedure of *A. tumefaciens*-mediated transformation using hypocotyl as the explant.

Following the development of an efficient protocol for *in vitro* regeneration of pepper cotyledons, Li et al. (2003) investigated the key factors affecting transformation and established a highly efficient genetic transformation system. All the genotypes of pepper tested presented a high differentiation efficiency (81.3% on average), elongation rate (61.5%), and rooting efficiency (89.5%). Polymerase chain reaction analysis results showed that 40.8% of the regenerated plantlets were transgenic plants. Then, in 2004, Lee et al. developed a system for selection by gathering shoots grown from calli that were induced from tissue cut of cotyledons and hypocotyls. Now, the difficulties of regeneration are overcome and the transformation is no longer a rare event; but is relatively frequent, even if frequencies are still low.

6.8   FUTURE DIRECTION

6.8.1   Genetic Resources: Moving the Priority from Collection to Evaluation and Data Diffusion

As for other species, one cannot ensure that the genetic variability of the *Capsicum* genus has been exhaustively collected. Most collections present numerous accessions of the domesticated species and their wild relatives (*C. annuum* complex including *C. frutescens* and *C. chinense*, *C. baccatum* and *C. pubescens* complexes), but much fewer accessions from the other wild species, so that collection efforts should now focus in these less known species in the primary diversification centers. Germplasm collection in the primary and secondary diversification centers has been encouraged in the past decades by national and international initiatives and resulted in the creation or strengthening of many local, national or international collections. In the second phase, the priority will be to move the investment toward the coordination of the different factors of pepper germplasm collection and to improve the availability of information and materials to the researchers and users.

IBPGR relayed by IPGRI remained essential promoters of the coordination initiatives at the international scale. USDA and the international centers of CATIE (Centro Agronomico Tropical de Investigacion y Ensenanza, Costa Rica) and AVRDC (Taiwan) have played an important role in the centralization and the coordination of material maintenance. In Europe, members of a network grouping the national germplasm collections recently coordinated their efforts and shared the passport data of all the maintained accessions in a single database. This permitted them to make available these data to the international pepper community and to direct the request to the right maintainer. Pepper is a major tropical crop, and many other collections are still unknown or only confidentially available in different countries in Africa and Asia, because of geographic or political isolation or the lack of public resources. Increased international initiatives must be encouraged because the local landraces are of generic interest, and it is desirable to maintain sources of genetic diversity.

A complementary effort to increase the phenotypic variation in *Capsicum* should involve induced variation populations created by mutagenesis. Although mutagenesis studies are not new to pepper, there has not been an attempt to produce a saturated mutagenesis population in which all or most genes have mutated in a common background. The construction of such populations will take advantage of reverse genetics technologies developed recently such as TILLING that will aid in the isolation of genes of interest.

Phenotypic and genetic characterization of the maintained germplasms is also a limiting factor for both users and collectors of pepper genetic resources. Knowledge of the genetic distances between the wild species may increase the number of genetic pools or may increase the number of species in the four characterized complexes, providing new potential for genetic improvement. Routes of pepper migration since post-Columbian times are complex and made difficult by genetic bottlenecks and the further secondary diversification. The representativeness of the present collections is unknown and should be the basis to direct further collections and evaluations. Pepper taxonomy made great progress during the 1970–1990 years thanks to excellent botanists. However, progress in this area has slowed in years. Genetic and molecular analysis provide highly pertinent information as shown by the recent studies of limited genetic pools and permitted to go further in the already acquired knowledge of the pepper genus organization. Two main directions in germplasm evaluation have to be increased. Evaluation of collections using neutral molecular markers is the first way to know the genetic structure of every germplasm and on a larger scale, the relationships between the different taxa of the genus. This will also provide key information for the representativeness and redundancy of local collections, sampling methods for phenotypic screening, and further studies in history of pepper migration. Non-neutral markers and phenotypic evaluations (using official descriptors) need to be developed to establish core collections for different traits. These will be valuable for breeding programs and for functional validation of cloned genes through phenotype/sequence polymorphism associations.

6.8.2 Breeding Objectives: Cumulating Previous Genetic Gain and Integrating the Environmental Impact of Cultivars as a Selection Criterium

Evolution of breeding objectives for pepper is close to the other plant and vegetable crops: recent genetic improvement permitted to increase the potential productivity and regularity of production of the different cultivars types. Most of the breeding objectives now aim at overcoming the yield-limiting factors, due, i.e., adaptation or resistance to pests, disease, and agro-climatic constraints that cannot be controlled with cultivation practices alone. If one considers the resistance to diseases, genetic resistances against the half of the diseases are available in commercial cultivars and many local populations have been improved or characterized for resistance against the remaining pathogens. A potential genetic control the majority of the mentioned parasitic constraints is available in the germplasm. Moreover, several resistance sources are reported for every disease and only a few gene combinations have been exploited.

The future breeding challenge is accessible and consists of improving both the control of epidemics in cultivated fields and in controlling the adaptation of pathogens to the newly cultivated genotypes. This means that breeding programs and screening criteria have to integrate the impact of the cultivated genotype on the environment: "sustainable breeding" is one primary component of sustainable agriculture. Considering durable resistance breeding, very few programs have considered the effect of resistance genes on the genetic evolution of the pathogen population. Pepper has some advantages in this field as polygenic resistances can control many pathogens. They were used with success against viruses such as the CMV or against the bacterial spot epidemics. Recent research was developed in pepper/potyvirus interaction and showed that the different resistance genes and QTLs presented different durabilities (ability to be overcome) in relation to the mutations required by the virus for resistance breakdown. This will permit researchers to choose the genes or gene combinations that will slow down the virus evolution. Such research should be extended to the emerging pepper pathogens, particularly for resistance to TSWV.

6.8.3   Breeding Methods: Genomics Information, Basis for Innovative Approaches

Optimizing the exploitation of the genetic variability, as illustrated for disease resistance, will depend on breeding methods and tools. Conventional pepper breeding methods were largely developed in a continuing effort to combine different traits. Many markers were recently made available to facilitate the selection of horticultural traits with late expression or that make experimentation or regulation problems (disease and pest resistance). Pepper breeders already use markers for Mendelian traits breeding and more recently, have experimented with marker-assisted selection for quantitative traits breeding. This practice was successful because it permitted to select the targeted genes and their phenotypic effects (foreground selection) and to accelerate the recovery of the genetic background on non-carrier chromosome. It also revealed two main limits of MAS, mainly due to technical constraints.

Limiting the linkage drag requires selection for recombination events flanking the targeted genes or QTLs. This necessitates additional "easy to use" markers in the genomic regions of interest, and for evaluating unfavorable linkages between different traits by integration of trait mapping data between crosses. The integrated molecular map offers users a large set of markers for a specific map position. These markers may be useful in different crosses or for saturating a genomic region with markers.

Most markers are biallelic and usable in pedigree or backcross selection in a constant genetic background, but breeders always need to introduce the new trait in several pepper cultivar types corresponding to different genetic backgrounds. Moreover, multiallelic screening in breeding populations with large genetic basis was already shown to be much more pertinent for the improvement of complex traits.

Development of diversified PCR-based markers is progressing in several laboratories, including SSR and SNPs that should progressively relay (or derive from) QTL- or gene-linked RFLPs and AFLPs. Fine mapping experiments and molecular characterization of loci are also crucial for breeding applications. The candidate gene approach already permitted the characterization of several loci in pepper and produced allele-specific markers that eliminated the risk due to crossovers between the gene and the marker, and that can be used in any genetic background. Other loci characterizations are underway. Genomic resources of pepper must be developed competitively and in synergy with the other strategic Solanaceae species (tomato and potato). Functional maps containing known function genes contributed to this task. EST libraries from pepper have already been produced (http://www.tigr.org/tigr-scripts/tgi/T_index.cgi?species=pepper) and permitted the identification of genes for fruit color and pungency; introgression libraries from interspecific and distant crosses are already underway with the long-term objective to identify pepper original genes and to transfer genomic data and knowledge from model genomes to pepper.

Past centuries of domestication and rapid conquest of the Old World by peppers gave evidence of its high nutritive value and of cultural attachment of the nations to this crop. World demand for peppers increases continuously, and world production has increased in parallel. Recent advances in science suggest that pepper can benefit from the genome model species, particularly within the Solanaceae synteny group, but *Capsicum* research has also delivered new discoveries of generic interest in human health, in plant metabolism (terpenoid biosynthesis) and in plant pathogen interaction. A happiness and beauty symbol in Inca civilization, the pepper will undoubtedly play an increasing role in the diversification and quality of our diet, in health improvement, and in knowledge progress.

6.9 SUMMARY

Pepper (*Capsicum* L.) is a member of the Solanaceae family that was domesticated 7,000 years ago in South and Central America. From the fifteenth to the eighteenth century, successive introductions occurred in Europe, Africa, and Asia, extending the cultivation of pepper around the world, with particular importance to countries in the tropical belt. Local selection for adaptation to these diverse environments, for various uses, and ethnic preferences increased the phenotypic variability of the genus, and generated new secondary diversification centers. Thousands of pepper cultivars and landraces are now cultivated worldwide with an extreme variability of plant types, fruit shape, and color. Cultivar types and cropping systems are specific for the various end-use products: from fresh vegetable to hot spice, food-dye, medicinal components, or ornamentals. The genus also offered new molecules to humankind. These include vitamin C, which was initially discovered in paprika, the red pigments capsanthin and capsorubin, frequently used as a natural food dye, and capsaicinoids, which confer pungency and are used in pharmaceutical research and industry.

Taxonomists presently recognize 25 species of *Capsicum*. Five species have been domesticated and are cultivated with *C. annuum* presenting the largest distribution. Part of the 20 strictly wild species have been related to the cultivated species and data from geographic distribution, numerical taxonomy, cross-compatibility experiments, biochemistry, cytogenetics, and molecular genetics agree to group these cultivated species and wild relatives in three main genetic pools that are separated by strong sexual barriers. More than half of the wild species remain to be studied and may be represent relatives of the previous groups or members of independent groups that were not subjected to domestication. Germplasm collections of various sizes and diversity are maintained by research or technical groups in more than 140 countries. Collection and maintenance of landraces and wild accessions was encouraged first by the International Board for Genetic Resources (IBPGR) and further by the International Plant Genetic Resource Institute (IPGRI) in national institutes of developing countries and international research centres. Priorities now move toward the phenotypic and molecular characterization of the accessions and the availability and interconnection of databases. This has been initiated within the international community with the European *Capsicum* database grouping collections of 14 countries.

Breeding pepper has become a very competitive activity due to the investment of many seed companies, high seed value of modern F_1 hybrids and globalization of the seed market. Breeding objectives are mainly directed toward fruit types and quality, adaptation to agro-climatic environment, and disease and pest resistance, particularly multi-virus and soilborne pathogen resistance. Natural genetic variability for these traits is very high so that conventional (sexual) breeding is predominant. Recurrent selection methods were developed for polygenic traits and pepper breeding took benefit of efficient haplomethods. Molecular markers were developed for Mendelian and quantitative traits and marker assisted selection has been integrated into breeding programs. Genetic transformation of pepper remains elusive. Transitory expression of genes was successful for functional genomic studies but stable transformation has not been successful enough to permit commercial exploitation. Pepper genomic research also took benefit from the advance of genomics in the Solanaceae synteny group (tobacco, potato, tomato), and from the recent development of pepper-specific genomic resources (BAC and EST libraries, introgression libraries, EMS mutants). Original genes were cloned from pepper and *Capsicum* proved to be a model plant for scientific research in host-pathogen interaction, fruit, and pigment components as well as a promising support for developing innovative breeding strategies for durable resistances. Genomic tools have to improve in pepper to integrate data from model species, but reciprocally, the genetic diversity of pepper already delivers knowledge of basic interest for the other crops.

ACKNOWLEDGMENTS

We gratefully acknowledge Dr. P. Bosland, Dr. I. Paran, and Dr. J. de Almeida Engler for constructive criticisms on the manuscript and critically reading the proof, and L. Pijarowski for helping in gathering bibliographic references.

REFERENCES

Ahamad, K., Gouse-Mohamed, M., and Murthy, N. S. R., Yield losses due to various pests in hot pepper, *Capsicum Newsl.*, 6, 83–84, 1987.

Ahmed, E.A. et al., Constructing multiresistant genotypes of sweet pepper for cultivation in the tropics, Paper presented at Eleventh Eucarpia Meeting on Genetics and Breeding of Capsicum and Eggplant, April 9–13, in Antalya, Turkey, p. 293, 2001.

Alcantara, T. P., Bosland, P. W., and Smith, D. W., Ethyl methanesulfonate-induced seed mutagenesis of C. annuum L, *J. Hered.*, 87, 239–241, 1996.

Andrews, J., *Pepper, the Domesticated Capsicum*, Austin, TX: University of Texas Press, 163 pp., 1984.

Arnedo-Andres, M. S., Gil-Ortega, R., Luis-Arteaga, M., and Hormaza, J. I., Development of rapid and scar markers linked to the pvr4 locus for resistance to pvy in pepper (*Capsicum annuum L.*), *Theor. Appl. Genet.*, 105, 1067–1074, 2002.

Arous, S., Boussaid, M., Marrakchi, M., Preliminary research about *in vitro* regeneration of Tunisian pepper *Capsicum annuum L*, Paper presented at Tenth Eucarpia Meeting on Genetics and Breeding of Capsicum and Eggplan, September 7–11, in Avignon, France, p. 187, 1998.

Arumuganathan, K. and Earle, E. D., Nuclear DNA content of some important plant species, *Plant Mol. Biol. Rep.*, 9(3), 208–218, 1991.

ASTA (American Spice Trade Association), *Official Analytical Methods of the American Spice Trade Association*, Englewood Cliffs, NJ: American Spice Trade Association, 1985.

Babaleye, T., Nematodes, an increasing problem in tropical agriculture, *Int. Pest Control*, 29(5), 114–119, 1987.

Baral, J. and Bosland, P. W., Genetic diversity of a Capsicum germplasm collection from Nepal as determined by randomly amplified polymorphic DNA markers, *J. Am. Soc. Horticult. Sci.*, 127(3), 318–324, 2002a.

Baral, J. B. and Bosland, P. W., An updated synthesis of the *Capsicum* genus, *Capsicum Eggplant Newsl.*, 21, 11–21, 2002b.

Baral, J. B. and Bosland, P. W., Unraveling the species dilemma *Capsicum frutescens* and *C. chinense* (*Solanaceae*): a multiple evidence approach using morphological, molecular analysis, and sexual compatibility, *J. Am. Soc. Hort.*, 129, 826–832, 2004.

Belletti, P., Marzachi, C., and Lanteri, S., Flow cytometric measurement of nuclear DNA content in *Capsicum* (Solanaceae), *Pl. Syst. Evol.*, 209(1–2), 85–91, 1998.

Ben Chaim, A. et al., QTL mapping of fruit-related traits in pepper (*Capsicum annuum*), *Theor. Appl. Genet.*, 102, 1016–1028, 2001a.

Ben Chaim, A. et al., Identification of quantitative trait loci associated with resistance to cucumber mosaic virus in *Capsicum annuum*, *Theor. Appl. Genet.*, 102, 1213–1220, 2001b.

Ben Chaim, A. et al., *Fs3.1*: a major fruit shape QTL conserved in *Capsicum*, *Genome*, 46, 1–9, 2003.

Berke, T., The AVRDC pepper-breeding unit, Lecture handout, Sixteenth RTP, ARC–AVRDC, Kasetsart University, Kamphaeng Saen, Thailand, 1997.

Besler, B., *Hortus Eystettensis: The Bishop's Garden: Hand-Colored Engraving*, Bauer, Konrad, Ed. 1st ed., Altdorf, Nuremburg, 1100 plant types in 367 color copperplate engravings, 463 pp., 1613.

Black, L. L., Hobbs, H. A., and Gatti, J. M., Tomato spotted wilt virus resistance in *Capsicum chinense* PI152225 and 159236, *Plant Dis.*, 75, 292–295, 1991.

Blum, E. et al., Molecular mapping of the *C* locus for presence of pungency in *Capsicum*, *Genome*, 45, 702–705, 2002.

Blum, E. et al., Molecular mapping of capsaicinoid biosynthesis genes and quantitative trait loci analysis for capsaicinoid content in Capsicum, *Theor. Appl. Genet.*, 108, 79–86, 2003.

Borovsky, Y. et al., The A locus that control anthocyanin accumulation in pepper encodes a MYB transcription factor homologous to Anthocyanin2 of Petunia, *Theor. Appl. Genet.*, 109, 23–29, 2004.

Borychowski, A., Niemirowicz-Szczytt, K. K., and Jedraszko, M., Plant regeneration from sweet pepper (*Capsicum annuum* L.) hypocotyls explants, *Acta Physiologiae Plantarum*, 24, 257–264, 2002.

Bosland, P. W., Chiles: a diverse crop, *Hort. Technol.*, 2(1), 6–10, 1992.

Bosland, P. W. and Votava, E. J., *Peppers: Vegetable and Spice Capsicum*, Wallingford, UK: CABI, 204 pp., 2000.

Boukema, I. W., Allelism of genes controlling resistance to TMV in Capsicum L, *Euphytica*, 29, 433–439, 1980.

Boukema, I. W., Resistance to a strain of TMV in Capsicum chacoense Hunz, *Capsicum Newsl.*, 1, 49–51, 1982.

Brouwer, D. J. and St. Clair, D. A., Fine mapping of three quantitative trait loci for late blight resistance in tomato using near isogenic lines (NILs) and sub-NILs, *Theor. Appl. Genet.*, 108(4), 628–638, 2004.

Brouwer, D. J., Jones, E. S., and St. Clair, D. A., QTL analysis of quantitative resistance to *Phytophthora infestans* (late blight) in tomato and comparisons with potato, *Genome*, 47(3), 475–492, 2004.

Cai, W. Q. et al., Development of CMV- and TMV-resistant chilli pepper: field performance and biosafety assessment, *Mol. Breed.*, 11(1), 25–35, 2003.

Caranta, C., A complementation of two genes originating from susceptible Capsicum annuum lines confers a new and complete resistance to pepper veinal mottle virus, *Phytopathology*, 86(7), 739–743, 1996.

Caranta, C., Lefebvre, V., and Palloix, A., Polygenic resistance of pepper to potyviruses consists of a combination of isolate-specific and broad-spectrum quantitative trait loci, *Mol. Plant–Microbe Interact.*, 10, 872–878, 1997a.

Caranta, C., Palloix, A., Lefebvre, V., and Daubèze, A. M., QTLs for a component of partial resistance to cucumber mosaic virus in pepper: restriction of virus installation in host-cells, *Theor. Appl. Genet.*, 94, 431–438, 1997b.

Caranta, C., Thabuis, A., and Palloix, A., Development of a CAPS marker for the Pvr4 locus: a tool for pyramiding potyvirus resistance genes in pepper, *Genome*, 42(6), 1111–1116, 1999.

Caranta, C. et al., QTLs involved in the restriction of cucumber mosaic virus (CMV) long-distance movement in pepper, *Theor. Appl. Genet.*, 104, 586–591, 2002.

Caranta, C., Ruffel, S., and Dussault, M. H., Gènes naturels de résistance aux virus chez les plantes: relations entre structure et fonction, *Virologie*, 7, 165–175, 2003.

Chaim, A. B., Borovsky, Y., de Jong, W., and Paran, I., Linkage of the A locus for the presence of anthocyanin and fs10.1, a major fruit-shape QTL in pepper, *Theor. Appl. Genet.*, 106, 889–894, 2003.

Chakraborty, S., Pandey, P. K., and Banerjee, M. K., Tomato leaf curl Gujarat virus, a new Begomovirus species causing a severe leaf curl disease of tomato in Varanasi, India, *Phytopathology*, 93(12), 1485–1492, 2003.

Chaney, W. E. et al., Insects and mites, In *University of California IPM Pest Management guidelines: Peppers*, ANR publication 3460, 45 pp., 2003.

Collins, M. D., Wasmund, L. M., and Bosland, P. W., Improved method for quantifying capsaicinoids in Capsicum using high performance liquid chromatography, *HortScience*, 30, 137–139, 1995.

Cook, A. A., Genetics of resistance in *Capsicum annuum* to two virus diseases, *Phytopathology*, 50, 364–367, 1960.

Cook, A. A. and Anderson, C. W., Multiple virus disease resistance in a strain of Capsicum annuum, *Phytopathology*, 49, 198–201, 1959.

Cook, A. A. and Guevara, Y. G., Hypersensitivity in Capsicum chacoense to race 1 of the bacterial spot pathogen of pepper, *Plant Dis.*, 68, 329–330, 1984.

Cook, A. A., Genetics of response in pepper to three strains of *Potato virus Y*, *Phytopathology*, 53, 720–722, 1963.

Csillery, G., New capsicum mutants found on seedling, growth type, leaf, flower and fruit, Paper presented at Fifth Eucarpia Meeting of the Capsicum and Eggplant working group, July 4–7 in Plovdiv, Bulgaria, pp. 127–130, 1983.

Csillery, G. and Rusko, J., The control of a new Tobamo virus strain by a resistance linked to anthocyanin deficiency in pepper (*Capsicum annuum* L.), Paper presented at Fourth Eucarpia Meeting on Genetics and Breeding of Capsicum, October 14–16 in Wageningen, The Netherlands, pp. 40–43, 1980.

Dabauza, M. and Pena, L., Response of sweet pepper (*Capsicum annuum* L.) genotypes to *Agrobacterium tumefaciens* as a means of selecting proper vectors for genetic transformation, *J. Hortic. Sci. Biotechnol.*, 78, 65–72, 2003.

Daubèze, A. M., Palloix, A., and Hennart, J. W., Resistance to *Leveillula taurica* in pepper (*Capsicum annuum*) is oligogeniccally controlled and stable in Mediterranean regions, *Plant Breed.*, 114, 327–332, 1995.

Daunay, M. C., Jullian, E., and Dauphin, F., Management of eggplant and pepper genetic resources in Europe: networks are emerging, Paper presented at Eleventh Eucarpia Meeting on Genetics and Breeding of Capsicum and Eggplant, April 9–13 in Antalya, Turkey, 2001.

Daunay, M. C., Maggioni, L., and Lipman, E., *Solanaceae* genetic resources in Europe, Report of two meetings, 21 September 2001, Nijmegen, The Netherlands/22 May 2003, Skierniewice, Poland, *International Plant Genetic Resources Institute*, Rome, Italy, 92 pp., 2003.

Davies, B. H., Matthews, S., and Kirk, J. T. O., The nature and biosynthesis of the carotenoids of different color varieties of *Capsicum annuum*, *Phytochemistry*, 9, 797–805, 1970.

DeWitt, D. and Bosland, P. W., *The Pepper Garden*, Berkeley: Ten Speed Press, 240 pp., 1993.

Di Vito, M. and Saccardo, F., Resistance of Capsicum species to Meloidogyne incognita, In *Root-knot Nematodes (Meloidogyne sp.), Systematics, Biology and Control*, Lamberti, F. and Taylor, E. E., Eds., London: Academic Press, pp. 455–456, 1979.

Di Vito, M. et al., Genetic of resistance to root-knot nematodes (Meloidogyne spp.) in Capsicum chacoense, C. chinense and C. frutescens, Paper presented at Eighth Eucarpia Meeting on Genetics and Breeding of Capsicum and Eggplant, September, 7–10, in Rome, Italy, pp. 205–209, 1992.

Djian-Caporalino, C. et al., Spectrum of resistance to root-knot nematodes and inheritance of heat-stable resistance in pepper (*Capsicum annuum L.*), *Theor. Appl. Genet.*, 99, 496–502, 1999.

Djian-Caporalino, C. et al., High-resolution genetic mapping of the pepper (*Capsicum annuum L.*) resistance loci Me_3 and Me_4 conferring heat-stable resistance to root-knot nematodes (*Meloidogyne spp.*), *Theor. Appl. Genet.*, 103, 592–600, 2001.

Djian-Caporalino, C., Fazari, A., Arguel, M. J., Vernie, T., VandeCasteele, C., Faure, I., and Brunoud, G., Root-knot nematode (*Meloidogyne* spp.) resistance genes *Me* in pepper (*Capsicum annuum* L.) are clustered on the chromosome p9. *Theor Appl Genet.*, Submitted for publication.

Doganlar, S. et al., Conservation of gene function in the Solanaceae as revealed by comparative mapping of domestication traits in eggplant, *Genetics*, 161, 1713–1726, 2002.

Dong, C. Z. et al., Transgenic tomato and pepper plants containing CMV sat-RNA cDNA, *Acta Hortic*, 402, 78–86, 1995.

Duffé, P., Gebre Selassie, K., Gognalons, P., and Grima, A., Semences de tomate. Procédés de désinfection efficaces contre virus et bactéries, *Etude et mise au point. PHM Revue Horticole*, 298, 59–63, 1989.

Dufour, O. et al., The distribution of Cucumber Mosaic Virus in resistant and susceptible plants of pepper, *Can. J. Bot.*, 67, 655–660, 1989.

Dumas de Vaulx, R. and Chambonnet, D., Influence of +35°C treatment and growth substances concentrations on haploid plant production through anther culture in Capsicum annuum, Paper presented at Fourth Eucarpia Meeting on Genetics and Breeding of Capsicum, October 14–16 1980, in Wageningen, The Netherlands, pp. 75–81, 1980.

Dumas de Vaulx, R. and Pitrat, M., Croisement interspécifique entre *C. annuum* et *C. baccatum*. Montfavet-Avignon, France, Paper presented at Third Eucarpia Meeting on Genetics and Breeding of Capsicum and Eggplant, July 5–8 July, in Montfavet-Avignon, France, 1977.

Dumas de Vaulx, R., Chambonnet, D., and Pochard, E., Culture *in vitro* d'anthères de piment (*Capsicum annuum* L.): amélioration des taux d'obtention de plantes chez différents génotypes par des traitements à 35°C, *Agronomie*, 1(10), 859–864, 1981.

Efrati, A., Eyal, Y., and Paran, I., Molecular mapping of the chlorophyll retainer (cl) mutation in pepper (Capsicum spp.) and screening for candidate genes using tomato ESTs homologous to structural genes of the chlorophyll catabolism pathway, *Genome*, 48(2), 347–351, 2005.

Emboden, W. A., A preliminary study of the crossing relationships of *Capsicum baccatum*, *Butler Univ. Bot. Studies*, 14, 1–5, 1961.

Erard, P., Ed, *Le poivron*, Paris: Centre Technique des Fruits et Légumes, 155 pp., 2002.

Escher, M. M., Fernandez, M. C. A., Iberia, R. L. D., and Preachy, A. L., Evaluation of Capsicum genotypes for resistance to the broad mite, *Hortic. Bras.*, 20(2), 217–221, 2002.

Eshbaugh, W. H., Genetic and biochemical systematic studies of chilli peppers (Capsicum–Solanaceae), *Bulletin Torrey Bota. Club.*, 102(6), 396–403, 1975.

Eshbaugh, W. H., The taxonomy of the genus Capsicum–Solanaceae, Montfavet-Avignon, France, Paper presented at Third Eucarpia Meeting on Genetics and Breeding of Capsicum and Eggplant, July 5–8, in Montfavet-Avignon, France, pp. 13–26, 1977.

Eshbaugh, W. H., Biosystematic and evolutionary study of the *Capsicum pubescens* complex, In *National Geographic Society Research Reports*, 1970 Projects, pp.143–162, National Geographic Society, Washington, DC, pp. 143–162, 1979.

Eshbaugh, W. H., The taxonomy of the genus *Capsicum (Solanaceae)*, *Phytologia*, 47, 153–166, 1980.

Eshbaugh, W. H., Variation and evolution in *Capsicum eximium* Hunz, *Baileya*, 21, 193–198, 1982.

Eshbaugh, W. H., History and exploitation of a serendipitous new crop discovery, In *New Crops*, Janick, J. and Simon, J. E., Eds., New York: Wiley, pp. 132–139, 1993.

Eshbaugh, W. H., Guttman, S. I., and McLeod, M. J., The origin and evolution of domesticated *Capsicum* species, *Ethnobiology*, 3, 49–54, 1983.

FAOSTAT data, Food and Agriculture Organization (FAO), Publishing Management Service, Information Division, Food and Agricultural Organization of the United Nations (FAO), Viale delle Terme di Caracalla, 00100 Rome, Italy, http://faostat.fao.org, 2004 (accessed on 20th December 2004).

Faria, J. C., Bezerra, I. C., and Zerbini, F. M., Current status of geminiviruses in Brazil, *Situacao atual das geminiviroses no Brasil, Fitopatologia Brasileira*, 25(2), 125–137, 2000.

Faten, G., Fekih-Hassen, I., and Nakhla, M. K., Molecular evidence of tomato yellow leaf curl virus-Sicily spreading on tomato, pepper and bean in Tunisia, *Phytopathologia Mediterranea*, 43(2), 177, 2004.

Federal Register of U.S. Environmental Protection Agency, Rules and regulations, Part II: Environmental Protection Agency 40 CFR, Part 82: Protection of Stratospheric Ozone: process for exempting critical uses from the phase out of methyl bromide, *Final Rule*, 69(246), 76981–77009, 2004.

Fegan, M. and Prior, P., How complex is the "Ralstonia solanacearum species complex", In *Bacterial Wilt: The Disease and the Ralstonia Solanacearum Species Complex*, Allen, C., Prior, P., and Hayward, C., Eds., St. Paul, MN: APS Press, pp. 449–461, 2005.

Fery, R. L. and Dukes, P. D., Southern root-knot nematode of pepper: studies on value of resistance, *Hort. Sci.*, 19(2), 211, 1984.

Fery, R. L. and Schalk, J. M., Resistance in pepper (*Capsicum annuum* L.) to western flower thrips (*Frankliniella occidentalis Pergrande*), *HortScience*, 26(8), 1073–1074, 1991.

Florini, D. A. and Zitter, T. A., Cucumber mosaic virus (CMV) in peppers (*C. annuum* L.) in New York and associated yield losses, *Phytopathology*, 77, 652, 1987.

Franceschetti, U., Natural cross pollination in pepper (*Capsicum annuum* L.), Paper presented at Eucarpia Meeting on Genetic and Breeding of Capsicum, September 16–18, in Turin, Italy, pp. 346–353, 1971.

Fuchs, L., Historia stirpium 1542, German translation in 1543 New Kreüterbuch, Taschen Ed., Köln Germany 960 p., 2001.

Garcilao de la Vega, Comentarios reales de los Incas, Carlos Aranibar, Ed. 1609. Lima, FCE, 1991.

G.N.I.S. (Interprofessional seeds and seedlings association), European Union Catalogue 2006, 24th intergral edition C275A, 8/11/2005, updated with the 3rd complement C132A, 7/06/2006, corresponding to cultivars registred up to the 31/12/2005, http://www.gnis.fr

Gonzalez de Leon, D. R., *Interspecific Hybridization and the Cytogenetic Architecture of Two Species of Chilli Pepper (Capsicum–Solanaceae)*, Ph.D. diss., University of Reading, 1986.

Govindarajan, V. S., Capsicum, Production, technology, chemistry and quality, *CRC Crit. Rev. in Food Sci. Nutr.*, 22, 109–176; 23, 207–288; 24, 245–355; 25, 158–282; 29, 435–474; 1985–1991.

Green, S. K. and Kim, J. S., *Characteristics and Control of Viruses Infecting Peppers: A Literature Review*, Asian Vegetable Research and Development Centre, Technical Bulletin, 18, 60 pp., 1991.

Greenleaf, W. H., Inheritance of resistance to tobacco-etch virus in *Capsicum frutescens* and in *Capsicum annuum*, *Phytopathology*, 46, 371–375, 1956.

Greenleaf, W. H., Pepper breeding, In *Breeding Vegetable Crops*, Bassett, M.J., Ed., Westport, CT: AVI, 584 pp., 1986.

Grube, R. C., Identification and comparative mapping of a dominant potyvirus resistance gene cluster in *Capsicum*, *Theor. Appl. Genet.*, 101(5–6), 852–859, 2000a.

Grube, R. C., Radwanski, E. R., and Jahn, M., Comparative genetics of disease resistance within the *Solanaceae*, *Genetics*, 155(2), 873–887, 2000b.

Hare, W. W., Resistance in pepper to *Meloidogyne incognita acrita*, *Phytopathology*, 46, 98–104, 1956.

Hartwell, J. L., Plants used against cancer, *Lloydia*, 34, 204–244, 1971.

Hayward, A. C., Biology and epidemiology of bacterial wilt caused by *Pseudomonas solanacearum*, *Annu. Rev. Phytopathol.*, 29, 65–87, 1991.

Hendy, H., Pochard, E., and Dalmasso, A., Identification de 2 nouvelles sources de résistance aux nématodes du genre *Meloidogyne* chez le piment *Capsicum annuum* L, *C.R. Acad. Agric., Fr.*, 817–822, 1983.

Hernandez-Verdugo, S., Guevara-Gonzalez, R. G., and Rivera-Bustamante, R. F., Screening wild plants of *Capsicum annuum* for resistance to pepper huasteco virus (PHV): presence of viral DNA and differentiation among populations, *Euphytica*, 122(1), 31–36, 2001.

Hibberd, A. M., Stall, R. E., and Bassett, M. J., Different phenotypes associated with incompatible races and resistance genes in bacterial leaf spot in pepper that is simply inherited, *Plant Dis.*, 7, 1075–1078, 1987.

Hoffman, P. G., Lego, M. C., and Galetto, W. G., Separation and quantification of red pepper major heat principles by reverse-phase high-pressure liquid chromatography, *J. Agric. Food Chem.*, 31, 1326–1330, 1983.

Holmes, F. O., Inheritance of resistance to tobacco mosaic disease in the pepper, *Phytopathology*, 27, 637–642, 1937.

Huang, S. W. et al., Development of pepper SSR markers from sequence databases, *Euphytica*, 117, 163, 2001.

Hugueney, P. et al., Metabolism of cyclic carotenoids: a model for the alteration of this biosynthetic pathway in *Capsicum annuum* chromoplasts, *Plant J.*, 8(3), 417–424, 1995.

Huh, J. H. et al., A candidate gene approach identified phytoene synthase as the locus for mature fruit color in red pepper (Capsicum spp.), *Theor. Appl. Genet.*, 102, 524–530, 2001.

Hurtado-Hernandez, H. and Smith, P. G., Inheritance of mature fruit color in *Capsicum annuum* L., *Heredity*, 76(3), 211–213, 1985.

Hussain, M., Mansoor, S., and Amin, I., First report of cotton leaf curl disease affecting chilli peppers, *Plant Pathol.*, 52(6), 809, 2003.

Hussain, M., Mansoor, S., and Iram, S., First report of Tomato leaf curl New Delhi virus affecting chilli pepper in Pakistan, *Plant Pathol.*, 53(6), 794, 2004.

IBPGR, Genetic resources of *Capsicum*, International Board for Plant Genetic Resources, AGPG/IBPGR/82/12, Rome, Italy, p. 49, 1983.

Inai, S., Ishikawa, K., Nunomura, O., and Ikehashi, H., Genetic analysis of stunted growth by nuclear–cytoplasmic interaction in interspecific hybrids of Capsicum by using RAPD markers, *Theor. Appl. Genet.*, 87, 416–422, 1993.

IPGRI, AVRDC, and CATIE, Descriptors for Capsicum (*Capsicum* spp.) AVRDC (Asian Vegetable Research and Development Center), CATIE (Tropical Agricultural Research and Higher Education Center), IPGRI (International Plant Genetic Resources Institute) (eds), 114 pp., 1995.

Jahn, M. et al., Genetic mapping of the *Tsw* locus for resistance to the Tospovirus tomato spotted wilt virus in *Capsicum* spp. and its relationship to the *Sw-5* gene for resistance to the same pathogen in tomato, *Mol. Plant–Microbe Interact.*, 13(6), 673–682, 2000.

Jensen, R. J., McLeod, M. J., Eshbaugh, W. H., and Guttman, S. I., Numerical taxonomic analyses of allozymic variation in *Capsicum* (*Solanaceae*), *Taxon.*, 28, 315–327, 1979.

Kang, B. et al., An interspecific (*Capsicum annuum* × *C. chinense*) F2 linkage map in pepper using RFLP and AFLP markers, *Theor. Appl. Genet.*, 102, 531–539, 2001.

Kang, B. C. et al., The *pvr1* locus in Capsicum encodes a translation initiation factor eIF4E that interacts with *Tobacco etch virus* VPg, *Plant J.*, 42, 392–405, 2005.

Kang, G. Q., Azzimonti, M. T., Marino, R., and Nervo, G., Factors affecting regeneration and transformation in Capsicum spp., Paper presented at the Tenth Eucarpia Meeting on Genetics and Breeding of Capsicum and Eggplant, September 7–11, in Avignon, France, p. 220, 1998.

Khan, M. W. and Haider, S. H., Comparative damage potential and reproduction efficiency of *Meloidogyne javanica* and races of *M. incognita* on tomato and eggplant, *Nematologica*, 37(3), 293–303, 1991.

Kim, B. D. et al., Construction of a molecular map and development of a molecular breeding technique, *J. Plant Biol.*, 40, 156–163, 1997.

Kim, B. S. and Kim, J. H., Evaluation of Phytophthora and Ralstonia multiple resistance selections of pepper for adaptability as rootstocks, Paper presented at the Twelfth Eucarpia Meeting on genetics and breeding of Capsicum and Eggplant, May 17–19, in Noordwijkerhout, The Netherlands, p. 185, 2004.

Kim, K. T. et al., Development of DNA markers linked to bacterial leaf spot resistance of chilli, *Acta Horticulturae*, 546, 597–601, 2001.

Kim, M. W., Kim, S. J., Kim, S. H., and Kim, B. D., Isolation of cDNA clones differentially accumulated in the placenta of pungent pepper by suppression subtractive hybridization, *Mol. Cells*, 11(2), 213–219, 2001.

Kim, S. J., Lee, S. J., Kim, B. D., and Paek, K. H., Satellite RNA mediated resistance to cucumber mosaic virus in transgenic plants of hot pepper, *Plant Cell Rep.*, 16, 825–830, 1997.

Knott, J. E. and Deanon, J. R., Jr., *Eggplant, Tomato and Pepper Vegetable Production in Southeast Asia*, Los Banos, Laguna, Philippines, University of the Philippines: Los Banos Press, pp. 99–109, 1967.

Kousik, C. S. and Ritchie, D. F., Response of bell pepper cultivars to bacterial spot pathogen races which individually overcome major resistance genes, *Plant Dis.*, 82(2), 181–186, 1998.

Kovacs, J. et al., Reaction of different Capsicum genotypes to Tobamoviruses and Cucumber mosaic virus, The Netherlands, Noordwijkerhout, Paper presented at the Twelfth Eucarpia Meeting on Genetics and Breeding of Capsicum and Eggplant, May 17–19, in Noordwijkerhout, The Netherlands, p. 186, 2004.

Kuhn, C. W., Nutter, F. W., and Padgett, G. B., Multiple levels of resistance to tobacco etch virus in pepper, *Phytopathology*, 79, 814–818, 1989.

Kyle, M. M. and Palloix, A., Proposed revision of nomenclature for potyvirus resistance genes in *Capsicum*, *Euphytica*, 97, 183–188, 1997.

Lafortune, D. et al., Partial resistance of pepper to bacterial wilt is under oligogenically controlled and efficient under hot and rainy season in tropical climate, *Plant Dis.*, 89, 501–506, 2005.

Lanteri, S., Lack of a caryotype class and skewed chromosome segregation in two backcross progenies of *Capsicum*, *J. Genet. Breed.*, 45(1), 51–57, 1991.

Lanteri, S. and Pickersgill, B., Chromosomal structural changes in *Capsicum annuum* L and *C chinense*, Jacq. *Euphytica*, 67(1–2), 155–160, 1993.

Lapidot, M. et al., Resistance to cucumber mosaic virus (CMV) in pepper: development of advanced breeding lines and evaluation for virus level, *Plant Dis.*, 81(2), 185–188, 1997.

Lecoq, H. et al., Durable resistance in plants trough conventional approaches: a challenge, *Virus Res.*, 100, 31–39, 2004.

Lee, C. J. et al., Non-pungent Capsicum contains a deletion in the Capsaicinoid Synthetase gene, which allows early detection of pungency with SCAR markers, *Mol. Cells*, 19, 262–267, 2005.

Lee, J. M., Nahm, S. H., Kim, Y. M., and Kim, B. D., Characterization and molecular genetic mapping of microsatellite loci in pepper, *Theor. Appl. Genet.*, 108, 619–627, 2004.

Lee, Y. H. et al., A new selection method for pepper transformation: callus-mediated shoot formation, *Plant Cell Rep.*, 23(1–2), 50–58, 2004.

Lefebvre, V., Molecular markers for genetics and breeding: development and use in pepper (*Capsicum* spp.), In *Biotechnology in Agriculture and Forestry, Molecular Marker Systems in Plant Breeding and Crop Improvement*, Lörz and Wenzel, Eds., Vol. 55, Berlin: Springer, pp. 189–214, 2005.

Lefebvre, V. and Palloix, A., Both epistatic and additive effects of QTLs are involved in polygenic induced resistance to disease, a case study, the interaction pepper-*Phytophthora capsici Leonian*, *Theor. Appl. Genet.*, 93, 503–511, 1996.

Lefebvre, V., Palloix, A., and Rives, M., Nuclear RFLP between pepper cultivars (*Capsicum annuum* L.), *Euphytica*, 71, 189–199, 1993.

Lefebvre, V., Palloix, A., Caranta, C., and Pochard, E., Construction of an intraspecific integrated linkage map of pepper using molecular markers and doubled-haploid progenies, *Genome*, 38(1), 112–121, 1995.

Lefebvre, V., Kuntz, M., Camara, B., and Palloix, A., The capsanthin–capsorubin synthase gene: a candidate gene for the y locus controlling the red fruit color in pepper, *Plant Mol. Biol.*, 36, 785–789, 1998.

Lefebvre, V. et al., Updated intraspecific maps of pepper, *Capsicum Eggplant Newsl.*, 16, 35–41, 1997.

Lefebvre, V. et al., Evaluation of genetic distances between pepper inbred lines for cultivar protection purposes: comparison of AFLP, RAPD and phenotypic data, *Theor. Appl. Genet.*, 102, 741–750, 2001.

Lefebvre, V. et al., Towards the saturation of the pepper linkage map by alignment of three intraspecific maps including known-function genes, *Genome*, 45, 839–854, 2002.

Lefebvre, V. et al., QTLs for resistance to powdery mildew in pepper under natural and artificial infections, *Theor. Appl. Genet.*, 107, 661–666, 2003.

Lefebvre, V. et al., Fine-mapping of a quantitative resistance locus to Phytophthora spp. and development of NIL–QTLs in pepper, Paper presented at the Twelfth Eucarpia Meeting on Genetics and Breeding of Capsicum and Eggplant, May 17–19, in Noordwijkerhout, The Netherlands, 2005.

Levesque, H., Vedel, F., Mathieu, C., and de Courcel, A. G. L., Identification of a short rDNA spacer sequence highly specific of a tomato line containing *Tm-1* gene introgressed from *Lycopersicon hirsutum*, *Theor. Appl. Genet.*, 80, 602–608, 1990.

Levy, A. et al., Carotenoid pigments and beta-carotene in paprika fruits (*Capsicum annuum*) with different genotypes, *J. Agric. Food Chem.*, 40, 2384–2388, 1995.

Li, D. et al., Establishment of a highly efficient transformation system for pepper (*Capsicum annuum* L.), *Plant Cell Rep.*, 21(8), 785–788, 2003.

Lim, Y. S. and Kim, B. S., Optimal conditions for resistance screening of *Cercospora* leaf spot by *Cercospora capsici* on pepper, *Res. Plant Dis.*, 9(3), 166–169, 2003.

Lindsey, D. L. and Clayshulte, M. S., Influence of initial population densities of Meloidogyne incognita on three chilli cultivars, *J. Nem.*, 14, 353–358, 1982.

Linné, C. von, *Species Plantarum*, 1st ed., Holmiae: L. Salvii, Stockholm, Sweden, 289 pp., 1753.

Lippert, L., Heterosis and combining ability in chilli peppers by diallel analysis, *Crop Sci.*, 15, 323–325, 1975.

Livingstone, K. D. et al., Genome mapping in *Capsicum* and the evolution of genome structure in the Solanaceae, *Genetics*, 152, 1183–1202, 1999.

Loaiza-Figueroa, F., Ritland, K., Laborde-Cancino, J. A., and Tanksley, S. D., Patterns of genetic variation of the genus *Capsicum* (*Solanaceae*) in Mexico, *Plant Syst. Evol.*, 165, 159–188, 1989.

Lotrakul, P., Valverde, R. A., and de la Torre, R., Occurrence of a strain of Texas pepper virus in tabasco and habanero pepper in Costa Rica, *Plant Dis.*, 84(2), 168–172, 2000.

Manoharan, M., Vidya, C. S. S., and Sita, G. L., *Agrobacterium*-mediated genetic transformation in hot chilli, *Plant Sci.*, 131(1), 77–83, 1998.

Marggaria, P., Ciuffo, M., and Turina, M., Resistance breaking strain of tomato spotted wilt virus (Tospovirus; Buniaviridae) on resistant pepper cultivars in Almeria, Spain, *Plant Pathol.*, 53, 795, 2004.

Maris, P. C., Joosten, N. N., Peters, D., and Goldbach, R. W., Thrips resistance in pepper and its consequences for the acquisition and inoculation of Tomato spotted wilt virus by the western flower thrips, *Phytopathology*, 93, 96–101, 2003.

Martelli, G. P. and Quacquarelli, A., The present status of tomato and pepper viruses, *Acta Hort.*, 127, 39–64, 1983.

McLeod, M. J., Eshbaugh, W. H., and Guttman, S. I., A preliminary biochemical systematic study of the genus *Capsicum, Solanaceae, The Biology and Taxonomy of the Solanaceae*, Hawkes, J. G., Lester, R. N., and Skelding, A. D., Eds., London: Academic Press, pp. 701–714, 1979.

McLeod, M. J., Guttman, S. I., Eshbaugh, W. H., and Rayle, R. E., An electrophoretic study of evolution in *Capsicum* (*Solanaceae*), *Evolution*, 37, 562–574, 1983.

McNeish, R. S., Ancient Mesoamerican civilization, *Science*, 143, 531–537, 1964.

Michelmore, R. W. and Meyers, B. C., Clusters of resistance genes in plants evolve by divergent selection and a birth-and-death process, *Genome Res.*, 8(11), 1113–1130, 1998.

Mihálka, V., Szasz, A., Fari, M., and Nagy, I., Gene transfer in pepper: comparative investigations of tissue culture factors and vectors systems, Paper presented at the Tenth Eucarpia Meeting on Genetics and Breeding of Capsicum and Eggplant, September 7–11, in, Avignon, France, 1998.

Mihálka, V. et al., Optimised protocols for efficient plant regeneration and gene transfer in pepper (*Capsicum annuum* L.), *J. Plant Biotechnol.*, 2, 143–149, 2000.

Mihálka, V., Balázs, E., and Nagy, I., Binary transformation systems based on "shooter" mutants of Agrobacterium tumefaciens: a simple, efficient and universal gene transfer technology that permits marker gene elimination, *Plant Cell Rep.*, 21(8), 778–784, 2003.

Milkova, L. and Daskalov, S., Ljulin-a hybrid cultivar of pepper based on induced male sterility, *Mutat. Breed Newsl.*, 24, 9, 1984.

Minsavage, C. V. et al., Gene-for-gene relationships specifying disease resistance in *Xanthomonas campestris* pv. *vesicatoria*–pepper interactions, *Mol. Plant–Microbe Interact.*, 3, 41–47, 1990.

Mityko, J., Andrasfalvy, A., Csillery, G., and Fari, M., Anther-culture response in different genotypes and F1 hybrids of pepper (*Capsicum annuum* L.), *Plant Breed*, 114, 78–80, 1995.

Morse, D., In the Hot-Sauce Biz, When You're Hot, It may Not be Enough, Wall Street Journal 15 May 2000.

Moury, B., Palloix, A., Gebre Selassie, K., and Marchoux, G., Hypersensitive resistance to tomato spotted wilt virus in three *Capsicum chinense* accessions is controlled by a single gene and is overcome by virulent strains, *Euphytica*, 94, 45–52, 1997.

Moury, B., Gebre Selassie, K., Marchoux, G., and Palloix, A., High temperature effects on hypersensitive resistance to Tomato Spotted Wilt Tospovirus (TSWV) in pepper (*Capsicum chinense* Jacq.), *Eur. J. Phytopathol.*, 104(5), 489–498, 1998a.

Moury, B., Palloix, A., Gebre Selassie, K., and Marchoux, G., L'émergence des tospovirus, *Virologie*, 2, 357–367, 1998b.

Moury, B. et al., A CAPS marker to assist selection of tomato spotted wilt virus (TSWV) resistance in pepper, *Genome*, 43(1), 137–142, 2000.

Moury, B. et al., Mutations in Potato virus Y genome-linked protein determine virulence toward recessive resistances in *Capsicum annuum* and *Lycopersicum hirsutum*, *Mol. Plant–Microbe Interact.*, 17, 322–329, 2004.

Nagy, I., Polley, A., and Ganal, M., Development and characterization of microsatellite markers in pepper, Paper presented at the Tenth Eucarpia Meeting on Genetics and Breeding of Capsicum and Eggplant, September 7–11, in Avignon, France, 1998.

Nagy, I. et al., Occurrence and polymorphism of microsatellite repeats in the sequence databases in pepper, Paper presented at the *Twelfth Eucarpia Meeting on Genetics and Breeding of Capsicum and Eggplant*, 17–19 May 2004, Noordwijkerhout, The Netherlands, 2004.

Nonnecke, L. L., *Vegetable Production*, New York: AVI Book, Van Nostrand Reinhold, 1989.

Nono-Womdim, R. et al., Study of multiplication of cucumber mosaic virus in susceptible and resistant Capsicum annuum lines, *Ann. Appl. Biol.*, 122, 49–56, 1993a.

Nono-Womdim, R., Palloix, A., Gebre-Selassie, K., and Marchoux, G., Partial resistance of Bell pepper to Cucumber Mosaic Virus Movement within plants: Field evaluation of its efficiency in southern France, *J. Phytopathol.*, 137, 125–132, 1993b.

Nuez, F., Gil Ortega, R., and Costa, J., Eds., *El Cultivo de Pimentos, Chile y Ajies*, Mundi-Prensa, 607 pp., 1996.

Odland, M. L. and Porter, A. M., A study of natural crossing in peppers (Capsicum frutescens), *Am. Soc. Hort. Sci. Proc.*, 38, 585–588, 1941.

Padgett, G. B., Nutter, F. W., Kuhn, C. W., and All, J. N., Quantification of disease resistance that reduces the rate of tobacco etch virus epidemics in bell pepper, *Phytopathology*, 80, 451–455, 1990.

Pakdeevaraporn, P., Wasee, S., Taylor, P., and Mongkolporn, O., QTL mapping of Anthracnose (Colletotrichum capsici) in Chilli, Paper presented at the Twelfth Eucarpia Meeting on Genetics and Breeding of Capsicum and Eggplant, May 17–19, in Noordwijkerhout, The Netherlands, p. 189, 2004.

Palloix, A., Pepper diseases and perspectives for genetic control, Paper presented at the Eighth Eucarpia Meeting on Genetics and Breeding of Capsicum and Eggplant, September 7–10, in Rome, Italy, Capsicum Newsletter, numero spécial, 1992.

Palloix, A., Daubèze, A. M., Phaly, T., and Pochard, E., Breeding transgressive lines of pepper for resistance to *Phytophthora capsici* in a recurrent selection system, *Euphytica*, 51(2), 141–150, 1990a.

Palloix, A., Pochard, E., Phaly, T., and Daubeze, A. M., Recurrent selection for resistance to *Verticillium dahliae* in pepper, *Euphytica*, 47(1), 79–89, 1990b.

Palloix, A. et al., Construction of disease resistance genotypes fitting cultivation conditions in pepper, *C.R. Acad. Agric. Fr.*, 83(7), 87–98, 1997.

Palloix A., et al., Breeding multiresistant bell peppers for intertropical cultivation conditions: the "Lira" program, Paper presented at the Tenth Eucarpia Meeting on Genetics and Breeding of Capsicum and Eggplant, September 7–11, in Avignon, France, 1998.

Palloix, A., Daubèze, A. M., and Pochard, E., Piments, In *Histoire de légumes. Des origines à l'orée du XXIe siècle*, Pitrat, M. and Foury, C., Eds., Paris: Edition INRA, pp. 278–290, 2004.

Panda, R. C., Kumar, O. A., and Rao, K. G. R., The use of seed protein electrophoresis in the study of phylogenetic relationship in Chile pepper (*Capsicum* L.), *Theor. Appl. Genet.*, 72, 665–670, 1986.

Paran, I., Aftergoot, E., and Shifriss, C., Variation in *Capsicum annuum* revealed by RAPD and AFLP markers, *Euphytica*, 99, 167–173, 1998.

Paran, I. et al., An integrated genetic linkage map of pepper (*Capsicum* spp.), *Mol. Breed.*, 13(3), 251–261, 2004.

Parkinson, J., *Theatrum Botanicum, The Theater of Plantes or An Universall and Compleate Herball*, London: T. Cotes, 1755 pp., 1640.

Parrella, G. et al., Recessive resistance genes against potyviruses are localized in colinear genomic regions of the tomato (*Lycopersicon* spp.) and pepper (*Capsicum* spp.) genomes, *Theor. Appl. Genet.*, 105, 855–861, 2002.

Pegard, A. et al., Histological characterization of resistance to different root-knot nematode species related to phenolics accumulation in *Capsicum annuum* L., *Phytopathology*, 95(2), 158–165, 2005.

Pernezny, K., Roberts, P. D., Murphy, J. F., and Goldberg, N. P., *Compendium of Pepper Diseases*, St. Paul, MN: APS Press, 63 pp., 2003.

Peterson, P. A., Cytoplasmically inherited male sterile in *Capsicum*, *Am. Nat.*, 92, 111–119, 1958.

Pflieger, S. et al., Disease resistance gene analogs as candidates for QTLs involved in pepper–pathogen interactions, *Genome*, 42(6), 1100–1110, 1999.

Pflieger, S., Lefebvre, V., and Causse, M., The candidate gene approach in plant genetics: a review, *Mol. Breed*, 7(4), 275–291, 2001a.

Pflieger, S. et al., Defense response genes co-localize with quantitative disease resistance loci in pepper, *Theor. Appl. Genet*, 103, 920–929, 2001b.

Pickersgill, B., The domestication of chilli pepper, In *The Domestication and Exploitation of Plants and Animals*, Ucko, P. J. and Dimbleby, G. W., Eds., London: Duckworth, 443 pp., 1969.

Pickersgill, B., Relationships between weedy and cultivated forms in some species in chilli peppers (genus *Capsicum*), *Evolution*, 25, 683–691, 1971.

Pickersgill, B., Migration of chile peppers, *Capsicum* spp, in the Americas, *Precolumbian plant migration*, Stone, D., Ed., Cambridge, MA: Haward Univ. Press, pp. 105–123, 1984.

Pickersgill, B., The genus *Capsicum*: a multidisciplinary approach to the taxonomy of cultivated and wild plants, *Biol. Zent.*, 107, 381–389, 1988.

Pickersgill, B., Cytogenetics and evolution of *Capsicum* L., In: *Chromosome Engineering in Plants, Genetics, Breeding and Evolution, Part B*, Tsuchiya, T. and Gupta, P. K., Eds., Amsterdam: Elsevier, pp. 139–160, 1991.

Pickersgill, B., Heiser, C. B., and McNeill, J., Numerical taxonomic studies on variation and domestication in some species of *Capsicum*, In *The Biology and Taxonomy of the Solanaceae*, Hawkes, J. G., Lester, R. N., and Skelding, A. D., Eds., London: Academic Press, pp. 679–700, 1979.

Pierre, M. et al., High-resolution genetic mapping of the pepper resistance locus *Bs3* governing recognition of the *Xanthomonas campestris* pv. *vesicatoria* AvrBs3 protein, *Theor. Appl. Genet.*, 101(1–2), 255–263, 2000.

Piton de Tournefort, J., Institutiones Rei herbariae, Typographia, Regia, Paris, 695 pp., 1700.

Pochard, E., Description des trisomiques de piment (*Capsicum annuum* L.) obtenus dans la descendance d'une plante haploïde, *Ann. Amélior. Plantes*, 20, 233–256, 1970.

Pochard, E., Localization of genes in *Capsicum annuum* L. by trisomic analysis, *Ann. Amélior. Plantes*, 27, 255–256, 1977a.

Pochard, E., Methods for the study of partial resistance to cucumber mosaic virus in the pepper, Paper presented at the Third Meeting on Genetics and Breeding of Capsicum and Eggplant, July 5–8, in Montfavet-Avignon, France, 1977.

Pochard, E. and Daubèze, A. M., Progressive construction of a polygenic resistance to cucumber mosaic virus in the pepper, Paper presented at the Seventh Eucarpia Meeting on Genetics and Breeding of Capsicum and Eggplant, June 27–30, in Smederevska-Palanka, Jugoslavia, 1989.

Pochard, E. and Dumas de Vaulx, R., Localization of vy2 and fa genes by trisomic analysis, *Capsicum Newsl.*, 1, 18–19, 1982.

Pochard, E., Palloix, A., and Daubeze, A. M., The use of androgenetic autodiploid lines for the analysis of complex resistance systems in the pepper, Paper presented at the Sixth Eucarpia on Genetics and Breeding of Capsicum and Eggplant, October 21–24 in Saragossa, Spain, 1986.

Popovsky, S. and Paran, I., Molecular genetics of the y locus in pepper: its relation to capsanthin–capsorubin synthase and to fruit color, *Theor. Appl. Genet.*, 101, 86–89, 2000.

Posh, A., van den Berg, B. M., Duranton, C., and Görg, A., Polymorphism of pepper (*Capsicum annuum* L.) seed proteins studied by two-dimensional electrophoresis with immobilized pH gradients: methodical and genetic aspects, *Electrophoresis*, 15(2), 297–304, 1994.

Poulos, J. M., Pepper breeding (*Capsicum* spp.): achievements, challenge and possibilities, *Plant Breed. Abstr.*, 64(2), 143–155, 1994.

Poulos, J. M., Reifschneider, F. J. B., and Coffman, W. R., Heritability and gain from selection for quantitative resistance to *Xanthomonas campestris* pv. *vesicatoria* in *Capsicum annuum* L, *Euphytica*, 56(2), 161–167, 1991.

Pozueta-Romero, J. et al., Enhanced regeneration of tomato and pepper seedling explants for *Agrobacterium*-mediated transformation, *Plant Cell, Tissue Org.*, 67(2), 173–180, 2001.

Prince, J. P., Loaiza-Figueroa, F., and Tanksley, S. D., Restriction fragment length polymorphism and genetic distance among Mexican accessions of *Capsicum*, *Genome*, 35, 726–732, 1992.

Prince, J. P., Pochard, E., and Tanksley, S. D., Construction of a molecular linkage map of pepper and a comparison of synteny with tomato, *Genome*, 36, 404–417, 1993.

Prince, J. P. et al., A survey of DNA polymorphism within the genus *Capsicum* and the fingerprinting of pepper cultivars, *Genome*, 38, 224–231, 1995.

Rao, G. U. and Paran, I., Polygalacturonase: a candidate gene for the soft flesh and deciduous fruit mutation in *Capsicum*, *Plant Mol. Biol.*, 51, 135–141, 2003.

Rao, G. U., Ben Chaim, A., Borovsky, Y., and Paran, I., Mapping of yield-related QTLs in pepper in an interspecific cross of *Capsicum annuum* and *C. frutescens*, *Theor. Appl. Genet.*, 106, 1457–1466, 2003.

Ritchie, D. F. and Dittapongpitch, V., Copper and streptomycin-resistant strains and host-differentiated races of *Xanthomonas campestris* pv. *vesicatoria* in North Carolina, *Plant Dis.*, 75(7), 733–736, 1991.

Rodriguez, J. M., Berke, T., Engle, L., and Nienhuis, J., Variation among and within *Capsicum* species revealed by RAPD markers, *Theor. Appl. Genet.*, 99, 147–156, 1999.

Roggero, P. et al., Two field isolates of tomato spotted wilt tospovirus overcome the hypersensitive response of a pepper (*Capsicum annuum*) hybrid with resistance introgressed from *C. chinense* PI152225, *Plant Dis.*, 83, 965, 1999.

Roggero, P., Masenga, V., and Tavella, L., Field isolates of tomato spotted wilt virus overcoming resistance in pepper and their spread to other hosts in Italy, *Plant Dis.*, 86, 950–954, 2002.

Ruffel, S., Recessive resistance genes against potyviruses in solanaceous crops and factors from the translation initiation complex, Ph.D. diss., University of Aix-Marseille II, France, 152 pp., 2004.

Ruffel, S. et al., A natural recessive resistance gene against potato virus Y in pepper corresponds to the eukaryotic initiation factor 4E (eIF4E), *Plant J.*, 32, 1067–1075, 2002.

Ruffel, S., Gallois, J. L., Lesage, M. L., and Caranta, C., The recessive potyvirus resistance gene *pot-1* is the tomato orthologue of the pepper *pvr2-eIF4E* gene, *Mol. Genet. Genomics*, 274(4), 346–353, 2005a.

Ruffel, S., et al., Simultaneous mutations in two distinct *eIF4E* genes are required to prevent *Pepper veinal mottle virus* infection in pepper, *J. Gen. Virol.*, 87, 2089–2098, 2006.

Saccardo, F. and Vitale, P., Mutations of practical value induced in pepper by gamete irradiation, *Capsicum Newsl.*, 1, 21–23, 1982.

Sasser, J. N., Worldwide dissemination and importance of the root-knot nematode, *Meloidogyne* spp., *J. Nematol.*, 22, 585–589, 1977.

SGN: Solanaceae Genomics Network database, URL: http://sgn.cornell.edu/, compendium first written in January 2004 by Christine Bunth of (PhD), researcher at Wageningen University, Laboratory for Bioinformatics, PO Box 8128, 6700 ET Wageningen, The Netherlands. The last major update of the compendium text was on March 2, 2004.

Sherf, A. F. and Macnab, A. A., *Vegetable Diseases and Their Control*, 2nd ed., New York: Wiley, 502 pp., 1986.

Shifriss, C., Male sterility in pepper (*Capsicum annuum* L.), *Euphytica*, 93, 83–88, 1997.

Shifriss, C., Breeding ornamental Capsicum, Paper presented at the Tenth Eucarpia Meeting on Genetics and Breeding of Capsicum and Eggplant, September 7–11, In Avignon, France, 1998.

Shin, R., Han, J. H., Lee, G. J., and Paek, K. H., The potential use of a viral coat protein gene as a transgene screening marker and multiple virus resistance of pepper plants coexpressing coat proteins of cucumber mosaic virus and tomato mosaic virus, *Transgen. Res.*, 11(2), 215–219, 2002.

Shivegowda, S. T., *In vitro* regeneration and transformation in chilli pepper (*Capsicum annuum* L.), *J. Hortic. Sci. Biotechnol.*, 77(5), 629–634, 2002.

Sibi, M., Dumas de Vaulx, R. D., and Chambonnet, D., Obtaining haploid plants by *in vitro* androgenesis in red pepper (*Capsicum annuum* L.), *Ann. Amélior Plantes*, 29(5), 583–606, 1979.

Singh, Y. and Sood, S., Screening of sweet pepper germplasm for resistance to bacterial wilt (*Ralstonia solanacearum*), *Capsicum Eggplant Newsl.*, 22, 117–120, 2003.

Siviero, P. and Gallerani, M., *La coltivazione del peperone*, Verona, Italia: L'Informatore Agrario, Caldiero, 217 pp., 1992.

Smith, C. E., Plant remains, *The Prehistory of the Tehuacan Valley, Environment and Subsistence*, Byers, D. S., Ed., Vol. 1, Austin, TX: University of Texas Press, pp. 220–225, 1967.

Smith, P. G. and Heiser, C. B., Taxonomy of *Capsicum chinense* Jacq. and the geographic distribution of the cultivated *Capsicum* species, *Bull. Torrey Bot. Club.*, 84, 413–420, 1957.

Somos, A., The paprika. Budapest: Akademiai Kiado, (1984).

Stewart, C., Jr. et al., The Pun1 gene for pungency in pepper encodes a putative acyltransferase, *Plant J.*, 42(5), 675–688, 2005.

Stirling, G. R., *Biological Control of Plant-Parasitic Nematodes*, Wallingford, UK: CAB International, 282 pp., 1991.

Szarka, J., Sardi, E., Szarka, E., and Csillery, G., General defence system in the plant kingdom, *Int. J. Horticult. Sci.*, 8(3–4), 45–54, 2002.

Tai, T. H. et al., High-resolution genetic and physical mapping of the region containing the Bs2 resistance gene of pepper, *Theor. Appl. Genet.*, 99, 1201–1206, 1999a.

Tai, T. H. et al., Expression of the *Bs2* pepper gene confers resistance to bacterial spot disease in tomato, *Proc. Natl Acad. Sci. U.S.A.*, 96(24), 14153–14158, 1999b.

Talekar, N. S., Berke, T., Breeding for pest resistance/tolerance in pepper, Paper presented at the Tenth Eucarpia Meeting on Genetics and Breeding of Capsicum and Eggplant, September 7–11, in Avignon, France, 1998.

Tanksley, S. D., Linkage relationships and chromosomal locations of enzyme-coding genes in pepper, *Capsicum annuum*, *Chromosoma*, 89(5), 352–360, 1984.

Tanksley, S. D., Bernatzky, R., Lapitan, N. L., and Prince, J. P., Conservation of gene repertoire but not gene order in pepper and tomato, *Proc. Natl Acad. Sci. U.S.A.*, 85, 6419–6423, 1988.

Tanksley, S. D. et al., High density molecular linkage maps of the tomato and potato genomes, *Genetics*, 132, 1141–1160, 1992.

Tewksbury, J. J. and Nabhan, G. P., Directed deterrence by capsaicin in chillies, *Nature*, 412, 403–404, 2001.

Thabuis, A., Construction de résistance polygénique assistée par marqueurs: application à la résistance quantitative du piment (*Capsicum annuum L.*) à Phytophthora capsici, Ph.D. diss, Paris, Grignon. Institut National Agronomique, 2002.

Thabuis, A. et al., Comparative mapping of *Phytophthora* resistance loci in pepper germplasm: evidence for conserved resistance loci across *Solanaceae* and for a large *genetic diversity*, *Theor. Appl. Genet.*, 106, 1473–1485, 2003.

Thabuis, A. et al., Phenotypic and molecular evaluation of a recurrent selection program for a polygenic resistance to *Phytophthora capsici* in pepper, *Theor. Appl. Genet.*, 109, 342–351, 2004a.

Thabuis, A. et al., Marker-assisted introgression of 4 *Phytophthora capsici* resistance QTL alleles into a bell pepper line: validation of additive and epistatic effects, *Mol. Breed.*, 14, 9–20, 2004b.

Thies, J. A., Mueller, J. D., and Fery, R. L., Effectiveness of resistance to southern root-knot nematode in "Carolina Cayenne" pepper (*Capsicum annuum* L.) in greenhouse, microplot, and field tests, *J. Amer. Soc. Hort. Sci.*, 122(2), 200–204, 1997.

Thorup, T. A. et al., Candidate gene analysis of organ pigmentation loci in the Solanaceae, *Proc. Natl Acad. Sci. U.S.A.*, 97(21), 11192–11197, 2000.

Toll, J. and van Sloten, D. H., *Directory of germplasm collections*, Rome: IBPGR, 187 pp., 1982.

Tong, N. and Bosland, P. W., *Capsicum tovarii*, a new member of the *Capsicum baccatum* complex, *Euphytica*, 109(2), 71–77, 1999.

Treaty of Tordesillas (1494). (Source: Davenport F. G., *European treaties bearing on the history of the United States and its dependencies to 1648*, Vol. 1, Washington, DC: The Carnegie Institution of Washington, pp. 75–78, 1917.)

UNEP (United Nations Environment Programme) (2004). The Vienna Convention for the Protection of the Ozone Layer & The Montreal Protocol on Substances that Deplete the Ozone Layer, adjusted and/or amended in Beijing 1999, Ozone Secretariat UNEP, PO Box 30552 Nairobi Kenya, http://www.unep.org/ozone.

Voorrips, R. E., Finkers, R., Sanjaya, L., and Groenwold, R., QTL mapping of anthracnose (*Colletotrichum* spp.) resistance in a cross between *Capsicum annuum* and *Capsicum chinense*, *Theor. Appl. Genet.*, 109, 1275–1282, 2004.

Walsh, B. M. and Hoot, S. B., Phylogenetic relationships of *Capsicum* (*Solanaceae*) using DNA sequences from two noncoding regions: the chloroplast atpB–rbcL spacer region and nuclear waxy introns, *Int. J. Plant Sci.*, 162, 1409–1418, 2001.

Wang, L. H. et al., Qtl analysis of fertility restoration in cytoplasmic male sterile pepper, *Theor. Appl. Genet.*, 109, 1058–1063, 2004.

Whitham, S. A. and Wang, Y., Roles for host factors in plant viral pathogenicity, *Curr. Opin. Plant Biol.*, 7(4), 365–371, 2004.

Wolf, D., Matzevitch, T., Steinitz, B., and Zelcer, A., Why is it difficult to obtain transgenic pepper plants? Tampere, Finland, Proceedings of the Fourth International Symposium on *In Vitro* Culture and Horticultural Breeding, July 2–7, 2000, in Tampere, Finland. *Acta Hortic.*, 560, 229–233, 2001.

Yamaguchi, M., *World Vegetables*, Van Nostrand Reinhold, CA, 1903.

Yoon, J. Y. et al., Pepper improvement for the tropics: Problems and the AVRDC approach, In *Tomota and Pepper Production in tropics, AVRDC*, Green, K., Griggs, T. D. and Mclean B. T., Eds., Taiwan, China: Shanhua, 86–98, 1989.

Young, N. D., Zamir, D., Ganal, M. W., and Tanksley, S. D., Use of isogenic lines and simultaneous probing to identify DNA markers tightly linked to the *Tm-2a* gene in tomato, *Genetics*, 120(2), 579–585, 1988.

Zhang, B. X., Huang, S. W., Yang, G. M., and Guo, J. Z., Two RAPD markers linked to a major fertility restorer gene in pepper, *Euphytica*, 113(2), 155–161, 2000.

Zhu, Y. X., OuYang, W. J., Zhang, Y. F., and Chen, Z. L., Transgenic sweet pepper plants from *Agrobacterium* mediated transformation, *Plant Cell Rep.*, 16, 71–75, 1996.

Zijlstra, S., Purimahua, C., and Lindhout, P., Pollen tube growth in interspecific crosses between *Capsicum* species, *HortScience*, 26, 585–586, 1991.

CHAPTER 7

Allium

Masayoshi Shigyo

CONTENTS

7.1 INTRODUCTION

The genus *Allium* is a diverse taxon with over 600 species (Traub 1968). *Allium* is assigned to the Amaryllidaceae, Liliaceae, and a distinct family, the Alliaceae (Hanelt 1990). Most *Allium* species are distributed in the Northern Hemisphere. Some of them such as bulb onion, shallot, and bunching onion have been cultivated in a wide geographical range from the temperate to the tropical zone. Of the edible alliums, *A. cepa* L. is one of the fastest-growing species in the food market. *Allium cepa* belongs to section *Cepa* of *Allium* and consists of three groups: the Common onion group, the Aggregatum group, and the Proliferum group (Jones and Mann 1963; Hanelt 1990). The Common onion and Aggregatum groups include two economically important cultivated crops, bulb onion and shallot, respectively. In both groups, large numbers of cultivars and strains have been developed and cultivated in many parts of the world. Recently, the Proliferum group is often excluded from *A. cepa* because of the interspecific nature strains of this group.

The bulb onion is the most important *Allium* crop; with a total world production in 2000 of approximately 52 Mt (FAO 2002). The gross world product of this crop is expected to increase by 30% for the decade from 1990 to 2000. China accounts for about 25% (11 Mt) of world production. The remaining leading producers are India (5.5 Mt), the United States (3.0 Mt), Turkey (2.3 Mt), and Japan (1.4 Mt). In contrast, the shallot is one of the more important crops in Southeast Asia. The shallot has the highest adaptability to tropical and sub-tropical zones of all edible alliums. However, due to the lack of suitable varieties, this plant is low yielding, has poor quality, and is susceptible to pests and pathogens. Understanding the inheritance of horticultural traits in the shallot is essential for efficient crop improvement and for the development of common onion cultivars suitable for low latitudes.

Morphological traits such as bulb formation, pathogen resistance, and other traits of economic importance differ between the cultivated *Allium* species and their wild relatives (Kik 2002). Interspecific crosses between the species have been made to transfer desirable features from one species to another. Genetic translocation via chromosome addition lines is one of the methods for introducing favorable genes on the extra chromosomes. In *Allium*, Shigyo et al. (1996) established eight monosomic additions, representing eight different chromosomes of the shallot in a background of Japanese bunching onion (*A. fistulosum*, $2n = 2x = 16$, genome FF). The monosomic additions ($2n = 2x + 1 = 17$, FF + nA) were used for genetic analysis of the shallot and for improving *A. fistulosum* cultivars. Genetic analyses identified 27 chromosome-specific genetic markers in shallot (Shigyo et al. 1996, 1997c). The effect of extra shallot chromosomes, on morphology and fertility of *A. fistulosum* were also identified (Shigyo et al. 1997b). The Faculty of Agriculture, Yamaguchi University, and the National Institute of Vegetable and Tea Science of Japan developed a novel breeding program for introducing disease resistance in *A. fistulosum* via the monosomic additions.

This review paper was prepared primarily to introduce the results so far obtained for the gene analyses of *A. cepa* using a complete set of *A. fistulosum*–shallot (*A. cepa* Aggregatum group) monosomic additions.

7.2 GENE POOLS FOR CULTIVATED ALLIUMS

The gene pool concept developed by Harlan and de Wet (1971) has played a pivotal role in the utilization of *Allium* germplasm resources for producing disease resistance cultivars, new CMS lines, etc. by conventional methods and by transformation technology (Figure 7.1).

The primary gene pools (GP-1) of bulb onion and Japanese bunching onion consist of local varieties and wild relatives. Primary gene pool A of bulb onion includes vegetatively propagated shallot that shows higher adaptability to tropical and sub-tropical zones than bulb onion. A wild relative, *A. vavilovii*, was placed in the primary gene pool B of bulb onion. Another wild relative, *A. altaicum*, was placed in the primary gene pool of Japanese bunching onion.

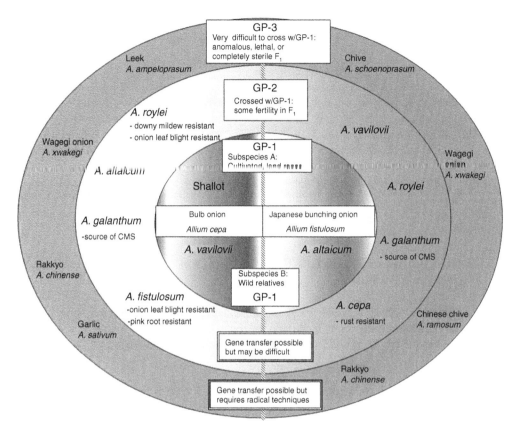

Figure 7.1 Gene pool concept in cultivated *Allium* species established based on hybridization.

The secondary gene pool (GP-2) included all species that can be crossed with GP-1 with at least some interspecific hybrid fertility. Gene transfer is possible, with difficulty. GP-2 for bulb onion and Japanese bunching onion were available as sources of disease resistance (Kik 2002 for *A. cepa*; Wako, T. personal communication 2005 for *A. fistulsoum*) and as new cytoplasmic sources of CMS lines (Havey 1999 for *A. cepa*; Yamashita and Tashiro 1999; Yamashita, Arita, and Tashiro 1999 for *A. fistulsoum*).

The tertiary gene pool (GP-3) consists of plants with marginal potential for genetic improvement. In most of the cases, the hybrids between the cultivated species of GP-1 and GP-3 can be generated through embryo culture. Pre- and post-zygotic abnormalities in the F_1s cause failure in the next hybridization process, inhibiting introgression between GP-1 and GP-3. The exploitation of distant relatives of bulb onion and Japanese bunching onion is often hampered by diminished compatibility, early embryo abortion, hybrid inviability, hybrid seedling lethality, and hybrid sterility due to low frequency of chromosome pairing. Technology to exploit GP-3 for broadening the genetic base of the two *Allium* cultivated crops has yet to be developed.

Allium germplasm collections, including cultivated and wild species, are maintained at Institute of Horticultural Research, Wellesbourne (UK), Plant Genetic Resources Unit, Geneva, NY (USA), the USDA-ARS Western Regional Plant Introduction Station (WRPIS), Pullman, WA (USA), the National Seed Storage Laboratory, Fort Collins, Colorado (USA), the National Institute of Vegetable and Tea Science, Ano, Mie (Japan), the Centre for Genetic Resources, Wageningen (The Netherlands), Institut für Pflanzengenetik und Kulturpflanzenforschung, Gatersleben (Germany), etc. The utilization of the germplasm collections may facilitate the use of molecular biology approaches for crop plant improvement.

7.3 CYTOGENETICS

7.3.1 Introduction

Allium has three basic chromosome numbers of $x=7$, 8, or 9 (Jones 1990). The majority of the species indigenous to Eurasia and the Mediterranean basin have $x=8$ (Ved Brat 1965). Species with $x=7$ (ca. 80) are largely confined to the New World. There are few species with $x=9$ (Ved Brat 1965). Most *Allium* species are diploid ($2n=2x=16$). Polyploidy is less common, but occurs among some of the cultivated forms. Generally speaking, the chromosomes of *Allium* species are very large and their complements show a high degree of symmetry and uniformity. Most of the chromosomes possess median or submedian centromeres and there is a gradation in size from the longest to the shortest member of the complement. Research emphasis had been placed on chromosome variation at mitosis and meiosis (Jones 1990). Development of *Allium* monosomic addition lines has provided new tools for the cytogenetic study of *Allium*.

7.3.2 Establishment of a Complete Set of *Allium fistulosum–Allium cepa* Monosomic Additions

A method for production of *A. fistulosum–A. cepa* alien additions is presented in Figure 7.2. Alien additions were selected from second backcross progenies (BC_2) of amphidiploid hybrids between *A. fistulosum* and shallot. The following characters were investigated especially for this purpose:

- Seed set in BC_2
- Chromosome numbers of somatic cells in BC_2
- Chromosome configurations at metaphase-I in pollen mother cells of BC_2

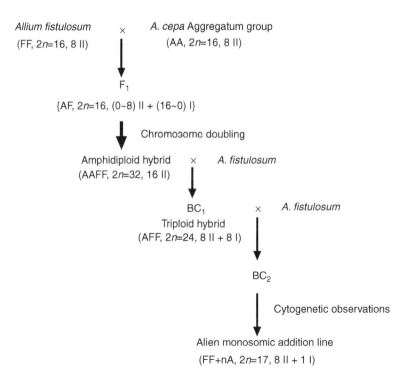

Figure 7.2 Procedure for producing Japanese bunching onion (*A. fistulosum*)–shallot (*A. cepa* Aggregatum group) monosomic additions. (From Shigyo, M. et al., *Genes Genet. Syst.*, 71, 363–371, 1996. With permission.)

Table 7.1 Seed Set, Seed Germination, and Seedling Survival in the Backcrossing of Triploid Hybrid (AFF) to *A. fistulosum* (FF)

Number of Flowers Pollinated	Number of Seeds Produced	Percentage of Ovules That Developed into Seeds[a]	Number of Seeds That Germinated	Percentage of Seeds That Germinated	Number of Seedlings That Survived	Percentage of Seedlings That Survived
6,870	1,122	2.7	404	36	274	67.8

[a] Percentage of ovules that developed into seeds $= \dfrac{\text{Number of seeds produced}}{\text{Number of flowers pollinated} \times \text{Number of ovules per flowers(6)}} \times 100$.

Source: From Shigyo, M. et al., *Genes Genet. Syst.*, 71, 363–371, 1996. With permission.

After backcrossing with *A. fistulosum*, the triploid hybrid AFF of BC_1 showed a very low seed set and germination rate (Table 7.1). Nevertheless, 274 BC_2 plants were ultimately obtained. The chromosome count using 253 BC_2 plants revealed that the chromosome numbers varied from $2n = 16$ to 24. Forty-seven plants had $2n = 17$ chromosomes. At metaphase-I, all PMCs of 43 plants formed $8II + 1I$. The PMCs of remaining three plants formed $8II + 1I$ and $7II + 3I$. The bivalents formed localized chiasmata characteristic of the meiotic chromosomes of *A. fistulosum*. Forty-one plants were selected as promising monosomic additions.

The monosomic additions were applied for elaborate karyotype observation with the calculations of relative chromosome length and centromeric index in each extra chromosome. The system of chromosome nomenclature for *A. fistulosum* and *A. cepa* was according to the Eucarpia 4th *Allium* Symposium (De Vries 1990). A set (eight different types) of the monosomic additions was found in the BC_2 population (Figure 7.3). Karyotype analysis at metaphase-I in PMCs confirmed the previous results. Frequencies of the monosomic additions with the extra chromosomes 1A–8A in the 41 BC_2 plants were as follows:

1A	5 plants
2A	3 plants
3A	5 plants
4A	9 plants
5A	4 plants
6A	2 plants
7A	11 plants
8A	2 plants

More monosomic additions had chromosomes 4A and 7A (22% + 27%) and fewer chromosomes with 6A and 8A (5% + 5%). It is possible that the presence of an extra chromosome affected the survival rate of the monosomic additions.

7.3.3 Application of Molecular Cytogenetics Observation

The application of the C-banding technique, as proposed by Kalkman (1984), to identify alien chromosomes of shallot was difficult. Few intercalary chromosome bands were present in the shallot. Therefore, an alien chromosome was identified by arm length and the position of the centromere. Genomic in situ hybridization (GISH) may be used for a quick and reliable identification of shallot chromosomes in Japanese bunching onion. GISH is a powerful tool for plant molecular cytogenetics (Jiang and Gill 1994). In *Allium*, GISH was used successfully by Hizume (1994), Khrustaleva and Kik (1998) to discriminate between the chromosome complements of *A. cepa* and *A. fistulosum* present in the interspecific hybrid between both species. Furthermore, this technique was used among others by Keller et al. (1996) and Friesen, Fritsch, and Bachmann (1997) to confirm the hybrid origin of presumed species hybrids.

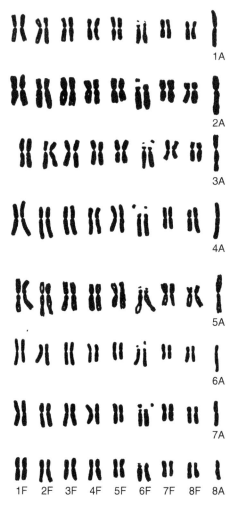

Figure 7.3 Somatic metaphase chromosomes in a complete set of *A. fistulosum* – shallot monosomic additions. (FF+1A, 130; FF+2A, 132; FF+3A, 5; FF+4A, 32; FF+5A, 71; FF+6A, 120; FF+7A, 23; FF+8A, 65). 1F–8F: Chromosomes from *A. fistulosum*. 1A–8A: Extra chromosomes from *A. cepa* Aggregatum group. (From Shigyo, M. et al., *Genes Genet. Syst.*, 71, 363–371, 1996. With permission.)

The GISH technique was applied to identify eight different alien chromosomes in a set of *A. fistulosum–A. cepa* monosomic additions (Shigyo et al. 1998). Metaphase chromosomes of each monosomic addition are presented in Figure 7.4. Probe hybridization sites were detected by FITC-conjugated avidin and anti-avidin antibodies that fluoresce green under blue light excitation. *Allium fistulosum* chromosomes showed very little hybridization. Shallot chromosomes were counterstained with propidium iodide and fluorescent orange–red with the same excitation wavelength. The green FITC label and orange–red counterstain interacted to give a yellow to yellow–green fluorescence. In a monosomic addition line with the chromosome 1A of shallot (FF+1A), one yellow–green fluorescent alien chromosome and 16 orange–red fluorescent chromosomes were observed. The alien chromosome was metacentric and much larger than any other *A. fistulosum* chromosome. This result with the conventional karyotype analysis conducted for the identification of alien chromosome 1A. Similarly, one yellow–green alien chromosome and 16 orange–red chromosomes were observed in the remaining seven types of the monosomic addition (FF+ 2A–FF+8A). Data on sizes and morphologies of the alien chromosomes (2A–8A) corresponded

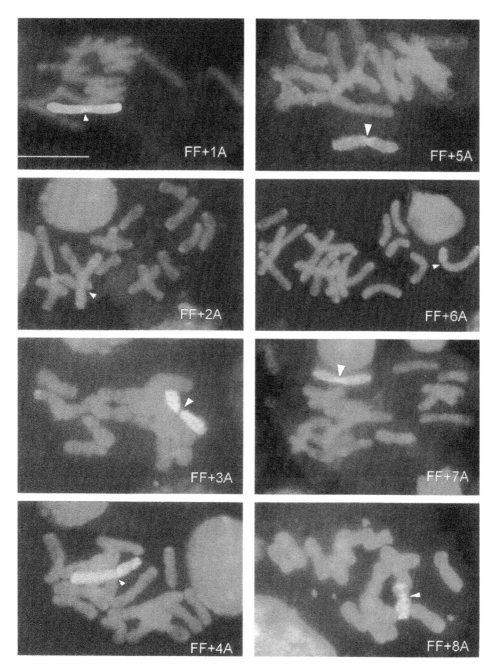

Figure 7.4 (See color insert following page 304.) Somatic metaphase cells of a complete set of *A. fistulosum* shallot monosomic additions (FF + 1A–FF + 8A) after genomic in situ hybridization. Each cell possesses one yellow–green alien chromosome of shallot (*A. cepa* Aggregatum group) and 16 orange–red stained chromosomes of *A. fistulosum*. A large-sized metacentric alien chromosome, namely 1A, is observed in FF + 1A; a large-sized sub-metacentric alien chromosome, i.e., 2A, in FF + 2A; a middle-sized sub-metacentric alien chromosome, i.e., 3A, in FF + 3A; a middle-sized sub-metacentric alien chromosome, i.e., 4A, in FF + 4A; a middle-sized metacentric alien chromosome, i.e., 5A, in FF + 5A; a middle-sized sub-telocentric alien chromosome, i.e., 6A, in FF + 6A; a small-sized metacentric alien chromosome, i.e., 7A, in FF + 7A; and a small-sized sub-metacentric alien chromosome, i.e., 8A, in FF + 8A. Each arrowhead indicates the centromere of the alien chromosome. (From Shigyo, M. et al., *Genes Genet. Syst.*, 73, 311–315, 1998. With permission.)

well to those shown in our previous study. Consequently, the alien chromosomes were successfully discriminated from the chromosome complement of *A. fistulosum* and identified as respective shallot chromosomes (1A–8A) by GISH.

No clear exchanges of chromosome segments between *A. fistulosum* and shallot were observed. This result shows that, in each addition line, an entire shallot chromosome was present in a diploid background of *A. fistulosum*. The frequency of homoeologous pairing and recombination is quite high in the meiosis of F_1 hybrids between these two species (Emsweller and Jones 1935; Maeda 1937; Levan 1941; Cochran 1950; Tashiro 1984; Peffley 1986) and implies a considerable degree of selection for entire chromosomes in the process of constructing the monosomic additions. Our monosomic additions were selected from the backcross progenies of interspecific triploids ($2 \times A.$ *fistulosum* $+ 1 \times$ shallot). At metaphase-I, almost all pollen mother cells of the triploids formed eight bivalents and eight univalents (Tashiro et al. unpublished data). This result suggests the occurrence of preferential pairing among *A. fistulosum* homologues, leaving shallot chromosomes univalents. Therefore, all shallot alien chromosomes may have been derived from a nonrecombinant chromatid in meiosis. In conclusion, GISH proved to be a valuable tool to identify alien *A. cepa* chromosomes in a diploid *A. fistulosum* background.

7.3.4 Morphological and Physiological Characteristics of Alien Monosomic Addition Lines

Extra chromosomes from the shallot may have different effects on the morphological and physiological characters of shoots and leaves in a vegetative period, morphological characters of flower organs, seed setting capacity, and pollen fertility of the Japanese bunching onion. Plant materials consisted of the 37 monosomic additions. *A. fistulosum* cv. "Kujyo-Hoso" was examined as a control.

In a previous report (Shigyo et al. 1997b), several morphological characters of the monosomic additions were found to be specific for the respective alien chromosomes from *A. cepa* Aggregatum group (Figure 7.5, Table 7.2). The most distinctive characteristics in each alien monosomic addition are listed in Table 7.3. The above morphological expressions are related to alien genes on extra chromosomes from the shallot. Further studies on the chromosomal locations of the genes controlling these chromosome-specific characters are necessary for gene manipulation in *Allium*.

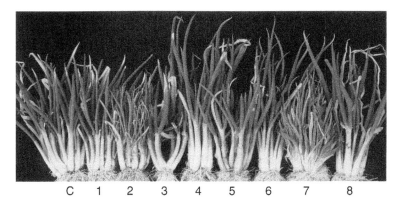

Figure 7.5 Plants of a complete set of *A. fistulosum*–shallot monosomic additions (1–8) and *A. fistulosum* (C) in vegetative stage. Each Arabic numeral (1–8) corresponds to the extra chromosomes (1A–8A) of the monosomic additions. X *ca.*1/12. (From Shigyo, M. et al., *Genes Genet. Syst.*, 72, 181–186, 1997b. With permission.)

Table 7.2 Morphological Data on Leaves and Flower Organs of Alien Monosomic Additions

| Plant Material | | | | | Leaf Length (cm)[b,c] | | | Leaf Sheath Diameter[b,c] (cm) | Leaf Color[a] | | Flower Stalk Length[b,c] (cm) | No. of Florets[b,c] | Pedicel Length[b,c] (mm) | Filament Length[b,c] (mm) | | Anther Size (mm)[b,c] | | Anther Color[b,c,f] | Ovary Color[b,c,g] |
Extra Chromosome	No. of Lines	No. of Plants Planted	Survival Rate (%)	No. of Tillers[b,c]	Sheath	Blade	Total		Sheath[d]	Blade[e,e]				Outer	Inner	Length	Width		
1A	5	82	46	5.7±0.6	10.9±0.3	40.4±1.1	51.3±1.4	5.3±0.1	−	3.4±0.8	53.3±1.1	440±20	11.4±0.5	9.6±0.4	8.3±0.3	2.0±0.0	1.0±0.0	6.2±1.0	3.6±0.8
2A	3	58	77	8.5±1.3	9.2±0.5	32.9±0.8	42.1±1.1	4.1±0.1	−	1.0±0.0	58.4±1.6	425	7.7±0.5	9.2±0.2	8.7±0.3	1.9±0.1	1.0±0.0	3.0±0.7	6.0±0.0
3A	4	22	28	7.6±3.6	11.0±0.4	29.2±2.1	40.2±2.3	5.1±0.5	−	2.0	43.8±0.9	317±36	10.3±0.4	10.5±0.1	10.7±0.3	1.8±0.1	1.0±0.0	5.5±0.8	3.6±1.2
4A	8	108	87	4.7±0.4	15.7±0.3	55.8±0.8	71.5±1.0	6.1±0.1	−	4.1±0.3	69.3±0.8	647±37	16.0±0.9	12.4±0.3	11.6±0.4	2.2±0.1	1.1±0.1	4.6±0.6	5.4±0.4
5A	3	55	80	8.3±0.8	10.8±0.5	36.5±1.0	47.2±1.4	5.1±0.1	+	4.0±0.7	60.1±1.6	386±36	15.8±1.2	10.0±0.3	9.4±0.3	1.9±0.1	1.0±0.0	3.3±0.4	4.2±1.2
6A	2	48	77	5.3±0.6	12.8±0.6	49.2±1.4	62.0±1.6	4.9±0.1	−	5.0	69.1±2.5	343	10.6±0.3	10.9±0.4	9.9±0.2	2.0±0.0	1.0±0.0	3.3	1.5
7A	10	160	83	9.0±0.6	12.4±0.3	43.3±0.5	55.7±0.7	5.3±0.1	−	1.1±0.1	59.7±0.7	486±23	14.1±0.5	9.8±0.2	9.4±0.2	2.1±0.0	1.1±0.0	5.8±0.5	3.1±0.4
8A	2	39	71	11.9±1.5	14.3±0.5	47.2±1.2	61.5±1.5	5.7±0.2	−	4.0	73.0±1.2	554	10.9±0.3	11.5±0.2	11.1±0.3	1.8±0.1	1.0±0.0	6.5	4.0
A. fistulosum	1	16	100	11.7±1.0	18.3±0.8	47.5±1.1	65.9±1.6	5.3±0.1	−	3.0	61.8±2.9	558	11.6±0.5	10.1±0.9	9.3±0.5	2.0±0.0	1.0±0.0	5.0	4.0

[a] Data from Table 2 in Shigyo M., et al. (1997b).
[b] Data from Table 3 in Shigyo M., et al. (1997b).
[c] Data are shown with mean ± SE.
[d] +, Reddish-yellow; −, White.
[e] Mean ± SE of color scales 1 (light green)–5 (deep green).
[f] Mean ± SE of color scales 1 (light yellow)–7 (deep yellow).
[g] Mean ± SE of color scales 1 (light green)–5 (deep green).
Source: From Shigyo, M. et al., *Genes Genet. Syst.*, 72, 181–186, 1997b. With permission.

Table 7.3   Morphological and Physiological Characteristics of Alien Monosomic Additions

Monosomic Additions	Characteristics
FF+1A	A little shorter leaf length, a little shorter flower stalk length, fast expansion of leaf, spheroidal spathe, low pollen and seed fertility
FF+2A	Short leaf length, short flower stalk length, slender leaf blade, light green leaf blade, bloom-less leaf blade, light yellow pollen, fast expansion of leaf, small floret, elongation of axillary bud from autumn to winter
FF+3A	Short leaf length, short flower stalk length, slow expansion of leaf, small spathe, very low seed fertility
FF+4A	Long leaf length, long flower stalk length, tough leaf blade, deep green leaf blade, few tillering, slow expansion of leaf, large floret, acuminate spathe, low seed fertility
FF+5A	A little shorter leaf length, a little shorter flower stalk length, slow expansion of leaf, reddish-yellow leaf sheath, low pollen fertility
FF+6A	A little slender leaf blade, deep green leaf blade, arch-like expansion of leaf
FF+7A	Fast expansion of leaf, elongation of axillary bud from autumn to winter, low pollen and seed fertility
FF+8A	Long leaf length, long flower stalk length, thick leaf blade, slow expansion of leaf, deep yellow anther, many tillering, high seed fertility

Source: From Shigyo, M. et al., *Genes Genet. Syst.*, 72, 181–186, 1997b. With permission.

7.3.5   Cheomosome Mapping by Cytogenetic and Molecular Methods

7.3.5.1   *Isozymes*

The isozymes are one of the most useful genetic markers because of co-dominance (Moore and Collins 1983). This marker has been used for the confirmation of hybridity, cultivar identification, detection of somatic variation, estimation of mating systems, genome identification, measurement of genetic variation, characterization of taxonomic relationships, purity testing of commercial seed lots, and for the confirmation of the introgression of the chromosome segments, including the loci-controlling morphological and economically important traits, etc. (Tanksley and Orton 1983). The chromosomal location of isozyme genes has been successfully determined by alien chromosomes (Hart and Tuleen 1983). The isozyme analyses were performed to determine the chromosomal location of enzyme genes in the shallot (Shigyo et al. 1995a, 1995b). Young expanding leaves were used for enzyme extraction. Polyacrylamide gel electrophoresis (PAGE) and starch gel electrophoresis (SGE) were employed for the isozyme analysis in our study. Nine different enzymes, including the GOT presented here, were examined.

Allium fistulosum and shallot showed different banding patterns. Bands 1 and 4 were detected in *A. fistulosum*, and bands 3 and 6 in shallot (Figure 7.6). In the amphidiploid hybrid AAFF and triploid hybrid AAF, two additional bands (bands 2 and 5) were detected. Band 2 migrated to an intermediate position between bands 1 and 3, and band 5 between bands 4 and 6. Dosage effects of the genes were clearly observed on the zymograms of the amphidiploid hybrid AAFF and triploid hybrid AFF. In the amphidiploid hybrid AAFF, band 2 was stained more intensely than bands 1 and 3, and band 5 more than bands 4 and 6. In the triploid hybrid AFF, bands 1 and 2 were stained more intensely than band 3, and bands 4 and 5 more intensely than band 6. These results indicate the following:

1. GOT is a dimeric enzyme.
2. *A. fistulosum* and shallot have two common gene loci (*Got-1*, *Got-2*).
3. There are two alleles (*Got-1^A* for shallot and *Got-1^F* for *A. fistulosum*) at *Got-1* and two alleles (*Got-2^A* for shallot and *Got-2^F* for *A. fistulosum*) at *Got-2*.

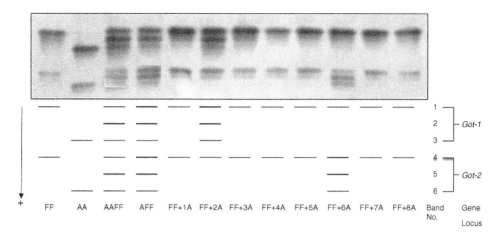

Figure 7.6 GOT zymograms. Schematic illustration is shown in the bottom half of the figure. From left to right lanes; *A. fistulosum* (FF), *A. cepa* Aggregatum group (AA), amphidiploid hybrid (AAFF), triploid hybrid (AFF), and a complete set of *A. fistulosum*–shallot monosomic additions (FF + 1A–FF + 8A).

The band patterns in the two plants of FF + 2A were identical with that of the triploid hybrid AFF at the gene locus *Got-1* and those of two plants of FF + 6A with that of the triploid AFF at the gene locus *Got-2*. In other monosomic additions, the band patterns were identical with that of *A. fistulosum* both at the two loci. These results reveal that the gene loci *Got-1* and *Got-2* are located on the chromosomes 2A and 6A, respectively. In the same way, the chromosomal locations of eight isozyme loci were determined in *A. cepa* (Table 7.4).

7.3.5.2 Ribosomal DNA (rDNA)

Ribosomal DNA (rDNA) has been recently used as a genetic marker (Uchimiya et al. 1983; Saul and Potrykus 1984; Rogers and Bendich 1987; Schwelzer et al. 1988; Honda and Hirai 1990; Koebner 1995; Sappal et al. 1995). In *Allium*, genetic analyses on several kinds of rDNA were applied to determine phylogenetic relationships (Havey 1992), determine the chromosomal locations (Ricroch, Peffley, and Baker 1992), identify the interspecific F_1 hybrids (Ohsumi et al. 1993), and to demonstrate the allodiploid nature of *A. ×wakegi* (Hizume 1994). Because eukaryotic genes coding for the 5S rDNA are present in multiple copies per genome and constitute a multigene family, it is relatively easy to determine the dispersed or tandem distribution of the repetitive DNA sequences on individual chromosomes (Jiang and Gill 1994). Determination of chromosomal

Table 7.4 Chromosome Markers of *A. cepa* Aggregatum Group Established by Isozyme and rDNA Studies

Chromosome	Chromosome Marker
1A	*Lap-1*
2A	*Got-1, 6-Pgdh-2*
3A	*Tpi-1*
4A	*Mdh-1*
5A	*Idh-1, Pgi-1*
6A	*Adh-1, Got-2*
7A	5S-Rdna-3
8A	*Gdh-1*

Source: From Shigyo, M. et al., *Genes Genet. Syst.*, 71, 363–371, 1996. With permission.

locations of 5S rDNA gene loci in *A. cepa* Common onion group by *in situ* hybridization resulted in detecting two gene loci at the interstitial regions of the short arm of the chromosome 7C (Hizume 1994). Polymerase chain reaction (PCR) is a rapid system for defining interspecific and intraspecific DNA sequence diversity of 5S rDNA in wheat (Cox, Bennett, and Dyer 1992). The 5S rDNA composed of highly conservative 5S rRNA sequence. The spacer of the 5S rDNA repeat unit was easily amplified in several plants by PCR using the universal primers (Cox, Bennett, and Dyer 1992). To determine the chromosomal location of the 5S rDNA locus in shallot, the analysis of 5S rDNA through PCR was carried out on the monosomic additions.

The electrophoretic profile of PCR products for 5S rDNA in *A. fistulosum* showed three bands 5S-Rdna-1 (approximating 1,000 bp), 5S-Rdna-2 (700 bp), and 5S-Rdna-4 (100 bp) (Figure 7.7). In shallot, the 5S-Rdna-3 band (approximating 500 bp), was detected in addition to the three bands observed in *A. fistulosum*. The amphidiploid hybrid AAFF and triploid hybrid AFF displayed the four bands observed in the parents. In both plants of the FF+7A, all four bands were observed. Other monosomic additions had the three products 5S-Rdna-1, 5S-Rdna-2, and 5S-Rdna-4, but did not have 5S-Rdna-3. These results confirm that 5S-Rdna-3 is on 7A. Hizume (1994) reported that onion had two 5S rDNA loci at the interstitial regions of the short arm of a small metacentric chromosome (7C). The close relationship of shallot and common onion is confirmed by the homology of chromosomes 7A and 7C, agreeing with the results of Hizume (1994).

7.3.5.3 *Random Amplified Polymorphic DNA (RAPD)*

Amplification of genomic DNA by PCR with the single primers of arbitrary sequence (10–12 nucleotides) reveals randomly amplified polymorphic DNA (RAPD). The RAPDs are useful genetic markers in plant genetics and breeding (Williams et al. 1990). The results of RAPD analyses using homozygous single chromosome recombinant lines and alien addition lines in wheat demonstrated that RAPD markers were effective in the analysis of genotypes (Devos and Gale 1992). In *Allium*, RAPDs were used to assist resistance breeding (De Vries et al. 1992), to examine variability among species (Wilkie, Isaac, and Slater 1993) and cultivars (Roxas and Peffley 1992), to assess integrities of inbreds (Bradeen and Havey 1995) and doubled haploid lines (Campion, Bohanec, and Javornik 1995), to confirm hybridities (Dubouzet et al. 1996), and to study phylogenetic relationships (Maaβ and Klaas 1995; Hong, Arisumi, and Etoh 1996). However, chromosomal locations of RAPD markers were not reported in *Allium*. RAPD analyses using a set of monosomic additions were needed to determine the chromosomal locations of the RAPD markers in shallot.

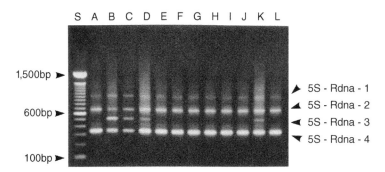

Figure 7.7 Amplification profiles for 5S rDNAs in *A. fistulosum* (A), *A. cepa* Aggregatum group (B), amphidiploid (C), triploid hybrid (D), monosomic addition FF+1A (E), FF+2A (F), FF+3A (G), FF+4A (H), FF+ 5A (I), FF+6A (J), FF+7A (K), and FF+8A (L). (S): Size marker (100-bp DNA ladder). (From Shigyo, M. et al., *Genes Genet. Syst.*, 71, 363–371, 1996. With permission.)

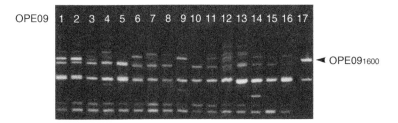

Figure 7.8   Representative RAPD profiles generated by a primer (OPE09) in a complete set of *A. fistulosum –* shallot monosomic additions (two plants of monosomic additions with chromosome 1A of *A. cepa* Aggregatum group, 1 and 2; 2A, 3 and 4; 3A, 5 and 6; 4A, 7 and 8; 5A, 9 and 10; 6A, 11 and 12; 7A, 13 and 14; 8A, 15 and 16) and *A. cepa* Aggregatum group (17). Arrowhead indicates a chromosome-specific band, $OPE09_{1600}$. (From Shigyo, M. et al., *Genes Genet. Syst.*, 72, 249–252, 1997c. With permission.)

Preliminary RAPD analyses using 108 primers (OPERON KITs A, E, and G; Wako DNA Oligomer sets A-1, C-4, F-4, and F-5) produced polymorphisms between *A. fistulosum* and shallot. The interspecific polymorphisms between these two species were observed for almost all the primers. Sixty-seven RAPD markers specific to shallot were obtained in 34 out of the 108 primers. Chromosomal locations of these markers were examined using the monosomic additions. Sixteen out of the 67 RAPD markers examined were detected in the monosomic additions (Figure 7.8). The RAPD marker $OPE09_{1600}$ was detected in two plants of the monosomic additions with chromosome 1A, but not in any of the others. In the same way, the locations of other RAPD markers on other chromosomes (2A–8A) were identified (Table 7.5). RAPD markers were assigned to all the chromosomes of shallot. The 16 RAPD bands detected in the monosomic additions have chromosome-specific unique sequences for shallot.

7.3.5.4   Flavonoid and Anthocyanin Genes

High-performance liquid chromatography (HPLC) analyses using a set of *A. fistulosum*–shallot monosomic additions were performed to determine the chromosomal locations of the genes for flavonoid and anthocyanin production in leaf sheaths of shallot (Shigyo et al. 1997a). In HPLC profiles at 360 and 520 nm exhibited several peaks in the shallot and monosomic additions with chromosome 5A from shallot, but no peak was observed in *A. fistulosum* or other monosomic additions (Figure 7.9). Four of the compounds observed at 360 nm were identified as known flavonoids, i.e., apigenin, kaempferol, quercetin, and rutin. Five out of the total 18 compounds at 520 nm were identified as known anthocyanins, i.e., cyanidin-3-glucoside, cyanidin-3-laminariobioside, peonidin-3-glucoside, cyanidin-3-malonylglucoside, and cyanidin-3-malonyl-laminariobioside. These results reveal that a group of the genes related to the flavonoid and anthocyanin production in the leaf sheath of shallot are located on chromosome 5A.

Table 7.5   Chromosome-Specific RAPD Markers of *A. cepa* Aggregatum Group

Chromosome	RAPD Marker
1A	$OPE09_{1600}$
2A	$OPE03_{600}$, $OPE18_{1400}$
3A	$OPA12_{700}$
4A	$OPA11_{950}$, $OPE18_{1500}$, $OPG10_{800}$, $WAA09_{1800}$, $WAF65_{900}$, $WAF83_{800}$
5A	$OPE17_{500}$, $OPE18_{2000}$
6A	$OPG08_{500}$
7A	$WAF81_{2800}$, $WAF83_{2200}$
8A	$WAC68_{500}$

Source: From Shigyo, M. et al., *Genes Genet. Syst.*, 72, 249–252, 1997c. With permission.

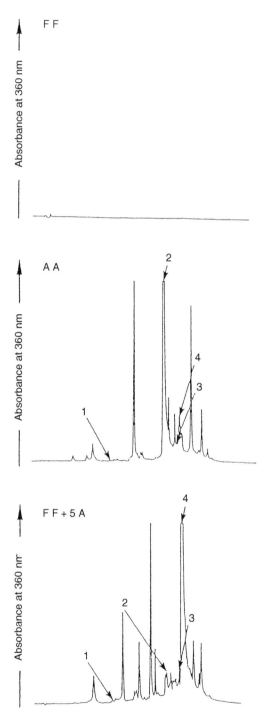

Figure 7.9   HPLC profiles of methanol extracts of *A. fistulosum* (FF), *A. cepa* Aggregatum group (AA) and *A. fistulosum*–shallot monosomic addition FF + 5A (detection: 360 nm). 1, rutin (Rt = 14.9); 2, quercetin (Rt = 23.0); 3, apigenin (Rt = 25.0); 4, kaempferol (Rt = 25.4). (From Shigyo, M. et al., *Genes Genet. Syst.* 72, 149–152, 1997a. With permission.)

7.3.6 Current Genetic Map of *Allium cepa*

Two genetic maps have been developed of *Allium cepa*. One is an AFLP linkage map using an interspecific backcross between *A. roylei* and *A. cepa* (van Heusden et al. 1999). The second is the RFLP, SSR and SNP linkage map of an intraspecific cross between onion inbreds "Brigham Yellow Globe 15–23" × "Alisa Craig 43" (King et al. 1998; Martin et al. 2006). The set of *A. fistulosum*–shallot monosomic additions is useful for assignment of the linkage maps to their respective chromosomes.

As shown in Figure 7.10, the *A. fistulosum*–shallot monosomic additions ($2n = 2x + 1 = 17$) and the AFLP linkage map were used to assign the eight linkage groups to chromosomes (Van Heusden et al. 2000). The 186 AFLP markers were present in *A. cepa* and not in *A. fistulosum*. Of these 186 AFLP markers, 51 were absent in *A. roylei* and were used to assign the eight onion linkage groups to chromosomes. Seven isozyme and three CAPS markers were also included. Two of the linkage groups were split because they included two sets of markers corresponding to different chromosomes. The monosomic additions were also applied to assign intraspecific onion RFLP linkage groups to chromosomes (Martin et al. 2006).

7.3.7 Other Types of Alien Addition Lines

Allium fistulosum contains several characteristics attractive to the breeding of *A. cepa* (bulb onion and shallot), such as resistance to pink root disease (Netzer, Rabinowitch, and Weintal 1985) (caused by the fungus *Pyrenochaeta terrestris*) and to onion leaf blight (Currah and Maude 1984) (caused by *Botrytis squamosa*). Among all the interspecific crosses in *Allium*, hybridization between *A. fistulosum* and *A. cepa* has been carried out the most extensively because *A. fistulosum* shows these desirable traits. Interspecific F_1 hybrids between these two species have often been produced and show resistance to several *A. cepa* diseases (Brewster 1994).

The interspecific hybrids are highly sterile because of chromosomal rearrangements between the two species (Havey 2002). The development of *A. fistulosum–A. cepa* monosomic additions via amphidiploid production, therefore, has proved to be a practical approach for introgressing genes from related species. Sears (1956) transferred genes for leaf rust resistance from *Aegilops* into wheat. Recently, rust resistance from *A. cepa* was tried to transferee into *A. fistulosum*, using the monosomic additions (Wako et al., unpublished data). While, *A. cepa–A. fistulosum* chromosome additions may be useful for the study of genome organization in *A. fistulosum* as well as for the practical breeding of *A. cepa*. Peffley et al. (1985) reported the production of four *A. cepa–A. fistulosum* monosomic additions ($2n = 17$) and several hypo-triploids ($2n = 20, 22, 25$). However, the monosomic additions still do not cover all the genomes of *A. fistulosum*. Shigyo, Tashiro, and Miyazaki (1994) proposed using hypo-allotriploids ($2n = 3x - 1 = 23$, AAF $- n$F) together with the *A. cepa–A. fistulosum* monosomic additions ($2n = 2x + 1 = 17$, AA $+ n$F) in allocating the genes and genetic markers to the *A. fistulosum* chromosome. Some trials have been conducted to find and characterize monosomic additions and hypo-allotriploids in the backcross progeny of amphidiploids between *A. cepa* and *A. fistulosum*.

The first and second backcrosses of amphidiploid hybrids ($2n = 4x = 32$, genomes AAFF) between shallot (*A. cepa* Aggregatum group) and *A. fistulosum* were conducted to produce *A. cepa–A. fistulosum* alien additions as shown in Figure 7.11 (Hang et al. 2004a). When shallot was used as a pollinator, the amphidiploids and allotriploids set viable BC_1 and BC_2 seeds. Of 237 BC_1 plants, 170 were allotriploids ($2n = 3x = 24$, AAF) and 42 were hypo-allotriploids possessing 23 chromosomes, i.e., single-alien deletions ($2n = 3x - 1 = 23$, AAF $- n$F). The single-alien deletions in the BC_1 progeny showed dwarfing and were discriminated from the allotriploids ($2n = 24$) and hyper-allotriploids ($2n = 25$) by means of flow cytometric analysis (Figure 7.12). The chromosome numbers of BC_2 seedlings varied from 16 to 24. Eight

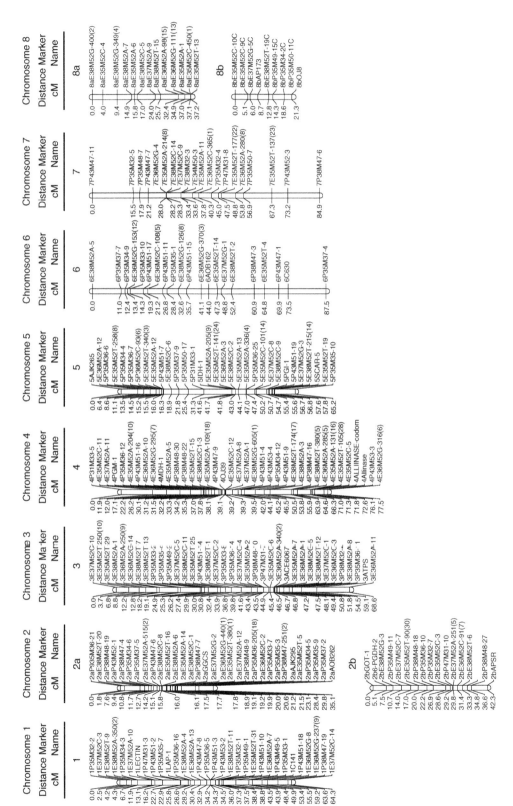

Figure 7.10 Linkage map of *Allium cepa*. (Revised and redrawn from van Heusden, A. W. et al., *Theor. Appl. Genet.*, 100, 480–486, 2000.)

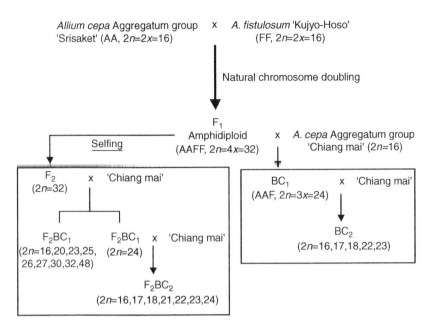

Figure 7.11 Pedigree of shallot (*A. cepa* Aggregatum group)–*A. fistulosum* chromosome additions. (From Hang, T. T. M. et al., *Genes Genet. Syst.*, 79, 263–269, 2004a. With permission.)

monosomic additions ($2n = 2x + 1 = 17$, AA + nF) and 20 single-alien deletions were found in these BC$_2$ seedlings. The *A. cepa–A. fistulosum* alien chromosome additions possessed different chromosome numbers ($2n = 17, 18, 20, 21, 22, 23$). A total of 79 aneuploids, including 62 single-alien deletions, were analyzed by a chromosome 6F-specific isozyme marker (*Got-2*). This analysis revealed that two out of 62 single-alien deletions did not possess 6F. One (AAF-6F) out of a possible eight single-alien deletions were identified (Figure 7.13). The present study is a first step toward the development of eight single-alien deletions, for the rapid chromosomal assignment of genes and genetic markers in *A. fistulosum*.

7.4 GERMPLASM ENHANCEMENT

7.4.1 Introduction

Only a few cases have been reported in which a complete set of monosomic additions in a diploid background was established (Singh 2003). Considerable attention has been devoted to maintaining these sets. However, the risk of losing this material by plant death is high, so it is important to maintain the set via other methods, such as generative reproduction.

7.4.2 Fertility and Transmission of the Extra Chromosome in *Allium* Monosomic Addition Lines

7.4.2.1 Fertility of Monosomic Addition Lines

Seed and pollen fertility in *A. fistulosum*–shallot monosomic additions was assessed to reveal the effect of the alien chromosomes on the fertility of the recipient species (Shigyo, Iino, and Tashiro 1999). Data for seed fertility were derived from three pollination methods: (open pollination,

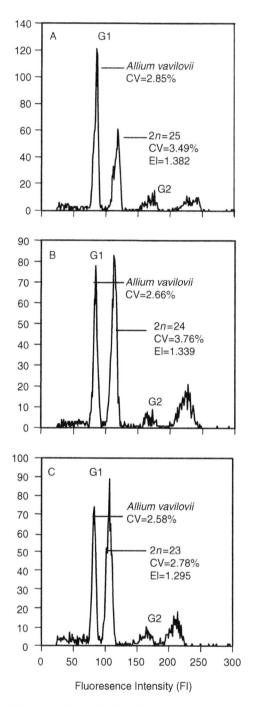

Figure 7.12 Representative histograms of nuclei isolated from mixtures of leaves of three plants with different numbers of chromosomes. (A) *Allium vavilovii*–hyper-allotriploid ($2n=3x+1=25$); (B) *A. vavilovii*–allotriploid ($2n=3x=24$); (C) *A. vavilovii*–hypo-allotriploid ($2n=3x-1=23$). (From Hang, T. T. M. et al., *Genes Genet. Syst.*, 79, 263–269, 2004a. With permission.)

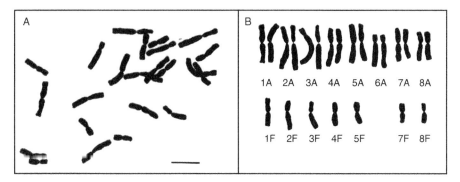

Figure 7.13 Somatic chromosome in hypo-allotriploid (2n=23, AAF-6F). Eight pairs of chromosomes, showing similar sizes and shapes, and seven solitary chromosomes are distinguished in B (an original picture is A). Scale bar=10 μm. (From Shigyo, M., Tashiro, Y., and Miyazaki, S., *Jpn. J. Genet.*, 69, 417–424, 1994. With permission.)

selfing, and backcrossing with *A. fistulosum* cv. Kujyo-Hoso). These methods yielded considerable variations among types of the monosomic additions. FF + 8A had high seed fertility, whereas FF + 3A had low (Table 7.6; open pollination). Pollen grains with the normal number and shape of nuclei were observed in all types of the monosomic additions, but pollen fertility varied among the types. FF + 4A expressed high pollen fertility (approx. 80%), whereas FF + 1A expressed low pollen fertility (approx. 31%). These results indicate that alien chromosomes influence both female and male fertility. A regression analysis (one-way ANOVA) revealed that there was no correlation between female and male fertility in the monosomic additions (data not shown), indicating that genes related to female and male fertility are located on different shallot chromosomes.

7.4.2.2 Transmission of Alien Chromosomes in A. fistulosum–A. cepa Monosomic Addition Lines

In a series of *A. fistulosum*–shallot monosomic additions, the female and male transmission rates of the extra chromosomes were assessed to examine the possibility of maintaining the series by seed propagation (Shigyo et al. 1999). Chromosome numbers of the seedlings obtained from reciprocal

Table 7.6 Seed Fertility, Seed Germination and Pollen Fertility in Opem-Pollinated Alien Monsomic Additions and *A. fistulosum*

Plant Material	FF + Extra Chromosomes	No. of Plants	Percentage of Ovules That Developed into Seeds (PODS)[a,b]	Percentage of Seeds That Germinated[b]	Pollen Fertility (%)[b]
Monosomic additions	FF + 1A	5	10.1 ± 4.5	4.1 ± 1.3	31.4 ± 8.9
of *A. fistulosum*	FF + 2A	2	14.1 + 6.4	36.0 + 3.8	68.0 + 2.3
	FF + 3A	4	3.1 ± 1.9	3.0 ± 1.7	62.7 ± 15.4
	FF + 4A	8	5.0 ± 1.0	31.6 ± 9.0	80.5 ± 5.9
	FF + 5A	3	18.8 ± 10.0	56.7 ± 17.9	50.1 ± 2.5
	FF + 6A	2	28.5 ± 6.3	70.5 ± 2.5	71.3 ± 11.9
	FF + 7A	10	8.4 ± 1.4	19.0 ± 7.1	49.0 ± 6.8
	FF + 8A	2	45.7 ± 8.0	74.5 ± 4.0	65.5 ± 20.1
A. fistulosum	FF	1	25.7	84.0	94.8

[a] Percentage of ovules that developed into seeds = $\frac{\text{Number of seeds produced}}{\text{Number of flowers pollinated} \times \text{Number of ovules per flowers(6)}} \times 100$.

[b] All data except *A. fistulosum* are shown with mean ± SE.

Source: From Shigyo, M., Iino, M., and Tashiro, Y., *J. Jpn. Soc. Hort. Sci.*, 68, 494–498, 1999. With permission.

crossings between the monosomic additions ($2n = 17$) and *A. fistulosum* ($2n = 16$) were 16 nd 17. The plants with 17 chromosomes in the seedlings obtained from the crossing FF (female)×FF+nA (male) and FF+nA (female)×FF (male) revealed the male and female transmission rates, respectively. The male transmission rates of extra chromosomes varied from 0 to 7.6% (mean = 2.6%), whereas the female transmission rates ranged from 6.1 to 40.4% (mean = 19.8%) (Figure 7.14). These results revealed the following:

1. Transmission rates are generally low.
2. Rates are variable among the extra chromosomes.
3. Female transmission rate is higher than the male transmission rate in all monosomic additions.
4. Transmission rate of the extra chromosome 8A were relatively high and 5A were low.

These results demonstrate that crossing with the monosomic additions as seed parents, are useful to maintain the monosomic additions by seed propagation.

Selfed progeny of a set of *A. fistulosum–A. cepa* monosomic additions were produced to examine the transmission rates of the alien chromosomes (Shigyo et al. 2003). All eight selfed monosomic additions set viable seeds. The numbers of chromosomes in the seedlings were 16, 17, or 18. The extra chromosomes varied in transmission rate from 9 (FF+2A) to 49% (FF+8A). The set of monosomic additions was reproduced successfully by self-pollination. A reliable way to maintain the monosomic additions used a combination of selfing and crossing as the female.

FF+8A produced two seedlings with 18 chromosomes (Figure 7.15). Cytogenetic analyses showed that these plants were disomic additions carrying two homologous chromosomes from

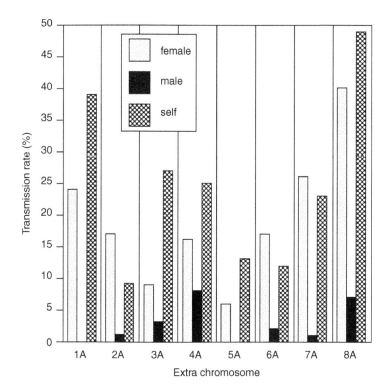

Figure 7.14 Female, male, and self transmission rates of extra chromosomes (1A–8A) in a complete set of *A. fistulosum*–shallot monosomic additions. (From Shigyo, M. et al., *J. Jpn. Soc. Hort. Sci.*, 68, 18–22, 1999a with data from Shigyo, M. et al., *Genome*, 46, 1098–1103, 2003. With permission.)

Figure 7.15   (See color insert following page 304.) Mitosis in a root tip cell of an *A. fistulosum* – shallot disomic addition FF+8A+8A. Eighteen somatic metaphase chromosomes of FF+8A+8A after GISH. Arrows indicate a pair of alien chromosomes 8A. Scale bar=10 μm.

A. cepa in a diploid background of *A. fistulosum*. Flow cytometry showed that a double dose of the alien 8A chromosome caused fluorescence intensity values spurring in DNA content, and isozyme analysis showed increased glutamate dehydrogenase activity at the gene locus *Gdh-1*.

7.4.3   Improvement of Chemical Composition in *Allium* by Chromosome Engineering

The effect of a single alien chromosome from shallot could reveal the chromosomal locations of shallot genes related to carbohydrate production in the leaf blade of bunching onion (*A. fistulosum*) (Hang et al. 2004b). In monosomic additions grown over two years in field plots at Yamaguchi University (34°N, 131°E), shallot chromosomes 2A and 8A altered carbohydrate contents in leaf-bunching onion (*A. fistulosum*). With the exception of FF+2A, every monosomic addition line accumulated nonreducing sugars in winter leaf blades. FF+2A produced low amounts of nonreducing sugar throughout the two-year study. FF+8A caused an increase in the amounts of nonreducing sugars in the winter. These results indicate those genes related to nonreducing sugar metabolisms are located on the 2A and 8A chromosomes.

7.5   FUTURE DIRECTION

The objectives of future studies should be to obtain genetic information on chromosomal location of genes related to *in vivo* production of several chemical compositions in shallot by using *Allium* alien-chromosome addition lines and to develop improved shallot varieties for cultivation in Southeast Asia. Our future research will focus on:

1. Analysis of chemical composition of A. *fistulosum*–shallot monosomic additions. Sets of A. *fistulosum*–shallot monosomic additions will be employed for analyzing qualitative traits, such as ascorbic acid, carbohydrates, chlorophyll, sulfides, anthocyanins, flavonoids, etc. Effects of shallot chromosome additions on chemical components in A. *fistulosum* will be evaluated. These analyses may define chromosomal locations of genes or QTLs for chemical composition production in shallot.

2. Production of A. *fistulosum*–shallot multiple alien addition lines. The triploid hybrid ($2n = 3x = 24$, AFF) previously derived, will be backcrossed again. Multiple alien addition lines of A. *fistulosum* containing shallot chromosomes ($2n = 18$–23, e.g., FF + 1A + 2A) will be selected by karyotype analysis and molecular marker studies. The interaction between shallot extra chromosomes at the production process of chemical compounds may be understood.

3. Breeding of new shallot cultivars suitable for Southeast Asia using chromosome engineering techniques. Another triploid hybrid ($2n = 3x = 24$, AAF) between A. *fistulosum* and shallot has been vegetatively maintained in my laboratory at Yamaguchi University. The triploid hybrid will be backcrossed to shallot and progeny examined for alien-chromosome addition lines of shallot with extra chromosomes from A. *fistulosum* ($2n = 17$–23, e.g., AA + 1F + 2F). Morphological and physiological characteristics of the addition lines will be determine the effect of A. *fistulosum* alien chromosomes on phenotypic expression of shallot, including production of chemical compositions. Subsequent lines will be evaluated for desirable agronomic traits. Field trials, at low latitudes, will be needed to select new shallot cultivars for Southeast Asia.

ACKNOWLEDGMENTS

I am deeply grateful to Professor Yosuke Tashiro of Saga University for his kind counsel, encouragement, and suggestions throughout the studies of *Allium* alien addition lines. Furthermore, I am indebted to Professor Naoki Yamauchi of Yamaguchi University for his valuable suggestions and critical comments especially on chemical composition analysis of the alien addition lines.

REFERENCES

Bradeen, J. M. and Havey, M. J., Randomly amplified polymorphic DNA in bulb onion and its use to assess inbred integrity, *J. Am. Soc. Hort. Sci.*, 120, 752–758, 1995.

Brewster, J. L., The genetics and plant breeding of *Allium* crops, *Onions and Other Vegetable Alliums*, Brewster, J. L., Ed., Melksham, UK: CAB International, Redwood Press, pp. 41–62, 1994.

Campion, B., Bohanec, B., and Javornik, B., Gynogenic lines of onion (*Allium cepa* L.): evidence of their homozygosity, *Theor. Appl. Genet.*, 91, 598–602, 1995.

Cochran, F. D., A study of the species hybrid, *Allium ascalonicum* × *Allium fistulosum* and its backcrossed progenies, *J. Am. Soc. Hort. Sci.*, 55, 293–296, 1950.

Cox, A. V., Bennett, M. D., and Dyer, T. A., Use of the polymerase chain reaction to detect spacer size heterogeneity in plant 5S-rRNA gene clusters and to locate such clusters in wheat (*Triticum aestivum* L.), *Theor. Appl. Genet.*, 83, 684–690, 1992.

Currah, L. and Maude, R. B., Laboratory tests for leaf resistance to *Botrytis squamosa* in onion, *Ann. Appl. Biol.*, 105, 277–283, 1984.

Devos, K. M., Gale, M. D. et al., The use of random amplified polymorphic DNA markers in wheat, *Theor. Appl. Genet.*, 84, 567–572, 1992.

De Vries, J. N., Onion chromosome nomenclature and homoeology relationships—workshop report, *Euphytica*, 49, 1–3, 1990.

De Vries, J. N. et al., RAPD markers assist in resistance breeding, *Prophyta*, 46, 50–51, 1992.

Dubouzet, J. G. et al., A diagnostic test to confirm interspecific *Allium* hybrids using random amplified polymorphic DNA from crude leaf DNA extracts, *J. Jpn. Soc. Hort. Sci.*, 65, 321–326, 1996.

Emsweller, S. L. and Jones, H. A., Meiosis in *Allium fistulosum*, *Allium cepa* and their hybrid, *Hilgardia*, 9, 277–288, 1935.

FAO, *FAO Production Yearbook*, 56, New York: Rome Free Press, p. 155, 2002.

Friesen, N., Fritsch, R., and Dadhmann, K., Hybrid origin of some ornamentals of *Allium* subgenus *Melanocrommyum* verified with GISH and RAPD, *Theor. Appl. Genet.*, 95, 1229–1238, 1997.

Hanelt, P., Taxonomy, evolution, and history, *Onions and Allied Crops*, Rabinowitch, H. D. and Brewster, J. L., Eds., Vol. 1, Boca Raton, FL: CRC Press, pp. 1–26, 1990.

Hang, T. T. M. et al., Production and characterization of alien chromosome additions in shallot (*Allium cepa* L. Aggregatum group) carrying extra chromosome(s) of Japanese bunching onion (*A. fistulosum* L.), *Genes Genet. Syst.*, 79, 263–269, 2004a.

Hang, T. T. M. et al., Effect of single alien chromosome from shallot (*Allium cepa* L. Aggregatum group) on carbohydrate production in leaf blade of bunching onion (*A. fistulosum* L.)., *Genes Genet. Syst.*, 79, 345–350, 2004b.

Harlan, J. R. and de Wet, J. M. J., Toward a rational classification of cultivated plants, *Taxon*, 20, 509–517, 1971.

Hart, G. E. and Tuleen, N. A., Introduction and characterization of alien genetic material, In *Isozymes in Plant Genetics and Breeding Part A*, Tanksley, S. D. and Orton, T. J., Eds., Amsterdam: Elsevier Science, pp. 339–362, 1983.

Havey, M. J., Restriction enzyme analysis of the chloroplast and nuclear 45s ribosomal DNA of *Allium* sections *Cepa* and *Phyllodolon* (*Alliaceae*), *Pl. Syst. Evol.*, 183, 17–31, 1992.

Havey, M. J., Seed yield, floral morphology, and lack of male-fertility restoration of male-sterile onion (*Allium cepa*) populations possessing the cytoplasm of *Allium galanthum*, *J. Am. Soc. Hort. Sci.*, 124, 626–629, 1999.

Havey, M. J., Genome organization in *Allium*. In *Allium Crop Science: Recent Advances*, Rabinowitch, H. D. and Currah, L., Eds., New York: CABI Publishing, pp. 59–79, 2002.

Hizume, M., Allodiploid nature of *Allium wakegi* Araki revealed by genomic in situ hybridization and localization of 5S and 18S rDNAs, *Jpn. J. Genet.*, 69, 407–415, 1994.

Honda, H. and Hirai, A., A simple and efficient method for identification of hybrids using nonradioactive rDNA as probe, *Jpn. J. Breed.*, 40, 339–348, 1990.

Hong, C., Arisumi, K., and Etoh, T., RAPD analysis of ornamental Alliums for phylogenetic relationship, *Mem. Fac. Agr. Kagoshima Univ.*, 32, 51–58, 1996.

Jiang, J. and Gill, B. S., Nonisotopic in situ hybridization and plant genome mapping: The first 10 years, *Genome*, 37, 717–725, 1994.

Jones, R. N., Cytogenetics. In *Onions and Allied Crops*, Rabinowitch, H. D. and Brewster, J. L., Eds., Vol. 1, Boca Raton, FL: CRC Press, pp. 199–214, 1990.

Jones, H. A. and Mann, L. K., *Onions and Their Allies*, London: Leonard Hill, 1963.

Kalkman, E. R., Analysis of the C-banded karyotype of *Allium cepa* L. standard system of nomenclature and polymorphism, *Genetica*, 65, 141–148, 1984.

Keller, E. R. J. et al., Interspecific crosses of onion with distant *Allium* species and characterization of the presumed hybrids by means of flow cytometry, karyotype analysis and genomic in situ hybridization, *Theor. Appl. Genet.*, 92, 417–424, 1996.

Khrustaleva, L. I. and Kik, C., Cytogenetical studies in the bridge cross *Allium cepa*×*A. fistulosum*×*A. roylei*, *Theor. Appl. Genet.*, 96, 8–14, 1998.

Kik, C., Exploitation of wild relatives for the breeding of cultivated *Allium* species. In *Allium Crop Science: Recent Advances*, Rabinowitch, H. D. and Currah, L., Eds., New York: CABI Publishing, pp. 81–100, 2002.

King, J. J. et al., A low-density genetic map of onion reveals a role for tandem duplication in the evolution of an extremely large diploid genome, *Theor. Appl. Genet.*, 96, 52–62, 1998.

Koebner, R. M. D., Generation of PCR-based markers for the detection of rye chromatinin a wheat background, *Theor. Appl. Genet.*, 90, 740–745, 1995.

Levan, A., The cytology of the species hybrid *Allium cepa*×*A. fistulosum* and its polyploid derivatives, *Hereditas*, 27, 253–272, 1941.

Maaβ, H. I. and Klaas, M., Infraspecific differentiation of garlic (*Allium sativum* L.) by isozyme and RAPD markers, *Theor. Appl. Genet.*, 91, 89–97, 1995.

Maeda, T., Chiasma studies in *Allium fistulosum*, *Allium cepa* and their F1, F2, and backcross hybrids, *Jpn. J. Genet.*, 13, 146–159, 1937.

Martin, J. M. et al., Genetic mapping of expressed sequences in onion and in silico comparisons with rice show scant colinearity, *Mol. Gen. Genomics*, 274, 197–204, 2006.

Moore, G. A. and Collins, G. B., New challenges confronting plant breeding. In *Isozymes in Plant Genetics and Breeding. Part A*, Tanksley, S. D. and Orton, T. J., Eds., Amsterdam: Elsevier Science, pp. 25–58, 1983.

Netzer, D., Rabinowitch, H. D., and Weintal, C., Greenhouse technique to evaluate pink root disease caused by *Pyrenochaeta terrestris*, *Euphytica*, 34, 385–391, 1985.

Ohsumi, C. et al., Interspecific hybrid between *Allium cepa* and *Allium sativum*, *Theor. Appl. Genet.*, 85, 969–975, 1993.

Peffley, E. B., Evidence for chromosomal differentiation of *Allium fistulosum* and *A. cepa*, *J. Am. Soc. Hort. Sci.*, 111, 126–129, 1986.

Peffley, E. B. et al., Electrophoretic analysis of *Allium* alien addition lines, *Theor. Appl. Genet.*, 71, 176–184, 1985.

Ricroch, A., Peffley, E. B., and Baker, R. J., Chromosomal location of rDNA in *Allium*: in situ hybridization using biotin- and fluorescein-labelled probe, *Theor. Appl. Genet.*, 83, 413–418, 1992.

Rogers, S. O. and Bendich, A. J., Ribosomal RNA genes in plants: variability in copy number and in the intergenic spacer, *Plant Mol. Biol.*, 9, 509–520, 1987.

Roxas, V. P. and Peffley, E. B., Short-day onion varietal identification using molecular (RAPD) markers, *Allium Improvement Newsl.*, 2, 15–17, 1992.

Sappal, N. P. et al., Restriction fragment length polymorphisms in polymerase chain reaction amplified ribosomal DNAs of three *Trichogramma* (*Hymenoptera*: *Trichogrammatidae*) species, *Genome*, 38, 419–425, 1995.

Saul, M. and Potrykus, I., Species-specific repetitive DNA used to identify interspecific somatic hybrids, *Plant Cell. Rep.*, 3, 65–67, 1984.

Schwelzer, G. et al., Species-specific repetitive DNA sequences for identification of somatic hybrids between *Lycopersicon esculentum* and *Solanum acaule*, *Theor. Appl. Genet.*, 75, 679–684, 1988.

Sears, E. R., The transfer of leaf-rust resistance from *Aegilops umbellutaca* to wheat, *Brookhaven Symp. Biol.*, 9, 1–22, 1956.

Shigyo, M., Tashiro, Y., and Miyazaki, S., Chromosomal locations of glutamate oxaloacetate transaminase gene loci in Japanese bunching onion (*Allium fistulosum* L.) and shallot (*A. cepa* L. Aggregatum group), *Jpn. J. Genet.*, 69, 417–424, 1994.

Shigyo, M. et al., Chromosomal locations of five isozyme gene loci (*Lap-1*, *Got-1*, *6-Pgdh-2*, *Adh-1* and *Gdh-1*) in shallot (*A. cepa* L. Aggregatum group), *Jpn. J. Genet.*, 70, 399–407, 1995a.

Shigyo, M. et al., Chromosomal locations of isocitrate dehydrogenase and phosphoglucoisomerase gene loci in shallot (*A. cepa* L. Aggregatum group), *Jpn. J. Genet.*, 70, 627–632, 1995b.

Shigyo, M. et al., Establishment of a series of alien monosomic addition lines of Japanese bunching onion (*Allium fistulosum* L.) with extra chromosomes from shallot (*A. cepa* L. Aggregatum group), *Genes Genet. Syst.*, 71, 363–371, 1996.

Shigyo, M. et al., Chromosomal locations of genes related to flavonoid and Anthocyanin production in leaf sheath of shallot (*Allium cepa* L. Aggregatum group), *Genes Genet. Syst.*, 72, 149–152, 1997.

Shigyo, M. et al., Morphological characteristics of a series of alien monosomic addition lines of Japanese bunching onion (*Allium fistulosum* L.) with extra chromosomes from shallot (*A. cepa* L. Aggregatum group), *Genes Genet. Syst.*, 72, 181–186, 1997.

Shigyo, M. et al., Assignment of randomly amplified polymorphic DNA markers to all chromosomes of shallot (*Allium cepa* L. Aggregatum group), *Genes Genet. Syst.*, 72, 249–252, 1997c.

Shigyo, M. et al., Identification of alien chromosomes in a series of *Allium fistulosum*–*A. cepa* monosomic addition lines by means of genomic in situ hybridization, *Genes Genet. Syst.*, 73, 311–315, 1998.

Shigyo, M. et al., Transmission rates of extra chromosomes in alien monosomic addition lines of Japanese bunching onion with extra chromosomes from shallot, *J. Jpn. Soc. Hort. Sci.*, 68, 18–22, 1999.

Shigyo, M., Iino, M., and Tashiro, Y., Fertility of alien monosomic addition lines of Japanese bunching onion with extra chromosomes from shallot (*A. cepa* L. Aggregatum group), *J. Jpn. Soc. Hort. Sci.*, 68, 494–498, 1999.

Shigyo, M. et al., Transmission of alien chromosomes from selfed progenies of a complete set of *Allium* monosomic additions: the development of a reliable method for the maintenance of a monosomic addition set, *Genome*, 46, 1098–1103, 2003.

Singh, R. J., Chromosome aberrations—Numerical chromosome changes, *Plant Cytogenetics*, Singh, R. J., Ed., Boca Raton, FL: CRC Press, pp. 157–276, 2003.

Tanksley, S. D. and Orton, T. J., *Isozymes in Plant Genetics and Breeding. Part A*, Amsterdam: Elsevier Science, 1983.

Tashiro, Y., Cytogenetic studies on the origin of *Allium wakegi* Araki, *Bull. Fac. Agr. Saga Univ.*, 56, 1–63, 1984.

Traub, H. P., The subgenera, sections and sub-sections of *Allium*, *Plant Life*, 24, 147–163, 1968.

Uchimiya, H. et al., Detection of two different nuclear genomes in parasexual hybrids by ribosomal RNA gene analysis, *Theor. Appl. Genet.*, 64, 117–118, 1983.

Van Heusden, A. W. et al., A genetic map of an interspecific cross in *Allium* based on amplified fragment length polymorphism (AFLP™) markers, *Theor. Appl. Genet.*, 86, 497–504, 1999.

Van Heusden, A. W. et al., AFLP linkage group assignment to the chromosomes of *Allium cepa* L. via monosomic addition lines, *Theor. Appl. Genet.*, 100, 480–486, 2000.

Ved Brat, S., Genetic systems in *Allium*. I. Chromosome variation, *Chromosoma*, 16, 486–499, 1965.

Wilkie, S. E., Isaac, P. G., and Slater, R. J., Random amplified polymorphic DNA (RAPD) markers for genetic analysis in *Allium*, *Theor. Appl. Genet.*, 86, 497–504, 1993.

Williams, J. G. K. et al., DNA polymorphisms amplified by arbitrary primers are useful as genetic markers, *Nucleic Acids Res.*, 18, 6531–6535, 1990.

Yamashita, K. and Tashiro, Y., Possibility of developing a male sterile line of shallot (*Allium cepa* L. Aggregatum group) with cytoplasm from *A. galanthum* Kar. et Kir, *J. Jpn. Soc. Hort. Sci.*, 68, 256–262, 1999.

Yamashita, K., Arita, H., and Tashiro, Y., Cytoplasm of a wild species, *Allium galanthum* Kar. et Kir. is useful for developing the male sterile line of *A. fistulosum* L., *J. Jpn. Soc. Hort. Sci.*, 68, 788–797, 1999.

CHAPTER 8

Cucurbits (Cucurbitaceae; *Cucumis* spp., *Cucurbita* spp., *Citrullus* spp.)

Aleš Lebeda, M. P. Widrlechner, J. Staub, H. Ezura, J. Zalapa, and E. Křístková

CONTENTS

8.1   INTRODUCTION

The Cucurbitaceae is a remarkable plant family, deserving of attention because of its economic, aesthetic, cultural, medicinal, and botanical significance. In the Old and New Worlds, cucurbits have been associated with human nutrition and culture for more than 12,000 years (Brothwell and Brothwell 1969; Lira-Saade 1995). Thus, the Cucurbitaceae, along with the Brassicaceae and Asteraceae, can be considered families of extraordinary importance to humans, and they follow cereals and legumes in their economic significance to human economy (Whitaker and Davis 1962; Nayar and More 1998).

From a lay perspective, the fruits of cultivated cucurbits characterize the family. The remarkable variation in size, shape, and color patterns are notable familial features. Equally impressive is the prodigious growth of the herbaceous, vining stems of some cucurbit species (Bates, Robinson, and Jeffrey 1990). The flesh of the fruits of many species is eaten as fruits or vegetables. Many cucurbits are important components of traditional medicines, and some may have modern pharmaceutical applications (Ng 1993). In certain cultures, the seeds, leaves, or shoots, roots, and flowers are also consumed. Moreover, edible and industrial oils can be extracted from the seeds. The name of the family, Cucurbitaceae, is likely derived from Latin, where the word "corbis" means "basket" or "bottle" (Pražák, Novotný, and Sedláček 1941), reflecting one of the ways that fruits are used. Mature fruits can also be strikingly ornamental or can serve as containers or as musical instruments. In addition, the fibers of the *Luffa* fruits can be used as sponges or in the production of shoes (Moravec, Lebeda, and Křístková 2004).

A striking feature of cultivated cucurbits is their adaptation to a wide variety of agricultural environments. Although they are often grown in extensive monoculture typical of crop production in developed economies, they are also grown in traditional small gardens typified by low external inputs. Many cucurbits are adapted to environments considered marginal for agriculture where some are gathered as food sources or for medicinal purposes (Bates, Robinson, and Jeffrey 1990).

Within this family, the genera *Cucumis, Cucurbita*, and *Citrullus* are considered to have high economic importance. Of about 3000 plant species used for human consumption, only 150 species are cultivated extensively, and of these, thirty provide the bulk of human food. Among these elite species, *Citrullus* is ranked twenty-fourth (Raven, Berg, and Johnson 1993).

China, India, Iran, Turkey, Egypt, and the U.S. are among the world's largest producers of cucurbits. China remains the world's leading producer of the major cucurbits, exporting fresh fruits, watermelon, and squash seeds (Maynard 2001). Production of these crops has dramatically increased during recent agricultural history. The cultivated area of *Cucumis, Cucurbita*, and *Citrullus*, as reported by the United Nations' Food and Agriculture Organization (FAO 2004), was 8.5×10^6 ha in the year 2004. More specifically, overall production reached 4×10^7 metric tons of cucumbers and gherkins; 2.7×10^7 metric tons of cantaloupes and melons; 1.9×10^7 metric tons of pumpkins, squashes, and gourds; and 9.3×10^7 metric tons of watermelons.

There are many cucumber market classes worldwide (e.g., U.S. processing (pickling), U.S. fresh market, European glasshouse, Mediterranean, Asian glasshouse). As noted above, world cucumber production ranks second among all cucurbits. Cucumber production is centered in Asia, where in 2004, almost 82% of world production occurred. China, Iran, Turkey, and the United States represented about 74% of the world's production (FAO 2004) with China accounting for almost 64%. European production followed well behind, representing about 10% of world production (Table 8.1). Extremely high yields of 7.1×10^6 kg/ha in the Netherlands, 4×10^6 kg/ha in Denmark, and 3.15×10^6 kg/ha in the United Kingdom are the result of intensive cultivation in glasshouses (FAO 2004). Other countries such as Japan, Spain, and Korea, also produce a significant volume of cucumbers in glasshouses and other protective structures (Rubatzky and Yamaguchi 1997). Much of this production utilizes parthenocarpic cultivars (seedless) for which cultivation is highly specialized (Rubatzky and Yamaguchi 1997).

Table 8.1   World Cucumber and Gherkin Production in 2004

Location	Production (metric tons)	Area Harvested (ha)	Mean Yield (kg/ha)
World	40,190,104	2,395,125	167,800
By Continent			
Africa	1,073,727	145,670	73,710
Asia	33,037,780	1,915,554	172,471
Australia	17,000	1,100	154,545
Europe	3,923,230	200,839	195,342
North and Central America	2,059,441	126,540	162,750
South America	76,430	5,237	145,942
By Nation			
China	25,558,000	1,502,900	170,058
Turkey	1,750,000	60,000	291,667
Iran	1,350,000	65,000	207,692
U.S.A.	1,046,960	72,000	145,411
Russian Federation	715,000	60,000	119,167

Source: From FAO, *FAOSTAT Agricultural Database*. http://apps.fao.org 2004.

Asia dominates the world production of cantaloupe and other melons by producing more than 71% of the total tonnage (Table 8.2) (FAO 2004). Europe, North America, and Central America follow, accounting for more than 20% of the global production. Among leading countries, China's production is the highest (52%) followed by Turkey, the United States, Iran, Romania, and Spain (Table 8.2). Extremely high yields have also been recorded in Puerto Rico (2.0×10^6 kg/ha) and Canada (1.2×10^6 kg/ha). Developed countries produce less than one-third of total production, suggesting that the potential for export to these countries is high and that international trade could increase. However, if one considers the large volume of home-grown melons, such statistics may significantly underestimate actual production (Rubatzky and Yamaguchi 1997).

As with cucumber, there are many melon market classes. For instance, the horticultural groups, Cantalupensis (i.e., cantaloupe) and Inodorus (i.e., honeydew), are of particular commercial importance in the United States, Europe, and Asia (McCreight, Nerson, and Grumet 1993). The importance of other melon groups depends on the region of production and usage (Rubatzky and Yamaguchi 1997).

The cultivated *Cucurbita* species, commonly referred to as squashes, pumpkins, and gourds, represent a very important source of nutrition, not only in Latin American countries, but also in many other regions worldwide. About 63% of the world's 2004 production was in Asia, which was led by China (29%) and India (18%) (Table 8.3). Extremely high yields have been reported from Netherlands (6×10^5 kg/ha), Spain (4.3×10^5 kg/ha), Israel (4.1×10^5 kg/ha), and France (4.1×10^5 kg/ha) (FAO 2004). Overall, production in developing countries is nearly four times that of the developed countries, which likely reflects the use of *Cucurbita* as a staple food in developing countries (Rubatzky and Yamaguchi 1997).

Watermelons are grown throughout the world in areas where a long, warm growing season prevails. In 2004, Asia was by far the most important watermelon production region with 79% of the world area in cultivation, resulting in 87% of global production (Table 8.4). China dominated Asian watermelon production with more than 75% of that continent's production area and more than 84% of the Asian harvest (Table 8.4).

Between three and four million ha are collectively devoted to watermelon production in Africa, North America, Central America, and Europe with a smaller area (and lower yields) found in South America (Table 8.4). Production yields in Europe are typically below the world average, which is likely due to less favorable growing conditions (Maynard 2001). Nevertheless, exceptionally high yields have been recorded in Cyprus (5.4×10^5 kg/ha) and Spain (4.1×10^5 kg/ha) (FAO 2004).

Table 8.2   World Cantaloupe and Other Melon Production (Excluding *Citrullus*) in 2004

Location	Production (metric tons)	Area Harvested (ha)	Mean Yield (kg/ha)
World	27,371,268	1,313,727	208,348
By Continent			
Africa	1,601,325	79,250	202,060
Asia	19,537,759	902,582	216,465
Australia	64,150	2,635	243,454
Europe	3,160,000	161,210	196,018
North and Central America	2,440,099	125,081	195,082
South America	565,400	42,800	132,103
By Nation			
China	14,338,000	558,500	256,723
Turkey	1,700,000	115,000	147,826
U.S.A.	1,240,000	46,000	269,565
Iran, Romania, Spain	1,000,000	70,000	142,857
Romania	1,000,000	52,000	192,308
Spain	1,000,000	39,100	255,754

Source: From FAO, *FAOSTAT Agricultural Database*. http://apps.fao.org 2004.

Table 8.3   World Pumpkin, Squash, and Gourd (*Cucurbita* spp.) Production in 2004

Location	Production (metric tons)	Area Harvested (ha)	Mean Yield (kg/ha)
World	19,015,901	1,468,434	129,498
By Continent			
Africa	1,782,414	226,010	78,864
Asia	12,121,615	890,753	136,083
Australia	93,226	6,584	141,595
Europe	2,177,300	118,510	183,723
North and Central America	120,811	160,710	120,811
South America	721,500	56,930	126,735
By Nation			
China	5,674,200	303,505	201,923
India	3,500,000	360,000	97,222
Ukraine	900,000	50,000	180,000
U.S.A.	740,000	35,010	211,368
Egypt	710,000	39,200	181,122

Source: From FAO, *FAOSTAT Agricultural Database*. http://apps.fao.org 2004.

In addition to the three genera noted above that are cultivated and traded worldwide, there are several other notable cucurbit genera of local or regional economic importance, including *Benincasa*, *Lagenaria*, *Luffa*, *Momordica*, and *Sechium*. General overviews of these secondary cucurbit crop species can be found in reports by Ng (1993), Bates, Merrick, and Robinson (1995), and Robinson and Decker-Walters (1997).

Extensive world production of cucurbits is based upon a broad array of genetic diversity, ranging from traditional landraces to elite F_1 hybrid cultivars. This diversity has evolved as a result of long periods of domestication designed to meet a wide range of human needs (e.g., various foods, medicines, processed, and other products) under vastly different environmental conditions and selection pressures. Taken collectively, this biodiversity comprises the body of cucurbit genetic resources. Many researchers have recognized the importance of the conservation, description, and effective use of these genetic resources, resulting in an extensive body of published research.

Table 8.4   World Watermelon Production in 2004

Location	Production (metric tons)	Area Harvested (ha)	Mean Yield (kg/ha)
World	93,481,266	3,461,023	270,097
By Continent			
Africa	3,799,605	182,604	208,079
Asia	81,156,597	2,658,635	305,257
Australia	110,955	1,335	255,952
Europe	3,954,683	351,392	112,543
North and Central America	3,135,114	127,181	246,508
South America	1,314,662	136,242	96,495
By Nation			
China	68,300,000	2,015,500	338,874
Turkey	4,000,000	140,000	285,714
Iran	1,900,000	90,000	211,111
U.S.A.	1,750,000	61,000	286,885
Egypt	1,600,000	60,000	266,667

Source: From FAO, *FAOSTAT Agricultural Database*. http://apps.fao.org 2004.

8.2 GERMPLASM COLLECTION, MAINTENANCE, CHARACTERIZATION, AND DISTRIBUTION

The efficient utilization of cucurbit germplasm for crop improvement relies upon the ability of the world's genebanks to conserve the breadth of genetic diversity present in cucurbitaceous crops and their wild and weedy relatives, to adequately characterize that diversity, and to make genetic resources and accurate information readily available to researchers. The following section summarizes the current status of cucurbit germplasm conservation and characterization, along with a brief examination of studies that investigate demand for, and usage of, cucurbit germplasm.

8.2.1 Gene Banks and Other Significant Germplasm Collections

The first global report of Cucurbitaceae genetic resources was compiled by Esquinas-Alcázar and Gulick (1983). The most recent compendium of cucurbit germplasm collections was prepared by Bettencourt and Konopka (1990). It provides general information about the holdings, maintenance conditions, availability, evaluation, and documentation of 68 of the world's collections, emphasizing national genebanks and citing important breeding collections. More recently, information about the holdings of the world's largest collections of *Citrullus*, *Cucumis*, and *Cucurbita* germplasm was summarized as part of the Food and Agriculture Organization's effort to present "The State of the World's Plant Genetic Resources for Food and Agriculture" (FAO 1998). In addition to these overviews, numerous articles have been published describing the current status of various genebank collections of cucurbit germplasm (Table 8.5).

In Europe, international cooperation among institutions holding germplasm collections of cucurbitaceous vegetables is coordinated by the International Plant Genetic Resources Institute (IPGRI), an autonomous international scientific organization, supported by the Consultative Group on International Agricultural Research (CGIAR). IPGRI's formal status is conferred under an establishment agreement which, by January 2002, had been signed and ratified by the governments of nearly 50 nations (http://www.ipgri.org).

Table 8.5 Publications Describing Important National Cucurbit Germplasm Collections

Nation	Genera	References
Brazil	*Citrullus*	DeQueiroz et al. (1996)
Bulgaria	*Citrullus, Cucumis, Cucurbita*	Krasteva (2000a, 2000b, 2002), Krasteva et al. (2002) and Todorova (1997)
China	*Cucumis*	Hou and Ma (2000)
Czech Republic	*Cucumis, Cucurbita*	Krístková and Lebeda (1995), Krístková (2002), and Krístková et al. (2003)
Hungary	Nine genera	Horváth (2002)
India	*Citrullus, Cucumis, Cucurbita, Lagenaria, Luffa, Momordica*	Mal et al. (1996), Seshadari (1988), and Sharma and Hore (1996)
Latvia	*Cucumis*	Lepse et al. (2000)
Netherlands	*Cucumis*	Visser and den Nijs (1980), and van Dooijeweert (2002, 2004)
Poland	*Citrullus, Cucumis, Cucurbita*	Kotlinska (1994)
Portugal	*Citrullus, Cucumis, Cucurbita*	Carnide (2002)
Russia	Fourteen genera	Piskunova (2002)
Spain	*Citrullus, Cucumis, Cucurbita, Lagenaria, Luffa, Momordica*	de la Cuadra and Varela (1996), Nuez et al. (2000a), and Picó et al. (2002)
Turkey	*Bryonia, Citrullus, Cucumis, Cucurbita, Ecballium, Lagenaria*	Küçük, Abak, and Sari (2002)
United States	*Cucumis, Cucurbita*	Clark et al. (1991)

In most European countries, research aimed at facilitating the long-term conservation and increased utilization of plant genetic resources operates under the aegis of The European Cooperative Programme for Crop Genetic Resources Networks (ECP/GR). The ECP/GR is entirely financed by the member countries, and it is coordinated by IPGRI. Within this program, working groups for various crop species have been established. The first *ad-hoc* meeting of the ECP/GR Informal Group on Cucurbits took place in January 2002 in Adana, Turkey (Díez, Picó, and Nuez 2002). The Cucurbits Working Group ECP/GR was officialy approved in 2003 (ECP/GR 2005; Thomas et al. 2005).

8.2.2 Acquisition and Exploration

Since the 1980s, considerable effort has been directed toward expanding the diversity of *ex situ* germplasm collections by collecting wild, weedy, and landrace populations of cucurbits from their centers of diversity and from the refuges of traditional agroecosystems throughout the world. Andres (2000) has published an excellent summary of exploration efforts for *Cucurbita* with special insights on the logistics of collection. No analogous overview has been published for other cucurbit genera, but many interesting and important exploration reports can be found in the literature (Table 8.6). Many more acquisition efforts have been described in less detail, often in reports on the evaluation of cucurbit landraces for their adaptive and horticultural characteristics (discussed in Section 8.6.2). In addition to the extensive collection and exploration programs developed in the United States (Williams 2005) at the beginning of the twenty-first century, similar activities are being undertaken in some European countries to help conserve landraces used in traditional farming systems [e.g., Poland (Kotlinska 1994) and Romania (Strajeru and Constantinovici 2004)].

An important research topic related to the collection of cucurbits from traditional agroecosystems is the investigation of in situ gene flow and delimitation of populations. Research on the extent

Table 8.6 Publications Describing Explorations to Collect Cucurbit Germplasm

Nation	Genera	References
Brazil	*Cucurbita*	Tasaki (1993)
Brazil	*Citrullus, Cucumis, Cucurbita*	Duarte and de Queiroz (2002)
China	*Citrullus, Cucumis, Luffa*	Wehner et al. (1996)
Cuba	*Citrullus, Cucumis, Cucurbita, Sechium*	Hammer et al. (1991)
Ecuador	*Cucurbita*	Andres (2000)
India	*Cucumis*	Anonymous (1993)
India	*Citrullus, Cucumis*	Pareek et al. (1999)
India	*Cucumis*	Pitchaimuthu and Dutta (1999)
Indonesia	*Lagenaria*	Plarre (1995)
Italy	*Citrullus, Cucumis, Cucurbita*	Laghetti et al. (1998)
Macedonia	*Cucurbita*	Ivanovska and Posimonova (2004)
Malawi	*Cucurbita*	Chigwe and Saka (1994)
Mexico	*Cucurbita*	Andres (2000)
Morocco	*Citrullus*	Prendergast et al. (1992)
Panama	*Cucurbita*	Andres (2000)
Romania	*Cucurbita*	Strajeru and Constantinovici (2004)
Spain	*Citrullus, Cucumis, Cucurbita*	Nuez et al. (1986a, 1986b, 1987, 1992)
Sudan	*Cucumis*	Mohamed and Taha (2004)
Tunisia	*Citrullus, Cucumis, Cucurbita, Lagenaria*	Pistrick et al. (1994)
United States	*Cucurbita*	Andres (2000)
Venezuela	*Cucurbita*	Segovia et al. (2000)
Zambia	Cucurbitaceae	Whitaker (1984)
Zimbabwe	*Citrullus, Cucumis*	Toll and Gwarazimba (1983)

of in situ gene flow has employed isozyme, morphological, and secondary chemical markers to document hybridization between cultivated populations and neighboring wild or weedy populations of *Cucurbita*, primarily in Mexico and Texas (Merrick and Nabhan 1984; Nabhan 1984; Kirkpatrick and Wilson 1988; Wilson 1990). Controlled experiments under more intensive, field-production conditions have also used genetic markers to document the frequency and distance of gene flow through pollen transfer between *Citrullus* (Rhodes, Adamson, and Bridges 1987) and *Cucumis* (Handel 1983; Handel and Mishkin 1984) populations in the absence of physical isolation other than distance.

8.2.3 Collection Regeneration and Maintenance

The ability of genebanks to conserve the genetic profiles of cucurbit germplasm relies on protocols that ensure the production of high-quality seed and minimize outcross contamination, genetic drift, and changes brought about through selective regeneration environments. Once produced, the seeds should be held under moisture and temperature conditions that prolong their longevity to increase the interval between regeneration events and reduce potential changes to genetic profiles.

Most cultivated cucurbits have large flowers that are attractive to pollinating insects. The ability to preserve their genetic identity relies upon the use of tents, cages, greenhouse isolation chambers (Figure 8.1 and Figure 8.2), and similar systems to ensure isolation and the manipulation of insects and/or human labor to effect pollination. Descriptions of the field cages and greenhouse chambers used to isolate cucurbit germplasm for seed production can be found in reports by Grewal and Sidhu (1979), Ellis et al. (1981), Ruszkowski and Biliński (1984), and Cox, Able, and Gustafson (1996).

Figure 8.1 Regeneration of wild *Cucumis* species in greenhouse cages. (Photo courtesy of L. Clark.)

Figure 8.2 Regeneration of *Cucurbita* species in field cages. (Photo courtesy of NCRPIS staff.)

Another key component in the regeneration system is to have effective, easily manipulated insect pollinators. The effective use of queenright nucleus hives of honeybees (*Apis mellifera*) in field cages of cucurbits was first described by Ellis et al. (1981) and was refined by Neykov (1997). Other bee genera, i.e., *Bombus*, *Megachile*, and *Pithitis* (Grewal and Sidhu 1979; Ruszkowski and Biliński 1984), and fly genera, i.e., *Lucillia*, *Musca*, and *Phomita* (Neykov 1997), have been used in field cages as alternatives to honeybees or hand pollination. Under greenhouse conditions, *Bombus ruderarius* and *B. terrestris* have been shown to be effective pollinators of *Cucumis* (Ruszkowski and Biliński 1984; Fisher and Pomeroy 1989). The pollination of *Cucumis* by *Megachile rotundata* in greenhouses has also been reported (Szabo and Smith 1970), but Ruszkowski and Biliński (1984) found *Megachile* to be somewhat inferior to *Bombus* as pollinators.

Widrlechner et al. (1992) used isozyme markers to compare genetic profiles between 157 pairs of genebank seedlots of *C. sativus* with one lot produced in a field cage with honeybees and the other being an older, open-pollinated sample of the same accession that the first lot replaced. As expected, the cage-pollinated samples were generally more homogeneous than the open-pollinated samples they replaced. In addition, certain rare isozyme alleles were found in the cage-pollinated samples, but not in the open-pollinated samples.

In addition to the preservation of genetic profiles, the production of disease-free, highly viable seeds with appropriate longevity for long-term storage must be an important consideration in any seed regeneration protocol (Sackville Hamilton and Chorlton 1997). Pertinent research reports on the quality of cucurbit seed production include investigations on optimal fruit maturity (Edwards, Lower, and Staub 1986; Oluoch and Welbaum 1996), the duration of seed fermentation (Nienhuis and Lower 1981) and washing (Oluoch and Welbaum 1996), and seed-drying methods (Zhang and Tao 1988; Ji, Guo, and Ye 1996; Kong and Zhang 1998; Shen and Qi 1998).

Once high-quality seeds are produced, they should be conserved under optimal storage conditions. Fortunately, typical storage conditions for germplasm conservation in base collections (-15 to $-18°C$) can preserve the viability of cucurbit seeds for many years. Roos and Davidson (1992) reported median viability periods (P_{50}s) for *C. melo* and *C. sativus* of 32–79 years and for *C. lanatus* of 34–48 years. Based on 11 years of storage at $-18°C$, Stoyanova (2001) calculated even more impressive P_{50} values of 297 years for *Cucurbita*, 506 years for *Cucumis melo*, and 785 years for *C. sativus*. Mean moisture contents for the samples in that study were typically in the range of 4–6%. Likewise, after 15 or more years of storage at −15°C, Specht et al. (1998) reported that all *C. sativus* and most *C. melo* samples tested were in the 71–100% germination class. Similar results were reported for *C. pepo* samples stored between 12 and 14 years.

Extended longevity for cucurbit seeds has also been noted in samples stored at above-freezing temperatures. Fan et al. (1989) reported that about half of 44 seed samples of *C. sativus* that had been stored for 23 years in a storehouse under low ambient humidity in western China retained at least 50% germination after storage. And, recently, Reitsma and Clark (2005) reported that the oldest *Cucumis* and *Cucurbita* seedlots of sufficient quality to distribute from the North Central Regional Plant Introduction Station's genebank were now more than 40 years old. Those seedlots have generally been held at 4°C and between 25 and 40% RH. These favorable results from using typical cold storage may help explain the overall lack of information about cryogenic seed storage of cucurbits. Only Chernova (1990) has reported on the successful use of liquid nitrogen vapor in storing *Citrullus* seeds.

Tissue culture and pollen storage can be valuable conservation tools as supplements or alternatives to seed storage, especially in cases where valuable germplasm accessions produce no viable seeds. Of the two methods, pollen storage may be the more intractable. An early report (Griggs, Vansell, and Iwakiri 1953) indicated that storage of *C. melo* pollen should be feasible at −18°C, but evidently did not lead to additional research or the refinement of protocols. Islam and Khan (1998) subsequently reported storage of *M. dioica* pollen up to 45 days at 0°C, but their determination of viability was based upon acetocarmine staining, which is not a vital stain. Wang and Robinson (1983) reported that *Cucurbita* pollen was extremely short-lived (typically living only a few hours) and noted that a variety of treatments, including the use of rapid freezing, a nitrogen atmosphere, and storage in organic solvents, could not extend its viability. On a more positive note, considerable progress has been made in Japan in the development of cucurbit pollen-storage methods, primarily for *Citrullus* (reviewed by Sugiyama, Morishita, and Nishino 2002b), with the best long-term results obtained by storing pollen in ethyl acetate at −20°C (Sugiyama, Morishita, and Nishino 2002b).

In contrast to pollen storage, tissue-culture methods for cucurbits are relatively well-developed, although organogensis and regeneration success can be strongly influenced by genotype (see Molina and Nuez 1996). Protocols for embryo and ovule culture were recently reviewed by Skálová, Lebeda, and Navrátilová (2004) and those for protoplast culture by Gajdová, Lebeda, and Navrátilová (2004). Somatic embryos of *C. melo* generated from callus culture have been successfully cryopreserved after desiccation (Shimonishi et al. 1991, 2000). From a germplasm conservation perspective, however, the most valuable culture system is one that produces and preserves apical and/or axillary meristems because intact meristems are generally more stable genetically than are many other cultured tissues (Withers 1989). A method to establish shoot-primordia cultures for *C. melo* was reported (Nagai, Nomura, and Oosawa 1989), but it is labor intensive, requiring repeated subculture. Ogawa et al. (1997) developed a slow prefreezing procedure to cryopreserve shoot primordia of a *C. melo* cultivar. However, to date, there are no reports of the application of this promising procedure to germplasm conservation.

8.2.4 Germplasm Characterization

It is crucial that genebank managers correctly establish the taxonomic identity of the collections they conserve (see Section 8.3 for an overview of taxonomic studies). Accurate characterization of germplasm should support its proper identification at both the species and cultivar levels, which facilitates both maintenance and utilization and reduces the frequency of unintended duplication or near-duplication (maintenance of closely related populations). A broad range of techniques has been employed to apply morphological, protein, and DNA markers to characterize germplasm collections (reviewed by Bretting and Widrlechner 1995), including those of cucurbits.

Studies solely employing traditional, morphological methods to characterize cucurbit germplasm are typically focused on revealing valuable horticultural traits and are discussed within that context in a review of research on screening for useful traits (see Section 8.6.2).

Polymorphic proteins have been widely used to characterize patterns of genetic diversity within, and among, cucurbit germplasm collections. Nearly all pertinent reports focus on enzyme or isozyme variation. The only notable exceptions are those made by Indian researchers who examined electrophoretic (SDS PAGE) variation in seed-storage proteins in *C. melo* (Singh, Shukla, and Tewari 1999; Sawant and More 2002) and in *C. sativus* (Singh and Ram 2001) and by Navot and Zamir (1987) who analyzed variation in seed-storage proteins in *Citrullus*.

Many of the initial studies of cucurbits that establish protocols for isozyme detection and analysis describe the genetic control of polymorphic banding patterns and examine evolutionary relationships were reviewed by Dane (1983), Doebley (1989), Puchalski and Robinson (1990), and Weeden and Robinson (1990). Two other key papers describing genetic control of and linkage relationships among loci responsible for isozyme variation are those of Knerr and Staub (1992) for *C. sativus* and Staub, Meglic, and McCreight (1998) for *C. melo*.

With well-developed arrays of isozyme polymorphisms available for *Citrullus*, *Cucumis*, *Cucurbita*, and *Momordica*, extensive surveys of isozyme variation have been used to elucidate genetic relationships among germplasm collections of those four genera. The most comprehensive of such surveys have been conducted for hundreds of accessions of *C. sativus* from throughout the world, emphasizing those conserved by the U.S. National Plant Germplasm System (NPGS). Results from this long-term survey were published as a series of papers with two overall assessments (Knerr et al. 1989; Meglic, Serquen, and Staub 1996) and more specialized studies of historic cultivars (Meglic and Staub 1996b) and collections from India (Staub, Serquen, and McCreight 1997b) and China (Staub et al. 1999). Similar surveys have been conducted on extensive samples of *Citrullus* (Navot and Zamir 1987) and *C. melo* germplasm (Meglic, Horejsi, and Staub 1994; McCreight et al. 2004) and more limited sets of *C. colocynthis* populations from Israel and Sinai (Zamir, Navot, and Rudich 1984); of landraces of *C. melo* from Spain (Esquinas-Alcázar 1981), India (Sujatha et al. 1991), and eastern and southern Asia (Akashi et al. 2002b; Kato et al. 2002); of wild *Cucumis* species (Frederick, and Marty 1987; Puchalski and Robinson 1990; Staub, Staub et al. 1992); of wild and domesticated *Momordica charantia* L. (Marr, Mei, and Bhattarai 2004); and of six genera of the Benincaseae (Walters et al. 1991). Peroxidase banding patterns were also used as one line of evidence in the correct assignment of the newly discovered Xishuangbanna cucumber to the species *C. sativus* (Qi, Yuan, and Li 1983).

Isozyme analyses have been used to evaluate traditional, morphologically defined taxa in *Cucurbita* with varying results. In an initial survey with three enzyme staining systems, Puchalski and Robinson (1978) reported general agreement with an earlier taxonomic treatment of relationships among *Cucurbita* species (Bemis et al. 1970). More extensive surveys of isozyme variation among *Cucurbita* species (Puchalski and Robinson 1990) and, more specifically, in wild, weedy, domesticated Mexican *Cucurbita* presented more complex views of genetic differentiation and gene flow among taxa (Wilson 1989). Isozyme analyses by Decker-Walters et al. (1990) did not support traditional infraspecific classifications for *C. maxima*, *C. argyosperma*, or *C. moschata*. The most thorough isozyme analyses, however, have been devoted to *C. pepo* in its wild, weedy, and domesticated forms (Ignart and Weeden 1984; Decker and Wilson 1987; Decker-Walters et al. 1993) and have been instrumental in clarifying patterns of variation and domestication within this species.

A recent analysis of *M. charantia* germplasm (Marr, Mei, and Bhattarai 2004) examined both morphometric and isozyme variation to describe patterns of crop evolution and overall diversity. Isozyme variation supported a single domestication event with extremely low levels of variation among domesticated accessions. However, this analysis could not identify the wild populations most closely related to the original domesticate.

The discovery of a tight linkage between an acid phosphatase locus and a locus controlling resistance to nematode infestation in *Lycopersicon* (Rick and Fobes 1974) encouraged other researchers to investigate possible linkages between enzyme loci and those controlling disease resistance and other useful, but difficult to evaluate, traits. In cucurbits, this led to the observation that peroxidase banding patterns could be used as a marker for resistance to *Pseudoperonospora cubensis* in *C. melo* (Reuveni, Shimoni, and Karchi 1990, 1992). In contrast, Lebeda and Doležal (1995) found no general relationship between peroxidase banding patterns and reaction to *P. cubensis* in *C. sativus*. Similarly, Kennard et al. (1994), Meglic and Staub (1996a), and Bradeen et al. (2001) found no useful linkage relationships between enzyme and disease resistance loci in *C. sativus*. In *Citrullus lanatus*, Benscher and Provvidenti (1991) conducted an initial study that suggested a possible linkage between a phosphoglucoisomerase locus and resistance to *Zucchini yellow mosaic virus* (ZYMV), but omission of this research from a later review of ZYMV resistance in *C. lanatus* by Provvidenti (1993) suggests that the proposed linkage could not be confirmed.

Many different classes of DNA markers have been used to characterize cucurbit germplasm, including both plastid and nuclear markers. Plastid markers are typically highly conserved, making them especially valuable for revealing phylogenetic relationships at or above the species level (Decker-Walters, Chung, and Staub 2004a). The first report investigating restriction endonuclease site polymorphisms in cucurbit chloroplast DNA was presented by Juvik and Palmer (1984) who conducted an initial survey of 12 accessions representing four genera: *Citrullus*, *Cucumis*, *Cucurbita*, and *Lagenaria*. A more extensive analysis of chloroplast DNA focusing on *Cucurbita* (Wilson, Doebley, and Duvall 1992) is valuable in elucidating relationships between domesticated taxa and their wild progenitors. More recently, mitochondrial DNA sequence analysis was conducted upon a similar set of *Cucurbita* accessions (Sanjur et al. 2002) built upon Wilson, Doebley, and Duvall (1992) findings and suggested at least six independent domestication events. Less-conserved, hypervariable regions within chloroplast DNA have also recently been evaluated as a source of markers, called consensus chloroplast simple sequence repeats (or cc SSRs), for the characterization of cucurbits by Chung and Staub (2003, 2004), but results of broad-based screening germplasm collections with cc SSRs have not been reported.

Of the many nuclear-marker classes available, the internal transcribed spacer (ITS) regions of nuclear ribosomal RNA genes evolve at a rate that gives them similar utility to many plastid gene markers in revealing phylogenetic relationships at or above the species level. For example, Jobst, King, and Hemleben (1998) and Garcia-Mas et al. (2004) conducted a phylogenetic analysis of the Cucurbitaceae based on ITS markers that clarified evolutionary relationships at the tribal level. Jarret and Newman (2000) evaluated ITS sequence variation to help construct a phylogeny of all known species of *Citrullus*.

Many other nuclear marker classes such as random amplified polymorphic DNA (RAPD) and simple sequence repeats (SSRs) are considerably more variable and are most useful for clarifying relationships at or below the species level. RAPD markers have been the most widely used to characterize intraspecific patterns of diversity among cucurbit germplasm collections. There are more than twenty publications reporting results of RAPD analyses of cucurbit germplasm, including analyses of *Citrullus*, *Cucumis*, *Cucurbita*, and *Lagenaria*. Key papers are summarized in Table 8.7. Notably, none of these studies (Table 8.7), either individually or collectively, provides a comprehensive analysis of diversity within any cucurbit crop in a manner analogous to the isozyme analyses of *C. sativus* described above.

The degree of repeatability and variation in genetic control of RAPD banding patterns can be a concern and has led some cucurbit researchers to employ other types of nuclear DNA markers. These marker classes are typically thought to be more repeatable and often display codominant product expression, theoretically supplying more genetic information. Of the codominant markers, early studies often used restriction fragment length polymorphisms (RFLPs) (see Dijkhuizen et al. 1996; Garcia-Mas et al. 2000), but once SSR loci and appropriate primers were established in

Table 8.7   Articles Describing Random Amplified Polymorphic DNA (RAPD) Variation among Cucurbit Germplasm Accessions

Taxon	Number and Type of Accessions	References
Citrullus colocynthis and *lanatus*	Forty-two germplasm accessions and five cultivars	Levi et al. (2001a)
Citrullus lanatus	Thirty-nine diverse cultivars	Lee et al. (1996)
Citrullus lanatus	Forty-six American cultivars and twelve germplasm accessions	Levi et al. (2001b)
Cucumis hystrix × *hytivus*, *melo*, *metuliferus*, and *sativus*	Twenty-two diverse accessions	Zhuang et al. (2004)
Cucumis melo	Thirty-two diverse breeding lines	Garcia et al. (1998)
Cucumis melo	Fifty-four germplasm accessions	Perl-Treves et al. (1998), Stepansky et al. (1999a)
Cucumis melo	Fifty-two Korean landraces and breeding lines	Mo et al. (1998)
Cucumis melo	Six diverse accessions	Garcia-Mas et al. (2000)
Cucumis melo	Forty-six germplasm accessions	Staub et al. (2000)
Cucumis melo	108 African germplasm accessions and eighteen reference lines	Mliki et al. (2001)
Cucumis melo	Fifteen Spanish landraces	López-Sesé et al. (2002)
Cucumis melo	114 East and South Asian landraces	Kato et al. (2002)
Cucumis melo	Forty-two New World wild populations, ten Old World wild populations, and fourteen cultivated accessions	Decker-Walters et al. (2002a)
Cucumis melo	One Turkish and ten Portuguese landraces	Carnide et al. (2004)
Cucumis melo	Nine Portuguese landraces and one French cultivar	Barroso et al. (2004)
Cucumis melo	Seventeem Greece landraces and inbred lines	Staub, Fanourakis, and López-Sesé (2004)
Cucumis melo	Sixty-seven Japanese cultivars	Nakata et al. (2005)
Cucumis sativus	Forty-four cultivars, thirty-nine breeding lines, and thirty-five germplasm accessions	Horejsi and Staub (1999)
Cucumis sativus	Fifty germplasm accessions	Zhang et al. (2002)
Cucumis sativus	Twenty-six African germplasm accessions and twenty-one reference lines	Mliki et al. (2003)
Cucumis sativus	Eight Greek landraces	Pavlikaki et al. (2004)
Cucurbita ficifolia, *maxima*, *moschata*, and *pepo* and *Lagenaria siceraria*	Nineteen germplasm accessions of *C. maxima* and eight related cucurbit accessions	Ferriol et al. (2003b)
Cucurbita moschata	Thirty-one African landraces	Gwanama et al. (2000)
Cucurbita pepo	Thirty-seven wild or weedy populations and sixteen cultivars	Decker-Walters et al. (2002b)
Lagenaria siceraria and *L. sphaerica*	Thirty-one landraces and forty-three cultivars of *L. siceraria* and one accession of *L. sphaerica*	Decker-Walters et al. (2001)
Lagenaria breviflora, *L. siceraria* and *L. sphaerica*	One accession of *L. breviflora*, one wild and three domesticated accessions of *L. siceraria*, and four accessions of *L. sphaerica*	Decker-Walters et al. (2004b)

cucurbits (Katzir et al. 1996; Jarret et al. 1997), SSRs have been the codominant marker class most extensively analyzed in cucurbits. Germplasm characterized for SSR variation includes *C. lanatus* (Jarret et al. 1997), *C. melo* (Staub et al. 2000; Akashi et al. 2001, 2002a; Decker-Walters et al. 2002a; López-Sesé et al. 2002), and *Cucurbita pepo* (Katzir et al. 2000; Paris et al. 2003).

There are many other polymerase chain reaction (PCR) based marker classes that have been used in recent years to characterize cucurbit germplasm, including amplified fragment length polymorphisms (AFLPs) (Garcia-Mas et al. 2000; Che et al. 2003; Paris et al. 2003; Ricciardi et al. 2003), cleaved amplified polymorphic sequences (CAPSs) (Akashi et al. 2001; Akashi et al. 2002a; Kato et al. 2002), inter simple sequence repeats (ISSRs) (Perl-Treves et al. 1998; Stepansky, Kovalski, and Perl-Treves 1999a; Katzir et al. 2000; Paris et al. 2003), restriction satellites (Helm and Hemleben 1997), and sequence-based amplified polymorphisms (SBAPs) (Ferriol, Picó, and Nuez 2003b) [also known as sequence-related amplified polymorphisms (SRAPs)] (Ferriol, Picó, and Nuez 2003a; Ferriol et al. 2004a).

A serious challenge to those who wish to evaluate results of molecular characterization or to initiate new characterization research is how to evaluate the appropriateness and efficiency of the various marker classes. Fortunately, studies have been conducted that compare multiple marker classes, both in relation to each other and in relation to morphological observations. One of the first comparative analyses was conducted by Dijkhuizen et al. (1996) who determined that RFLP marker data revealed differences between *C. sativus* var. *sativus* and var. *hardwickii* when isozyme data were not as conclusive.

Since then, many molecular studies in cucurbits have been conducted via RAPD analysis, so it should not be surprising that considerable knowledge has been gained about the comparative utility and characteristics of that marker class. In terms of coefficients of variation (CV) per band, RAPDs exhibited lower CVs than did either isozymes or SSRs in *Cucumis* (Staub et al. 1997a, 2000) with moderate congruence between results generated by each marker class. In contrast, Zhuang et al. (2004) reported a high degree of congruence between genetic relationships revealed by RAPDs and SSRs among *Cucumis* taxa.

In a three-way comparison, measurements of genetic distance from RAPDs more closely resembled those from RFLPs than those from isozymes for *C. sativus* (Horejsi and Staub 1999). In addition, RAPDs were superior to quantitative analysis of morphological traits in *C. melo* in calculating genetic distances in comparison to known pedigree data (Garcia et al. 1998). A three-way comparison of AFLP, RAPD, and RFLP markers to describe patterns of genetic variation among six diverse accessions of *C. melo* gave similar results for each marker class (Garcia-Mas et al. 2000), but the authors indicated that AFLPs were most efficient in detecting polymorphisms. In addition, Ferriol, Picó, and Nuez (2003b) reported that SBAPs were superior to RAPDs in revealing patterns of genetic diversity in *C. maxima*, noting that the results of RAPD analyses conformed neither to groupings by morphology nor geographic origin, and, similarly, there was no correlation between RAPD analyses and morphological or geographical groupings of *C. melo* landraces from Portugal (Carnide et al. 2004) or Spain (López-Sesé, Staub, and Gomez-Guillamon 2003). Ferriol, Picó, and Nuez (2003a) and Ferriol et al. (2004a) also evaluated other *Cucurbita* species with SRAPs and AFLPs and noted some interesting distinctions. SRAPs were generally more concordant with morphological variation and agronomic traits, consistent with this marker class's connection to open reading frames; however, AFLPs grouped *C. maxima* accessions by geographic origin, reflecting the bottleneck that occurred with the introduction of this species to the Old World (Ferriol, Picó, and Nuez 2004b).

Katzir et al. (2000) noted that SSR data supported a dendrogram of *C. pepo* germplasm that had been developed by ISSR marker analysis, and, in a more comprehensive study of *C. pepo* germplasm, Paris et al. (2003) confirmed a high degree of congruence among AFLP, ISSR, and SSR marker data that they then used in support of existing botanical and horticultural classifications.

Ideally, characterization based on multiple lines of evidence, including molecular markers and morphological, phenological, and biochemical information such as that reported by Paris et al.

(2003) for *C. pepo*, should shed the greatest light upon relationships among germplasm accessions. In *C. sativus*, Horejsi and Staub (1999) reported that patterns of variation revealed by RAPD marker analysis were consistent with geographic origin and morphological characters. Shortly thereafter, Staub and Ivandic (2000) reported on a combined analysis of genetic variation in *C. sativus* including both isozyme and RAPD markers. For *C. melo*, Perl-Treves et al. (1998), Stepansky, Kovalski, and Perl-Treves (1999a) presented results of a comprehensive analysis of morphological markers, sugar composition, and RAPD and ISSR markers to test an infraspecific classification scheme proposed by Munger and Robinson (1991). From a more limited analysis of fifteen Spanish landraces of *C. melo*, López-Sesé et al. (2002) reported that patterns of genetic variation revealed by RAPDs and SSRs were generally congruent, and bulk sampling within an accession was appropriate for the assessment of large collections. Another recent study combining multiple lines of evidence, in this case, from AFLP data and morphological types, was published by Che et al. (2003) for *C. lanatus*.

Beyond simple characterization, molecular-marker data can be analyzed to help create core collections, as defined by Frankel (1984), that can serve as a management tool for the efficient sampling of genetic diversity from large germplasm collections. The most progress in the establishment of core collections for cucurbit germplasm has been made following the suggestion of Staub (1994a, 1994b), which ultimately led to the selection of a core collection for *C. sativus* from the holdings of the U.S. NPGS based on past isozyme analyses, historical and geographical information, and disease-evaluation data (Staub et al. 2002).

From a practical perspective, just as work has been conducted to search for close associations between isozyme loci and those that control useful traits (reviewed above), researchers have searched for such associations between DNA markers and useful-trait loci. These efforts can be facilitated through the development of comprehensive linkage maps (see Kennard et al. 1994; Périn et al. 2002b; Levi et al. 2004b; Zraidi and Lelley 2004), but also can emerge from germplasm characterization studies. Examples linking RAPD banding patterns with useful trait loci in *Citrullus* have been reported by Lee et al. (1996) for high levels of free fruit sugars and by Levi et al. (2001a) for resistance to gummy stem blight (*Didymella bryoniae*) and *Fusarium* wilt. And the development of a linkage map for *C. melo* by Périn et al. (2002b) has been used to locate loci controlling tolerance to sulfur application (Perchepied et al. 2004) and to map two disease-resistance loci (Brotman et al. 2004). A second linkage map for *C. melo* has been used to locate loci controlling resistance to cotton aphid (*Aphis gossypii*) and powdery mildew caused by *P. xanthii* (syn. *S. fuliginea*) (Fukino et al. 2002). Additional examples of exploitable linkage relationships are discussed in Section 8.5.2 and Section 8.5.3.

8.2.5 Germplasm Descriptors

Standardized lists of evaluation descriptors and protocols are often used to provide structure and comparability for germplasm characterization and evaluation. Internationally recognized descriptor lists have typically been published by the International Board for Plant Genetic Resources (IBPGR) and its successor organization, IPGRI. However, only a single international list for cucurbits has been published, one for *C. melo* (IPGRI 2003). Development of international descriptor lists for other important cucurbits was identified as a crucial task by the newly developed Working Group on Cucurbitaceous Vegetables in Adana, Turkey (Díez, Picó, and Nuez 2002). As a preliminary step, sets of the most significant descriptors, called minimum descriptor lists, were elaborated (Thomas et al. 2004). Descriptor lists for *Cucumis* (Křístková et al. 2003) and cultivated *Cucurbita* species (Křístková et al. 2005) have been developed for the national gene bank collections of the Czech Republic (Table 8.8 and Figure 8.3). A recent discussion of the development of new descriptor systems for *Cucumis* and *Cucurbita* was recently presented by Vinter et al. (2004).

Table 8.8 Morphological Descriptors for Cultivated *Cucurbita* Species

Number	Descriptor Name	Scale	Descriptor State	Explanation	Note
1; Morphological descriptors					
1.4; Leaf					
1.4.1.4; * S (I)	Leaf blade— shape of apex of terminal lobe	1 2 3 4 5 6 7	acute subacute obtuse mucronate rounded truncate sinuate	Figure 8.3	Fully developed leaf from the middle part of plant at botanical maturity

Note: *, Highly discriminating descriptor; S = descriptor characterizing species; I = descriptor discriminating infraspecific variation; ( ), letter in parentheses is on a secondary significance.
Source: From Křístková, E., Křístková, A. Vinter, V., and Lebeda, A., *Hort Sci.,* Prague., in press.

Although descriptor lists, whether from international standards or individual genebanks, can be used to guide morphological characterization *per se* (the verification of botanical and horticultural identity and clarification of genetic relationships among collections), as exemplified in *Cucumis melo* by Costa et al. (1989) and in *Cucurbita* by Nuez et al. (2000b) and by Křístková, Křístková, and Vinter (2004), they are more often applied to the evaluation of germplasm to uncover valuable traits (see Section 8.6.2 for an in-depth treatment of this topic). In addition to the reports cited above, an extensive project to characterize germplasm accessions of *C. pepo*, primarily to elucidate patterns of variation, has been conducted by researchers at the Newe Ya'ar Research Center in Israel (Nerson, Paris, and Paris 2000; Paris 2001a; Paris and Nerson 2003).

8.2.6 Germplasm Distribution Analysis and Its Incorporation into Collection Management

Knowledge of the patterns of distribution of germplasm accessions from genebanks, especially as related to the extent and intended uses of distributions, is central to efficient genebank management and can help curators meet future demand (Widrlechner and Burke 2003). Unfortunately, to date, the body of published research on patterns of cucurbit germplasm distribution or how well it meets users' needs is extremely limited. Without such information, it is difficult for curators to apply international guidelines for seed regeneration (Sackville Hamilton and Chorlton 1997) that require good projections of future germplasm demand.

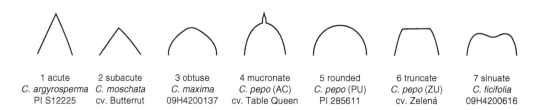

1 acute	2 subacute	3 obtuse	4 mucronate	5 rounded	6 truncate	7 sinuate
C. argyrosperma	*C. moschata*	*C. maxima*	*C. pepo* (AC)	*C. pepo* (PU)	*C. pepo* (ZU)	*C. ficifolia*
PI S12225	cv. Butterrut	09H4200137	cv. Table Queen	PI 285611	cv. Zelená	09H4200616

Figure 8.3 Descriptor 1.4.1.4. Leaf blade of *Cucurbita* species—shape of apex of terminal; lobe (example, From Křístková, E., Křístková, A., Vinter, V., and Lebeda, A., *Hort Sci.,* Prague, in press.) Morphotypes of *C. pepo* (According to Paris, H. S., *Econ. Bot.,* 43, 13–43, 1989.): AC = acorn; PU = pumpkin; ZU = zucchini.Origin of accessions: PI 285611, PI 512225 (Plant Introduction Station, Iowa State University, Ames, U.S.A.); 09H4200137, 09H4200616 (Research Institute of Crop Production, Gene Bank Workplace in Olomouc, Czech Republic); *C. moschata* cv. Butternut = Stokes, U.S.A.; *C. pepo* cv. Table Queen = Ball Seeds, Canada; *C. pepo* cv. Zelená = Nohel, Czech Republic.

Clark et al. (1991) presented a simple distribution analysis, classifying the types of uses that were made of the cucurbit germplasm conserved by the North Central Regional Plant Introduction Station (NCRPIS) in Iowa. Gao et al. (2000, 2001) conducted an extensive user survey to classify important characteristics of researchers requesting samples and of the germplasm that had been distributed, how those germplasm accessions were being used, and factors limiting utilization. *C. sativus* was included among the ten major crops in their analysis, and, as part of that study, Gao et al. (2001) presented a summary of how *C. sativus* germplasm accessions were important in the development of new cultivars in China. In a somewhat less formal fashion, Queiroz et al. (2004) described how local *Citrullus* germplasm has been successfully integrated into varietal development in Brazil.

Widrlechner and Burke (2003) analyzed distribution patterns for *C. melo* and *sativus* and *C. pepo* germplasm accessions from the NCRPIS over a 12-year period. They noted an overall decline in distribution rate for all three species, which they attributed to three possible causes: (1) declines experienced after the completion of large-scale evaluation projects (which was true for *C. sativus*); (2) those caused by the gradual incorporation of valuable germplasm into users' research and crop improvement programs; and (3) those resulting from an increase in targeting of germplasm distributions as increasing amounts of evaluation data are made available to users. Circumstantial evidence was presented in support of increased targeting as average shipment size declined over time. The distributions of cucurbit germplasm on a per-accession basis were normally distributed, simplifying the prediction of demand among individual accessions, which can then be applied to diverse managerial tasks (Sackville Hamilton and Chorlton 1997; Widrlechner and Burke 2003).

8.3 TAXONOMY

8.3.1 Introduction

Taxonomy is both the master and servant of biology: *master* in that the results of all other fields of biological research contribute to its database, and *servant* in that it provides name assignment opportunities for all practitioners of biology, without which repeatability and, therefore, application of the scientific method in biology would be possible. This dichotomy is also reflected by differences in philosophical and methodological approaches: one termed *empiricist,* which explores phylogenetic methods and cladistics often with the goal of accurately depicting evolutionary history, and the other, *instrumentalist* that is more closely connected to phenetic methods of analysis whose goal is the production of practical taxonomic treatments (Jeffrey 1990).

Organisms, as a result of evolutionary processes, possess basic genetic information about ordered and evolved biological patterns. Thus, a key task for systematics is to characterize and apply such information to plant classification. There is, however, little certainty in taxonomy. Scientists generally choose among competing taxonomic treatments solely on the basis of their ability to explain observations of organismal properties under consideration.

Issues that complicate the adoption of taxonomic treatments arise from attempts to adhere to the dynamic nomenclatural provisions of the International Code of Botanical Nomenclature (ICBN) (Greuter et al. 2000). Such immediate procedural issues, when combined with a legacy of complex taxonomic history, often result in valid taxonomic treatments that may or may not be functional in the long term. Such adherence to taxonomic precedents is, nevertheless, important because it ensures that for any given taxonomic circumscription, a taxon will have only one name by which it is known (Jeffrey 1990).

The assignment of a unique name is particularly important for cucurbits because their long history of cultivation in diverse regions has resulted in an abundance of common names where identical names have been applied to different species (Rubatzky and Yamaguchi 1997). However, just as International Code of Nomenclature for Cultivated Plants (ICNCP) establishes the rules for botanical names (Brickell et al. 2004), defines rules for the proper description and nomenclature of cultivated plants at the cultivar and group levels. These rules have been increasingly applied to the description and classification of cultivated cucurbits (Jeffrey 2001).

8.3.2 Taxonomic Relationships at the Familial Level

Cucurbitaceae is taxonomically isolated from other genetically related families and is best referred to a monotypic order, the Cucurbitales (Jeffrey 1990). Although, to date, its sole putative relatives are members of the Begoniaceae and Datiscaceae of the order Begoniales, wider genetic affinities among three families are somewhat obscure (Jeffrey 1990).

Members of the Cucurbitaceae are predominantly tropical with 90% of the species found in three main areas: Africa and Madagascar, Central and South America, and Southeast Asia and Malaysia. There are about 118 extant genera and 825 species (Jeffrey 1990). Geographic distribution of cucurbit diversity is illustrated in Figure 8.4 that maps the cumulative natural ranges of 120 cucurbit genera with patterns of generic diversity corresponding well to those at the specific level.

The family is divided into two unequally sized subfamilies. The Zanonioideae, with only 18 genera and 80 species of little economic value, is characterized by small, tricolporate, striate pollen grains; two or three free styles; and bifid tendrils spiralling both below and above the branching point. The subfamily Cucurbitoideae includes about 100 genera with styles united into a single column, branched tendrils with a non-spiralling basal part, and various, but not striate, pollen grains. The following tribes (and genera) of Cucurbitoideae are of major economic importance: Benincaseae (*Benincasa, Citrullus, Coccinia, Lagenaria, Luffa*), Melothrieae (*Cucumeropsis, Cucumis*), Cucurbiteae (*Cucurbita*), Cyclanthereae (*Cyclanthera*), Joliffieae (*Momordica, Telfairia*), Sicyoeae (*Sechium*), Schizopeponeae (*Schizopepon*), and Trichosantheae (*Hodgsonia, Trichosanthes*) (Jeffrey 1990; Rubatzky and Yamaguchi 1997).

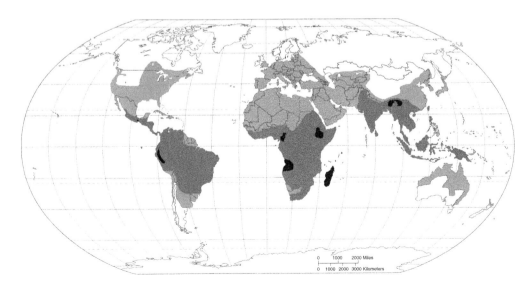

Figure 8.4 Composite map of the native ranges of the world's cucurbit genera. Light gray = 1–9 native genera; dark gray = 10–19 native genera; black = 20 + native genera. (Created by: M.P. Widrlechner.)

8.3.3 General Characterization

The Cucurbitaceae is represented by annual and perennial herbs, semi-shrubs, aculeateous shrubs, and a few rare succulent trees. Trailing or climbing vine growth with nodal branching is typical of many species. However, primary stem growth of some *Cucurbita* and *Cucumis* cultivars is substantially reduced, giving plants a "bush" habit (Rubatzky and Yamaguchi 1997). In *Cucurbita*, plants generally develop a strong, fairly long taproot (1–2 m long) and a highly branched network of shallow, secondary roots. Some perennial species develop large storage roots, and in moist environments after a period of drought, adventitious roots develop at the nodes of some *Cucurbita* species. In *Cucumis* species, plant roots are diffuse and reside in the upper 25 cm of the soil horizon.

Cucurbit leaves have bi-collateral vein fascicles and generally contain phytoliths (Piperno et al. 2002). Leaves of most cucurbits are alternate, and their blade is entire, toothed, or palmately lobed. Among species, foliage differs in size, texture, and shape such as the differences in the orientation and depth of lobes. Tendrils are borne in leaf axils and may be simple, highly spiralled, or branched. They are usually absent in bush-type *Cucurbita*.

Floral morphology is often poorly differentiated within a genus, but flowers vary widely among the various genera. Commonly, plants are monoecious, but other forms of sex expression occur or can be induced (Kalloo 1988). In contrast, sex expression in cucumber ranges from androecious to gynoecious to include hermaphroditic flower types. Most cucurbit fruits are fleshy berries or hard-rinded pepos, but extreme variation occurs in fruit shape, color, and size. Cucurbit seeds do not contain endosperm, but their large cotyledons contain amounts of carbohydrates and lipids sufficient for embryo development (Rubatzky and Yamaguchi 1997). A detailed morphological description of the family is given by Robinson and Decker-Walters (1997).

Any comprehensive discussion of taxonomic characterization must go beyond an examination of gross morphology and its comparative analysis. Modern taxonomic treatments of Cucurbitaceae also incorporate evidence from biochemistry such as the production of toxic cucurbitacins from terpene metabolism, cytology, cytogenetics, molecular genetics, taxon crossing ability, and coevolution with insects and pathogens in delimiting taxa (Bates, Robinson, and Jeffrey 1990; Kirkbride 1993). Data obtained by classical means (e.g., anatomical studies of phytoliths) are also considered (Piperno et al. 2002). The following sections briefly summarize the current state of taxonomic literature for economically important cucurbit genera.

8.3.4 *Cucumis*

In the most recent comprehensive biosystematic monograph of the genus *Cucumis*, Kirkbride (1993) recognized 32 species (Table 8.9, Figure 8.5). Of these, in addition to the two economically important species, cucumber (*C. sativus* L.) and melon (*C. melo* L.), two wild species *C. anguria* L. (West Indian gherkin), and *C. metuliferus* E. Meyer ex Naudin (African horned cucumber or jelly melon) are also commercially exploited for fruit production (Morton 1987; Baird and Thieret 1988). Other wild species originating mostly from arid and/or semi-arid regions of Africa are cultivated as ornamental plants, e.g., *C. dipsaceus* Ehrenberg ex Spach (hedgehog gourd) and *C. myriocarpus* Naudin (gooseberry gourd) (Kirkbride 1993).

Based on their geographic distributions, morphological characteristics, chromosome numbers, isozyme banding patterns, chloroplast DNA profiles, and patterns of interspecific hybridization, the species of genus *Cucumis* fall into two principal groups, generally recognized as subgenera (Jeffrey 1980; Raamsdonk, den Nijs, and Jongerius 1989; Kirkbride 1993). Sometimes these groups have been treated as distinct genera (as *Cucumis* and *Melo*) (Pangalo 1951), but that classification has not been commonly accepted.

Subgenus *Cucumis* represents a compact and isolated group of two species, *C. sativus* ($2n=2x=14$; Indian subcontinent origin) and *C. hystrix* Chakr. ($2n=2x=24$; Chinese origin)

Table 8.9   Organization of the Genus *Cucumis*

Cucumis spp.	Chromosome Number (n)	Distribution
Subgenus Cucumis		Asia
C. sativus L.	7	India, Sri Lanka, Burma, China,
C. sativus L. var. hardwickii (Royle) Alef.	7	India
C. hystrix Chakravarty	12 (Chen et al. 1997b)	India, Burma, China, Thailand
Subgenus Melo (Miller) **C. Jeffrey** **Section Aculeatosi Kirkbride** **Series Myriocarpi Kirkbride**		
C. myriocarpus Naudin subsp. myriocarpus	12	Lesotho, Mozambique, S. Africa, Zambia
subsp. leptodermis (Schweickerdt) Jeffrey & Halliday	12	S. Africa, Lesotho
C. africanus L.	12	S. Africa, Angola, Botswana, Namibia Zimbabwe
C. quintanilhae R. Fernandes & A. Fernandes	?	S. Africa, Botswana
C. heptadactylus Naudin	24	S. Africa
C. calahariensis Meeuse	?	Botswana, Namibia
Series Angurioidei Kirkbride		
C. anguria L. var. anguria	12	both vars: Angola, Botswana, Cape Verde Islands, Malawi, Mozambique, Namibia, S. Africa, Sierra Leone, Swaziland, Tanzania, Zaire, Zambia
var. longaculeatus Kirkbride	12	Zimbabwe
C. sacleuxii Paillieux & Bois	12	Kenya, Masdagascar, Tanzania, Uganda, Zaire
C. carolinus Kirkbride	?	Ethiopia, Kenya
C. dipsaceus Ehrenberg ex Spach	12	Ethiopia, Kenya, Somalia, Tanzania, Uganda
C. prophetarum L. subsp. prophetarum	12	Egypt, Mali, Mauritania, Nigeria, Senegal, Somalia, Sudan, Iran, Iraq, Israel, Oman, Qatar, Saudi Arabia, Yemen, Socotra, Syria, United Arab Emirates, Jordan
subsp. dissectus (Naud.) C. Jeffrey		Chad, Egypt, Ethlopia, Kenya, Mauritania, Niger, Rwanda, Somalia, Tanzania, Uganda, Saudi Ararbia, Yemen
C. pubituberculatus Thulin	?	Somalia
C. zeyheri Sonder	12 (24)	Lesotho, Mozambique, S. Africa, Swaziland, Zambia, Zimbabwe
C. prolatior Kirkbride	?	Kenya
C. insignis C. Jeffrey	12?	Ethiopia
C. globosus C. Jeffrey	12?	Tanzania
C. thulinianus Kirkbride	?	Somalia
C. ficifolius A. Richard	12 (24)	Ethiopia, Kenya, Rwanda, Tanzania, Uganda, Zaire
C. aculeatus Cogniaux	24	Ethiopia, Kenya, Rwanda, Tanzania, Uganda, Zaire
C. pustulatus Naudin ex Hooker	12, 48, 72	Ethiopia, Chad, Kenya, Niger, Nigeria, Sudan, Tanzania, Uganda, Saudi Arabia, Yemen
C. meeusei C. Jeffrey	24	S. Africa, Botswana, Namibia

continued

Table 8.9   Continued

Cucumis spp.	Chromosome Number (n)	Distribution
C. jeffreyanus Thulin	?	Ethiopia, Kenya, Somalia
C. hastatus Thulin	?	Somalia
C. rigidus E. Meyer ex Sonder	?	S. Africa, Namibia
C. baladensis Thulin	?	Somalia
Series Metuliferi Kirkbride		
C. metuliferus E. Meyer ex Naudin	12	Angola, Botswana, Ethiopia, S. Africa, Kenya, Malawi, Mozambique, Namibia, Senegal, Sudan, Swaziland, Tanzania, Uganda, Zaire, Zambia, Zimbabwe, Cameroon, Central African Republic, Liberia, Burkina, Yemen
C. rostratus Kirkbride	?	Ivory Coast, Nigeria
Section Melo (Miller) Kirkbride		
Series Hirsuti Kirkbride		
C. hirsutus Sonder	12	S. Africa, Angola, Botswana, Burundi, Congo, Kenya, Malawi, Mozambique, Sudan, Swaziland, Tanzania, Zaire, Zambia, Zimbabwe
Series Humifructosi Kirkbride		
C. humifructus Stent	12	S. Africa, Ethiopia, AngolaKenya, Namibia, Zaire, Zambia, Zimbabwe
Series Melo (Miller) Kirkbride		
C. melo L.		
subsp. melo	12	Africa, Iran, Afghanistan, Burma, China, India, Japan, Pakistan, Malesia, New Guinea, Australia, Fiji Islands, Papua New Guinea
subsp. agrestis (Naudin) Pangalo	12	Africa, Saudi Arabia, Yemen, China, Burma, India, Japan, Korea, Nepal, Pakistan, Sri Lanka, Thailand, Malesia, Indonesia, New Guinea, Philippines, Australia, Guam, Papua New Guinea, Samoa, Solomon Islands, Tonga Islands
C. sagittatus Peyritsch	12	S. Africa, Angola, Namibia

Source: From Kirkbride, Jr., J. H., *Biosystematic Monograph of the Genus Cucumis (Cucurbitaceae)*, Boone, NC: Parkway Publishers, 1993.

(Kirkbride 1993; Chen et al. 1997b). Molecular-marker analyses have recently confirmed that these two species form a coherent group in relation to subgenus *Melo* (Zhuang et al. 2004). This is also reflected in the development of a stable amphidiploid hybrid, *C.×hytivus* Chen & Kirkbride, derived from the artificial hybridization of these two species (Chen and Kirkbride 2000).

The foothills of the Himalaya Mountains in India are the mostly likely center of origin for the domestication of *C. sativus*, which fits well with levels of genetic diversity observed in Indian cucumber landraces (Staub, Serquen, and McCreight 1997b). Interpretation of the domestication of the cucumber depends on the status of *C. sativus* var. *hardwickii* (i.e., if this taxon is truly wild or feral or if it represents both wild and feral populations) (Bates and Robinson 1995; for more details, see Section 8.3.5). Cucumber has been cultivated for about 3,000 years in India, and soon after was

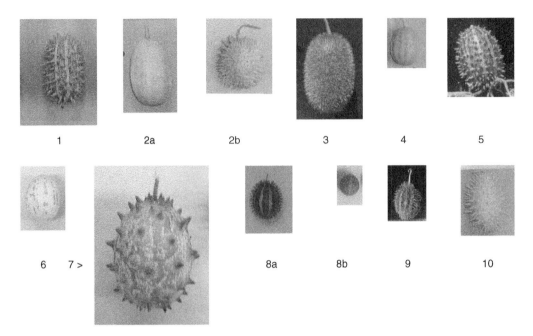

Figure 8.5 **(See color insert following page 304.)** Wild *Cucumis* species fruit variation (composite photo: NCRPIS staff) 1. *C. africanus* (PI 542127), 2a. *C. anguria* var. *anguria* (PI 196477), 2b. *C. anguria* var. *longaculeatus* (Ames 23548), 3. *C. dipsaceus* (PI 236468), 4; *C. ficifolius* (PI 196844), 5. *C. heptadactylus* (PI 282446), 6. *C. melo* (Ames 24297), 7. *C. metuliferus* (Ames 23535), 8a. *C. myriocarpus* (Ames 23537), 8b. *C. myriocarpus* subsp. *leptodermis* (PI 282447), 9. *C. prophetarum* (PI 193967), 10. *C. zeyheri* (PI 532629).

disseminated to the south and east of the Himalayas. From India, cucumber was brought to Greece and Italy, and later, China (Bisognin 2002). Historical records have confirmed cucumber cultivation in France and the Great Moravian Empire in central Europe in the ninth century (Moravec, Lebeda, and Křístková 2004).

The subgenus *Melo* is essentially African in its distribution and can be divided into three to six groups (Singh 1990; Kirkbride 1993). Most species are diploids with $2n = 2x = 24$, but some species have $2n = 48$ or even $2n = 72$ (Table 8.9) (den Nijs and Custers 1990; Singh 1990; Bates and Robinson 1995). The natural range of *C. melo* has yet to be conclusively determined. Prior to domestication, it may have been limited to Africa or may have reached the Near East or perhaps farther east to Asia (Bates and Robinson 1995). The center of diversity and perhaps of the origin of the principal melons of world commerce (i.e., the *C. melo* Inodorus and Cantalupensis Groups) is located in the Near East and adjacent central Asian regions (Jeffrey 1980). The Indian subcontinent may have been the original home of the Conomon Group and other local variants and may have been the scene of redomestication of feral forms (Bates and Robinson 1995). Melon was introduced to Central America in 1516 (Ware and McCollum 1980) and quickly expanded widely in the New World.

A recent provisionary infraspecific classification of cultivated *C. melo*, based on fruit morphology, plant biology, and geographic distribution, was proposed by Pitrat, Hanelt, and Hammer (2000). Decker-Walters et al. (2002a) conducted a complementary analysis of feral, New World *C. melo* populations in relation to those found in the Old World. Although extensive assessment of genetic variation at the protein and DNA level has occurred (Staub and Ivandic 2000), a similar comprehensive assessment of morphological characteristics within the species *C.*

sativus remains to be completed. The description of cultivar groups by Křístková et al. (2003), based on fruit shape and geographic distribution, however, serves as a good starting point. As new cucumber cultivars are being developed with new combinations of traits, new groupings within this species may be needed.

8.3.5 *Cucurbita*

The genus *Cucurbita* is exclusively native to the New World (Figure 8.4 and Figure 8.6), but has been widely cultivated in the Old World since the 1500s (Paris 1989, 2001a, 2001b). It is not closely related to other cucurbit genera (Merrick 1995). The basic chromosome number of all *Cucurbita* species is $2n = 2x = 40$, and karyotypes suggest that these species are of allopolyploid origin (Singh 1979; Weeden and Robinson 1990). Results from electrophoretic analyses also helped confirm this genus' polyploid (Kirkpatrick, Decker, and Wilson 1985) or, more specifically, allotetraploid (Weeden 1984) origin.

Whitaker and Bemis (1975) recognized 27 *Cucurbita* species. After careful analysis of type specimens, synonymy and additional taxonomic evidence (including coevolutionary studies involving the specialized pollinators, *Peponapis* and *Xenoglossa*; Hurd, Linsley, and Whitaker 1971), the genus is now thought to consist of between 12 and 15 species with five regularly cultivated (Lira-Saade 1995; Jeffrey 2001; Sanjur et al. 2002) (Table 8.10, Figure 8.7 through Figure 8.9). The genus can be divided into two groups based on ecological adaptation. The first group consists of mesophytic annuals or short-lived perennials with fibrous root systems. It includes all five major cultivated species. Wild taxa within this group occur from the southeastern United States south to central Argentina, typically below 1300 m above sea level. A recent sequence analysis of an intron from the mitochondrial gene, *nad1*, indicated that *C. ficifolia* Bouché was basal to all other taxa in this group (Sanjur et al. 2002).

The second group, the xerophytic, long-lived perennial species, are characterized by the presence of fleshy storage roots. They are adapted to arid zones or high-elevation regions from the southwestern United States to southern Mexico (Merrick 1995) and, based on chloroplast DNA

Figure 8.6 *Cucurbita argyrosperma* subsp. *sororia* in native habitat in Manzanillo, Colima, Mexico, November 1982. (Photo courtesy of L.C. Merrick.)

Table 8.10 Organization of the Genus *Cucurbita*

Cucurbita spp.	Distribution
Group Argyrosperma	
C. argyrosperma C. Huber[a]	Mexico, Central America, southwestern U.S.A.
subsp. *argyrosperma*	
subsp. *sororia* (L.H. Bailey) L. Merrick & D.M. Bates	Pacific Coast from Mexico to Nicaragua
Group Ficifolia	
C. ficifolia Bouché[a]	From Mexico highlands south to northern Chile and Argentina
Group Maxima	
C. maxima Duchesne[a]	Argentina, Bolivia, Chile
subsp. *maxima*	
subsp. *andreana* (Naudin) Filov	Argentina, Bolivia
C. moschata Duchesne[a]	Lowlands of Mexico, Central America
Group Pepo	
C. pepo L.[a]	Northern Mexico and southern U.S.A.
subsp. *fraterna* (L.H. Bailey) Filov	Northeastern Mexico
subsp. *ovifera* (L.) Harz	
subsp. *ozarkana* D.S. Decker	South-central U.S.A.
subsp. *pepo*	
subsp. *texana* (Scheele) Filov	Texas and southeastern U.S.A.
C. ecuadorensis H.C. Cutler & Whitaker	Pacific coast of Ecuador
Group Okeechobeensis	
C. okeechobeensis (J.K. Small) L.H. Bailey	
subsp. *okeechobeensis*	Palm Beach County, Florida, U.S.A.
subsp. *martinezii* (L.H. Bailey) T.W. Walters & D.S. Decker	Veracruz, Mexico
C. lundelliana L.H. Bailey	Mexico lowlands of Yucatan, Guatemala, Belize
Group Digitata	
C. digitata A. Gray[b]	New Mexico and Arizona, U.S.A.
C. cylindrata L.H. Bailey[b]	Baja California, Mexico
C. palmata S. Watson[b]	Southern California and Arizona, U.S.A.
Group Foetidissima	
C. foetidissima Kunth[b]	Western U.S.A. and Mexico
C. pedatifolia L.H. Bailey[b]	Central Mexico
C. × *scabridifolia* L.H. Bailey[b]	Northeastern Mexico
C. radicans Naudin[b]	Mexico

[a] Cultivated species.
[b] Perennials.
Source: Adapted from Lira-Saade, R., *Cucurbita* L., In: Estudios Taxonómicos y Ecogeográphicos de las Cucurbitaceae Latinoamericanas de Importancia Económica; Systematics and Ecogeographic Studies on Crop Genepools, No; 9; IPGRI, Rome, 1995; Jeffrey, C. Cucurbitaceae; In *Mansfeld's Encyclopedia of Agricultural and Horticultural Crops,* Vol; 3, Ed; P; Hanelt, Springer-Verlag, Berlin, 1510–1557, 2001; U.S. National Plant Germplasm System's Germplasm Resources Information Network database, National Germplasm Resources Laboratory, 2005.

Figure 8.7 Fruits of cultivated *Cucurbita* species 1 = *C. argyrosperma* PI 512225 (Plant Introduction Station, Iowa State University, Ames, U.S.A.); 2 = *C. ficifolia* CZE09H4200616 (Research Institute of Crop Production, Gene Bank Workplace Olomouc, Czech Republic); 3 = *C. maxima* CZE09H4200137 (Research Institute of Crop Production, Gene Bank Workplace Olomouc, Czech Republic); 4 = *C. moschata* cv. Butternut, Stokes, U.S.A.; 5 = *C. pepo* PI 285611 (Plant Introduction Station, Iowa State University, Ames, U.S.A.). (Composite photo courtesy of A. Křístková.)

analyses, are believed to be ancestral to members of the first group (Wilson, Doebley, and Duvall 1992). Of this second group, there has been interest in domesticating buffalo gourd, *C. foetidissima* (De Veaux and Schultz 1985). Its storage roots contain starch, and the lipids extracted from its seeds could serve as edible oil (Gathman and Bemis 1990).

The current state of knowledge on the origin and evolution of the domesticated *Cucurbita* species has been summarized by Merrick (1995) and Sanjur et al. (2002). The five domesticated *Cucurbita* species differ significantly in terms of their early distribution in the Americas. Current research suggests that each species is likely to have been domesticated independently from the others in distinct regions: (1) *C. maxima* Duchesne (Figure 8.10) in southern South America, *C. ficifolia* perhaps in the northern or central South American highlands; (2) *C. moschata* Duchesne in the southern Central American or northern South American lowlands; (3) *C. argyrosperma* C. Huber in southern Mexico (Figure 8.6); and (4) *C. pepo* L. (Figure 8.11) in northern Mexico and most probably also in the south-central United States (Sanjur et al. 2002). There is strong evidence for two domestication events in *C. pepo*, but the precise locations of those events remains obscure.

Figure 8.8 Leaf variation of cultivated *Cucurbita* species 1 = *C. argyrosperma* PI 512225 (Plant Introduction Station Iowa State University, Ames, U.S.A.); 2 = *C. ficifolia* CZE09H4200616 (Research Institute of Crop Production, Gene Bank Workplace Olomouc, Czech Republic); 3 = *C. maxima* CZE09H4200137(Research Institute of Crop Production, Gene Bank Workplace Olomouc, Czech Republic); 4 = *C. moschata* cv. Butternut (Stokes, U.S.A.); 5 = *C. pepo* cv. Ghada (Royal Sluis, The Netherlands); Herbarium specimens from collection of Department of Botany, Palacký University, Czech Republic. (Composite photo courtesy of A. Křístková.)

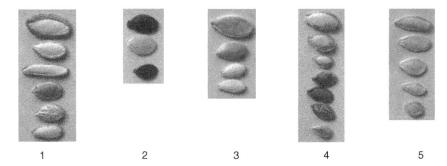

| 1 | 2 | 3 | 4 | 5 |

Figure 8.9   (See color insert following page 304.) *Cucurbita* species seed variation (composite photo: L.C. Merrick) 1 = *C. argyrosperma*, 2 = *C. ficifolia*, 3 = *C. maxima*, 4 = *C. moschata*, 5 = *C. pepo*.

The wild taxon *C. maxima* subsp. *andreana* (Naudin) Filov has been considered the progenitor of *C. maxima* by many authors yet sometimes has been considered as a feral escape and not the wild ancestor (Nee 1990). Recently, Sanjur et al. (2002) presented evidence confirming the close relationship between these two taxa. The progenitor of *C. ficifolia* has not been identified (Nee 1990; Sanjur et al. 2002).

The ancestor of cultivated *C. moschata* also remains unknown, but will probably be found among wild cucurbits in northern Colombia (Nee 1990; Sanjur et al. 2002; Andres 2004b). *Cucurbita lundelliana* L.H. Bailey had been considered to be its probable ancestor by Whitaker and others (Merrick 1995). However, Merrick (1991) demonstrated that *C. argyrosperma* subsp. *sororia* (L.H. Bailey) L. Merrick & D.M. Bates has a closer genetic affinity to *C. moschata* than does any other known wild *Cucurbita* taxon (Merrick 1995). *Cucurbita argyrosperma* subsp. *sororia* is also the most likely progenitor of cultivated forms of *C. argyrosperma* (Merrick 1995; Sanjur et al. 2002) with data providing interesting geographic correlations between the center of origin of this domesticate and that of *Zea mays* L. (Doebley 1990; Sanjur et al. 2002). Finally,

Figure 8.10   (See color insert following page 304.) Examples of *Cucurbita maxima* morphotypes (classification according to Whitaker, T. W. and Davis, G. N., Cucurbits: Botany, Cultivation, and Utilization, New York: Interscience Publishers, 1962. With permission.) 1 = hubbard, 2 = field pie pumpkin, 3 = turban, 4 = banana. (Composite photo courtesy of E. Křístková.)

Figure 8.11 **(See color insert following page 304.)** Morphotypes of *Cucurbita pepo* (classification according to Paris, H. S., Historical records, origins, and development of the edible cultivar groups of *Cucurbita pepo* (Cucurbitaceae), *Econ. Bot.*, 43, 13–43, 1989. With permission.) 1=pumpkin, 2=acorn, 3= vegetable marrow, 4=cocozelle, 5=zucchini, 6=straighneck, 7=scallop, 8=crookneck, 9=ornamental gourd. (Composite photo courtesy of E. Křístková.)

C. pepo was domesticated independently in two areas, most likely in the south-central U.S., from wild *C. pepo* subsp. *texana* (Scheele) Filov, and in Mexico, from the wild taxon, *C. pepo* subsp. *fraterna* (L.H. Bailey) Filov (Decker 1988; Wilson, Doebley, and Duvall 1992; Sanjur et al. 2002).

A recent critical summary of the infraspecific classification of cultivated *Cucurbita* species, based primarily on the fruit morphology, was given by Jeffrey (2001) with in-depth discussions of classification within *C. moschata*, including cultivar origin and domestication presented by Andres (2004a, 2004b) and within *C. pepo* presented by Paris and his colleagues (Paris 1986, 1989, 2001a, 2001b; Paris and Nerson 2003; Paris et al. 2003).

8.3.6 *Citrullus*

The domestication of the watermelon, *C. lanatus* (Thunb.) Matsum & Nakai, is still somewhat unclear. All *Citrullus* species originated in tropics and subtropics of the Old World, primarily in Africa. Watermelon is an important crop in warmer parts of Russia and other parts Asia Minor, the Near East, China, and Japan. It was brought to the New World by Spanish explorers (Robinson and Decker-Walters 1997). By one interpretation (Bates and Robinson 1995), the edible cultivated watermelon was derived from *C. colocynthis* (L.) Schrad., a closely related species endemic to northern Africa, southwestern Asia and the eastern Mediterranean with a long archaeological history (Zohary and Hopf 2000). A more recent synthesis by Robinson and Decker-Walters (1997) indicated that wild populations of *C. lanatus* var. *citroides* (L.H. Bailey) Mansf., which are common in central Africa, probably gave rise to the domesticate, subsp. *lanatus*. This is supported by viewpoint that domesticated watermelons were derived from indigenous African populations of *C. lanatus* in or near the Kalahari Desert of Namibia and South Africa (Jeffrey 2001). It has been hypothesized that the initial site of intentional breeding and wide diversification of *C. lanatus* was in southwestern central Asia where it had spread from Africa as a wild or weedy plant (Filov 1959, in Sinskaja 1969).

Four species of *Citrullus* have been generally recognized, all sharing a chromosome number of $2n=2x=22$ (Table 8.11). In addition to the two widely distributed species noted above, the genus includes two species native to the desert regions of Namibia: the perennial vine, *C. ecirrhosus* Cogn., and the annual, *C. rehmii* De Winter (De Winter 1990; Jarret and Newman 2000; Levi et al. 2000). Jarret and Newman (2000) presented evidence from a comparative study of ITS sequences

Table 8.11 Organization of the Genus *Citrullus*

Citrullus spp.	Distribution
C. colocynthis (L.) Schrad.	Northern Africa, southwestern Asia, eastern Mediterranean
C. ecirrhosus Cogn.	Namibia and South Africa
C. lanatus (Thunb.) Matsum; & Nakai	Kalahari of Namibia, southern Africa, cultivated elsewhere
subsp. *lanatus*	As wild plant in the Kalahari
subsp. *vulgaris* (Schrad; ex Eckl; & Zeyh.)	Semicultivated forms in Sahara, Sudan, Egypt; warmer areas
subsp. *mucosospermus* Fursa	Western Africa (e.g., Senegal, Mali, Guinea, Ghana, Niger)
C. rehmii De Winter	Namibia

Source: Adapted from Jeffrey, C. Cucurbitaceae; In *Mansfeld's Encyclopedia of Agricultural and Horticultural Crops*, Vol. 3, Hanelt, P. Ed. Springer-Verlag, Berlin, 1510–1557, 2001; U.S. National Plant Germplasm System's Germplasm Resources Information Network database, National Germplasm Resources Laboratory, 2005.

that the annual species, *C. lanatus* and *C. rehmii*, were both derived from perennial ancestors, that *C. colocynthis* was more remotely related to *C. lanatus* than was *C. ecirrhosus*, and that *C. lanatus* var. *citroides* is likely the immediate ancestor of cultivated watermelons. Their data supports the views of Robinson and Decker-Walters (1997) and the results of molecular studies based on SSRs by Jarret et al. (1997) who divided *C. lanatus* into two basic groups (*C. lanatus* var. *citroides* and *C. lanatus* var. *lanatus*). However, a recent summary of the classification of cultivated *Citrullus* species (Jeffrey 2001; Table 8.11) partly differs from this concept.

An alternative classification that treats cultivated forms of *C. lanatus* was presented by Jeffrey (2001) who subdivided the species into three subspecies. These subepecies are subsp. *lanatus* that includes: (1) wild populations and one cultivar group, Group Citroides (including fodder melon, citron, and preserving melon); (2) subsp. *vulgaris* (Schrad. ex Eckl. and Zeyh.) with an inexactly defined Group Cordophanus with bitter and non-bitter bland forms and a Group Dessert, including sweet edible cultivars; and (3) subsp. *mucosospermus* Fursa that includes a single cultivar group, Group Mucosospermus, with large, soft-coated seeds rich in oil and protein.

Two additional species, closely related to *Citrullus* and once included therein, are *Praecitrullus fistulosus* (Stock) Pang., which is cultivated in India and Pakistan for its edible fruits and *Acanthosicyos naudinianus* (Sond.) C. Jeffrey, a wild species native to southern Africa (Robinson and Decker-Walters 1997; Levi et al. 2000).

8.3.7 Other Cultivated Genera

A useful summary of modern taxonomic understanding of secondary cucurbit crops, including *Benincasa*, *Lagenaria*, *Luffa*, and *Sechium*, was compiled by Bates, Merrick, and Robinson (1995). Noteworthy specific examples of biosystematic studies of these crops include an assessment of cultivar groups in *Benincasa hispida* (Walters and Decker-Walters 1989) and broader investigations of taxonomic relationships among members of tribe Benincaseae (Walters et al. 1991; Chung, Decker-Walters, and Staub 2003) and the evolution of the Cucurbitaceae (Decker-Walters, Chung, and Staub 2004a). In addition, Decker-Walters et al. (2001) combined information about geographic origins and fruit and seed types with RAPD data to investigate patterns of evolution and genetic diversity among landraces and cultivars of *L. siceraria*, which provided base-line data in support of a remarkable, recent report of a wild lineage of *L. siceraria* from Zimbabwe (Decker-Walters et al. 2004b).

Examining the evolution of another domesticate, Marr, Mei and Bhattarai (2004) investigated patterns of morphological and isozyme variation among wild and cultivated populations of *Momordica charantia*, and they observed comparable levels of morphological variation in wild and cultivated populations but a great reduction in isozyme polymorphisms among cultivated populations. Their results can be placed into the larger context of variation within the genus *Momordica* in southeast Asia as a result of de Wilde and Duyfjes's (2002) recent taxonomic treatment.

8.4 GENE POOLS AND GENETIC DIVERSITY

8.4.1 General Characterization

Gene pools serve as a tool for conceptualizing the ability of plant populations to cross with conspecific populations and those of other species, usually of the same genus (Harlan and deWet 1971). The primary gene pool is represented by interfertile populations, generally of a specific biological species and may include other species that are fully cross-compatible. The primary gene pool typically encompasses a crop and any fully interfertile wild progenitor(s). The secondary gene pool is represented by all other populations that can be crossed with the crop; the gene flow is possible but is connected with a reduction of fertility within hybrid generations. Species from the tertiary gene pool cannot be crossed with the crop species except through special biotechnological approaches such that the resulting hybrids often express abnormalities and often are lethal or completely sterile. Until the advent of transformation technologies for cucurbits (Clough and Hamm 1995), access to the genetic diversity present in tertiary gene pools by breeding programs was severely limited by sexual incompatibility (Gepts and Papa 2003).

In its most basic sense, however, a gene pool can also be thought of as consisting of a collection of gametes with the potential of intermating. The different types and frequencies of gametes that occur in a gene pool ultimately depend upon the pool's original parental genotypes and selection at the gametic level. If alleles having a positive overall fitness are present in the parental genotypes, they will frequently be predominant in the gene pool and, consequently, in genotypes within the resulting population.

Vavilov (1926, 1997), in his work on the origin of cultivated plants, used the concept of "centres of diversity" to predict where crop species were initially domesticated. This system works well for a high percentage of crops (Hancock 2004), including cucurbits (Smartt and Simmonds 1995). Secondary centers of diversity can be described from parts of the world where centers of diversity occur distinct from centers of origin. Such secondary centers can be created by the migration and continued domestication of genotypes beyond their original centers of origin.

Genetic diversity analyses provide estimates of allelic frequencies for the determination of genetic associations among populations within gene pools and to help determine the relationships of such gene pools to centers of origin. Such analyses are based upon the characterization of genetic diversity through the measurement of heritable phenotypic traits and molecular markers, which may not always be correlated (Reed and Frankham 2001). This situation can be attributed principally to weak linkages between molecular markers (mainly neutral) and genes coding for phenotypic traits (which often confer selective advantage), differences in gene action and heritability among marker types, and mutation rates and mutation input (Gepts and Papa 2003).

In cucurbit crop species, genetic diversity analysis has been rigorously applied to cucumber and melon to determine the nature and structure of their gene pools and to examine hypotheses that may provide insights into the nature of their domestication. The following section describes gene pools in cucumber and melon in relationship to their evolution, migration, and differentiation.

8.4.2 *Cucumis*

8.4.2.1 *Gene Pools of C. sativus and C. melo*

Considerable research has been conducted to help delimit primary, secondary, and tertiary gene pools within the genus *Cucumis*. However, results are insufficient to support a complete synthesis of domestication of *Cucumis* species. The nature of these relationships is important for the application of appropriate technologies to transfer desirable genes from wild *Cucumis* species to cucumber and melon and between cucumber and melon (Bates and Robinson 1995). Many researchers have attempted to make interspecific crosses in *Cucumis*, and data resulting from initial efforts was summarized by Raamsdonk, den Nijs, and Jongerius (1989). Complex experiments on crossing ability within the genus have been conducted by den Nijs and Custers (1990) with the aim of transferring valuable characters from wild species from Africa cultivated *C. sativus* and *C. melo*. Their results confirmed the division of the genus into two distinct gene pools that can be treated as subgenera, *Cucumis* and *Melo*, and the further subdivision of subgenus *Melo* (den Nijs and Custers 1990). Although young embryos have been recovered from crosses between *C. sativus* and *C. melo*, these consistently abort at an early stage (Bates and Robinson 1995).

The species *C. heptadactylus*, *C. humifructus*, *C. melo*, and *C. sativus* have never been successfully crossed with any other species of *Cucumis* to directly produce a fertile F_1 progeny (Kirkbride 1993).

The primary gene pool of *C. sativus* consists solely of that species and its two interfertile varieties, the domesticate, var. *sativus*, and the wild or feral type, var. *hardwickii* (Royle) Gabaev. Its secondary gene pool includes *C. hystrix* Chakr. (Bates and Robinson 1995) because fertile interspecific hybrid progeny have been synthetically made through F_1 embryo rescue with subsequent chromosome manipulation (Chen and Adelberg 2000). *Cucumis hystrix* has been collected only in the Yunnan Province of China. *C. sativus* var. *hardwickii* (R.) Alef. grows in the foothills of the Himalayan Mountains sympatrically with var. *sativus* and is used by native peoples of northern India as a laxative (Deakin, Bohn, and Whitaker 1971). It possesses a multiple fruiting and branching habit not present in *C. sativus* var. *sativus* (commercial cucumber; Horst and Lower 1978), suggesting considerable potential for increasing the genetic diversity available for improvement of commercial cucumber Staub and Kupper 1985). There is evidence for the relative recent introgression of *C. sativus* var. *hardwickii* genes into some *C. sativus* var. *sativus* landraces in northern India (Horejsi and Staub 1999). Another biological variation with respect to prefertilization barriers in crosses with *C. sativus* was also found in *C. zeyheri* (den Nijs and Custers 1990).

Gene pools of *C. melo* remain poorly defined. Analyses of chloroplast DNA (Perl-Treves et al. 1985; Perl-Treves and Galun 1985) clearly grouped *C. melo* and *C. sagittatus* together, distinct from the rest of the genus with just two mutations separating them. Fruit set is stimulated in *C. melo* and *C. sagittatus* by members of *Cucumis* ser. *Angurioidei*, but not by *C. humifructus* or *C. metuliferus* (Raamsdonk, den Nijs, and Jongerius 1989). Promising results were obtained by den Nijs and Custers (1990) after crossing *C. sativus* var. *hardwickii* accession IVT Gbn 1811A with *C. melo* because hybrid embryos developed much further than usual in this combination and appeared to accept pollen of both *C. metuliferus* and diploid *C. zeyheri* (den Nijs and Custers 1990). A hybrid between *C. melo* and *C. metuliferus* has been reported but not confirmed (Puchalski and Robinson 1990).

Isozyme studies provide two interpretations of the position of *C. melo* within subgenus *Melo*. In one interpretation, it constitutes a sister group with *C. humifructus* and *C. sagittatus* by sharing a common ancestor (Perl-Treves et al. 1985); in the second, it is a sister group with *C. humifructus* and *C. hirsutus* (Puchalski and Robinson 1990). In contrast, cpDNA results suggest different relationships among theses taxa (Bates and Robinson 1995).

8.4.2.2 Genetic Diversity of C. sativus and C. melo

Cucumis sativus

Genetic relationships among cucumber accessions in relation to their geographic origins are consistent with generally accepted, historic dispersal patterns (Horejsi and Staub 1999). Distinct

C. sativus var. *sativus* gene pools exist in Europe (Dijkhuizen et al. 1996; Horejsi and Staub 1999), Africa (Mliki et al. 2003), North America (Dijkhuizen et al. 1996), and Asia (Staub, Serquen, and McCreight 1997b, 1999; Horejsi and Staub 1999).

Genetic markers (morphological and biochemical) have been employed for the characterization of genetic diversity present in cucumber (Knerr et al. 1989; Meglic, Serquen, and Staub 1996; Staub and Ivandic 2000). Assessment of genetic diversity in *C. sativus* var. *sativus* and var. *hardwickii* using isozymes, restriction fragment length polymorphisms (RFLPs), and random amplified poly-morphic DNAs (RAPDs) indicated that diversity in *C. sativus* var. *sativus* is relatively low (3–8%) when compared with other allogamous *Cucumis* species (10–25%) (Dane 1976, 1983; Esquinas-Alcazar 1977; Knerr et al. 1989; Dijkhuizen et al. 1996; Horejsi and Staub 1999). Levels of allelic polymorphism in *C. sativus* var. *hardwickii* (17–25%) are predictably higher than in *C. sativus* var. *sativus* (Dijkhuizen et al. 1996; Meglic, Serquen, and Staub 1996; Horejsi and Staub 1999) as is the case for many pairs of wild progenitors and domesticates. These data lend support to the hypothesis that cucumber originated in India (Leppik 1966). Chinese (China is a putative secondary center for cucumber diversity; Leppik 1966; Staub et al. 1999) and Indian germplasm accessions, when examined collectively, were distinct from each other and from all other *C. sativus* var. *sativus* and *C. sativus* var. *hardwickii* accessions examined by Staub et al. (1999). Chinese (Staub et al. 1999) and Indian (Staub, Serquen, and McCreight 1997b) accessions represent the two most diverse subsets of the primary gene pool present in nature and in the world's major genebanks such as the U.S. NPGS (Staub et al. 2002).

It is thought that northern and southern Chinese cucumber cultivars have different origins (Leppik 1966; Staub et al. 1999), and it has been hypothesized that the genetic variation present in cucumber germplasm in southern China is endemic to that region and/or has been historically augmented by infrequent introductions of germplasm from northern Indian sources via ancient Himalayan trade routes (Staub et al. 1999). Southern Chinese germplasm has been isolated by the Himalayas and the region's social structure. In contrast, it is thought that the genetic diversity of northern Chinese cucumber germplasm has been a direct beneficiary of the Silk Road where germplasm (genetic variation) has been continually shared across central Asia and the Near East based on observed protein and DNA polymorphisms (Staub et al. 1999; Horejsi and Staub 1999). Such findings (Staub et al. 1999) suggest that northern and southern Chinese cultivars may be more similar than previously thought. The fact that the Indian and Chinese accessions examined in several studies are genetically different (Meglic, Serquen, and Staub 1996; Staub, Serquen, and McCreight 1997b, 1983) supports the hypothesis that cucumber germplasm exchanged between these countries has been limited, allowing for unimpeded genetic changes through divergent selection resulting from different growing conditions and unique human needs.

Test arrays constitute artificially constructed subsamples of gene pools based on unique charac-teristics (e.g., specific disease-resistance alleles; Bretting and Widrlechner 1995) and, ideally, also on broad genetic diversity as defined by geography and marker-based genetic distance. The vari-ation observed in cucumber germplasm has allowed for the construction of test arrays for reaction to angular leafspot (*Pseudomonas syringae* pv. *lachrymans*), anthracnose (*Colletotrichum orbicu-lare*), downy mildew (*Pseudoperonospora cubensis*), rhizoctonia fruit rot (*Rhizoctonia solani*), target leafspot (*Corynespora cassiicola*), and water and heat stress and for the designation of a core collection consisting of 115 accessions from those test arrays and 32 more accessions to assist in encompassing the global genetic diversity in the NPGS collection (Staub et al. 2002). At the time of its selection, the core collection of 147 accessions (115 + 32) represented about 11% of the total collection's size (1352).

Provisional classification of cultivar groups has been proffered for *C. melo* by Pitrat, Hanelt, and Hammer (2000) and Jeffrey (2001). Similar to melon, horticultural market classes exist for cucumber; however, a similar subdivision of *C. sativus* with regard to morphological and evolution-ary aspects has not yet been elaborated, and a core collection for melon has not been proposed.

Cucumis melo

Melon is a morphologically diverse outcrossing species (Kirkbride 1993). Since the time of its domestication, melon cultivation has been expanding from points of origin along well-defined trade routes (e.g., the Silk Road and international shipping routes) (McCreight, Nerson, and Grumet 1993; Robinson and Decker-Walters 1997).

Based on vegetative and fruit variation, Munger and Robinson (1991) defined melon morphotypes of *C. melo* as botanical varieties [treated herein as essentially synonymous with Groups as defined by Brickell et al. (2004) in the ICNCP] to include Agrestis, Flexuosus, Conomon, Cantalupensis, Inodorus, Chito, Dudaim, and Momordica. The economically important *C. melo* morphotypes are commonly partitioned into market classes according to their culinary attributes (Staub et al. 2000). For instance, Group Cantalupensis includes marker classes Earl's, House, Galia, Charentais, and Ogen; Group Inodorus includes Honeydew and Casaba; and Group Conomon includes Oriental with each having a different fruit morphology valued for their unique aromas and flavors and variable shelf life. While Groups Cantalupensis and Inodorus are of commercial importance in the United States and Europe as well as in Mediterranean and Asian countries, Group Conomon types have their origin and are widely grown in Asia (McCreight, Nerson, and Grumet 1993; Robinson and Decker-Walters 1997).

Recently, Pitrat, Hanelt, and Hammer (2000) assessed earlier studies and conducted a morphological analysis of melon germplasm and provisionally classified 6 *C. melo* morphotypes into five groups in subspecies *agrestis* (Naudin) Pangalo (i.e., wild, weedy, and free-living types) and 11 groups in subspecies *melo* (to include diverse commercial types having specific consumer-based culinary attributes). A more rigorous taxonomic treatment of this provisionary classification may allow for the formal designation of these morphotypes as Groups, consistent with the ICNCP (Brickell et al. 2004) as discussed by Spooner et al. (2003).

Attempts to clarify interspecific taxonomic relationships within *C. melo* and evaluate its patterns of genetic diversity, assess market–class relationships, and determine likely centers of origin for the various groupings have employed a diverse array of molecular markers, including isozymes (Esquinas-Alcazar 1981; Perl-Treves et al. 1985; Staub, Frederick, and Marty 1987; Staub et al. 2000; Akashi et al. 2002b; McCreight et al. 2004), RFLPs (Neuhausen 1992; Silberstein et al. 1999), RAPDs (Staub et al. 1997a, 2000; Garcia et al. 1998; Silberstein et al. 1999; Stepansky, Kovalski, and Perl-Treves 1999a; Mliki et al. 2001; López-Sesé et al. 2002, López-Sesé, Staub, and Gomez-Guillamon 2003), SSRs (Katzir et al. 1996; Staub et al. 2000; Danin-Poleg et al. 2001; Decker-Walters et al. 2002a; López-Sesé et al. 2002; López-Sesé, Staub, and Gomez-Guillamon 2003; Monforte, Garcia Mas, and Arus 2003), ISSRs (Danin-Poleg et al. 1998b; Perl-Treves et al. 1998; Stepansky, Kovalski, and Perl-Treves 1999a), and AFLPs (Garcia-Mas et al. 2000). Although both morphological and DNA (molecular-marker) variation has been used to define melon taxonomic groups and market classes, results derived from different lines of evidence do not always agree (López-Sesé, Staub, and Gomez-Guillamon 2003).

Melons have been the subject of numerous ethnobotanical studies (Jacquat and Berstossa 1990; Robinson and Decker-Walters 1997; Staub et al. 2000; Goldman 2002), and it is thought that the species is of African origin (Robinson and Decker-Walters 1997). However, Bates and Robinson (1995) noted the possibility of multiple origins in that at least some melons may have been domesticated in Asia. Typical of many vegetable crops, the paucity of archaeological remains of melon (because of its perishable tissues) does not allow for intense scrutiny of its domestication (Zohary and Hopf 2000). The only important exception to this limitation can be found in Egypt where, because of its unique climatic conditions and cultural and agricultural history, domestication theories can be postulated. A combination of archaeological evidence, interpretations of mural paintings, and genetic analyses supports the hypothesis that wild ancestors of cultivated melon originated in Africa and were domesticated in multiple areas of secondary diversity to include the Middle and Near East and India (Staub et al. 1992; Robinson and Decker-Walters 1997; Mliki et al. 2001). These diverse geographic areas house several related, but differentiated, gene pools.

Cultivation of *C. melo* in India began as early as 2000 BC where early forms of Groups Momordica, Agrestis, and Flexuosus were cultivated and selected for local culinary preferences and medicinal purposes (Robinson and Decker-Walters 1997). These non-sweet melon types are, based on phenotype and fruit morphology, clearly distinct from Group Cantalupensis and Inodorus types, which are cultivated primarily in Europe and the Middle East (Stepansky, Kovalski, and Perl-Treves 1999a). Thus, it is likely that Indian subsp. *melo* types were developed independently from those in Europe and the Middle East (López-Sesé, Staub, and Gomez-Guillamon 2003; Staub, Fanourakis, and López-Sese 2004).

The Oriental Asian melons collectively form Group Conomon and comprise a distinct germplasm pool (Nakata et al. 2005). However, Pitrat, Hanelt, and Hammer (2000) described how Group Conomon morphotypes could be further partitioned into vars. *makuwa* Makino and *conomon* Thunberg. According to Kitamura (1950), var. *makuwa* melons were established in northern China whereas var. *conomon* melons were solely cultivated in southern China. Seed size and isozyme analyses by Akashi et al. (2002b) suggested that ancestral types of vars. *makuwa* and *conomon* melons were distinct lineages before their introduction to China. Although the introduction of these melons to China may not have been directly from India, they are, based on their small seed size, most closely related to the melons of eastern India. Thus, Akashi et al. (2002b) hypothesized that melons grown in China originated in central India, and then, through progressive domestication, were transported eastward to Laos and eastern China. Alternatively, melon types could have been introduced to western China via the Silk Road (from Baghdad to Iran to Kashmir and then to China; ca. 700–1000 AD; Kitamura 1951).

Relationships among the gene pools of Groups Flexuosus and Inodorous and African melons have not yet been clearly defined. But according to Stepansky, Kovalski, and Perl-Treves (1999a) and Silberstein et al. (1999), Group Conomon and African melons (no distinct market classes) possess closer genetic affinities than are shared between Group Conomon and Groups Flexuosus and Inodorus. In fact, the hypothesized close genetic relationship between African landrace melons and Group Conomon Oriental market class melons was confirmed by Mliki et al. (2001), suggesting a common origin. These landraces are, in many respects, morphologically distinct from Indian populations of Group Agrestis (McCreight and Staub 1993; Mliki et al. 2001), and their progenitor(s) may have reached Africa from southeast or western Asia relatively late (ca. 1600 AD) by sea-trade routes.

The history of European melons' domestication is also obscure, confounding an understanding of relationships among European market classes. The variation in fruit morphology of European melons is, however, broad and dramatic. Genetic-marker studies have documented differences among market classes, creating unique gene pools (García et al. 1998; Stepansky, Kovalski, and Perl-Treves 1999a; Staub et al. 2002). For example, the Group Cantalupensis market classes of central European origin clearly differ from Group Inodorus Spanish landraces as do the Spanish landraces from Asian Group Conomon and Cantalupensis melons (López-Sesé, Staub, and Gomez-Guillamon 2003; Nakata et al. 2005). These differences suggest divergent selection for culturally based culinary attributes (Staub et al. 2000, López-Sesé et al. 2002; López-Sese, Staub, and Gomez-Guillamon 2003). Within Group Inodorus, Casaba types may have originated in the Middle East and were then transported to sentral and southern Europe where they were subsequently refined to form a distinct gene pool designated as modern Casaba melons (López-Sesé, Staub, and Gomez-Guillamon 2003; Staub, Fanourakis, and López-Sesé 2004).

Based on fruit morphology, the Group Flexuosus and Inodorus accessions from Greece are similar to west Asian and Mediterranean market classes, respectively (Staub, Fanourakis, and López-Sesé 2004). Their RAPD-based genetic similarity to reference accessions from Japan and Europe argues that this may be the case (Table 8.12). The Japanese reference accessions employed in

Table 8.12 Statistical Measures of Genetic Variation as Measured by RAPD Markers for Melon Accessions (*Cucumis melo* L.) Grouped by Origin and Botanical Groups (Population)

Population Origin (Botanical Group)[a]	N[b]	na[c]	ne[d]	H[e]	I[f]	% Polym.[g]
Greece (Cantalupensis)	2	1.3	1.2	0.11	0.16	26.5
Greece (Flexuosus)	8	1.6	1.4	0.20	0.30	52.9
Greece (Inodorus)	7	1.5	1.3	0.16	0.25	47.1
Japan (Cantalupensis)	9	1.7	1.5	0.27	0.40	67.6
Japan (Conomon)	2	1.4	1.3	0.17	0.25	41.2
Japan (Inodorus)	8	1.5	1.3	0.18	0.26	47.1
African (landraces)	15	1.9	1.6	0.34	0.50	85.3
Spain (Inodorus)	14	1.7	1.4	0.22	0.33	67.7
RA (Cantalupensis)	11	1.6	1.4	0.24	0.36	55.9
RA (Inodorus)	6	1.6	1.4	0.23	0.34	55.9

[a] RA = U.S. and European market reference array.
[b] N = Number of accessions in each population (bulk of 15 individuals per accession); only populations with more than one accession were considered for statistical purposes.
[c] na = Average observed number of alleles.
[d] ne = Average effective number of alleles.
[e] H = Nei's gene diversity (Nei 1973).
[f] I = Shannon's Information index.
[g] Percentage of polymorphic loci.
Source: From Staub, J. E., Fanourakis, N., and López-Sesé A. I., *Euphytica*, 136, 151–166, 2004.

such studies typified variation in common Japanese market classes (i.e., House, Earl's, and Oriental) (Nakata et al. 2005). Moreover, genetic affinities between the Greek accessions and Japanese accessions reported by Staub, Fanourakis, and López-Sesé (2004) might have been predicted because Japanese market classes were developed in part from European germplasm beginning in the middle to late 1800s (Nakata et al. 2005).

Fruits of many of the Greek accessions examined by Staub, Fanourakis, and López-Sesé (2004) resembled the Group Inodorus Casaba market class. In fact, the RAPD data they presented indicated that accessions from Crete also share genetic affinities with that market class. Upon closer inspection of relationships among Spanish melon landraces, López-Sesé, Staub, and Gomez-Guillamon (2003) detected strong genetic affinities between Spanish landraces originating from southwestern and central growing regions (Extremadura and Andalucia) and Group Flexuosus accessions from Greece. Moreover, some Greek Group Inodorus landraces (e.g., Agiou Vassiliou and Argous) are similar in fruit shape, size, and color to that of Piel de Sapo and other green Spanish melon cultivar types (López-Sesé, Staub, and Gomez-Guillamon 2003; Staub, Fanourakis, and López-Sesé 2004). Intentional, strict selection among Spanish melon cultivars for specific morphological characteristics and their culinary uniqueness (distinctive textures and specialized tastes) may have prevented the introgression of genes from additional germplasm sources of diverse origin despite the recent lack of geographic isolation (Esquinas-Alcázar 1977). Differences in melon fruit morphology and plant phenotype are controlled by relatively few genes (Pitrat 1994); therefore, these genetic relationships may have been forged either by common ancestral origins or through gene introgression from one or multiple sources.

The proximity of Greece and Crete to ancient trade routes might partially explain the genetic relationships detected by Staub, Fanourakis, and López-Sesé (2004). Immigrants from Asia Minor first inhabited Crete during the Neolithic period (6th millennium BC). Crete and mainland Greece became important crossroads (by sea and land), linking the continents of Asia, Africa and Europe with vital local and international commerce during the Minoan (2600–1100 BC), Achaean (1450 BC), and Dorian (1100 BC) cultural periods (Zohary and Hopf 2000; Staub, Fanourakis, and López-Sesé 2004).

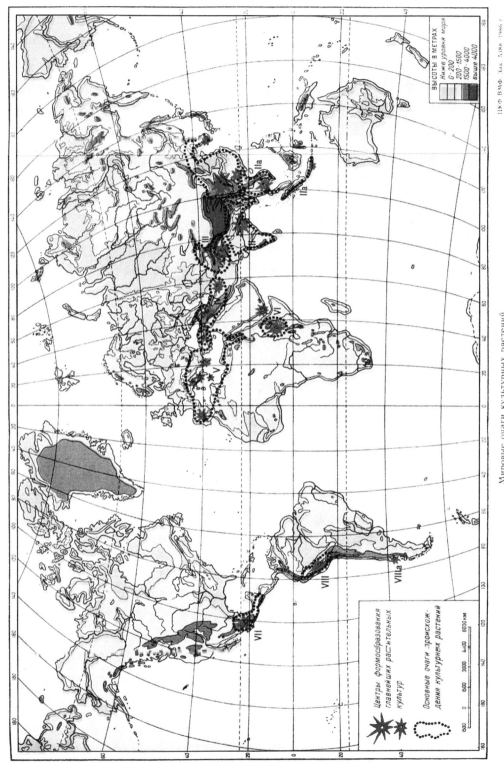

Мировые очаги культурных растений

Figure 1.1

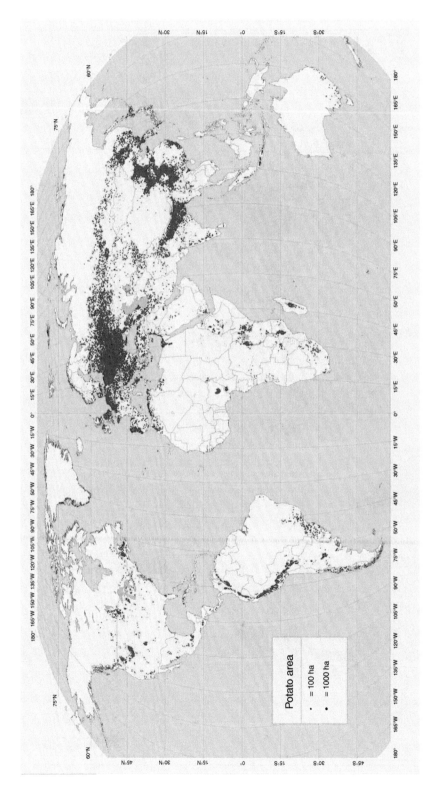

Figure 2.1

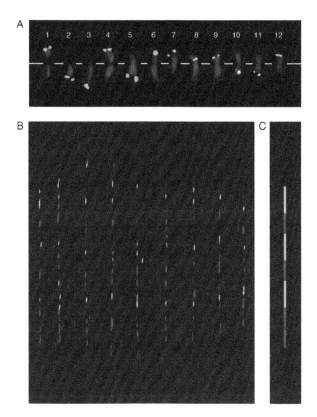

Figure 2.3

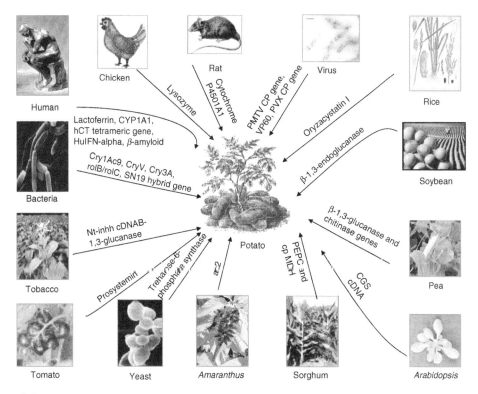

Figure 2.7

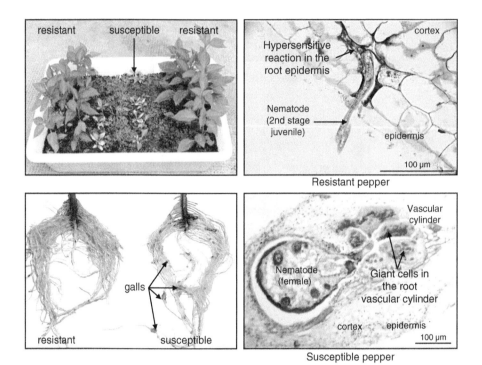

resistant   susceptible   resistant

cortex

Hypersensitive reaction in the root epidermis

Nematode (2nd stage juvenile)

epidermis

100 μm

Resistant pepper

galls

resistant   susceptible

Vascular cylinder

Nematode (female)

Giant cells in the root vascular cylinder

cortex   epidermis

100 μm

Susceptible pepper

Figure 6.6

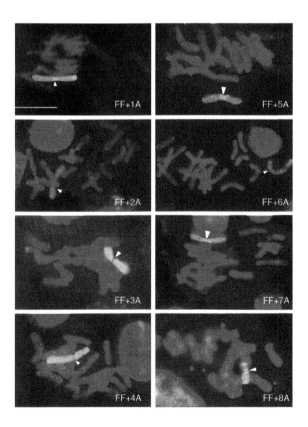

FF+1A

FF+5A

FF+2A

FF+6A

FF+3A

FF+7A

FF+4A

FF+8A

Figure 7.4

Figure 7.15

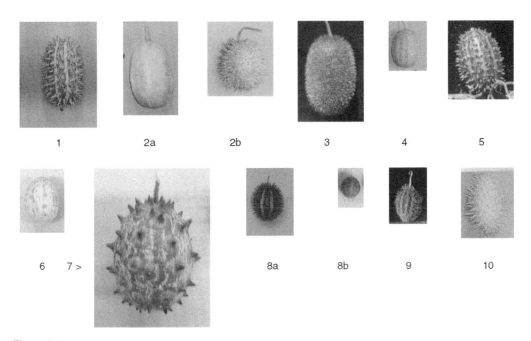

| 1 | 2a | 2b | 3 | 4 | 5 |

| 6 | 7 > | | 8a | 8b | 9 | 10 |

Figure 8.5

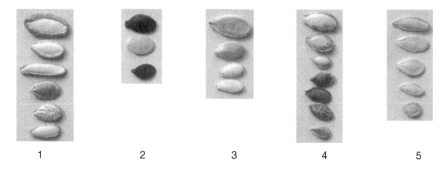

Figure 8.9

Figure 8.10

Figure 8.11

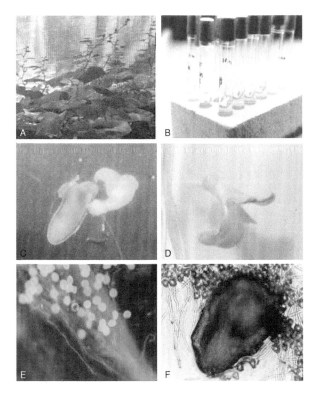

Figure 8.14

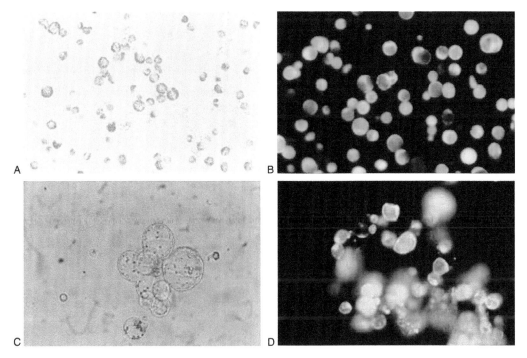

Figure 8.15

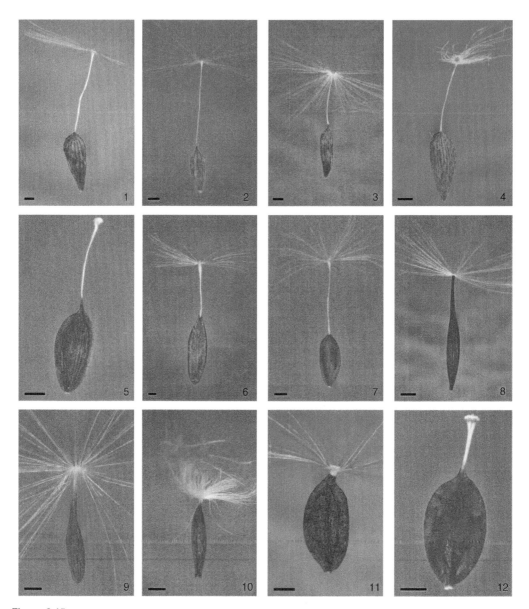

Figure 9.15

Figure 9.16

Figure 9.19

Greeks consumed melon in the third century BC. Mediterranean trade routes were well established by the second century AD, and by the third century AD, Rome was importing melons from Armenia in Asia Minor. The King's Highway (ca. AD 400) was a well-travelled route of agricultural commerce, linking Asia with the Mediterranean and extending from Egypt across the Sinai Peninsula, through Jordan into Syria, through Damascus ending at the Euphrates River. Somewhat later, the early Islamic diffusion of the eighth to eleventh centuries AD also allowed for the westward transport of Indian and south Asian crops, including several fruit tree species, condiments, and vegetables (Zohary and Hopf 2000). This Arabic expansion purportedly facilitated the dissemination of unique melon germplasm westward through the Mediterranean region to Spain. Such trade routes afforded Spain and Crete doors of commercial opportunity during these periods and the avenues for gene exchange to produce distinct gene pools.

Most accessions from Crete studied by Staub, Fanourakis, and López-Sesé (2004) shared genetic similarities with the original Galia hybrid accession from Israel as well as Galia cultivars derived from it (Staub et al. 2000). The pedigree of Galia is complex, and its relatively recent development from Ogen (Italian) and Charentais (French) market class types and a line of Russian origin now represent a unique market class (released in 1974 by Zvi Karchi, Agricultural Research Organization). Ogen market class melons, in fact, have strong genetic marker-based affinities with Galia melon types (Staub et al. 2000) and with Spanish Group Inodorus Casaba landrace melons (López-Sesé, Staub, and Gomez-Guillamon 2003). The melon accessions of Greek origin examined by Staub, Fanourakis, and López-Sesé (2004) were cultivated on the mainland, Crete, and other islands of the Aegean Sea (Samos) and Ionian Sea (Zakynthos) before the release of the original Galia. Because genetic affinities were detected between Greek accessions and Charentais but not Ogen accessions, it is likely that western European germplasm is in the pedigree of the Greek accessions examined by Staub, Fanourakis, and López-Sesé (2004), or that Charentais shares common ancestors with the Greek landraces.

8.4.3 *Cucurbita*

8.4.3.1 *Gene Pools of Cultivated Cucurbita Species*

The five cultivated species of *Cucurbita* are believed to have been derived from mesophytic progenitors. The xerophytic, long-lived perennial species of *Cucurbita* (including the buffalo gourd) are considered by Merrick (1995) to be terminal evolutionary lineages distantly related to the mesophytic species, yet Sanjur et al.'s (2002) phylogeny indicated that at least one of the xerophytic species (*C. foetidissima*) is basal to four of the mesophytic species.

Generally, the domesticated *Cucurbita* species are reproductively isolated from one another. The primary gene pools of each species are represented by their landraces and commercial cultivars as well as by their infraspecific taxa (Table 8.13).

Experimental crosses among them can be made with difficulty, and interspecific progenies are usually either sterile or sparingly fertile (Merrick 1995). Spontaneous crosses between the cultivated *Cucurbita* species are uncommon, but hybrids interspecific natural can occasionally be detected in landraces mostly from Mexico (Decker-Walters et al. 1990; Merrick 1990, 1991). In addition, Křístková (1991) reported a spontaneous hybrid of *C. maxima* by *C. pepo* under central European field conditions.

However, despite the high degree of genetic differentiation within *Cucurbita*, none of the genus's species is completely reproductively isolated from the others in term of barriers to hybridization. Of the domesticates, *C. moschata* is considered to be the extant species with the most ancestral-like genome and displays wide cross-compatibility (Merrick 1995). The domesticated species *C. argyrosperma*, *C. pepo*, and *C. maxima* can be crossed with their wild or feral relatives (Lira-Saade 1995; Merrick 1995). In the Americas, pairs of closely related domesticated and wild

Table 8.13 Gene Pools of Cultivated *Cucurbita* Species

| *Cucurbita* spp. | Gene Pools | | |
	Primary	Secondary	Tertiary
C. argyrosperma	*C. argyrosperma* subsp. *sororia* subsp. *argyrosperma (sensu lato)*	*C. moschata*	*C. pepo* *C. maxima* *C. foetidissima*
C. ficifolia	*C. ficifolia*	*C. pedatifolia* *C. foetidissima*	*C. lundelliana* *C. maxima* *C. pepo*
C. maxima	*C. maxima* subsp. *maxima* subsp. *andreana*	*C. ecuadorensis*	*C. lundelliana* *C. argyrosperma* *C. ficifolia* *C. pepo*
C. moschata	*C. moschata*	*C. argyrosperma*	*C. lundelliana* *C. maxima* *C. pepo*
C. pepo	*C. pepo* subsp. *pepo (sensu lato)* subsp. *fraterna* subsp. *texana*	*C. argyrosperma* *C. okeechobeensis* *C. moschata* *C. ecuadorensis*	*C. lundelliana* *C. ficifolia* *C. maxima*

Source: Adapted from Lira-Saade, R., *Cucurbita* L., In: *Estudios Taxonómicos y Ecogeográphicos de las Cucurbitaceae Latinoamericanas de Importancia Económica; Systematics and Ecogeographic Studies on Crop Genepools*, No. 9. IPGRI, Rome 1995.

species can occur sympatrically, and genetic interchange between them takes place, providing a natural source of variation within populations (Decker 1988; Wilson 1990; Merrick 1991).

Hybridization experiments (and some field observations) involving *C. argyrosperma* and other wild and cultivated *Cucurbita* taxa have revealed (Lira-Saade 1995) that, among the cultivated species, *C. moschata* has the highest degree of compatibility with *C. argyrosperma*, placing it into its secondary gene pool (Table 8.13). The next level of cross-compatibility involves the wild and cultivated taxa of *C. pepo,* some cultivars of *C. maxima*, and the wild perennial species *C. foetidissima*, which collectively represent the tertiary gene pool. The wild species that have shown some degree of compatibility with *C. argyrosperma s.l.* possess genes of resistance to some viral diseases that have a high incidence in the cultivated species (Lira-Saade 1995).

There are strong hybridization barriers between *C. ficifolia* and other species of this genus, making the definition of its gene pools problematical. Some interspecific hybrids have been obtained from crosses with *C. pedatifolia*, *C. foetidissima*, and *C. lundelliana*, but they often lack the capacity to produce an F_2 generation (Lira-Saade 1995).

The secondary gene pool of *C. maxima* is represented by *C. ecuadorensis*; and its tertiary gene pool includes *C. lundelliana*, *C. argyrosperma*, *C. ficifolia*, and *C. pepo* (Lira-Saade 1995).

The secondary gene pool of *C. moschata* is represented by *C. argyrosperma* as both species possess very close evolutionary relations. The tertiary gene pool is formed by *C. lundelliana* and some taxa of the "groups" Maxima and Pepo (Lira-Saade 1995).

The primary gene pool of *C. pepo* is formed by its various edible and ornamental cultivars (Figure 8.11) as well as populations of the wild taxa, var. *fraterna* and var. *texana*, until recently considered as distinct species or subspecies of *C. pepo*. There are a great many commercial cultivars with particular characteristics that, together with local landraces (grown mainly in Mexico), constitute an extraordinary genetic stock. However, in contrast to other *Cucurbita* species, this diversity does not represent an important source of resistance genes to pests and diseases because *C. pepo s.l.* is probably the one species with the greatest susceptibility to the most important viral diseases that attack cultivated *Cucurbita* species (Whitaker and Robinson 1986; Provvidenti 1990; Lira-Saade 1995). Populations that could be considered as part of its secondary gene pool are scarce as most attempts at hybridizing *C. pepo* with other wild or cultivated species have required special techniques such as embryo culture (see Section 8.6.3).

8.4.3.2   Genetic Diversity of Cultivated Cucurbita Species

Studies of isozyme variation in *Cucurbita* by Andres (1990), Decker (1988), Decker-Walters et al. (1990, 1993), and Merrick (1991) revealed moderate amounts of genetic differentiation within *C. pepo* and *C. moschata* while, in marked contrast, little genetic differentiation has occurred within *C. argyrosperma* and *C. ficifolia* (and perhaps in *C. maxima* as well, although the survey of the latter species has been less comprehensive than those of the others; Merrick 1995). Similar analyses to survey the variation present in DNA markers have been conducted, but on a somewhat more limited scale, with the most comprehensive work accomplished within *C. pepo* (Katzir et al. 2000; Decker-Walters et al. 2002b; Ferriol, Picó, and Nuez 2003a; Paris et al. 2003).

The diversity of *C. argyrosperma* is less than that found within the other three most widely cultivated domesticates, *C. pepo*, *C. moschata*, and *C. maxima*. Local varieties have evolved in a rather restricted geographic range including the southeastern United States, Mexico and Central America (Lira-Saade 1995). The most important variation observed within cultivated morphotypes corresponds to fruit size, shape, and color pattern and to seed morphology (Lira-Saade 1995). Three major lineages which comprise the domesticated forms of *C. argyrosperma* have been recognized at the varietal level by Merrick (1990): vars. *argyrosperma*, *stenosperma*, and *callicarpa*. The different degrees of variation in the nutritionally important parts of the three cultivated varieties suggest a strong association with human interests. The relatively large seed size of var. *argyrosperma* indicates that it was mainly selected to obtain seeds for human consumption while the great diversity of shapes, colors, and sizes of the fruits and seeds of vars. *stenosperma* and *callicarpa* indicate that selection had two aims: to obtain flesh as well as seeds (Lira-Saade 1995).

The wide range of elevations at which *C. moschata* is cultivated within the New World suggest that this species has evolved diverse adaptations to various environmental conditions (Andres 2004b). The species is highly polymorphic (Andres 2004a) with considerable morphological diversity of its seeds and fruit (color, shape, thickness, and durability of the fruit's skin). The existence of varieties with life cycles of differing phenology as well as the breadth of its numerous cultivars developed in other parts of the world (see Gwanama, Labuschagne, and Botha 2000) and of local varieties with excellent horticultural characteristics strongly suggest that its collective genetic variation is very extensive (Filov 1959; Lira-Saade 1995).

Some interesting Latin American landraces have been noted in the traditional agroecosystems of the Yucatán Peninsula where two types with distinct life cycles are grown and in the Mexican states of Guanajuato and Chiapas where landraces with resistance to some viral diseases have been noted (Lira-Saade 1995). Among the landraces from Yucatán, the short-cycle type commonly grown in Mayan vegetable gardens is of great interest because it was a likely progenitor of the most commercially important variety in the region. And, notably, selected landraces from Guanajuato and Chiapas were incorporated into genetic improvement programs (Lira-Saade 1995).

With regard to the range of variation of *C. moschata* populations developed outside its center of origin, the best example may be that of a landrace native to Nigeria, which represents the only source of resistance to certain viral diseases (Provvidenti 1993). Another part of the gene pool of *C. moschata* is represented by the numerous commercial cultivars that have been mainly developed in the United States and, to a lesser extent, in Brazil. Some of these commercial cultivars also have different levels of resistance and/or susceptibility to certain diseases, indicating a wide genetic variation of this species (Lira-Saade 1995). The possibilities of hybridization that *C. moschata* has shown with other cultivated species (e.g., *C. maxima*) suggest that that there are good prospects for the improvement of these other species as well (Lira-Saade 1995).

Cucurbita maxima has been shown to encompass a high degree of diversity for both morphological and molecular characters (Ferriol, Picó, and Nuez 2004b). Its variation includes local varieties and numerous commercial cultivars with different plant habits, fruit (Figure 8.10) and seed shapes and colors, and levels of host-plant resistance to viral diseases (Lira-Saade 1995). Also, the variability in the duration of its life cycle and, in some cases, its adaptation to marginal ecological conditions are remarkable (Filov 1959; Lira-Saade 1995). On the basis of morphological variation primarily in Spanish landraces, Ferriol, Picó, and Nuez (2004b) defined eight different groups, yet these were not congruent with concomitant molecular analyses of AFLP and SRAP variation. In that same study, Ferriol, Picó, and Nuez (2004b) noted that the high degree of genetic variability present in their small sample of New World populations (as compared to their larger sample of Spanish accessions) suggests that much work remains to be done to understand the full range of variation found in this species.

The diversity of *C. pepo* is comparable to or higher than that found in *C. maxima* (Lira-Saade 1995). Two major evolutionary lineages that collectively comprise the cultivated forms of *C. pepo* were recognized at the subspecific level by Decker (1988): subsp. *pepo* and subsp. *ovifera*. More recently, Decker-Walters et al. (2002b) used RAPD polymorphisms to characterize three distinct, New World populations of wild and weedy *C. pepo*. Local varieties and commercial cultivars have been categorized into at least ten cultivar groups (Figure 8.11) related to their fruit shape and quality and reflecting their evolutionary histories (Paris 1989; Paris et al. 2003).

Morphological and genetic variation is quite limited in *C. ficifolia* (Andres 1990) with a limited range of landraces and no commercial cultivars. Its scant morphological variation is consistent with its limited degree of isozyme variation. Yet, there may be cryptic variation for environmental adaptation because it is cultivated over a wide geographic range at various elevations in both agroecosystems with high competition, such as in maize fields in high rainfall areas, and in those with less competition and more intensive cultivation (Lira-Saade 1995).

8.4.4 *Citrullus*

Species of the genus *Citrullus* share a common chromosome number, and all its taxa are cross-compatible with each other to varying degrees. *Citrullus lanatus* and *C. ecirrhosus* are evidently more closely related to each other than either is to *C. colocynthis* (Navot and Zamir 1987; Jarret and Newman 2000). Yet even *C. lanatus* and *C. colocynthis* can be crossed experimentally and also cross spontaneously in nature to produce partially or fully fertile hybrids (Bates and Robinson 1995), indicative either of a single, primary gene pool or of a secondary gene pool encompassing two, poorly differentiated primary pools.

Considerable research has been conducted to describe the intra- and interspecific variation in *Citrullus* within this broad gene pool. However, a comprehensive synthesis of patterns of variation remains to be completed. The nature of phylogenetic and geographic relationships may guide the application of appropriate technologies to improve the efficiency of transferring desirable genes from wild *Citrullus* species to cultivated watermelon (Bates and Robinson 1995).

Watermelon fruits vary in size and shape, rind and seed color, and color and density of their flesh. Many of these variant forms were recorded by the European Middle Ages (Sturtevant 1919) and led to the description of a moderately complex array of species and botanical varieties, which for the most part, could be placed in cultivar groups (Bates and Robinson 1995). Complicating the situation is the enormous diversity of forms, developed and cultivated in southeastern countries of the former Soviet Union, that were described by Filov (1954). A recent overview of cultivar groups in *C. lanatus* was given by Jeffrey (2001), which addressed Fursa and Filov's (1982) recognition of thirty-one cultivar types based on fruit form and color and their parallel description of ten ecological–geographical groups.

Levi et al. (2001a) analyzed a group of watermelon cultivars, representing a wide range of horticultural traits (fruit size and shape, flesh texture and color, and firmness), but RAPD-based genetic variation among them was relatively low. That study did demonstrate, however, that molecular markers can be useful in evaluating groups based on their genetic similarity values. Such a classification can enable the selection of representatives from each group of closely related accessions for evaluation of such traits as disease or pest resistance.

Levi et al. (2001b) noted that low genetic diversity among watermelon cultivars bred in the U.S., which are widely grown throughout the world, increases the need to expand the genetic base of elite cultivars. Their study also indicated a higher genetic diversity within the wild subspecies *C. lanatus* var. *citroides* than in *C. lanatus* var. *lanatus* although accessions of *C. lanatus* var. *lanatus* are preferred in watermelon breeding programs because of their horticultural qualities, their close proximity to elite lines, and their reported resistance to anthracnose (Boyhan et al. 1994) and *Watermelon mosaic virus* (WMV) (Gillaspie and Wright 1993). The wild species *C. colocynthis*, which has the widest natural geographic distribution, also has the highest genetic diversity among *Citrullus* species. Wide genetic diversity among *C. colocynthis* plant introductions (PIs; Levi et al. 2000) indicated that this species may possess various genes that could confer pest resistance to cultivated watermelon. Beyond the traditional use of watermelon flesh, the gene pool of wild *C. colocynthis* can be explored as a source of genes for developing watermelon as an edible oilseed (Schafferman et al. 1998).

8.5 GENETIC MAPPING AND GENOMICS

8.5.1 Introduction

The primary utility of genetic maps in plant improvement is their deployment in marker-assisted selection (MAS) and breeding. The predictive value of genetic markers used in MAS depends upon their inherent repeatability, map position, and linkage with economically important traits (quantitative or qualitative) (Staub, Serquen, and Gupta 1996). While genetic mapping experiments have been well documented in *Cucumis* species, there are comparatively few examples for other cucurbit genera and species. The creation of genetic maps for watermelon, *C. lanatus* var. *lanatus*, has been hampered by lack of genetic diversity (Navot, Sarfatti, and Zamir 1990; Jarret et al. 1997; Levi et al. 2001a), few mapable, co-dominant markers (Hawkins et al. 2001; Levi et al. 2001a), and distored marker segregation patterns (Levi et al. 2001c). This has necessitated the use of wide intervarietal crosses (*C. lanatus* var. *lanatus* × *C. lanatus* var. *citroides/colocynthis*) for the construction of relatively unsaturated maps (Levi et al. 2002, 2004a; Hashizume, Shimamoto, and Hirai 2003; Zhang et al. 2004). Genetic diversity analysis (Gwanama, Labuschagne, and Botha 2000; Ferriol et al. 2004a; Ferriol, Picó, and Nuez 2004b) and map construction (Brown and Myers 2002) is also in its infancy in *Cucurbita*. Given the paucity of information on other cucurbit genera, the following discussion on genetic mapping will focus solely on *Cucumis* species.

8.5.2 *Cucumis sativus*

8.5.2.1 *Mapping*

Cucumber linkage maps have been constructed using phenotypic markers (Fanourakis and Simon 1987; Pierce and Wehner 1990; Vakalounakis 1992), isozymes (Knerr and Staub 1992), isozymes and phenotypic markers (Meglic and Staub 1996a, 1996b), and combinations of molecular (RFLP, RAPD, isozyme) and phenotypic markers (Kennard et al. 1994; Serquen, Bacher, and Staub 1997a). The maps of Fanourakis and Simon (1987), Knerr and Staub (1992a),

Vakalounakis (1992), Kennard et al. (1994), Meglic and Staub (1996a,b), Serquen, Bacher, and Staub (1997a), and Park et al. (2000) spanned 168, 166, 95, 766 (narrow-based) and 480 (wide-based), 584, 600, and 816 cM, respectively (750–1000 cM total genome length). These reports have been augmented more recently with markers associated with basic cell functions (Xie, Wehner, and Conkling 2002; Xie et al. 2003) and cytoplasmic factors controlling economically important traits (Chung and Staub 2003).

Park et al. (2000) employed 347 RAPD, RFLP, and AFLP marker loci and those conditioning virus resistances to construct a map with 12 linkage groups (LOD $\leq$ 3.5) and a mean marker interval of 4.2 cM. Resistances to *Papaya ringspot virus* (PRV) and ZYMV were closely linked (2.2 cM) to each other and were also tightly linked (~5.2 cM) to three AFLP markers. A map constructed by Serquen, Bacher, and Staub (1997a) defined nine linkage groups and spanned ca. 600 cM with an average distance between RAPD markers of 8.4 cM. Information from the Serquen, Bacher, and Staub (1997a) map was recently merged with other maps (Fanourakis and Simon 1987; Knerr and Staub 1992; Kennard et al. 1994; Meglic and Staub 1996a, 1996b; Horejsi, Staub, and Thomas 2000) to synthesize a consensus map containing 255 markers, including morphological traits, disease resistance loci, isozymes, RFLPs, RAPDs, and AFLPs spanning ten linkage groups (Bradeen et al. 2001). The mean marker interval in this consensus map was 2.1 cM, spanning a total length of 538 cM. More recently, Fazio, Stauba, and Stevens (2003) constructed a map containing 14 SSR, 24 sequence characterized amplified region (SCAR), 27 AFLP, 62 RAPD, one single nucleotide polymorphism (SNP), and three morphological markers (131 total markers) spanning seven linkage groups (the theoretical number based on the haploid chromosome number) by using recombinant inbred lines (RILs). This map spanned 706 cM with a mean marker interval of 5.6 cM. Because the map of Fazio, Staub, and Stevens (2003) contains several anchor markers common to the maps of Serquen, Bacher, and Staub (1997a) and Bradeen et al. (2001), it is possible that these maps and that of Park et al. (2000) could be merged for synteny comparison with melon (*C. melo*) using anchor markers (Danin-Poleg et al. 2000a).

8.5.2.2 QTL Analysis

The wide cross [GY-14 (U.S. elite processing) $\times$ PI 432860 (China)] used by Kennard et al. (1994) was employed to study QTL associated with fruit quality by using different mating designs (Kennard and Havey 1995). The two-year, single-location study identified five, three, three, and two QTL for fruit length, diameter, seed-cavity size, and color, respectively. More recently, Serquen, Bacher, and Staub (1997a; 100 F2:3 progeny), Fazio, Staub, and Stevens (2003; 170 RILs) characterized and mapped important yield components in U.S. adapted cucumber populations developed from the same genetically distinct parental lines (i.e., GY7 (synom. G421) and H-19) (Figure 8.12). These components included earliness (days to anthesis), sex expression [*F* locus for gynoecy (femaleness)], multiple lateral branching (MLB; branching on the main stem in the first seven leaf nodes), fruit number and weight, and fruit size [measured as length:diameter (L:D) ratio; commercially acceptable ratio is 3:1]. All of these characters are under the genetic control of relatively few QTL (2–6 for each trait) (Serquen, Bacher, and Staub 1997a, 1997b). Four significant QTL [logarithmic odds (LOD) $>$ 3] for sex expression were detected as well as four for multiple lateral branching (MLB), two for earliness, and two for fruit length. Cumulative QTL effects (R^2) explained at least 25% (~45% for gynoecy) of the observed variation for any particular trait.

The RIL-based QTL analysis by Fazio, Staub, and Stevens (2003) re-examined those traits studied by Serquen, Bacher, and Staub (1997a, 1997b). Fazio, Staub, and Stevens (2003; four-location study) confirmed the QTL detected by Serquen, Bacher, and Staub. (1997a, 1997b; two-location study). In general, Fazio, Staub, and Stevens (2003) mapped these QTL to smaller intervals than those identified by Serquen, Bacher, and Staub (1997a). The highest R^2 value indicated that the major genes controlling *F* (female flowering) and determinate (*de*; dwarf plant type) (Figure 8.12)

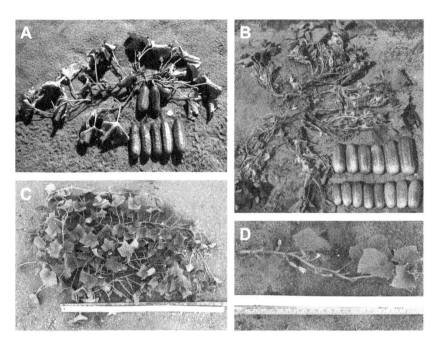

Figure 8.12   Phenotypes described in the mapping of yield components in cucumber (*Cucumis sativus*); Panel A = multiple lateral determinate (*de*), Panel B = multiple lateral compact (*cp*), Panel C = multiple lateral indeterminate (*De*; e.g., H-19), and Panel D = unilateral determinate [e.g., GY-7 (synom; G421)]. (Composite photo courtesy of J.E. Staub.)

were linked to SSR loci CSWCT28 and CSWCTT14 at distances of 5.0 cM and 0.8 cM, respectively. Moreover, Fazio, Staub, and Stevens (2003) revealed four, location-independent factors that cumulatively explained 42% of the observed phenotypic variation for MLB. QTL conditioning lateral branching (mlb1.1), fruit length/diameter ratio (ldr1.2), and sex expression (sex1.2) were associated with *de*, confirming the results of Serquen, Bacher, and Staub (1997a). Six QTL were identified for sex expression, five for MLB, two for earliness, and five for fruit length. Sex expression was influenced by three genomic regions corresponding to *F* and *de*, both on Linkage Group 1 and a third locus (sex6.1) on Linkage Group 6 (Figure 8.13). The potential value of these marker-trait associations with other yield components for plant improvement has been documented (Serquen, Bacher, and Staub 1997a, 1997b; Fazio, Staub, and Stevens 2003; Fazio and Staub 2003). Their efficacy is portended by the relatively high LOD scores (2.6–13.0) and associated R^2 values (1.5%–32.4%) resulting from comparatively few genetic factors (perhaps three to ten). These and other QTL-based data (Dijkhuizen and Staub 2003) confirm the reliability and consistency of these QTL-marker associations over multiple environments, populations, and various experimental designs.

8.5.2.3   *Application of Marker-Assisted Selection*

A test of the efficiency of MAS in cucumber was conducted in concert with efforts to modify MLB (Figure 8.2), a metric trait controlled by at least five effective factors (Serquen, Bacher, and Staub 1997b; Fazio, Staub and Stevens 2003). Yield increase in processing cucumber is positively correlated with increased number of fruit-bearing branches (Cramer and Wehner 2000). Fazio and Staub (2003) compared three breeding schemes for MLB in a backcross population: Phenotypic selection under open-field conditions (PHE), random intermating without selection (RAN), and MAS, employing five markers for the selection of greenhouse-grown plants. Their study

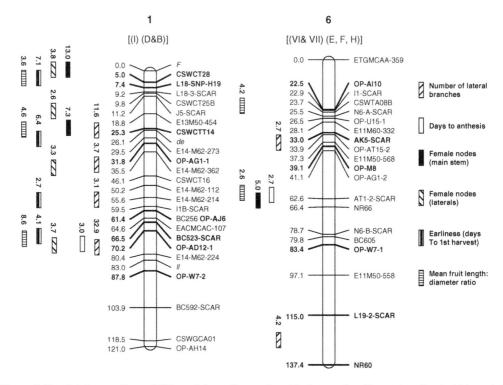

Figure 8.13   Relative positions of QTL on linkage Groups 1 and 6 of cucumber (*Cucumis sativus*) which contain
RAPD, SCAR, AFLP, and morphological markers (italicized) (Modified from Fazio, G., Staub, J. E.,
and Stevens, M. L., *Theor. Appl. Genet*, 107, 864–874, 2003a.); Linkage groups are designated by
numbers (1 and 6) above Roman numerals and letters corresponding to linkage groups in maps by
Bradeen et al. (2001) and Serquen, Bacher, and Staub (1997a) (From Serquen F. C., Bacher, J., and
Staub J. E., *Mol. Breed.*, 3, 257–268, 1997a.), respectively; RAPDs are identified by the preceding
letters OP and BC according to Serquen, Bacher, and Staub (1997a), SSR by the preceding letters
CS, CM and NR, AFLP by E_M_, and SCARs by the designation SCAR according to Fazio (2001);
The vertical bars to the left of each linkage group represent the QTL regions detected with their
respective LOD score; Markers associated with and/or bracketing economically important regions
are given in bold.

employed two SSRs, two RAPDs, and one SNP marker to independently confirm previously
determined linkages and their utility in MAS. No significant differences ($p < 0.001$) were detected
between the mean values of MLB from PHE and MAS. However, values for both PHE and MAS
populations were significantly higher than for the RAN control. Because MAS BC populations were
produced in one year and phenotypic selection required three years to complete, markers linked to
MLB increased overall breeding efficiency.

8.5.3   *Cucumis melo*

The first melon gene list was published in 1979 and has been updated every four years since 1986
(Pitrat 2002). Initially, this list included genes for simply inherited traits such as disease resistance
and morphological characters (Pitrat 1990). The most recent melon gene list, however, included
162 genes (forty-seven mapped unto linkage groups as phenotypic markers), with seven QTL for
disease resistance, forty-three QTL for fruit-quality traits, and forty-six cloned genes with complete
sequences (Pitrat 2002).

8.5.3.1 Mapping of Disease and Pest Resistance

Molecular markers (e.g., AFLPs, CAPs, ISSRs, RAPDs, RFLPs, SCARs, SNPs, and SSRs) have been used to identify and map disease and pest-resistance genes in melon. Marker–trait associations have been identified for host reaction to fusarium wilt (*fom1* and *fom2*; Wechter et al. 1995; Wechter, Thomas, and Dean 1998; Wang, Thomas, and Dean 1998, 2000; Zheng et al. 1999; Karsies, Dean, and Thomas 2000; Zheng and Wolff 2000; Garcia-Mas et al. 2001; Brotman et al. 2004), ZYMV (*Zym 2* and *Zym 3*; Danin-Poleg et al. 2000b, 2002), PRV (*Prv*; Garcia-Mas et al. 2001; Brotman et al. 2002, 2004), powdery mildew (*Pm*; Fukino et al. 2002), *Melon necrotic spot virus* (MNSV) (*Nsv*; Garcia-Mas et al. 2004; Morales et al. 2004a, 2004b), and for resistance to the transmission of viruses by aphids (*Vat*; Brotman et al. 2002).

8.5.3.2 Linkage Maps

The genome size of melon has been estimated to be $4.5–5.0 \times 10^8$ bp, about three times that of *Arabidopsis* (Arumuganathan and Earle 1991). In 1984, Pitrat initiated a systematic characterization of linkage relationships among phenotypic markers and developed the first linkage map for melon (Pitrat 1984, 1991). Initial reports were followed by a map consisting of 28 loci partitioned into eight linkage groups (Pitrat 1994). Some of these linkage groups contained genes for economically important traits, including disease resistance (*Fom-2*, *Nsv*, *Pm*, *Pvr*, and *Zym*), floral characters (*a* and *ms*), and plant architecture (*si* and *lmi*).

The development of molecular markers has led to the construction of numerous linkage maps in melon. However, most of the initial maps were relatively unsaturated and identified the position of relatively few traits of economic importance. Baudracco-Arnas and Pitrat (1996) constructed the first DNA-based map based on 218 F_2 progeny from the cross Vedrantais × Songwhan Charmi (PI 161375). The 103-point map consisted of 34 RFLPs, 63 RAPDs, one isozyme, four disease-resistance markers (*Fom-1*, *Fom-2*, *Nsv*, and *Vat*), and one morphological marker (*p*, conditioning carpel number). The total length of this map was 1,390 cM, where marker loci were fairly evenly dispersed among 14 linkage groups at an average distance of 17.7 cM. Wang, Thomas, and Dean (1997) constructed a 204-point map based on 66 BC_1 lines derived from a cross between line MR-1 and Anas Yokneam. That map consisted of 14 major and six minor linkages groups composed of 197 AFLPs, six RAPDs, and one SSR marker, spanning 1942 cM, with an average distance between markers of 11 cM (Wang, Thomas, and Dean 1997, 1998). Similarly, Liou et al. (1998) used 64 F_2 progeny (Makuna #SLK-V-052 × "Sky Rocket") and 125 RAPD markers to construct a map consisting of twenty-nine linkage groups spanning 1348 cM, with markers spaced at an average distance of 10.8 cM. Staub, Meglic, and McCreight (1998) also used F_2 and BC_1 progenies originating from strategic crosses among 400 USDA PIs to construct two linkage groups consisting of 11 isozyme loci spaced at an average distance of 9 cM.

These initial efforts were followed by the publication of more robust maps with higher degrees of saturation by French, Spanish, and Israeli research groups. Périn et al. (1998) reported the use of 122 RILs derived from the Ved161 population used by Baudracco-Arnas and Pitrat (1996) for the creation of a map consisting of 354 markers (294 AFLP, 46 ISSR, and14 morphological markers). That map consisted of twelve major and five minor linkage groups covering 1,366 cM with an average marker interval of 3.9 cM (Périn et al. 1998). Dogimont et al. (2000) subsequently constructed an improved 527-point map based on the backbone map of Périn et al. (1998). That map consisted of 12 linkage groups spanning 1583 cM where the average marker interval was 3 cM. Périn et al. (2000) continued this mapping effort by using 120 RILs from Ved161 and 60 RILs from Ved414 (Vedrantais × PI 414723) to construct a composite map spanning 1590 cM. This consensus map consisted of 777 markers, including 608 AFLP, 25 RFLP, 128 ISSR, 16 RAPD, and 14 morphological markers evenly dispersed in 12 linkage groups. More recently, Périn et al. (2002b) developed a 668-point map by using 163 RILs from Ved161 and 63 RILs from Ved414 and AFLP, RAPD, RFLP, SSR, and morphological markers, which were grouped into 12 linkage groups spanning 1654 cM (Périn et al. 2002a, 2002b, 2002c).

Spanish researchers published several reports that culminated initially in the creation of 411-point map (Oliver et al. 1998, 2000, 2001). This map was constructed using 93 F_2 progeny from a cross between Songwhan Charmi and Pinyonet Piel de Sapo consisting of 234 RFLP, 94 AFLP, 47 RAPD, 29 SSR, five ISSR, two isozyme, and one phenotypic marker (p). These markers were randomly distributed across 12 linkage groups that spanned 1197 cM where the average marker interval was 3.1 cM (Oliver et al. 2001). Gonzalo et al. (2005) more recently published a more saturated map by adding 26 SSRs to the extensive array used by Oliver et al. (2001). The 287-point map consisted of 12 linkage groups and spanned 1240 cM with an average marker interval of 4.3 cM. In addition, Gonzalo et al. (2005) mapped 173 polymorphic markers to 12 linkage groups using 77 dihaploid (DH) lines derived from the F_2 mapping population of Oliver et al. (2001). This mapping effort included 33 previously described SSRs (Katzir et al. 1996; Danin-Poleg et al. 2000a, 2001; Fazio, Staub, and Chung 2002), 41 newly developed SSRs, 79 RFLPs, 16 EST-SSR, three SNPs (Morales et al. 2004b), and the *Nsv* locus (Morales et al. 2004a). The map consisted of 12 linkage groups spanning 1,223 cM with a mean marker interval of 7 cM. These two maps were merged (Gonzalo et al. 2005), resulting in a consensus map consisting of 327 loci (226 RFLPs, 97 SSRs, 3 SNPs, and the *Nsv* locus) distributed over 12 linkage groups spanning 1,021 cM, with an average distance between loci of 3.1 cM.

Important contributions for the development of molecular markers and genetic maps in melon have also been made by Israeli scientists. Initially, Katzir et al. (1995, 1996) and Danin-Poleg et al. (1996, 2001) developed SSR markers, which lead to the construction of three SSR-based maps designed to assess the utility of this marker class for map construction and merging (Danin-Poleg et al. 1998a, 2000a; 2002). The first map (Danin-Poleg et al. 1998a, 2000a) was constructed from 60 F_2-derived F_3 bulks randomly chosen from the population used by Baudracco-Arnas and Pitrat (1996). Their initial 123-point map (138 RFLPs, 65 RAPDs, one isozyme, *Fom-1*, *Fom-2*, *Nsv*, and *Vat*, and 14 SSRs) consisted of 13 linkage groups spanning 1716 cM. A second 82-point map (Danin-Poleg et al. 1998a, 2000a) spanning 1130 cM and consisting of 12 linkage groups was constructed using 93 F_2 progeny from a Songwhan Charmi × Pinyonet Piel de Sapo mating (Oliver et al. 2001). A third map (Danin-Poleg et al. 2000a, 2002) in this initial series was developed using 122 F_2 progeny from the cross PI 414723 × Dulce consisting of 22 SSR, 46 RAPD, 2 ISSR, and four phenotypic markers (*a*, *Fom-1*, *st*, and *pH*) in 14 linkage groups, spanning 610 cM. A common set of SSR markers resident in these three maps allowed for marker-order confirmation and synteny assessement. More recently, maps have been developed based on 113 F_2 individuals derived from the cross between PI 414723 × Top-Mark (Brotman et al. 2000; Silberstein et al. 2003). Progeny segregation allowed for the construction of a 179-point consensus map using AFLPs, RAPDs, ISSRs, SSRs, RFLPs, and two phenotypic markers (*vat* and *a*), consisting of 24 linkage groups spanning 1421 cM (Silberstein et al. 2003).

8.5.3.3 QTL Analysis and Map Merging

At this time, there are only four known studies involving the molecular dissection of QTL controlling horticulturally important traits in melon. These include the characterization of QTL for *Cucumber mosaic virus* (CMV) resistance (Dogimont et al. 2000), ethylene production during fruit maturation (Périn et al. 2002a), and ovary and fruit shape, fruit weight, sugar content, external color, and flesh color (Périn et al. 2002c; Monforte et al. 2004).

The paucity of QTLs studies in melon is in part due to a lack of linkage maps constructed from populations designed for the extensive replication required for effective QTL analysis. Most melon maps have been developed from F_2 or BC_1 populations, which are not particularly well-suited for extensive replicated, multi-location evaluation (Baudracco-Arnas and Pitrat 1996;

Liou et al. 1998; Wang, Thomas, and Dean 1998; Oliver et al. 2001; Danin-Poleg et al. 2002; Silberstein et al. 2003). Although immortalized populations (e.g., DH and RIL populations) are ideal for analyzing complex trait QTL, saturated melon maps constructed from such populations are relatively new (Périn et al. 2002a,b,c; Gonzalo et al. 2005).

The development of more densely saturated maps and the use of DH or RIL populations will likely lead to the placement of economically important traits conditioned by QTL. However, many initial melon maps included many dominant markers (Baudracco-Arnas and Pitrat 1996; Liou et al. 1998; Wang, Thomas, and Dean 1998; Danin-Poleg et al. 2002; Silberstein et al. 2003) with information that is usually not readily transferable to other populations (Périn et al. 2002a). Codominant markers (e.g., SSRs) are often syntenic and can be used in map comparison and merging experiments (Danin-Poleg et al. 2001; Gonzalo et al. 2005). Such markers will likely be important for interspecific synteny analysis (melon and cucumber) and the identification, characterization, and mapping of QTL for complex traits (Katzir et al. 1996; Danin-Poleg et al. 2000a, 2001, 2002; Oliver et al. 2001; Périn et al. 2002a, 2002b, 2002c; Monforte, Garcia Mas, and Arus 2003; Silberstein et al. 2003; Fukino et al. 2004; Gonzalo et al. 2005).

8.6   GERMPLASM ENHANCEMENT

8.6.1   Introduction

The process of genetic improvement often involves the incorporation of genetic material [gene(s)] from one or more exotic accessions (Walters and Wehner 1994a; Wehner and St. Amand 1995) into a germplasm pool (Wehner, Elsey, and Kennedy 1985, Wehner et al. 1989; Wehner, Shetty, and Clark 2000a; Wehner, Shetty, and Wilson 2000b; Wehner and Shetty 1997, 2000; Shetty and Wehner 2002) with subsequent cycles of enhancement of the broadened pool followed by testing of derived populations or lines for hybrid or *per se* performance (Rubino and Wehner 1986; Cramer and Wehner 1999a). This general strategy has resulted in important public germplasm releases in cucumber (Peterson et al. 1982, 1986b; Peterson, Staub, and Palmer 1986; Wehner et al. 1996b; Walters and Wehner 1997, Wehner 1998a, 1998b), melon (Jagger and Scott 1937; McCreight, Kishaba, and Bohn 1984; Thomas 1986), and squash (Coyne and Hill 1976; Mutschler and Pearson 1987).

The selection scheme used in cucurbits depends on the objectives of the enhancement/breeding project, the inheritance patterns and heritability of the characters to be selected, the available germplasm, and the species selection history. Cucurbits are typically genetically diverse and allogamous, but tolerant to some degree of inbreeding, making them amenable to a myriad of selection procedures. Selection procedures often evolve over time because breeders react to new information about the genetics of their target species and its key traits and to changes in economic conditions (e.g., market target), management practices (e.g., hand vs. machine harvesting operations), and environmental factors that affect abiotic and biotic stresses. For traits that are conditioned by few genes (such as many disease resistances), segregating progeny can be tested against simple Mendelian ratios. In contrast, response of metric traits conditioned by multiple allelic arrays (such as yield and quality components) to selection is measured by changes in population means and variances that can be used to define heritabilities (Wehner 1984; Strefeler and Wehner 1986; Serquen, Bacher, and Staub 1997b; Cramer and Wehner 1998a; St. Amand and Wehner 2001a, 2001b) and develop optimal recurrent selection procedures (Wehner and Cramer 1996a, 1996b; Cramer and Wehner 1998b, 1999b).

8.6.2   Gene Pools: Evaluation and Utilization

Gathering appropriate germplasm is a critical step in any enhancement program. Plants can be visualized on a continuum, based on degree genetic improvement and local adaptation, from exotic, unadapted types, which may possess specific traits of interest to elite, locally-adapted materials that are uniform, possessing many commerically acceptable traits. Exotic cucurbit germplasm typically includes genetic diversity absent in elite types. Because of this and the fact that exotic germplasm usually also exhibits undesirable characteristics (e.g., lack of disease resistance, bitterness, relatively small fruit), refinement of exotic populations often requires a substantial, long-term effort. Therefore, the types and proportions of adapted and exotic germplasm that are selected to form a base population that maximizes the genetic diversity for desirable traits is dictated by program objectives (e.g., long vs. short-term), time (reproduction cycles/year) and cost considerations (e.g., molecular-aided vs. conventional), and potential benefits. Strategic intercrossing of the germplasm forming the base population is needed when the breeder requires that the population eventually utilized is random-mating and at linkage equilibrium (e.g., F_2, three-way, mass intercrossing). Often, where the objective of the enhancement program is the development of an elite population for further improvement, a large array of diverse parental stocks based on their performance (disease resistance, yield heterosis) is initially chosen for intermating, followed by selection over seasons and years for a range of specific characterisitics (e.g., sex expression, flowering date, disease resistance, and fruit number and quality; e.g., Wehner et al. 1989; Wehner and Cramer 1996b).

The breeding of cucurbits has primarily focused on improving the production and quality of fruits by increasing resistance to a wide range of pathogens and diseases and by modifying plant architecture and sex expression. Until the end of twentieth century, interspecific hybridization, even when aided by embryo culture and other biotechnologies, had not been yet used to breed cucumber and melon cultivars (Bates and Robinson 1995; Robinson and Decker-Walters 1997). Principal attention was focused on the search for valuable characters within cultivated species and on basic genetic studies (i.e., inheritance of useful features and identification of genes). Gene lists for cucurbitaceous vegetables summarize many of the results of above mentioned approaches (Pitrat 2002).

The identification of genetic variation within cucurbit germplasm collections to overcome the challenges of abiotic and biotic stresses and reduce the cost and/or increase the value of production has motivated considerable research. This body of research encompasses both the commonly cultivated taxa and, to a lesser extent, their wild and weedy relatives as evaluated for many different traits. These include horticultural and agronomic characteristics, biochemical and quality traits, and reaction to biotic and abiotic stresses. Given such an extensive body of literature, the following section will focus on published literature reviews (where such exist) and on recent findings.

8.6.2.1 *Horticultural and Agronomic Characteristics*

An excellent example of a long-term project where standardized morphological evaluations of cucurbit germplasm were used to identify superior horticultural forms for breeding and/or direct introduction has been conducted by the Institute for Introduction and Plant Genetic Resources in Bulgaria, primarily for *C. sativus* (Neykov 1994, 1998; Neykov and Alexandrova 1997) but also for *Citrullus*, *C. melo*, and *Cucurbita* (Stefanova, Neykov, and Todorova 1994; Krasteva 2002). Brazilian researchers have also directed considerable attention to the evaluation of local cucurbit (*Citrullus* and *Cucurbita*) germplasm with the goal of identifying valuable accessions for breeding (Choer 1999; Romão et al. 1999; Ramos et al. 2000; Queiroz et al. 2004). *Cucurbita* germplasm has received similar attention by Chinese researchers (Chen 1993; Zhou et al. 1995; Ren 1998). Other reports on horticultural evaluations directed toward the identification of superior genotypes among local landraces include studies of *Cucurbita* landraces from Cuba (Ríos Labrada, Fernández Almirall, and Casanova Galarraga 1998) and Jordan (Kasrawi 1995) and of *C. melo* landraces from India (Seshadri and More 2002), Italy, and Albania (Ricciardi et al. 2003).

While horticultural production of cucurbits in Europe and North America focuses on *Citrullus*, *C. melo*, *C. sativus*, and *Cucurbita*, worldwide production is much more diverse, including many genera and species of local importance (Ng 1993). The body of literature on horticultural evaluation includes a number of reports on these less widely cultivated genera and species. For example, two uncommon species of *Cucumis* have been evaluated for specialty fruit production. Marsh (1993) evaluated 26 accessions of *C. metuliferus* under field conditions in Missouri, U.S., and Dhaliwal (1997) evaluated 48 accessions of *C. anguria* under field conditions in Punjab, India. Both of these projects identified superior lines for potential commercial production. Evaluations of *Sechium edule* germplasm have been reviewed by Newstrom (1990), of *Lagenaria siceraria* germplasm have been reported by Ram et al. (1996) and Upadhyay, Ram, and Singh (1997–1998), and of *Momordica charantia* germplasm by Xu and Huang (1995), Ram et al. (1996), and Marr, Mei, and Bhattarai (2004).

As noted above, there are many papers that report the results of general horticultural and morphological evaluations. However, there is also a body of research that has investigated particular plant traits that contribute to fruit yield and marketability; on the basis of such work, in-depth evaluations of specific traits have also been conducted. This has been true especially for *C. sativus* where germplasm has been evaluated for variation in root growth (Grumet et al. 1992; Walters and Wehner 1994a), postharvest storage (Wehner, Shetty, and Wilson 2000b), combining ability for yield as measured via testcrosses with a common inbred tester (Wehner, Shetty, and Clark 2000a), specific yield components such as fruit number and weight (Shetty and Wehner 2002), and precocious staminate flowering (Walters and Wehner 1994b). In addition, Sugiyama (1998) evaluated *Citrullus lanatus* germplasm for variation in the number of pistillate flowers and its relationship to fruit yield.

8.6.2.2 Biochemical and Quality Traits

Variation in soluble solids, individual sugars, and the enzymes that control sugar production has received special attention especially in *C. melo* (Stepansky et al. 1999b; Burger et al. 2004), and to a lesser extent, in *C. sativus* (Robinson 1987). Burger et al. (2004) also surveyed variation in ascorbic acid (vitamin C) production in *C. melo* germplasm.

The evaluation of seed composition, especially of fatty acids, has been conducted in the domesticated oilseed forms of *Cucurbita pepo* (reviewed by Teppner 2004) and in both *Cucurbita foetidissima* (Scheerens et al. 1978; De Veaux and Schultz 1985) and *Citrullus colocynthis* (Schafferman et al. 1998) with the goal of developing them as new, arid-land perennial oilseed crops. Research to develop *Citrullus colocynthis* as a new crop in India has been summarized by Mal, Rana, and Joshi (1996). Thompson, Dierig, and White (1992) presented a succinct review of research that had been conducted in an unsuccessful attempt to domesticate *Cucurbita digitata* and *Cucurbita foetidissima* in the United States between the 1940s and 1990.

Information on the challenges and opportunities related to the domestication of *C. foetidissima* has been reported by Gathman and Bemis (1990). The substantial production of cucurbitacins, a class of unusual tetracyclic triterpenoids by *C. foetidissima*, has potential application in the production of insect attractants (baits). The complex roles of cucurbitacins as vertebrate toxins, insecticides, and attractants has been reviewed by Metcalf and Rhodes (1990). This review included papers that summarized the evaluation of cucurbit germplasm for variation in the production of these distinctive compounds.

8.6.2.3 Reaction to Biotic and Abiotic Stresses

The discovery of valuable genes that allow plants to resist pathogens and pests and to tolerate abiotic stresses and their subsequent incorporation into commercial crops are key justifications for the conservation of *ex situ* germplasm collections. As described below, cucurbit germplasm collections have received considerable attention for evaluation, both as sources of pathogen and pest resistance and for their responses to a wide range of abiotic stresses.

Viral Pathogens

Provvidenti (1982, 1986, 1989, 1993) published three review papers that summarize the body of knowledge on screening cucurbit germplasm for its reaction to viral pathogens. In addition, a single, more specialized review focused exclusively on the genus *Cucurbita* (Provvidenti 1990). Munger (1993) also reviewed the genetic control of viral resistance in relation to cucurbit breeding. Many germplasm viral pathogen evaluations have been published since these early reviews.

Three destructive viruses have received much of the attention for screening: CMV, ZYMV, and *Watermelon mosaic virus-2* (WMV-2) (Lebeda and Křístková 1996; Křístková and Lebeda 2000a; Paris and Cohen 2000). Evaluations discovered new sources of resistance in both *C. maxima* and *C. pepo*. New sources of resistance to ZYMV infection have been reported for both *Citrullus* (Boyhan et al. 1992; Guner and Wehner 2004) and *C. melo* (Herrington and Prytz 1990). *Cucumis. hystrix*, which is sexually compatible with *C. sativus*, evidently also harbors some degree of resistance to CMV and ZYMV and comparatively strong resistance to PRV (Chen et al. 2004).

Horváth (1993a) screened a set of 67 accessions of 12 *Cucumis* species for the resistance to seven viruses [*Cucumber green mottle mosaic virus* (CGMMV), *Cucumber leaf spot virus* (CLSV), CMV, MNSV, *Melon yellow fleck virus* (MYFV), WMV-2, ZYMV]. An immune reaction was found within cultivated *C. melo* and *C. sativus* genotypes, but potentially valuable sources of resistance were also identified within *C. africanus* (CGMMV, CLSV, CMV, WMV-2, ZYMV), *C. anguria* (CLSV, CMV, WMV-2, ZYMV), *C. ficifolius* (CGMMV, MNSV, WMV-2, ZYMV), *C. figarei* (MNSV), *C. meusii* (CGMMV, WMV-2), *C. myriocarpus* (CLSV, CMV, WMV-2), *C. melo* var. *agrestis* (CMV), and *C. zeyheri* (WMV-2). Horváth (1993b) conducted a similar screening of four *Cucurbita* species.

Other viral pathogens that have received special attention for germplasm evaluation include the *Cucurbit yellowing stunting disorder virus* in *Citrullus* and *Cucumis* (Hassan et al. 1990, 1991; López-Sesé and Gómez-Guillamón 2000) and *Cucurbit aphid-borne yellows luteovirus* (Dogimont et al. 1996), *Cucumber vein yellowing virus* (CVYV) (Montoro et al. 2004), and *Kyuri green mottle mosaic virus* (KGMMV) (Daryono, Somowiyarjo, and Natsuaki 2004) in *C. melo*.

Bacterial Pathogens

The major bacterial diseases of cucurbits have been described by Zitter, Hopkins, and Thomas (1996). Important germplasm evaluation reports include those by Barry, Burnside, and Myers (1976) for reaction to bacterial wilt (*Erwinia tracheiphila*) in *Cucumis* by Sowell and Schaad (1979), Hopkins and Thompson (2002) for reaction to bacterial fruit blotch (*Acidovorax avenae* subsp. *citrullii*) in *Citrullus* and by Kůdela and Lebeda (1997) and Olczak-Woltman et al. (2004) for reaction to angular leaf spot (*Pseudomonas syringae* pv. *lachrymans*) in wild *Cucumis* species and *C. sativus*, respectively. Earlier publications describing reactions of *C. sativus* germplasm to angular leaf spot were reviewed by Kůdela and Lebeda (1997) and Olczak-Woltman et al. (2004) with the exception of an extensive screening project conducted on *C. sativus* germplasm by Staub et al. (1989).

Fungal Pathogens

Any comprehensive reviews of research to evaluate cucurbit germplasm for reaction to the diverse array of known fungal pathogens (Zitter, Hopkins, and Thomas 1996) has not been found. However, considerable work has been conducted on screening for reaction to downy and powdery mildews and, to a lesser extent, on many other fungal pathogens. Much of this work is complicated by variation in pathogen isolates and screening protocols.

Downy Mildew. Downy mildew (*Pseudoperonospora cubensis*) damages many cucurbit species with pathotypes differing in host range and virulence. Lebeda and Widrlechner (2003) reviewed germplasm evaluation reports for all cucurbits as part of the establishment of a comprehensive set of differential lines that discriminate pathotypes. In addition to evaluations reviewed by Lebeda and Widrlechner (2003), extensive screening of more than 2000 accessions of *C. melo* has been conducted by Thomas and Jourdain (1992), Thomas (1999), More (2002) and More, Dhakare, and Sawant (2002) evaluated 368 accessions of *C. melo*; Staub and Palmer (1987) evaluated seedling response in nine wild species of *Cucumis*; and Lebeda (1992a, 1992b) and Lebeda and Prášil (1994) evaluated large sets of wild *Cucumis* species and *C. sativus* germplasm. However, to date, efficient and highly effective sources of resistance in *Cucumis* species are unavailable; only significant differences in field resistance have been recorded (Lebeda 1999). Recently, Lebeda and Widrlechner (2004) published the results of downy mildew screening on wild and weedy accessions of *Cucurbita*. Cultivated *C. pepo*, represented by eight groups of morphotypes, expressed significant differences in resistance/susceptibility to *P. cubensis* and both powdery mildews (*G. cucurbitacearum* and *P. xanthii*). Generally, there was an inverse relationship detected in resistance to the two groups of mildews. While zucchini, cocozelle, and vegetable marrow were highly resistant to *P. cubensis*, they had relatively high powdery mildew sporulation. Cultivars with the fruit type acorn, straightneck, and ornamental gourd were quite susceptible to *P. cubensis*; however, they were considered resistant to powdery mildew in laboratory and field evaluations (Lebeda and Křístková 2000).

Powdery Mildew. Another widespread foliar disease of cucurbits is powdery mildew, that is caused by *Podosphaera xanthii* (syn. *Sphaerotheca fuliginea*) and *Golovinomyces cucurbitacearum* (syn. *Erysiphe cichoracearum*). Jahn, Munger, and McCreight (2002) published a comprehensive review of sources and genetic control of resistance to powdery mildew in cucurbits, which included some references to past germplasm evaluations. Two more recent extensive evaluations for reaction of *C. sativus* germplasm to inoculation by *P. xanthii* have been reported by Morishita et al. (2003) and by Block and Reitsma (2005). Among *C. sativus* PI accessions, Lebeda and Křístková (1997) found significant differences in field resistence to both powdery mildews.

For *C. melo*, Pitrat, Dogimont, and Bardin (1998) reviewed the literature on screening germplasm for powdery mildew resistance. Since that time, Fanourakis, Tsekoura, and Nanou (2000) reported results of an evaluation of Greek landraces of *C. melo* for reaction to *P. xanthii*. The only report of a comprehensive screening of wild *Cucumis* germplasm to *P. xanthii* and *G. cucurbitacearum* was contributed by Lebeda (1984) who identified the best resistance in *C. anguria, dinteri, ficifolius*, and *sagittatus*. Recently, Thomas, Levi, and Caniglia (2005) presented the results of screening 266 *C. lanatus* accessions for reaction to race 2 *P. xanthii*, identifying 23 accessions with intermediate levels of resistance. Evaluations of *Cucurbita* germplasm for resistance to powdery mildew have been well summarized by Jahn, Munger, and McCreight (2002). One additional report, examining differences in damage to leaf surfaces and stems and petioles of *C. pepo*, germplasm was published by Křístková and Lebeda (2000b).

Other Foliar Pathogens. Cucurbit germplasm has been evaluated for reaction to other important foliar pathogens, including scab (*Cladosporium cucumerinum*) in wild *Cucumis* species (Staub and Palmer 1987), *C. sativus* (Lebeda 1985), and *C.* species (Strider and Konsler 1965), anthracnose (*Colletotrichum orbiculare*) in *C. sativus* (Staub et al. 1989; Wehner and St. Amand 1995) and *C. lanatus* (Wang and Liu 1989), and target leaf spot (*Corynespora cassiicola*) in *C. sativus* (Staub et al. 1989). Gummy stem blight (*Didymella bryoniae*) damages both foliar and vascular tissues. The results of three extensive evaluations for reaction to gummy stem blight, emphasizing *C. melo* germplasm, have been published (Zhang et al. 1997; Sakata et al. 2000; Wako et al. 2002) that also include reviews of past work. And more recently, Chen, Moriarty, and Jahn (2004b) reported finding resistance to *D. bryoniae* in *C. hystrix*. Higher levels of resistance to *D. bryoniae* were also found in some wild *Cucumis* species (*C. ficifolius, C. melo* var. *agrestis*, and *C. myriocarpus*) (Lhotský, Lebeda, and Zvára 1991). Also inbred pickling cucumber with resistence to *D. bryoniae* was released (Wehner, St. Amand, and Lower 1996).

Fusarium Wilt. Fusarium wilt (*Fusarium oxysporum*) can cause significant economic losses in many cucurbits. Results of germplasm screening have been summarized for *Citrullus* landraces from Botswana (Wang and Zhang 1988) and for a broader sampling of more than 100 *Citrullus* accessions, including *C. lanatus* var. *lanatus*, *C. lanatus* var. *citroides*, and *C. colocynthis* (Huh, Om, and Lee 2002), for *C. sativus* germplasm challenged by isolates representing three different races (Armstrong, Armstrong, and Netzer 1978), for a diverse sampling of *C. melo* cultivars tested against a single race (Zink, Gubler, and Grogan 1983), for selfed progeny of landraces and cultivars against two races (Pitrat et al. 1996), and for Iberian landraces of *C. melo* tested against various local races (Alvarez and González-Torres 1996). Vine decline in *C. melo* can also be caused by *Acremonium cucurbitacearum* and *Monosporascus cannonballus*. Crosby (2001) has evaluated *C. melo* var. *agrestis* for resistance to *M. cannonballus* in Texas, and de C.S. Dias et al. (2001) conducted a field assay of a diverse sampling of *C. melo* accessions and related species in southeastern Spain where highly aggressive isolates of both fungi had been isolated.

Belly Rot, Phytophthora Fruit Rot. Screening has also been conducted to evaluate resistance to two fungal pathogens that can cause fruit damage: belly rot (*R. solani*) and Phytophthora fruit rot (*Phytophthora capsici*). Sloane, Wehner, and Jenkins (1983, 1984) described the most resistant of 1063 *C. sativus* accessions they evaluated for reaction to *R. solani* in field and laboratory tests. Lopes, Brune, and Henz (1999) presented results from an inoculation of 150 *Cucurbita* accessions with *P. capsici*.

Nematodes, Insects, and Mites. Nematodes are serious pests of cucurbits in both field and greenhouse production with the root-knot nematodes (*Meloidogyne* spp.) among the most destructive. Thies (1996) presented an overview of nematode damage on cucurbits and reviewed key evaluations of germplasm conducted to identify sources of resistance to *Meloidogyne*. Two evaluations, unnoted by Thies (1996), have been published. Zhang, Qian, and Liu (1989) presented the results of an extensive screening of *Citrullus* germplasm to three species of *Meloidogyne*, and Dalmasso, Dumas de Vaulx, and Pitrat (1981) published a brief summary of research on wild species of *Cucumis* and *Cucurbita*. In addition, Walters, Wehner, and Barker (1997) characterized a recessive gene for resistance to *Meloidogyne javanica* from *C. sativus* var. *hardwickii*, and resistance has also been reported from *C. hystrix* (Chen and Lewis 2000; Chen et al. 2001).

Elsey (1989) and Robinson (1992) assembled comprehensive reviews of research conducted to evaluate cucurbit germplasm for reaction to insect and mite pests. However, in the intervening years, results from more recent evaluations for reaction to destructive insects and mites have been published. For example, extensive screening of *C. melo* germplasm to identify sources of host-plant resistance to whiteflies (*Bemisia* spp.) has been conducted with promising results (McCreight 1995; Simmons and McCreight 1996). In addition, resistance to whiteflies has been identified among *Citrullus colocynthis* germplasm accessions (Simmons and Levi 2002). *C. melo* germplasm has also been evaluated for tolerance to the melon aphid, *Aphis gossypii* (Bohn, Kishaba and McCreight 1996), and to melonworm, *Diaphania hyalinata* (Guillaume and Boissot 2001), where field tests identified moderate levels of tolerance in some *C. melo* lines and higher levels of resistance in *C. metuliferus* and *C. pustulatus*. Western flower thrips, *Frankliniella occidentalis*, is a serious pest of cucurbits under greenhouse conditions. Balkema-Boomstra and Mollema (1996) identified sources of resistance to this pest within a geographically diverse sample of *C. sativus* germplasm.

Early reports by Leppik (1968), Tulisalo (1972), and Knipping et al. (1975) suggested that wild species of *Cucumis* might harbor valuable sources of resistance to spider mites (*Tetranychus* spp.). To that end, Lebeda (1996) evaluated a broad sampling of wild *Cucumis* accessions and identified useful levels of resistance to *Tetranychus urticae* in *C. africanus* and *C. zeyheri*. Very broad variation for resistance to glasshouse whitefly (*Trialeurodes vaporariorum*) has also been found in wild *Cucumis* germplasm (Láska and Lebeda 1989). Even after the completion of a diverse array of insect and mite evaluations by the 1990s, however, Webb (1998) found only a few cucurbit breeding lines and one cultivar with insect resistance among releases from that time period.

Abiotic Stresses. Extensive evaluations of *C. sativus* germplasm have been conducted to identify sources of tolerance to the herbicides, clomazone (Staub et al. 1991) and atrazine (Werner and Putnam 1977), and to high temperatures (Staub and Krasowska 1990) and moisture deficits (Wann 1992). Wehner (1982) evaluated the ability of *C. sativus* accessions to germinate under low-temperature conditions and determined the heritability of this characteristic based on parent-progeny regressions (Wehner 1984). And for *C. melo* accessions, Hutton and Loy (1985) compared rates of seedling development at 15 and 30°C. Ríos Labrada, Fernández Almirall, and Casanova Galarraga (1998) evaluated Cuban landraces of *Cucurbita moschata* for tolerance to high temperatures and drought, and they developed a stress-tolerance index as part of a larger project to identify germplasm to breed new cultivars for low-input, tropical agriculture.

Sulfur dusting is used to control powdery mildew in *C. melo*, but it can induce severe foliar damage. Perchepied et al. (2004) recently evaluated more than 200 *C. melo* accessions for reaction to sulfur and reported interesting geographic patterns of variation in the distribution of resistant germplasm. They subjected one source of resistance to QTL analysis through an examination of recombinant inbred lines and detected one major and two minor QTL.

8.6.3 Cross-Compatible Breeding Strategies

8.6.3.1 *Interspecific Hybridization: Conventional Approaches*

Transfer of economically important characteristics such as host-plant resistance to diseases and pests or stress tolerance from wild cucurbit species to their cultivated counterparts is one of the most important challenges for breeders. Substantial differences can be found among the three major genera of cultivated Cucurbitaceae (*Cucumis*, *Cucurbita*, and *Citrullus*) with regard to their cross-compatiblities with wild relatives. They exhibit a wide range of crossing barriers at both the presyngamic (failure of pollen tube growth) and postsyngamic (breakdown of embryo development) developmental phases. Many techniques, including embryo and callus culture, bridge crosses, bud pollination, repeated pollination, regulating ploidy levels, and the use of growth regulators, have been employed in attempts to improve the success of interspecific hybridization (Kalloo 1988).

To date, conventional crosses between *C. sativus* and other *Cucumis* species, with the exception of *C. hystrix*, have not been successful; however, somatic hybridization via protoplast fusion, especially asymmetrical fusion, has been suggested as a solution to this problem (Tatlioglu 1993).

Melon, *C. melo*, is not easily crossable with a majority of wild *Cucumis* species in spite of sharing the same base chromosome number and ploidy level with many of those species. In some cases, *in vitro* culture of hybrid embryos (discussed in Section 8.6.5.3) has been successfully employed for their recovery (Sauton 1987).

Similarly, among *Cucurbita* species, conventional interspecific pollination followed by *in vitro* culture of hybrid embryos can result in the production of fertile hybrids (Šiško, Ivančič, and Bohanec 2003). Information related to hybridization among *Cucurbita* species and techniques to overcome crossing barriers and hybrid sterility has been summarized by Lira-Saade (1995).

As the cultivated watermelon, *C. lanatus* var. *lanatus*, is cross-compatible with its congeneric wild relatives, biotechnological methods (discussed in Section 8.6.5) have been applied to primarily increase the fertility of obtained hybrids. Although unique crossing approaches are used for the creation of triploid seedless cultivars (Kihara 1951), such approaches have not been needed for overcoming crossing barriers. Specific examples documenting the application of various conventional techniques to improve the success of wide hybridization follow.

The power of the bridge cross emanates from its ability to exploit the natural potential of crossing ability among species. If two species are not crossable directly, the breeder searches for a third species that is compatible with both. Within the genus *Cucurbita*, *C. moschata* is crossable with many wild and cultivated species, and it has been exploited as a bridge species (Provvidenti, Robinson, and Munger 1978). Three-way crosses have been used to transfer resistance genes for powdery mildew and CMV from *C. martinezii* to *C. pepo*. Initially, *C. martinezii* was crossed with *C. moschata*, then the resulting hybrid was crossed to *C. pepo* (Whitaker and Robinson 1986).

Bridging species can also be used to circumvent sterility in interspecific crosses. Whitaker (1959) found that *C. lundelliana* could be crossed with each of the cultivated species of *Cucurbita*. Rhodes (1959) developed an interbreeding population from crosses involving *C. lundelliana* and the cultivated species: *C. pepo*, *C. argyrosperma*, *C. moschata*, and *C. maxima*. Thus, *C. lundelliana* served as a bridge to transfer genes between species that would otherwise be difficult to cross (Whitaker and Robinson 1986).

Cross-compatibility also depends upon the particular accessions selected to represent a given species in interspecific hybridization. Cultivars of the same species may vary greatly in crossability. For example, crosses with *C. moschata* Butternut were more easily made with Scallop than with other *C. pepo* cultivars (Whitaker and Robinson 1986). In the genus *Cucumis*, there is ample variation within two species (*C. anguria* and *C. zeyheri*) that carry CGMMV resistance. This variability has, in fact, been exploited to overcome hybridization barriers in wild African *Cucumis* species (Visser and den Nijs 1983). These examples and many others (see Raamsdonk, den Nijs, and Jongerius 1989) suggest that biosystematic and gene-pool studies can guide the selection of genotypes to increase the probability of successful interspecific hybridization.

Common problems that occur in the F_1 and early succeeding generations of distant crosses are sterility and poor seed development. Often the embryo does not abort, but the nutritive tissue of the seed fails to develop normally. Embryo culture may be required in such cases (see Section 8.6.5.3). Poorly developed seeds from interspecific hybrids or their progeny often germinate better when their seed coats are removed (Whitaker and Robinson 1986). F_1 hybrid seedlings so obtained from a cross between *C. pepo* Black Jack and *C. martinezii* were resistant to *G. cichoracearum* and CMV as was the male parent (*C. martinezii*) (Metwally, Haroun, and El-Fadly 1996). This method has also been recommended for use when making compatible crosses in *Cucurbita* species (Hong and Hyo-Guen 1994). Interspecific hybrids of *C. moschata* × *C. maxima* are known as Kabocha and have been offered by Sakata Seed Company (Japan) under different cultivar names (e.g., Alguri, Kikusui, Tetsakabuto). Fruits are reported to combine favorable characters from both species, and there is considerable heterosis for degree of female sex expression and for yield (Whitaker and Robinson 1986).

Researchers have also explored the ability of mentor pollen to help overcome presyngamic crossing barriers. In such cases, the stigma is pollinated with a mixture of pollen grains from a compatible species with those from a typically incompatible species possessing important traits. The presence of compatible pollen can facilitate pollen-tube growth. This approach improved hybrid seed set in the cross *C. metuliferus* × *C. africanus* (Kho, den Nijs, and Franken 1980). In *Cucumis*, success has been achieved when mentor pollen combined with chemical treatment was followed by embryo culture (den Nijs and Oost 1980; den Nijs, Custers, and Kooistra 1980).

The use of irradiated pollen has been investigated for various purposes. One experimental approach was based on the assumption that irradiation leads to the selective elimination of chromosomes of *C. melo* in the early stages after pollination. This approach was attempted by Custers and Bergervoet (1984) in crossing *C. sativus* with *C. melo*. However, they found that with increasing doses of irradiation, there was a decrease in the number and size of embryos and endosperm.

Irradiated pollen has also been used to produce haploid plants. Pollen can be irradiated at dosages that retain its capacity for fertilization but destroy its potential to deliver viable chromosomal material to offspring. This approach (pollen irradiated by gamma rays) has been applied to *C. melo* for the induction of parthenogenetic ovule development. Haploid embryos excised at the globular or heart stage were cultured *in vitro* (Sauton 1988), producing haploid melon plants where there was no spontaneous doubling of chromosomes, but there are other reports of the recovery of both haploid and dihaploid plants (Sauton and de Vaulx 1987; Savin et al. 1988; Dirks and van Buggenum 1991).

Hybridization barriers can be overcome by changing ploidy levels, either to restore fertility or to match ploidy levels between two species (Kalloo 1988). Different chromosome numbers in *C. sativus* and other *Cucumis* species are the main reason of their non-crossability (Dane 1991). Polyploidization could favorably influence the constitution of chromosome bivalents in the first stage of the zygotic cell division and to promote subsequent growth and development (Colijn-Hooymans et al. 1994; Lebeda, Křístková, and Kubaláková 1996). Colchicine has been used to restore fertility in sterile *C. maxima* × *C. moschata* progeny (Pearson, Hopp, and Bohn 1951; Whitaker and Robinson 1986). Moreover, Bemis (1973) reported that the original interspecific crosses of *C. moschata* × *C. palmata* were successful at the tetraploid level.

Another potential method to overcome either presyngamic or postsyngamic crossing barriers is the application of growth regulators (AVG, BAP, IBA, 4-CPA) to the peduncles of female flowers. This technique was first applied in crosses among *Cucumis* species (Deakin, Bohn, and Whitaker 1971), but it was without the expected positive results (Chatterjee and More 1991; Lebeda, Křístková, and Kubaláková 1996).

8.6.3.2 *Developing Unique Genetic Stocks*

Bulk segregant analysis (Michelmore, Paran, and Kesseli 1991), in combination with near isogenic lines (NILs) and other unique genetic stocks, has been used in cucumber to identify markers linked to disease resistance (Horejsi, Staub, and Thomas 2000; Park et al. 2000) and to study sex-expression genes (Trebitsh, Staub, and O'Neill 1997; Witkowicz, Urbanczyk-Wochniak, and Przybecki 2003). The cucumber RILs developed by Fazio, Staub, and Stevens (2003) are useful in mapping markers and QTL, but they are not, by themselves, immediately useful to plant breeding programs. Genes for yield components are dispersed throughout these RILs, and the lines themselves are not entirely homozygous (Fazio, Staub, and Stevens 2003). Therefore, it would be desirable to create genetic stocks containing QTL in useful homozygous genetic backgrounds for use as introgression lines. One method to introgress QTL into breeding lines, called "advanced backcross QTL analysis" (AB-QTL analysis), has been proposed by Fulton et al. (2000). This method involves backcross breeding in combination with MAS to create NILs that vary in useful donor alleles in the elite parent background. Once created, these NILs could be used for the efficient introgression of economcially important QTL into elite germplasm (Tanksley et al. 1996; Fulton et al. 1997; Bernacchi et al. 1998a, 1998b). For instance, NILs (BC_2S_3) strategically created in cucumber (e.g., Gy-7 × H19; Serquen, Bacher, and Staub 1997a; Fazio, Staub, and Stevens 2003) could carry single-gene donor introgressions for desirable QTL-alleles (i.e., branching; fruit length) derived from the donor parent (i.e., H-19) but transferred to the genetic background of the elite recurrent parent (i.e., Gy-7). Such genetic stocks would allow for the evaluation of epistatic interactions to determine efficacy of gene action under different environments.

In addition to NILs, Ezura, Kikuta, and Oosawa (1994) have produced aneuploid plants by crossing tetraploid and triploid melon lines followed by *in vitro* culture. From these aneuploid plants, trisomic lines could be developed, which have long served as useful tools in genetic analysis (reviewed by Khush 1973).

8.6.3.3 *Broadening Genetic Diversity*

Genetic diversity is essential for the continued development of commercial cultivars. The NPGS cucumber core collection (147 accessions; Staub et al. 2002) effectively samples genetic diversity at molecular loci to allow for the incorporation genetically diverse exotic accessions into the core

(Horesji and Staub 1999). However, genetic analysis of the NPGS cucumber collection (~ 1350 accessions of worldwide origin) and elite commercial germplasm (> 150 accessions developed between 1846 and 1995) documented reduced diversity among commercial germplasm ($\sim 3\%$ polymorphism in elite commercial germplasm and $\sim 12\%$ across the entire primary gene pool) (Meglic and Staub 1996b). Thus, rigorous genomic characterization of the elite lines suffers from a paucity of genetic variation (Dijkhuizen et al. 1996; Staub et al. 2002), highlighting the need for the genetic enrichment of elite cucumber germplasm to ensure continued potential for plant improvement.

The inbred backcross line (IBL) breeding method and analytical procedures, as originally described by Wehrhahn and Allard (1965), could be applied in combination with MAS for broadening the genetic base of commerical cucumber germplasm. The IBL procedure applies no selection (as opposed to AB-QTL analysis) in the production of BC_2S_3 lines in distinct genetic backgrounds (e.g., U.S., European, and Asian cucumber) with extensive replicated evaluation. A modification of this procedure involves the initial cross of a donor (exotic germplasm) to a recurrent (elite germplasm) parent in a target genetic background followed by a backcross to the recurrent parent to produce the BC_1 (this procedural modification of IBL proposed initially herein). In this modified IBL procedure, germplasm at various stages of introgression (i.e., BC_1 and BC_2) are genotyped by using a standard marker array (López-Sesé, Staub, and Gomez-Guillamon 2003; Mliki et al. 2003) that has proven useful across various gene pools to maximize genetic diversity. For example, the BC_1 progeny could be genotyped by molecular markers and about 200 of the most genetically diverse BC_1 would then be crossed to the recurrent parent to produce 200 BC_2 progeny (either marker-genotyped or chosen at random), which can then be self-pollinated thrice after marker genotyping to obtain BC_2S_3 lines prior to replicated evaluation in appropriate target environments (e.g., European greenhouse cucumber under controlled environments and U.S. processing under open-field conditions) for qualitative and quantitative traits of interest. The best lines could then be further refined (e.g, inbreeding or backcrossing) or used directly in breeding programs (e.g., evaluated for combining ability).

8.6.4 Cross-Incompatible Breeding Strategies

The identification and incorporation of resistance to economically important pests such as root-knot nematode (Thies 1996) and various foliar pathogens (den Nijs and Custers 1990) has been required to sustain cucumber production. However, resistance genes for several crop-limiting pests and pathogens have not been found within the primary gene pool of *C. sativus*.

The incorporation of wild *Cucumis* species to broaden the genetic base of cucumber has been a goal of plant geneticists and breeders for over one hundred years (Robinson and Decker-Walters 1997). Meeting this goal has been daunting because nearly all wild *Cucumis* species are cross-incompatible with *C. sativus* and *C. melo* (Staub et al. 1992). However, these species are potentially important for plant improvement programs as sources of resistance to economically important pathogens such as powdery mildew, downy mildew, anthracnose, and fusarium wilt (Leppik 1966; Lower and Edwards 1986; Kirkbride 1993). In the last 60 years, conventional and biotechnological approaches for overcoming species crossing barriers in *Cucumis* have often been applied without success (Whitaker 1930; Batra 1953; Smith and Venkat Ram 1954; Deakin, Bohn, and Whitaker 1971; Fassuliotis and Nelson 1988). Until recently, repeated attempts at interspecific hybridization of wild *Cucumis* species (e.g., *C. metuliferus*, *C. melo*) with cucumber were either unsuccessful or irreproducable (Fassuliotis and Nelson 1988). Thus, it had been concluded that mating of *C. sativus* with any other *Cucumis* species possessing a chromosome series of $2n = 24$ or 48 would be unsuccessful (Křístková and Lebeda 1995).

However, in 1995, a successful interspecific hybridization was accomplished between *C. hystrix* Chakr. ($2n = 2x = 24$), native to Yunnan Province, China (Chen et al. 1995), and

C. sativus var. *sativus* (Chen et al. 1997b). *Cucumis hystrix* appears to possess economically important traits such as root-knot nematode resistance (Chen and Lewis 2000; Chen et al. 2001), gummy stem blight (*D. bryoniae*) resistance, downy mildew (*P. cubensis*) resistance, unique nutritional qualities, and tolerance to growth under low irradiance and temperature (Chen et al. 2003b; Chen, Moriarty, and Jahn 2004b).

The cross was attempted based on *Cucumis* phylogenetic relationships as revealed by isozyme (Chen et al. 1995, 1997a) and RAPD/SSR (Zhuang et al. 2004) analyses. Because the initial F_1 hybrids ($2n=19$, with 12 and 7 chromosomes contributed by *C. hystrix* (H) and *C. s.* var. *sativus* (C), respectively) were completely heterologous, chromosome-doubling experiments were initiated to restore fertility (Chen and Staub 1997). This doubling of chromosome number in F_1 interspecific hybrid progeny was accomplished by monitoring *in vitro* cultures for somaclonal variation (Chen et al. 1998, 2003b; Chen and Adelberg 2000). Putative tetraploid somaclonal variants were initially identified by their unique morphological characters (e.g., serrated leaf edge) and then by chromosome counts (Chen et al. 2003a, 2003b). Reciprocal differences have been observed between F_1 progeny from *C. sativus* var. *sativus* × *C. hystrix* matings, indicating maternal and paternal inheritance for certain morphological and molecular (RAPD) characters (Chen et al. 2004c). The resulting amphidiploid (HHCC, $2n=4x=38$) was designated as a new synthetic nothospecies (*C.* × *hytivus* Chen) and has been reproduced in subsequent experiments, ultimately resulting in an array of amphidiploids that have been self-pollinated to produce viable seeds (Chen and Kirkbride 2000; Chen et al. 2003b). More recently, hybridization experiments have produced allotriploids ($2n=3x=26$; genome designated as HCC) from amphidoploid × diploid (*C. sativus* var. *sativus*) matings as well as diploid derivatives ($2n=2x=14$), which are fully fertile with *C. sativus* var. *sativus* (Chen et al. 2004a; Qian et al. 2005). The amphidiploid and its allotriploid and diploid derivatives provide a species bridge in *Cucumis*, and a source for broadening the genetic base of *C. s.* var. *sativus* and perhaps eventually also of *C. melo* (Chen et al. 2003a; Zhuang et al. 2004; Qian et al. 2005).

An attempt to transfer resistance to *P. cubensis* from *C. melo* (line MR-1) to *C. sativus* combined the conventional pollination of intact *C. sativus* flowers with *C. melo* pollen with *in vitro* culture of excised young seeds and embryos (Lebeda, Křístková, and Kubaláková 1996), resulting in the regeneration of seven embryos. Five embryos developed small roots and/or shoot meristems. However, only callus formation was observed during the culture of these embryos that is likely due to an increase in ploidy level (Lebeda, Křístková, and Kubaláková 1996). Isozyme analyses of callus derived from superior embryos verified the presence of hybrid zones. Two embryos developed into flowering plants after which their hybrid origin was confirmed through isozyme analyses. Although the morphology of these plants resembled their maternal (*C. sativus*) parent (Lebeda et al., 1999), increased resistance to *P. cubensis* was observed (Lebeda et al., 1999). Lebeda et al. (1999) concluded that the regenerated plants were not symmetrical interspecific hybrids, but they were more likely asymmetrical hybrids (Lebeda et al. 1999). More detailed studies of the mechanism and methodology of interspecific crossing between *C. melo* and *C. sativus* are ongoing (Ondřej, Navrátilová, and Lebeda 2000, 2001, 2002; Skálová, Lebeda, and Navrátilová 2004).

8.6.5 Biotechnological Approaches

8.6.5.1 Haploid Technology

Cucumber (*Cucumis sativus*)

The first cucumber haploids were obtained by stimulating parthenogenesis through pollination with irradiated pollen (Troung-Andre 1988; Niemirowicz-Szczytt and Dumas de Vaulx 1989;

Sauton 1989) and have been characterized by Przyborowski and Niemirowicz-Szczytt (1994). Three weeks after pollination, 0.49–1.70 embryos per fruit were isolated, but only 3.3–7.7% of the haploid embryos developed into haploid plants. Some plants showed abnormal development such as lack of a primary meristem, short internodes, rosette growth habit, or thick curled leaves. To help optimize the efficiency of haploid production, Faris, Nikolova, and Niemirowicz-Szczytt (1999) examined the effects of radiation dosage. Of the three conditions they tested, a dose of 0.1 kGy from a gamma ray (^{60}Co) source stimulated development of the largest number of haploid embryos.

Haploid cucumber plants are infertile and typically do not undergo spontaneous diploidization. Thus, haploids from four different genotypes were treated with colchicine to induce doubled haploids (DH) (Nikolova and Niemirowicz-Szczytt 1996). The following procedure was used in the production of haploid plants: (1) an apical shoot-meristem treatment, (2) the soaking of shoot explants, and (3) the culture of shoot explants on a medium containing colchicine. Optimal results (20.9% DH) were obtained with repeated treatment of apical shoot meristems with colchicine. Many chimeras (28.5%) as well as haploids and tetraploids resulted from this treatment protocol. The DH plants were fertile and gave uniform progeny. Chimeras had decreased fertility when compared to untreated control plants and showed disturbances in meiotic divisions. DH lines have also been occasionally produced by regenerating spontaneous diploid shoots from tissue cultures (Faris et al. 1997; Sztanger et al. 2004).

Haploid cucumber plants have also been produced via *in vitro* gynogenesis with limited success (Gémes-Juhász, Vencze, and Balogh 1996; Gémes-Juhász et al. 2002). By optimizing culture conditions, including a special heat treatment, gynogenic plants of cucumber were successfully produced *in vitro* from unpollinated ovules, including both haploids and spontaneous DH plants directly through embryogenesis. The highest frequency of gynogenesis obtained was 18.4% with rates of plant regeneration of up to 7.1%. Analysis with flow cytometry indicated that 87.7% of the regenerants were haploid.

In vitro conservation and storage of haploid cucumber lines has been attempted by Niemirowicz-Szczytt et al. (2000) who reported that lines at optimal developmental stages could be maintained through timely subculturing for two years.

Melon (Cucumis melo)

Rapid production of homozygous lines is extremely valuable for commercial hybrid melon breeding. However, several generations of inbreeding are typically required to produce such lines. The significance of haploid and DH lines in accelerating inbred-line development has been recognized by melon breeders for many years. Spontaneous haploid production can occur via parthenogenesis, but at a very low frequency (Bhojwani and Razdan 1983).

The recovery of gynogenetic haploid melon plants obtained from an interspecific cross between *C. melo* and *C. ficifolius* was first reported by Dumas de Vaulx (1979). However, efficiency was quite low. *In situ* haploid gynogenesis has been induced by using irradiated pollen (Sauton and Dumas de Vaulx 1987; Cuny et al. 1992) treated with either gamma radiation or soft x-rays (Sauton 1989). The frequency of haploid production was usually less than a few percent of the melon ovules cultured *in vitro*, although frequency was influenced by season and genotypes (Sauton 1988; Ficcadenti et al. 1995) with the best results obtained in summer. Most of plants recovered from gynogenesis were haploid, in contrast to DH plants in other species (see Antoine Michard and Beckert 1997; Gu, Zhou, and Hagberg 2003). Melon haploids are sterile, and chromosome doubling must be induced to obtain fertile homozygous lines. Colchicine is commonly used for this purpose (Yashiro et al. 2002; Lotfi et al. 2003; Yetisir and Sari 2003).

Yashiro et al. (2002) treated lateral shoots of 31 haploid melon plants grown in a greenhouse with a 0.1% colchicine solution. Eventually, 68% of the treated haploid plants were doubled by this method. Lotfi et al. (2003) treated haploid melon plants with colchicine *in vitro*. From 167 micropropagated haploid shoots, 10 diploid and 100 mixoploid plants were produced.

Yetisir and Sari (2003) compared both *in vitro* and *in situ* methods for the diploidization of haploid melon plants. Immersion of *in vitro* plantlets or of single-node explants in a colchicine solution was compared to immersion of shoot tips of greenhouse-grown plants. In addition, single drops of colchicine were applied to lateral buds of greenhouse-grown plants. The diploidization rate achieved by the immersion of shoot tips was about 89%, a rate three times greater than that attained by the treatment of *in vitro* plantlets or single-node explants.

Haploid technology has been put into practice by melon breeders; however, two challenges limit haploidization's progress. Haploid production's frequency is still low with the number of expected haploids per plant ranges from one to five. Recently, Kuzuya et al. (2003) investigated whether resistance to powdery mildew (*P. xanthii*) could be manifest in haploids derived two disease-resistant lines, PMR 45 and WMR 29. Encouragingly, the haploids responded to powdery mildew identically to their diploid progenitors.

Squash (Cucurbita pepo)

The production of haploids from unfertilized ovules of summer squash (*C. pepo*) has been reported by Dumas de Vaulx and Chambonnet (1986), Metwally et al. (1998a). Ovaries from squash plants were collected one day before anthesis and exposed to low temperature (4°C) for zero, two, four, or eight days. Ovules were then cultured on MS medium, supplemented with four concentrations of 2,4-D. After incubation at 25 ± 1°C under 16-h photoperiod for four weeks, the ovules were transferred to an MS medium lacking 2,4-D for four weeks. The highest rate of plant recovery was from ovules that did not receive a cold treatment when cultured on MS medium supplemented with 1 or 5 mg/L 2,4-D. Of the plants recovered, one third of those examined were haploid and the others were DH.

Haploid squash plants have also been derived from anther culture by Metwally et al. (1998b) who excised anthers at the mid or late uninucleate microspore stage from sterilized buds and plated them on twenty different induction media. The most plantlets resulted from an induction medium supplemented with 150 g/L sucrose and 5 mg/L 2,4-D. Root tips from twenty plantlets were cytologically examined, revealing that ten were diploid plants and ten plants were haploid.

In addition, haploids have been obtained by stimulating parthenogenesis through pollination with irradiated pollen (Kurtar, Sari, and Abak 2002). Haploid embryos and plants were obtained with production strongly influenced by gamma ray dosage, embryo development, and maternal genotypes with gamma ray doses of 25 and 50 Gy giving the greatest response. Different shapes and stages of embryos were derived from seeds extracted from fruits harvested four to five weeks after pollination. All of the pointed globular arrow-tip and stick-shaped embryos developed into haploid plants. However, only 53.8% of torpedo and 23.1% of heart-shaped embryos resulted in haploid plants. In contrast, cotyledonary embryos or those with amorphous shapes produced only diploid plantlets.

No publications describing successful haploid production in other species of *Cucurbita* are currently known.

Watermelon (Citrullus lanatus)

Parthenogenetic haploid embryos of Crimson Sweet, Halep Karasi, Sugar Baby, and Panonia F_1 watermelon were obtained after pollination with gamma-irradiated (200 or 300 Gy) pollen (Sari et al. 1994). Some globular and heart-shaped embryos were observed in fruit harvested two to five weeks after pollination. The number of embryos per 100 seeds was highest for Halep Karasi. After *in vitro* culture, Sari et al. (1994) obtained seventeen haploid plants, and DH lines were generated after chromosome doubling with colchicine.

Both direct (chromosome counting) and indirect (flow cytometry, stomatal size, chloroplast number of the guard cells, and morphological observations) methods were used by Sari, Abak, and Pitrat (1999) to determine the most efficient method(s) for ploidy level identification. Their results revealed that all the techniques tested could be successfully used. Although counting chromosomes is cumbersome, producing plants for morphological observations requires considerable time, and flow cytometry is expensive. On the other hand, measurement of stomata and chloroplast counting methods are relatively simple and less labor intensive than cytological or morphological inspection and are recommended as practical alternatives (Sari, Abak, and Pitrat 1999).

8.6.5.2 Polyploid Technology

Cucumber (Cucumis sativus)

Tetraploid cucumbers were discovered among plants regenerated via somatic embryogenesis by Custers, Zijlstra, and Jansen (1990) who noted that tetraploid comprised 11% of the recovered regenerants. More recently, Plader et al. (1998), Ladyzynski, Burza, and Malepszy (2002) have reported even higher frequencies of tetraploid generation after *in vitro* culture with some cases reaching over 50%.

Triploid plants have been regenerated from embryos isolated from crosses between diploid and tetraploid plants (Mackiewicz et al., 1998; Malepszy et al., 1998). *In vitro* culture of late heart-stage triploid embryos isolated from reciprocal crosses of tetraploid and diploid forms of three cucumber lines produced variable results, depending on genotype and the direction of the cross. Five main obstacles that decrease the probability of obtaining triploid plants have been reported (Malepszy et al. 1998): (1) inability of embryos to germinate; (2) non-rooting plantlets; (3) albinism; (4) mixoploid chimeras; and (5) death after transplanting to soil. The lowest failure rate occurred when the tetraploid was used as the female parent. Depending on the line, 11.7–45.8% of embryos obtained from maternal tetraploids developed into mature plants. Surprisingly, triploid plants did not show visible abnormalities. Flow-cytometry analyses of DNA content revealed that normal plants were triploid while poorly growing and abnormal plants were hexaploid or mixoploid (Malepszy et al. 1998). Triploid plants of various origins were characterized (Mackiewicz et al. 1998) where the parthenocarpy and seedlessness, larger floral corollas and fruits, increased production of female flowers, new classes of pollen size, and morphology, and decreased pollen fertility were reported as unique traits. Notably among these triploid plants, parthenocarpic fruit set ranged between 13 and 41% without pollination and a 9–11-fold increase in fruit set as a result of pollination by diploid plants.

Melon (Cucumis melo)

Melon has a basic chromosome number of $x = 12$, and cultivated melons are mostly diploid ($2n = 24$), but triploids ($2n = 36$) have also been reported (Dane 1991) with the triploids resulting from hybridization between diploid and tetraploid ($2n = 48$) lines. Tetraploid melons have been discovered spontaneously growing in field and greenhouse plantings (Suzuki 1958; Nugent and Ray 1992; Nugent 1994) and have also been induced by colchicine treatment of germinating seedlings (Batra 1952) although their incidence is generally quite low. In addition, tetraploids can occur as a result of adventitious-shoot organogenesis (Bouabdallah and Branchard 1986) and among plants regenerated via other *in vitro* culture systems, including somatic embryogenesis and shoot-primordial aggregates (Ezura et al. 1992b). The frequency of tetraploidy generated by somatic embryogenesis and adventitious shoot organogenesis can be remarkably high, up to 30% of regenerated plants (Ezura et al. 1992a, 1992b; Adelberg et al. 1994), which is favored by selective regeneration conditions (Ezura and Oosawa 1994a, 1994b). Tetraploid plants have been characterized by low fertility, large male and female flowers, thickened leaves, short internodes, round seeds, and flat, lighter fruits, and no significant improvement over diploid fruits have been displayed (Ezura et al. 1992a).

Although triploid melon plants are produced by crossing tetraploid and diploid lines, their frequency is relatively low (Suzuki 1959, 1960). However, when seeds produced by crossing diploid and tetraploid plants are sown *in vitro*, triploid plants can be obtained at comparatively high frequencies (Ezura, Amagai, and Oosawa 1993). Pollen fertility in triploid melons is low. Moreover, while triploid lines cannot set fruit by simple self-pollination, they are reported to set fruit by self-pollination following growth-regulator treatment (Ezura, Amagai, and Oosawa 1993) or when pollinated by adjacent diploid pollen sources (Adelberg et al. 1995). The fruits of triploids also have shown no significant improvement over those of diploids.

Watermelon (Citrullus lanatus)

Seedless watermelon fruits have been produced either by induction of triploid lines, as reviewed by Compton, Barnett, and Gaba (2004), or by using pollen irradiated with soft X-rays on diploid parents (Sugiyama and Morishita 2000a, 2000b; Sugiyama, Morishita, and Nishino 2002a). Triploid seeds are produced by crossing diploid and tetraploid plants (Kihara 1951). Traditionally, tetraploid plants have been generated by treating diploid seedlings with colchicine (Kihara 1951). However, that method is difficult; thus, only a few tetraploid breeding lines are currently used to produce commercial triploid hybrids.

As with other cucurbits, tetraploid watermelons can be identified among plants regenerated *in vitro* from diploids (Compton, Barnett, and Elmstrom 1994; Zhang, Rhodes, and Adelberg 1994a), and a method to identify tetraploids *in vitro* has been developed (Compton, Gray, and Elmstrom 1996; Compton, Barnett, and Gray 1999) based on differences in the number of chloroplasts per guard-cell pair in leaves (Compton, Gray, and Elmstrom 1996). The mean number of chloroplasts for tetraploid regenerants was 19.1, whereas diploids averaged 11.2. An enhanced method was developed by painting the lower epidermis of intact *in vitro*-derived leaves with fluorescein diacetate and observing fluorescence of guard cell chloroplasts under a microscope with UV light in comparison to fluorescence from known diploid cultivars and tetraploid breeding lines (Compton, Barnett, and Gray 1999).

Plant regeneration protocols in watermelon have been improved and more reliable protocols to produce tetraploids developed to expand the genetic diversity of tetraploid breeding lines (Compton 1999, 2000; Chaturvedi and Bhatnagar 2001). By germinating embryos in darkness, Compton (1999) significantly improved the number of explants that then produced harvestable shoots during a six-week incubation period on a shoot-regeneration medium under a 16-h photoperiod. The organogenic competence of different explant sizes and sources from young watermelon seedlings was determined by calculating the percentage of various cotyledon explants that produced adventitious shoots (Compton 2000). Although about 52% of explants prepared from the proximal region of cotyledons formed shoots, only about 6% of distal explants did so. Shoot formation was limited to the proximal end of basal explants but was not restricted to any specific region on distal ones. This study indicated that watermelon lines which respond poorly to *in vitro* procedures may have fewer cells competent for shoot regeneration, requiring special care during explant preparation.

8.6.5.3 Embryo Rescue

Cucumis

To overcome post-zygotic barriers to interspecific hybridization, immature zygotic embryos can be removed from developing seeds and cultured *in vitro* (Figure 8.14). Recent progress, problems, and future trends in the development of embryo-rescue and ovule culture methods for *Cucumis* and their

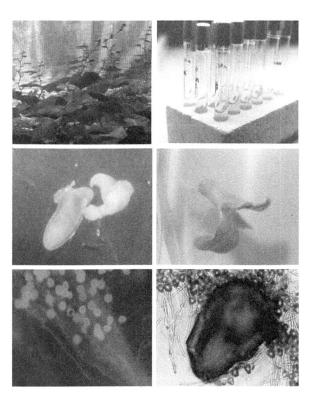

Figure 8.14   (See color insert following page 304.) Embryo culture in the genus *Cucumis* (adapted from Skálová, D., Lebeda, A., and Navrátilová, B., In *Progress in Cucurbit Genetics and Breeding Research, Proceedings of Cucurbitaceae 2004, the 8th EUCARPIA Meeting on Cucurbit Genetics and Breeding*, A., Lebeda and H.S., Paris, eds., Olomouc, Czech Republic: Palacký University in Olomouc, pp. 415–430, 2004.); A=mother flowers (*Cucumis sativus*) in the glasshouse; B= isolated embryos in the test-tubes in the cultivation room; C=germinated embryo (*Cucumis melo*, cv. Solartur) 2 weeks after self-pollination; D=embryo (*Cucumis melo*) after 3 weeks; E= *in vitro* pollination in the cross-pollination between *Cucumis sativus* (female flowers) and *Cucumis melo* (male flowers); F=germinated pollen tubes (*Cucumis melo*, colored FDA).

utilization in interspecific hybridization were summarized above and discussed in detail by Chen and Adelberg (2000) and Skálová, Lebeda, and Navrátilová (2004). Key developments in the utilization of embryo rescue for interspecific hybridization in the genus *Cucumis* are summarized in Table 8.14.

Cucurbita

Several attempts have been made to produce interspecific *Cucurbita* hybrids with the assistance of embryo rescue, following techniques reviewed by Rakoczy-Trojanowska and Malepszy (1989a) and Rakoczy-Trojanowska, Plęder, and Malepszy (1992). By using this strategy, one could expect novel traits to be introgressed or the generation of hybrid genotypes with entirely new characteristics. Limited success in obtaining fully developed seeds that germinated and produced viable plants has been reported for a few interspecific combinations, including *C. andreana* × *C. ficifolia* (Whitaker 1954), *C. lundelliana* × *C. moschata* (Whitaker 1959), and *C. maxima* × *C. ecuadorensis* (Paran, Shifriss, and Raccah 1989). Embryo rescue was employed in transferring resistance to powdery mildew, CMV, and WMV from *C. ecuadorensis* to *C. pepo* (de Vaulx and Pitrat 1980). However, the successful rescue of interspecific *Cucurbita* hybrid embryos has also been reported for many other combinations, including *C. pepo* × *C. moschata* (Wall 1954; de Oliveira et al. 2003;

Table 8.14   Key Examples of the Use of Embryo Rescue or Embryos Production in Conjunction with Interspecific Hybridization in *Cucumis*

Species	Results	Reference
C. sativus × *C. melo*	Embryos (5 callus formation, 2 flowering plants)	Lebeda, Krístková, and Kubaláková (1996), Lebeda et al. (1997, 1999); Ondrej, Navrátilová, and Lebeda (2001)
C. sativus × *C. hystrix*	Plants 2*n* = 19 (sterile), *Cucumis* × *hytivus* 2*n* = 38 (amphidiploid, fertile)	Chen and Staub (1996), Chen and Kirkbride (2000); Chen et al. (2003a b)
C. metuliferus × *C. africanus*	Fruit set with embryos	Custers, den Nijs, and Riepma (1981)
	Embryos, plants	Wehner, Cade, and Locy (1990)
C. metuliferus × *C. melo*	Embryos	Fassuliotis (1977)
	Fertile F$_1$	Norton and Granberry (1980)
C. metuliferus × *C. zeyheri*	Fruit set, some embryos	Custers and den Nijs (1986)
C. melo × *C. metuliferus*	Embryos, plants	Wehner, Cade, and Locy (1990)
	Pollen tubes penetrated into the upper part of the style	Beharav and Cohen (1994)
	Fruit set (irradiation pollen, high irradiation dose increases fruit set)	Beharav and Cohen (1995)
C. melo × *C. ficifolius*	Embryos	Kho, Franken, and den Nijs (1981)
C. hystrix × *C. sativus*	Fertile plants (4*n*)	Chen et al. (1998)
C. sagittatus × *C. melo*	Embryos	Deakin, Bohn, and Whitaker (1971)

Source: Based on Skálová, D., Lebeda, A., and Navrátilová, B., In *Progress in Cucurbit Genetics and Breeding Research, in Proceedings of Cucurbitaceae 2004, the 8th EUCARPIA Meeting on Cucurbit Genetics and Breeding*, A. Lebeda and H. S. Paris, Eds., Placký University in Olomouc, Oloumouc, Czech Republic, 415–430, 2004.

Šiško, Ivančič, and Bohanec 2003), *C. pepo* × *C. martinezii* (DeVaulx and Pitrat 1979; Metwally, Haroun, and El-Fadly 1996), *C. maxima* × *C. pepo* (Rakoczy-Trojanowska and Malepszy 1986, 1989b; Šiško, Ivančič, and Bohanec 2003), *C. moschata* × *C. maxima* (Kwack and Fujieda 1987), *C. ficifolia* × *C. maxima* (El-Mahdy, Metwally, and El-Fadly 1991; Šiško, Ivančič, and Bohanec 2003), *C. maxima* × *C. foetidissima* (Rakoczy-Trojanowska, Plęder, and Malepszy 1992), and *C. argyrosperma* × *C. moschata* (Šiško, Ivančič, and Bohanec 2003).

8.6.5.4   *Protoplast Culture and Somatic-Cell Fusion*

In melon, protoplasts were first prepared and cultured from leaves (Moreno et al. 1986), cotyledons (Roig et al. 1986b; Fellner and Lebeda 1998), and hypocotyls (Fellner and Lebeda 1998). Since that time, there have been successful reports of plant regeneration from cultured protoplasts (Li et al. 1990; Debeaujon and Branchard 1992; Tabei, Nishio, and Kanno 1992), but reported regeneration frequencies remained low, Clearly, efficient regeneration protocols from protoplasts are needed for protoplast culture and for somatic-cell fusion to be of practical importance (Figure 8.15 and Figure 8.16). Recent progress and future trends in protoplast culture for *Cucumis* and *Cucurbita* species were reviewed and discussed in detail by Gajdová, Lebeda, and Navrátilová (2004).

In cucumber, Burza and Malepszy (1995) described a protocol for the isolation and culture of protoplasts from embryogenic calli and embryogenic suspension cultures (ESC). From these isolated protoplasts, they observed a high degree of direct embryogenesis with up to ca. 5,000 embryo structures/g tissue. Some embryos developed into plants after six to eight weeks in culture. ESC-derived plants, when transferred into the glasshouse, flowered normally and set seed. Later,

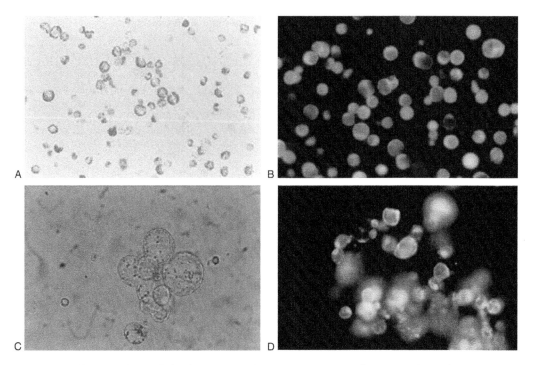

Figure 8.15   (See color insert following page 304.) Protoplast culture of *Cucumis melo* (line MR-1) (adapted from Gajdová, J., Lebeda, A., and Navarátilová, B., In *Progress in Cucurbit Genetics and Breeding Research, Proceedings of Cucurbitaceae 2004, the 8th EUCARPIA Meeting on Cucurbit Genetics and Breeding*, A. Lebeda and H.S. Paris, Eds., Czech Republic: Palacký University, Olomouc, pp. 441–454, 2004.) A = mesophyll protoplasts immediately after isolation; B = viability of callus protoplasts after isolation (fluorescence staining with FDA); C = dividing hybrid cell—product of protoplasts fusion between *Cucumis melo* and *Cucumis metuliferus*, 7 days after fusion; D = viability of mesophyll protoplasts (13 days after isolation, FDA staining).

various protocols and enzyme mixtures were described for protoplast isolation from cotyledons and hypocotyls of cucumber (Fellner and Lebeda 1998).

With the culture systems described by Burza and Malepszy (1995), various attempts have been made to fuse protoplasts and create somatic hybrids. Roig et al. (1986a) attempted to fuse protoplasts of *C. melo* and *C. myriocarpus* and obtained putative hybrid plants, but hybridity was not confirmed. Yamaguchi and Shiga (1993) fused protoplasts of melon (*C. melo*) and pumpkin (interspecific hybrid *C. maxima* × *C. moschata*), and they also obtained putative somatic hybrid plants. At an early developmental stage, the regenerated plants included a chromosomal segment from *Cucurbita*, but it was evidently eliminated at a later stage of growth. Jarl, Bokelmann, and DeHaas (1995) attempted asymmetric protoplast fusion between *C. melo* and *C. sativus*, but recovered only callus. Molecular analysis of the callus showed the incorporation of genetic materials from cucumber to melon.

Cotyledonary protoplasts from an albino *C. melo* mutant were electrofused with protoplasts of *Cucumis anguria var. longipes* (Dabauza et al. 1998). Selection of putative somatic hybrids was based on competence of the albino cells to grow and regenerate shoots when combined with the photosynthetic wild-type cells. The ITS region of *C. anguria* ribosomal DNA was sequenced to design species-specific primers, which were used to distinguish between parental lines and fusion products by PCR amplification. All the organogenic lines characterized by this method proved to be somatic hybrids. Three of 16 selected lines produced shoots with

Figure 8.16   Protoplast culture of *Cucumis sativus* (line 6514) (Adapted from Gajdová, J., Lebeda, A., and Navrátilová, B., In *Progress in Cucurbit Genetics and Breeding Research, Proceedings of Cucurbitaceae 2004, the 8th EUCARPIA Meeting on Cucurbit Genetics and Breeding*, A., Lebeda, H. S., Eds., Palacký University, Olomouc, Czech Republic, pp. 441–454, 2004. With permission.) A= mesophyll protoplasts immediately after isolation; B=first division and viability of mesophyll protoplasts, 11 days after isolation (fluorescence staining with FDA); C=regenerated microcallus, 62 days after isolation; D=leaf-derived callus, material for protoplast isolation; E=*in vitro* plants used as a source of protoplasts.

albino and green sectors. Eleven lines remained green, but shoots developed abnormally and did not produce roots *in vitro*. Two hybrid lines regenerated normal shoots, but with a limited ability to produce roots *in vitro* or *in vivo*.

Cotyledonary protoplasts from an albino *C. melo* mutant were also electrofused with protoplasts of *C. myriocarpus* (Bordas et al. 1998). Putative somatic hybrids were selected on the basis of albinism complementation and the differential behavior of protoplast-derived cells of each of the parental species in a defined sequence of culture media. Somatic hybrids were characterized at a molecular level, and the green calli recovered after fusion were shown to be interspecific hybrids. RFLP patterns that identified both the *C. melo* and *C. myriocarpus* genomes could be obtained by digestion of PCR-derived DNA fragments containing the 18S–25S ribosomal DNA (rDNA) spacer region with the endonuclease Sau3A. The selected calli recovered after fusion contained part of the *C. myriocarpus* genome. Because these calli also showed the RFLP melon pattern, they were clearly a combination of both parental genomes.

8.6.5.5 Somaclonal Variation

The selection of useful phenotypes, known as somaclonal variants, created by revealing cryptic, somatic genetic variation through *in vitro* culture or through the *de novo* generation of variation as part of the culture process became a prominent topic for research beginning in the late 1970s, in part because of successes reviewed by Larkin and Scowcroft (1981). The following paragraphs summarize efforts to document and, in one case, to exploit somaclonal variation in cucurbits.

Cucumis sativus

Plader et al. (1998) examined somaclonal variation in a cucumber line subjected to five different *in vitro* culture systems: (1) micropropagation (MP); (2) direct leaf-callus regeneration (DLR); (3) leaf-callus regeneration (LCR); (4) recurrent leaf-callus regeneration (RLCR); and (5) direct proto- plast regeneration (DPR). They measured the frequency of new phenotypes that appeared in R_1 lines and the stability of the rDNA region by using five probes. MP was not subject to change while DLR and LCR caused only infrequent changes. The highest frequencies of change arose through DPR (90% of lines) and RLCR (43%). Tetraploids were produced only through LCR (4.7%) and RLCR (28%).

The effects of seven *in vitro* culture protocols on somaclonal variation were also compared in an inbred cucumber line by Ladyzynski, Burza, and Malepszy (2002). They regenerated plants by using the following systems: 12- and 18-month-old liquid culture of meristematic clumps (LMC 12 and 18); 10-month-old, embryogenic, cytokinin-dependent suspension (CDS); 18-month-old, embryogenic, cytokinin-dependent suspension in medium with a modified NH_4^+/NO_3^- ratio (CDS 1.7); 12-month-old, embryogenic, auxin-dependent suspension (ADS); 36-month-old, embryogenic, auxin-dependent suspension in medium with a modified NH_4^+/NO_3^- (ADS 1.7); and recurrent leaf-callus regeneration (RLC). Differences were noted among the various systems for ploidy levels in R_0 plants and by the segregation of new morphological traits in the R_1 gener- ation and the germination ability of R_1 seeds. While segregating R_1 families with new phenotypes were most numerous from CDS (62.5%) and LMC18 (57.9%), less so from CDS1.7 (35.7%), the smallest number was found from LMC12 (11.1%) and RLC (3.4%). Among the new phenotypes, Ladyzynski, Burza, and Malepszy (2002) noted some truly novel traits, including ginkgo like leaf (*gll*), yellow-green chlorophyll mutants (*y-gc*), and flowers with serrate corolla margins (*smc*).

In addition, tetraploid and mixoploid plants were regenerated from ADS1.7 and ADS (100%), but only tetraploids were regenerated from CDS and RLC. There were no changes of ploidy after LMC12, LMC18, or CDS1.7.

Cucumis melo

High frequencies of somaclonal variation among melon plants regenerated after *in vitro* culture have been reported (Moreno and Roig 1990; Ezura et al. 1992a, 1992b). Selection among soma- clonal variants to identify desirable traits at the germination or seedling stage can be performed effectively as demonstrated by Ezura et al. (1995). They were able to select somaclonal variants with low-temperature germinability from melon plants regenerated via somatic embryogenesis.

Selected variants also set larger fruits than did the original plants used for somatic embryogenesis. By using those mutants as a source of parental lines, a commercial F$_1$ hybrid cultivar was created.

8.6.5.6 Genetic Transformation

Cucumber (Cucumis sativus)

Transgenic cucumber plants were first produced by *Agrobacterium rhizogenes*-mediated transformation via somatic embryogenesis (Trulson, Simpson, and Shahin 1986). Subsequent reports of the production of transgenic cucumbers mediated by *A. tumefaciens* via somatic embryogenesis include those by Chee (1990) and Tabei et al. (1994). Transgenic cucumber plants were also obtained from hypocotyl explants inoculated with *A. tumefaciens,* via organogenesis (Nishibayash, Kaneko, and Hayakawa 1996b), in experiments where acetosyringone enhanced the transformation efficiency. Recently, Rajagopalan and Perl-Treves (2005) reported an increased efficiency of transformation resulting from special explant wounding methods and selection protocols.

Biolistic transformation of cucumber has also been reported (Chee and Slighton 1992; Schulze et al. 1995). Neomycin phosphotransferase (*nptII*) and the reporter, glucoronidase *GUS* genes were used in bombarding a highly embryogenic cell-suspension culture by using a particle gun, and transgenic plants were regenerated via embryogenesis.

Improvement of virus resistance by the introduction of the CMV coat protein (CP) gene was the first example of transgenic modification of a useful trait in cucumber (Chee and Slighton 1991). Nishibayashi, Hayakawa, and Nakajima (1996a) also introduced a CMV-O CP gene into cucumber plants by using a Ti-*Agrobacterium*-mediated transformation. Progeny of a cross between cv. Sharp 1 and transgenic plants possessing the CP gene displayed strong resistance to cotyledonary inoculation by a CMV-Y strain; although, both the control (Sharp 1) and segregating CP($-$) plants displayed foliar disease symptoms 5–6 days after CMV-Y inoculation. However, the transgenic cucumber plants displayed no direct resistance to ZYMV, yet transgenic plants did show a reduced degree of disease symptom development following a double inoculation of CMV and ZYMV.

Various chitinase-encoding genes have also been introduced into cucumber to enhance resistance to fungal pathogens (Punja and Raharjo 1996; Raharjo et al. 1996; Tabei et al. 1998). For example, Tabei et al. (1998) introduced rice chitinase cDNA (*RCC2*) driven by the CaMV 35S promoter into cucumber. Sixty elongated shoots were examined for the presence of the integrated *RCC2* gene. Of these, 20 were tested for resistance against gray mold (*Botrytis cinerea*) by conidial infection; 15 exhibited a higher level of resistance than did the non-transgenic plants. The three transgenic cucumber strains exhibiting the highest resistance against *B. cinerea* completely inhibited disease transmission. Different patterns of gene expression were observed among the highly resistant strains. One strain inhibited appressoria formation and hyphal penetration. Another permitted hyphal penetration but restricted further hyphal invasion. These transgenic cucumber strains have been further characterized by Ishimoto et al. (2002), Koga-Ban et al. (2004).

Szwacka et al. (2002) produced transgenic cucumber plants with stable integrated constructs consisting of the cauliflower mosaic virus 35S promoter and thaumatin II cDNA, which codes for a unique protein with a sweet flavor. Inter- and intra-transformant variability in the expression of the thaumatin II gene was observed. Observed variation was independent of integrated copy number. Variation in thaumatin II protein accumulation levels in ripe fruits and a lack of correlation between protein and mRNA levels were both noted, suggesting that protein production may be controlled both during transcription and translation. Transgenic fruits accumulating thaumatin II exhibited a sweet phenotype, and a positive correlation between thaumatin accumulation levels and the intensity of perceived sweetness was reported.

A novel use for transgenic cucumber fruits was proposed by Lee et al. (2003a), who introduced the CuZnSOD cDNA (*mSOD1*) gene from cassava to produce superoxide dismutase (SOD) for human cosmetic applications (putative anti-aging agent). They employed *Agrobacterium*-mediated transformation with an ascorbate oxidase promoter that gives high levels of gene expression in fruits. Southern blot analysis confirmed that the *mSOD1* gene was properly integrated into the nuclear genomes of three cucumber plants. It was highly expressed in transgenic cucumber fruits; whereas, it was expressed at a low level in transgenic leaves. SOD-specific activity (units/mg protein) in transgenic fruits was approximately three times higher than were those of non-transgenic plants, suggesting that cucumber fruits with elevated levels of SOD could serve as a functional cosmetic material.

Endogenous ethylene levels have been correlated with different floral sex phenotypes in cucumber (Perl-Treves 1999). Rajagopalan et al. (2004) recently transformed cucumber plants for overexpression of an ethylene receptor, encoded by the cucumber *CS-ERS1* gene, and produced two families segregating for plants with more female flowers produced earlier in plant development. These plants also exhibited accelerated yellowing of detached leaves. This approach could ultimately lead to earlier, high-yielding cultivars.

Field tests of these transgenic cucumber have been conducted in the United States (29 records) (Anonymous 2005). Most of these reports involve plants transformed with CP genes from CMV, ZYMV, and WMV-2, but they also include genes that enhance salt tolerance.

Melon (Cucumis melo)

Various *Agrobacterium*-mediated and particle-bombardment transformation procedures have reported in melon (reviewed by Guis et al. 1998), and *Agrobacterium*-mediated methods have been practically used for producing transgenic melon plants. The first successful *Agrobacterium*-mediated transformation event produced conferred kanamycin resistance in melon by introducing the *NPT-II* gene (Fang and Grumet 1990). The introduction and expression of the reporter gene, *GUS*, has also been noted (Dong et al. 1991; Valles and Lasa 1994).

However, from those early studies, two problems in melon transformation became clear. First, many of the transgenic melon plants were found to be tetraploid. In response, Guis et al. (2000) developed a simple and efficient regeneration system, facilitating the production of diploid transformants at a high rate.

A second problem was the low overall frequency of transformation events because of the presence of escapes (Guis et al. 1998). In initial studies, transgenic plants were generated via adventitious-shoot organogenesis. To reduce the frequency of escapes, an alternative regeneration system was needed. Several groups reported the production of somatic embryos from melon cell-suspension cultures (Oridate and Oosawa 1986). Although somatic embryogenesis can sometimes lead to problems such as abnormal embryos and hyperhydricity, the liquid-culture system is considered useful for efficient selection of transformed tissues as whole explants absorb antibiotics more easily when suspended in liquid media than when cultured on solidified media (Ezura, unpubl. results). Recently, Akasaka-Kennedy, Tomita, and Ezura (2004) published an efficient transformation and plant regeneration system via somatic embryogenesis for two melon types of *C. melo* subsp. *melo* Group Cantalupensis (i.e., vars. *cantalupensis* and *reticulata*). By following their protocol, transgenic plants were successfully produced at a rate greater than 2.3%, sufficient for practical use.

Other approaches to improve the frequency of transformation events have been proposed. Galperin (2003) screened melon genotypes for ease of transformation and regeneration and noted variation between 0.4 and 1.5 transgenic shoots per explant. Ezura et al. (2000) noted that during *Agrobacterium* inoculation, explants produced ethylene. By adding an ethylene biosynthesis inhibitor, AVG, to the co-cultivation medium, they reduced ethylene production from the explants, resulting in increased transformation efficiency. And Atarés et al. (2004) recently described a protocol that employs polyethylene glycol to increase direct (non-*Agrobacterium* mediated) transformation efficiency of melon protoplasts.

The first useful trait in melon modified by genetic transformation was virus resistance conferred by the expression of viral CP genes. A CP gene from CMV was introduced into a commercial melon variety (Yoshioka et al. 1992; Gonsalves et al. 1994), and the transgenic plants showed improved resisitance to CMV (Yoshioka et al. 1993; Gonsalves et al. 1994). This CP-mediated strategy has been applied to improve the resistance to other cucurbit viruses in melon. Transgenic melon plants with a CP gene from ZYMV showed resistance to ZYMV and other potyviruses (Fang and Grumet 1993). Transgenic melon plants over-expressing both CPs of ZYMV and WMV-2 were produced and subjected to extensive field trials (Clough and Hamm 1995). A significant reduction in disease incidence in the transgenic lines was observed. In addition to CP-mediated virus resistance, ribozyme-mediated resistance has been used to produce virus-resistant melons. Transgenic plants with a polyribozyme construct showed resistance to a natural CMV infection (Plages 1997) and to WMV (Huttner et al. 2001). Gaba, Zelcher, and Gal-On (2004) has summarized the results from many field trials of these transgenic lines with virus resistance.

Recently, three genes, *Vat, Nsv,* and *Fom-2,* which are responsible for major disease resistance in melon, have been isolated by map-based cloning. *Vat* confers a double resistance, resistance to plant colonization by the important pest, melon/cotton aphid, *Aphis gossypii,* and to virus transmission by *A. gossypii* (Pauquet et al. 2004). Functional validation was obtained by the stable transformation of susceptible melons with *Vat. Nsv* confers resistance to MNSV (Garcia-Mas et al. 2004), a single-stranded RNA virus that infects cucurbits grown under glass.

Transformation for disease resistance has not been restricted to viruses and their vectors. The third gene, *Fom-2,* confers resistance to races 0 and 1 of soil-borne fungal pathogen, *Fusarium oxysporum* f. sp. *melonis* (Joobeur et al. 2004), which causes significant losses in cultivated melon.

Another major fungal pathogen, *Pseudoperonospora cubensis,* the causal agent of downy mildew in cucurbits (i.e., cucumber, melon, watermelon, and squash), can be controlled through transgenic manipulation. The wild melon line, PI 124111F, is highly resistant to *P. cubensis,* through its enhanced expression of the resistance genes, *At1* and *At2,* which encode glyoxylate aminotransferases. These enzymes are important in photorespiration. Transgenic melon plants overexpressing either *At1* or *At2* displayed enhaced glyoxylate aminotransferase activity and remarkable resistance against *P. cubensis* (Taler et al. 2004).

The use of genetic transformation to improve tolerance to abiotic stresses has also received attention. The *Saccharomyces cerevisiae HAL1* gene was introduced to two melon cultivars (Bordas et al. 1997). It confers salt tolerance to yeast by increasing intracellular K + and decreasing Na$^+$ levels by an unknown mechanism (Rios et al. 1997). Primary transformants carrying the *35S:HAL1* construct exhibited improved rooting in a NaCl-containing medium with respect to the control, but no clear differences were observed in aboveground plant parts.

Modification of postharvest fruit characteristics has been another target of genetic engineering in melon. Postharvest physiology has been extensively studied in melon (Martinez-Madrid et al. 1999; Yang and Oetiker 1998), and key genes have been isolated and characterized. For example, the aminocyclopropane-1-carboxylic acid (ACC) oxidase gene (*Cm-ACO1*) encodes the ACC-oxidase emzyme, which catalyizes the conversion of ACC to ethylene (Balague et al. 1993), a critical phytohormone responsible for fruit ripening. The antisense version of *Cm-ACO1* was introduced to Charentais type melon (Ayub and Guis 1996). By supressing the expression of *Cm-ACO1,* ethylene production was greatly reduced in transgenic fruit, and ripening was delayed as expected. Transgenic fruits could be stored for at least two weeks without over-ripening while the wild-type fruits became decomposed under the same conditions.

As previously noted for cucumber, melon lines also display a range of sex phenotypes that are subject to ethylene control (Rudich 1990). Direct evidence of the importance of endogenous ethylene production in floral sex determination was presented by Papadopoulou et al. (2002).

They transformed the melon cultivar, Hale's Best Jumbo, with the *ACS* gene encoding ACC synthase, responsible for the first step in ethylene biosynthesis, and with an *Arabidopsis etr1-1* gene that blocks ethylene perception. The *ACS* transformants showed increased ethylene production and increased femaleness, and the *Etr1-1* melons exhibited phenotypes associated with ethylene insensitivity and failed to produce pistillate flowers. In practice, the use of such genes may facilitate the development of melon lines with earlier fruit production.

Field tests of transgenic melons have been conducted (143 records; Anonymous 2005). The reports have predominantly been of transgenic plants exhibiting virus resistance through the expression of CP genes, but they have also included transformants exhibiting altered fruit ripening via the expression of S-adenosylmethione hydrolase from *E. coli* and those exhibiting male sterility from the expression of phosphinothricin acetyl transferase or glucanase.

Squash (Cucurbita maxima and C. pepo)

The production of transformants in *Cucurbita* species has been limited by the development of appropriate tissue-culture regeneration protocols. Juretic and Jelaska (1991) reported the induction of embryogenic callus derived from summer squash (*C. pepo*) hypocotyl segments, and Chee (1991) reported that summer squash cultivars were regenerated via somatic embryos produced from calli derived from shoot apices. *Cucurbita pepo* cultivars have also been regenerated via somatic embryogenesis from cotyledonary explants (Gonsalves, Xue, and Gonsalves 1995). An efficient regeneration protocol using cotyledonary explants of winter squash (*C. maxima*) has also been developed (Lee, Chung, and Ezura 2002, 2003b). Squash regeneration via organogenesis has also recently been reported (Anathakrishnan et al. 2003).

Methods to produce transformants of *Cucurbita* containing viral CP genes via *Agrobacterium*-mediated transfer have been described by Pang et al. (2000) based on systems initially developed by Tricoli et al. (1995). By using various protocols, both those published in the scientific literature and others found only in patent documentation, transgenic *Cucurbita* lines exhibiting CP-mediated resistance to CMV, ZYMV, and WMV-2 have all been produced. Field tests of these transgenic squash have been conducted (68 records; Anonymous 2005). There are two deregulated transgenic squash lines, ZW-20 and CWZ-3, in the United States permitted for commercial sale. The transgenic line ZW-20 exhibits resistance to WMV-2 and ZYMV while CWZ-3 is resistant to CMV, WMV-2, and ZYMV.

Watermelon (Citrullus lanatus)

Genetic transformation in watermelon was first reported by Choi, Soh, and Kim (1994) who used an *Agrobacterium*-mediated protocol to introduce a *35S:GUS* fusion construct to yield transformants created by direct shoot organogenesis from cotyledon explants. Reed et al. (2001) described an *in vitro* method to select transformants via the production of phosphomannose isomerase (PMI). PMI catalyzes the reversible interconversion of mannose 6-phosphate and fructose-6-phosphate. Plant cells lacking this enzyme are incapable of surviving on synthetic media containing only mannose as a carbon source. Selection by PMI resulted in an average transformation frequency of 2% (Reed et al. 2001).

A more efficient and reproducible *Agrobacterium*-mediated protocol suitable for transferring interesting genes into different watermelon cultivars was reported by Ellul et al. (2003). Their protocol resulted in transformation efficiencies ranging from 2.8 to 5.3% depending on cultivar. Using this protocol, the *HAL1* gene (as described in the above melon section) related to salt tolerance was introduced into watemelon cultivars. Transgenic plants expressing the *HAL1* gene performed better than non-transformed plants did under salt-stress conditions.

Novel techniques for introducing foreign genes into intact plant organs have also been reported for watermelon. Chen et al. (1998) injected a phosphorylated GUS plasmid containing the CaMV35S promoter and DNA from a *Cucurbita* cultivar resistant to *Fusarium oxysporum* f.sp. *niveum* into the ovaries of the F_1 hybrid watermelon, Pink Orchid, after hand pollination. Transformation events were confirmed through staining, Southern blot analysis, and RAPD analysis. A transformation frequency of 0.5% (i.e., 10 of 200 transformants) yielded a low number of *Fusarium* plants. More recently, Hema, Prasad, and Vani (2004) introduced a chimeric *GUS* gene following electroporation of zygotic embryos and nodal buds. The stable integration of this chimeric gene in progeny was confirmed by dot and Southern blot analyses.

Genetic transfromation of a wild relative of watermelon, *Citrullus colocynthis*, has also been reported (Dabauza, Bordas, and Salvador 1997). *C. colocynthis* exhibits resistance to various diseases, including to *Yellow-stunting disorder*, ZYMV, CMV, and WMV-2. A GUS gene was introduced into *C. colocynthis* by *Agrobacterium*-mediated protocol, and the transformation efficiency was 14.2%, which was considerably higher than that reported for cultivated types.

Using the above protocols, transgenic watermelon lines have been mainly generated by commercial companies and have been field tested in the United States and Europe (thirteen records; Anonymous 2005). Most of the transgenic watermelon lines field tested were for virus resistance conferred by the expression of CP genes from CMV, ZYMV, and WMV-2. However, no transgenic watermelon has been yet released for commercial use.

8.7 CONCLUSIONS

The Cucurbitaceae is a remarkable plant family; the fruits of wild and cultivated cucurbits displays enormous variation in size, shape, and color patterns (Rubatzky and Yamaguchi 1997). A few cucurbit genera (e.g., *Cucumis, Cucurbita,* and *Citrullus*) are widely cultivated and have extraordinary human importance for many reasons reviewed (economic, aesthetic, cultural, medicinal, and botanical). Cucurbits are among the oldest domesticated plants, emerging as some of the first vegetable crops 8,000–12,000 years ago, both in the Old and New Worlds (Hancock 2004). However, the current state of knowledge about the origins, domestication processes, post-domestication diversification, and geographic expansion of important cucurbits is still relatively limited. Gaps in current understanding present many research opportunities to collect additional historical, geographical, biological, and genetical evidence related to the domestication and diversification of these crops.

The family Cucurbitaceae is taxonomically well-defined and treated as the only member of the monotypic order, Cucurbitales. Its putative relatives are members of the Begoniaceae and Datiscaceae in the Begoniales (Jeffrey 1990). About 90% of cucurbit species occur in three tropical regions: Africa (south of the Sahara, including Madagascar), Central and South America, and southeast Asia and Malaysia. There are about 118 extant genera and 825 species (Jeffrey 1990, 2001). From the perspective of life form, the family is represented by annual and perennial herbs, semi-shrubs, aculeateous shrubs, and a few rare succulent trees. Trailing or climbing vine growth with nodal branching is typical of many species. Recent taxonomic treatments of the family build upon traditional morphological and anatomical studies by incorporating evidence from biochemistry, cytology, cytogenetics, molecular genetics, crossing barriers, and coevolution with insects and pathogens in delimiting taxa (Bates, Robinson, and Jeffrey 1990; Kirkbride 1993). Since 1990, comprehensive biosystematic and evolutionary monographs and/or detailed studies have been published for the genera *Cucumis* (Kirkbride 1993), *Cucurbita* (Merrick 1995; Sanjur et al. 2002) and *Citrullus* (Jarret and Newman 2000; Levi et al. 2000), and an overview of the classification of cultivated cucurbits was provided by Jeffrey (2001). However, careful comprehensive studies to elaborate taxonomic and phylogenetic relationships among most wild and cultivated cucurbits are still required, and information about biogeography and ecobiology of wild relatives is still rather limited.

Germplasm collections of cucurbits are among the oldest and the most extensive of horticultural plant germplasm collections. Recent information about the holdings of the world's largest collections of *Citrullus*, *Cucumis*, and *Cucurbita* germplasm was summarized as part of the "The State of the World's Plant Genetic Resources for Food and Agriculture" (FAO 1998). The world's largest cucurbits collections are represent by the U.S. National Plant Germplasm System and the Vavilov Institute in Russia. Among European countries, research and long-term conservation of cucurbits germplasm is coordinated by the Working Group on Cucurbits (Díez, Picó, and Nuez 2002), which operates under the aegis of The European Cooperative Programme for Crop Genetic Resources Networks (ECP/GR). In the areas of acquisition and exploration, since the 1980s, considerable effort has been directed toward expanding the diversity of *ex situ* germplasm collections by collecting wild, weedy, and landrace populations of cucurbits from their centers of diversity. For selected species such as crops in the genus *Cucurbita*, an excellent summary of exploration efforts has been published (Andres 2000); however, no analogous overview has been published for other important cucurbit genera. This summary of published exploration reports (Table 8.6) could serve as a starting point in the development of such overviews. An important topic related to the collection of cucurbits from traditional agroecosystems is the investigation of *in situ* gene flow and delimitation of populations. The most significant progress on this topic has been made with New World *Cucurbita* spp. (cf. Wilson 1990).

Substantial progress has been achieved in seed regeneration and maintenance protocols of cucurbits from both the perspective of controlled pollination through the manipulation of insects and from that of seed physiology. In addition to seed storage, tissue-culture methods for cucurbits are relatively well developed (see Molina and Nuez 1996), and pollen-storage methods, primarily for *Citrullus*, are being developed (Sugiyama, Morishita and Nishino 2002b).

Germplasm characterization can be considered a crucial component of effective germplasm management. Without accurate characterization, it is impossible to establish or verify taxonomic identity. This is not only important for meeting the germplasm needs of researcher, but it is also for efficient management of collections (e.g., fingerprinting for quality assurance, identification of duplicates, and gaps in collections) (Bretting and Widrlechner 1995). During the last two decades, diverse techniques to assess highly heritable phenotypic features (morphological, anatomical, biochemical, pest and pathogen reactions, polymorphic proteins) have been employed to characterize cucurbit germplasm, often in concert with analyses of polymorphic DNA markers. Nevertheless, in many germplasm collections, basic characterization data are incomplete, limiting the utility of the collections.

Given the importance of germplasm characterization, it is surprising that only a single internationally approved descriptor list for cucurbits has been published, one for *C. melo* (IPGRI 2003). The development of international descriptor lists for other important cucurbits has been identified as a crucial task (Díez, Picó, and Nuez 2002). Recently, however, detailed descriptor lists (*Cucumis* and *Cucurbita* spp.) have been developed for national gene bank collections (Křístková et al. 2003, 2005).

During the last two decades, considerable research has been directed toward the exploitation of wild relatives in cucurbit breeding (Robinson and Decker-Walters 1997). Closely related to this topic, various authors have applied the concept of gene pools (sensu Harlan and deWet 1971) to cucurbits. In cucurbit crop species, genetic diversity analysis has been rigorously applied primarily to *C. sativus* and *C. melo* to determine the nature and structure of their gene pools and to examine hypotheses that may provide insights into the nature of their domestication (Bates and Robinson 1995). These gene pools are now relatively well described in relation to their evolution, migration, and differentiation (e.g., Raamsdonk, den Nijs, and Jongerius 1989; Horejsi and Staub 1999). The concept of gene pools is also rather well elaborated for *Cucurbita* spp. (e.g., Lira-Saade 1995). Species of the genus *Citrullus* share a common chromosome number, and all its taxa are cross-compatible (Bates and Robinson 1995), indicative either of a single primary gene pool or of a secondary gene pool encompassing two poorly differentiated primary pools.

The development of polymorphic protein and DNA markers and their use in germplasm characterization created fertile ground for application of genetic mapping and genomic technologies. Genetic mapping has progressed quickly for *C. sativus* and *C. melo*, but lags for other cucurbit genera. Genetic diversity analysis and map construction (Brown and Myers 2002) is in its infancy in *Cucurbita*, and the creation of genetic maps for watermelon, *C. lanatus* var. *lanatus*, has been hampered by insufficient genetic diversity (Levi et al. 2001a, 2001c).

Cucumber linkage maps have been constructed by using combinations of molecular and phenotypic markers (Kennard et al. 1994; Serquen, Bacher, and Staub 1997a) and more recently with markers associated with basic cell functions (Xie, Wehner, and Conkling 2002; Xie et al. 2003) and cytoplasmic factors controlling economically important traits (Chung and Staub 2003). During the last decade, important contributions toward the development of molecular markers and genetic maps in melon (*C. melo*) have also been made (Baudracco-Arnas and Pitrat 1996; Katzir et al. 1996; Périn et al. 2002b; Gonzalo et al. 2005). The development of linkage maps for cucumber have spurred research on quantitative trait loci (QTL) associated with fruit quality (Kennard and Havey 1995) and important yield components (e.g., earliness, sex expression, multiple lateral branching, fruit number and weight, and fruit size) (Fazio, Staub, and Stevens 2003), leading to the application of MAS (Fazio and Staub 2003). In the related area of the development of unique genetic stocks, bulk segregant analysis (BSA) (Michelmore, Paran, and Kesseli 1991), in combination with near isogenic lines (NILs) and other similar techniques, has been used in cucumber to identify markers linked to disease resistance (Horejsi, Staub, and Thomas 2000) and to study sex-expression genes (Witkowicz, Urbanczyk-Wochniak, and Przybecki 2003). NILs could be also used for the efficient introgression of economically important QTL into elite germplasm (e.g., Bernacchi et al. 1998a, 1998b).

The breeding of cucurbits has primarily focused on improving the production and quality of fruits by increasing resistance to a wide range of pathogens and diseases and by modifying plant architecture and sex expression (Robinson and Decker-Walters 1997). Toward those ends, many different selection schemes have been used depending on the objectives of the enhancement/ breeding project, the inheritance patterns and heritability of the characters to be selected, the available germplasm, and the species' selection history.

Broadening the genetic diversity is considered to be essential for the continued development of commercial cucurbit cultivars. However, genetic analysis of the U.S. National Plant Germplasm System's cucumber collection showed reduced diversity among commercial germplasm in relation to the entire primary gene pool ($\sim 3\%$ polymorphism in elite commercial germplasm vs. $\sim 12\%$ across the entire collection; Meglic and Staub 1996b), highlighting the need for the genetic enrichment of elite cucumber germplasm to ensure continued potential for plant improvement (Staub et al. 2002).

To this end, considerable progress has been made in the areas of germplasm enhancement and introduction of novel traits into various cultivated cucurbits. This typically involves the incorporation of genes from exotic accessions into a more elite germplasm pool (Shetty and Wehner 2002). This strategy has resulted in important public germplasm releases in cucumber (Walters and Wehner 1997), melon (Thomas 1986), and squash (Mutschler and Pearson 1987). A crucial step in the enhancement process is an evaluation of available germplasm diversity for key phenotypic characters. Exotic and wild cucurbit germplasm often includes genetic diversity absent in elite types.

Germplasm evaluation projects have often focused on horticultural and agronomic characteristics. Many horticultural evaluations have been directed toward the identification of superior genotypes among local landraces of *C. sativus*, *C. melo*, *Cucurbita* spp., and *C. lanatus* with some of the most detailed evaluations of specific traits (e.g., root growth, postharvest storage, combining ability for yield, specific yield components) conducted on *C. sativus* (Wehner, Shetty, and Clark 2000a; Wehner, Shetty, and Wilson 2000b; Shetty and Wehner 2002). The characterization of biochemical and quality traits can also contribute significantly to crop marketability. Variation in soluble solids, ascorbic acid, individual sugars, and the enzymes that control sugar production have received special attention, especially in *C. melo* (e.g., Burger et al. 2004), and to a lesser extent, in *C. sativus* (Robinson 1987). Substantial progress has been made in the evaluation of seed composition (especially fatty acids) in *Cucurbita pepo* (Teppner 2004), *C. foetidissima* (Scheerens et al. 1978), and *C. colocynthis* (Schafferman et al. 1998). There is also interest in the roles of cucurbitacins as vertebrate toxins, insecticides, and attractants, which has led to evaluations for cucurbitacin composition and concentration (Metcalf and Rhodes 1990).

Cucurbit germplasm collections have received considerable attention for evaluation for pathogen, pest, and abiotic stress resistance. Clearly, the discovery and incorporation of resistance genes into commercial crops are key justifications for the conservation of *ex situ* germplasm collections. Comprehensive reviews have been published to summarize extensive research on the evaluation of cucurbit germplasm for its reaction to viral pathogens, including information on the genetics of resistance (Munger 1993; Provvidenti 1993). Since the publication of those reviews, many additional germplasm evaluations to different viral pathogens have been conducted and new sources of resistance identified (e.g., Horváth 1993a; Lebeda and Křístková, 1996; Křístková and Lebeda 2000b; López-Sesé and Gómez-Guillamón 2000). In contrast, reports on the screening of cucurbit germplasm for resistance to bacterial pathogens are relatively scarce.

A diverse array of known fungal pathogens has been reported to infect cucurbits (Zitter, Hopkins, and Thomas 1996); however, of this extensive list, only a smaller number are considered as economically important. Considerable progress has been achieved on research and screening of germplasm for reaction to key pathogens such as the downy and powdery mildews. Recent progress in pathotype discrimination, cucurbit germplasm evaluation, and resistance breeding against downy mildew (*Pseudoperonospora cubensis*) was reviewed by Lebeda and Widrlechner (2003). Cucurbit powdery mildew, caused by *Podosphaera xanthii* and *Golovinomyces cichoracearum*, is a widespread foliar disease. Jahn, Munger, and McCreight (2002) published a comprehensive review of sources and genetic control of resistance to powdery mildew in cucurbits, which included some references to past germplasm evaluations. More recently, additional data were published on the reactions of *C. sativus* germplasm to inoculation by *P. xanthii* (Morishita et al. 2003; Block and Reitsma 2005). Results of extensive cucurbit germplasm evaluations has been published for reactions to other important foliar pathogens as well [e.g., scab (*Cladosporium cucumerinum*), anthracnose (*Colletotrichum orbiculare*), target leaf spot (*Corynespora cassiicola*), gummy stem blight (*Didymella bryoniae*)]. Of the fungal root-rot diseases, the most damaging include fusarium wilt (*Fusarium oxysporum*) and vine decline (caused by *Acremonium cucurbitacearum* and *Monosporascus cannonballus*) in *C. melo*. Research to evaluate melon germplasm to vine decline is in progress (Crosby 2001; Dias et al. 2001).

Substantial progress has been made in the study of cucurbits resistance against nematodes, insects, and mites. Thies (1996) reviewed key evaluations of cucurbit germplasm conducted to identify sources of resistance to root-knot nematodes (*Meloidogyne* spp.). Robinson (1992) comprehensively reviewed research conducted to evaluate cucurbit germplasm for reaction to insect and mite pests. However, many additional results from more recent evaluations have been published and are reviewed herein (e.g., Bohn, Kishaba, and McCreight 1996; Simmons and McCreight 1996; Simmons and Levi 2002).

Abiotic stresses are often as important as are pests and pathogens in limiting the production of cucurbit crops. Extensive germplasm evaluations have been conducted to identify sources of tolerance to herbicides (Staub et al. 1991) and other chemical treatments such as sulphur dusting (Perchepied et al. 2004), high temperatures and moisture deficits (Staub and Krasowska 1990; Wann 1992; Ríos Labrada, Fernández Almirall, and Casanova Galarraga 1998), and the ability to germinate under low-temperature conditions (Wehner 1984).

Transfer of economically important traits from wild cucurbit species to their cultivated counterparts continues to be one of the most important challenges facing breeders. This is most crucial in cases where resistance genes for several crop-limiting pathogens and pests have not been found within the primary gene pools such as resistance to *P. cubensis* in

C. sativus. Substantial differences in ease of interspecific hybridization have been recognized among the three major genera of cultivated Cucurbitaceae (*Cucumis*, *Cucurbita*, and *Citrullus*; Robinson and Decker-Walters 1997). Many techniques, including embryo and callus culture, somatic hybridization via protoplast fusion, bridge crosses, bud pollination, repeated pollination, regulation of ploidy levels, and growth regulators use, have been developed in attempts to improve the success of interspecific hybridization (e.g., Kalloo 1988; den Nijs and Custers 1990; Tatlioglu 1993; Lebeda, Křístková, and Kubaláková 1996; Lebeda et al. 1999; Chen and Adelberg 2000; Gajdová, Lebeda, and Navrátilová 2004; Skálová, Lebeda and Navrátilová 2004). More specifically, progress in overcoming post-zygotic barriers has been achieved in *Cucumis* by using embryo rescue (Lebeda, Křístková, and Kubaláková 1996; Lebeda et al. 1999; Chen and Adelberg 2000; Skálová, Lebeda, and Navrátilová 2004), and both haploid and polyploid technologies (e.g., Ezura and Oosawa 1994a, 1994b) have been developed and used in *C. sativus, C. melo, C. pepo, and C. lanatus.* The successful rescue of interspecific *Cucurbita* hybrid embryos has also been reported for many combinations (Šiško, Ivančič, and Bohanec 2003). Recent progress and future trends in protoplast culture and somatic-cell fusion in *Cucumis* and some *Cucurbita* species have been reviewed and discussed in detail by Gajdová, Lebeda, and Navrátilová (2004). Despite the large body of technical protocols to help optimize wide crosses, careful biosystematic and gene-pool studies will continue to guide the selection of genotypes that increase the probability of successful interspecific hybridization (see Raamsdonk, den Nijs, and Jongerius 1989).

Genetic transformation is one of the most quickly developing and expanding technologies being harnessed in the improvement of cucurbits. To date, transgenic plants of *C. sativus* (Truson, Simpson, and Shahin 1986; Chee 1990), *C. melo* (Guis et al. 1998), *Cucurbita maxima* and *C. pepo* (Pang et al. 2000), and *C. lanatus* (Choi, Soh, and Kim 1994) were released. Various methodologies, including *Agrobacterium rhizogenes* and *A. tumefaciens*-mediated transformation via somatic embryogenesis and organogenesis and biolistic transformation, have been used for production of these plants. A wide range of genes, including ones coding for chitinase, conferring resistance to viral and fungal pathogens, modifying fruit ripening, and influencing many other physiological processes, have now been introduced and expressed *in vivo*.

The array of biotechnological tools that are being brought to bear on the improvement of cucurbits in extensive and rapidly evolving. While the next generation of such tools cannot be accurately predicted, current advancements in functional genomics and metabolomics will play key roles in their development. What can be predicted with confidence is that the successful application of today's technologies and those yet to come (no matter how promising or broadly applicable) will continue to rely upon researchers' *sound* understanding of the organism and its inherent diversity. This comprehensive review of recent cucurbit research may help provide that understanding, connecting today's researchers with their colleagues' work and allowing those that follow to build upon it.

ACKNOWLEDGMENTS

A. Lebeda's work on this chapter was supported by the following projects: (1) MSM 6198959215 (Ministry of Education, Youth and Sports, Czech Republic); and (2) QD 1357 (NAZV, Ministry of Agriculture, Czech Republic). The authors wish to thank the following colleagues for contributing information to this review: Charles Block, Lucinda Clark, Laura Merrick, Kathleen Reitsma, and John Wiersema. We are also grateful for assistance with the making of the world composite range map provided by Jeff Carstens, David Kovach, and Ashlee Langholz.

REFERENCES

Adelberg, J., Rhodes, B., Skorupska, H., and Bridges, W., Explant origin affects the frequency of tetraploid plants from tissue cultures of melon, *HortSci.*, 29, 689–692, 1994.

Adelberg, J., Rhodes, B., Zhang, X., and Skorupska, H., Fertility and fruit characters of hybrid triploid melon, *Breeding Sci.*, 45, 37–43, 1995.

Akasaka-Kennedy, Y., Tomita, K. O., and Ezura, H., Efficient plant regeneration and *Agrobacterium*-mediated transformation via somatic embryogenesis in melon (*Cucumis melo* L.), *Plant Sci.*, 166, 763–769, 2004.

Akashi, Y., Ezura, H., Kubo, Y., Masuda, M., and Kato, K., Microsatellite and CAPS markers for ethylene-related genes, 1-aminocyclopropane-1-carboxylic acid (ACC) synthase and ACC oxidase genes, and their variation in melon (*Cucumis melo* L.), *Breeding Sci.*, 51, 107–112, 2001.

Akashi, Y., Ezura, H., Kubo, Y., Masuda, M., and Kato, K., Varietal variation in microsatellite and CAPS for ethylene-related genes and its possible association with agronomic characters in melon (*Cucumis melo*), *Acta Hort.*, 588, 313–315, 2002a.

Akashi, Y., Ezura, H., Kubo, Y., Masuda, M., and Kato, K., Genetic variation and phylogenetic relationships in East and South Asian melons *Cucumis melo* L., based on the analysis of five isozymes, *Euphytica*, 125, 385–396, 2002b.

Alvarez, J. and González-Torres, R., Resistance to physiological races of *Fusarium oxysporum* f.sp. *melonis* in Iberian melon genotypes, In *Cucurbits towards 2000. Proceedings of the Sixth Eucarpia Meeting on Cucurbit Genetics and Breeding*, Gómez-Guillamón, M. L. et al., Eds., Algarrobo, Málaga, Spain: Estación Experimental "La Mayora", pp. 217–222, 1996.

Ananthakrishnan, G., Xia, X., Elman, C., Singer, S., Paris, H., Gal-On, A., and Gaba, V., Shoot production in squash (*Cucurbita pepo*) by *in vitro* organogenesis, *Plant Cell Rep.*, 21, 739–746, 2003.

Andres, T. C., Biosystematics, theories on the origin, and breeding potential of *Cucurbita ficifolia*, In *Biology and Utilization of the Cucurbitaceae*, Bates, D. M. et al., Eds., Ithaca, NY: Comstock Publishing Associates, pp. 102–119, 1990.

Andres, T. C., Searching for *Cucurbita* germplasm: collecting more than seeds, *Acta Hort.*, 510, 191–198, 2000.

Andres, T. C., Diversity in tropical pumpkin (*Cucurbita moschata*): a review of infraspecific classifications, In *Progress in Cucurbit Genetics and Breeding Research, Proceedings of Cucurbitaceae 2004, the 8th EUCARPIA Meeting on Cucurbit Genetics and Breeding*, Lebeda, A. and Paris, H. S., Eds., Olomouc, Czech Republic: Palacký University in Olomouc, pp. 107–112, 2004a.

Andres, T. C., Diversity in tropical pumpkin (*Cucurbita moschata*): cultivar origin and history, In *Progress in Cucurbit Genetics and Breeding Research, Proceedings of Cucurbitaceae 2004, the 8th EUCARPIA Meeting on Cucurbit Genetics and Breeding*, Lebeda, A. and Paris, H. S., Eds., Olomouc, Czech Republic: Palacký University in Olomouc, pp. 113–118, 2004b.

Anonymous, India. Activities at NBPGR, *IBPGR Newsletter for Asia, the Pacific and Oceania* 11, 5–7, 1993.

Anonymous. *Information Systems for Biotechnology Website*, http://www.isb.vt.edu/cfdocs/fieldtests1.cfm (accessed January 2005), 2005.

Antoine Michard, S. and Beckert, M., Spontaneous versus colchicine-induced chromosome doubling in maize anther culture, *Plant Cell Tiss. Organ Cult.*, 48, 203–207, 1997.

Armstrong, G. M., Armstrong, J. K., and Netzer, D., Pathogenic races of the cucumber-wilt *Fusarium*, *Plant Dis. Rep.*, 62, 824–828, 1978.

Arumuganathan, K. and Earle, E. D., Nuclear DNA content of some important plant species, *Plant Mol. Biol. Rep.*, 9, 211–215, 1991.

Atarés, A., García-Sogo, B., Pineda, B., Ellul, P., and Moreno, V., Transformation of melon via PEG-induced direct DNA uptake into protoplasts, *Progress in Cucurbit Genetics and Breeding Research. In Proceedings of Cucurbitaceae 2004, the 8th EUCARPIA Meeting on Cucurbit Genetics and Breeding*, Lebeda, A., Paris, H. S. et al., Eds., Olomouc, Czech Republic: Palacky University in Olomouc, pp. 465–469, 2004.

Ayub, R., Guis, M., Ben Amor, M., Gillot, L., Roustan, J. P., Latche, A., Bouzayen, M., and Pech, J. C., Expression of ACC oxidase antisense gene inhibits ripening of cantaloupe melon fruits, *Nature Biotechnol.* 14, 862–866, 1996.

Baird, J. R. and Thieret, J. W., The bur gherkin (*Cucumis anguria* var. *anguria*, Cucurbitaceae), *Econ. Bot.*, 42, 447–451, 1988.

Balague, C., Watson, C. F., Turner, A. J., Rouge, P., Picton, S., Pech, J. C., and Grierson, D., Isolation of a ripening and wound-induced cDNA from *Cucumis melo* L. encoding a protein with homology to the ethylene-forming enzyme, *Eur. J. Biochem.*, 212, 27–34, 1993.

Balkema-Boomstra, A. G. and Mollema, C., Resistance to western flower thrips in cucumber, In *Cucurbits towards 2000, Proceedings of the Sixth Eucarpia Meeting on Cucurbit Genetics and Breeding*, Gómez-Guillamón, M. L. et al., Eds., Algarrobo, Málaga, Spain: Estación Experimental "La Mayora", pp. 340–343, 1996.

Barroso, M. R., Martins, S., Vences, F. J., Sáenz de Miera, L. E., and Carnide, V., Comparative analysis of melon landraces from South Portugal using RAPD markers, In *Progress in Cucurbit Genetics and Breeding Research, Proceedings of Cucurbitaceae 2004, the 8th EUCARPIA Meeting on Cucurbit Genetics and Breeding*, Lebeda, A. and Paris, H. S., Eds., Olomouc, Czech Republic: Palacký University in Olomouc, pp. 143–150, 2004.

Barry, B. D., Burnside, J. A., and Myers, H. S., Cucumis species resistance to striped cucumber beetle seedling feeding and bacterial wilt, *USDA-Agric. Res. Service. Rep*, ARS-NC-46, 1976.

Bates, D. M. and Robinson, R. W., Cucumbers, melons and water-melons, In *Evolution of Crop Plants*, Smartt, J. and Simmonds, N. W., Eds., 2nd ed., Harlow, Essex, U.K.: Longman Scientific, pp. 89–96, 1995a.

Bates, D. M., Robinson, R. W., and Jeffrey, C., Eds., *Biology and Utilization of the Cucurbitaceae*, Ithaca, NY: Comstock, 1990.

Bates, D. M., Merrick, L. C., and Robinson, R. W., Minor cucurbits, In *Evolution of Crop Plants*, Smartt, J. and Simmonds, N. W., Eds., 2nd ed., Harlow, Essex, U.K.: Longman Scientific, pp. 105–111, 1995b.

Batra, S., Induced tetraploidy in muskmelon, *J. Hered.*, 43, 141–148, 1952.

Batra, S., Interspecific hybridization in the genus *Cucumis*, *Sci. Cult.*, 18, 445–446, 1953.

Baudracco-Arnas, S. and Pitrat, M., A genetic map of melon (*Cucumis melo* L.) with RFLP, RAPD, isozyme, disease resistance and morphological markers, *Theor. Appl. Genet.*, 93, 57–64, 1996.

Beharav, A. and Cohen, Y., The crossability of *Cucumis melo* and *C. metuliferus*, an investigation of *in vivo* pollen tube growth, *Cucurbit Genet. Coop. Rep.*, 17, 97–100, 1994.

Beharav, A. and Cohen, Y., Attempts to overcome barrier of interspecific hybridization between *Cucumis melo* L., and *C. metuliferus*, *Isr. J. Plant Sci.*, 43, 113–123, 1995.

Bemis, W. P., Interspecific aneuploidy in *Cucurbita*, *Genet. Res.*, 21, 221–228, (Cambridge) 1973.

Bemis, W. P., Rhodes, A. M., Whitaker, T. W., and Carmer, S. G., Numerical taxonomy applied to *Cucurbita* relationships, *Am. J. Bot.*, 57, 404–412, 1970.

Benscher, D. and Provvidenti, R., Allozyme diversity at the *Pgi-2* locus in landraces of *Citrullus lanatus* from Zimbabwe, *Cucurbit Genet. Coop. Rep.*, 14, 104–106, 1991.

Bernacchi, D., Beck-Bunn, T., Emmatty, D., Eshed, Y., Inai, S., Lopez, J., Petiard, V. et al., Advanced backcross QTL analysis of tomato II. Evaluation of near-isogenic lines carrying single-donor introgressions for desirable wild QTL-alleles derived from *Lycopersicon hirsutum* and *L. pimpinellifolium*, *Theor. Appl. Genet.*, 97, 170–180, 1998a.

Bernacchi, D., Beck-Bunn, T., Eshed, Y., Lopez, J., Petiard, V., Uhlig, J., Zamir, D., and Tanksley, S. D., Advanced backcross QTL analysis in tomato. I. Identification of QTLs for traits of agronomic importance from *Lycopersicon hirsutum*, *Theor. Appl. Genet.*, 97, 381–397, 1998b.

Bettencourt, E. and Konopka, J., *Directory of Germplasm Collections 4. Vegetables Abelmoschus, Allium, Amaranthus, Brassicaceae, Capsicum, Cucurbitaceae, Lycopersicon, Solanum and Other Vegetables*, Rome: IBPGR, 1990.

Bhojwani, S. S. and Razdan, M. K., Haploid production, In *Plant Tissue Culture: Theory and Practice*, Amsterdam, Netherlands: Elsevier, pp. 113–141, 1983.

Bisognin, D. A., Origin and evolution of cultivated cucurbits, *Cienc. Rural*, 32, 715–723, 2002.

Block, C. C. and Reitsma, K. R., Powdery mildew resistance in the U.S. national plant germplasm system cucumber collection, *HortSci.*, 40, 416–420, 2005.

Bohn, G. W., Kishaba, A. N., and McCreight, J. D, A survey of tolerance to *Aphis gossypii* Glover in part of the world collection of *Cucumis melo* L., In *Cucurbits towards 2000. Proceedings of the Sixth Eucarpia Meeting on Cucurbit Genetics and Breeding*, Gómez-Guillamón, M. L. et al., Eds., Algarrobo, Málaga, Spain: Estación Experimental "La Mayora", pp. 334–339, 1996.

Bordas, M., Gonzalez-Candelas, L., Dabauza, M., Romon, D., and Moreno, V., Somatic hybridization between an albino *Cucumis melo* L. mutant and *Cucumis myriocarpus* Naud, *Plant Cell Rep.*, 132, 179–190, 1998.

Bordas, M., Montesinos, C., Dabauza, M., Salvador, A., Roig, L. A., Serrano, R., and Moreno, V., Transfer of the yeast salt tolerance gene *HAL1* to *Cucumis melo* L. cultivars and *in vitro* evaluation of salt tolerance, *Transgenic Res.*, 6, 41–50, 1997.

Bouabdallah, L. and Branchard, M., Regeneration of plants from callus cultures of *Cucumis melo* L., *Z. Pflanzenzucht*, 96, 82–85, 1986.

Boyhan, G. E., Norton, J. D., Abrahams, B. R., and Wen, H. H., A new source of resistance to anthracnose (race 2) in watermelon, *HortSci.*, 29, 111–112, 1994.

Boyhan, G., Norton, J. D., Jacobsen, B. L., and Abrahams, B. R., Evaluation of watermelon and related germ plasm for resistance to zucchini yellow mosaic virus, *Plant Dis.*, 76, 251–252, 1992.

Bradeen, J. M., Staub, J. E., Wyse, C., Antonise, R., and Peleman, J., Towards an expanded and integrated linkage map of cucumber (*Cucumis sativus* L.), *Genome*, 44, 111–119, 2001.

Bretting, P. K. and Widrlechner, M. P., Genetic markers and plant genetic resource management, *Plant Breeding Rev.*, 13, 11–86, 1995.

Brickell, C. D., Baum, B. R., Hetterscheid, W. L. A., Leslie, A. C., McNeill, J., Trehane, P., Vrugtman, F., and Wiersema, J. H., International code of nomenclature for cultivated plants, 7th ed., *Acta Hort.*, 647(i–xxi), 1–123, 2004.

Brothwell, D. and Brothwell, P., *Food in Antiquity: A Survey of the Diet of Early Peoples*, New York: Fredrick A. Praeger Publishers, 1969.

Brotman, Y., Kovalski, I., Dogimont, C., Pitrat, M., Katzir, N., and Perl-Treves, R., Molecular mapping of the melon *Fom-1/Prv* locus, In *Progress in Cucurbit Genetics and Breeding Research, Proceedings of Cucurbitaceae 2004, the 8th EUCARPIA Meeting on Cucurbit Genetics and Breeding*, Lebeda, A. and Paris, H. S., Eds., Olomouc, Czech Republic: Palacký University in Olomouc, pp. 485–489, 2004.

Brotman, Y., Silberstein, L., Kovalski, I., Perin, C., Dogimont, C., Pitrat, M., Klinger, J., Thompson, G. A., and Perl-Treves, R., Resistance gene homologies in melon are linked to genetic loci conferring disease and pest resistance, *Theor. Appl. Genet.*, 104, 1055–1063, 2002.

Brotman, Y., Silberstein, L., Kovalski, I., Thompson, G., Katzir, N., and Perl-Treves, R., Linkage groups of *Cucumis melo*, including resistance gene homologies and known genes, *Acta Hort.*, 510, 441–448, 2000.

Brown, R. N. and Myers, J. R., A genetic map of squash (*Cucurbita* sp.) with randomly amplified polymorphic DNA markers and morphological markers, *J. Am. Soc. Hort. Sci.*, 127, 568–575, 2002.

Burger, Y., Yeselson, Y., Saar, U., Paris, H. S., Katzir, N., Tadmor, Y., and Schaffer, A. A., Screening of melon (*Cucumis melo*) germplasm for consistently high sucrose content and for high ascorbic acid content, In *Progress in Cucurbit Genetics and Breeding Research*, Lebeda, A. and Paris, H. S., Eds., Olomouc, Czech Republic: Palacký University in Olomouc, pp. 151–155, 2004.

Burza, W. and Malepszy, S., *In vitro* culture of *Cucumis sativus* L. 18. Plant from protoplasts through direct somatic embryogenesis, *Plant Cell Tissue Organ Cult.*, 41, 259–266, 1995.

Carnide, V., Genetic resources of Cucurbitaceae in Portugal, In *Cucurbit Genetic Resources in Europe*, Díez, M. J., Picó, B., and Nuez, F., Eds., Rome: International Plant Genetic Resources Institute, *Ad hoc Meeting*, 19 January 2002, Adana, Turkey, p. 36, 2002.

Carnide, V., Martins, S., Vences, F. J., Sáenz de Miera, L. E., and Barroso, M. R., Evaluation of Portuguese melon landraces conserved on farm by morphological traits and RAPDs, In *Progress in Cucurbit Genetics and Breeding Research, Proceedings of Cucurbitaceae 2004, the 8th EUCARPIA Meeting on Cucurbit Genetics and Breeding*, Lebeda, A. and Paris, H. S., Eds., Olomouc, Czech Republic: Palacký University in Olomouc, pp. 135–142, 2004.

Chatterjee, M. and More, T. A., Techniques to overcome barrier of interspecific hybridization in *Cucumis*, *Cucurbit Genet. Coop. Rep.*, 14, 66–68, 1991.

Chaturvedi, R. and Bhatnagar, S. P., High-frequency shoot regeneration from cotyledon explants of watermelon cv. sugar baby, *In Vitro* Cell. Devel. Biol. Plant, 37, 255–258, 2001.

Che, K. P., Liang, C. Y., Wang, Y. G., Jin, D. M., Wang, B., Xu, Y., Kang, G. B., and Zhang, H. Y., Genetic assessment of watermelon germplasm using the AFLP technique, *HortSci.*, 38, 81–84, 2003.

Chee, P. P., Transformation of *Cucumis sativus* tissue by *Agrobacterium tumefaciens* and the regeneration of transformed plants, *Plant Cell Rep.*, 9, 245–248, 1990.

Chee, P. P., Somatic embryogenesis and plant regeneration of squash *Cucurbita pepo* L cv. YC 60, *Plant Cell Rep.*, 9, 620–622, 1991.

Chee, P. P. and Slighton, J. L., Transfer and expression of cucumber mosaic virus coat protein gene in the genome of *Cucumis sativus*, *J. Am. Soc. Hort. Sci.*, 116, 1098–1102, 1991.

Chee, P. P. and Slighton, J. L., Transfer of cucumber tissues by microprojectile bombardment: identification of plants containing functional and non-functional transferred genes, *Gene*, 118, 255–260, 1992.

Chen, J. F. and Adelberg, J., Interspecific hybridization in *Cucumis*—progress, problem, and perspectives, *HortSci.*, 35, 11–15, 2000.

Chen, J. F., Adelberg, J. W., Staub, J. E., Skorupska, H. T., and Rhodes, B. B., A new synthetic amphidiploid in *Cucumis* from *C. sativus* × *C. hystrix* Chakr. F₁ interspecific hybrid, In *Cucurbitaceae '98 Evaluation and Enhancement of Cucurbit Germplasm*, McCreight, J., Ed., Alexandria, VA: ASHS Press, pp. 336–339, 1998a.

Chen, J. F., Isshiki, S., Tashiro, Y., and Miyazaki, S., Studies on a wild cucumber from China (*Cucumis hystrix* Chakr.). I. Genetic distances between *C. hystrix* and two cultivated *Cucumis* species (*C. sativus* L. and *C. melo* L.) based on isozyme analysis, *J. Jpn. Soc. Hort. Sci.*, 64(suppl. 2), 264–265, 1995.

Chen, J. F., Isshiki, S., Tashiro, Y., and Miyazaki, S., Biochemical affinities between *Cucumis hystrix* Charkr. and two cultivated *Cucumis* species (*C. sativus* L. and *C. melo.* L.) based on isozyme analysis, *Euphytica*, 97, 139–141, 1997a.

Chen, J. F. and Kirkbride, J. H., A new synthetic species *Cucumis* (Cucurbitaceae) from interspecific hybridization and chromosome doubling, *Brittonia*, 52, 315–319, 2000.

Chen, J. F. and Lewis, S., New source of nematode resistance was identified in *Cucumis*, *Cucurbit Genet. Coop. Rep.*, 23, 32–35, 2000.

Chen, J. F., Lin, M. S., Qian, C. T., Zhuang, F. Y., and Lewis, S., Identification of *Meloidogyne incognita* (Kofoid & White) Chitwood resistance in *Cucumis hystrix* Chakr. and the progenies of its interspecific hybrid with cucumber (*C. sativus* L.), *J. Nanjing Agric. Univ.*, 24, 21–24, 2001.

Chen, J. F., Luo, X. D., Qian, C. T., Jahn, M. M., Staub, J. E., Zhuang, F. Y., Lou, Q. F., and Ren, G., *Cucumis* monosomic alien addition lines: morphological, cytological, and genotypic analyses, *Theor. Appl. Genet.*, 108, 1343–1348, 2004a.

Chen, J. F., Luo, X. D., Staub, J. E., Qian, C. T., Zhuang, F. Y., and Ren, G., An allotriploid derived from an amphidiploid × diploid mating in *Cucumis* I: production, micropropagation and verification, *Euphytica*, 131, 235–241, 2003a.

Chen, J. F., Moriarty, G., and Jahn, M., Some disease resistance tests in *Cucumis hystrix* and its progenies from interspecific hybridization with cucumber, In *Progress in Cucurbit Genetics and Breeding Research, Proceedings of Cucurbitaceae 2004, the 8th EUCARPIA Meeting on Cucurbit Genetics and Breeding*, Lebeda, A. and Paris, H. S., Eds., Olomouc, Czech Republic: Palacký University in Olomouc, pp. 189–196, 2004b.

Chen, J. F. and Staub, J. E., Regeneration of interspecific hybrids of *Cucumis sativus* L. × *C. hystrix* Char. by direct embryo culture, *Cucurbit Genet. Coop. Rep.*, 19, 34–35, 1996.

Chen, J. F. and Staub, J. E., Attempts at colchicine doubling of an interspecific hybrid of *Cucumis sativus* × *C. hystrix*, *Cucurbit Genet. Coop. Rep.*, 20, 24–26, 1997.

Chen, J. F., Staub, J. E., Adelberg, J., Lewis, S., and Kunkie, B., Synthesis and preliminary characterization of a new species (amphidiploid) in *Cucumis*, *Euphytica*, 123, 315–322, 2002.

Chen, J. F., Staub, J. E., Qian, C. T., Jiang, J. M., Luo, X. D., and Zhuang, F. Y., Reproduction and cytogenetic characterization of interspecific hybrids derived from *Cucumis hystrix* Chakr. × *C. sativus* L., *Theor. Appl. Genet.*, 106, 688–695, 2003b.

Chen, J. F., Staub, J. E., Tashiro, Y., Isshiki, S., and Miyazaki, S., Successful interspecific hybridization between *Cucumis sativus* L. and *Cucumis hystrix* Chakr, *Euphytica*, 96, 413–419, 1997b.

Chen, J. F., Zhuang, F. Y., Liu, X. A., and Qian, C. T., Reciprocal differences of morphological and DNA characters in interspecific hybridization in *Cucumis*, *Can. J. Bot.*, 82, 16–21, 2004c.

Chen, W.-S., Chiu, C.-C., Liu, H.-Y., Cheng, J.-T., Lin, C.-C., Wu, Y.-J., and Chang, H.-Y., Gene transfer via pollen-tube pathway for anti-Fusarium wilt in watermelon, *Biochem. Molec. Biol. Internat.*, 46, 1201–1209, 1998b.

Chen, Y. M., Preliminary study on local germplasm of pumpkins of inner Mongolia (in Chinese), *Crop Genet. Res.*, 2, 13–14, 1993.

Chernova, L. V., Influence of cryogenic seed treatment on the development and yield of watermelon in an unheated plastic greenhouse, *Nauchno Tekhnicheskii Byulleten Vsesoyuznogo Ordena Lenina I Ordena Druzhby Narodov Nauchno*, 199, 57–60, 1990.

Chigwe, C. F. B. and Saka, V. W., Collection and characterization of Malawi pumpkin germplasm. Zimbabwe, *J. Agric. Res*, 32, 139–147, (actual publication date 1996) 1994.

Choer, E., Avaliação morfologica de acessos de *Cucurbita* spp., *Agropecuaria Clima Temperado*, 2, 151–158, 1999.

Choi, P. S., Soh, W. Y., and Kim, Y. S., Genetic transformation and plant regeneration of watermelon using *Agrobacterium tumefaciens*, *Plant Cell Rep.*, 13, 344–348, 1994.

Chung, S.-M., Decker-Walters, D. S., and Staub, J. E., Genetic relationships within the Cucurbitaceae as assessed by ccSSR marker and sequence analysis, *Can. J. Bot.*, 81, 814–832, 2003.

Chung, S.-M. and Staub, J. E., The development and evaluation of consensus chloroplast primer pairs that possess highly variable sequence regions in a diverse array of plant taxa, *Theor. Appl. Genet*, 107, 757–767, 2003.

Chung, S.-M. and Staub, J. E., Consensus chloroplast primer analysis: a molecular tool for evolutionary studies in Cucurbitaceae, In *Progress in Cucurbit Genetics and Breeding Research, Proceedings of Cucurbitaceae 2004, the 8th EUCARPIA Meeting on Cucurbit Genetics and Breeding*, Lebeda, A. and Paris, H. S., Eds., Olomouc, Czech Republic: Palacký University in Olomouc, pp. 477–483, 2004.

Clark, R. L., Widrlechner, M. P., Reitsma, K. R., and Block, C. C., Cucurbit germplasm at the north central regional plant introduction station, Ames, Iowa, *HortSci.*, 26, (326), 450–451, 1991.

Clough, G. H. and Hamm, P. B., Coat protein transgenic resistance to watermelon mosaic and zucchini yellow mosaic virus in squash and cantaloupe, *Plant Dis.*, 79, 1107–1109, 1995.

Colijn-Hooymans, C. M., Hakkert, J. C., Jansen, J. C., and Custers, J. B. M., Competence for regeneration of cucumber cotyledons is restricted to specific developmental stages, *Plant Cell Tiss. Org. Cult.*, 39, 211–217, 1994.

Compton, M. E., Dark pretreatment improves adventitious shoot organogenesis from cotyledons of diploid watermelon, *Plant Cell Tiss. Org. Cult.*, 58, 185–188, 1999.

Compton, M. E., Interaction between explant size and cultivar affects shoot organogenic competence of watermelon cotyledons, *HortSci.*, 35, 749–750, 2000.

Compton, M. E., Barnett, N., and Gray, D. J., Use of fluorescein diacetate (FDA) to determine ploidy of *in vitro* watermelon shoots, *Plant Cell Tiss. Org. Cult.*, 58, 199–203, 1999.

Compton, M. E., Gray, D. J., and Elmstrom, G. W., Regeneration of tetraploid plants from cotyledons of diploid watermelon, *Proc. Fla. State Hort. Soc.*, 107, 107–109, 1994.

Compton, M. E., Gray, D. J., and Elmstrom, G. W., Identification of tetraploid regenerants from cotyledons of diploid watermelon cultured *in vitro*, *Euphytica*, 87, 165–172, 1996.

Compton, M. E., Gray, D. J., and Gaba, B., Use of tissue culture and biotechnology for the genetic improvement of watermelon, *Plant Cell Tiss. Org. Cult.*, 77, 231–243, 2004.

Costa, J., Catala, M. S., Cortez, C., Nuez, F., Abadia, J., and Cuartero, J., Evaluación de la variabilidad en los principales tipos de melon cultivados en España, *Investigación Agraria Producción Y Protección Vegetales*, 4, 43–57, 1989.

Cox, R. L., Abel, C., and Gustafson, E., A novel use for bees: controlled pollination of germplasm collections, *Am. Bee J.*, 136, 709–712, 1996.

Coyne, D. P. and Hill, R. M., "Butternut Patriot" squash, *HortSci.*, 11, 618, 1976.

Cramer, C. S. and Wehner, T. C., Performance of three selection cycles from four slicing cucumber populations hybridized with a tester, *J. Am. Soc. Hort. Sci.*, 123, 396–400, 1998a.

Cramer, C. S. and Wehner, T. C., Fruit yield and yield component means and correlations of four slicing cucumber populations improved through six to ten cycles of recurrent selection, *J. Am. Soc. Hort. Sci.*, 123, 388–395, 1998b.

Cramer, C. S. and Wehner, T. C., Testcross performance of three selection cycles from four pickling cucumber populations, *J. Am. Soc. Hort. Sci.*, 124, 257–261, 1999a.

Cramer, C. S. and Wehner, T. C., Little heterosis for yield and yield components in hybrids of six cucumber inbreds, *Euphytica*, 110, 99–108, 1999b.

Cramer, C. S. and Wehner, T. C., Path analysis of the correlation between fruit number and plant traits of cucumber populations, *HortSci.*, 35, 708–711, 2000.

Crosby, K. M., Screening *Cucumis melo* L. *agrestis* germplasm for resistance to *Monosporascus cannonballus*, *Subtropic. Plant Sci.*, 53, 24–26, 2001.

Cuny, F., Dumas de Vaulx, R., Longhi, B., and Siadous, R., Analysis of muskmelon plants (*Cucumis melo* L.) obtained after pollination with gamma-irradiated pollen: effect of different doses, *Agronomie*, 12, 623–630, 1992.

Custers, J. B. M. and Bergervoet, J. H. W., *In vitro* adventitious bud formation on seedling and embryo plants of *Cucumis sativus* L, *Cucurbit Genet. Coop. Rep.*, 7, 2–4, 1984.

Custers, J. B. M. and den Nijs, A. P. M., Effect of aminoethoxyvinylglycine (AVG), environment, and genotype in overcoming hybridization barriers between *Cucumis* species, *Euphytica*, 35, 639–647, 1986.

Custers, J. B. M., den Nijs, A. P. M., and Riepma, A. W., Reciprocal crosses between *Cucumis africanus* L.f. and *C. metuliferus* Naud. Effects of pollination aids, physiological condition and genetic constitution of the maternal parent on crossability, *Cucurbit Genet. Coop. Rep.*, 4, 50–52, 1981.

Custers, J. B. M., Zijlstra, S., and Jansen, J., Somaclonal variation in cucumber (*Cucumis sativus* L.) plants regenerated via somatic embryogenesis, *Acta Bot. Neerl.*, 39, 153–161, 1990.

Dabauza, M., Bordas, M., and Salvador, A., Plant regeneration and *Agrobacterium*-mediated transformation of cotyledon explants of *Citrullus colocynthis* (L.) Schrad, *Plant Cell Rep.*, 16, 888–892, 1997.

Dabauza, M., Gonzalez-Candelas, L., Bordas, M., Roig, L. A., Ramon, D., and Moreno, V., Regeneration and characterisation of *Cucumis melo* L.(+) *Cucumis anguria* L. var. *longipes* (Hook. fil.) Meeuse somatic hybrids, *Plant Cell Tissue Org. Cult.*, 52, 123–131, 1998.

Dalmasso, A., Dumas de Vaulx, R., and Pitrat, M., Response of some *Cucurbita* and *Cucumis* accessions to three *Meloidogyne* species, *Cucurbit Genet. Coop. Rep.*, 4, 53–54, 1981.

Dane, F., Evolutionary studies in the genus *Cucumis*, Ph.D. diss., Fort Collins, CO.: Colorado State Univ., 1976.

Dane, F., Cucurbits, In *Isozymes in Plant Genetics and Breeding, Part B*, Tanksley, S. D. and Orton, T. J., Eds., Amsterdam: Elsevier, pp. 369–390, 1983.

Dane, F., Cytogenetics of the genus *Cucumis*, In *Chromosome Engineering in Plants: Genetics, Breeding and Evolution, Part B*, Tsuchiya, T. and Gupta, P. K., Eds., Amersterdam: Elsevier, 1991.

Danin-Poleg, Y., Reis, N., Baudracco-Arnas, S., Pitrat, M., Staub, J. E., Oliver, M., Arus, P., de Vincente, C. M., and Katzir, N., Simple sequence repeats in *Cucumis* mapping and map merging, *Genome*, 43, 963–974, 2000a.

Danin-Poleg, Y., Reis, N., Tzuri, G., and Katzir, N., Simple sequence repeats as reference points in *Cucumis* mapping, In *Cucurbitaceae 98: Evaluation and Enhancement of Cucurbit Germplasm*, McCreight, J. D., Ed., Alexandria, VA: ASHS Press, pp. 349–353, 1998a.

Danin-Poleg, Y., Reis, N., Tzuri, G., and Katzir, N., Development and characterization of microsatellite in *Cucumis*, *Theor. Appl. Genet.*, 102, 61–72, 2001.

Danin-Poleg, Y., Tadmor, Y., Tzuri, G., Reis, N., Hirschberg, J., and Katzir, N., Construction of a genetic map of melon with molecular markers and horticultural traits, and localization of genes associated with ZYMV resistance, *Euphytica*, 125, 373–384, 2002.

Danin-Poleg, Y., Tzuri, G., Karchi, Z., Cregan, P. B., and Katzir, N., Length polymorphism and homologies of microsatellites in several cucurbitacea species, In *Cucurbits towards 2000, Proceedings of the Sixth Eucarpia Meeting on Cucurbit Genetics and Breeding*, Gómez-Guillamón, M. L. et al., Eds., Algarrobo, Málaga, Spain: Estación Experimental "La Mayora", pp. 179–186, 1996.

Danin-Poleg, Y., Tzuri, G., Reis, N., Karchi, Z., and Katzir, N., Search for molecular markers associated with resistance to viruses in melon, *Acta Hort.*, 510, 399–404, 2000b.

Danin-Poleg, Y., Tzuri, G., Reis, N., and Katzir, N., Application of inter-SSR markers in melon (*Cucumis melo* L.), *Cucurbit Genet. Coop. Rep.*, 21, 25–28, 1998b.

Daryono, B. S., Somowiyarjo, S., and Natsuaki, K. T., Detection of resistant melons to the Indonesian isolate of KGMM, In *Progress in Cucurbit Genetics and Breeding Research*, Lebeda, A. and Paris, H. S., Eds., Olomouc, Czech Republic: Palacký University, pp. 213–217, 2004.

Deakin, J. R., Bohn, G. W., and Whitaker, T. W., Interspecific hybridization in *Cucumis*, *Econ. Bot.*, 25, 195–211, 1971.

Debeaujon, I. and Branchard, M., Induction of somatic embryogenesis and caulogenesis from cotyledon and leaf protoplast-derived colonies of melon (*Cucumis melo* L.), *Plant Cell Rep.*, 12, 37–40, 1992.

Decker, D. S., Origin(s), evolution, and systematics of *Cucurbita pepo* (Cucurbitaceae), *Econ. Bot.*, 42, 4–15, 1988.

Decker, D. S. and Wilson, H. D., Allozyme variation in the *Cucurbita pepo* complex: *C. pepo* var. *ovifera* vs. *C. texana*, *Syst. Bot.*, 12, 263–273, 1987.

Decker-Walters, D. S., Chung, S. M., and Staub, J. E., Plastid sequence evolution: a new pattern of nucleotide substitutions in the Cucurbitaceae, *J. Molec. Evol.*, 58, 606–614, 2004a.

Decker-Walters, D., Staub, J., Lopez-Sese, A., and Nakata, E., Diversity in landraces and cultivars of bottle gourd (*Lagenaria siceraria*; Cucurbitaceae) as assessed by random amplified polymorphic DNA, *Genet. Resour. Crop Evol.*, 48, 369–380, 2001.

Decker-Walters, D. S., Chung, S. M., Staub, J. E., Quemada, H. D., and López-Sesé, A. I., The origin and genetic affinities of wild populations of melon (*Cucumis melo* Cucurbitaceae) in North America, *Plant Syst. Evol.*, 233, 183–197, 2002a.

Decker-Walters, D. S., Staub, J. E., Chung, S. M., Nakata, E., and Quemada, H. D., Diversity in free-living populations of *Cucurbita pepo* (Cucurbitaceae) as assessed by random amplified polymorphic DNA, *Syst. Bot.*, 27, 19–28, 2002b.

Decker-Walters, D., Staub, J., López-Sesé, A., and Nakata, E., Diversity in landraces and cultivars of bottle gourd (*Lagenaria siceraria*, Cucurbitaceae) as assessed by random amplified polymorphic DNA, *Genet. Res. Crop Evol.*, 48, 369–380, 2002c.

Decker-Walters, D. S., Walters, T. W., Cowan, C. W., and Smith, B. D., Isozymic characterization of wild populations of *Cucurbita pepo*, *J. Ethnobiol.*, 13, 55–72, 1993.

Decker-Walters, D. S., Walters, T. W., Posluszny, U., and Kevan, P. G., Genealogy and gene flow among annual domesticated species of *Cucurbita*, *Can. J. Bot.*, 68, 782–789, 1990.

Decker-Walters, D. S., Wilkins-Ellert, M., Chung, S.-M., and Staub, J. E., Discovery and genetic assessment of wild bottle gourd [*Lagenaria siceraria* (Mol.) Standley. Cucurbitaceae] from Zimbabwe, *Econ. Bot.*, 58, 501–508, 2004b.

de C.S. Dias, R., Picó, B., Adalid, A. M., Herraiz, J., Espinós, A., and Nuez, F., Field resistance to melon vine decline in wild accessions of *Cucumis* spp. and in a Spanish accession of *Cucumis melo*, *Cucurbit Genet. Coop. Rep.*, 24, 23–25, 2001.

de la Cuadra, C. and Varela, F., Cucurbitaceae genetic resource collection in base bank of CRF-INIA, In *Cucurbits towards 2000. Proceedings of the Sixth Eucarpia Meeting on Cucurbit Genetics and Breeding*, Gómez-Guillamón, M. L. et al., Eds., Algarrobo, Málaga, Spain: Estación Experimental La Mayora, pp. 88–96, 1996.

den Nijs, A. P. M. and Custers, J. B. M., Introducing resistances into the cucumber by interspecific hybridization, In *Biology and Utilization of the Cucurbitaceae*, Bates, D. M. et al., Eds., Ithaca, NY: Comstock Publishing Associates, pp. 382–396, 1990.

den Nijs, A. P. M., Custers, J. M. R., and Kooistra, A. J., Reciprocal crosses between *Cucumis africanus* L. and *Cucumis metuliferus* Naud. I. Overcoming barriers to fertilization by mentor pollen and AVG, *Cucurbit Genet. Coop. Rep.*, 3, 60–62, 1980.

den Nijs, A. P. M. and Oost, E. N., Effect of mentor pollen on pistil-pollen incongruities among species of *Cucumis*, *Euphytica*, 29, 267–271, 1980.

de Oliveira, A. C. B., Maluf, W. R., Pinto, J. E. B. P., and Azevedo, S. M., Resistance to papaya ringspot virus in summer squash *Cucurbita pepo* L. introgressed from an interspecific *C. pepo* $\times$ *C. moschata* cross, *Euphytica*, 132, 211–215, 2003.

DeQueiroz, M. A., Romão, R. L., De C.S. Dias, R., De A. Assis, J. G., Borges, R. M. E., Da F. Ferreria, M. A. J., Ramos, S. R. R., Costa, M. S. V., and da C. C. L. Moura, M., Watermelon germplasm bank for the Northeast of Brazil. An integrated approach, M.L. Gómez-Guillamón et al., Eds., *Cucurbits towards 2000. Proceedings of the Sixth Eucarpia Meeting on Cucurbit Genetics and Breeding*, 97–103, 1996.

de Vaulx, R. D. and Pitrat, M., Interspecific cross between *Cucurbita pepo* and *C. martinezii*, *Cucurbit Genet. Coop. Rep.*, 2, 35, 1979.

de Vaulx, R. D. and Pitrat, M., Realization of the interspecific hybridization (F_1 and BC_1) between *Cucurbita pepo* and *C. ecuadorensis*, *Cucurbit Genet. Coop. Rep.*, 3, 42, 1980.

DeVeaux, J. S. and Schultz, E. B., Jr, Development of buffalo gourd (*Cucurbita foetidissima*) as a semiaridland starch and oil crop, *Econ. Bot.*, 39, 454–472, 1985.

de Wilde, W. J. J. O. and Duyfjes, B. E. E., Synopsis of *Momordica* (Cucurbitaceae) in SE Asia and Malesia, *Botanicheskii Zhurnal*, 87(3), 132–148, 2002.

De Winter, B., A new species of *Citrullus* (Benincaseae) from the Namib desert, Namibia, *Bothalia*, 20, 209–211, 1990.

Dhaliwal, M. S., Evaluation of gherkin germplasm, *Cucurbit Genet. Coop. Rep.*, 20, 60–62, 1997.

Díez, M. J., Picó, B., and Nuez, F., compil, *Cucurbit Genetic Resources in Europe, Ad hoc* meeting, 19 January 2002, Adana, Turkey, Rome, Italy: IPGRI, 2002.

Dijkhuizen, A., Kennard, W. C., Havey, M. J., and Staub, J. E., RFLP variation and genetic relationships in cultivated cucumber, *Euphytica*, 90, 79–87, 1996.

Dijkhuizen, A. and Staub, J. E., Effects of environment and genetic background on QTL affecting yield and fruit quality traits in a wide cross in cucumber [*Cucumis sativus* L. ×*Cucumis hardwickii* (R.) Alef.], *J. New Seeds*, 4, 1–30, 2003.

Dirks, R. and van Buggenum, M., Regeneration of plants from mesophyll protoplasts of haploid and diploid *Cucumis sativus, Physiol. Plant* 82, A18, (Poster Abstr.), 1991.

Doebley, J. F., Isozymic evidence and the evolution of crop plants, In *Isozymes in Plant Biology*, Soltis, D. E. and Soltis, P. S., Eds., Portland, OR: Dioscorides Press, pp. 165–191, 1989.

Doebley, J., Molecular evidence and the evolution of maize, *Econ. Bot.*, 44(3 suppl.), 6–27, 1990.

Dogimont, C., Bussemakers, A., Slama, S., Martin, J., Lecoq, H., and Pitrat, M., Diversity of resistance sources of cucurbit aphid-borne yellows luteovirus in melon and genetics of resistance, In *Cucurbits towards 2000. Proceedings of the Sixth Eucarpia Meeting on Cucurbit Genetics and Breeding*, Gómez-Guillamón, M. L. et al., Eds., Algarrobo, Málaga, Spain: Estación Experimental La Mayora, pp. 328–333, 1996.

Dogimont, C., Leconte, L., Périn, C., Thabuis, A., Lecoq, H., and Pitrat, M., Identification of QTLs contributing to resistance to different strains of cucumber mosaic cucumovirus in melon, *Acta Hort.*, 510, 391–398, 2000.

Dong, J. Z., Yang, M. Z., Jia, S. R., and Chua, N. H., Transformation of melon (*Cucumis melo* L.) and expression from the cauliflower mosaic virus 35S promoter in transgenic melon plants, *Bio/Technol.*, 9, 858–863, 1991.

Duarte, R. L. R. and de Queiroz, M. A., Coleta e caracterização de germoplasma de cucurbitaceas do Piauí, *Revista Científica Rural*, 7(1), 1–5, 2002.

Dumas de Vaulx, R., Production of haploid plants in melon (*Cucumis melo* L.) after pollination by *Cucumis ficifolius* A. Rich, *Comptes Rendus Hebdomadaires des Seances de l' Academie des Sciences, Série 3*, 289, 875–878, 1979.

Dumas de Vaulx, R. and Chambonnet, D., Obtention of embryos and plants from *in vitro* culture of unfertilized ovules of *Cucurbita pepo*, In *Genetic Manipulation in Plant Breeding*, Horn, W., Jensen, C. J., Odenback, W., and Schieder, O., Eds., Berlin: De Gruyter, pp. 295–297, 1986.

ECP/GR, *Vegetables, Medicinal and Aromatic Plants Network*, http://www.ecpgr.cgiar.org/Networks/Vegetables/vegetables.htm (accessed on May 2005).

Edwards, M. D., Lower, R. L., and Staub, J. E., Influence of seed harvesting and handling procedures on germination of cucumber seeds, *J. Am. Soc. Hort. Sci.*, 111, 507–512, 1986.

Ellis, M. D., Jackson, G. S., Skrdla, W. H., and Spencer, H. C., Use of honey bees for controlled interpollination of plant germplasm collections, *HortSci.*, 16, 488–491, 1981.

Ellul, P., Rios, G., Atares, A., Roig, L. A., and Serrano, R., The expression of the *Saccharomyces cerevisiae* HAL1 gene increases salt tolerance in transgenic watermelon (*Citrullus lanatus* (Thumb.) Matsun. & Nakai.), *Theor. Appl. Genet.* 107, 462–469, 2003.

El-Mahdy, I., Metwally, E. I., and EL-Fadly, G. A., *In vitro* differentiation of the immature interspecific embryo derived from the crosses between *Cucurbita pepo* and *Cucurbita ficifolia*, In *Proc. 4th Nat. Conf. of Pest. & Dis. of Veg. & Fruits in Egypt*, Vol. 2, pp. 598–606, 1991.

Elsey, K. D., Insect resistance in the cucurbits; status and potential In *Proceedings of Cucurbitaceae 89: Evaluation and Enhancement of Cucurbit Germplasm*, Thomas, C. E., Ed., USDA/ARS, Charleston, SC, pp. 49–59, 1989.

Esquinas-Alcázar, J. T., Alloenzyme variation and relationships in the genus *Cucumis*, Ph.D. Diss., Davis, CA: Univ. of California, 1977.

Esquinas-Alcázar, J. T., Alloenzyme variation and relationships among Spanish land-races of *Cucumis melo* L, *Kulturpflanze*, 29, 337–352, 1981.

Esquinas-Alcázar, J. T. and Gulick, P. J., *Genetic Resources of Cucurbitaceae: A Global Report*, Rome: IBPGR, 1983.

Ezura, H., Amagai, H., Kikuta, I., Kobota, M., and Oosawa, K., Selection of somaclonal variants with low-temperature germinability in melon (*Cucumis melo* L.), *Plant Cell Rep.*, 14, 684–688, 1995.

Ezura, H., Amagai, H., and Oosawa, K., Efficient production of triploid melon plants by *in-vitro* culture of abnormal embryos excised from dried seeds of diploid×tetraploid crosses and their characteristics, *Jpn. J. Breed.*, 43, 193–199, 1993.

Ezura, H., Amagai, H., Yoshioka, K., and Oosawa, K., Efficient production of tetraploid melon (*Cucumis melo* L.) by somatic embryogenesis, *Jpn. J. Breed.*, 42, 137–144, 1992a.

Ezura, H., Amagai, H., Yoshioka, K., and Oosawa, K., Highly frequent appearance of tetraploidy in regenerated plants, an universal phenomenon in tissue cultures of melon (*Cucumis melo* L.), *Plant Sci*, 85, 209–213, 1992b.

Ezura, H., Kikuta, I., and Oosawa, K., Production of aneuploid melon plants following *in vitro* culture of seeds from a triploid×diploid cross, *Plant Cell Tiss. Org. Cult.*, 38, 61–63, 1994.

Ezura, H. and Oosawa, K., Selective regeneration of plants from diploid and tetraploid cells in adventitious shoot cultures of melon (*Cucumis melo* L.), *Plant Tiss. Cult. Lett.*, 11, 26–33, 1994a.

Ezura, H. and Oosawa, K., Ploidy of somatic embryos and the ability to regenerate plantlets in melon (*Cucumis melo* L.), *Plant Cell Rep.*, 14, 107–111, 1994b.

Ezura, H., Yuhashi, K. I., Yasuta, T., and Minamisawa, K., Effect of ethylene on *Agrobacterium tumefaciens*-mediated gene transfer to melon, *Plant Breeding*, 119, 75–79, 2000.

Fan, K., Qiao, X., Gu, L., and Lin, C., A study on the viability of long-term stored cucumber seed and the effects of this storage on plant characteristics, In *Proceedings of Cucurbitaceae 89: Evaluation and Enhancement of Cucurbit Germplasm*, Thomas, C. E., Ed., USDA/ARS, Charleston, SC, Charleston, SC, pp. 92–93, 1989.

Fang, G. and Grumet, R., *Agrobacterium tumefaciens* mediated transformation and regeneration of muskmelon plants, *Plant Cell Rep.*, 9, 160–164, 1990.

Fang, G. and Grumet, R., Genetic engineering of potyvirus resistance using constructs derived from the zucchini yellow mosaic virus coat protein gene, *Mol. Plant-Microbe Interact.*, 6, 358–367, 1993.

Fanourakis, N. E. and Simon, P. W., Analysis of genetic linkage in cucumber, *J. Hered.*, 78, 238–242, 1987.

Fanourakis, N., Tsekoura, Z., and Nanou, E., Morphological characteristics and powdery mildew resistance of *Cucumis melo* land races in Greece, *Acta Hort.*, 510, 241–245, 2000.

FAO, *The State of the World's Plant Genetic Resources for Food and Agriculture*, Food and Agriculture Organization of the United Nations, Rome, 1998.

FAO, *FAOSTAT Agricultural Database*, Food and Agriculture Organization of the United Nations, Rome, Italy, http://apps.fao.org (accessed on March 2005).

Faris, N. M., Burza, W., Malepszy, S., and Niemirowicz-Szczytt, K., Direct regeneration from leaf explant of cucumber (*Cucumis sativus* L.) haploids, *Sci. Pap. Agric. Univ. Cracow*, 50, 249–252, 1997.

Faris, N. M., Nikolova, V., and Niemirowicz-Szczytt, K., The effect of gamma irradiation dose on cucumber (*Cucumis sativus* L.) haploid embryo production, *Acta Physiol. Plant.*, 21, 391–396, 1999.

Fassuliotis, G., Self-fertilization of *Cucumis metuliferus* Naud. and its cross compatibility with C. *melo* L, *J. Am. Soc. Hort. Sci.*, 102, 336–339, 1977.

Fassuliotis, G. and Nelson, B. V., Interspecfic hybrids of *Cucumis metuliferus*×*C. anguria* obtained through embryo cuture and somatic embryogenesis, *Euphytica*, 37, 53–60, 1988.

Fazio, G., Comparative study of marker-assisted and phenotypic selection and genetic analysis of yield components in cucumber, PhD Thesis, University of Wisconsin, Department of Horticulture, Madison, 2001.

Fazio, G. and Staub, J. E., Comparative analysis of response to phenotypic and marker-assisted selection for multiple lateral branching in cucumber (*Cucumis sativus* L.), *Theor. Appl. Genet.*, 107, 875–883, 2003.

Fazio, G., Staub, J. E., and Chung, S. M., Development and characterization of PCR markers in cucumber (*Cucumis sativus* L.), *J. Am. Soc. Hort. Sci.*, 127, 545–557, 2002.

Fazio, G., Staub, J. E., and Stevens, M. L., Genetic mapping and QTL analysis of horticultural traits in cucumber (*Cucumis sativus* L.) using recombinant inbred lines, *Theor. Appl. Genet.*, 107, 864–874, 2003.

Fellner, M. and Lebeda, A., Callus induction and protoplast isolation from tissues of *Cucumis sativus* L. and C. *melo* L. seedlings, *Biol. Plant.*, 41, 11–24, 1998.

Ferriol, M., Picó, B., de Córdova, P. F., and Nuez, F., Molecular diversity of a germplasm collection of squash (*Cucurbita moschata*) determined by SRAP and AFLP markers, *Crop Sci.*, 44, 653–664, 2004a.

Ferriol, M., Picó, B., and Nuez, F., Genetic diversity of a germplasm collection of *Cucurbita pepo* using SRAP and AFLP markers, *Theor. Appl. Genet.*, 107, 271–282, 2003a.

Ferriol, M., Picó, B., and Nuez, F., Morphological and molecular diversity of a collection of *Cucurbita maxima* landraces, *J. Am. Soc. Hort. Sci.*, 129, 60–69, 2004b.

Ferriol, M., Picó, B., and Nuez, F., Genetic diversity of some accessions of *Cucurbita maxima* from Spain using RAPD and SBAP markers, *Genet. Res. Crop Evol.*, 50, 227–238, 2003b.

Ficcadenti, N., Veronese, P., Sestili, S., Crino, P., Lucretti, S., Schiavi, M., and Saccardo, F., Influence of genotype on the induction of haploidy in *Cucumis melo* L. by using irradiated pollen, *J. Genet. Breed.*, 49, 359–364, 1995.

Filov, A. I., *Rukovodstvo po aprobacii selskochozijajstvennych kultur, Tom 6, Bachčevyje kultury*, Gosudarst-vennoje Izdatelstvo Selskochozijajstvennoj Literatury, Moskva, Leningrad, 1954.

Filov, A. I., *Bachčevodstvo*, Gosudarstvennoje Izdatelstvo Selskochozijajstvennoj Literatury, Moskva, 1959.

Fisher, R. M. and Pomeroy, N., Pollination of greenhouse muskmelons by bumble bees (Hymenoptera: Apidae), *J. Econ. Entomol.*, 82, 1061–1066, 1989.

Frankel, O. H., Genetic perspectives of germplasm conservation, In *Genetic Manipulation: Impact on Man and Society*, Arber, W. K. et al., Eds., Cambridge: Cambridge University Press, pp. 161–170, 1984.

Fukino, N., Kuzuya, M., Kunishia, M., and Matsumoto, S., Characterization of a simple sequence repeats (SSR) and development of SSR markers in melon (*Cucumis melo*), In *Progress in Cucurbit Genetics and Breeding Research*, Lebeda, A. and Paris, H. S., Eds., Olomouc, Czech Republic: Palacký University, pp. 503–506, 2004.

Fukino, N., Taneishi, M., Saito, T., Nishijima, T., and Hirai, M., Construction of a linkage map and genetic analysis for resistance to cotton aphid and powdery mildew in melon, *Acta Hort.*, 588, 283–286, 2002.

Fulton, T. M., Beck-Bunn, T., Emmatty, D., Eshed, Y., Lopez, J., Petiard, V., Uhlig, J., Zamir, D., and Tanksley, S. D., QTL analysis of an advanced backcross of *Lycopersicon peruvianum* to the cultivated tomato and comparisons with QTLs found in other wild species, *Theor. Appl. Genet.*, 95, 881–894, 1997.

Fulton, T. M., Grandillo, S., Beck-Bunn, T., Fridman, E., Frampton, A., Lopez, J., Petiard, V., Uhlig, J., Zamir, D., and Tanksley, S. D., Advanced backcross QTL analysis of a *Lycopersicon esculentum* × *L. parviflorum* cross, *Theor. Appl. Genet.*, 100, 1025–1042, 2000.

Fursa, T. B., and Filov, A. I., Tykvennye (arbuz, tykva), In *Kulturnaja Flora SSSR, 21*, Kolos, Moskva, 1982.

Gaba, V., Zelcher, A., and Gal-On, A., Cucurbit biotechnology: the importance of virus resistance, *In vitro* Cell. Dev. Biol. Plant., 40, 346–358, 2004.

Gajdová, J., Lebeda, A., and Navrátilová, B., Protoplast cultures of *Cucumis* and *Cucurbita* spp., In *Progress in Cucurbit Genetics and Breeding Research, Proceedings of Cucurbitaceae 2004, the 8th EUCARPIA Meeting on Cucurbit Genetics and Breeding*, Lebeda, A. and Paris, H. S., Eds., Olomouc, Czech Republic: Palacký University, pp. 441–454, 2004.

Galperin, M., A melon genotype with superior competence for regeneration and transformation, *Plant Breed.*, 122, 66–69, 2003.

Gao, W., Fang, J., Zheng, D., Li, Y., Lu, X., Rao, R. V., Hodgkin, T., and Zhang, Z., Utilization of germplasm conserved in Chinese national genebanks—a survey, *Plant Genet. Resour. Newslet.*, 123, 1–8, 2000.

Gao, W., Fang, J., Zheng, D., Li Y., Lu, X., Rao, V. R., Hodgkin, T., and Zhang, Z., The utilization of germplasm conserved in Chinese national genebanks, In *Plant Genetic Resource Conservation and Use in China. Proceedings of National Workshop on Conservation and Utilization of Plant Genetic Resources*, Gao W., Rao, V.R., and Zhou, M., Eds., Beijing: Chinese Academy of Agricultural Sciences and IPGRI Office for East Asia, pp. 40–52, 25–27 October 1999, Beijing, China, 2001.

Garcia, E., Jamilena, M., Alvarez, J. I., Arnedo, T., Oliver, J. L., and Lozano, R., Genetic relationships among melon breeding lines revealed by RAPD markers and agronomic traits, *Theor. Appl. Genet.*, 96, 878–885, 1998.

Garcia-Mas, J., van Leeuwen, H., Monforte, A. J., de Vincente, M. C., Puigdomenech, P., and Arus, P., Mapping of resistance gene homologues in melon, *Plant Sci.*, 161, 165–172, 2001.

Garcia-Mas, J., Morales, M., van Leeuwen, H., Monforte, A. J., Puigdomenech, P., Arus, P., Nieto, C. et al., A physical map covering the *Nsv* locus in melon, In *Progress in Cucurbit Genetics and Breeding Research*, Lebeda, A. and Paris, H. S., Eds., Olomouc, Czech Republic: Palacký University, pp. 209–212, 2004.

Garcia-Mas, J., Oliver, M., Gómez-Paniagua, H., and de Vicente, M. C., Comparing AFLP, RAPD and RFLP markers for measuring genetic diversity in melon, *Theor. Appl. Genet.*, 101, 860–864, 2000.

Gathman, A. C. and Bemis, W. P., Domestication of the buffalo gourd, *Cucurbita foetidissima*, In *Biology and Utilization of the Cucurbitaceae*, Bates, D. M. et al., Eds., Ithaca, NY: Comstock Publishing Associates, pp. 335–348, 1990.

Gémes-Juhász, A., Balogh, P., Ferenczy, A., and Kristof, Z., Effect of optimal stage of female gametophyte and heat treatment on *in vitro* gynogenesis induction in cucumber (*Cucumis sativus* L.), *Plant Cell Rep.*, 21, 105–111, 2002.

Gémes-Juhász, A., Venczel, G., and Balogh, P., Haploid plant induction in zucchini (*Cucurbita pepo* L. convar, Giromontiina Duch) and in cucumber (*Cucumis sativus* L.) lines through *in vitro* gynogenesis, *Acta Hort.*, 447, 623–625, 1996.

Gillaspie, A. G. and Wright, J. M., Evaluation of *Citrullus* sp. germplasm for resistance to watermelon mosaic virus-2, *Plant Dis.*, 77, 352–354, 1993.

Gepts, P. and Papa, R., Possible effects of (trans)gene flow from crops on the genetic diversity from landraces and wild cultivars, *Environ. Biosafety Res.*, 2, 89–100, 2003.

Goldman, A., *Melons for the Passionate Grower*, New York: Artisan/Workman Publishing, 2002.

Gonzalo, M. J., Oliver, M., Garcia-Mas, J., Monfort, A., Dolcet-Sanjuan, R., Katzir, N., Arus, P., and Monforte, A. J., Simple-sequence repeat markers used in merging linkage maps of melon (*Cucumis melo* L.), *Theor. Appl. Genet.*, 110, 337–345, 2005.

Gonsalves, C., Xue, B., and Gonsalves, D., Somatic embryogenesis and regeneration from cotyledon explants of six squash cultivars, *HortSci.*, 30, 1295–1297, 1995.

Gonsalves, C., Xue, B., Yepes, M., Fuch, M., Ling, K., Namba, S., Chee, P., Slightom, J. L., and Gonsalves, D., Transferring cucumber mosaic virus-white leaf strain coat protein gene into *Cucumis melo* L. and evaluating transgenic plants for protection against infections, *J. Am. Soc. Hort. Sci.*, 119, 345–355, 1994.

Greuter, W., McNeill, J., Barrie, F. R., Burdet, H. M., DeMoulin, V., Filgueira, T. S., Nicolson, D. H. et al., *International Code of Botanical Nomenclature (St. Louis Code). Regnum Vegetabile, Volume 138*, Königstein, Germany: Koeltz Scientific Books, 2000.

Grewal, G. S. and Sidhu, A. S., Note on the role of bees in the pollination of *Cucurbita pepo*, *Ind. J. Agric. Sci.*, 49, 386–388, 1979.

Griggs, W. H., Vansell, G. H., and Iwakiri, B. T., Pollen storage: high viability of pollen obtained after storage in home freezer, *Califor. Agric.*, 7(7), 12, 1953.

Grumet, R., Barczak, M., Tabaka, C., and Duvall, R., Aboveground screening for genotypic differences in cucumber root growth in the greenhouse and field, *J. Am. Soc. Hort. Sci.*, 117, 1006–1011, 1992.

Gu, H. H., Zhou, W. J., and Hagberg, P., High frequency spontaneous production of doubled haploid plants in microspore cultures of *Brassica rapa* subsp. *chinensis*, *Euphytica*, 134, 239–245, 2003.

Guillaume, R. and Boissot, N., Resistance to *Diaphania hyalinata* (Lepidoptera: Crambidae) in *Cucumis* species, *J. Econ. Entomol.*, 94, 719–723, 2001.

Guis, M., Amor, M. B., Latché, A., Pech, J. C., and Roustan, J. P., A reliable system for the transformation of cantaloupe charentais melon (*Cucumis melo* L. var. *cantalupensis*) leading to a majority of diploid regenerants, *Sci. Hort.*, 84, 91–99, 2000.

Guis, M., Roustan, J. P., Dogimont, C., Pitrat, M., and Pech, J. C., Melon biotechnology, *Biotech. Genet. Engineer. Rev.*, 15, 289–311, 1998.

Guner, N. and Wehner, T. C., Resistance to a severe strain of zucchini yellow mosaic virus in watermelon, In *Progress in Cucurbit Genetics and Breeding Research*, Lebeda, A. and Paris, H. S., Eds., Olomouc, Czech Republic: Palacký University in Olomouc, 2004.

Gwanama, C., Labuschagne, M. T., and Botha, A. M., Analysis of genetic variation in *Cucurbita moschata* by random amplified polymorphic DNA (RAPD) markers, *Euphytica*, 113, 19–24, 2000.

Hammer, K., Esquivel, M., and Carmona, E., Plant genetic resources in Cuba. Report of a collecting mission, February–March 1990, *FAO/IBPGR Plant Genet. Resour. Newslet.*, 86, 21–27, 1991.

Hancock, J. F., *Plant Evolution and the Origin of Crop Species*, 2nd ed., Wallingford, U.K.: CABI Publishing, 2004.

Handel, S. N., Contrasting gene flow patterns and genetic subdivision in adjacent populations of *Cucumis sativus* (Cucurbitaceae), *Evolution*, 37, 760–771, 1983.

Handel, S. H. and Mishkin, J. L., Temporal shifts in gene flow and seed set: evidence from an experimental population of *Cucumis sativus*, *Evolution*, 38, 1350–1357, 1984.

Harlan, J. R. and deWet, J. M. J., Toward a rational classification of cultivated plants, *Taxon*, 20, 509–517, 1971.

Hashizume, T., Shimamoto, I., and Hirai, M., Construction of a linkage map and QTL analysis of horticultural traits for watermelon [*Citrullus lanatus* (Thunb.) Matsum & Nakai] using RAPD, RFLP and ISSR markers, *Theor. Appl. Genet.*, 106, 779–785, 2003.

Hassan, A. A., Obaji, U. A., Wafi, M. S., Quuronfilan, N. E., Al-Masry, H. H., and Al-Rays, M. A., Evaluation of domestic and wild *Cucumis melo* germplasm for resistance to the yellow-stunting disorder in the United Arab Emirates, *Egypt. J. Hort.*, 17, 181–199, 1990.

Hassan, A. A., Quronfilah, N. E., Obaji, U. A., Al-Rays, M. A., and Wafi, M. S., Evaluation of domestic and wild *Citrullus* germplasm for resistance to the yellow-stunting disorder in the United Arab Emirates, *Egypt. J. Hort.*, 18, 11–21, 1991.

Hawkins, L. K., Dane, F., Kubisiak, T. L., Rhodes, B. B., and Jarret, R. L., Linkage mapping in a watermelon population segregating for fusarium wilt resistance, *J. Am. Soc. Hort. Sci.*, 126, 344–350, 2001.

Helm, M. A. and Hemleben, V., Characterization of a new prominent satellite DNA of *Cucumis metuliferus* and differential distribution of satellite DNA in cultivated and wild species of *Cucumis* and in related genera of Cucurbitaceae, *Euphytica*, 94, 219–226, 1997.

Hema, M. V., Prasad, D. T., and Vani, A., Transient expression and stable integration of chimeric Gus gene in watermelon following electroporation, *J. Hort. Sci. Biotechnol.*, 79, 364–369, 2004.

Herrington, M. E. and Prytz, S., Further sources of resistance to ZYMV in *Cucumis melo* L, *Cucurbit Genet. Coop. Rep.*, 13, 25–26, 1990.

Hong, K. H. O. and Hyo-Guen, Y. H. P., Interspecific hybridization between *Cucurbita pepo* and *C. moschata* through ovule culture, *J. Kor. Soc. Hort. Sci.*, 35, 438–448, 1994.

Hopkins, D. L. and Thompson, C. M., Evaluation of *Citrullus* sp. germplasm for resistance to *Acidovorax avenae* subsp. *citrulli*, *Plant Dis.*, 86, 61–64, 2002.

Horejsi, T. and Staub, J. E., Genetic variation in cucumber (*Cucumis sativus* L.) as assessed by random amplified polymorphic DNA, *Genet. Res. Crop Evol.*, 46, 337–350, 1999.

Horejsi, T., Staub, J. E., and Thomas, C., Linkage of random amplified polymorphic DNA markers to downy mildew resistance in cucumber (*Cucumis sativus* L.), *Euphytica*, 115, 105–113, 2000.

Horst, E. K. and Lower, R. L., *Cucumis hardwickii*: a source of germplasm for the cucumber breeder, *Cucurbit Genet. Coop. Rep.*, 1, 5, 1978.

Horváth, J., Reactions of sixty-seven accessions of twelve *Cucumis* species to seven viruses, *Acta Phytopathol. Entomol. Hung.*, 28, 403–414, 1993a.

Horváth, J., Reactions of thirty-nine accessions of four *Cucurbita* species from different origin to seven viruses, *Acta Phytopathol. Entomol. Hung.*, 28, 415–425, 1993b.

Horváth, J., Status of the national cucurbit collection in Hungary, In *Cucurbit Genetic Resources in Europe*, Díez, M. J., Picó, B., and Nuez, F., Eds., Rome, Italy: International Plant Genetic Resources Institute, *Ad hoc Meeting*, 19 January 2002, Adana, Turkey, pp. 30–32, 2002.

Hou, F. and Ma, D., Cucumber germ plasm study and breeding in China, *Acta Hort.*, 523, 221–227, 2000.

Huh, Y. C., Om, Y. H., and Lee, J. M., Utilization of *Citrullus* germplasm with resistance to fusarium wilt (*Fusarium oxysporum* f.sp. *niveum*) for watermelon rootstocks, *Acta Hort.*, 588, 127–132, 2002.

Hurd, P. D., Jr, Linsley, E. G., and Whitaker, T. W., Squash and gourd bees (*Peponapis, Xenoglossa*) and the origin of the cultivated *Cucurbita*, *Evolution*, 25, 218–238, 1971.

Huttner, E., Tucker, W., Vermeulen, A., Ignart, F., Sawyer, B., and Birch, R., Ribozyme genes protecting transgenic melon plants against potyviruses, *Curr. Issues Mol. Biol.*, 3, 27–34, 2001.

Hutton, M. G. and Loy, J. B., Cold germinability of *Cucumis melo*, *Cucurbit Genet. Coop. Rep.*, 8, 11–13, 1985.

Ignart, F. and Weeden, N. F., Allozyme variation in cultivars of *Cucurbita pepo* L, *Euphytica*, 33, 779–785, 1984.

IPGRI, *Descriptors for Melon (Cucumis melo L.)*, Rome: International Plant Genetic Resources Institute, 2003.

Ishimoto, K., Nishizawa, Y., Tabei, Y., Hibi, T., Nakajima, M., and Akutsu, K., Detailed analysis of rice chitinase gene expression in transgenic cucumber plants showing different levels of disease resistance to gray mold (*Botrytis cinerea*), *Plant Sci.*, 162, 655–662, 2002.

Islam, M. S. and Khan, S., Pollen viability of *Momordica dioica* Roxb. as affected by storage period and temperature, *Bangladesh J. Bot.*, 27, 153–155, 1998.

Ivanovska, S. and Posimonova, G., Initial activities for agrobiodiversity conservation in Macedonia (FYR), *IPGRI Newsletter for Europe*, 28, 10, 2004.

Jacquat, C. and Berstossa, G., *Plants from the Markets of Thailand*, Bangkok, Thailand: Editions Duang Kamol, 1990.

Jagger, I. C. and Scott, G. W., Development of powdery mildew resistant cantaloup no. 45, *USDA Circ.*, 441, 6, 1937.

Jahn, M., Munger, H. M., and McCreight, J. D., Breeding cucurbit crops for powdery mildew resistance, In *The Powdery Mildews, A Comprehensive Treatise*, Bélanger, R. R. et al., Eds., American Phytopathological Society: St. Paul, MN, pp. 239–248, 2002.

Jarl, C. I., Bokelmann, G. S., and DeHaas, J. M., Protoplast regeneration and fusion in *Cucumis*: Melon × cucumber, *Plant Cell Tiss. Org. Cult.*, 43, 259–265, 1995.

Jarret, R. L., Merrick, L. C., Holms, T., Evans, J., and Aradhya, M. K., Simple sequence repeats in watermelon (*Citrullus lanatus* (Thunb.) Matsum & Nakai), *Genome*, 40, 433–441, 1997.

Jarret, R. L. and Newman, M., Phylogenetic relationships among species of *Citrullus* and the placement of *C. rehmii* De Winter as determined by Internal Transcribed Spacer (ITS) sequence heterogeneity, *Genet. Res. Crop Evol.*, 47, 215–222, 2000.

Jeffrey, C., A review of the Cucurbitaceae, *Bot. J. Linnean Soc.*, 81, 233–247, 1980.

Jeffrey, C., Systematics of the Cucurbitaceae: an overview, In *Biology and Utilization of the Cucurbitaceae*, Bates, D. M. et al., Eds., Ithaca, NY: Comstock Publishing Associates, pp. 3–7, 1990.

Jeffrey, C., Cucurbitaceae, In *Mansfeld's Encyclopedia of Agricultural and Horticultural Crops*, Hanelt, P., Ed., Vol. 3, Berlin: Springer-Verlag, pp. 1510–1557, 2001.

Ji, Z. X., Guo, C. G., and Ye, Y. F., Study on ultra-dry storage in four melon seeds (in Chinese), *Acta Agric. Zhejiangensis*, 8, 50–53, 1996.

Jobst, J., King, K., and Hemleben, V., Molecular evolution of the internal transcribed spacers (ITS1 and ITS2) and phylogenetic relationships among species of the family Cucurbitaceae, *Molec. Phylogenet. Evol.*, 9, 204–219, 1998.

Joobeur, T., King, J. J., Nolin, S. J., Thomas, C. E., and Dean, R. A., The fusarium wilt resistance locus *Fom-2* of melon contains a single resistance gene with complex features, *Plant J.*, 39, 283–297, 2004.

Juretic, B. and Jelaska, S., Plant development in long-term embryogenic callus lines of *Cucurbita pepo*, *Plant Cell Rep.*, 9, 623–626, 1991.

Juvik, J. A. and Palmer, J. D., Potential of restriction endonuclease analysis of chloroplast DNA for the determination of phylogenetic relationships among members of the Cucurbitaceae, *Cucurbit Genet. Coop. Rep.*, 7, 66–68, 1984.

Kalloo, G., *Vegetable Breeding*, Vol. I, Boca Raton, FL: CRC Press, Inc., 1988.

Karsies, T., Dean, R A., and Thomas, C. E., Toward the development of molecular markers linked to race 2 Fusarium wilt resistance in melon (*Cucumis melo* L.), *Acta Hort.*, 510, 415–419, 2000.

Kasrawi, M. A., Diversity in landraces of summer squash from Jordan, *Genet. Res. Crop Evol.*, 42, 223–230, 1995.

Kato, K., Akashi, Y., Tanaka, K., Wako, T., and Masuda, M., Genetic characterization of East and South Asian melons *Cucumis melo*, by the analysis of molecular polymorphisms and morphological characters, *Acta Hort.*, 588, 217–222, 2002.

Katzir, N., Dannin-Poleg, Y., Tzuri, G., Karchi, Z., Lavi, U., and Cregan, P. B., Application of RADP and SSR analyses to the identification and mapping of melon (*Cucumis melo* L.) varieties, In *Proceedings Cucurbitaceae 94: Evaluation and Enhancement of Cucurbit Germplasm*, Lester, G. E. and Dunlap, J. R., Eds., Edinburg, TX: Gateway Printing and Office, p. 196, 1995.

Katzir, N., Danin-Poleg, Y., Tzuri, G., Karchi, Z., Lavi, U., and Cregan, P. B., Length polymorphism and homologies of microsatellites in several Cucurbitaceae species, *Theor. Appl. Genet.*, 93, 1282–1290, 1996.

Katzir, N., Tadmor, Y., Tzuri, G., Leshzeshen, E., Mozes-Daube, N., Danin-Poleg, Y., and Paris, H. S., Further ISSR and preliminary SSR analysis of relationships among accessions of *Cucurbita pepo*, *Acta Hort.*, 510, 433–439, 2000.

Kennard, W. C. and Havey, M. J., Quantitative trait analysis of fruit quality in cucumber: QTL detection, confirmation, and comparison with mating-design variation, *Theor. Appl. Genet.*, 91, 53–61, 1995.

Kennard, W. C., Poetter, K., Dijkhuizen, A., Meglic, V., Staub, J. E., and Havey, M. J., Linkages among RFLP, RAPD, isozyme, disease-resistance, and morphological markers in narrow and wide crosses of cucumber, *Theor. Appl. Genet.*, 89, 42–48, 1994.

Kho, Y. O., den Nijs, A. P. M., and Franken, J., Interspecific hybridization in II. The crossability of species, an investigation of *in vivo* pollen tube growth and seed set *Cucumis*, *Euphytica*, 29, 661–672, 1980.

Kho, Y. O., Franken, J., and den Nijs, A. P. M., Species crosses under controlled temperature condition, *Cucurbit Genet. Coop. Rep.*, 4, 56–57, 1981.

Khush, G. S., *Cytogenetics of Aneuploids*, New York: Academic Press, 1973.

Kihara, H., Triploid watermelons, *Proc. Am. Soc. Hort. Sci.*, 58, 217–230, 1951.

Kirkbride, J. H., Jr, *Biosystematic Monograph of the Genus Cucumis (Cucurbitaceae)*, Boone, NC: Parkway Publishers, 1993.

Kirkpatrick, K. J., Decker, D. S., and Wilson, H. D., Allozyme differentiation in the *Cucurbita pepo* complex: *C. pepo* var. *medullosa* vs. *C. texana*, *Econ. Bot.*, 39, 289–299, 1985.

Kirkpatrick, K. J. and Wilson, H. D., Interspecific gene flow in *Cucurbita*: *C. texana* vs. *C. pepo*, *Am. J. Bot.*, 75, 519–527, 1988.

Kitamura, S., Notes on *Cucumis* of the Far East, *Acta Phytotax. Geobot.*, 14, 41–44, 1950.

Kitamura, S., The origin of cultivated plants of China, *Acta Phyotax. Geobot.*, 14, 81–86, 1951.

Knerr, L. D. and Staub, J. E., Inheritance and linkage relationships of isozyme loci in cucumber (*Cucumis sativus* L.), *Theor. Appl. Genet.*, 84, 217–224, 1992.

Knerr, L. D., Staub, J. E., Holder, D. J., and May, B. P., Genetic diversity in *Cucumis sativus* L. assessed by variation at 18 allozyme coding loci, *Theor. Appl. Genet.*, 78, 119–128, 1989.

Knipping, P. A., Patterson, C. G., Knavel, D. E., and Rodriguez, J. G., Resistance of cucurbits to twospotted spider mite, *Environ. Entomol.*, 4, 507–508, 1975.

Koga-Ban, Y., Tabei, Y., Ishimoto, M., Nishizawa, Y., Tsuchiya, K., Imaizumi, N., Nakamura, H., Kayano, T., and Tanaka, H., Biosafety assessment of transgenic plants in the greenhouse and the field—a case study of transgenic cucumber, *JARQ*, 38, 167–174, 2004.

Kong, X. H. and Zhang, H. Y., The effect of ultra-dry methods and storage on vegetable seeds, *Seed Sci. Res.*, 8(suppl. 1), 41–45, 1998.

Kotlinska T., Vegetable crops genetic resources conservation and utilization, In *Integration of Conservation Strategies of Plant Genetic Resources in Europe, Proceedings of an International Symposium on Plant Genetic Resources in Europe*, Begemann, F. and Hammer, K., Eds., Bonn, Germany: ZADI and Gatersleben, Germany: IPK, Gatersleben, Germany, December 6–8, pp. 195–201, 1993.

Krasteva, L., Organization of melon plant genetic resources in Bulgaria, *Acta Hort.*, 510, 247–251, 2000a.

Krasteva, L., Watermelon genetic resources in Bulgaria, *Acta Hort.*, 510, 253–256, 2000b.

Krasteva, L., Evaluation, use and conservation of the *Cucumis melo* L. collection in Bulgaria, In *Cucurbit Genetic Resources in Europe*, Díez, M. J., Picó, B., and Nuez, F., Eds., Rome: International Plant Genetic Resources Institute, *Ad hoc Meeting*, 19 January 2002, Adana, Turkey, pp. 12–17, 2002.

Krasteva, L., Lozanov, I., Neykov, S., and Todorova, T., Cucurbitaceae genetic resources in Bulgaria, In *Cucurbit Genetic Resources in Europe*, Díez, M. J., Picó, B., and Nuez, F., Eds., Rome: International Plant Genetic Resources Institute, *Ad hoc Meeting*, 19 January 2002, Adana, Turkey, pp. 8–11, 2002.

Křístková, E., Lze křížit dýni s cuketou (It is possible to cross pumpkin with squash), Záhradníctvo, 16, 159, 1991.

Křístková, E., The Czech national collection of cucurbitaceous vegetables, In *Cucurbit Genetic Resources in Europe*, Díez, M. J., Picó, B., and Nuez, F., Eds., Rome: International Plant Genetic Resources Institute, *Ad hoc Meeting*, 19 January 2002, Adana, Turkey, pp. 18–29, 2002.

Křístková, E., Křístková, A., and Vinter, V., Morphological variation of cultivated *Cucurbita* species, In *Progress in Cucurbit Genetics and Breeding Research, Proceedings of Cucurbitaceae 2004, the 8th EUCARPIA Meeting on Cucurbit Genetics and Breeding*, Lebeda, A. and Paris, H. S., Eds., Olomouc, Czech Republic: Palacký University in Olomouc, pp. 119–128, 2004.

Křístková, E., Křístková, A., Vinter, V., and Lebeda, A., Genetic resources of the cultivated *Cucurbita* species (*C. argyrosperma, C. ficifolia, C. maxima, C. moschata, C. pepo*) and their morphological description (English-Czech version), *Hort. Sci.*, Prague, in press, 2005.

Křístková, E. and Lebeda, A., Genetic resources of vegetable crops from the family Cucurbitaceae, *Zahradnictví (Hort. Sci.*, Prague), 22, 123–128, 1995.

Křístková, E. and Lebeda, A., Resistance in *Cucurbita pepo* and *Cucurbita maxima* germplasm to watermelon mosaic potyvirus-2, *Plant Genet. Resour. Newslet.*, 121, 47–52, 2000a.

Křístková, E. and Lebeda, A., Powdery mildew field infection on leaves and stems of *Cucurbita pepo* accessions, *Acta Hort.*, 510, 61–66, 2000b.

Křístková, E., Lebeda, A., Vinter, V., and Blahoušek, O., Genetic resources of the genus *Cucumis* and their morphological description (English-Czech version), *Hort. Sci., Prague*, 30, 14–42, 2003.

Kubaláková, M., Doležel, J., and Lebeda, A., Ploidy instability of embryogenic cucumber (*Cucumis sativus* L.) callus culture, *Biol. Plant.*, 38, 475–480, 1996.

Küçük, A., Abak, K., and Sari, N., Cucurbit genetic resources collections in Turkey, In *Cucurbit Genetic Resources in Europe*, Díez, M. J., Picó, B., and Nuez, F., Eds., Rome: International Plant Genetic Resources Institute, *Ad hoc Meeting*, 19 January 2002, Adana, Turkey, pp. 46–51, 2002.

Kůdela, V. and Lebeda, A., Response of wild *Cucumis* species to inoculation with *Pseudomonas syringae* pv. *lachrymans*, *Genet. Res. Crop Evol.*, 44, 271–275, 1997.

Kurtar, E. S., Sari, N., and Abak, K., Obtention of haploid embryos and plants through irradiated pollen technique in squash (*Cucurbita pepo* L.), *Euphytica*, 127, 335–344, 2002.

Kuzuya, M., Hosoya, K., Yashiro, K., Tomita, K., and Ezura, H., Powdery mildew (*Sphaerotheca fuliginea*) resistance in melon is selectable at the haploid level, *J. Exper. Bot.*, 54, 1069–1074, 2003.

Kwack, S. N. and Fujieda, K., Seed abortion and techniques for obtaining hybrids in interspecific crosses of *Cucurbita*, *J. Jpn. Soc. Hort. Sci.*, 55, 455–460, 1987.

Ladyzynski, M., Burza, W., and Malepszy, S., Relationship between somaclonal variation and type of culture in cucumber, *Euphytica*, 125, 349–356, 2002.

Laghetti, G., Hammer, K., Olita, G., and Perrino, P., Crop genetic resources from Ustica Island (Italy): collecting and safeguarding, *Plant Genet. Resour. Newslet*, 116, 12–17, 1998.

Larkin, P. J. and Scowcroft, W. R., Somaclonal variation—a novel source of variability from cell cultures for plant improvement, *Theor. Appl. Genet.*, 60, 197–214, 1981.

Láska, P. and Lebeda, A., Resistance in wild *Cucumis* species to the glasshouse whitefly (*Trialeurodes vaporariorum*), *Archiv für Züchtungsforschung*, 19, 89–93, 1989.

Lebeda, A., Screening of wild *Cucumis* species for resistance to cucumber powdery mildew (*Erysiphe cichoracearum* and *Sphaerotheca fuliginea*), *Sci. Hort.*, 24, 241–249, 1984.

Lebeda, A., Resistance of *Cucumis sativus* cultivars to *Cladosporium cucumerinum*, *Sci. Hort.*, 26, 9–15, 1985.

Lebeda, A., Screening of wild *Cucumis* species against downy mildew (*Pseudoperonospora cubensis*) isolates from cucumbers, *Phytoparasitica*, 20, 203–210, 1992a.

Lebeda, A., Susceptibility of accessions of *Cucumis sativus* to *Pseudoperonospora cubensis*, *Tests of Agrochem. and Cult.*, 13(Ann. Appl. Biol., 120 suppl.), 102–103, 1992b.

Lebeda, A., Resistance in wild *Cucumis* species to twospotted spider mite (*Tetranychus urticae*), *Acta Phytopathol. Entomol. Hung.*, 31, 247–252, 1996.

Lebeda, A., *Pseudoperonospora cubensis* on *Cucumis* spp. and *Cucurbita* spp.—resistance breeding aspects, *Acta Hort.*, 492, 363–370, 1999.

Lebeda, A. and Doležal, K., Peroxidase isozyme polymorphism as a potential marker for detection of field resistance in *Cucumis sativus* to cucumber downy mildew (*Pseudoperonospora cubensis* (Berk. et Curt.) Rostov.), *Zeitschrift Pflanzenkrankh. Pflanzenschtz.*, 102, 467–471, 1995.

Lebeda, A. and Křístková, E., Resistance in *Cucurbita pepo* and *Cucurbita maxima* germplasms to cucumber mosaic virus, *Genet. Resour. Crop Evol.*, 43, 461–469, 1996.

Lebeda, A. and Křístková, E., Evaluation of *Cucumis sativus* L. germplasm for field resistance to the powdery mildew, *Acta Phytopath. Entom. Hung.*, 32, 299–303, 1997.

Lebeda, A. and Křístková, E., Interaction between morphotypes of *Cucurbita pepo* and obligate biotrophs (*Pseudoperonospora cubensis*, *Erysiphe cichoracearum* and *Sphaerotheca fuliginea*), *Acta Hort.*, 510, 219–225, 2000.

Lebeda, A., Křístková, E., and Kubaláková, M., Interspecific hybridization of *Cucumis sativus* × *Cucumis melo* as potential way to transfer resistance to *Pseudoperonospora cubensis*, In *Cucurbits towards 2000*, GómezGuillamón, M. L., et al., Ed., *Proceedings of the sixth Eucarpia Meeting on Cucurbit Genetics and Breeding*, Algarrobo, Malaga, Spain Estación Experimental, La Mayora, pp. 31–37, 1996.

Lebeda, A., Kubaláková, M., Křístková, E., Navrátilová, B., Doležal, K., Doležel, J., and Lysák, M., Inter-specific hybridization of *Cucumis sativus* L.×*C. melo* L., In *Biological and Technical Development in Horticulture*, Kobza, F., Pidra, M., and Pokluda, R., Eds., Brno and Lednice na Moravě, Czech Republic: Mendel Agricultural and Forestry University, pp. 49–52, 1997.

Lebeda, A., Kubaláková, M., Křístková, E., Navrátilová, B., Doležal, K., Doležel, J., and Lysák, M., Morpho-logical and physiological characteristics of plants issued from an interspecific hybridization of *Cucumis sativus*×*Cucumis melo*, Acta Hort., 492, 149–155, 1999.

Lebeda, A. and Prášil, J., Susceptibility of *Cucumis sativus* cultivars to *Pseudoperonospora cubensis*, Acta Phytopath. Ent. Hung., 29, 89–94, 1994.

Lebeda, A. and Widrlechner, M. P., A set of Cucurbitaceae taxa for the differentiation of *Pseudoperonospora cubensis* pathotypes, J. Plant Dis. Protect., 110, 337–349, 2003.

Lebeda, A. and Widrlechner, M. P., Response of wild and weedy *Cucurbita* L. to pathotypes of *Pseudopero-nospora cubensis* (Berk. & Curt.) Rostov. (cucurbit downy mildew), In *Advances in Downy Mildew Research, Vol. 2*, Spencer-Phillips, P. T. N. and Jeger, M., Eds., Dordrecht, The Netherlands: Kluwer Academic Publishers, pp. 203–210, 2004.

Lee, H. S., Kwon, E. J., Kwon, S. Y., Jeong, Y. J., Lee, E. M., Jo, M. H., Kim, H. S. et al., Transgenic cucumber fruits that produce elevated level of an anti-aging superoxide dismutase, Mol. Breed., 11, 213–220, 2003a.

Lee, K. G., Chung, W. I., and Ezura, H., Efficient plant regeneration via organogenesis in winter squash (*Cucubita maxima* Duch.), Plant Sci., 164, 413–418, 2003b.

Lee, S. J., Shin, J. S., Park, K. W., and Hong, Y. P., Detection of genetic diversity using RAPD-PCR and sugar analysis in watermelon [*Citrullus lanantus* (sic) (Thunb.) Mansf.] germplasm, Theor. Appl. Genet., 92, 719–725, 1996.

Lee, Y. K., Chung, W. I., and Ezura, H., Plant regeneration via organogenesis in the Korean and Japanese winter squash (*Cucurbita maxima*), Acta Hort., 588, 299–302, 2002.

Leppik, E. E., Searching gene centers of the genus *Cucumis* through host-parasite relationship, Euphytica, 15, 323–328, 1966.

Leppik, E. E., Relative resistance of *Cucumis* introductions to diseases and insects, Adv. Front. Plant Sci., 19, 43–51, 1968.

Lepse, L., Baumane, M., and Rashal, I., Cucumber genetic resources of Latvian origin, Acta Hort., 510, 257–262, 2000.

Levi, A., Thomas, C. E., Joobeur, T., Zhang, X., and Davis, A., A genetic linkage map for watermelon derived from a testcross population: (*Citrullus lanatus* var. *citroides*×*C. lanatus* var. *lanatus*)×*Citrullus colo-cynthis*, Theor. Appl. Genet., 105, 555–563, 2002.

Levi, A., Thomas, C. E., Keinath, A. P., and Wehner, T. C., Estimation of genetic diversity among *Citrullus* accessions using RAPD markers, Acta Hort., 510, 385–390, 2000.

Levi, A., Thomas, C. E., Keinath, A. P., and Wehner, T. C., Genetic diversity among watermelon (*Citrullus lanatus* and *Citrullus colocynthis*) accessions, Genet. Resour. Crop Evol., 48, 559–566, 2001a.

Levi, A., Thomas, C. E., Newman, M., Reddy, O. U. K., Zhang, X., and Xu, Y., SSR and AFLP markers differ among American watermelon cultivars with limited genetic diversity, J. Am. Soc. Hort. Sci., 129, 553–558, 2004a.

Levi, A., Thomas, C. E., Thies, J., Simmons, A., Xu, Y., Zhang, X., Reddy, O. U. K. et al., Developing a genetic linkage map for watermelon: polymorphism, segregation and distribution of markers, In *Progress in Cucurbit Genetics and Breeding Research, Proceedings of Cucurbitaceae 2004, the 8th EUCARPIA Meeting on Cucurbit Genetics and Breeding*, Lebeda, A. and Paris, H. S., Eds., Olomouc, Czech Republic: Palacký University in Olomouc, pp. 515–523, 2004b.

Levi, A., Thomas, C. E., Wehner, T. C., and Zhang, X., Low genetic diversity indicates the need to broaden the genetic base of cultivated watermelons, HortSci., 36, 1096–1101, 2001b.

Levi, A., Thomas, C. E., Zhang, X., Joobeur, T., Dean, R. A., Wehner, T. C., and Carle, B. R., A genetic linkage map for watermelon based on randomly amplified polymorphic DNA markers, J. Am. Soc. Hort. Sci., 126, 730–737, 2001c.

Lhotský, B., Lebeda, A., and Zvára, J., Resistance of wild *Cucumis* species to gummy stem blight (*Didymella bryoniae*), Acta Phytopath. Ent. Hung., 26, 303–306, 1991.

Li, R., Sun, Y., Zhang, L., and Li, X., Plant regeneration from cotyledon protoplasts of Xinjian muskmelon, Plant Cell Rep., 9, 199–203, 1990.

Liou, P. C., Chang, Y. M., Hsu, W. S., Cheng, Y. H., Chang, H. R., and Hsiao, C. H., Construction of a linkage map in *Cucumis melo* (L.) using random amplified polymorphic DNA markers, *Acta Hort.*, 461, 123–132, 1998.

Lira-Saade, R., *Cucurbita* L. *Estudios Taxonómicos y Ecogeográphicos de las Cucurbitaceae Latinoamericanas de Importancia Económica; Systematics and Ecogeographic Studies on Crop Genepools*, Vol. 9, Rome: IPGRI, 1995.

Lopes, J. F., Brune, S., and Henz, G. P., Metodologia de avaliação da resistência de germoplasma de abóboras e morangas a *Phytophthora capsici*, *Hort. Brasil.*, 17(Suppl.), 30–32, 1999.

López-Sesé, A. I. and Gómez-Guillamón, M. L., Resistance to Cucurbit Yellowing Stunting Disorder Virus (CYSDV) in *Cucumis melo* L., *HortSci.*, 35, 110–113, 2000.

López-Sesé, A. I., Staub, J. E., and Gomez-Guillamon, M. L., Genetic analysis of Spanish melon (*Cucumis melo* L.) germplasm using a standardized molecular marker array and reference accessions, *Theor. Appl. Genet.*, 108, 41–52, 2003.

López-Sesé, A. I., Staub, J. E., Katzir, N., and Gómez-Guillamón, M. L., Estimation of between and within accession variation in selected Spanish melon germplasm using RAPD and SSR markers to assess strategies for large collection evaluation, *Euphytica*, 127, 41–51, 2002.

Lotfi, M., Alan, A. R., Henning, M. J., Jahn, M. M., and Earle, E. D., Production of haploid and doubled haploid plants of melon (*Cucumis melo* L.) for use in breeding for multiple virus resistance, *Plant Cell Rep.*, 21, 1121–1128, 2003.

Lower, R. L. and Edwards, M. D., Cucumber breeding, In *In Breeding Vegetable Crops*, Basset, M. J., Ed., Westport, CT: AVI, pp. 173–207, 1986.

Mackiewicz, H. O., Malepszy, S., Sarreb, D. A., and Narkiewicz, M., Triploids in cucumber: II. Characterization of embryo rescue plants, *Gartenbauwissenschaft*, 63, 125–129, 1998.

Mal, B., Rana, R. S., and Joshi, V., Status of research on new crops in India, In *In Proceedings of the 9th International Conference on Jojoba and its Uses and of the 3rd International Conference on New Industrial Crops and Products, Catamarca, Argentina*, Princen, L. H. and Rossi, C., Eds., Peoria, IL: Association for the Advancement of Industrial Crops, pp. 198–204, 1996.

Malepszy, S., Sarreb, D. A., Mackiewicz, H. O., and Narkiewicz, M., Triploids in cucumber: I. Factors influencing embryo rescue efficiency, *Gartenbauwissenschaft*, 63, 34–37, 1998.

Marr, K. L., Mei, X. Y., and Bhattarai, N. K., Allozyme, morphological and nutritional analysis bearing on the domestication of *Momordica charantia* L. (Cucurbitaceae), *Econ. Bot.*, 58, 435–455, 2004.

Marsh, D. B., Evaluation of *Cucumis metuliferus* as a specialty crop for Missouri, In *New Crops, In Proceedings of the 2nd National Symposium: New Crops, Exploration, Research and Commercialization*, Janick, J. and Simon, J. E., Eds., New York: Wiley, New York, pp. 558–559, 1993.

Martinez-Madrid, M. C., Martinez, G., Pretel, M. T., Serrano, M., and Romojaro, F., Role of ethylene and abscisic acid in physicochemical modifications during melon ripening, *J. Agric. Food Chem.*, 47, 5285–5290, 1999.

Maynard, D. N., Ed., *Watermelons. Characteristics, Production and Marketing*, Alexandria, VA: ASHS Press, 2001.

McCreight, J. D., Screening of melons for silverleaf whitefly resistance, *Cucurbit Genet. Coop. Rep.*, 18, 45–47, 1995.

McCreight, J. D., Kishaba, N. N., and Bohn, G. W., AR Hale's Best Jumbo, AR 5, and AR Topmark, melon aphid-resistant muskmelon breeding lines, *HortSci.*, 19, 309–310, 1984.

McCreight, J. D., Nerson, H., and Grumet, R., Melon, *Cucumis melo* L., In *In Genetic Improvement of Vegetable Crops*, Kalloo, G. and Bergh, B. O., Eds., New York: Pergamon Press, pp. 267–294, 1993.

McCreight, J. D. and Staub, J. E., Indo-U.S. *Cucumis* germplasm expedition, *HortSci.*, 28, 467, 1993.

McCreight, J. D., Staub, J. E., López-Sesé, A., and Chung, S. M., Isozyme variation in Indian and Chinese melon (*Cucumis melo* L.) germplasm collections, *J. Am. Soc. Hort. Sci.*, 129, 811–818, 2004.

Meglic, V. V., Horejsi, T. F., and Staub, J. E., Genetic diversity, and inheritance and linkage of isozyme loci in melon (*Cucumis melo* L.), *HortSci.*, 29, 449, (Abstract) 1994.

Meglic, V., Serquen, F., and Staub, J. E., Genetic diversity in cucumber (*Cucumis sativus* L.): I. A reevaluation of the U.S. germplasm collection, *Genet. Resour. Crop Evol.*, 43, 533–546, 1996.

Meglic, V. and Staub, J. E., Inheritance and linkage relationships of allozyme and morphological loci in cucumber (*Cucumis sativus* L.), *Theor. Appl. Genet.*, 92, 865–872, 1996a.

Meglic, V. and Staub, J. E., Genetic diversity in cucumber (*Cucumis sativus* L.): II. An evaluation of selected cultivars released between 1846 and 1978, *Genet. Resour. Crop Evol.*, 43, 547–558, 1996b.

Merrick, L. C., Systematics and evolution of a domesticated squash, *Cucurbita argyrosperma*, and its wild and weedy relatives, In *In Biology and Utilization of the Cucurbitaceae*, Bates, D. M. et al., Eds., Ithaca: Comstock Publishing Associates, pp. 77–95, 1990.

Merrick, L. C., Systematics, evolution, and ethnobotany of a domesticated squash. *Cucurbita argyrosperma*, PhD diss., Ithaca, NY: Cornell University, 1991.

Merrick, L. C., Squashes, pumpkins and gourds, In *Evolution of Crop Plants*, Smartt, J. and Simmonds, N. W., Eds., 2nd ed., London: Longman Scientific and Technical, pp. 97–105, 1995.

Merrick, L. C. and Nabhan, G. P. Natural hybridization of wild *Cucurbita sororia* group and domesticated *C. mixta* in Southern Sonora, Mexico, *Cucurbit Genet. Coop. Rep.*, 7, 73–75, 1984.

Metcalf, R. L. and Rhodes, A. M., Coevolution of the Cucurbitaceae and Luperini (Coleoptera:Chrysomelidae): basic and applied aspects, In *Biology and Utilization of the Cucurbitaceae*, Bates, D. M. et al., Eds., Ithaca, NY: Comstock Publishing Associates, pp. 167–182, 1990.

Metwally, E. I., Haroun, S. A., and El-Fadly, G. A., Interspecific cross between *Cucurbita pepo* L. and *Cucurbita martinezii* through *in vitro* embryo culture, *Euphytica*, 90, 1–7, 1996.

Metwally, E. I., Moustafa, S. A., El-Sawy, B. I., Haroun, S. A., and Shalaby, T. A., Production of haploid plants from *in vitro* culture of unpollinated ovules of *Cucurbita pepo*, *Plant Cell Tiss. Organ Cult.*, 52, 117–121, 1998a.

Metwally, E. I., Moustafa, S. A., El-Sawy, B. I., and Shalaby, T. A., Haploid plantlets derived by anther culture of *Cucurbita pepo*, *Plant Cell Tiss. Organ Cult.*, 52, 171–176, 1998b.

Michelmore, R. W., Paran, I., and Kesseli, R. V., Identification of markers linked to disease-resistance genes by bulked segregant analysis: a rapid method to detect markers in specific genome regions by using segregating populations, *Proc. Natl. Acad. Sci. (U.S.A.)*, 88, 9828–9832, 1991.

Mliki, A., Staub, J. E., Zhangyoung, S., and Ghorbel, A., Genetic diversity in melon (*Cucumis melo* L.): an evaluation of African germplasm, *Genet. Resour. Crop Evol.*, 48, 587–597, 2001.

Mliki, A., Staub, J. E., Zhangyoung, S., and Ghorbel, A., Genetic diversity in African cucumber (*Cucumis sativus* L.) provides potential for germplasm enhancement, *Genet. Resour. Crop Evol.*, 50, 461–468, 2003.

Mo, S. Y., Im, S. H., Go, G. D., Ann, C. M., and Kim, D. H., RAPD analysis for genetic diversity of melon species (in Korean), *Korean J. Hort. Sci. Technol.*, 16, 21–24, 1998.

Mohamed, E. I. and Taha, M., Indigenous melons (*Cucumis melo* L.) in Sudan: a review of their genetic resources and prospects for use as sources of disease and insect resistance, *Plant Genet. Resour. Newslet.*, 138, 36–42, 2004.

Molina, R. V. and Nuez, F., The inheritance of organogenic response in melon, *Plant Cell Tiss. Organ Cult.*, 46, 251–256, 1996.

Monforte, A. J., Garcia Mas, J., and Arus, P., Genetic variability in melon based on microsatellite variation, *Plant Breeding*, 122, 153–157, 2003.

Monforte, A. J., Oliver, M., Gonzalo, M. J., Alvarez, J. M., Dolcet-Sanjuan, R., and Arus, P., Identification of quantitative trait loci involved in fruit quality traits in melon (*Cucumis melo* L.), *Theor. Appl. Genet.*, 108, 750–758, 2004.

Montoro, T., Sánchez-Campos, S., Camero, R., Marco, C. F., Corella, P., and Gómez-Guillamón, M. L., Searching for resistance to cucumber vein yellowing virus in *Cucumis melo*, In *Progress in Cucurbit Genetics and Breeding Research*, In *Proceedings of Cucurbitaceae 2004, the 8th EUCARPIA Meeting on Cucurbit Genetics and Breeding*, Lebeda, A. and Paris, H. S., Eds., Olomouc, Czech Republic: Palacký University in Olomouc, pp. 197–202, 2004.

Morales, M., Luis-Arteaga, M., Alvares, J. M., Dolcet-Sanjuan, R., Monforte, A. J., Arus, P., and Garcia-Mas, J., Marker saturation of the region flanking the gene *Nsv* conferring resistance to the melon necrotic spot *Carmovirus* (MNSV) in melon, *J. Am. Soc. Hort. Sci.*, 127, 540–544, 2004a.

Morales, M., Roig, E., Monforte, A. J., Arus, P., and Garcia-Mas, J., Single-nucleotide polymorphisms detected in expressed sequence tags of melon (*Cucumis melo* L.), *Genome*, 47, 352–360, 2004b.

Moravec, J., Lebeda, A., and Křístková, E., History of growing and breeding of cucurbitaceaous vegetables in Czech Lands, *Progress in Cucurbit Genetics and Breeding Research*, In *Proceedings of Cucurbitaceae 2004, the 8th EUCARPIA Meeting on Cucurbit Genetics and Breeding*, Lebeda, A. and Paris, H. S., Eds., Olomouc, Czech Republic: Palacký University in Olomouc, pp. 21–38, 2004.

More, T. A., Enhancement of muskmelon resistance to disease via breeding and transformation, *Acta Hort.*, 588, 205–211, 2002.

More, T. A., Dhakare, B. B., and Sawant, S. V., Identification of downy mildew resistant sources in muskmelon genotypes, *Acta Hort.*, 588, 241–245, 2002.

Moreno, V. and Roig, L. A., Somaclonal variation in cucurbits, In *Biotechnology in Agricultural and Forestry, Vol. 11, Somaclonal Variation in Crop Improvement I*, Bajaj, Y. S. P., Ed., Berlin: Springer-Verlag, pp. 435–464, 1990.

Moreno, V., Zubeldia, L., Garcia-Sogo, B., Nuez, F., and Roig, L. A., Somatic embryogenesis in protoplast-derived cells of *Cucumis melo* L., In *Genetic Manipulation in Plant Breeding*, Horn, W., Jensen, C. J., Odenback, W., and Schieder, O., Eds., Berlin: De Gruyter, pp. 491–493, 1986.

Morishita, M., Sugiyama, K., Saito, T., and Sakata, Y., Powdery mildew resistance in cucumber, *JARQ*, 37, 7–14, 2003.

Morton, J. F., The horned cucumber, alias "Kiwano" (*Cucumis metuliferus*, Cucurbitaceae), *Econ. Bot.*, 41, 325–327, 1987.

Munger, H. L., Breeding for viral disease resistance in cucurbits, In *Resistance to Viral Diseases of Vegetables: Genetics & Breeding*, Kyle, M. M., Ed., Portland, OR: Timber Press, pp. 44–60, 1993.

Munger, H. L. and Robinson, R. W., Nomenclature of *Cucumis melo* L., *Cucurbit Genet Coop. Rep.*, 14, 43–44, 1991.

Mutschler, M. A. and Pearson, O. H., The origin, inheritance, and instability of butternut squash (*Cucurbita moschata* Duchesne), *Hort. Sci.*, 22, 535–539, 1987.

Nabhan, G. P., Evidence of gene flow between cultivated *Cucurbita mixta* and a field edge population of wild *Cucurbita* at Onavas Sonora, *Cucurbit Genet. Coop. Rep.*, 7, 76–77, 1984.

Nagai, T., Nomura, Y., and Oosawa, K., Induction of shoot primordial and plantlet formation in melon, *J. Jpn. Soc. Hort. Sci.*, 58(supplement 1), 208–209, 1989.

Nakata, E., Staub, J. E., López-Sesé, A., and Katzir, N., Genetic diversity in Japanese melon (*Cucumis melo* L.) as assessed by random amplified polymorphic DNA and simple sequence repeat markers, *Genet. Resour. Crop Evol.*, 52, 403–417, 2005.

National Germplasm Resources Laboratory (Database Management Unit). Germplasm Resources Information Network—National Plant Germplasm System database, http://www.ars-grin.gov/npgs (accessed January 2005), 2005.

Navot, N., Sarfatti, M., and Zamir, D., Linkage relationships of genes affecting bitterness and flesh color in watermelon, *J. Hered.*, 81, 162–165, 1990.

Navot, N. and Zamir, D., Isozyme and seed protein phylogeny of the genus *Citrullus* (Cucurbitaceae), *Plant Syst. Evol.*, 156, 61–67, 1987.

Nayar, N. M. and More, T. A., *Cucurbits*, Enfield, NH: Science Publishers, 1998.

Nee, M., The domestication of *Cucurbita* (Cucurbitaceae), *Econ. Bot.*, 44(3, Supplement), 56–68, 1990.

Nei, M., Analysis of gene diversity in subdivided populations, *Proc. Nat. Acad. Sci. (USA)*, 70, 3321–3323, 1973.

Nerson, H., Paris, H. S., and Paris, E. P., Fruit shape, size and seed yield in *Cucurbita pepo*, *Acta Hort.*, 510, 227–230, 2000.

Neuhausen, S. L., Evaluation of restriction fragment length polymorphism in *Cucumis melo*, *Theor. Appl. Genet*, 83, 379–384, 1992.

Newstrom, L. E., Origin and evolution of chayote, *Sechium edule*, In *Biology and Utilization of the Cucurbitaceae*, Bates, D. M. et al., Eds., Ithaca, NY: Comstock Publishing Associates, pp. 141–149, 1990.

Neykov, S., Characteristics and potential breeding use of introduced cucumber samples, *Plant Genet. Resour. Newslet.*, 99, 1–2, 1994.

Neykov, S., Reproduction methods of cucumber accessions for conservation and exchange in the genebank, *Acta Hort.*, 462, 773–776, 1997.

Neykov, S., Evaluation of introduced and local salad cucumber cultivars in Bulgaria, *Plant Genet. Resour. Newslet.*, 114, 45–46, 1998.

Neykov, S. and Alexandrova, M., Characteristics of the Bulgarian National Collection of cucumber cultivars (*Cucumis sativus* L.) in relation to breeding, *Acta Hort.*, 462, 217–223, 1997.

Ng, T. J., New opportunities in the Cucurbitaceae, *New Crops*, In *Proceedings of the 2nd National Symposium: New Crops, Exploration, Research and Commercialization*, Janick, J. and Simon, J. E., Eds., New York: Wiley, pp. 538–546, 1993.

Niemirowicz-Szczytt, K. and Dumas de Vaulx, R., Preliminary data on haploid cucumber (*Cucumis sativus* L.) induction, *Cucurbit Genet. Coop. Rep.*, 12, 24–25, 1989.

Niemirowicz-Szczytt, K., Faris, N. M., Rucinska, M., and Nikolova, V., Conservation and storage of a haploid cucumber (*Cucumis sativus* L.) collection under *in vitro* conditions, *Plant Cell Rep.*, 19, 311–314, 2000.

Nienhuis, J. and Lower, R. L., The effects of fermentation and storage time on germination of cucumber seeds at optimal and suboptimal temperatures, *Cucurbit Genet. Coop. Rep.*, 4, 13–15, 1981.

Nikolova, V. and Niemirowicz-Szczytt, K., Diploidization of cucumber (*Cucumis sativus* L) haploids by colchicine treatment, *Acta Soc. Bot. Poloniae*, 65, 311–317, 1996.

Nishibayashi, S., Hayakawa, T., and Nakajima, T., CMV protection in transgenic cucumber plants with introduced CMV-O cp gene, *Theor. Appl. Genet.*, 93, 672–678, 1996a.

Nishibayashi, S., Kaneko, H., and Hayakawa, T., Transformation of cucumber (*Cucumis sativus* L.) plants using *Agrobacterium tumefaciens* and regeneration from hypocotyl explants, *Plant Cell Rep.*, 15, 809–814, 1996b.

Norton, J. D. and Granberry, D. M., Characteristics of progeny from an interspecific cross *Cucumis melo* with *C. metuliferus*, *J. Am. Soc. Hort. Sci.*, 105, 174–180, 1980.

Nuez, F., Anastasio, G., Cortés, C., Cuartero, J., Gómez-Guillamón, M. L., and Costa, J., Germplasm resources of *Cucumis melo* L. from Spain, *Cucurbit Genet. Coop. Rep.*, 9, 60–63, 1986a.

Nuez, F., Ayuso, M. C., Molina, R. V., Costa, J., and Cuartero, J., Germplasm resources of *Cucumis sativus* L. from Spain, *Cucurbit Genet. Coop. Rep.*, 9, 10–11, 1986b.

Nuez, F., Díez, M. J., Palomares, G., Ferrando, C., Cuartero, J., and Costa, J., Germplasm resources of *Citrullus lanatus* [(Thunb.) Matsum and Nakai], *Cucurbit Genet. Coop. Rep.*, 10, 64–65, 1987.

Nuez, F., Fernández de Córdova, P., and Díez, M. J., Collecting vegetable germplasm in the Iberian Peninsula, *FAO/IBPGR Plant Genet. Resour. Newslet.*, 90, 31–33, 1992.

Nuez, F., Fernández de Córdova, P., Valcárcel, M., Ferriol, J. V., Picó, B., and Diez, M. J., *Cucurbita* spp. and *Lagenaria siceraria* collection at the Center for Conservation and Breeding of Agricultural Biodiversity (CCMAV), Polytechnical University of Valencia, *Cucurbit Genet. Coop. Rep.*, 23, 60–61, 2000a.

Nuez, F., Ruiz, J. J., Valcárcel, J. V., and Fernández de Córdova, P., *Colección de semillas de calabaza del Centro de Conservación y Mejora de la Agrodiversidad Valenciana*, Monografías INIA, Instituto Nacional de Investigación y Tecnología Agraria y Alimentaria, Madrid, 2000b.

Nugent, P. E., Tetraploid "Planters Jumbo" melon lines C883-m6-4x and 67-m6-1004x, *HortSci.*, 29, 48–49, 1994.

Nugent, P. E. and Ray, D. T., Spontaneous tetraploid melons, *HortSci.*, 27, 47–50, 1992.

Ogawa, R., Ishikawa, M., Niwata, E., and Oosawa, K., Cryopreservation of shoot primordial cultures of melon using a slow prefreezing procedure, *Plant Cell Tissue Org. Cult.*, 49, 171–177, 1997.

Olczak-Woltman, H., Bękowska, M., Schollenberger, M., and Niemirowicz-Szczytt, K., Cucumber screening for resistance to angular leaf spot, *Progress in Cucurbit Genetics and Breeding Research. Proceedings of Cucurbitaceae 2004, the 8th EUCARPIA Meeting on Cucurbit Genetics and Breeding*, Lebeda, A. and Paris, H. S., Eds., Olomouc, Czech Republic: Palacký University in Olomouc, pp. 245–249, 2004.

Oliver, M., Garcia-Mas, J., Arroyo, M., Morales, M., Dolcet-Sanjuan, R., de Vicente, M. C., Gomez, H. et al., The Spanish melon genome project: construction of a saturated genetic map, *Acta Hort.*, 510, 375–378, 2000.

Oliver, M., Garcia-Mas, J., Cardus, M., Pueyo, N., López-Sesé, A. I., Arroyo, M., Gomez-Paniagua, H., Arus, P., and de Vincente, M. C., Construction of a reference linkage map for melon, *Genome*, 44, 836–845, 2001.

Oliver, M., Garcia-Mas, J., López-Sesé, A. I., Gomez, H., and de Vincente, M. C., Towards a sturdy map of melon (*Cucumis melo* L.), In *Cucurbitaceae 98: Evaluation and Enhancement of Cucurbit Germplasm*, McCreight, J. D., Ed., Alexandria, VA: ASHS Press, pp. 362–365, 1998.

Oluoch, M. O. and Welbaum, G. E., Effect of postharvest washing and post-storage priming on viability and vigour of six-year-old muskmelon (*Cucumis melo* L.) seeds from eight stages of development, *Seed Sci. Technol.*, 24, 195–209, 1996.

Ondřej, V., Navrátilová, B., and Lebeda, A., Embryo cultures of wild and cultivated species of the genus *Cucumis* L., *Acta Hort.*, 510, 409–414, 2000.

Ondřej, V., Navrátilová, B., and Lebeda, A., Determination of the crossing barriers in hybridization of *Cucumis sativus* and *Cucumis melo*, *Cucurbit Genet. Coop. Rep.*, 24, 1–5, 2001.

Ondřej, V., Navrátilová, B., and Lebeda, A., Influence of GA$_3$ on the zygotic embryogenesis of *Cucumis* species *in vitro*, *Biologia (Bratislava)*, 57, 523–525, 2002.

Oridate, T. and Oosawa, K., Somatic embryogenesis and plant regeneration from suspension callus culture in melon (*Cucumis melo* L.), *Jpn. J. Breed.*, 36, 424–428, 1986.

Pang, S.-Z., Jan, F. J., Tricoli, D. M., Russell, P. F., Carney, K. J., Hu, J. S., Fuchs, M., Quemada, H. D., and Gonsalves, D., Resistance to squash mosaic comovirus in transgenic squash plants expressing its coat protein genes, *Mol. Breed.*, 6, 87–93, 2000.

Pangalo, K. I., Melon as an independent genus *Melo* Adans, *Bot. Zhur.*, 36, 571–580, 1951.

Papadopoulou, E., Hammar, S., Little, H., and Grumet, R., Effect of modified endogenous ethylene production and perception on sex expression in melon (*Cucumis melo* L.), In *Cucurbitaceae 2002*, Maynard, D. N., Ed., Alexandria, VA: ASHS Press, pp. 157–164, 2002.

Paran, I., Shifriss, C., and Raccah, B., Inheritance of resistance to zucchini yellow mosaic virus in the interspecific cross *Cucurbita maxima* × *Cucurbita ecuadorensis*, *Euphytica*, 42, 227–232, 1989.

Pareek, O. P., Vashishtha, B. B., and Samadia, D. K., Genetic diversity in drought hardy cucurbits from hot arid zone of India, *IPGRI Newsletter for Asia, the Pacific and Oceania*, 28, 22–23, 1999.

Paris, H. S., A proposed subspecific classification for *Cucurbita pepo*, *Phytologia*, 61, 133–138, 1986.

Paris, H. S., Historical records, origins, and development of the edible cultivar groups of *Cucurbita pepo* (Cucurbitaceae), *Econ. Bot.*, 43, 13–43, 1989.

Paris, H. S., Characterization of the *Cucurbita pepo* collection at the Newe Ya'ar Research Center, Israel, *Plant Genet. Resour. Newslet.*, 126, 41–45, 2001a.

Paris, H. S., History of the cultivar-groups of *Cucurbita pepo*, *Hort. Rev.*, 25, 71–170, 2001b.

Paris, H. S. and Cohen, S., Oligogenic inheritance for resistance to *Zucchini yellow mosaic virus* in *Cucurbita pepo*, *Ann. Appl. Biol.*, 136, 209–214, 2000.

Paris, H. S. and Nerson, H., Seed dimensions in the subspecies and cultivar-groups of *Cucurbita pepo*, *Genet. Resour. Crop Evol.*, 50, 615–625, 2003.

Paris, H. S., Yonash, N., Portnoy, V., Mozes-Daube, N., Tzuri, G., and Katzir, N., Assessment of genetic relationships in *Cucurbita pepo* (Cucurbitaceae) using DNA markers, *Theor. Appl. Genet.*, 106, 971–978, 2003.

Park, Y. H., Senory, S., Wye, C., Antonise, R., Peleman, J., and Havey, M. J., A genetic map of cucumber composed of RAPDs, RFLPs, AFLPs, and loci conditioning resistance to papaya ringspot and zucchini yellow mosaic viruses, *Genome*, 43, 1003–1010, 2000.

Pauquet, J., Burget, E., Hagen, L., Chovelon, V., Le Menn, A., Valot, N., Desloire, S., Caboche, M., Rousselle, P., Pitrat, P., Bendahmane, M., and Dogimont, C., Map-based cloning of the *Vat* gene from melon conferring resistance to both aphid colonization and aphid transmission of several viruses, In *Progress in Cucurbit Genetics and Breeding Research, Proceedings of Cucurbitaceae 2004, the 8th EUCARPIA Meeting on Cucurbit Genetics and Breeding*, Lebeda, A. and Paris, H. S., Eds., Olomouc, Czech Republic: Palacký University in Olomouc, pp. 325–329, 2004.

Pavlikaki, H., Ponce Navarro, C., and Fanourakis, N., Genetic relationships of different Greek landraces of cucumber (*Cucumis sativus*) as assessed by RAPDs, In *Progress in Cucurbit Genetics and Breeding Research, Proceedings of Cucurbitaceae 2004, the 8th EUCARPIA Meeting on Cucurbit Genetics and Breeding*, Lebeda, A. and Paris, H. S., Eds., Olomouc, Czech Republic: Palacký University in Olomouc, pp. 101–106, 2004.

Pearson, O. H., Hopp, R., and Bohn, G. W., Notes on species crosses in *Cucurbita*, *Proc. Am. Soc. Hort. Sci.*, 57, 310–322, 1951.

Perchepied, L., Périn, C., Giovinazzo, N., Besombes, D., Dogimont, C., and Pitrat, M., Susceptibility to sulfur dusting and inheritance in melon, In *Progress in Cucurbit Genetics and Breeding Research, Proceedings of Cucurbitaceae 2004, the 8th EUCARPIA Meeting on Cucurbit Genetics and Breeding*, Lebeda, A. and Paris, H. S., Eds., Olomouc, Czech Republic: Palacký University in Olomouc, pp. 353–357, 2004.

Périn, C., Gomez-Jimenez, M., Hagen, L., Dogimont, C., Pech, J. C., Latche, A., Pitrat, M., and Lelievre, J. M., Molecular and genetic characterization of a non-climacteric phenotype in melon reveals two loci conferring altered ethylene response in fruit, *Plant Physiol.*, 129, 300–309, 2002a.

Périn, C., Hagen, L. S., de Conto, V., Katzir, N., Danin-Poleg, Y., Portnoy, V., Baudracco-Arnas, S., Chadoeuf, J., Dogimont, C., and Pitrat, M., A reference map of *Cucumis melo* based on two recombinant inbred line populations, *Theor. Appl. Genet.*, 104, 1017–1034, 2002b.

Périn, C., Hagen, L., Dogimont, C., de Conto, V., Lecomte, L., and Pitrat, M., Construction of a reference genetic map of melon, *Acta Hort.*, 510, 367–374, 2000.

Périn, C., Hagen, L., Dogimont, C., de Conto, V., and Pitrat, M., Construction of a genetic map of melon with molecular markers and horticultural traits, In *Cucurbitaceae 98: Evaluation and Enhancement of Cucurbit Germplasm*, McCreight, J. D., Ed., ASHS Press: Alexandria, VA, pp. 370–376, 1998.

Périn, C., Hagen, L. S., Giovinazzo, N., Besombes, D., Dogimont, C., and Pitrat, M., Genetic control of fruit shape acts prior to anthesis in melon (*Cucumis melo* L.), *Mol. Genet. Genom.*, 266, 933–941, 2002c.

Perl-Treves, R., Male to female conversion along the cucumber shoot: Approaches to study sex genes and floral development in *Cucumis sativus*, In *Sex Determination in Plants*, Ainsworth, C. C., Ed., Oxford: BIOS Scientific Publisher, pp. 189–206, 1999.

Perl-Treves, R. and Galun, E., The *Cucumis* plastome: physical map, intrageneric variation and phylogenetic relationships, *Theor. Appl. Genet.*, 71, 417–429, 1985.

Perl-Treves, R., Stepansky, A., Schaffer, A. A., and Kovalski, I., Intraspecific classification of *Cucumis melo*: how is the morphological and biochemical variation of melons reflected at the DNA level?, In *Cucurbitaceae 98: Evaluation and Enhancement of Cucurbit Germplasm*, McCreight, J. D., Ed., Alexandria, VA: ASHS Press, pp. 310–319, 1998.

Perl-Treves, R., Zamir, D., Navot, N., and Galun, E., Phylogeny of *Cucumis* based on isozyme variability and its comparison with plastome phylogeny, *Theor. Appl. Genet.*, 71, 430–436, 1985.

Peterson, C. E., Staub, J. E., and Palmer, M. J., Wisconsin 5207, a mutliple disease resistant cucumber population, *HortSci.*, 21, 333–335, 1986a.

Peterson, C. E., Staub, J. E., Williams, P. H., and Palmer, M. J., Wisconsin 1983 cucumber, *HortSci.*, 21, 1082–1083, 1983.

Peterson, C. E., Williams, P. H., Palmer, M. J., and Louward, P., Wiscosin 2757 cucumber, *HortSci.*, 17, 268, 1982.

Picó, B., Díez, M. J., Ferriol, M., Fernández de Córdoba, P., Valcárcel, J. V., and Nuez, F., Status of the cucurbit collection at COMAV, Spain, In *Cucurbit Genetic Resources in Europe*, Díez, M. J., Ed., Rome: International Plant Genetic Resources Institute, *Ad hoc* meeting, 19 January 2002, Adana, Turkey, pp. 39–45, 2002.

Pierce, L. K. and Wehner, T. C., Review of genes and linkage groups in cucumber, *HortSci.*, 25, 605–615, 1990.

Piperno, D. R., Holst, I., Wessel-Beaver, L., and Andres, T. C., Evidence for the control of phytolith formation in *Cucurbita* fruits by the hard rind (*Hr*) genetic locus: archaeological and ecological implications, *Proc. Nat. Acad. Sci. (U.S.A.)*, 99, 10923–10928, 2002.

Piskunova, T., Status of the cucurbit collections in Russia, In *Cucurbit Genetic Resources in Europe*, Díez, M. J. et al., Eds., Rome: International Plant Genetic Resources Institute, *Ad hoc meeting*, 19 January 2002, Adana, Turkey, pp. 37–38, 2002.

Pistrick, K., Loumerem, M., and Haddad, M., Field studies of plant genetic resources in South Tunisia, *Plant Genet. Resour. Newslet.*, 98, 13–17, 1994.

Pitchaimuthu, M. and Dutta, O. P., Collecting and evaluation of muskmelon germplasm in Andhra Pradesh, *IPGRI Newsletter for Asia, the Pacific and Oceania*, 29, 21–22, 1999.

Pitrat, M., Linkages studies in muskmelon, *Cucurbit Genet. Coop. Rep.*, 7, 51–63, 1984.

Pitrat, M., Gene list for melon, *Cucurbit Genet. Coop. Rep.*, 13, 58–68, 1990.

Pitrat, M., Linkage groups in *Cucumis melo* L., *J. Hered.*, 82, 406–411, 1991.

Pitrat, M., Linkage groups in *Cucumis melo* L., *Cucurbit Genet. Coop. Rep.*, 17, 148–149, 1994.

Pitrat, M., Gene list for melon, *Cucurbit Genet. Coop. Rep.*, 25, 76–93, 2002.

Pitrat, M., Dogimont, C., and Bardin, M., Resistance to fungal diseases of foliage in melon, In *Cucurbitaceae 98: Evaluation and Enhancement of Cucurbit Germplasm*, McCreight, J. D., Ed., Alexandria, VA: ASHS Press, pp. 167–173, 1998.

Pitrat, M., Hanelt, P., and Hammer, K., Some comments on infraspecific classification of cultivars of melon, *Acta Hort.*, 510, 29–36, 2000.

Pitrat, M., Riser, G., Bertrand, F., Blancard, D., and Lecoq, H., Evaluation of a melon collection for disease resistance, In *Cucurbits towards 2000, Proceedings of the VIth Eucarpia Meeting on Cucurbit Genetics and Breeding*, Gómez-Guillamón, M. L. et al., Eds., Algarrobo, Málaga, Spain: Estación Experimental "La Mayora", CSIC, pp. 49–58, 1996.

Plader, W., Malepszy, S., Burza, W., and Rusinowski, Z., The relationship between the regeneration system and genetic variability in the cucumber (*Cucumis sativus* L.), *Euphytica*, 103, 9–15, 1998.

Plages, J. N., L'avenir des variétés génétiquement modifiées pour la résistance aux virus (un exemple développé par Limagrain), *Comptes Rendus Acad. Agric. Francaise*, 83, 161–164, 1997.

Plarre, W., Evolution and variability of special cultivated crops in the highlands of West New Guinea (Irian Jaya) under present Neolithic conditions, *Plant Genet. Resour. Newslet.*, 103, 1–13, 1995.

Pražák, J. M., Novotný, F., and Sedláček, J., *Latin-Czech Dictionary*, Prague: Czech-Slovak Graphical Union, 1941.

Prendergast, H. D. V., Birouk, A., and Tazi, M., International team collects a rich trove of wild species in Morocco, *Diversity*, 8(3), 16–19, 1992.

Provvidenti, R., Sources of resistance and tolerance to viruses in accessions of *Cucurbita maxima*, *Cucurbit Genet. Coop. Rep.*, 5, 46–47, 1982.

Provvidenti, R., Viral diseases of cucurbits and sources of resistance, *ASPAC Food & Fertilizer Technology Center Technical Bulletin* (Taipei) No. 93,1986.

Provvidenti, R., Sources of resistance to viruses in cucumber, melon, squash, and watermelon, In *Proceedings of Cucurbitaceae 89: Evaluation and Enhancement of Cucurbit Germplasm*, C. E., Thomas, Ed., Charleston, SC, pp. 29–36, 1989.

Provvidenti, R., Viral diseases and genetic sources of resistance in *Cucurbita* species, In *Biology and Utilization of the Cucurbitaceae*, Bates, D. M. et al., Eds., Ithaca, NY: Comstock, pp. 427–435, 1990.

Provvidenti, R., Resistance to viral diseases of cucurbits, In *Resistance to Viral Diseases of Vegetables: Genetics & Breeding*, Kyle, M. M., Ed., Portland, OR: Timber Press, pp. 8–43, 1993.

Provvidenti, R., Robinson, R. W., and Munger, H. M., Resistance in feral species to six viruses infecting *Cucurbita*, *Plant Dis. Rep.*, 62, 326–329, 1978.

Przyborowski, J. and Niemirowicz-Szczytt, K., Main factors affecting cucumber (*Cucumis sativus* L.) haploid embryo development and haploid plant characteristics, *Plant Breeding*, 112, 70–75, 1994.

Puchalski, J. T. and Robinson, R. W., Comparative electrophoretic analysis of isozymes in *Cucurbita* species, *Cucurbit Genet Coop. Rep.*, 1, 28, 1978.

Puchalski, J. T. and Robinson, R. W., Electrophoretic analysis of isozymes in *Cucurbita* and *Cucumis* and its application for phylogenetic studies, In *Biology and Utilization of the Cucurbitaceae*, Bates, D. M. et al., Eds., Ithaca, NY: Comstock, pp. 60–76, 1990.

Punja, Z. K. and Raharjo, S. H. T., Response of transgenic cucumber and carrot plants expressing different chitinase enzymes to inoculation with fungal pathogens, *Plant Dis.*, 80, 999–1005, 1996.

Qi, C., Yuan, Z., and Li, Y., A new type of cucumber—*Cucumis sativus* L. var. *xishuangbannanesis* (in Chinese with English abstract), *Acta Hort. Sinica*, 10, 259–263, 1983.

Qian, C. T., Jahn, M. M., Staub, J. E., Lou, X. D., and Chen, J. F., Meiotic chromosome behaviour in an allotriploid derived from an amphidiploid×diploid mating in *Cucumis*, *Plant Breeding*, 124, 272–276, 2005.

Queiroz, M. A., Silva, M. L., Silveira, L. M., Dias, R. C. S., Ferreira, M. A. J. F., Ramos, S. R. R., Romão, R. L., Assis, J. G. A., Souza, F. F., and Moura, M. C. C. L., Pre-breeding in the watermelon germplasm bank of the Northeast of Brazil, In *Progress in Cucurbit Genetics and Breeding Research*, Lebeda, A. and Paris, H. S., Eds., Olomouc, Czech Republic: Palacký University in Olomouc, 2004.

Raamsdonk, L. W. D., den Nijs, A. P. M., and Jongerius, M. C., Meiotic analyses of *Cucumis* hybrids and an evolutionary evaluation of the genus *Cucumis* (Cucurbitaceae), *Plant Syst. Evol.*, 163, 133–146, 1989.

Raharjo, S. H. T., Hernandez, M. O., Zhang, Y. Y., and Punja, Z. K., Transformation of pickling cucumber with chitinase-encoding genes using *Agrobacterium tumefaciens*, *Plant Cell Rep.*, 15, 591–596, 1996.

Rajagopalan, P. A. and Perl-Treves, R., Improved cucumber transformation by a modified explant dissection and selection protocol, *HortSci.*, 40, 431–435, 2005.

Rajagopalan, P. A., Saraf-Levy, T., Lizhe, A., and Perl-Treves, R., Increased femaleness in transgenic cucumbers that over-express an ethylene receptor, In *Progress in Cucurbit Genetics and Breeding Research, Proceedings of Cucurbitaceae 2004, the 8th EUCARPIA Meeting on Cucurbit Genetics and Breeding*, Lebeda, A. and Paris, H. S., Eds., Olomouc, Czech Republic: Palacký University in Olomouc, pp. 525–531, 2004.

Rakoczy-Trojanowska, M. and Malepszy, S., Obtaining of hybrids within family Cucurbitaceae by *in vitro* culture of immature embryos. I. Characteristics of hybrids from *Cucurbita maxima* × *Cucurbita pepo*, *Genet. Polonica*, 27, 259–271, 1986.

Rakoczy-Trojanowska, M. and Malepszy, S., A method for increased plant regeneration from immature F1 and BC1 embryos of *Cucurbita maxima* Duch. × *C. pepo* L. hybrids, *Plant Cell Tiss. Org. Cult.*, 18, 191–194, 1989a.

Rakoczy-Trojanowska, M. and Malepszy, S., Obtaining of hybrids within the family Cucurbitaceae by *in vitro* culture of immature embryos, II. Characteristics of BC1 and F2 hybrids between *Cucurbita maxima* × *C. pepo*, *Genet. Polonica*, 30, 67–73, 1989b.

Rakoczy-Trojanowska, M., Pleder, W., and Malepszy, S., Embryo rescue hybrids between various Cucurbitaceae, In *Proceedings of Fifth Eucarpia Cucurbitaceae Symposium*, Doruchowski, R. W. et al., Eds., Skierniewice, Poland: Research Institute for Vegetable Crops, pp. 91–94, 1992.

Ram, H. H., Singh, D. K., Tripathi, P. C., and Rai, P. N., Indigenous germplasm resource in cucurbits, *Recent Hort.*, 3(1), 70–75, 1996.

Ramos, S. R. R., de Queiroz, M. A., Casali, V. W. D., and Cruz, C. D., Divergencia genetica em germoplasma de abóbora procedente de diferentes areas do nordeste, *Hort. Brasil.*, 18, 195–199, 2000.

Raven, P. H., Berg, L. R., and Johnson, G. B., *Environment*, Ft. Worth, TX: Saunders College Publishing, 1993.

Reitsma, K. and Clark, L, Vegetables, in USDA/ARS North Central Regional Plant Introduction Station NC7 Annual Report, January 1–December 31, 2004, http://www.ars-grin.gov/ars/MidWest/Ames/Reports_New/Reports.html (accessed on December 2005).

Reed, D. H. and Frankham, R., How closely correlated are molecular and quantitative measures of genetic variation? A meta-analysis, *Evolution*, 55, 1095–1103, 2001.

Reed, J., Privalle, L., Powell, M. L., Meghiji, M., Dawson, J., Dunder, E., Suttie, J. et al., Phosphomannose isomerase: an efficient selectable marker for plant transformation, *In Vitro* Cell. Develop. Biol.-Plant., 37, 127–132, 2001.

Ren, J. J., Germplasm resources of *Cucurbita pepo* L. in Heilongjiang Province (in Chinese), *Crop Genet. Res.*, 1, 49, 1998.

Reuveni, R., Shimoni, M., and Karchi, Z., A rapid assay for monitoring peroxidase activity in melon as a marker for resistance to *Pseudoperonospora cubensis*, *J. Phytopathol.*, 129, 333–338, 1990.

Reuveni, R., Shimoni, M., Karchi, Z., and Kuc, J., Peroxidase activity as a biochemical marker for resistance of muskmelon (*Cucumis melo*) to *Pseudoperonospora cubensis*, *Phytopathol.*, 82, 749–753, 1992.

Rhodes, A. M., Species hybridization and interspecific gene transfer in the genus, *Cucurbita. Proc. Am. Soc. Hort. Sci.*, 74, 546–552, 1959.

Rhodes, B. B., Adamson, W. C., and Bridges, W. C., Outcrossing in watermelons, *Cucurbit Genet. Coop. Rep.*, 10, 66–68, 1987.

Ricciardi, L., De Giovanni, C., Dell'Orco, P., Marcotrigiano, A. R., and Lotti, C., Phenotypic and genetic characterization of *Cucumis melo* L. landraces collected in Apulia (Italy) and Albania, *Acta Hort.*, 623, 95–105, 2003.

Rick, C. M. and Fobes, J. F., Association of an allozyme with nematode resistance, *Rep. Tomato Genet. Coop.*, 24, 25, 1974.

Rios, G., Ferrando, A., and Serrano, R., Mechanism of salt tolerance conferred by overexpression of the *HAL1* gene in *Saccharomyces cerevisiae*, *Yeast*, 13, 515–528, 1997.

Ríos Labrada, H., Fernández Almirall, A., and Casanova Galarraga, E., Tropical pumpkin (*Cucurbita moschata*) for marginal conditions: breeding for stress interactions, *Plant Genet. Resour. Newslet.*, 113, 4–7, 1998.

Robinson, R. W., Genetic variation in soluble solids of cucumber fruit, *Cucurbit Genet. Coop. Rep.*, 10, 9, 1987.

Robinson, R. W., Genetic resistance in the Cucurbitaceae to insects and spider mites, *Plant Breed. Rev.*, 10, 309–360, 1992.

Robinson, R. W. and Decker-Walters, D. S., *Cucurbits Crop Production Science in Horticulture Series*, Wallingford, U.K.: CAB International, 1997.

Roig, L. A., Roche, M. V., Orts, M. C., Zubeldia, L., Garcia-Sogo, B., and Moreno, V., Obtencion de hibridos *Cucumis melo* × *C. myriocarpus* mediante fusion de protoplastos, *Actas II Congreso S.E.C.H.*, Cordoba, Spain, 1986a.

Roig, L. A., Roche, M. V., Orts, M. C., Zubeldia, L., and Moreno, V., Plant regeneration from cotyledons protoplasts of *Cucumis melo* L. cultivar Cantaloup charentais, *Cucurbit Genet. Coop. Rep.*, 9, 74–77, 1986b.

Romão, R. L., de Queiroz, M. A., Martins, P. S., and Cordeiro, C. M. T., Caracterização morfologica de acessos de melancia do banco ativo de germoplasma (BAG) de cucurbitaceas para o nordeste do Brasil, *Hort. Brasil.*, 17(suppl.), 23–25, 1999.

Roos, E. E. and Davidson, D. A., Longevities of vegetable seeds in storage, *HortSci.*, 27, 393–396, 1992.

Rubatzky, V. E. and Yamaguchi, M., *World Vegetables Principles, Production and Nutritive Values*, 2nd ed., New York: Chapman & Hall, International Thompson Publishing, 1997.

Rubino, D. B. and Wehner, T. C., Effect of inbreeding on horticultural performance of lines developed from an open-pollinated pickling cucumber population, *Euphytica*, 35, 459–464, 1986.

Rudich, J., Biochemical aspects of hormonal regulation of sex expression in Cucurbits, In *Biology and Utilization of the Cucurbitaceae*, Bates, D. M. et al., Eds., Ithaca, NY: Comstock, pp. 269–280, 1990.

Ruszkowski, A. and Biliński, M., Próba wykorzystania trzmieli (*Bombus* Latr.) i miesiarki lucernówki (*Megachile rotundata* F.) do zapylania materiału hodowlanego ogórków, *Pszczelnicze Zeszyty Naukowe*, 28, 175–184, 1984.

Sackville Hamilton, N. R. and Chorlton, K. H., *Regeneration of Accessions in Seed Collections: A Decision Guide, Handbook for Genebanks No. 5*, Rome: IPGRI, 1997.

Sakata, Y., Wako, T., Sugiyama, M., and Morishita, M., Screening melons for resistance to gummy stem blight, *Acta Hort.*, 510, 171–177, 2000.

Sanjur, O. I., Piperno, D. R., Andres, T. C., and Wessel-Beaver, L., Phylogenetic relationships among domesticated and wild species of *Cucurbita* (Cucurbitaceae) inferred from a mitochondrial gene: implications for crop plant evolution and areas of origin, *Proc. Nat. Acad. Sci. (U.S.A.)*, 99, 535–540, 2002.

Sari, N., Abak, K., and Pitrat, M., Comparison of ploidy level screening methods in watermelon: *Citrullus lanatus* (Thunb.), Matsum. and Nakai, *Sci. Hort.*, 82, 265–277, 1999.

Sari, N., Abak, K., Pitrat, M., Rode, J. C., and DeVaulx, R. D., Induction of parthenogenetic haploid embryos after pollination by irradiated pollen in watermelon, *HortSci.*, 29, 1189–1190, 1994.

Sauton, A., Obtention of embryos and plants from *in vitro* culture of fertilized ovules of *Cucumis melo* 5 days after pollination, *Cucurbit Genet. Coop. Rep.*, 10, 62, 1987.

Sauton, A., Effect of season and genotype on gynogenetic haploid production in muskmelon, *Cucumis melo*. *Sci. Hort.*, 35, 71–75, 1988.

Sauton, A., Haploid gynogenesis in *Cucumis sativus* L. induced by irradiated pollen, *Cucurbit Genet. Coop. Rep.*, 12, 22–23, 1989.

Sauton, A. and Dumas de Vaulx, R., Production of haploid plants in melon (*Cucumis melo* L.) as a result of gynogenesis induced by irradiated pollen, *Agronomie*, 7, 141–147, 1987.

Savin, F., Decomble, V., Le Couviour, M., and Hallard, J., The X-ray detection of haploid embryos arisen in muskmelon (*Cucumis melo* L.) seeds, and resulting from a parthenogenetic development induced by irradiated pollen, *Cucurbit Genet. Coop. Rep.*, 11, 39–42, 1988.

Sawant, S. V. and More, T. A., Electrophoretic variation for seed protein of muskmelon genotypes, *Acta Hort.*, 588, 247–254, 2002.

Schafferman, D., Beharav, A., Shabelsky, E., and Yaniv, Z., Evaluation of *Citrullus colocynthis*, a desert plant native to Israel, as a potential source of edible oil, *J. Arid Environ.*, 40, 431–439, 1998.

Scheerens, J. C., Bemis, W. P., Dreher, M. L., and Berry, J. W., Phenotypic variation in fruit and seed characteristics of buffalo gourd, *J. Am. Oil Chem. Soc.*, 55, 523–525, 1978.

Schulze, J., Balko, C., Zellner, B., Koprek, T., Hänsch, R., Nerlich, A., and Mendel, R. R., Biolistic transformation of cucumber using embryogenic suspension cultures: long-term expression of reporter genes, *Plant Sci.*, 112, 197–206, 1995.

Segovia, V., Fuenmayor, F., and Mazzani, E., Recursos fitogenéticos de interés agrícola de la Orinoquia venezolana, *Plant Genet. Resour. Newslet.*, 122, 7–12, 2000.

Serquen, F. C., Bacher, J., and Staub, J. E., Mapping and QTL analysis of a narrow cross in cucumber (*Cucumis sativus* L.) using random amplified polymorphic DNA markers, *Mol. Breed.*, 3, 257–268, 1997a.

Serquen, F. C., Bacher, J., and Staub, J. E., Genetic analysis of yield components in cucumber (*Cucumis sativus* L.) at low plant density, *J. Am. Soc. Hort. Sci.*, 122, 522–528, 1997b.

Seshadari, V.S., Genetic resources and their utilization in vegetable crops, In *Plant Genetic Resources—Indian Perspective, Proceedings of the National Symposium on Plant Genetic Resources, New Delhi, India, March 3–6, 1987*, Paroda, R. S., Arora, R. K., and Chandel, K.P.S., Eds., New Delhi, India: National Bureau of Plant Genetic Resources, pp. 335—343, 1988.

Seshadri, V. S. and More, T. A., Indian land races in *Cucumis melo*, *Acta Hort.*, 588, 187–193, 2002.

Sharma, B. D. and Hore, D. K., Indian cucumber germplasm and challenges ahead, *Genet. Resour. Crop Evol.*, 43, 7–12, 1996.

Shen, D. and Qi, X., Short- and long-term effects of ultra-drying on germination and growth of vegetable seeds, *Seed Sci. Res.*, 8(suppl. 1), 47–53, 1998.

Shetty, N. V. and Wehner, T. C., Screening the cucumber germplasm collection for fruit yield and quality *Crop Sci.*, 42, 2174–2183, 2002.

Shimonishi, K., Ishikawa, M., Suzuki, S., and Oosawa, K., Cryopreservation of melon somatic embryos by desiccation method, *Jpn. J. Breed.*, 41, 347–351, 1991.

Shimonishi, K., Ishikawa, M., Suzuki, S., and Oosawa, K., Cryopreservation of melon somatic embryos by desiccation method, In *Cryopreservation of Tropical Plant Germplasm: Current Research Progress and Application*, Engelmann, F. and Takagi, H., Eds., Rome, Italy: International Research Center for Agricultural Sciences, Tsukuba, Japan, and International Plant Genetic Resources Institute, pp. 167–171, 2000.

Silberstein, L., Kovalski, I., Brotman, Y., Perin, C., Dogimont, C., Pitrat, M., Klingler, J. et al., Linkage map of *Cucumis melo* including phenotypic traits and sequence-characterized genes, *Genome*, 46, 761–773, 2003.

Silberstein, L., Kovalski, I., Huang, R. G., Anagnostu, K., Jahn, M. M. K., and Perl-Treves, R., Molecular variation in melon (*Cucumis melo* L.) as revealed by RFLP and RAPD markers, *Sci. Hort.*, 79, 101–111, 1999.

Simmons, A. M. and Levi, A., Sources of whitefly (Homoptera: Aleyrodidae) resistance in *Citrullus* for the improvement of cultivated watermelon, *HortSci.*, 37, 581–584, 2002.

Simmons, A. M. and McCreight, J. D., Evaluation of melon for resistance to *Bemisia argentifolii* (Homoptera: Aleyrodidae), *J. Econ. Entomol.*, 89, 1663–1668, 1996.

Singh, A. K., Cucurbitaceae and polyploidy, *Cytologia*, 74, 897–905, 1979.

Singh, A. K., Cytogenetics and evolution of the Cucurbitaceae, In *Biology and Utilization of the Cucurbitaceae*, Bates, D. M. et al., Eds., Ithaca, NY: Comstock Publishing Associates, pp. 10–14, 1990.

Singh, D. K., Shukla, D., and Tewari, D., Electrophoretic characterization of indigenous germplasm lines of muskmelon [*Cucumis melo* (L.)], *Vegetable Sci.*, 26, 174–175, 1999.

Singh, D. K. and Ram, H., Characterization of indigenous germplasm lines of cucumber (*Cucumis sativus* L.) through SDS PAGE, *Vegetable Sci.*, 28, 22–23, 2001.

Sinskaja, E. N., *Istoričeskaja Geografia Kulturnoj Flory; Na Zare Zemledelija (Historical Geography of Cultivated Plants; In the Beginning of Agriculture)*, Leningrad, USSR: Izdatelstvo Kolos, 1969.

Šiško, M., Ivančič, A., and Bohanec, B., Genome size analysis in the genus *Cucurbita* and its use for determination of interspecific hybrids obtained using the embryo-rescue technique, *Plant Sci.*, 165, 663–669, 2003.

Skálová, D., Lebeda, A., and Navrátilová, B., Embryo and ovule culture in *Cucumis* species and their utilization in interspecific hybridization, *Progress in Cucurbit Genetics and Breeding Research,* In *Proceedings of Cucurbitaceae 2004, the 8th EUCARPIA Meeting on Cucurbit Genetics and Breeding*, Lebeda, A. and Paris, H. S., Eds., Olomouc, Czech Republic: Palacký University in Olomouc, pp. 415–430, 2004.

Sloane, J. T., Wehner, T. C., and Jenkins, S. F. Jr., Screening cucumber for resistance to belly rot caused by *Rhizoctonia solani*, *Cucurbit Genet. Coop. Rep.*, 6, 29–31, 1983.

Sloane, J. T., Wehner, T. C., and Jenkins, S. F. Jr., Evaluation of screening methods and sources for *Rhizoctonia* fruit rot in cucumber, *Cucurbit Genet. Coop. Rep.*, 7, 23–24, 1984.

Smartt, J. and Simmonds, N. W., *Evolution of Crop Plants*, 2nd ed., Harlow, U.K.: Longman Scientific & Technical, 1995.

Smith, P. G. and Venkat Ram, B. R., Interspecific hybridization between melon and cucumber, *J. Hered.*, 45, 24, 1954.

Sowell, G., Jr. and Schaad, N. W., *Pseudomonas pseudoalcaligenes* subsp. *citrulli* on watermelon: seed transmission and resistance of plant introductions, *Plant Dis. Rep.*, 63, 437–441, 1979.

Specht, C. E., Freytag, U., Hammer, K., and Börner, A., Survey of seed germinability after long-term storage in the Gatersleben genebank (part 2), *Plant Genet. Resour. Newslet.*, 115, 39–43, 1998.

Spooner, D. M., van den Berg, R. G., Hetterscheid, W. L. A., and Brandenburg, W. A., Plant nomenclature and taxonomy: an horticultural and agronomic perspective, *Hort. Rev.*, 28, 1–60, 2003.

St. Amand, P. C. and Wehner, T. C., Heritability and genetic variance estimates for leaf and stem resistance to gummy stem blight in two cucumber populations, *J. Am. Soc. Hort. Sci.*, 126, 90–94, 2001a.

St. Amand, P. C. and Wehner, T. C., Generation means analysis of leaf and stem resistance to gummy stem blight in cucumber, *J. Am. Soc. Hort. Sci.*, 126, 95–99, 2001b.

Staub, J. E., A core collection for cucumber: to be or not to be, *Cucurbit Genet. Coop. Rep.*, 17, 1–5, 1994a.

Staub, J. E., A core collection for cucumber: a starting point, *Cucurbit Genet. Coop. Rep.*, 17, 6–11, 1994b.

Staub, J. E. and Kupper, R. S., Results of the use of *Cucumis sativus* var. *Hardwickii* germplasm backcrossing with *Cucumis sativus* var. *sativus*, *HortSci.*, 20, 436–438, 1985.

Staub, J., Barczynska, H., Van Kleinwee, D., Palmer, M., Lakowska, E., Dijkhuizen, A., Clark, R., and Block, C., Evaluation of cucumber germplasm for six pathogens, In *Proceedings of Cucurbitaceae 89: Evaluation and Enhancement of Cucurbit Germplasm* Thomas, C. E., Ed., USDA/ARS, Charleston, SC, pp. 149–153, 1989.

Staub, J. E., Box, J., Meglic, V., Horejsi, T. F., and McCreight, J. D., Comparison of isozyme and random amplified polymorphic DNA data for determining intraspecific variation in *Cucumis*, *Genet. Resour. Crop. Evol.*, 44, 257–269, 1997a.

Staub, J., Crubaugh, L., Baumgartner, H., and Hopen, H., Screening of the cucumber germplasm collection for tolerance to clomazone herbicide, *Cucurbit Genet. Coop. Rep.*, 14, 22–24, 1991.

Staub, J. E., Dane, F., Reitsma, K., Fazio, G., and López-Sesé, A. I., The formation of test arrays and a core collection in (*Cucumis sativus* L.) using phenotypic and molecular marker data, *J. Am. Soc. Hort. Sci.*, 127, 558–567, 2002.

Staub, J. E., Danin-Poleg, Y., Fazio, G., Horejsi, T., Reis, N., and Katzir, N., Comparative analysis of cultivated melon groups (*Cucumis melo* L.) using random amplified polymorphic DNA and single sequence repeat markers, *Euphytica*, 115, 225–241, 2000.

Staub, J. E., Fanourakis, N., and López-Sesé, A. I., Genetic diversity in melon (*Cucumis melo* L.) landraces from the island of Crete as assessed by random amplified polymorphic DNA and simple sequence repeat markers, *Euphytica*, 136, 151–166, 2004.

Staub, J. E., Frederick, L., and Marty, T. C., Electrophoretic variation in cross-compatible wild diploid species of *Cucumis*, *Can. J. Bot.*, 65, 792–798, 1987.

Staub, J. E. and Ivandic, V., Genetic assessment of the United States national cucumber collection, *Acta Hort.*, 510, 113–121, 2000.

Staub, J. E., Knerr, L. D., Holder, D. J., and May, B., Phylogenetic relationships among several African *Cucumis* species, *Can. J Bot.*, 70, 509–517, 1992.

Staub, J. E. and Krasowska, A., Screening of the U.S. germplasm collection for heat stress tolerance, *Cucurbit Genet. Coop. Rep.*, 13, 47, 1990.

Staub, J. E., Meglic, V., and McCreight, J. D., Inheritance and linkage relationships of melon (*Cucumis melo* L.) isozymes, *J. Am. Soc. HortSci.*, 123, 264–272, 1998.

Staub, J. E. and Palmer, M. J., Resistance to downy mildew [*Pseudoperonospora cubensis* (Berk & Curt.) Rostow.] and scab (spot rot) [*Cladosporium cucumerinum* Ellis & Arthur] in *Cucumis* spp., *Cucurbit Genet. Coop. Rep.*, 10, 21–23, 1987.

Staub, J. E., Serquen, F., and Gupta, M., Genetic markers, map construction and their application in plant breeding, *HortSci.*, 31, 729–741, 1996.

Staub, J. E., Serquen, F. C., Horejsi, T., and Chen, J. F., Genetic diversity in cucumber (*Cucumis sativus* L.) IV; an evaluation of Chinese germplasm, *Genet. Resour. Crop Evol.*, 46, 297–310, 1999.

Staub, J. E., Serquen, F. C., and McCreight, J. D., Genetic diversity in cucumber (*Cucumis sativus* L): III. an evaluation of Indian germplasm, *Genet. Resour. Crop Evol.*, 44, 315–326, 1997b.

Stefanova, L., Neykov, S., and Todorova, T., Genetic diversity in the cucurbit family, *Plant Genet. Resour. Newslet.*, 99, 3–4, 1994.

Stepansky, A., Kovalski, I., and Perl-Treves, R., Intraspecific classification of melons (*Cucumis melo* L.) in view of their phenotypic and molecular variation, *Plant Syst. Evol.*, 217, 313–332, 1999a.

Stepansky, A., Kovalski, I., Schaffer, A. A., and Perl-Treves, R., Variation in sugar levels and invertase activity in mature fruit representing a broad spectrum of *Cucumis melo* genotypes, *Genet. Resour. Crop Evol.*, 46, 53–62, 1999b.

Stoyanova, S. D., *Ex situ* conservation in the Bulgarian genebank: 1. Effect of storage, *Plant Genet. Resour. Newslet*, 128, 68–76, 2001.

Strajeru, S. and Constantinovici, D., Romania's on-farm conservation project completed, *IPGRI Newsletter for Europe*, 28, 10, 2004.

Strefeler, M. S. and Wehner, T. C., Estimates of heritabilities and genetic variances of three yield and five quality traits in three fresh-market cucumber populations, *J. Am. Soc. Hort. Sci.*, 111, 599–605, 1986.

Strider, D. L. and Konsler, T. R,, An evaluation of the *Cucurbita* for scab resistance, *Plant Dis Rep* 49 388–391, 1965.

Sturtevant, E. L., Notes on edible plants, In *Report of the New York Agricultural Experiment Station for the Year 1919*, Vol. 2, Part II, Hedrick, U. P., Ed., Albany, NY: J.B. Lyon, 1919.

Sugiyama, K., Varietal differences in female flower bearing ability and evaluation method in watermelon, *JARQ*, 32, 267–273, 1998.

Sugiyama, K. and Morishita, M., Production of seedless watermelon using soft-X-irradiated pollen, *Scientia Hort.*, 84, 255–264, 2000a.

Sugiyama, K. and Morishita, M., Fruit and seed characteristics of diploid seedless watermelon (*Citrullus lanatus*) cultivars produced by soft-X-irradiated pollen, *J. Jpn. Soc. Hort. Sci.*, 69, 684–689, 2000b.

Sugiyama, K., Morishita, M., and Nishino, E., Seedless watermelons produced via soft-X-irradiated pollen, *HortSci.*, 37, 292–295, 2002a.

Sugiyama, M., Sakata, Y., Kitadani, E., Morishita, M., and Sugiyama, K., Pollen storage for production of seedless watermelon (*Citrullus lanatus*) using soft-X-irradiated pollen, *Acta Hort.*, 588, 269–272, 2002b.

Sujatha, V. S., Seshadri, V. S., Srivastava, K. N., and More, T. A., Isozyme variation in muskmelon (*Cucumis melo* L.), *Indian J. Genet Plant Breed.*, 51, 438–444, 1991.

Suzuki, E., Studies on muskmelon (*Cucumis melo* var. *reticulatus* NAUDIN).I. On the spontaneous tetraploid mutants, *Bull. Fac. Educ. Shizuoka Univ.*, 9, 169–176, 1958.

Suzuki, E., Studies on muskmelon (*Cucumis melo* var. *reticulatus* NAUDIN). III. On the triploid cross 4x×2x, *Bull. Fac. Educ. Shizuoka Univ.*, 10, 218–224, 1959.

Suzuki, E., Studies on muskmelon (*Cucumis melo* var. *reticulatus* NAUDIN); IV.On the reciprocal cross 2x× 4x, *Bull. Fac. Educ. Shizuoka Univ.*, 11, 176–186, 1960.

Szabo, T. I. and Smith, M. V., The use of *Megachile rotundata* for the pollination of greenhouse cucumbers, *Arkansas Agric. Exten. Service Misc. Publ.*, 127, 95–104, 1970.

Sztanger, J., Wronka, J., Galecka, T., and Korzeniewska, A., Cucumber (*Cucumis sativus*) haploids developed from parthenocarpic hybrids, In *Progress in Cucurbit Genetics and Breeding Research. In Proceedings of Cucurbitaceae 2004, the 8th EUCARPIA Meeting on Cucurbit Genetics and Breeding*, Lebeda, A. and Paris, H. S., Eds., Olomouc, Czech Republic: Palacký University in Olomouc, pp. 411–414, 2004.

Szwacka, M., Krzymowska, M., Osuch, A., Kowalczyk, M. E., and Malepszy, S., Variable properties of transgenic cucumber plants containing the thaumatin II gene from *Thaumatococcus daniellii*, *Acta Physiol. Plant.*, 24, 173–185, 2002.

Tabei, Y., Kitade, S., Nishizawa, Y., Kikuchi, N., Kayano, T., Hibi, T., and Akutsu, K., Transgenic cucumber plants harboring a rice chitinase gene exhibit enhanced resistance to gray mold (*Botrytis cinerea*), *Plant Cell Rep.*, 17, 159–164, 1998.

Tabei, Y., Nishio, T., and Kanno, T., Shoot regeneration from cotyledonary protoplast of melon (*Cucumis melo* L. cultivar Charentais), *J. Jpn. Soc. Hort. Sci.*, 61, 317–322, 1992.

Tabei, Y., Nishio, T., Kurihara, K., and Kanno, T., Selection of transformed callus in a liquid-medium and regeneration of transgenic plants in cucumber (*Cucumis sativus* L.), *Breeding Sci.*, 44, 47–51, 1994.

Taler, D., Galperin, M., Benjamin, I., Cohen, Y., and Kenigsbuch, D., Plant *eR* genes that encode photo-respiratory enzymes confer resistance against disease, *Plant Cell*, 16, 172–184, 2004.

Tanksley, S. D., Grandillo, S., Fulton, T. M., Zamir, D., Eshed, Y., Petiard, V., Lopez, J., and Beck-Bunn, T., Advanced backcross QTL analysis in a cross between an elite processing line of tomato and its wild relative *L. pimpinellifolium*, *Theor. Appl. Genet.*, 92, 213–224, 1996.

Tasaki, S., Collection and evaluation of germplasm of indigenous *Cucurbita* species in Brazil, *JICA Techn. Assist. Activ. Genet. Res. Project Newslet.*, 16, 1993.

Tatlioglu, T., Cucumber *Cucumis sativus* L., In *Genetic Improvement of Vegetable Crops*, Kalloo, G. and Bergh, B. O., Eds., Tarrytown, NY: Pergamon Press, pp. 197–234, 1993.

Teppner, H., Notes on *Lagenaria* and *Cucurbita* (Cucurbitaceae)—review and new contributions, *Phyton*, 44, 245–308, 2004.

Thies, J. A., Diseases caused by nematodes, In *Compendium of Cucurbit Diseases*, Zitter, T. A. et al., Eds., St. Paul, MN: APS Press, pp. 56–58, 1996.

Thomas, C. E., Downy and powdery mildew resistant muskmelon breeding line MR-1, *HortSci.*, 21, 329, 1986.

Thomas, C. E., Additional evaluations of *Cucumis melo* L. germplasm for resistance to downy mildew, *HortSci.*, 34, 920–921, 1999.

Thomas, C. E. and Jourdain, E. L., Evaluation of melon germplasm for resistance to downy mildew, *HortSci.*, 27, 434–436, 1992.

Thomas, C. E., Levi, A., and Caniglia, E., Evaluation of U.S. plant introductions of watermelon for resistance to powdery mildew, *HortSci.*, 40, 154–156, 2005.

Thomas, G., Astley, D., Boukema, I., Daunay, M. C., Del Greco, A., Diez, M. J., van Dooijeweert, W., et al., Eds., *Report of a Vegetables Network and ad hoc Leafy Vegetables Meeting* with and adhoc group on Leafy Vegetables, 22–24 May 2003, Skierniewice, Poland. International Plant Genetic Resources Institute, Rome, Italy, 154 pp. 2005.

Thompson, A. E., Dierig, D. A., and White, G. A., Use of plant introductions to develop new industrial crop cultivars, In *Use of Plant Introductions in Cultivar Development, Part 2*, Shands, H. L. and Weisner, L. E., Eds., Madison, WI: CSSA Special Publication 20. Crop Science Society of America, pp. 9–48, 1992.

Todorova, T., Status of the national vegetable crops collection in Bulgaria, *Plant Genet. Resour. Newslet*, 110, 55–56, 1997.

Toll, J. and Gwarazimba, V., Collecting in Zimbabwe, *FAO/IBPGR Plant Genet. Resour. Newslet.*, 53, 2–5, 1983.

Trebitsh, T., Staub, J. E., and O'Neill, S. D., Identification of an 1-aminocyclopropane-1-carboxylate synthase gene linked to the *Female* gene (*F*) that determines female sex expression in cucumber (*Cucumis sativus* L.), *Plant Physiol.*, 113, 987–995, 1997.

Tricoli, D. M., Carney, K. J., Russell, P. F., McMaster, J. R., Groff, D. W., Hadden, K. C., Himmel, P. T., Hubbard, J. P., Boeshore, M. L., and Quemada, H. D., Field evaluation of transgenic squash containing single or multiple virus coat protein gene constructs for resistance to cucumber mosaic virus, watermelon mosaic virus 2, and zucchini yellow mosaic virus, *Bio/Technol.*, 13, 1458–1465, 1995.

Troung-Andre, I., *In vitro* haploid plants derived from pollination by irradiated pollen of cucumber, In *Proceedings of Eucarpia Meeting on Cucurbits*, France: Avignon-Monfavet, pp. 143–144, 1988.

Truson, A. J., Simpson, R. B., and Shahin, E. A., Transformation of cucumber (*Cucumis sativus* L.) plants with *Agrobacterium rhizogenes*, *Theor. Appl. Genet.*, 73, 11–15, 1986.

Tulisalo, U., Resistance to the two-spotted spider mite; *Tetranychus urticae* Koch (Acarina, Tetranychidae) in the genera *Cucumis* and *Citrullus* (Cucurbitaceae), *Ann. Entomol. Fennici*, 38, 60–64, 1972.

Upadhyay, R., Ram, H. H., and Singh, D. K., Breeding potential of bottle gourd strains collected from Uttar Pradesh, *Recent Hort.*, 4, 159–162, 1997–1998.

Vakalounakis, D. J., Heart leaf, a recessive leaf shape marker in cucumber: linkage with disease resistance and other traits, *J. Hered.*, 83, 217–221, 1992.

Valles, M. P. and Lasa, J. M., *Agrobacterium*-mediated transformation of commercial melon (*Cucumis melo* L., cv. Amarillo Oro), *Plant Cell Rep.*, 13, 145–148, 1994.

van Dooijeweert, W., Status of the cucumber (*Cucumis sativus*) collection of CGN, In *Cucurbit Genetic Resources in Europe*. Ad hoc Meeting, 19 January 2002, Adana, Turkey, Díez, M. J. et al., Eds., Rome: International Plant Genetic Resources Institute, pp. 33–35, 2002.

van Dooijeweert, W., The status of the cucumber (*Cucumis sativus*) collection of CGN, In *Progress in Cucurbit Genetics and Breeding Research, Proceedings of Cucurbitaceae 2004, the 8th EUCARPIA Meeting on Cucurbit Genetics and Breeding*, Lebeda, A. and Paris, H. S., Eds., Olomouc, Czech Republic: Palacký University in Olomouc, pp. 91–94, 2004.

Vavilov, N. I., *Studies on the Origin of Cultivated Plants*, Leningrad, USSR: Institute of Applied Botany and Plant Breeding, 1926.

Vavilov, N. I., *Five Continents*, Rome: IPGRI, 1997.

Vinter, V., Křístková, A., Lebeda, A., and Křístková, E., Descriptor lists for genetic resources of the genus *Cucumis* and cultivated species of the genus *Cucurbita*, In *Progress in Cucurbit Genetics and Breeding Research, Proceedings of Cucurbitaceae 2004, the 8th EUCARPIA Meeting on Cucurbit Genetics and Breeding*, Lebeda, A. and Paris, H. S., Eds., Olomouc, Czech Republic: Palacký University in Olomouc, pp. 95–99, 2004.

Visser, D. L. and den Nijs, A. P. M., The *Cucumis* species collection at the IVT, *Cucurbit Genet. Coop. Rep.*, 3, 68–69, 1980.

Visser, D. L. and den Nijs, A. P. M., Variation for interspecific crossability of *Cucumis anguria* L. and *Cucumis zeyheri*, *Cucurbit Genet. Coop. Rep.*, 6, 100–101, 1983.

Wako, T., Sakata, Y., Sugiyama, M., Ohara, T., Ishiuchi, D., and Kojima, A., Identification of melon accessions resistant to gummy stem blight and genetic analysis of the resistance using an efficient technique for seedling test, *Acta Hort.*, 588, 161–164, 2002.

Wall, J. R., Interspecific hybrids of *Cucurbita* obtained by embryo culture, *Proc. Am. Soc. Hort. Sci.*, 63, 427–430, 1954.

Walters, T. W. and Decker-Walters, D. S., Systematic re-evaluation of *Benincasa hispida* (Cucurbitaceae), *Econ. Bot.*, 43, 274–278, 1989.

Walters, T. W., Decker-Walters, D. S., Posluszny, U., and Kevan, P. G., Determination and interpretation of comigrating allozymes among genera of the Benincaseae (Cucurbitaceae), *Syst. Bot.*, 16, 30–40, 1991.

Walters, S. A. and Wehner, T. C., Evaluation of the U.S. cucumber germplasm collection for root size using a subjective rating technique, *Euphytica*, 79, 39–43, 1994a.

Walters, S. A. and Wehner, T. C., Evaluation of the U.S. cucumber germplasm collection for early flowering, *Plant Breeding*, 112, 234–238, 1994b.

Walters, S. A. and Wehner, T. C., 'Lucia', 'Manteo,' and 'Shelby' root-knot nematode-resistant cucumber inbred lines, *HortSci.*, 32, 1301–1303, 1997.

Walters, S. A., Wehner, T. C., and Barker, K. R., A single recessive gene for resistance to the root-knot nematode (*Meloidogyne javanica*) in *Cucumis sativus* var. *hardwickii*, *J. Hered.*, 88, 66–69, 1997.

Wang, M. and Liu, L., Studies on anthracnose resistance of watermelon germplasm at the seedling stage, In *Cucurbitaceae 89: Evaluation and Enhancement of Cucurbit Germplasm*, Thomas, C. E., Ed., USDA/ARS, Charleston, SC, pp. 154–156, 1989.

Wang, M. and Zhang, X., Evaluation and utilization of the valuable African watermelon germplasm, *Cucurbit Genet. Coop. Rep.*, 11, 69, 1988.

Wang, Y. H., Thomas, C. E., and Dean, R. A., A genetic map of melon (*Cucumis melo* L.) based on amplified fragment length polymorphism (AFLP) markers, *Theor. Appl. Genet.*, 95, 791–798, 1997.

Wang, Y. H., Thomas, C. E., and Dean, R. A., Genome mapping of melon (*Cucumis melo* L.) for localizing disease resistance genes, In *Cucurbitaceae 98: Evaluation and Enhancement of Cucurbit Germplasm*, McCreight, J. D., Ed., Alexandria, VA: ASHS Press, pp. 354–361, 1998.

Wang, Y. H., Thomas, C. E., and Dean, R. A., Genetic mapping of a fusarium wilt resistance gene (*Fom-2*) in melon (*Cucumis melo* L.), *Mol. Breed.*, 6, 379–389, 2000.

Wang, Y. J. and Robinson, R. W., Influence of temperature and humidity on longevity of squash pollen, *Cucurbit Genet. Coop. Rep.*, 6, 91, 1983.

Wann, E. V., Evaluation of the U.S; cucumber germplasm collection for tolerance to soil moisture deficit, *Cucurbit Genet. Coop. Rep.*, 15, 1–3, 1992.

Ware, G. W. and McCollum, J. P., *Producing Vegetable Crops*, Danville, IL: Interstate, 1980.

Webb, S. E., Insect resistance in cucurbits: 1992–1998, In *Cucurbitaceae 98, Evaluation and Enhancement of Cucurbit Germplasm*, McCreight, J. D., Ed., Alexandria, VA: ASHS Press, 1998.

Wechter, W. P., Thomas, C. E., and Dean, R. A., Development of sequence specific primers which amplify a 1.5 kb DNA marker for race 1 Fusarium-wilt resistance in *Cucumis melo*, *Hort. Sci.*, 33, 291–292, 1998.

Wechter, W. P., Whitehead, M. P., Thomas, C. E., and Dean, R. A., Identification of a randomly amplified polymorphic DNA marker linked to the Fom 2 Fusarium wilt resistance gene in muskmelon MR-1, *Phytopathology*, 85, 1245–1249, 1995.

Weeden, N. F., Isozyme studies indicate that the genus *Cucurbita* is an ancient tetraploid, *Cucurbit Genet. Coop. Rep.*, 7, 84–85, 1984.

Weeden, N. F. and Robinson, R. W., Isozyme studies in *Cucurbita*, Bates, D. M. et al., Ed., In *Biology and Utilization of the Cucurbitaceae*, pp. 51–59, 1990.

Wehner, T. C., Genetic variation for low-temperature germination ability in cucumber, *Cucurbit Genet. Coop. Rep.*, 5, 16–17, 1982.

Wehner, T. C., Estimates of heritabilities and variance components for low-temperature germination ability in cucumber, *J. Am. Soc. Hort. Sci.*, 109, 664–667, 1984.

Wehner, T. C., Two special cucumber populations: NCH1 and NCBA1, *HortSci.*, 33, 766–768, 1998a.

Wehner, T. C., Three pickling cucumber populations: NCWBP, NCMBP and NCEP1, *HortSci.*, 32, 941–944, 1998b.

Wehner, T. C., Cade, R. M., and Locy, R. D., Cell, tissue and organ culture techniques for genetic improvement of cucurbits, In *Biology and Utilization of the Cucurbitaceae*, Bates, D. M. et al., Eds., Ithaca, NY: Comstock, pp. 367–381, 1990.

Wehner, T. C. and Cramer, C. S., Gain for pickling cucumber yield and fruit shape using recurrent selection, *Crop Sci.*, 36, 1538–1544, 1996a.

Wehner, T. C. and Cramer, C. S., Ten cycles of recurrent selection for fruit yield, earliness, and quality in three slicing cucumber populations, *J. Am. Soc. Hort. Sci.*, 121, 362–366, 1996b.

Wehner, T. C., Elsey, K. D., and Kennedy, G. G., Screening for cucumber antibiosis to pickleworm, *Hort. Sci.*, 20, 1117–1119, 1985.

Wehner, T. C., Lower, R. L., Staub, J. E., and Tolla, G. E., Convergent-divergent selection for cucumber fruit yield, *HortSci.*, 24, 667–669, 1989.

Wehner, T. C., McCreight, J. D., Rhodes, B., and Zhang, X., Mutually beneficial cucurbit expedition to the People's Republic of China leads to continued collaboration with the United States, *Diversity*, 12(1), 13, 1996.

Wehner, T. C. and Shetty, N. V., Downy mildew resistance of the cucumber germplasm collection in North Carolina field tests, *Crop Sci.*, 37, 1331–1340, 1997.

Wehner, T. C. and Shetty, N. V., Screening the cucumber germplasm collection for resistance to gummy stem blight in North Carolina field tests, *HortSci.*, 35, 1132–1140, 2000.

Wehner, T. C., Shetty, N. V., and Clark, R. L., Screening the cucumber germplasm collection for combining ability for yield, *HortSci.*, 35, 1141–1150, 2000a.

Wehner, T. C., Shetty, N. V., and Wilson, L. G., Screening the cucumber germplasm collection for fruit storage ability, *HortSci.*, 35, 699–707, 2000b.

Wehner, T. C. and St. Amand, P. C., Anthracnose resistance of the cucumber germplasm collection in North Carolina field tests, *Crop Sci.*, 35, 228–236, 1995.

Wehner, T. C., St. Amand, P. C., and Lower, R. L., 'M 17' gummy stem blight resistant pickling cucumber inbred, *HortSci.*, 31, 1248–1249, 1996.

Wehrhahn, C. and Allard, R. W., The detection and measurement of the effects of individual genes involved in the inheritance of a quantitative character in wheat, *Genetics*, 51, 109–119, 1965.

Werner, G. M. and Putnam, A. R., Triazine tolerance in *Cucumis sativus* L, *Proceedings of the North Central Weed Control Conference*, 32, 26, 1977.

Whitaker, T. W., Chromosome number in cultivated cucurbits, *Am. J. Bot.*, 17, 1033–1040, 1930.

Whitaker, T. W., A cross between an annual species and a perennial species of *Cucurbita*, *Madrono*, 12, 213–217, 1954.

Whitaker, T. W., An interspecific cross in *Cucurbita* (*C. lundelliana* Bailey × *C. moschata* Duch.), *Madrono*, 15, 4–13, 1959.

Whitaker, T. W., A collection of wild and cultivated Cucurbitaceae from Zambia, *Cucurbit Genet. Coop. Rep.*, 7, 89–90, 1984.

Whitaker, T. W. and Bemis, W. P., Origin and evolution of the cultivated *Cucurbita*, *Bull. Torrey Bot. Club*, 102, 362–368, 1975.

Whitaker, T. W. and Davis, G. N., *Cucurbits: Botany, Cultivation, and Utilization*, New York: Interscience Publishers, 1962.

Whitaker, T. W. and Robinson, R. W., Squash breeding, In *Breeding Vegetable Crops*, Bassett, M. J., Ed., Westport, CT: AVI Publishing Co., Inc., pp. 209–242, 1986.

Widrlechner, M. P. and Burke, L. A., Analysis of germplasm distribution patterns for collections held at the North Central Regional Plant Introduction Station, Ames, Iowa, U.S.A., *Genet. Resour. Crop Evol.*, 50, 329–337, 2003.

Widrlechner, M. P., Knerr, L. D., Staub, J. E., and Reitsma, K. R., Biochemical evaluation of germplasm regeneration methods for cucumber, *Cucumis sativus* L., *FAO/IBPGR Plant Genet Resour. Newslet.*, 88/89, 1–4, 1992.

Williams, K. A., An overview of the U.S. National Plant Germplasm System's exploration program, *HortSci.*, 40, 297–301, 2005.

Wilson, H. D., Discordant patterns of allozyme and morphological variation in Mexican, *Cucurbita*, *Syst. Bot.*, 14, 612–623, 1989.

Wilson, H. D., Gene flow in squash species, *BioScience*, 40, 449–455, 1990.

Wilson, H. D., Doebley, J., and Duvall, M., Chloroplast DNA diversity among wild and cultivated members of *Cucurbita* (Cucurbitaceae), *Theor. Appl. Genet.*, 84, 859–865, 1992.

Withers, L. A., *In vitro* conservation and germplasm utilization, In *The Use of Plant Genetic Resources*, Brown, A. H. D. et al., Eds., Cambridge: Cambridge University Press, pp. 309–334, 1989.

Witkowicz, J., Urbanczyk-Wochniak, E., and Przybecki, Z., AFLP marker polymorphism in cucumber (*Cucumis sativus* L.) near isogenic lines differing in sex expression, *Cell. Mol. Biol. Lett.*, 8, 375–381, 2003.

Xie, J., Wehner, T. C., and Conkling, M. A., PCR-based single-strand conformation polymorphism (SSCP) analysis to clone nine aquaporin genes in cucumber, *J. Am. Soc. Hort. Sci.*, 127, 925–930, 2002.

Xie, J., Wehner, T. C., Wollenberg, K., Purugganan, M. D., and Conkling, M. A., Intron and polypeptide evolution of conserved NPA to NPA motif regions in plant aquaporins, *J. Am. Soc. Hort. Sci.*, 128, 591–597, 2003.

Xu, D. and Huang, Y., Study on Hunan local bitter gourd (*Momordica* (sic) *charantia* L.) variety resource, *Acta Hort.*, 402, 329–333, 1995.

Yamaguchi, J. and Shiga, T., Characteristics of regenerated plants via protoplast electrofusion between melon (*Cucumis melo*) and pumpkin (interspecific hybrid *Cucurbita maxima* × *Cucurbita moschata*), *Jpn. J. Breed.*, 43, 173–182, 1993.

Yang, S. F. and Oetiker, J. H., Molecular biology of ethylene biosynthesis and its application in horticulture, *J. Jpn. Soc. Hort. Sci.*, 67, 1209–1214, 1998.

Yashiro, K., Hosoya, K., Kuzuya, M., Tomita, K., and Ezura, H., Efficient production of doubled haploid melon plants by modified colchicines treatment of parthenogenetic haploids, *Acta Hort.*, 588, 335–338, 2002.

Yetisir, H. and Sari, N., A new method for haploid muskmelon (*Cucumis melo* L.) dihaploidization, *HortSci.*, 98, 277–283, 2003.

Yoshioka, K., Hanada, K., Harada, T., Minobe, Y., and Oosawa, K., Virus resistance in transgenic melon plants that express the cucumber mosaic virus coat protein gene and in their progeny, *Jpn. J. Breed.*, 43, 629–634, 1993.

Yoshioka, K., Hanada, K., Nakazaki, Y., Minobe, Y., Yakuwa, T., and Oosawa, K., Successful transfer of the cucumber mosaic virus coat protein gene to *Cucumis melo* L., *Jpn. J. Breed.*, 42, 277–285, 1992.

Zamir, D., Navot, N., and Rudich, J., Enzyme polymorphism in *Citrullus lanatus* and *C. colocynthis* in Israel and Sinai, *Plant Syst. Evol.*, 146, 163–170, 1984.

Zhang, G. P., Xiang, C. P., and Li, X. X., Analysis of 50 cucumber accessions by RAPD markers, *Hunan Agric. Sci. Technol. Newsl.*, 3, 10–14, 2002.

Zhang, R. B., Xu, Y., Vi, K., Zhang, H. Y., Liu, L. G., Yi, G. G., and Levi, A., A genetic linkage map for watermelon derived from recombinant inbred lines, *J. Am. Soc. Hort. Sci.*, 129, 237–243, 2004.

Zhang, X. P., Rhodes, B. B., and Adelberg, J. W., Shoot regeneration from immature cotyledons off watermelon, *Cucurbit Genet. Coop. Rep.*, 17, 111–115, 1994.

Zhang, X. W., Qian, X. L., and Liu, J. W., Evaluation of the resistance to root-knot nematode of watermelon germplasm and its control (in Chinese), *J. Fruit Sci.*, 6, 33–38, 1989.

Zhang, X. Y. and Tao, K. L., Silica gel seed drying for germplasm conservation—practical guidelines, *FAO/IBPGR Plant Genet. Resour. Newslet.*, 75/76, 1–5, 1988.

Zhang, Y., Kyle, M., Anagnostou, K., and Zitter, T. A., Screening melon (*Cucumis melo*) for resistance to gummy stem blight in the greenhouse and field, *HortSci.*, 32, 117–121, 1997.

Zheng, X. Y. and Wolff, D. W., Randomly amplified polymorphic DNA markers linked to fusarium wilt resistance in diverse melons, *HortSci.*, 35, 716–721, 2000.

Zheng, X. Y., Wolff, D. W., Baudracco-Arnas, S., and Pitrat, M., Development and utility of cleaved amplified polymorphic sequences (CAPS) and restriction fragment length polymorphisms (RFLPs) linked to the *Fom-2* fusarium wilt resistance gene in melon (*Cucumis melo* L.), *Theor. Appl. Genet.*, 99, 453–463, 1999.

Zhou, S. K., Qiu, Z. H., Li, G. X., Li, Z., and Li, X. R., Study on the germplasm resources of *Cucurbita* for seed and its utilization (in Chinese), *Crop Genet. Res.*, 2, 13–15, 1995.

Zhuang, F. Y., Chen, J. F., Staub, J. E., and Qian, C. T., Assessment of genetic relationships in *Cucumis* species by SSR and RADP analysis, *Plant Breeding*, 123, 167–172, 2004.

Zink, F. W., Gubler, W. D., and Grogan, R. G., Reaction of muskmelon germ plasm to inoculation with *Fusarium oxysporum* f. sp. *melonis* race 2, *Plant Dis.*, 67, 1251–1255, 1983.

Zitter, T. A., Hopkins, D. L., and Thomas, C. E., Eds., *Compendium of Cucurbit Diseases*, St. Paul, MN: APS Press, 1996.

Zohary, D. and Hopf, M., *Domestication of Plants in the Old World; The Origin and Spread of Cultivated Plants in West Asia, Europe, and the Nile Valley*, Oxford: Oxford University Press, 2000.

Zraidi, A. and Lelley, T., Genetic map for pumpkin *Cucurbita pepo* using random amplified polymorphic DNA markers, In *Progress in Cucurbit Genetics and Breeding Research, Proceedings of Cucurbitaceae 2004, the 8th EUCARPIA Meeting on Cucurbit Genetics and Breeding*, Lebeda, A. and Paris, H. S., Eds., Olomouc, Czech Republic: Palacký University in Olomouc, pp. 507–514, 2004.

CHAPTER 9

Lettuce (Asteraceae; *Lactuca* spp.)

Aleš Lebeda, E. J. Ryder, R. Grube, I. Doležalová, and E. Křístková

CONTENTS

9.1 INTRODUCTION

Lettuce (*Lactuca sativa* L.) is a member of the family Asteraceae, the largest of the dicotyledonous families (Judd et al. 1999). The genus includes about 100 species (Lebeda, Doležalová, and Astley, 2004a, 2004b). Nearly all of these are not compatible with cultivated lettuce. However, three species: *L. serriola, L. saligna*, and *L. virosa,* are compatible, to different degrees (Zohary 1991). *L. serriola* (prickly lettuce, compass lettuce or common wild lettuce), is found worldwide (Lebeda, Pink, and Mieslerová, 2001c, 2004a). It is completely compatible with lettuce and is considered by some to be the same species and a direct progenitor of cultivated lettuce (de Vries 1997; Koopman et al. 1998). *L. saligna* is primarily a Mediterranean species, occurring in Europe, northern Africa, and the Mideast (Zohary 1991; Lebeda et al. 2004b). It is less closely related to

L. sativa than is *L. serriola*. Crosses are most easily made when *L. saligna* is used as the female parent. *L. virosa* is also a Mediterranean species. In Europe it is distributed more towards the south and west than *L. saligna* (Feráková 1977; Lebeda, Doležalová, and Astley 2004a). It is the least closely related of the three to *L. sativa*. Crosses with cultivated lettuce are very difficult and F_1 plants are highly sterile.

Several other taxonomic groups which are closely related to *L. sativa* or *L. serriola* are: *L. aculeata*, *L. altaica*, *L. augustana*, *L. georgica*, *L. dregeana*, and *L. azerbaijanica* (Lebeda and Astley 1999). They are very similar taxonomically and some have been successfully crossed with *L. sativa* or *L. serriola* (Lebeda 1998; Lebeda and Astley 1999).

Lettuce belongs to the core group of crops as one of the earliest domesticated vegetable crops (8,000 to 4,000 years before present) (Hancock 2004). The earliest known artifacts indicating that lettuce was used as a food are tomb decorations from Ancient Egypt. However, it seems more likely that the center of diversity for the related *Lactuca* species is in southwestern Asia in the Tigris–Euphrates region, and that lettuce may have become a food plant there (Zohary and Hopf 1993). If so, it evidently found its way to the Nile Valley, and our first knowledge of its use comes from the tomb decorations first created about 4,500 B.C. (Keimer 1924; Harlan 1986; de Vries 1997) (Figure 9.1 through Figure 9.3). These depict plants similar to modern stalk or stem lettuce. The oilseed lettuces also evolved at about the same time, or perhaps earlier, because they are wilder in appearance. Variations leading to consumption of the leaves rather than the stems may have occurred before or during the movement of lettuce into ancient Greek and Roman cultures. Romaine lettuce, likely introduced from Italy, was grown in southern France in the early fifteenth century and head type lettuce was first noted in the sixteenth century. The stem type lettuce traveled to China, perhaps from Persia, possibly between 600 and 900 A.D. Lettuce was brought to the Americas early; probably during the second voyage of Christopher Columbus in 1494 (Hedrick 1972; Ryder 1986, 1999a).

Lettuce is a variable species and occurs in seven distinct types: crisphead, cos (romaine), butterhead, leaf, Latin, stem, and oilseed (de Vries 1997; Ryder 1999a). Crisphead has two important subtypes, iceberg and batavia. Crisphead lettuce forms a head with crisp textured leaves. The iceberg subtype is the larger of the two, both in weight and volume. It forms a dense spherical head of overlapping white or yellowish leaves. The head is partially enclosed near the bottom by large wrapper leaves. Batavia is smaller and less dense and the transition between head leaves and wrappers is less obvious. Some crisphead cultivars are intermediate in appearance.

Cos lettuce forms an erect, elongated, or loaf-shaped head of leaves that are not as broad as leaves of the crisphead type. They are also crisp and somewhat rough in texture. There is little enclosure of inner by outer leaves, and more of them retain green color than those of the crisphead type. Butterhead lettuce also forms a head, essentially a flattened sphere, which is considerably smaller than the crisphead and cos types. The leaves are more broad than long, and have a soft oily texture. The plants weigh about 1/4 to 1/3 as much as iceberg lettuces. The fourth major type includes the leaf or cutting lettuces. They are a highly variable group, having in common the lack of a heading or enclosure stage. The leaves vary in shape (broad, elongated, lobed, curled); size; texture (crisp, soft); and color (red, green; dark, light). Because the rosette remains open, all leaves have some green color.

The other types are less commonly grown. Latin, or grassé, lettuce forms a small head, slightly elongated but shorter than romaine. The leaves look like butterhead leaves, but are smaller and have a crisp texture. It is grown primarily in Europe. Stalk (stem) lettuce has a thick elongated stalk and narrow leaves. The stalk is tender and is the part that is consumed, raw in Egypt and cooked in China. Oilseed lettuce is a primitive type with larger seeds than other lettuces. The seeds are crushed to produce an oil that is used for cooking, especially in Egypt.

The importance of lettuce (*L. sativa* L.) rests on the fact that it is the most popular vegetable, from the group of leafy vegetables, used in salads in most parts of the world. It is almost exclusively

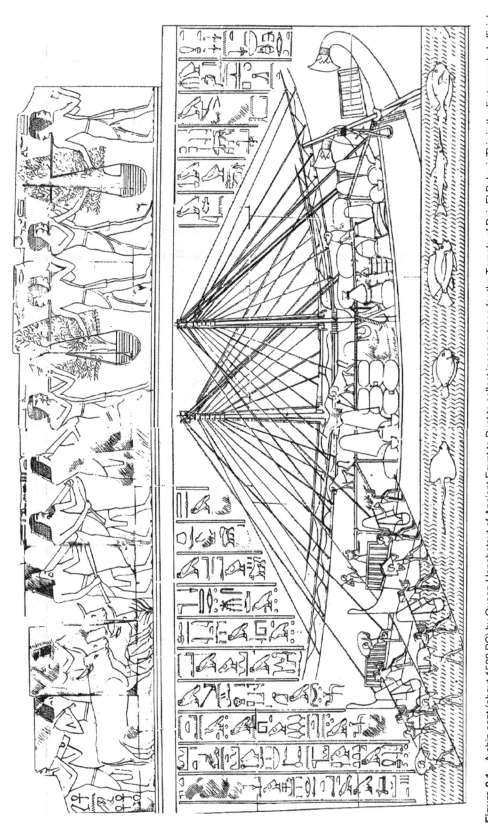

Figure 9.1 A ship sent (about 1500 BC) by Queen Hapshetsut of Ancient Egypt to Punt to collect incense trees for the Temple of Deir El Bahari. This is the first recorded official expedition to collect plants from another county. (From Harlan, J. R., *Crops and Man*, 2nd ed., Crop Science Society of America, Madison, WI, 1992. With permission.)

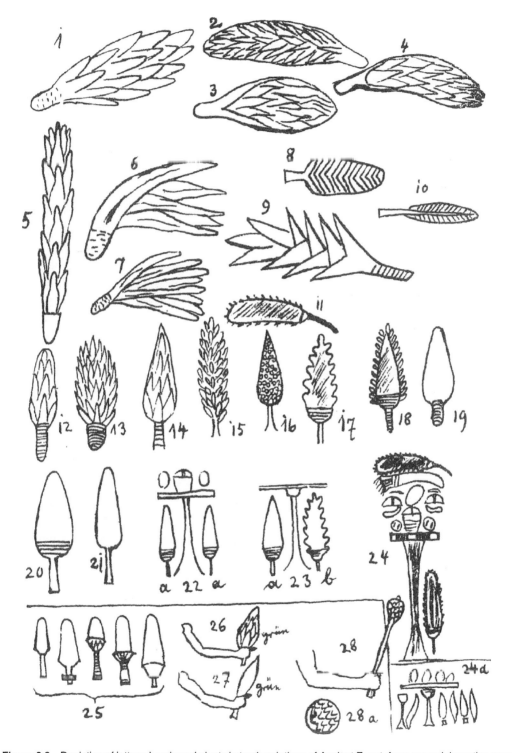

Figure 9.2 Depiction of lettuce heads and plants in tomb paintings of Ancient Egypt, from several dynasties over many years. From top to bottom, drawings change from natural to highly stylized over time. (From Keimer, L., *Die Gartenpflanzen in Alten Aegypten*, Hoffmann und Campe, Hamburg, 1924.)

Figure 9.3 Picture of several plants left as offerings in a tomb near Meir, Egypt. Lettuce is at the top, near the center, and appears to be a type of stem lettuce. (From Harlan, J., *Econ. Bot.* 40, 4–15, 1986.)

used as a fresh, uncooked product with moderate dietetic value (Table 9.1). It occurs in several forms, all except two of which are consumed as a raw vegetable. One type of stalk lettuce is used in China as a cooked vegetable. In Egypt, the seeds of a primitive form of lettuce are crushed to produce an oil used for cooking (Rubatzky and Yamaguchi 1997; Ryder 1999a).

Lettuce is produced commercially in many countries worldwide and is grown as a garden vegetable in most places. It is especially important as a commercial crop in Asia, North and Central America, and Europe. China, United States, Spain, Italy, India and Japan are among the world's largest producers of lettuce (Table 9.2).

The U.S. has the most intensive lettuce production in the world. Nearly all production is in two states: California and Arizona; with small amounts grown in Florida, Colorado, and New Jersey (Table 9.3). Until recently, several other states contributed to lettuce production, but they no longer grow and market significant amounts (USDA 2004).

Different regions of the world have been associated with different types of lettuce (Rubatzky and Yamaguchi 1997; Ryder 1999a). Of the major types, cos lettuce became well established and remains popular in the Mediterranean Basin, in the countries of northern Africa, southwest Asia, and southern Europe. In northern Europe, butterhead lettuce became highly popular both in the western and eastern sections. In the United States, early cultivars were of several types (Tracy 1904), but in the late nineteenth and early twentieth centuries, iceberg lettuce became overwhelmingly popular. More recently it has become popular in other countries, such as Spain, Australia, Japan, Sweden, and the United Kingdom. Still more recently, cos and leaf lettuces have increased in use in the United States and in other countries.

Table 9.1 Nutrient Values for Several Lettuce Types Compared to Other Selected Vegetables, Content in 100 g Portions (Rubatzky and Yamaguchi, 1997)

Vegetable	Vitamins				Minerals (g)			
	A (I.U.)	C (g)	Ca	P	K	Na	Mg	Fe
Lettuce								
Crisphead	470	7	22	26	166	7	11	1.5
Butter	1,065	8	35	26	260	7	11	1.8
Romaine	1,925	22	44	35	277	9	9	1.3
Leaf	1,900	18	68	25	264	9	11	1.4
Endivo	2,110	0	66	41	304	50	13	1.3
Onion	40	9	26	33	156	6	11	0.4
Celery	184	8	52	32	313	107	17	0.4
Cabbage	150	50	49	29	272	17	16	0.5
Broccoli	3,150	109	101	77	389	23	23	1.2
Spinach	7,045	50	107	57	605	110	92	2.7
Tomato	900	25	12	26	244	3	14	0.5
Potato	20	18	12	51	420	10	27	0.7

Source: From Rubatzky, V. E. and Yamaguchi, M., *World Vegetables, Principles, Production, and Nutritive Values*, 2nd Ed., Chapman & Hall, New York, 1997.

Traditionally, lettuce was harvested and marketed as a whole head, with only a few lower leaves trimmed off. Three changes occurred in the latter quarter of the twentieth century (Ryder 1999a). The first change was the development of the wrap pack for crisphead lettuce, in which all but one or two outer leaves are trimmed off and the head is wrapped in plastic. Shortly afterwards, a light processing procedure was developed. The heads were chopped or shredded and packed in plastic bags. At first the bags were large and used only for institutional packs. Later, the product was packed in small, consumer-sized packages. Over the years, many combinations were devised, with several kinds of lettuce, or other salad vegetables, in one package. The third innovation is called mesclun or salad mix. It consists mostly of immature leaves of several types of lettuce, as well as spinach, beet tops, endive, tat soi, arugula, and other leafy vegetables (Rubatzky and Yamaguchi 1997). As many as ten

Table 9.2 World Lettuce Production in 2004

Location	Production (metric tons)	Area Harvested (ha)	Mean Yield (kg/ha)
World	21,902,397	1,017,419	21,527
By Continent			
Africa	309,421	15,830	19,547
Asia	12,673,289	687,977	18,421
Australia	121,508	6,134	19,809
Europe	3,216,116	140,979	22,813
North & Central America	5,358,640	150,134	35,692
South America	101,731	15,005	12,778
By Nation			
China	10,505,000	500,250	20,999
U.S.A.	4,976,880	131,280	37,910
Spain	1,044,200	37,500	27,845
Italy	830,580	43,564	19,066
India	790,000	120,000	6,583
Japan	550,000	22,000	25,000
France	462,917	17,237	26,856
Turkey	340,000	18,700	18,182

Source: From FAO, *FAOSTAT Agricultural Database*, http://apps.fao.org, 2004.

Table 9.3   Production of Lettuce in Leading Producing States of the United States, 2003

State	Area (ha)	Production (Tonnes)
Iceberg lettuce		
California	54,660	2,142,000
Arizona	19,640	771,000
Colorado	810	23,600
New Jersey	360	7,200
Romaine lettuce		
California	20,650	995,400
Arizona	680	220,400
Leaf and butter lettuce		
California	9,000[a]	453,500
Arizona	3,440	74,400

[a] Estimate.
Source: From USDA, personal communication (2004).

leafy vegetables may be used in a specific mix, depending upon availability. A total of 25–30 may have potential for use in a mixture.

9.2   GERMPLASM COLLECTION, MAINTENANCE, EVALUATION, AND DISTRIBUTION

Wild plant germplasm is defined as that which is not grown as a crop but also differs from other related wild ancestors by various characters (Lenné and Wood 1991; Dinoor and Eshed 1997). The exploitation of the wild gene pool of plants is not yet a common practice in the breeding of many crops. Rational exploitation of wild germplasm or wild relatives is based on, and conditioned by, the following: (1) identification of the wild gene pool of the crop; (2) availability of sufficient material for screening and evaluation; and (3) the use of appropriate methods for gene transfer (Ladizinsky 1989). The utilization of lettuce and wild *Lactuca* species germplasm for research and breeding relies upon three things in the world's germplasm collections: (1) to conserve the breadth of genetic diversity present in *L. sativa* and their primitive, wild and weedy relatives; (2) to adequately characterize that diversity; and (3) to make genetic resources and accurate information readily available to researchers and breeders. The following section summarizes the current status of lettuce germplasm conservation, characterization and usage.

9.2.1   Genebanks and Other Significant Germplasm Collections

Limited information is available about lettuce germplasm collections (Boukema, Hazekamp, and van Hintum 1990; McGuire et al. 1993; Soest and Boukema 1997; Lebeda 1998; Hintum and Boukema 1999; Lebeda and Astley 1999; Ryder 1999a; Lebeda and Boukema 2001, 2005; Thomas et al. 2005). These sources provide general information about the holdings, maintenance conditions, availability, evaluation, and documentation of the most important of the world's collections; emphasizing national genebanks and working collections. More recently, information about the holdings of the world's largest collections of leafy vegetable germplasm was summarized as part of the Food and Agriculture Organization's effort to present "The State of the World's Plant Genetic Resources for Food and Agriculture" (FAO 1998) and by Cross (1998), and in Europe by Lebeda and Boukema (2001). Table 9.4 lists examples of the world's most important *Lactuca* species genetic resources collections.

Table 9.4   Important Lettuce (*Lactuca sativa*) and Wild *Lactuca* Species Germplasm Collections

Name	Location	Material	References
Europe			
Aegean Agricultural Research Institute (AARI)	Izmir, Turkey	Lr, Ws	Lebeda and Boukema (2001)
Agricultural Institute of Slovenia, Department of Crop and Seed Production	Ljubljana, Slovenia	Cv, Lr	Lebeda and Boukema (2001)
AgriFood Research Institute, Germplasm Bank	Zaragoza, Spain	Cv, Lr	Valcárcel et al. (2005)
Genetic Resources Unit, Warwick HRI	Wellesbourne, U.K.	Cv, Lr, Ws	Pink and Astley (2005)
INRA, Unité de Génétique et d'Amelioration des Fruits et Légumes	Montfavet, France	Cv, Lr, Ws	Lebeda and Boukema (2001)
Institute for Agrobotany	Tápioszele, Hungary	Cv, Ws	Lebeda and Boukema (2001)
Institute of Plant Genetics and Crop Plant Research (IPK), Department of Genebank	Gatersleben, Germany	Cv, Lr, Ws	Lebeda and Boukema (2001)
Lettuce Collection, Centre for Genetic Resources (CGN)	Wageningen, the Netherlands	Cv, Lr, Ws	Boukema and de Groot (2005)
N.I. Vavilov Institute of Plant Industry	St. Petersburg, Russia	Cv, Lr, Ws	Lebeda and Boukema (2005)
Polytechnic University of Valencia, Centre for the Conservation and Agrodiversity	Valencia, Spain	Cv	Valcárcel et al. (2005)
Research Institute of Crop Production, Gene Bank Department	Olomouc, Czech Republic	Cv, Lr, Gs, Ws	Křístková and Chytilová (2005)
Research Institute for Vegetable Crops, Plant Genetic Resources Laboratory	Skierniewice, Poland	Cv, Lr, Ws	Kotlińska (2005)
Working Collection of the Genus *Lactuca*, Department of Botany, Palacky University	Olomouc, Czech Republic	Cv, Lr, Gs, Ws	Lebeda, Doležalová, and Vondráková (2005)
United States			
Horticultural Sciences Department, Cornell University	Geneva, New York, U.S.A.	Cv, Lr, Bl, Gs, Ws	McGuire et al. (1993)
Lactuca Genetic Resources Collection, University of California (UC)	Davis, California, U.S.A.	Cv, Lr, Bl, Gs, Ms, Ws	McGuire et al. (1993)
USDA–ARS Regional Plant Introduction Station	Pullman, Washington, U.S.A.	Lr, Ws	McGuire et al. (1993)
USDA *Lactuca* Germplasm Collection	Salinas, California, U.S.A.	Cv, Lr, Bl, Gs, Ws	McGuire et al. (1993)
U.S. National Seed Storage Laboratory	Fort Collins, Colorado, U.S.A.	Cv	McGuire et al. (1993)

Note: For U.S. collections, see also GRIN database: http://www.ars-grin,gov/npgs/searchgrin.html.
Cv, cultivars; Lr, landraces; Bl, breeding lines; Gs, genetic stocks; Ms, molecular stocks; Ws, wild species.

   In Europe, international cooperation among institutions holding germplasm collections of vegetable crops, including leafy vegetables and lettuce, is coordinated by the International Plant Genetic Resources Institute (IPGRI) (Maggioni 2004), an autonomous international scientific organization, and supported by the Consultative Group on International Agricultural Research (CGIAR). IPGRI's formal status is conferred under an "establishment agreement" that, by January 2002, had been signed and ratified by the governments of nearly 50 nations (http://www.ipgri.org).

In most European countries, research aimed at facilitating the long-term conservation and increased utilization of plant genetic resources operates under the aegis of The European Cooperative Programme for Crop Genetic Resources Networks (ECP/GR). The ECP/GR is entirely financed by the member countries and is coordinated by IPGRI (Maggioni 2004). Within this program, working groups for various vegetable crops have been established. The first ad hoc meeting of the ECP/GR Informal Group on leafy vegetables took place in May 2003 in Skierniewice, Poland (Lebeda and Boukema 2005). The Leafy Vegetables Working Group ECP/GR was officialy approved in 2004 (ECP/GR, 2005) and the first meeting of this group held in October 2005 in Olomouc, Czech Republic.

In the United States, germplasm research regarding conservation, evaluation, and utilization of resources is overseen by Crop Germplasm Committees (CGC) under the National Plant Germplasm System of USDA-ARS. The committee for lettuce is the Leafy Vegetables CGC, that also deals with chicory, endive, spinach and celery.

9.2.2   Acquisition and Exploration

Biological diversity is a global asset of tremendous value to present and future generations (Anonymous 1994). During the last two decades, the protection of genetic diversity within plant species has become a priority for conservation efforts (Huenneke 1991). Good and complex knowledge of genetic resources (Lebeda et al. 1999, 2001a) is an important key for their conservation and utilization, including the breeding of lettuce (Lebeda and Astley 1999).

Wild relatives and progenitors of cultivated plants represent a strategic part of the germplasm collections of economically utilized plant species (Guarino et al. 1995). Landraces of *L. sativa* and wild *Lactuca* species that may be in the secondary and tertiary gene pools (see Section 9.4) are valuable sources of different traits utilized in lettuce breeding (Pink and Keane 1993; Ryder 1999a) and their practical role has increased dramatically since the end of the 1980s (Lebeda 1998; Lebeda, Pink, and Astley 2002a, 2002b). There is also increasing interest in sampling, conservation, evaluation, and utilization of germplasm collections in *L. sativa* and wild *Lactuca* species (Lebeda et al. 2001b, 2004a). However, out of 100 *Lactuca* species described (for details see Section 9.3), there are only about 20 species maintained in genebank collections (Lebeda, Doležalová, and Astley 2004a). A recent comparative study about natural distributions of wild *Lactuca* species that are represented in world genebank collections is summarized in the "International *Lactuca* database" (ILDB; for details see Section 9.2.5). It shows surprising data. A total of 27 wild *Lactuca* species are

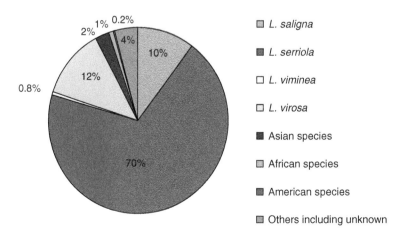

Figure 9.4   Representation of wild *Lactuca* species in genebank collections according to the most common species and geographic groups of species. (From Lebeda, A., Doležalová, I., and Astley, D., *Genet. Resour. Crop Evol.*, 51, 167–174, 2004.)

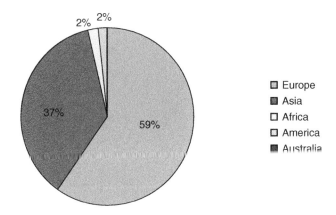

Remark:
Only one accession of *L. serriola* was collected in Australia

Figure 9.5 Representation of wild *Lactuca* species in genebank collections according to the continents of their origin. (From Lebeda, A., Doležalová, I., and Astley, D., *Genet. Resour. Crop Evol.*, 51, 167–174, 2004. With permission.)

reported in world genebank collections in ILDB. However, due to incorrect taxonomic determination, the real number of species is lower (about 20 species). Over 90% of the collections are represented by only three species (*L. serriola, L. saligna, L. virosa*; Figure 9.4). And from a geographic viewpoint, they are mostly European in origin (Figure 9.5). The autochthonous species originating from other continents (Asia, Africa, America), which form about 83% of known *Lactuca* species richness (Figure 9.6), are very poorly represented in the collections (only about 3%). It is significant that genetic resource collections do not hold some species (e.g., *L. azerbaijanica, L. georgica, L. scarioloides*) (Lebeda, Doležalová, and Astley 2004a), which are considered to be primary progenitors of cultivated lettuce (Zohary 1991; de Vries 1997).

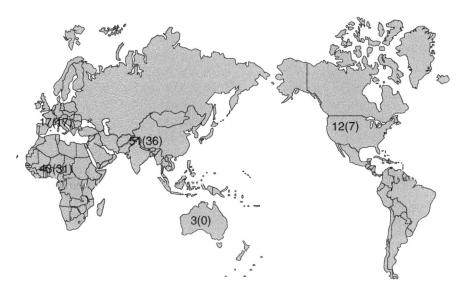

Figure 9.6 Generalized map of the geographical distribution of wild *Lactuca* species numbers of autochthonous species for each continent are in parentheses. (From Lebeda, A., Doležalová, I., Feráková, V., and Astley, D., *Bot. Rev.*, 70, 328–356, 2004. With permission.)

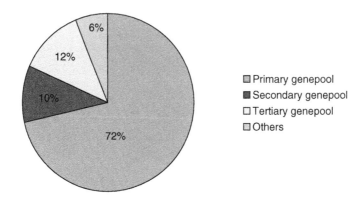

Categorization and representatives of lettuce genepools:
Primary genepool: *L. serriola, L. altaica, L. aculeata, L. georgica, L. scarioloides* and *L. dregeana;*
Secondary genepool: *L. saligna;*
Tertiary genepool: *L. virosa.*

Figure 9.7   Representation of wild *Lactuca* species in genebank collections according to the categorization into gene pools. (From Lebeda, A., Doležalová, I., and Astley, D., *Genet. Resour. Crop Evol.*, 51, 167–174, 2004.)

It is also evident that the more localized species (for details see Lebeda et al. 2004b) are completely absent from genebank collections, with the exception of *L. aculeata*. The majority of accessions originate from Europe (about 60%) and are from the primary gene pool of *L. sativa* (about 70%; Figure 9.7). Thus, the global biodiversity of *Lactuca* spp. germplasm is represented very poorly and is biased in genebank collections (Lebeda, Doležalová, and Astley 2004a).

Considerable effort must be directed toward expanding the diversity of *ex situ* germplasm collections by collecting wild, weedy, and landrace populations of *Lactuca* species from their centers of diversity, especially Africa and Asia, and from the refuges of traditional agroecosystems throughout the world (Lebeda et al. 2004b). In recent years only a limited number of exploration and acquisition missions have been realized (Pistrick and Malćev 1998; Soest et al. 1998; Lebeda and Boukema 2001, 2005). An extensive collection and exploration program was developed during the 1990s in the Czech Republic (Lebeda and Křístková 1995; Křístková and Lebeda 1999; Lebeda et al. 1999, 2001b, 2005a, 2005b; Doležalová, Lebeda, and Křístková 2001, 2002b; Křístková, Lebeda, and Doležalová 2001).

As was pointed out by Lebeda, Doležalová, and Astley, (2004a), the collection and maintenance of wild *Lactuca* species and their accessions must be considered as a significant target for future progress in research areas such as taxonomy, species diversification, spatial population, structure and relationships in the genepool, and genetic variability of host–pathogen interactions. It was concluded (Lebeda, Doležalová, and Astley 2004a) that future collecting strategies must be more intensively oriented to the hotspots of *Lactuca* species biodiversity (Lebeda et al. 2004b; e.g., Asia, central and southern Africa, North America). Broadening the collecting activities in natural habitats and traditional agroecosystems of these areas is valuable not only from the viewpoint of biodiversity conservation, but also for increasing our knowledge of taxonomy, biogeography, ecobiology, stress biology, weed–crop interactions, population biology and genetics, gene flow, and delimitation of populations.

9.2.3 Collection Regeneration and Maintenance

Regeneration protocols of *Lactuca* species germplasm accessions (Figure 9.8) should ensure the production of high-quality seed and minimize outcross contamination, genetic drift, and changes

Figure 9.8 Regeneration of wild *Lactuca* species germplasm in a greenhouse. (Photo courtesy of I. Doležalová.)

brought about through selective regeneration environments. The seeds produced should be held under moisture and temperature conditions that prolong their viability to increase the interval between regeneration events and reduce potential changes to genetic profiles.

Standards for regeneration of genetic resources of the genus *Lactuca* are elaborated for each national collection in holding countries. An overview of basic regeneration protocols summarized for the most important European genebanks is given by Lebeda and Boukema (2005). One of the goals of the Leafy Vegetables Working Group ECP/GR established by IPGRI in 2003 (Thomas et al. 2005) is to harmonize these protocols with regard to the germplasm quality expected and the local technical conditions for regeneration.

9.2.3.1 Isolation

Regeneration protocols are based on the biology of *Lactuca* species. *L. sativa* and the majority of wild species are obligate self-fertilizing species, but the possibility of pollen transmission by insects cannot be excluded. Also interspecific hybrids of *L. serriola* (maternal component) with

L. sativa (source of pollen) created under natural conditions have been identified (Lebeda et al. unpublished).

In the production of seeds of lettuce cultivated (*Lactuca sativa* L.) for commercial purposes, up to 5% cross-pollination has been observed. In most areas lettuce is regarded as a self-pollinated crop and only a physical barrier (e.g., adjacent sections of greenhouses) or a minimum of 2 m between different cultivars is recommended (George 1999).

Generally, isolation is not strictly recommended for regeneration of accessions of self-pollinated *Lactuca* species. However, the official method accepted by the National Council for Plant Genetic Resources of the Czech Republic in 2004 recommends the regeneration of *L. sativa* accessions in insect-free isolation cages, to avoid potential cross-pollination and virus transmission (*LMV*) by insects (Chytilová et al. 2004; Křístková et al. 2005).

Because most wild *Lactuca* species are obligate self-fertilizing, the accessions are not isolated during their regeneration. The accessions of outbreeding species (*L. perennis* and *L. viminea* subsp. *chondrilliflora*) are isolated and manually pollinated (Doležalová et al. 2002a).

9.2.3.2   Plant Numbers for Regeneration

From the strict genetic point of view, the most important goal for consideration during the regeneration of plant genetic resource accessions is to maintain all alleles included in the accession, i.e., to maintain the initial diversity of the accession. So, the frequency of alleles in accessions should play a fundamental role in establishing the number of plants to be regenerated.

This number is primarily influenced by flower and pollination biology, i.e., whether the species is self- or cross-pollinated, by the character (type) of the original sample of the accession, and by the number of seeds needed for long term storage and to maintain viability in seeds provided to requestors. Recent experience with long-term storage of lettuce seeds shows that good germination can be maintained for 20 years of storage at $-18°C$ ($-20°C$) (Křístková 2005, unpublished data). A sufficient amount of seed obtained in one regeneration, can minimize the need for repeated regenerations and thus reduce the potential risk of genetic drift, contamination, and other changes caused by the environment. Generally, if the accession represents a uniform, advanced, and well described variety (cultivar), the plant number should be at a lower level of about 8 to 15 plants. Heterogeneous samples of cultivated lettuce and of wild *Lactuca* species demand a higher number, at least 16 individual plants.

An overview of basic regeneration protocols, with the number of plants used, has been summarized for the most important European gene banks by Lebeda and Boukema (2005). Plant numbers range between 8 and 60; generally the average number of plants is between 20 and 24.

9.2.3.3   Regeneration Protocols

Under conditions in central Europe, seeds are sown at the beginning of March. After pre-cultivation, the plantlets are transplanted to the field, soil in isolation cages, or garden soil in plastic pots in the first half of April. The accessions are regenerated under long-day conditions.

To obtain seeds in one growing season, all wild species are vernalized for six weeks at 1°C, when the seeds are just germinating. Heading types of *L. sativa* are treated with gibberellic acid (20 ppm GA_3) at a young stage to expedite bolting and to prevent the plants from rotting (Boukema, Hazekamp, van Hintum, 1990). Further approaches, recommended for commercial seed production and given by George (1999), should also be applied in germplasm regeneration.

The species *L. indica* and *L. schweinfurthii* (or *L. longespicata*), which flower under short-day conditions, are kept in a vegetative state till late autumn. The biennial species (*L. canadensis, L. biennis, L. virosa*) and perennial (*L. viminea, L. perennis, L. tatarica, L. tenerrima*) overwinter at

5–7°C in a glasshouse at the basal rosette stage (Doležalová et al. 2002a). Basic information about regeneration of wild *Lactuca* species was given by Boukema, Hazekamp, and van Hintum (1990) and Hintum and Boukema (1999).

9.2.3.4 Seed Harvest, Conservation Protocol

Seeds of wild *Lactuca* species typically ripen continuously, so the most efficient approach to harvest seeds from single plants is by shaking their heads into canvas bags or sacks. If it is done every two to three days, the maximum amount of seed is collected.

The seed harvest of cultivated lettuce should follow the method used for commercial seed production, summarized by George (1999). If the seed producer waits for the development and ripening of the seeds from later ripening capitula, the earlier ripened highest quality seeds will probably have been lost. It is general practice to cut the upper part of the stalk when an estimated 50% of the seeds are ready on a typical sample plant. The stage of ripeness at which the pappus is fully developed and dry is referred to as "feathering" (George 1999).

The cut materials are placed on canvas and left in windrows for several days, and when the plants are dry, they are crushed. The seeds are separated from most of the debris, and the seeds with small pieces of plant debris are cleaned.

9.2.3.5 Seed Drying and Storage

After cleaning, seeds are either dried by using a silica gel or lyolab device, or for several days in a special room at a temperature of 20°C and relative humidity of 28–30%. Seeds with moisture content of 5–6% are stored in hermetically closed jars or special bags at −18°C (−20°C). Recent experiments showed that a storage temperature of −5°C also maintains long-term viability of seeds (Mgr. Iva Fáberová, Gene Bank RICP in Praha–Ruzyně, CR, 2005, pers. commun.). The composition of oils and other substances in lettuce seeds allows the use of a technique of "ultra dry seed" storage. This procedure is based on seed drying under a temperature of +30°C to a moisture content of 3%. Then seeds are packed hermetically and can be stored at a laboratory temperature of about +25°C for the long term. This procedure is as effective as storage at −20°C. During the ultra-drying procedure, the critical degree of moisture content in the seeds must not be exceeded, to prevent irreversible degradations of seed components (proteins, etc.). Exact procedural conditions should be studied and established for individual species and varieties.

During storage, the seed quality (viability) is periodically monitored and in case of a germination decrease, the next regeneration is recommended.

9.2.4 Germplasm Characterization

Accurate characterization of lettuce and *Lactuca* germplasm should support its proper taxonomic identification and specification on a cultivar level. A broad range of techniques has been employed to apply morphological, protein, and DNA markers to characterize germplasm collections (Breuing and Widrlechner 1995).

A discipline aimed at the classification of species, and understanding of phylogeny and evolutionary relationships is also considered as a useful tool for genebank management (Koopman 1999). Correct establishment of the taxonomic identity of *Lactuca* species (see Section 9.3 for an overview of taxonomic studies) and their conserved accessions must be considered as a crucial step of germplasm characterization. As was stressed elsewhere in this book (Chapter 8, Section 2) the proper identification at both the species and cultivar levels facilitates both maintenance and utilization of germplasms, and reduces the frequency of unintended duplications (Lebeda et al. 2006). Plant species maintained in genebanks are characterized to enhance the understanding of the collections for both collection management and for potential users (Doležalová et al. 2003b). Recently the recording of classical morphological features (keystone of descriptive databases) has been complemented with records on biochemical and molecular features (Waycott and Fort 1994; Ayad et al. 1997; Hintum 2003). However, the absence of or poor quality data on taxonomy and morphology can result in misinterpretation when using material for scientific purposes. Moreover, the paucity of data may limit the elucidation of relationships between individual species (accessions) within a given taxonomic group (Doležalová et al. 2003b). Recently, it has been shown repeatedly that many wild *Lactuca* species accessions maintained in world genebank collections are not correctly determined (Lebeda, Doležalová, and Astley 2004a) and that the incorrect determination of taxonomic identity has led to incorrect passport data (Lebeda et al. 1999, 2001b; Doležalová et al. 2002b, 2004). It was concluded that taxonomic validation (at the level of species and subspecies) is urgently needed and such work must be primarily based on classical taxonomy and the use of original herbarium specimens (Lebeda, Doležalová, and Astley 2004a). Only coordinated research, which includes classical taxonomic studies in tandem with a molecular approach, can solve these taxonomic questions objectively (Doležalová et al. 2003b, 2004).

Storing of germplasm seeds in genebanks must be done efficiently to avoid accidental loss of diversity. This raises many questions concerning which samples to include in a collection; how to determine the redundancies; how to multiply samples without loss of diversity due to contamination, inadvertent selection, or genetic drift (Hintum 2003). One of the most important problems of efficient genebanking is duplication (Hintum and Knüpfer 1995).

The appropriate maintenance of *Lactuca* collections according to international standards is space and time consuming, and hampered by limited financial resources (Lebeda and Boukema 2001). There is now a need for detecting the existence of duplicate accessions on a national and international level. (Sretenović-Rajičić et al. 2006). It is well documented that *Lactuca* genetic resource collections consist of a large amount of duplicated material (Hintum and Boukema 1999). Comparison of four large collections (CGN, WRPIS, IPK, and HRI), which represent about 30% of the world's accessions, has shown a high level of overlapping between collections. The analysis clearly demonstrates that only about 40% of accessions are not duplicated, while 60% of accessions are duplicated once, twice, or in all studied collections. In this study it was concluded that the extent of duplication among the world's lettuce germplasm collections is enormous. The main reasons for duplication are: (1) total lack of information, (2) poor documentation (basic passport data), (3) typing errors in descriptions of names of landraces and old cultivars, and (4) use of different languages for descriptions of the same material (Hintum and Boukema 1999). There are different ways to solve this problem. At least three individual approaches, or their combination, could be used: (1) obtaining basic missing information on material, (2) morphological comparison of accessions within and among collections based on very good knowledge of the material, and (3) use of biochemical and molecular markers for exact genetic discrimination of germplasm. Recently substantial progress was made in all these approaches, as for example the development of: (1) efficient information technologies and databases related to *Lactuca* collections (e.g., International Lactuca Database: ILDB (Stavelikova, Boukema, and van Hintum 2002); GRIN database: http://www.ars-grin.gov/npgs), (2) elaboration of precise descriptors for lettuce (Křístková et al. 2005) and wild *Lactuca* species (Doležalová et al. 2002a, 2003a), and (3) development of fingerprinting technologies for germplasm characterization (Hintum 1999, 2003; Michelmore et al. 2003; Dziechciarková et al. 2004a, 2004b). Recent results clearly showed that combining these approaches could be very efficient not only for detection of duplicates, but also for verification of the taxonomic status of *Lactuca* species (Sretenović-Rajičić et al. 2006).

The genus *Lactuca* is extremely variable from the morphological and phenological viewpoints. However, only a relatively limited number of macroscopically visible morphological features (e.g., leaf morphology and rosette character, inflorescence, panicle, ligules, capitula, achenes and beak, stem character) are frequently used in classical taxonomy and descriptions of *Lactuca* species variation (Lebeda et al. 1999). Substantial progress was achieved in this area by the development of detailed descriptors of lettuce (Křístková et al. 2006) and wild *Lactuca* species (Doležalová et al. 2002a, 2003a; for details see Section 9.2.5). Recently, for basic germplasm characterization, minimum descriptors were developed (Lebeda and Boukema 2005). Nevertheless, there are many other important underutilized features (e.g., stomata, indumentum, pollen grains), which are important for species characterization (Lebeda et al. 1999, 2001a).

Detailed research on morphologic variation and evolutionary relationships, defining the intraspecific and interspecific variation of *L. sativa* cultivars in comparison with *L. serriola, L. saligna, L. virosa*, and the study of the formal basis of distinction between *L. sativa* and *L. serriola* was carried out by de Vries and van Raamsdonk (1994). Analysis of data (vegetative and generative characteristics) based on multivariate methods (principal component analysis) clearly distinguished four groups corresponding with the four studied *Lactuca* species. The greatest intraspecific variation was demonstrated for *L. sativa* resulting from many years of breeding for various vegetative traits. Analyses showed that lettuce (*L. sativa*) cultivar classification based on morphological differences (Rodenburg 1960) is more or less valid (for details see Section 9.4.5). The butterhead, stalk, and crisphead groups were very homogeneous. Cos lettuce appeared to be homogeneous and was intermediate between the stalk and butterhead groups. Latin lettuce was heterogeneous and could be split into two separate subgroups. Cutting lettuce was also heterogeneous and could possibly be split into two or three subgroups. According to these results, a global division into two supergroups was suggested (for details see Section 9.4.5). It was concluded that on the basis of this data, it should be possible to choose a standard lettuce cultivar for each morphotype group. However, morphological cultivar classification should be flexible as new cultivars and new combinations resulting from breeding may be added to the system (de Vries and van Raamsdonk 1994). From the taxonomical viewpoint, a very important conclusion was that morphological differences between *L. sativa* and *L. serriola* are too large to consider them as one species (de Vries and van Raamsdonk 1994). This is in contradiction to some later conclusions based on karyotype and molecular studies (for details see Section 9.3.4 and Section 9.4.2).

Until recently, research on morphology and phenophases was very limited and focused only on few wild *Lactuca* species (i.e., *L. serriola, L. saligna, L. virosa, L. viminea, L. quercina*) (Feráková 1977; de Vries and van Raamsdonk 1994). Preliminary studies of phenology in some wild *Lactuca* germplasm showed enormous variation in various developmental stages (e.g., seedling stage, formation of basal rosettes, bolting, occurrence of first flowers, time of full flowering, first achenes etc.) (Lebeda et al. 1999). It is evident that these characters are related to ecogeography and ecological adaptations of individual species (Lebeda et al. 2004b). Detailed intraspecific study of phenologic variation among *L. serriola* accessions originating from four European countries (Czech Republic, Germany, the Netherlands and United Kingdom) clearly demonstrated geographic differences (Lebeda et al. 2004c, 2005c; Doležalová et al. 2005), some preliminary conclusions were also made on an interspecific level (Lebeda et al. 1999). Altogether 50 *L. serriola* populations were morphologically characterized following the descriptor list of wild *Lactuca* species germplasm (Doležalová et al. 2003a) by growing the plants in the glasshouse in standard conditions. Assessment included 26 quantitative and qualitative characters of the stem (e.g., stem length), rosette and cauline leaves (e.g., depth of incisions), and the inflorescence and flower (e.g., anthocyanin coloration on bracts) (Lebeda et al. 2004c, 2005c; Doležalová et al. 2005). Mean length of stem ranged between 1.46 m for plants originated from Germany and 1.54 m for plants from the U.K. As for morphology of rosette and cauline leaves, only *L. serriola* f. *serriola* was recorded in the Czech Republic, while 91% of plants originated from Germany were form *serriola*; both form *serriola* and form *integrifolia* were represented about equally in the Netherlands populations; and in populations originated from the U.K., 98% plants were form *integrifolia* (Figure 9.9). The distribution of anthocyanin was in the apex of the bracts in the majority of cases. Plants from the Czech Republic and Germany possessed anthocyanin in the apex of bracts in 34.3 and 32.1% of individuals, respectively. Plants from the Netherlands populations possessed anthocyanin in 68.9% of the individuals and plants from U.K. had 99.3% (Figure 9.10).

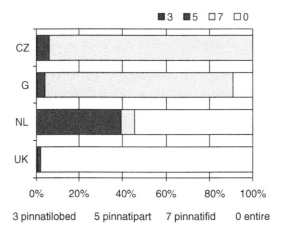

Figure 9.9 Variation in the shape of rosette leaves in European populations of *L. serriola*.

A corymbose panicle was characteristic for the majority of plants collected in Germany, while the majority of plants from the U.K. displayed a pyramidal panicle (Lebeda et al. 2004c; Doležalová et al. 2005).

Among the developmental characteristics, substantial differences in the time of flowering were recorded between accessions originating from individual countries (Lebeda et al. 2004c, 2005c; Doležalová et al. 2005). Flowering stage (number of days from sowing) was divided into three categories: early (109–134), medium (135–164) and late (165–223). The median time of flowering was recorded for the majority of plants. Early flowering was characteristic for plants originating from Germany and the Czech Republic (50.2 and 34.1% of plants, respectively). *L. serriola* plants from the U.K. and the Netherlands showed late flowering in 21.5 and 13.1% of individuals (Figure 9.11).

It was concluded that there is a shift in this developmental feature from eastern to western Europe, and *L. serriola* populations in the U.K. differ substantially from those in continental Europe (Lebeda et al. 2005c).

Research on specificity of interactions between plants and pathogens is not possible without germplasm collections of the host and the pathogens, to gain basic knowledge of their biology and evolution. The specificity of plant–pathogen interactions could be considered at various levels of biological hierarchy and from different viewpoints (Lebeda 1998). There has been a general tendency in plant breeding programs toward the rapid elimination of genetic variability, based on the notion that strictly uniform crop populations are the universal ideal. However, as uniform

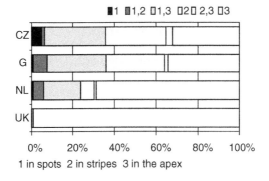

Figure 9.10 Variation in distribution of anthocyanin in European populations of *L. serriola*.

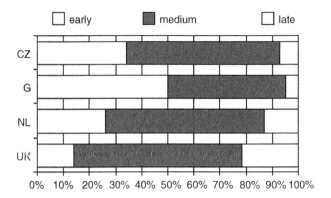

Figure 9.11   Variation in the time of flowering in European populations of *L. serriola.*

cultivars have been grown over wider areas, their vulnerability to diseases has increased (Lenné and Wood 1991). The need to broaden the genetic base of crops by use of wild germplasm resources has been widely recognized and there is a continuing requirement that the genetic diversity of crop species and their wild relatives, including lettuce and wild *Lactuca* species, be conserved as a resource for future programs of genetic improvement; including resistance to lettuce diseases and pests (Crute 1992a, 1992b). Wild relatives of crop plants have proved to be fruitful sources of resistance genes against different pathogens (Burdon and Jarosz 1989; Lenné and Wood 1991; Dinnor and Eshed 1997). Also in lettuce research and breeding programs there is increasing interest in the study of resistance variation and the use of wild *Lactuca* germplasm, especially as sources of new and efficient resistance (Alconero 1988; Kalloo 1988; Pink and Keane 1993; Lebeda 1998; Reinink 1999; Ryder 1999b; Lebeda, Pink, and Astley 2002; Ryder et al. 2003). Details about recent research progress in this area are summarized in Section 9.6.

Nuclear DNA content and genome size (C-value) are important biodiversity characters, whose study provides a strong unifying element in plant biology with practical and predictive uses. It is well known that DNA content varies among and within the plant species. However, intraspecific variation remains one of the most controversial topics regarding plant genome size (Bennett and Leitch 2005a).

Absolute nuclear DNA content is known only in about 3% of angiosperm species (Bennett and Leitch 2005b). Knowledge of variation of DNA content in *Lactuca* germplasm is rather limited. Until the end of the twentieth century, nuclear DNA content variation analyses have been performed on only a limited number of *Lactuca* species (*L. sativa*, *L. serriola*, *L. saligna*, *L. virosa*; Bennett and Leitch 1995; Koopman and de Jong 1996; Koopman 1999, 2000). Using flow cytometry (measurement of relative and absolute DNA content) for *Lactuca* species may be a tool for distinguishing at least: (1) taxonomic status and evolutionary relationships between the species, (2) intraspecific variability, (3) identification of ploidy level, and (4) karyotype analysis in relationship to DNA content.

Flow cytometry was tested for its reliability as a tool to distinguish some *Lactuca* species (Koopman 1999, 2000). The measurement of its relative DNA amount showed that *L. serriola* content was significantly higher than in *L. saligna*, but significantly lower than in *L. virosa*. The mean relative DNA amount of *L. sativa* did not differ from that of *L. serriola*. These tests revealed that individual plants of *L. serriola, L. saligna,* or *L. virosa* can be identified by flow cytometry (Koopman 2000). Doležalová et al. (2002b) analyzed fifty accessions of 25 *Lactuca* species (including hybrid *L. serriola* × *L. sativa* and *Mycelis muralis*) for chromosome number and relative DNA content variation using flow cytometry (DAPI staining) and showed that 2C DNA content ranged from 2.02 ru (relative units) in *L. capensis* to 17.96 ru in *L. canadensis*. Statistical and cluster analysis of data corresponded well with the recently accepted taxonomic classification of the genus *Lactuca*. New data of Koopman, Hadam, and Doležel (2002) showed that *Lactuca* s.l. species, relative to other angiosperms, have low 2C DNA contents, ranging from 1.913 pg (picograms) in *L. tenerrima* to 13.068 pg in *L. indica*, and high AT contents (base composition), ranging from 61.4% in *L. virosa* to 64.2% in *L. perennis*. It was concluded that interspecific variation is not generally applicable for species identification.

Koopman (2000) demonstrated significant differences in relative DNA amount between the accessions within the *Lactuca* species studied (*L. sativa, L. serriola, L. saligna, L. virosa*), showing the presence of intraspecific variation. Intraspecific variation of relative DNA amounts was significantly higher for *L. serriola* than for *L. saligna* accessions, but significantly lower than for *L. virosa* accessions. Later, Koopman, Hadam, and Doležel (2002) showed that five *Lactuca* species (*L. viminea, L. virosa, L. serriola, L. sativa*, and *L. sibirica*) have significant intraspecific variation in DNA content, but it was concluded that only the variation within *L. virosa* seems to have evolutionary significance. Recent studies focused on intraspecific differences in relative DNA content in *L. serriola* germplasm, originating from 12 European countries, which showed significant variation among individual accessions (Lebeda et al. 2004d, 2005d). In the set studied, the mean of 2C DNA content ranged from 4.34 to 6.85 ru. Statistical analyses divided the studied set into 21 groups differing in relative DNA content. Simplification of this analysis reduced the number of groups to five, characterized by the following ranges of DNA content: (1) 4.34–4.54; (2) 4.55–4.81; (3) 4.82–5.05; (4) 5.06–5.76; (5) 5.77–6.85. In the categories with extremely low (1) or high (5) DNA content, 5 and 11% of the accessions, respectively, occurred. There was no clear relationship between the subspecific taxa of *L. serriola* (f. *serriola* and f. *integrifolia*) originating from various countries and relative DNA content. It was concluded that there is great variation in relative DNA content within *L. serriola* and this feature could be used as one of the tools for germplasm characterization (Lebeda et al. 2004d) and future detailed research of population variability (Lebeda et al. 2005d).

Ploidy level by flow cytometry was not identifiable in detail. In some plant genera, genome size was found to correlate with the number of chromosomes and ploidy level (Doležel 1997). This was partly confirmed also in *Lactuca* species (Doležalová et al. 2002b). However, it was concluded that there is no broad clear relationship between relative DNA content and chromosome number. Karyotype analysis and relative DNA content were used for characterization of *L. sativa, L. serriola, L. saligna*, and *L. virosa* and their evolutionary relationships (Koopman and de Jong 1996). No significant differences were found between *L. sativa* and *L. serriola*. However, *L. saligna* differed from *L. sativa/L. serriola* for all parameters. The largest differences were found between *L. saligna* and *L. virosa*. Both species have asymmetric karyotypes compared to *L. sativa/ L. serriola*. Because asymmetric karyotypes are considered, in relation to phylogeny and evolution, to be derived (Stebbins 1971), it was concluded that *L. saligna* and *L. virosa* are advanced species that evolved in different directions. On the other hand, the domestication process of lettuce is probably not reflected in the karyotype (Koopman and de Jong 1996).

Protein and molecular marker technologies substantially contributed to the development and understanding of taxonomy, phylogeny, genetic variation, germplasm maintenance and characterization, and practical breeding (Bretting and Widrlechner 1995; Ayad et al. 1997; Karp, Isaac, and Ingram, 1998; Henry 2005). These techniques have been also broadly applied in the genus *Lactuca* (Vosman 1997; Dziechciarková et al. 2004a) for different purposes. In the following paragraphs, we primarily focused on studies related to *Lactuca* germplasm variation and related topics. Detailed information about application of these methods in taxonomy, genetic diversity studies, genetic mapping, and breeding are summarized elsewhere (for details, see Section 9.3 through Section 9.6).

Polymorphic proteins have been widely used to characterize patterns of genetic diversity within and among plants (Soltis and Soltis 1989); including lettuce germplasms (Lebeda 1998). Enzyme and isozyme variation has been used for over 60 years for different purposes in biology; e.g., to delineate phylogenetic relationships, to estimate genetic variation, to study population genetics and developmental biology, and to direct utilization in plant genetic resources management and plant breeding (Bretting and Widrlechner 1995; Staub and Serquen 1996). Only a few articles have been published focusing on the study of *Lactuca* species using isozyme analysis (Table 9.5). For isozyme variation in *Lactuca* germplasm different parts of the plants were used, mostly seeds (Mejia and McDaniel 1986; Cole, Sutherland, and Riggall 1991; de Vries 1996) or leaf tissue (Roux et al. 1985; Kesseli and Michelmore 1986; Dziechciarková et al. 2004b). These studies described the application of isozyme techniques for the identification of genetic variability among germplasms of lettuce and wild *Lactuca* species and the determination of the genetic and phylogenetic relationships of *Lactuca* species. The analysis of results showed a lower level of intraspecies than interspecies diversity in *Lactuca*. Nevertheless, it was concluded that isozyme markers display a high level of polymorphism in *Lactuca* species, and can be useful for the characterization of variability and the determination of taxonomic relationships and species identity (Dziechciarková et al. 2004a). However, the polymorphism of closely related species was relatively low, thus limiting the resolution of some problems related to the species relationships (Kesseli, Ochoa, and Michelmore 1991). Therefore, there has been a tendency to use new, more sensitive methods to eliminate the disadvantages of isozyme techniques.

Many different molecular methods and classes of DNA markers have been used to characterize lettuce germplasm (Vosman 1997; Dziechciarková et al. 2004a). A survey of molecular marker methods and studies used for lettuce and wild *Lactuca* species germplasm characterization is summarized in Table 9.6. From this survey it is evident that a relatively broad spectrum of recently available molecular techniques and approaches has been used for various goals related to genetic variability and genetics of *Lactuca* germplasm. The following were studied: (1) relationships between the closely related genera and *Lactuca* species; (2) phylogenetic relationships between various sections and groups of *Lactuca* species; (3) intraspecific variability of wild *Lactuca* species; (4) polymorphism in lettuce varieties; (5) variety identification and their genetic homogeneity; (6) construction of lettuce genetic maps; (7) determination of spatial diversity in wild *Lactuca* species populations; and (8) analysis of gene flow in the lettuce crop–weed complex. Various techniques and approaches were applied: RFLP (restriction fragment length polymorphism), RAPD (random amplified polymorphic DNA), AFLP (amplified fragment length polymorphism), SSRs (simple sequence repeats), SAMPL (selectively amplified microsatellite polymorphic locus), STMS (sequence-tagged microsatellite site), SSAP (sequence-specific amplified polymorphism) and NBS (nucleotide-binding site) profiling (Dziechciarková et al. 2004a; van de Wiel et al. 2004).

There are various advantages and disadvantages of each of these methods, and possibilities of usage for different goals. RFLP and RAPD markers show similar levels of polymorphism. RFLP markers can easily distinguish between most accessions of *L. sativa*, except for sister lines of the same breeding population (Landry et al. 1987; Kesseli, Ochoa, and Michelmore 1991, 1994; Vermeulen et al. 1994). RAPDs appear to be able to distinguish between nearly identical germplasm accessions of *L. sativa* (Waycott and Fort 1994). However, RAPDs have been shown to be poorly reproducible (Karp et al. 1997a, 1997b), and are therefore less useful for routine identification. AFLP markers are useful for measurement of genetic diversity and the determination of genetic relationships within and among *Lactuca* species (Koopman, Zevenbergen, and van den Berg 2001). There is also indication about the presence of phylogenetic signals in the AFLP data sets of *Lactuca s.l.* (Koopman 2005). Microsatellites have been used for characterization, lettuce variety identification, and screening of diversity of *Lactuca* germplasm collections (van de Wiel, Arens, and Vosman 1998, 1999; van Hintum 2003).

Another approach recently used for plant genetic diversity studies is the assessment of variation in retrotransposable elements (Kumar and Hirochika 2001). SSAP is one of the methods used for this purpose and is based on retrotransposons occurring ubiquitously in plant genomes. This methodology is used for assessment of the level of gene flow between cultivated lettuce and *L. serriola* (van de Wiel et al. 2004). Diversity of genomic regions of cultivated and wild

Table 9.5 Survey of Protein Marker Methods and Isozyme Systems Used for *Lactuca* Species Characterization

Method	*Lactuca* spp.	Enzyme Systems	Goal of Work	References
Electrofocusing	*L. sativa, L. serriola, L. saligna, L. aculeata, L. virosa*	EST	Characterization of *L. sativa* and related species by electrofocusing of esterases	Roux et al. (1985)
Isozymes	*L. sativa, L. serriola, L. saligna, L. virosa*	ACO, ACP, ADH, ALD, ALO, ALP, AMY, ASO, CAT, DIA, ENP, EST, FDP, FUM, GALD, GPD, GPT, GDH, GOT, GLS, GAPD, HXK, ICD, LAC, LAD, LED, LAP, MDH, ME, MDR, PER, PGD, PGI, PGM, PHP, SHDH, SCD, SOD, TPI, TYR, XDH	Genetic variability and phylogenetic study	Kesseli and Michelmore (1986)
Isozymes	*L. sativa*	ACP, ADH, EST, GOT, ICD, LAP, MDH, ME, PGD	Electrophoretic characterisation of lettuce cultivars	Mejia and McDaniel (1986)
Isozymes	*L. sativa, L. serriola, L. saligna, L. virosa*	ADH, DIA, EST, GOT, IDH, LAP, MDH, PGD, PGI, SHDA	Characterization of wild populations of four *Lactuca* species by ten enzyme systems by using polyacrylamide gradient gel electrophoresis	Cole et al. (1991)
SDS-electrophoresis	*L. sativa, L. serriola, L. saligna, L. virosa*		Characterization and identification of *L. sativa* cultivars and wild relatives	de Vries (1996)
Isozymes	*L. sativa, L. serriola, L. aculeata, L. altaica, L. canadensis, L. dregeana, L. indica, L. perennis, L. saligna, L. sativa, L. serriola, L. taraxacifolia, L. tatarica, L. tenerrima, L. virosa, L. viminea*	ACP, ADH, EST, PER	Characterization germplasm collection of *Lactuca* species by four enzyme systems, confirmation of taxonomy determination	Lebeda et al. (1999, 2001a), Doležalová et al. (2003b)
Isozymes	*L. sativa, L. serriola, L. saligna, L. virosa, L. indica*	ADH, EST, DIA, FDH, GDH, GOT, ICD, MDH, ME, PGI, PGM, PGD, SADH, SOD	Characterization of polymorphism, segregation analysis of the F_2 progeny between *L. sativa* and *L. serriola*, linkage analysis, genetic markers for lettuce breeding	Mizutani and Tanaka (2003)
Isozymes	*L. serriola* f. *serriola* and *L. serriola* f. *integrifolia*	DIA, EST, GOT, PGM, GPI, LAP, MDH, NADH, NADH DH, SHDH, 6-PGDH	Characterization of polymorphisms in European populations	Dziechciarková et al. (2004b)

Note: ACO (aconitase); ACP (acid phosphatase); ADH (alcohol dehydrogenase); ALD (aldolase); ALO (aldehyde oxidase); ALP (alkaline phosphatase); AMY (amylase); ASO (ascorbate oxidase); CAT (catalase); DIA (diaphorase); ENP (endopeptidase); EST (esterase); FDH (formate dehydrogenase); FDP (fructose-1,6-diaphorase); FUM (fumarase); GALD (galactose dehydrogenase); GPD (glucose-6-phosphate dehydrogenase); GPT (glucose-1-phosphate transferase); GDH (glutamate dehydrogenase); GOT (glutamate oxaloacetate transaminase); GLS (glutamate syntetase); GAPD (glyceraldehyde-3-phosphate dehydrogenase NAD, glyceraldehyde-3-phosphate dehydrogenase NADP); HEX (hexokinase); ICD (isocitrate dehydrogenase); LAC (laccase); LAD (lactose dehydrogenase); LED (leucine dehydrogenase); LAP (leucine aminopeptidase); MDH (malate dehydrogenase NAD$^+$; malate dehydrogenase NADP$^+$); ME (malic enzyme NAD$^+$); MDR (menidione reductase); PER (peroxidase); PGD (6-phosphogluconate dehydrogenase); PGI (phosphoglucoisomerase); PGM (phosphoglucomutase); PHP (phosphorylase); SADH (shikimic acid dehydrogenase); SHDG (shikimate dehydrogenase); SCD (sucinate dehydrogenase); SOD (superoxide dismutase); TPI (triosephosphate isomerase); TYR (tyrosinase); XDH (xanthinine dehydrogenase).

Source: Adapted from Dziechciarková, M., Lebeda, A., Doležalová, I., and Astley, D., *Plant Soil Environ.*, 50, 49–60, 2004.

Table 9.6 Survey of Molecular Marker Methods Used for *Lactuca* Species Characterization

Method	*Lactuca* Species	Goal of Work	References
RFLP	*L. sativa* (Calmar×Kordaat cross)	A construction of lettuce genetic map	Landry et al. (1987)
	L. sativa, L. saligna, L. serriola, L. virosa, L. perennis, L. indica	Study of relationships between cultivated lettuce and five related wild *Lactuca* species; Study on the origin of cultivated lettuce (*L. sativa*)	Kesseli et al. (1991)
	L. sativa, L. serriola, L. saligna, L. virosa, L. alpina, L. perennis, L. tatarica	Study of relationships among *Cichorium* species and related genera of the tribe *Lactuceae*	Vermeulen et al. (1994)
	L. sativa (Calmar×Kordaat cross)	A construction of genetic map of *L. sativa*	Kesseli et al. (1994)
RAPD	*L. sativa* (Calmar×Kordaat cross)	A construction of genetic map of *L. sativa*	Kesseli et al. (1994)
	L. sativa	DNA polymorphisms in 12 lettuce (*L. sativa*) varieties were identified by amplification with 21 arbitrary RAPD primers	Yamamoto et al. (1994)
AFLP	*L. sativa, L. serriola, L. saligna, L. virosa, L. perennis, L. indica*	Study genetic relationships in *Lactuca* species	Hill et al. (1996)
	L. sativa, L. saligna	An integrated interspecific AFLP map of lettuce based on two *L. sativa* x *L. saligna* F$_2$ populations	Jeuken et al. (2001)
	L. sativa, L. serriola, L. dregeana, L. altaica, L. aculeata, L. saligna, L. virosa, L. tenerrima, L. perennis, L. tatarica, L. sibirica, L. quercina, L. viminea, L. indica	Study of species relationships in *Lactuca* species	Koopman et al. (2001)
Microsatellites	*L. sativa, L. serriola, L. saligna, L. virosa, L. perennis, L. indica*	Identification, genetic localization, and allelic diversity SAMPL in lettuce and wild relatives (*Lactuca* species)	Witsenboer et al. (1997)
	L. sativa, L. serriola, L. saligna, L. virosa	Variety identification in lettuce cultivars (*L. sativa*) and discriminate between cultivated lettuce and wild relatives	Van de Wiel et al. (1998)
	L. sativa, L. serriola, L. saligna, L. virosa	Distinguishing lettuce cultivars and screening diversity of genetic resources	Van de Wiel et al. (1999)
ITS-1 DNA sequence	*L. sativa, L. serriola, L. aculeata, L. dregeana, L. saligna, L. altaica, L. virosa, L. tenerrima, L. perennis, L. tatarica, L. sibirica, L. quercina, L. viminea, L. indica.* Other related species: *Mycelis muralis, Steptorhampus tuberosus, Cicerbita plumieri, C. alpina, Prenanthes purpurea, Chondrilla juncea,*		

continued

Table 9.6   Continued

Method	*Lactuca* Species	Goal of Work	References
	Taraxacum officinale, Sonchus asper, Cichorium intybus	Phylogenetic relationships among *Lactuca* species and related genera; specification position of *L. altaica;* delimitation of genus *Lactuca;* the taxonomic position of *Cichorium*	Koopman et al. (1998)
SSAP	*L. sativa, L. serriola*	Gene flow between cultivated and wild lettuce	van de Wiel (2004)
NBS-profiling	*L. sativa, L. serriola*	Gene flow between cultivated and wild lettuce	van de Wiel (2004)
	L. sativa, L. serriola, L. saligna, L. virosa, L. augustana, L. indica, L. perennis	Genomic diversity of cultivated and wild lettuce	Sicard et al. (1999)

Source: Adapted from Dziechciarková, M., Lebeda, A., Doležalová, I., and Astley, D., *Plant Soil Environ.*, 50, 49–60, 2004.

Lactuca germplasm was also analyzed by using molecular markers derived from resistance genes of the NBS-LRR (leucine-rich repeat) type (Sicard et al. 1999; van de Wiel et al. 2004). NBS profiles target variation in and around the most prominent family of disease resistance genes; the NBS-LRR family (Young 2000). All types of molecular marker methods were also used for the character-ization of different resistance genes in lettuce (Irwin et al. 1999; Sicard et al. 1999; Chin et al. 2001; Jeuken et al. 2001; Grube et al. 2003; Hand et al. 2003; Maisonneuve 2003; Mou and Ryder 2003; Nicaise et al. 2003b; Ryder et al. 2003).

9.2.5   Germplasm Descriptors

An increasing need for international cooperation in the area of leafy vegetable genetic resources, including lettuce, was stressed during the Eucarpia Leafy Vegetables Conference in the Czech Republic in 1999 (Lebeda and Křístková, 1999), at the meetings of the European Cooperative Programme (ECP/GR) Network Coordinating Group on Vegetables in Portugal in 2000 (Lebeda and Boukema 2001), and in Skierniewice in 2003 (Lebeda and Boukema 2005). The necessity for a more structured characterization of wild lettuce species was discussed, as well as an elaboration of basic morphological descriptors for wild *Lactuca* species.

A first step for coordinating international activities was achieved by the development of the International *Lactuca* Database (Boukema, Stavelíková, and van Hintum, 2001; Stavelikova, Boukema, and van Hintum, 2002; Boukema 2005), which contains passport data of almost all lettuce accessions from world germplasm collections. It is available at the URL web site: www.plant.wageningen-ur.nl/cgn/ildb.

The descriptors and codes for cultivated lettuce (*Lactuca sativa* L.) (Křístková et al. 2006, in press) were developed as a basic rule for documentation of the characterization and evaluation of the Czech collection of *L. sativa*. Traits were scored in multiple trials within the framework of the National Programme of Conservation and Utilization of Plant Genetic Resources and will be used for purposes of central documentation of plant genetic resources EVIGEZ (http://genbank.vurv.cz/genetic/resources). As primary sources for creation of recent Czech descriptors, the descriptor lists of the Centre for Genetic Resources; Centre for Plant Breeding Research (recently Plant Research

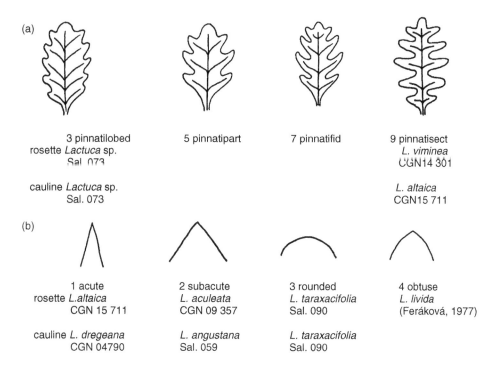

Figure 9.12 (a) Descriptors 1.3.3. and 1.3.7. Divided rosette (cauline) leaf—depth of incisions. (b) Descriptors 1.3.4 and 1.3.8. Rosette (cauline) leaf—shape of apex. (Example from Doležalová, I., Křístková, E., Lebeda, A., Vinter, V., Astley, D., and Boukema, I. W., *Lactuca* spp. *Plant Genet. Resour. Newsl.*, 134, 1–9, 2003.)

International); Wageningen (The Netherlands) (Boukema, Hazenkamp, and van Hintum 1990); the Western Regional Plant Introduction Station; Pullman; Washington (U.S.A.) (McGuire et al. 1993); and Guidelines UPOV for the conduct of tests for distinctness, homogeneity, and stability of lettuce (UPOV 1981) were used.

A description of morphological characters of wild *Lactuca* L. species genetic resources (English–Czech version) (Doležalová et al. 2002a) was also primarily created for characterization and evaluation of the *Lactuca* National Gene Bank Collection of the Czech Republic, and the international Gene-Mine project of the Fifth Framework Programme of the European Union (Jansen 2001; Doležalová et al. 2003a; Figure 9.12 and Figure 9.13). This list is also proposed for the international gene bank community. Recently, a minimum descriptor list for leafy vegetables has been developed for a basic and quick description of *L. sativa* and wild *Lactuca* spp. (*L. serriola* and related species from the primary gene pool; Lebeda and Boukema 2005).

9.3 TAXONOMY

9.3.1 Evolutionary and Taxonomic Relationships at the Familial Level

The Asteraceae (Compositae) is one of the largest plant families. One of every 10 flowering plant species is in this family (Funk et al. 2005). The Asteraceae contains many important crop plants revealing a remarkable morphological diversity, from trees up to 30 m tall to small herbs barely

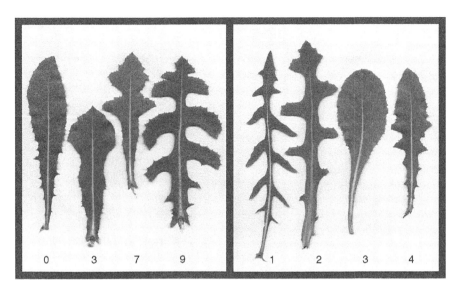

Figure 9.13   Variability of rosette leaves in *L. serriola* (09H5800723) 0. entire; 3. pinnatilobed; 7. pinnatifid; 9. pinnatisect. Shapes of leaf apices in *Lactuca* species 1. acute—*L. saligna* (LSA/6); 2. subacute— *L. serriola* (09H5800723); 3. rounded—*L. virosa* (CGN 04681); 4. obtuse—*L. serriola* (09H5800722).

1 cm high. It comprises approximately 23,000 species in 1,535 currently accepted genera (Judd et al. 1999).

Ideas about classification and diversification of the Asteraceae remained largely unchanged from Bentham (1873) to Cronquist (1955, 1977). However, recent results of molecular studies produced a meta-supertree formed by the tribal trees onto a base supermatrix tree (Funk et al. 2005). There are indications that the origin of extant members of the Asteraceae lies in southern South America; and that subsequent radiation in Africa gave rise to most extant tribes. The African radiation contains clades from Asia, Eurasia and Australia. The origin and diversification of the Heliantheae *s.l.* is likely to have occurred in North America, subsequent to separation from Gondwana (Funk et al. 2005).

The species within the Asteraceae have often been arranged into three subfamilies: Asteroideae, Barnadesioideae, and Cichorioideae representing in total 17 tribes (Bremer et al. 1994). Bremer (1996) has recently modified the classification and placed the Cardueae, formerly treated as a tribe, in a fourth subfamily: Carduoideae Cass. ex Sweet. The tribe Lactuceae from the subfamily Cichorioideae, formerly known as the Cichorieae, is perhaps the best known and most easily recognized tribe of the family (Tomb 1977). It is characterized by distinctive autapomorphic florets, and is divided into ten monophyletic subtribes, including the Lactucinae *s.s.*, with about 100 genera and over 1,550 species (Gemeinholzer and Kilian 2005). The subtribe Lactucinae itself comprises 17 genera, including the genus *Lactuca*, and about 270 species (Bremer et al. 1994). Recent phylogenetic studies confirm the origin of the tribe Lactuceae in Eurasia (Gemeinholzer and Kilian 2005).

9.3.2 General Characterization of the Tribe Lactuceae

Although the classification of the Lactuceae into genera and subtribes is in a much better stage than is classification of most other tribes, the generic limits of the genus *Lactuca* are still problematic. Based on the available literature, the genus *Lactuca* L. comprises approximately 100 wild species. However, the specific number of *Lactuca* taxa differs from author to author (Feráková, 1977;

Meusel and Jäger 1992; Bremer et al. 1994; Lebeda 1998; Lebeda and Astley 1999). The disagreements are due to the broad biodiversity within the genus and the occurrence of spontaneous hybridization, as well as the fact that the large *Lactuca* genus includes chorologically (spatially) strongly divergent groups of species.

There have been 17 European, 40 Asian, 33 African and 10 North American wild *Lactuca* species reported, confined mostly to temperate and warm regions of the northern hemisphere (Feráková 1977; Rulkens 1987; Lebeda and Astley 1999), of which only 54 species have been included in a world index of plant distribution maps (Lundqvist and Jäger 1995). Some taxa (e.g., *L. serriola*, *L. saligna*) are also naturalized in Australia and Tasmania (Burbinge and Gray 1970). The most recent survey of available literature has shown that at least 98 wild *Lactuca* spp. are distributed around the world (Table 9.7 and Figure 9.6); 17 species in Europe, 51 in Asia, 3 in Australia, 43 in Africa and 12 in America (mostly North America; Lebeda et al. 2004b; Figure 9.14). These species form three groups, based on different chromosome numbers (Table 9.8).

9.3.3 Delimitation and Characterization of the Genus *Lactuca*

Frietema (1994) considered the common (vernacular) name of the species belonging to the genus *Lactuca* to be "lettuce." To characterize the taxonomic treatment of *Lactuca*, Feráková (1977) used the eloquent words Babcock (1947) applied to the genus *Crepis* as an "oscillation between splitting and lumping." This is valid even after almost 30 years from the publication of the revision of *Lactuca* in Europe, as the generic delimitation still varies.

Of the most important characters of *Lactuca*, the following are most frequently stressed: cylindrical involucre of several rows of upright and rigid or reflexed bracts, receptacle without scales, corolla tube glabrous, collection hairs on style arms long and prominent, homomorphic, distinctly but moderately compressed and many-ribbed beaked (rarely unbeaked) achenes, pappus of many fine simple smooth or scabrid bristles without (exceptionally with) an outer ring of very short, smooth hairs. According to Kilian (2001) in the circumscription of *Lactuca* in particular, three features have been strongly emphasized: (1) the presence or absence of an outer row of minute pappus hairs; (2) the presence or absence of a beak (Figure 9.15); and (3) the number of flowers per capitulum (Figure 9.16).

The three major generic concepts of *Lactuca* were elaborated by Stebbins (1937), Tuisl (1968), and Feráková (1977). Stebbins (1937) defines the genus broadly and includes *Mulgedium* Cass., *Lactucopsis* Schultz-Bip. ex Vis et Panč., *Phaenixopus* Cass., *Mycelis* Cass., and part of *Cicerbita* Wallr. (*C. alpina*, with a coarse pappus and nearly columnar, slightly compressed achenes, is

Table 9.7 Geographical Distribution of Wild *Lactuca* Species

Continent	Number of *Lactuca* Species
Europe	17
North America	12
Africa (incl. Madagascar)	43
Asia	51
Australia	3

Source: From Lebeda, A., Biodiversity of the interactions between germplasm of wild *Lactuca* spp. and related genera and lettuce downy mildew (*Bremia lactucae*), Report on research programme OECD Biological Resource Management for Sustainable Agricultural Systems, HRI, Wellesbourne, U.K., 1998; Lebeda, A., Doležalová, I., Feráková, V., and Astley, D., *Bot. Rev.*, 70, 328–356, 2004.

Figure 9.14   Wild *Lactuca* species. (Composite photo courtesy of Doležalová, I., Lebeda, A., and Dreiseitl, A.) 1. *L. aculeata* (origin of accession—Israel), 2. *L. tatarica* (CGN 09390), 3. *L. perennis* (CGN 09323), 4. *L. saligna* (stony habitat, Adana, Turkey), 5. *L. virosa* (stony habitat, Mount St. Helen, Washington, U.S.A.), 6. *L. virosa* (swamp habitat, Redwoods, California, U.S.A.), 7. *L. quercina* (original habitat near village Velemín, Czech Republic), 8. *L. viminea* (ruderal, Avignon, France).

Table 9.8   Chromosomal Groups of *Lactuca* Species

Group	Chromosome Number (n)	Geographic Location
1	8	Mountain species from Europe and the Himalayas
2	9	Majority of European species, Indian and Mediterranean species African species
3	17	North American species (from Canada to Florida)

Source.   From Lebeda, A., Biodiversity of the interactions between germplasm of wild *Lactuca* spp. and related genera and lettuce downy mildew (*Bremia lactucae*), Report on research programme OECD Biological Resource Management for Sustainable Agricultural Systems, HRI, Wellesbourne, U.K., 1998.

excluded). Tuisl (1968) in Flora Iranica defined the genus in a narrow sense. He divided *Lactuca* s.l. into the following six genera *Mulgedium* Cass., *Scariola* F.W. Schmidt (=*Phaenixopus* Cass.), *Cicerbita* Wallr., *Cephalorrhynchus* Boiss., *Steptorhamphus* Bunge, and *Lactuca* L. on the basis of morphological and anatomical studies of fruit, flower, involucre, and pappus. The narrow generic concept of *Lactuca* has been supported among others; by Soják (1961, 1962), who accepted *Scariola* and treated *Lactuca* subg. *Mulgedium* (Cass.); C.B. Clarke on a generic level as *Lagedium* Soják (a genus of an intermediate position between *Lactuca* and *Mulgedium*); and by Jeffrey (1975) in Flora of Turkey.

Feráková (1977) suggested a classification, at least for European *Lactuca* species, by considering two main possibilities: (1) to define the genus in a narrow sense according to Tuisl (1968) with the separation of *Lactucopsis*; and (2) to treat it in a broad sense as explained by Stebbins (1937) and supported by Vuilleumier (1973). She takes an intermediate position by including *Mulgedium, Lactucopsis,* and *Phaenixopus* (*Scariola*) in the genus as sections *Mulgedium* (Cass.) C.B. Clarke, *Lactucopsis* (Schultz-Bip. ex Vis et Panč.) Rouy, and *Phaenixopus* (Cass.) Benth., while she treats *Mycelis, Steptorhamphus, Cicerbita,* and *Cephalorrhynchus* as separate genera.

A more recent revision of *Lactuca* is that of Shih (1988b). He restricted the genus to those species having seven to twenty-five yellow ligular florets and one to ten longitudinal ribs on each side of the achene, with an acute to filiform beak at its apex. Such a definition limits the genus to the *serriola*-like species from the sect. *Lactuca* subsect. *Lactuca,* according to Feráková (1977), excepting *L. virosa* and *L. livida,* species with broadly elliptical narrowly winged achenes.

Intergeneric transfers concern mainly species of the related genera: from the subtribe *Lactucinae* Dumort (*Cephalorrhynchus* Boiss., *Steptorhamphus* Bunge); from the *Sonchinae* (*Launaea* Cass., *Prenanthes* L., *Sonchus* L., delimited by Boulos (1972, 1973, 1974a, 1974b); and from the *Crepidinae* (*Ixeris* Cass., *Youngia* Cass.). Some genera such as *Scariola* F.W. Schmidt, *Mulgedium* Cass., *Cicerbita* Wallr., rarely *Mycelis* Cass., are sometimes kept separate from *Lactuca* L., and sometimes are included.

Bremer et al. (1994), in the list of genera belonging to the tribe *Lactuceae*, subtribe *Lactucinae* Dumort, specified as separate genera *Scariola* F.W. Schmidt (ten species), *Cicerbita* Wallr. (35 species), *Mulgedium* Cass. (15 species), *Mycelis* Cass. (1 species), *Lagedium* Soják (1 species), as well as Asian genera, described in the last twenty years with the majority of representatives in China: *Notoseris* C. Shih 1987, *Pterocypsela* C. Shih 1988, *Paraprenanthes* Chang in Shih 1988, *Lactucella* É.A. Nazarova 1990, *Chaetoseris* C. Shih 1991, and *Stenoseris* C. Shih 1991 (Shih 1987, 1988a, 1988b, 1991, Nazarova 1990). In newer regional floras of the central European countries (e.g., Adler, Oswald, and Fischer (1994), Rothmaler et al. (1999), Kubát et al. (2002), and Grulich (2004)) the broader generic conception of *Lactuca* (incl. *Scariola* and *Lagedium*) has been accepted.

Figure 9.15 **(See color insert following page 304.)** Wild *Lactuca* species achene variation 1. *L. aculeata* (CGN 09357), 2. *L. saligna* (PI 261653), 3. *L. serriola* (09H5801249), 4. *L. serriola*—oil seed (PI 236396), 5. *L. virosa* (CGN 4956), 6. *L. perennis* (CGN 09323), 7. *L. tenerrima* (CGN 13351), 8. *L. viminea* subsp. *chondrilliflora* (CGN 14301), 9. *l. tatarica* (CGN 09390), 10. *L. taraxacifolia* (Sal 090), 11. *L. indica* (CGN 14316), 12. *L. canadensis* (CGN 14308). (Composite photo courtesy of I. Doležalová.)

Recent investigations have found that the complex of *Lactuca* species originating from other continents is not very well elucidated from a taxonomic viewpoint. A classification of these species, based on taxonomic and geographic concepts, was elaborated by Lebeda (1998) and Lebeda and Astley (1999). In this view the genus is divided into seven sections and two (African and North American) geographic groups.

In the decades after the treatment of African taxa by Jeffrey (1966), European species by Feráková (1977), both European and Asian ones in the former U.S.S.R. by Kirpicznikov (1964),

Figure 9.16   **(See color insert following page 304.)** Variation in flowers of *Lactuca* species. 1. *L. aculeata* (CGN 09357), 2. *L. saligna* (PI 261653), 3. *L. serriola* (09H5801249), 4. *L. serriola*—oil seed (PI 236396), 5. *L. virosa* (CGN 4956), 6. *L. perennis* (CGN 09323), 7. *L. viminea* subsp. *chondrilliflora* (CGN 14301), 8. *L. tatarica* (CGN 09390), 9. *L. indica* (CGN 13393), 10. *L. capensis* (Sal 060), 11. *L. biennis* (09H5800889), 12. *L. canadensis* (CGN 14308). (Composite photo courtesy of I. Doležalová.)

those of Turkey by Jeffrey (1975), and of Iran by Tuisl (1968), the new approaches have been used to elucidate taxonomical status, and as well as relationships within *Lactuca*. Chromosome numbers were established—e.g., for *L. longidentata* (Arrigoni and Mori 1976), *L. palmensis* (Ortega and Navarro 1977), *L. viminea* subsp. *chondrilliflora* (Brullo and Pavone 1978), *L. viminea* subsp. *alpestris* (Montmollin 1986; Tzanoudakis 1986), *L. livida* (Blanca López and Cueto Romero 1984), *L. perennis* subsp. *granatensis,* and *L. viminea* subsp. *ramosissima* (Mejías 1993), and chromosome banding patterns were studied (Koopman, de Jong, and de Vries 1993) while new data on biochemical and molecular variability—e.g., isozyme markers in selected species—were obtained by Kesseli and Michelmore (1986). The results did not significantly influence the generic and infrageneric divisions. Recently, the genus concept was discussed in detail by Koopman, Hadam, and Doležel (2002), who applied molecular markers to elucidate the relationships within *Lactuca*. Koopman et al. (1998) in their paper on phylogenetic relationships among *Lactuca* species and related genera based on ITS-1 DNA sequences, stated that the molecular data support the broader generic concept of Stebbins (1937), with the inclusion of *Cicerbita alpina*. They also demonstrated that delimitation of the genus *Lactuca* based on absence or presence of an outer pappus, a beak, or the number of flowers per capitulum clearly conflicts with molecular data. They do not confirm the position of *Lactuca* subsect. *cyanicae* Dc. The taxa belonging to this taxonomic unit seem to be relatively unrelated to cultivated lettuce and the nominate subsection. Greuter (2003) emphasized that molecular data have not borne out the traditional, narrowly defined generic concepts of Lactucineae. All Euro-Mediterranean Lactucinae genera such as *Cephalorrhynchus*, *Cicerbita*, *Mycelis* and *Steptorhamphus* (all recognized as distinct in Flora Europea) are merged with *Lactuca* as well as *Mulgedium*, *Scariola* and three other SW Asian *Prenanthes* species. According to Doležalová et al. (2002b) the data on relative nuclear DNA content more or less correspond well with the recently accepted taxonomical classification of *Lactuca*. Whether a narrower or broader concept of *Lactuca* is more appropriate will have to be shown by new methodological approaches. Molecular and biochemical studies have shown that flavonoids can be particularly useful tools as taxonomic markers (Harborne 1967). Unfortunately, for the present, flavonoid data has been available for only 15 species including a number of cultivars of cultivated lettuce (Bohm and Stuessy 2001). For a revised delimitation of *Lactuca* further research is needed, involving species from *Lactuca* and from the related genera as well.

Kilian (2001) suggested a wide concept of *Lactuca*, *sensu* Stebbins (1937) with inclusion of Stebbins' redefinition of *Cicerbita* (Stebbins 1937), to consequently keep the various tropical African species in *Lactuca*, as has been done by Jeffrey (1966), Dethier (1982), Pope (1992), and Jeffrey and Beentje (2000). At this time we are following the formal classification of *Lactuca* and the subgeneric division suggested by Lebeda (1998), by whom seven sections and two heterogeneous geographical groups (from Africa and North America) are recognized (Table 9.9), until more facts on the natural relationship between individual representatives of the genus are collected.

Table 9.9   The Taxonomy of *Lactuca* L.: Differentiation to Sections and Groups, Respectively

Sections (Groups) and Subsections of the Genus *Lactuca* L.
Lactuca L.
subsect. *Lactuca* L.
subsect. *Cyanicae* DC.
Phaenixopus (Cass.) Bentham
Mulgedium (Cass.) C.B. Clarke
Lactucopsis (Schultz Bip. ex Vis. et Pančić) Rouy
Tuberosae Boiss.
Micranthae Boiss.
Sororiae Franchet
North American Group
African Group

Source: From Lebeda, A., Biodiversity of the interactions between germplasm of wild *Lactuca* spp. and related genera and lettuce downy mildew (*Bremia lactucae*). Report on research programme OECD Biological Resource Management for Sustainable Agricultural Systems, HRI, Wellesbourne, U.K., 1998.

9.3.4   Sections and Groups of *Lactuca* Species

In this part, description and characterization of different *Lactuca* species follow the recent general view on the taxonomy of *Lactuca* L. (Table 9.9 and Table 9.10), but with some simplification concerning mostly Asian species. European *Lactuca* species are included in sections Lactuca, Phaenixopus, Mulgedium, and Lactucopsis (Feráková 1977).

9.3.4.1   Section Lactuca

The division of this section into two subsections *Lactuca* and *Cyanicae* is based on the life cycle of their representatives and on their different chromosome counts. (Feráková 1977). The subsection Lactuca comprises annual, winter annual, and biennial herbs with rich inflorescences. The capitulum has from 10 to 30 (50) yellow florets. The achene is obovate with many ribs, narrowed in its upper part into a slender pale beak usually at least as long as the body. This subsection includes the most common and widespread wild species *L. serriola*, *L. saligna,* and *L. virosa*, representing the primary, secondary and tertiary gene pools, respectively, of a popular leafy vegetable: cultivated lettuce (*L. sativa*). Southwest Asian elements *L. aculeata*, *L. azerbaijanica*, *L. georgica*, *L. scarioloides,* and *L. altaica* constitute the primary gene pool as well, thus are of primary importance in relation to cultivated lettuce. Species belonging to subsection Cyanicae are perennial herbs with capitula composed of not more than 22 florets of blue or lilac color and 1–3 ribbed achenes.

9.3.4.2   Section Phaenixopus

Most of the species included in this section are prevalent in the Mediterranean region (Crete, Greece, Iberian Peninsula, and Sardinia) and some of them are endemic (*L. longidentata*, *L. viminea* subsp. *alpestris*). *Lactuca longidentata* and *L. viminea* are biennial, while *L. acanthifolia* and the Asiatic species *L. orientalis* are perennial (Lebeda 1998). Representatives of this section are marked by decurrent leaves. The capitulum, solitary or in fascicles, has from 5 to 6 florets (in the upper part almost glabrous) forming a densely branched panicle. The achene is 5–11 ribbed, oblong-elliptical, and contracted into a concolorous beak not longer than the body.

9.3.4.3   Section Mulgedium

This section is represented by the perennial species *L. tatarica* and *L. sibirica*, occurring in northern areas of Europe and Asia. *Lactuca taraxacifolia* is a newly described species from the mountains of central Asia (Altay, Pamir) (Chalkuziev 1974). The inflorescence of species included in this section comprises few capitula on ascending branches. Florets are numerous; blue, lilac, and rarely white in color. The achene is slightly compressed, marked by a very short beak of the same color as the body.

9.3.4.4   Section Lactucopsis

To this section belong the biennial species *L. quercina* and the perennial *L. aurea*, which are found in woodland and scrub areas of Europe and Asia. The perennial *L. watsoniana* is endemic to the Azores, occupying volcanic craters (Feráková 1976). The inflorescence of this species is usually corymbose with capitula of 6–15 florets. The achene is oblong-elliptic with 2–10 ribs and a concolorous beak 1/4 to 1/2 as long as the body.

Table 9.10 The Taxonomy of *Lactuca* L. Species and Examples of Their Categorization to Different Sections, Subsections and Groups

Section/Subsection (*n*=Chromosomes Number)	*Lactuca* Species[a] (Synonyms)
Lactuca L. (*n*=9)	
Lactuca L.	*L. aculeata* Boiss. et Kotschy
	L. altaica Fisch. et C.A. Mey.
	L. azerbaijanica Rech. f. (*n*=?)
	L. dregeana DC.
	L. georgica Grossh.
	L. saligna L.
	L. sativa L.
	L. scarioloides Boiss. (*n*=?)
	L. serriola L.
	(syn. *L. scariola* L., *L. augustana* All., *L. laciniata* Roth.)
	f. *serriola*
	f. *integrifolia* (S.F. Gray) S.D. Prince et R.N. Carter
	L. virosa L.
	L. livida Boiss. et Reut.
	(syn. *L. virosa* subsp. *livida* (Boiss. et Reuter) Ladero et Velasco)
Cyanicae DC.	*L. intricata* Boiss.
	L. perennis L.
	subsp. *perennis*
	subsp. *granatensis* Charpin et Fernandez Casas
	L. tenerrima Pourr. (*n*=8)
Phoenixopus (Cass.) Bentham (*n*=9)	
	L. acanthifolia (Willd.) Boiss.
	(syn. *L. amorgina* Heldr. et Orph. ex Halácsy)
	L. longidentata Moris (*n*=8)
	L. orientalis Boiss. (Boiss) (*n*=?)
	L. viminea (L.) J. et C. Presl
	subsp. *alpestris* (Gand.) Feráková
	subsp. *chondrilliflora* (Boreau) Bonnier
	subsp. *ramosissima* (All.) Bonnier
	subsp. *viminea*
Mulgedium (Cass.) C.B. Clarke (*n*=9)	
	L. sibirica (L.) Benth. ex Maxim
	L. tatarica (L.) C.A. Mey.
	(syn. *L. pulchella* Pursch DC.)
	subsp. *tatarica*
	subsp. *pulchella* (Pursh) Stebbins
	L. taraxacifolia Schum et Thonn.
Lactucopsis (Schultz Bip. ex Vis. et Pančić) Rouy	
	L. aurea (Sch. Bip. ex Vis. et Panč.) Stebbins (*n*=8)
	(syn. *L. sonchifolia* Pančić)
	L. quercina L. (*n*=9)
	subsp. *quercina*
	(syn. *L. sagittata* Waldst. et Kit., *L. stricta* Waldst. et Kit., *L. altissima* Bieb.)
	subsp. *wilhelmsiana* (Fischer et C.A. Meyer ex DC.)
	L. watsoniana Trel. (*n*=?)

continued

Table 9.10   Continued

Section/Subsection (*n*=Chromosomes Number)	*Lactuca* Species[a] (Synonyms)
Tuberosae Boiss. (*n*=9)	
	L. indica L.
	(syn. *L. laciniata* (Houtt.) Makino,
	L. squarrosa var. *laciniata* O. Kuntze,
	L. amurensis Regel.)
	L. formosana Maxim.
	L. graciliflora DC. (*n*=8)
	L. Iɔɔɔᴄɾʈɪᴀɾɪɑ (DO.) O.D. Claɾkᴇ (*n*=8)
	L. raddeana Maxim. (*n*=?)
	(syn. *L. alliariaefolia* Lév. et Van't.)
	L. triangulata Maxim.
Micranthae Boiss. (*n*=8)	
	L. auriculata DC. (*n*=?)
	L. dissecta D. Don
	L. glauciifolia Boiss. (*n*=?)
	L. rosularis Boiss. (*n*=?)
	L. undulata Ledeb. (*n*=?)
Sororiae Franchet (*n*=9)	
	L. sororia Miq.
	(syn. *L. polypodiifolia* Franch.)
African group	
	L. capensis Thunb. (*n*=8)
	L. dregeana DC. (*n*=9)
	L. glandulifera Hook. f. (*n*=?)
	L. homblei De Wild. (*n*=?)
	L. imbricata Hiern (*n*=?)
	L. praecox R. E. Fr. (*n*=?)
	L. schweinfurthii Oliver et Hiern (*n*=9?)
	L. schulzeana Buettn. (*n*=8)
	L. tysonii (Phillips) C. Jeffrey (*n*=?)
North American Group (*n*=17)	
	L. biennis (Moench) Fernald
	L. canadensis L.
	L. floridana (L.) Gaertn.
	L. graminifolia Michx.
	L. hirsuta Muhl. ex Nutt. (*n*=?)
	L. ludoviciana (Nutt.) Riddell
	L. terrae-novae Fernald (*n*=?)

[a] Modified according to IPNI (International Plant Names Index; data downloaded April 2003 from http://www.ipni. org/index.html) and Lebeda et al. (2004b).

Source: From Lebeda, A., Biodiversity of the interactions between germplasm of wild *Lactuca* spp. and related genera and lettuce downy mildew (*Bremia lactucae*). Report on research programme OECD Biological Resource Management for Sustainable Agricultural Systems, HRI, Wellesbourne, U.K., 1998; Lebeda, A. and Astley, D., World genetic resources of *Lactuca* spp., their taxonomy and biodiversity, In *Eucarpia Leafy Vegetables '99, Proceedings of the Eucarpia Meeting on Leafy Vegetables Genetics and Breeding*, edited by Lebeda A. and Křístková, E. pp. 81–94, Palacký University in Olomouc, Czech Republic, 1999.

9.3.4.5   *Asian Species*

The Asian species are mostly representatives of the sections *Tuberosae*, *Micranthae*, and *Sororiae* (Table 9.10), except for the species from the section *Lactuca* (Lebeda 1998; Lebeda and Astley 1999). Altogether there are 51 *Lactuca* species recorded from Asia, which represent about 52% of the total number of species in the genus (Lebeda et al. 2004b). However, only a few species are distributed throughout the continent. The greatest species richness is recognized in Iran, India and Pakistan (15, 18, and 23 species, respectively), while in some other countries (e.g., Mongolia, Israel, Lebanon, Syria) only a few *Lactuca* species (3–7) are found (Lebeda et al. 2004b).

L. aculeata is restricted to the Near East. It is an Irano-Turanian steppe element occurring mostly among rocks and in uncultivated fields (Tuisl 1968; Jeffrey 1975). *L. scarioloides* is distributed over eastern Turkey, Iraqi Kurdistan, Iran, and Afghanistan on high plateaus (2,000–3,000 m altitude) (Tuisl 1968; Jeffrey 1975; Rechinger 1977). *L. azerbaijanica* is an endemic species known from a single location in Iranian Azerbaijan (Rechinger 1977). The distribution of *L. georgica* is restricted to the wet Euxinian–Hyrcanian region of southwest Asia (Caucasus, northeast Anatolia, and northern Iran) (Kirpicznikov 1964; Jeffrey 1975).

The section *Tuberosae* comprises annual or biennial herbs with fusiform roots and morphologi-caly variable leaves and inflorescences. Florets are yellow, blue, or lilac in color. Achenes are distinctly flattened, elliptic, black with short pale beaks (Ohwi 1965). The most common species belonging to the section *Tuberosae* are *L. indica*, *L. raddeana,* and *L. triangulata* (Lebeda 1998; Lebeda and Astley 1999). *L. indica* var. *laciniata* occupies grassy places in lowlands of Japan, Korea, China, Taiwan, and Indonesia; nevertheless, this species is original in Africa (South Africa, Mozam-bique, Madagascar, Mauritius, Seychelles; Jeffrey 1966). *L. raddeana* is distributed in the far east of Russia (Sakhalin, Kurilen), Japan, China, Taiwan, Indochina, and Korea. *L. triangulata* occurs on mountain slopes of Japan, Sakhalin, Manchuria, and Korea (Kirpicznikov 1964; Ohwi 1965).

The section *Micranthae* includes annual or biennial species with violet or purple florets. The achene is 1–3 ribbed with a beak 2–4 times longer than the body. *Lactuca auriculata*, *L. dissecta*, *L. glauciifolia*, and *L. undulata* are the representatives of the section *Micranthae* (Lebeda 1998; Lebeda and Astley 1999) and are widely distributed. *Lactuca auriculata* occurs in different habitats (among rocks, along rivers, as a weed, mostly over 2,500 m) of central Asia (Iran, India, former U.S.S.R.) (Pavlov 1966). *Lactuca dissecta* and *L. glauciifolia* are distributed in Iran, Pakistan, Afghanistan, India, China and the former U.S.S.R. (Rechinger 1977). *Lactuca undulata* is an Irano–Turanian element occurring in the eastern Mediterranean, Iraq, Pakistan, central Asia, and western China (Jeffrey 1975).

The section *Sororiae* is closely related to the genus *Prenanthes* and comprises perennial species with purplish corollas and barely flattened achenes with fusiform beaks (Lebeda and Astley 1999). *Lactuca sororia* is a relatively common wild lettuce of the section *Sororiae* occurring in Japan (Honshu, Kyushu, Shikoku) and China in deciduous forests (Iwatsuki et al. 1995); it also grows in grasslands at lower elevations in the north of Taiwan (Hui-Lin et al. 1978).

9.3.4.6 African Species

The African species were summarized by Jeffrey (1966) who reported a total of 33 species. Lebeda and Astley (1999) pointed out that during the last two decades some new species have been recorded or determined. The most recent inventory was done by Lebeda et al. (2004b), who numbered at least 43 *Lactuca* species from the African continent; thus this is the second largest geographic group of *Lactuca* species. About 75% of *Lactuca* species (in total 32) occurring in Africa can be considered as autochthonous and are restricted to limited areas; e.g., *L. schulzeana* (Cameroun, Congo, Uganda), *L. homblei* (Congo, Zambia), *L. dregeana,* and *L. tysonii* (South Africa). Some others are more widely distributed (*L. capensis, L. glandulifera, L. imbricata, L. lasiorhiza, L. schweinfurthii*). Only eight species have been reported in northern Africa. Higher numbers of species are recorded from eastern Africa (15 species) and tropical west Africa (10 species), while the greatest species richness is evident in central (23 species) and southern (23 species) Africa. Jeffrey (1966) outlined a new system for the tribe Lactuceae comprising 5 groups, 8 subgroups and 18 series. This system is still used in part.

9.3.4.7 American Species

A recent survey showed that 12 *Lactuca* species have been recorded on the American continent (Lebeda et al. 2004b). This group of species, classified as the North American group (Lebeda and

Astley 1999), includes species originating and distributed in North America (from Canada to Florida), as well as species which are synanthropic and cosmopolitan (*L. serriola*, *L. saligna*, *L. virosa*; Steyermark 1963; Nessler 1976; Strausbaugh and Core 1978; Cronquist 1980; McGregor et al. 1986). Seven *Lactuca* species in North America have been reported as autochthonous (Figure 9.17). These species, genetically distant from other *Lactuca* species, are characterized by the consistent presence of $2n = 34$ chromosomes (Babcock, Stebbins, and Jenkins 1937) and

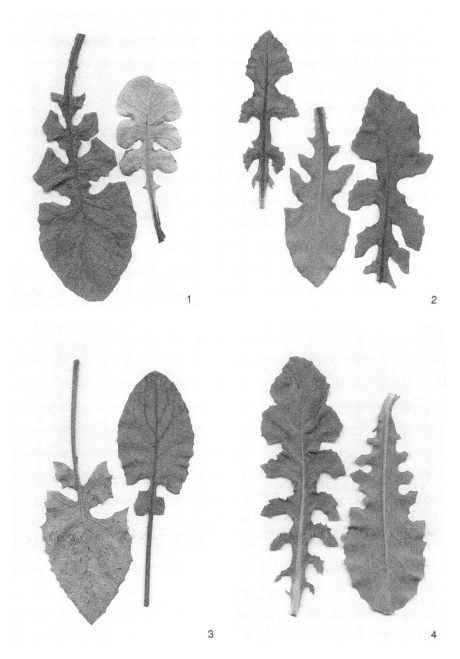

Figure 9.17 Variation in rosette leaves in North American *Lactuca* species. 1. *L. biennis*, 2. *L. canadensis*, 3. *L. floridana*, 4. *L. ludoviciana*. (Composite photo courtesy of I. Doležalová.)

a completely different relative DNA content (Lebeda et al. 2001a; Doležalová et al. 2002b). *Lactuca canadensis*, *L. graminifolia*, *L. biennis*, and *L. intybacea* are the most common species in this area. *Lactuca canadensis* occurs throughout a great part of the northeastern and northcentral United States and southern Canada in thickets, at the edges of woods and in forest clearings (Fernald 1950; Cronquist 1980; McGregor et al. 1986; Hickman 1993; Kartesz 1994). *Lactuca graminifolia* is widespread in sandy fields, open woods, and clearings mainly in North and South Carolina, Arizona, Texas, Florida, and Mexico (Cronquist 1980); it is also recorded from Guatemala (Nash 1976). *Lactuca biennis* is scattered throughout the Great Plains (U.S.A.) in forest clearings, along streams, and lake shores (McGregor et al. 1986). *Lactuca intybacea* is a characteristic species for Central America (Belize, Guatemala, Cuba, Bahamas) (Alain 1964; Nash 1976; Balick, Nee, and Atha 2000) and northern parts of South America (Venezuela, Peru) (Lasser 1964; Brako and Zarucchi 1993). It grows mainly on open banks or in dry thickets, and sometimes along sandy stream beds. Information about the occurrence of other *Lactuca* species in Central and South America is rather rare. In South America, *L. serriola* and *L. saligna* are reported only from Argentina (Lebeda et al. 2004b).

9.3.5 Genera Related to *Lactuca* L.

Following the Feráková (1977) concept, closely related genera are: *Mycelis*, *Steptorhamphus*, *Cicerbita*, and *Cephalorrhynchus*. Frequent synonymy related to *Lactuca* species occurs also in genera *Ixeris* and *Youngia* (Lebeda 1998; Table 9.11); however, these genera are not closely related to *Lactuca* L.

9.3.5.1 *Cephalorrhynchus Boiss.*

These are biennial or perennial herbs, often with tuberous rootstocks and variable, entire, denticulate, runcinate, pinnatifid, or lyrate-pinnatisect leaves. Capitula are in corymbose-paniculate synflorescences. Involucral bracts are in many rows. Florets are yellow, whitish, blue, violet, or purplish. Achenes are slightly compressed, fusiform-ellipsoid to elliptic-oblong, many ribbed, glabrous, with distinct, slender to filiform beaks. The pappus has many fine, white, smooth bristles with an outer row of very short, fine hairs. About 15 species are known from southeastern Europe, Turkey, Middle East, Iran, Afghanistan, Pakistan, the Himalayas, and central Asia.

9.3.5.2 *Cicerbita Wallr.*

These are perennial herbs with almost entire to coarsely dentate, runcinate, pinnatifid, or lyrate-pinnatisect leaves. Capitula are in corymbose-paniculate to racemose synflorescens. Florets are blue, violet, purplish, or rarely yellow. Achenes are slightly compressed; fusiform-ellipsoid to elliptic-oblong; glabrous; apically constricted; without or sometimes with short, indistinct beaks. The pappus has many fine, white, smooth, or scabrid bristles and an outer row of very short, fine hairs. The genus includes about 35 species in Europe, southwest and central Asia, the Himalayan region, and China.

9.3.5.3 *Mycelis Cass.*

Mycelis is a monotypic Eurasian genus with some populations occasionally occurring in northern Africa. It comprises perennial herbs with solitary branched stems reaching up to 40–90 cm in height. Leaves are lyrate-pinnatisect. Flowers are bright yellow. Achenes are flattened, elliptic-obovate, many ribbed, with distinct, rather short and stout beaks. The pappus has many fine, white, scabrid bristles with an outer row of very short, fine hairs. *M. muralis* (Wall lettuce) is a common species, growing as a ruderal plant in waste places, around roads, and sometimes in walls.

Table 9.11   Genera Related to the Genus *Lactuca* L. and Their Synonyms

Genus	Species[a]	Synonym (*Lactuca* spp.)
Cicerbita Wallroth	*C. alpina* Wallr.	*L. alpina* (L.) Gray (*Sonchus alpinus* L.)
	C. bourgaei Beauverd	*L. bourgaei* Boiss. (*Mulgedium bourgaei* Boiss.)
	C. plumieri (L.) Kirschl.	*L. plumieri* Schulz Bip. (*Sonchus plumieri* L.)
	C. macrophylla Wallr.	*L. macrophylla* (Willd.) A. Gray
	C. racemosa Beauverd	*L. albana* C. A. Mey.
Mycelis Cass. (Cass.)	*M. muralis* Dumort.	*L. muralis* (L.) Gaertn.
Steptorhamphus Bunge	*S. tuberosus* (L.) Grossh	*L. tuberosa* Jacq. (*L. cretica* Desf.)
Ixeris Cass.	*I. dentata* Nakai	*L. dentata* (Thunb.) Robins. (*L. thunbergii* (A. Gray) Maxim.)
	I. chinensis (Thunb.) Nakai	*L. chinensis* Makino
	I. chinensis (Thunb.) Nakai var. *strigosa* (Lev et Van't) Ohwi	*L. strigosa* Lev et Van't
	I. japonica Nakai	*L. debilis* (Thunb.) Benth.
	I. laevigata (Blume) Yamamoto	*L. stenophylla* (Makino)
	I. makinoana (Kitam.) Kitam.	*L. thunbergii* var. *angustifolia* Makino (*L. dentata* var. *angustifolia* (Makino) Makino, *L. makinoana* Kitam.)
	I. nipponica Nakai	*L. nipponica* Nakai ex Makino et Nemoto
	I. polycephala Cass.	*L. polycephala* (Cass.) Bruth. et Hook.
	I. repens A. Gray	*L. repens* (L.) Benth.
	I. stolonifera A. Gray	*L. stolonifera* (A. Gray) Benth.
	I. tamagawaensis (Makino) Kitam.	*L. tamagawaensis* Makino (*L. versicolor* var. *arenicola* Makino)
Youngia Cass.	*Y. chelidoniifolia* (Makino) Stebbins	*L. chelidoniifolia* Makino (*I. chelidoniifolia* (Makino) Stebbins)
	Y. denticulata (Houtt.) Nakai ex Stebbins	*L. denticulata* (Houtt.) Maxim. (*I. denticulata* (Houtt.) Nakai)
	Y. yoshinoi (Makino et Nakai) Kitam.	*L. yoshinoi* (Makino) Makino et Nakai (*L. denticulata* var. *yoshinoi* Makino)

[a] Modified according to IPNI (International Plant Names Index; data downloaded November 2005 from http://www.ipni.org/index.html).

Source: Modified by Lebeda, A., Biodiversity of the interactions between germplasm of wild *Lactuca* spp. and related genera and lettuce downy mildew (*Bremia lactucae*), Report on research programme OECD Biological Resource Management for Sustainable Agricultural Systems, HRI, Wellesbourne, U.K., 1998.

9.3.5.4   *Steptorhamphus Bunge*

These are perennial, somewhat succulent, herbs with swollen roots. Stems are solitary, erect, simple, or branched. Leaves are entire to pinnatified and amplexicaul. Ligulate capitula have rather wide involucral bracts, with 15–50 yellow or lilac ligules. Achenes are homomorphic, compressed and winged, with very long, slender beaks. The pappus has two rows of simple fine bristles, with an outer row of very short hairs. *S. tuberosus* is found in the southern parts of the Balkan peninsula, Aegean region and Krym. The genus includes seven species occurring in southeast Europe, Turkey, the Caucasus, the Middle East, Iran, Afghanistan, and central Asia.

9.4 GENE POOLS AND GENETIC DIVERSITY

9.4.1 General Characterization

Harlan and de Wet (1971) proposed a concept involving three gene pool levels to classify crop species and their wild relatives. The primary gene pool includes the cultivated crop species and their wild ancestors, which can be intercrossed regularly and produce fertile hybrids. The cultivated gene pool comprises the commercial cultivars of the crop and landraces, and the wild species closely related to the crop that do not show crossing barriers. The secondary pool includes wild taxa that exchange genes with the crop to a limited degree. The tertiary gene pool consists of wild relatives with barriers making crosses difficult with the primary gene pool species.

9.4.2 Gene Pools and Species Concept of *Lactuca sativa*

During the last sixty years, the genus *Lactuca* together with the species of related genera have been intensively studied with respect to exploitation of wild relatives in commercial lettuce breeding (Lebeda and Křístková 1999; Lebeda, Doležalová, and Astley 2004a, 2004b). There has not yet been any general agreement among taxonomists about the classification and delimitation of *Lactuca* (Koopman et al. 1998; Lebeda et al. 2004b). In general, scientists working in this field have largely accepted genus delimitation according to Feráková (1977). She proposed an intermediate position between a broad genus definition by Stebbins (1937) and the view of Tuisl (1968) who takes the genus in a narrow sense. Following the Feráková (1977) concept, the species closely related to cultivated lettuce belong to the section *Lactuca* subsection *Lactuca* (see Section 9.3). Except for *L. sativa*, this section includes the three best known and most common European wild species, all with the same chromosome number ($2n = 18$): *L. serriola, L. saligna,* and *L. virosa* as well as an Euro-Asiatic element *L. altaica*.

The lettuce primary gene pool is represented mainly by *L. sativa* with its enormous morphological variation, numerous cultivars and many landraces (Rodenburg 1960; Phillips and Rix 1993; Stickland 1998). Despite the diversity of *L. sativa*, all the various lettuce types are interfertile and form the primary gene pool together with the wild species *L. serriola*. The species *L. aculeata* and *L. dregeana* can also be included in this group, because they are fully sexually compatible with *L. sativa* and *L. serriola* (Zohary 1991). According to McGuire et al. (1993) and Zohary (1991), *L. saligna* is included in the secondary gene pool. Because of very difficult crossing barriers, *L. virosa* belongs in the tertiary gene pool (Figure 9.18).

Among wild species, *L. serriola* is the closest to cultivated lettuce. It freely hybridizes with *L. sativa* in nature and both are fully cross-compatible and fully interfertile (Thompson et al. 1941; Lindqvist 1960a; Whitaker 1969). In addition to morphologic and genetic resemblance (Zohary 1991; Koopman, de Jong, and de Vries 1993; de Vries and Raamsdonk 1994; Koopman and de Jong 1996), the close relationship has been revealed by comparative studies of different biochemical and molecular marker systems (Kesseli, Ochoa, and Michelmore 1991; de Vries 1996; Koopman 1998, 2001, 2002; van de Wiel, Arens, and Vosman 1998, 1999; Dziechciarková et al. 2004a, 2004b). Although de Vries and Raamsdonk (1994) have regarded them as separate species, some authors, based on the above mentioned analyses have considered *L. sativa* conspecific with *L. serriola* (Frietema de Vries 1996; Koopman et al. 1998, 2001; van de Wiel, Arens, and Vosman 1998, 1999). Frietema de Vries (1996) proposed distinction on the subspecific level: *L. sativa* subsp. *serriola* and *L. sativa* subsp. *sativa*. This is in contrast to Koopman, Zevenbergen, and van den Berg, (2001) who stated

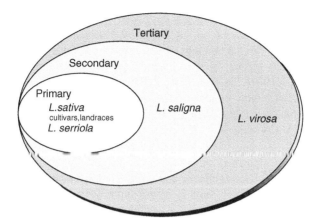

Figure 9.18 Categorization of *Lactuca* species into gene pools following the McGuire et al. (1993) concept.

that taxa so similar should not be considered as subspecies. He proposed the name *L. sativa* for both *L. sativa* and *L. serriola*.

It is very difficult to consider *L. sativa* and *L. serriola* as a single species. In general, crop species are considered dynamic populations of taxons characterized by many different specific attributes (e.g., morphology, geographic distribution, ecology, breeding system etc.; Judd et al. 1999). *L. sativa* is a typical cultivated and highly domesticated species (Hancock 2004). It has no natural distribution in nature, no specific area of distribution or adaptation to "wild" conditions, different ecological parameters, differing performance patterns, substantial differences in morphology, and extremely broad variation in different characters, etc. Some plant taxonomists consider these two taxa as independent species (Feráková 1977; Grulich 2004), but with very close evolutionary or domestication relationships. The long history of lettuce (*L. sativa*), with cultivation, selection, and breeding in various parts of the world, has resulted in enormous variation and loss of "wild" traits characteristic of *L. serriola*. In the specific case of lettuce, we must rather strictly apply the principles of the International Code of Botanical Nomenclature (ICBN; Greuter et al. 2000), which establishes the rules for botanical names; and the International Code of Nomenclature for Cultivated Plants (ICNCP; Brickell et al. 2004), which defines rules for the proper description and nomenclature of cultivated plants at the cultivar and cultivar group levels.

The seven wild *serriola*-like species (*L. serriola, L. aculeata, L. altaica, L. azerbaijanica, L. georgica, L. scarioloides,* and *L. dregeana*), which are taxonomically closely related to cultivated lettuce, were studied by Zohary (1991). After analysis of detailed combined taxonomic and genetic evidence, he stated that *L. aculeata, L. altaica, L. azerbaijanica, L. georgica, L. scarioloides,* and *L. dregeana* indicate very similar interfertility with the crop species and constitute its primary gene pool. Because taxonomically more distant *L. saligna* is partly cross-fertile with the crop, it is possible, but not compelling for it to be considered in the lettuce gene pool as well. According to the taxonomic survey by Feráková (1977), the central west Asian wild lettuce *L. altaica* occupies an intermediate position between *L. serriola* and *L. saligna*. Lindqvist (1960b) also suggested that *L. altaica* may be a hybrid between *L. saligna* and *L. serriola*. However, the analysis of DNA ITS-1 sequences showed a closer relationship to *L. serriola* than to *L. saligna* (Koopman et al. 1998). He states, that except for morphological similarity, no evidence was found to support a close relationship between *L. altaica* and *L. saligna*. To elucidate the taxonomic position of *L. altaica*, the available literature and herbarium specimens were searched and plants collected in Uzbekistan (van Soest 1998) were intensively studied using numerical morphologic analysis and molecular technique (AFLP; Jacobs 2001). After the comparison of the original protoloque with a designated lectotype and living plant material, it appeared that analyzed

L. altaica accessions matched the species described in floras (Kirpicznikov 1964, 2000). However, they did not fit the lectotype suggested by Kirpicznikov (Jacobs 2001). Thus, the taxonomical position of *L. altaica* remains still questionable until certain discrepancies have been resolved. A different theory on *serriola*-like species, in contrast with earlier concepts (Feráková 1977; Zohary 1991), has been set forth by Koopman, Zevenbergen, and van den Berg (2001). His cladistic and phenetic analyses of AFLP results showed the group of *serriola*-like species clustered together in accordance with the previous ITS-1 sequence study (Koopman et al. 1998). Consequently, Koopman et al. (1998) regarded *serriola*-like species *L. sativa*, *L. serriola*, *L. altaica,* and *L. dregeana* as conspecific. It is very difficult to accept conclusions made only through the "molecular view" on this question. Many other parameters must be seriously considered (see arguments and discussion of the authors of this Section above).

Although *L. serriola*, *L. saligna,* and *L. virosa* have been intensively studied by breeders and plant evolutionists, the categorization to secondary and tertiary gene pools has remained open to question. The comparative study of achene protein variability (de Vries 1996) has shown that *L. virosa* is more closely related to *L. sativa* and *L. serriola* than to *L. saligna*. On the other hand, the AFLP analysis of leaf DNA demonstrated that *L. saligna* is not so distinct from *L. sativa* and *L. virosa* (Hill et al. 1996). Koopman, Zevenbergen, and van den Berg (2001) discussed the position of *L. saligna* and *L. virosa* in the subsection *Lactuca*. He summarized all available data about plant morphology, crossability, karyotypes, chromosome banding patterns, and isozyme and molecular analyses, which showed different possibilities for the position of *L. saligna* and *L. virosa* in relation to the *serriola*-like species. He stated that AFLP data were inconclusive to fix their position within the subsection *Lactuca*, and the relationship to *serriola*-like species remains unanswered.

A completely different view from the previous conception of the lettuce gene pool was proposed by Koopman et al. (1998). Based on analysis of DNA ITS-1 sequences, supported with data from crossing experiments (Thompson et al. 1941; Chupeau et al. 1994; Maisonneuve et al. 1995; Mazier et al. 1999), he adjusted genus limitation to coincide with the lettuce gene pool. He stated that the species in subsection *Cyanicae* do not belong to the lettuce gene pool and therefore should be excluded from *Lactuca*. He proposes that section *Lactuca* subsection *Lactuca* comprises the primary and secondary gene pool, while the sections *Phaenixopus*, *Mulgedium*, and *Lactucopsis* comprise the tertiary gene pool.

9.4.3 Interspecific Hybridization and Gene Pools

Since the 1930s many crossing experiments within *Lactuca* species have been made, mainly between *Lactuca* and closely related species. The available information about this topic was summarized by Rulkens (1987), de Vries (1990), de Vries and Raamsdonk (1994), and Ryder (1999a). In commercial plant breeding, including lettuce breeding, interspecific hybridization is considered a major tool for increasing the genetic diversity of crops and the transfer of required new genes (Sanchez-Monge and Garcia-Olmedo 1978; Pink and Keane 1993; Ladizinsky 1998; Lebeda 1998; Lebeda, Pink, and Astley 2002; Hancock 2004). The crossing experiments and practical utilization of wild *Lactuca* germplasm were carried out mostly with species that form the section *Lactuca*, especially with the primary pool (*L. serriola*) and also the secondary pool (*L. saligna*). The tertiary gene pool (*L. virosa*) has also been utilized. Details about this topic are summarized in Section 9.6.

9.4.4 The Origin of Cultivated Lettuce (*Lactuca sativa*)

In general, domestication is a process which is based on genetic shift in various populations; making them better adapted to the environment created by cultivation, and at the same time ill adapted to their original habitat in the wild (Ladizinsky 1998). The only domesticated *Lactuca* species, *Lactuca sativa*, in its various forms, is the world's most important salad crop (Rubatzky and Yamaguchi 1997). In the last few decades, studies on *L. sativa* origin and evolution have resulted in several hypotheses.

The original theory proposed that cultivated lettuce arose from wild forms of *L. sativa*. Lindqvist (1960b) implied that such forms could previously have existed, but have gone out of cultivation. Therefore the theory could not be confirmed because there is no actual evidence at this point. The second theory assumed that *L. sativa* originated directly from *L. serriola*, because nearly all the variations in cultivated lettuce are present in *L. serriola*, except for the extreme forms of head formation (Ryder and Whitaker 1995). Several theories based on the concept of hybrid speciation indicate the various possible directions of this process: (a) *L. sativa* and *L. serriola* have descended from heterogeneous hybrid populations from one crop–weed complex. Then *L. sativa* developed through human selection, and *L. serriola* through its the adaptation to man-made waste habitats; (b) Progenitors of *L. sativa* were hybrids between *L. serriola* and some other *Lactuca* species; (c) *L. serriola* is as a product of hybridization between cultivated forms of *L. sativa* and other *Lactuca* species. This last is improbable, because *L. serriola* possess many unique alleles and does not occupy an intermediate position between *L. sativa* and the other related taxa (Kesseli, Ochoa, and Michelmore 1991).

According to de Vries and van Raamsdonk (1994), *L. sativa* is probably of a polyphyletic origin and was selected out from a gene pool of *L. serriola* with simultaneous introgression of genes from another species. This source could be a crop–weed complex (de Vries 1990, 1997; Koopman, de Jong, and de Vries 1993). The concept of a crop–weed complex, which considered *L. sativa* as a cultivated species and *L. serriola* as a camp-following weed, was suggested by Lindqvist (1960b) and has been accepted in general by other authors (de Vries 1990, 1997; Koopman, de Jong, and de Vries 1993; de Vries and Raamsdonk 1994). The basic biosystematic concept of crop–weed complexes was developed by Pickersgill (1981) who formulated three different possibilities of interactions between crop plants and weedy or wild relatives: (1) crops and weeds arise simultaneously from a wild ancestor; (2) wild plants give rise to weeds and crops evolved from these weeds; (3) escaped individuals of a crop became weeds. Van der Maesen (1994) defined the crop–weed complexes as composed of cultivated and related wild or weedy plants, growing together and influencing each other through introgression. Introgression could be important if crops and wild relatives are interfertile; these crosses may result also in weedy intermediates as a part of "crop–weed complexes."

L. serriola is a worldwide distributed *Lactuca* species (Lebeda et al. 2004b) and one of the most common weed plants in Europe (Lebeda et al. 2001b; Figure 9.19). It is interfertile with *L. sativa*, but introgression may be limited because each has an autogamous breeding system. Nevertheless, cross-pollination has been confirmed. Thompson (1933) and Thompson et al. (1958) have shown that cross-pollination to the extent of 2–3% occurred between lettuce cultivars grown in close proximity. Significant differences in the amounts of natural cross-pollination were found at different periods during the flowering cycle. These authors confirmed that natural hybrids between *L. sativa* cultivars frequently occur in commercial plantings. Bohn and Whitaker (1951), Lindqvist (1960b), and Feráková (1977) mentioned that spontaneous hybridization takes place occasionally between *L. sativa* and *L. serriola*. Frietema (1994) discussed the possibility that the recent spread of *L. serriola* could be linked to the cultivation of lettuce. Recently an attempt was made to assess the level of gene flow between cultivated and wild forms of lettuce by comparing two molecular marker systems, AFLP (amplified fragment length polymorphism) and SSAP (sequence-specific amplified polymorphism; van de Wiel et al. 2004).

Although Lindqvist (1960b), in one of the earliest studies on the origin of *L. sativa*, did not consider *L. serriola* as a direct progenitor of *L. sativa*, recent analyses of protein (Kesseli and Michelmore 1986; de Vries 1996) and molecular markers (Kesseli, Ochoa, and Michelmore 1991; de Vries and Raamsdonk 1994; Hill et al. 1996; Witsenboer, Vogel, and Michelmore 1997;

Figure 9.19   (See color insert following page 304.) Examples of *L. serriola* habitats. 1. *L. serriola* f. *integrifolia*,
Rhône river (France), 2. *L. serriola* f. *serriola*, ruderal in town Dvůr Králové n. Labem (Czech
Republic), 3. *L. serriola* f. *serriola*, stony wall in village Hostkovice (Czech Republic), 4. *L. serriola*
f. *serriola*, ruderal—broken bricks, Torremolines (Spain), 5. *L. serriola* f. *serriola*, stony slope, Adana
(Turkey), 6. *L. serriola* f. *serriola*, beet field near village Ústín (Czech Republic), 7. *L. serriola* f.
serriola, pavement in town Olomouc (Czech Republic), 8. *L. serriola* f. *serriola*, aluvium of Gail
river (Austria). (Composite photo courtesy of Doležalová, I. and Lebeda, A.)

Koopman, Hadam, and Doležel 1998, 2001, 2002) have indicated a polyphyletic origin of cultivated lettuce including *L. serriola* as a direct ancestor. Unfortunately it is not known which other
species were involved in the domestication of lettuce. However, de Vries (1997) concluded that
the other Asian species from the section *Lactuca* subsection *Lactuca*: *L. scarioloides*,
L. azerbaijanica, and *L. altaica*, and even the South African *L. dregeana*, may have also
played a role in the origin of lettuce. He proposed the possibility of their involvement in the

origin because, especially in the center of origin (southwest Asia), the breeding system is more allogamous and hybrids could more easily have arisen. The hypothesis supports the occurrence of spontaneous hybridization of some southwest Asian *Lactuca* species in their natural habitats (Zohary 1991) as well as the findings of Koopman et al. (1998) who reported (nearly) identical ITS-1 DNA sequences of *L. dregeana*, *L. aculeata*, and *L. altaica* as those of cultivated lettuce (*L. sativa*). Koopman et al. (1998) further expects that the remaining species from the subsection *Lactuca*, e.g., *L. scarioloides*, *L. azerbaijanica*, and *L. georgica* will show the similar ITS-1 DNA sequences and crossing behavior. The close relationship of *serriola*-like species *L. serriola*, *L. dregeana*, *L. altaica*, and *L. aculeata* to *L. sativa* showed in results of the AFLP fingerprints (Koopman, Zevenbergen, and van den Berg, 2001) and in establishment of DNA content and base composition in *Lactuca* (Koopman, Hadam, and Doležel 2002).

To conclude, *L. sativa* probably has a polyphyletic origin and was selected out from a gene pool of *L. serriola*, probably *L. serriola* f. *integrifolia* (Lebeda 1998), with more or less simultaneous introgression from another species. The current model includes the possible influence of *L. saligna* (de Vries and Raamsdonk 1994; Lebeda 1998; Koopman, Hadam, and Doležel 2002) and perhaps other southwest Asian *Lactuca* species. The species *L. scarioloides*, *L. azerbaijanica*, *L. georgica*, *L. altaica*, and *L. dregeana* have not yet been investigated cytogenetically. Information is still missing about their genetic affinities to cultivated lettuce. Further detailed research is necessary to elucidate their role in the origin of cultivated lettuce.

Estimations of the center and time of the origin of *Lactuca sativa* vary from author to author. According to Lindqvist (1960b), cultivated lettuce originated in Egypt where the paintings of tall plants with pointed leaves in Egyptian tombs were found. The paintings have been interpreted as lettuces and their age have been dated at about 2500 B.C. (Keimer 1924; Harlan 1986). These forms, which look like stalk or stem lettuces, and are similar in appearance to cos lettuce, as well as forms of primitive oil seed lettuce, are still grown in this area (Phillips and Rix 1993; de Vries 1997). According to Ryder and Whitaker (1995) and Ladizinsky (1998) the forms similar to cos have lanceolate leaves with pointed apices and resemble modern cultivars of asparagus lettuce. Ladizinsky (1998) mentioned that lettuce was not formerly used as it is used today, as a leafy vegetable, but rather as an oil seed crop and medicinal plant based on its production of milky latex. The forms with tall large narrow-leaves spread from Egyptian provinces to Rome and Greece. The stalk lettuces probably moved to China around 600–900 A.D. (Boukema, Hazekamp, van Hintum, 1990). Crisp lettuce was introduced to America by the first settlers (about 1,500) and probably the modern American crisphead types descended from the French batavia types (Boukema, Hazekamp, van Hintum, 1990). De Vries (1997) cited numerous references, primarily herbals, where the popularity of lettuce, not only as a vegetable crop but also as soporific medicine during the Middle Ages in Europe, is recorded. Use of lettuce as a frequently grown vegetable crop in Europe from the sixteenth century is depicted on paintings from that time (Zeven and Brandenburg 1986).

However, several other places of origin other than Egypt have been proposed. Ryder (1986) and Ryder and Whitaker (1995) mentioned the eastern Mediterranean area as a place of lettuce origin. This is also the conclusion of Harlan (1992) and Rubatzky and Yamaguchi (1997), who linked the Mediterranean types to the near-eastern complex. Boukema, Hazekamp, and van Hintum (1990) assumed that lettuce was domesticated in southwest Asia, in the region between Egypt and Iran. According to Rulkens (1987) and de Vries (1997) the Kurdistan–Mesopotamia region, rather than Egypt, is thought to be the most probable primary center for lettuce because a large number of related species occur there, and the area between the Euphrates and Tigris rivers is also regarded as the region of the ancient origin of agriculture itself (Vavilov 1995). Zohary and Hopf (1993) considered cultivated lettuce as a west Asiatic and Mediterranean element later grown in ancient Egypt, Greece, and Rome. Accepting *L. serriola* as a one of the direct progenitors of cultivated lettuce, human effort in the process of domestication probably involved selection against leaf spines, latex content, and the bitter taste of leaves, and selection for late bolting and flowering.

Domestication also has led to shortening of internodes, bunching of leaves, increase in seed size, and nonshattering seeds.

9.4.5 Variability of *L. sativa*

L. sativa is an annual glabrous herb with a thin tap root and an erect stem, 30–100 cm high, branched in the upper part. Leaves are spirally arranged, forming a dense rosette or a head before bolting. Their shape is oblong to transverse elliptic, orbicular to triangular, undivided to pinnatisect. The leaf margin is entire to setose dentate, often curly. Stem leaves are oblong elliptic, with a cordate base. The inflorescence is composed of 7–15 (35) yellow ligulate florets in a corymbose densely bracted panicle. The anthocyanin can be distributed on the cotyledons, true leaves, stems, and ligules. The involucre is 10–15 mm long, cylindrical, with involucral bracts that are broadly to narrowly lanceolate, light green, with white margins, and erect. The fruit (achene) has 5–7 setose ribs on each side, a beak, and a white pappus. Its length (including beak) is 6–8 mm and the color is white, cream to yellow, gray, brown, or black (Dostál 1989; Rubatzky and Yamaguchi 1997).

Morphologically, lettuce is the most diverse species of the genus *Lactuca*. The broad variability of cultivated *L. sativa* is caused by its polyphyletic origin and a complicated process of domestication proceeding through an accumulation of small changes over a long period of time (Kesseli,

Figure 9.20 Drawing of six representatives of the six *Lactuca* cultivar groups according to Rodenburg (1960). (Example Vries de, I. M., and van Raamsdonk, L. W. D. *Plant Syst. Evol.* 193, 125–141, 1994.), 1. Hilde (Butterhead), 2. Great Lakes (Crisphead), 3. Bella (Latin), 4. Oakleaf (Cutting), 5. Celtuce (Stalk), 6. Verte Maraîchères (Cos).

Ochoa, and Michelmore, 1991). A survey of lettuce cultivars and classification of types were provided by Rodenburg (1960). The recent most comprehensive overview of taxonomical and morphological analyses of lettuce cultivars was summarized by de Vries and van Raamsdonk (1994) and de Vries (1997). The crop comprises seven main groups of cultivars (including oil seed lettuce) differing phenotypically; they are usually described as morphotypes (Figure 9.20 and Figure 9.21).

1. *Butterhead lettuce* (var. *capitata* L. *nidus tenerrima* Helm) (Kopfsalat, Laitue pommé): Heading type with soft and tender leaves, eaten raw. It is most popular in England, France, the Netherlands, and other western and central European countries (Ryder 1986). In recent decades many cultivars have been bred and grown in the United States (Ryder 1999a).
2. *Crisphead lettuce* (var. *capitata* L. *nidus jäggeri* Helm) (Iceberg type, Eissalat, Batavia): Heading type with thick crispy leaves and flabellate leaf venation, eaten raw. It is mainly cultivated in the United States, where it was introduced after discovery of the continent (de Vries 1997). However, it is also grown now in the western and central European countries, including the Netherlands, the United Kingdom, France, Spain, Belgium, Germany, Poland, and Czech Republic, as well as in Japan, China, and Australia.

Figure 9.21 Variability of lettuce (*L. sativa*) morphotypes. 1. Achát (Butterhead), 2. Dubáček (Cutting), 3. Pennlake (Crisphead—Iceberg), 4. Romana (Cos), 5. Karmína (Cutting), 6. Dubared (Cutting). (Composite photo courtesy of I. Doležalová.)

3. *Cos lettuce* (*L. sativa* var. *longifolia* Lam., var. *romana* hort. in Bailey; Römischer Salat, Laitue romaine): Plants with tall loose heads that are sometimes tied up, oblong rigid leaves with a prominent midrib running almost to the apex, eaten raw or cooked. The name of the morphotype is taken from the Greek island Cos (Kos) where the type has been cultivated for a long time. Cos lettuce is the most common in the Mediterranean countries of Europe, western Asia, and northern Africa (Ryder 1986). According to Boukema, Hazekamp, and van Hintum (1990) many landraces maintained at the CGN gene bank collection originate mainly from Egypt, Iran, Turkey, and Syria.

4. *Cutting lettuce* (var. *acephala* Alef., syn. var. *secalina* Alef., syn.var. *crispa* L.) (Gathering lettuce, Loose-leaf, Picking lettuce, Schnittsalat, Laitue à couper): Nonheading type. Harvested as whole rosettes, occasionally as separate leaves, eaten raw. Cutting lettuces have been very popular in the United States, Italy, France, and Czech and Slovak Republics (de Vries 1997). The group of morphotypes is extremely heterogeneous. Cultivars may have leaves with entire, curled or fringed leaves, straight or deeply incised margins, elongated or broad leaves, various shades of green, and various distribution and intensities of anthocyanin. The Greeks and Romans cultivated cutting lettuces. Boukema, Hazekamp, and van Hintum (1990) stated that CGN gene bank landraces of this type come from Turkey and Greece.

5. *Stalk (Asparagus) lettuce* (var. *angustana* Irish ex Bremer, syn. var. *asparagina* Bailey, syn. *L. angustana* hort. in Vilm.; Stem lettuce, Stengelsalat, Laitue-tige): Plants with swollen stalks, which are eaten raw or cooked like asparagus. Leaves can be eaten raw in a very young stage or cooked like spinach (Lebeda and Křístková 1995).

   According to Lindqvist (1960b) there are two types recognized within this group. The Chinese cultivars have light gray leaves resembling cos lettuce leaves; the second type has long lanceolate leaves with pointed apices. According to Helm (1954) stalk lettuce originated from Tibet, which would account for its extensive cultivation in China, in the Pamirs, and India (Rodenburg 1960, de Vries 1997). However, the lettuce shown in Egyptian tombs is stalk lettuce and dates back to about 2,500 B.C. If lettuce originated in Mesopotamia, it is even older in the Middle East. Both asparagus types and cos-like types are found in Egypt. We think it is more likely that the original types migrated to the Far East overland, showing up there up to 1,500–2,000 years later. It is possible that Helm (1954) was referring to *L. indica*, which is common in the Far East. Stalk lettuce material collected in Afghanistan appeared to be an intermediate between cos and stalk lettuces and is sometimes used as a food for livestock (Boukema, Hazekamp, and van Hintum 1990).

6. *Latin lettuce* (without scientific name). Plants have loose heads with thick leathery leaves, dark green color, and are eaten raw. It is mainly cultivated in the Mediterranean countries, including northern Africa, and in South America (Rodenburg 1960).

7. *Oil seed lettuce*. Because of the bitter taste of the leaves it is not eaten as a vegetable. Oilseed lettuce is characterized by a high percentage (35%) of oil in the seeds, which is used for cooking. The oil contains vitamin E, which is used in human medicine (Boukema, Hazekamp, and van Hintum 1990). In Egypt, cultivation of oil-producing forms has continued to the present time (Ryder 1986). Boukema, Hazekamp, and van Hintum (1990) mentioned that some of its forms may be either *L. serriola* or *L. sativa* or intermediate types between two species. de Vries and van Raamsdonk (1994), from a detailed comparison based on a multivariate analysis of vegetative characters of lettuce cultivars, proposed two supergroups: (1) Butterhead, Crisphead (Iceberg and Cabbage) and Latin groups; (2) Cos, Cutting and Stalk (Stem, Asparagus) groups.

9.4.6 Genetic Diversity of *L. sativa* and Wild *Lactuca* Species

Genetic diversity of plants is frequently measured by molecular markers which are routinely used to track genotypes and deduce degrees of relatedness between individuals in diversity and evolutionary studies, and genetic mapping. Several marker types have been used to study genetic and phylogenetic relationships within *Lactuca* (Vosman 1997; Dziechciarková et al. 2004a). These include isozyme (Kesseli and Michelmore 1986; Dziechciarková et al. 2004b), RFLP (Kesseli, Ochoa, and Michelmore 1991; Kesseli, Paran, and Michelmore 1994), RAPD (Waycott and Fort 1994; Yamamoto, Nishikawa, and Oeda 1994), AFLP (Hill et al. 1996; Koopman, Zevenbergen, and van den Berg 2001) and microsatellite (Witsenboer, Vogel, and Michelmore 1997; Wiel van de, Arens, and Vosman 1999) markers, as well as internal transcribed spacer (ITS-1) sequences (Koopman et al. 1998). Use of the earliest marker systems (e.g., isozyme, RFLP) was limited by cost and a low number of available markers. Technologies that rapidly produce large numbers of markers at a relatively low cost have now permitted definitive phenetic analyses of the genus. Marker types vary for several characteristics, including polymorphism and reproducibility (Staub et al. 1996). More conserved markers, such as ITS-1 sequences, are ideal for studying more distantly related species. In contrast, highly polymorphic markers (e.g., AFLP) are most appropriate for resolving relationships among close relatives, such as varieties or types within *L. sativa*.

In general, molecular studies have confirmed the systematic relationships proposed earlier, based on morphology and cytogenetic traits (Lindqvist 1960a). These aspects are discussed from the viewpoint of taxonomy and classification in more detail in this Section above. In general, the low number of accessions and variability within the less common species probably makes their classification less certain. Precise taxonomic determination of *Lactuca* germplasm requires complex research (Lebeda et al. 1999, 2001a), including classical taxonomic studies in tandem with a molecular approach, to solve complex taxonomic and biodiversity questions objectively (Doležalová et al. 2003b).

Within *L. sativa*, markers have also been used to study the diversity and relationships between different cultivated types. Studies using different marker types were generally in good agreement (Kesseli, Ochoa, and Michelmore, 1991; Waycott and Fort 1994; Hill et al. 1996; Witsenboer, Vogel, and Michelmore 1997). Slight variation among phenograms developed for *L. sativa* cultivars using RAPD, RFLP, AFLP, and SAMPL markers might be due to the specific markers chosen and the relatively small numbers of markers used. Molecular diversity measured using SAMPL and RFLP markers was greatest for butterhead cultivars, intermediate for other types, and least for crisphead cultivars (Witsenboer, Vogel, and Michelmore 1997). One possible use for markers within *L. sativa* would include hybrid identification in the context of a backcross breeding program. Most cultivars were differentiated using a low number of markers in each study; however, highly polymorphic markers will be required to differentiate more closely related genotypes. Relatively conserved RFLP markers were ineffective at identifying sister lines from a breeding program (Kesseli, Ochoa, and Michelmore 1991), but hypervariable TCT microsatellite fingerprints differentiated closely related cultivars as well as seedlots of a heterogeneous cultivar (Wiel van de, Arens, and Vosman 1998). Markers have also been proposed as a way to define the degree of relatedness that makes a variety "essentially derived" from another variety. Intellectual property protection conferred by a plant variety patent (PVP) extends to very close relatives of the protected variety, or those that are considered to be "essentially derived." Currently, an informal guideline based upon the breeding pedigree or number of backcrosses is used to determine the breadth of a PVP in lettuce.

Molecular data may have great use in germplasm management (Vosman 1997). Within *Lactuca*, markers have been used to show that assigned species names were incorrect for some accessions from public germplasm collections (Kesseli, Ochoa, and Michelmore 1991). Marker data can help curators characterize existing collections, as well as minimize duplication and prioritize accessions to be maintained. The primary limitation to this use is the need for a set of markers that are unbiased (provide genotypic information for the entire genome) and that are low cost and highly reproducible. Some additional data related to the topic of lettuce and wild *Lactuca* germplasm variability are given in Section 9.2.4.

9.5 GENETIC MAPPING AND GENOMICS

9.5.1 Recent Status

The status of genetic mapping in lettuce was reviewed by Michelmore, Kesseli, and Ryder (1994). The first molecular maps of lettuce were developed using RAPD and RFLP markers in intraspecific populations (Landry et al. 1987; Kesseli, Paran, and Michelmore 1994; Waycott et al. 1999). Molecular maps have now been constructed for intraspecific (*L. sativa* × *L. sativa*) F2 and inter-specific (*L. sativa* × *L. serriola* and *L. sativa* × *L. saligna*) F2 and recombinant-inbred line (RIL) populations in lettuce (Table 9.12). Using common markers, several of the maps have now been integrated and combined. An integrated interspecific map was constructed using AFLP and SSR markers for two *L. saligna* × *L. sativa* populations (Jeuken et al. 2001). Another interspecific map was created using AFLP markers from a *L. sativa* × *L. serriola* F_2 population (Johnson et al. 2000). The current consensus map was developed using F_6 RILs from the same *L. sativa* × *L. serriola* population, and now has over 2,600 markers on the nine linkage groups that correspond to the nine *Lactuca* chromosomes (Argyris et al. 2005; http://www.compositdb.ucdavis.edu; R. Michelmore; personal communication).

These studies have not provided any evidence for major chromosomal rearrangements within *L. sativa* or between *L. sativa* and *L. saligna* or *L. serriola*, and conserved marker order facilitated alignment of multiple maps (Johnson et al. 2000; Jeuken et al. 2001; Kesseli, Paran, and Michelmore 1994). Even so, fine QTL mapping in the *L. saligna* × *L. sativa* populations was limited due to sterility and segregation distortion in some genomic regions. Due to the great interest in accessing resistance genes from *L. saligna* without these problems, a backcross-inbred line (BIL) population with *L. saligna* introgressions in an *L. sativa* genetic background was created (Jeuken and Lindhout 2004). The inbreeding nature of *Lactuca* makes developing F_2 or RIL populations simpler than backcrossing or testcrossing, but AFLP marker positions from earlier studies (Jeuken et al. 2001) permitted the final backcross generation of the BIL development.

9.5.2 Qualitative Genes

Over 70 qualitative genes known to control disease resistances and morphological traits have been identified through classical genetic analysis of lettuce (summarized by Michelmore, Kesseli, and Ryder, 1994). A portion of these have been localized on one or more of the lettuce molecular maps (Kesseli, Paran, and Michelmore 1994; Waycott et al. 1999; Moreno-Vazquez et al. 2003; Grube et al. 2005a, 2005b). Most of the mapped genes in lettuce are disease resistance genes, the majority being *Dm* genes that control downy mildew (*Bremia lactucae*).

As in other plant species, lettuce disease resistance genes have been found to occur in clusters (Farrara, Ilott, and Michelmore, 1987; Kesseli, Paran, and Michelmore 1994; Witsenboer et al. 1995). There are at least four known clusters of resistance genes in lettuce, each of which contains at least one *Dm* gene (Table 9.13). The gene *Dm3* has been cloned and isolated from the lettuce genome, and was shown to be one member of a complex gene family (Meyers et al. 1998a, 1998b). *Dm* gene sequences from several accession of *L. sativa* and wild relatives have been used to develop models about the evolution of plant disease resistance genes (Michelmore and Meyers 1998; Sicard et al. 1999; Chin et al. 2001; Kuang et al. 2004).

Table 9.12   Populations Used for Molecular Mapping in Lettuce

Population	Type	No. Linkage Groups	No. Markers	Marker Types	Reference
L. sativa "Calmar" × *L. sativa* "Kordaat"	Intraspecific F_2	13	319	RAPD, RFLP	Landry et al. (1987) and Kesseli et al. (1994)
L. sativa "87-25-1M" × *L. sativa* "87-109M"	Intraspecific F_2	20	108	RAPD	Waycott et al. (1999)
L. sativa "dwarf-2" × *L. sativa* "Caffior"	Intraspecific F_2				
L. sativa "Olaf" × *L. saligna* "CGN5271"	Interspecific F_2	9	488	AFLP, SSR	Jeuken et al. (2001)
L. saligna "CGN11341" × *L. sativa* "Norden"	Interspecific F_2				
L. sativa "Olaf" × *L. saligna* "CGN5271"	Interspecific BC_4 BILs	9	757	AFLP, SSR	Jeukent and Lindhout (2004)
L. sativa "Salinas" × *L. serriola* "UC96US23"	Interspecific F_6 RILs	9	>2000	AFLP, RFLP, SSR, RAPD	http://www.compositdb.ucdavis.edu, R. Michelmore personal communcation, Argyris et al. (2005)
L. sativa "Salinas" × *L. serriola* "UC92G489"	Interspecific F_2	10	513	AFLP	Johnson et al. (2000)

9.5.3   Quantitative Traits

The existence of detailed molecular maps for lettuce has permitted mapping of minor genes that control quantitative traits; quantitative trait loci (QTL). In lettuce, interspecific RIL mapping populations involving *L. sativa* and either *L. serriola* or *L. saligna* have been used for most QTL mapping studies. Interspecific populations ensure sufficient genetic variation within the population, and are also well suited for studying agronomically important domestication-related traits. These traits include seed and seedling traits relating to germination and early plant growth (Argyris et al. 2005) and root architecture (Johnson et al. 2000). These techniques have also been applied to study disease resistance genes that confer partial or quantitative resistance, such as the downy mildew resistance QTL from *L. saligna* (Jeuken and Lindhout 2002) and *L. sativa* (Hand et al. 2003). Studies are also underway to map QTL for resistance to *Sclerotinia minor* (Grube unpublished data).

Table 9.13   Disease Resistance Gene Clusters in Lettuce

Cluster	Genes
1	*Dm5, Dm8, Dm10, Tu, plr*
2	*Dm1, Dm2, Dm3, Dm6, Dm14, Dm15, Dm16, Dm18, Ra*
3	*Dm13, cor*
4	*Dm4, Dm7, Dm11, mo1*

Note: *Dm*, downy mildew (*Bremia lactucae*); *Tu*, Turnip mosaic virus; *plr*, *Plasmopara lactucae-radicis*; *Ra*, root aphid (*Pemphigus bursarius*); *cor*, corky root (*Rhizomonas suberifaciens*); *mo1*, Lettuce mosaic virus-1.

9.5.4 Genomic Manipulation

Other technologies that facilitate genomic manipulation are also available in lettuce, including a bacterial artificial chromosome (BAC) library (Frijters et al. 1997) and a facile transformation system (see below). Genomics approaches have been taken to study the relationships between members of the Compositae (sunflower, lettuce) and the model plant species *Arabidopsis thaliana* (http://compgenomics.ucdavis.edu; Plocik, Layden, and Kesseli 2004). For lettuce, an extensive database of over 135,000 expressed sequence tag (EST) sequences has also been generated from two genotypes, *L. sativa* "Salinas" and *L. serriola* "UC96US23" (http://cgpdb.ucdavis.edu). By using bioinformatic approaches to compare EST sequences from lettuce with other organisms, ESTs that represent candidate genes for phenotypic traits of interest (e.g., disease resistance, stress responses, plant development) can be readily identified. Coupled with whole-plant physiological and expression studies, it is hoped that genetic variability could then more rapidly and more accurately be coupled with corresponding phenotypic effects.

9.6 GERMPLASM ENHANCEMENT

9.6.1 Introduction and Goals

Germplasm enhancement may be defined as the introduction of desirable traits or novel combinations thereof into cultivated germplasm. The genetic diversity that is required may be found within cultivated *Lactuca sativa* and *Lactuca* species with no fertility barriers (the primary gene pool), within wild *Lactuca* species with minor fertility barriers (the secondary gene pool), within wild species with marginal sexual compatibility (the tertiary gene pool), or within other organisms (the quaternary gene pool). For lettuce, nearly all germplasm enhancement has targeted traits from the primary, secondary and tertiary gene pools, although recent developments have made access to more distant gene pools a possibility.

For cultivated lettuce, desirable traits include disease and insect resistances, early vegetative development but slow bolting, novel leaf colors and shapes, nutritional value, uniform development and harvest, and broad environmental adaptation. Optimum horticultural characteristics differ greatly for different types of lettuce (e.g., crisphead, romaine, leaf), different end uses within a type (e.g., hearts versus whole heads for romaine), and different production systems (e.g., protected vs. open field).

Regional production of specialized types of lettuce provides significant genetic variation within *L. sativa*. For several traits, even more genetic variation is found within exotic germplasm such as landraces and wild relatives. The primary objective of commercial breeding programs is to rapidly develop marketable finished cultivars. As a result, most private sector programs have utilized traits that are already found within the primary gene pool, or even more narrowly, within the specific morphological type of interest (e.g., crisphead, romaine). In contrast, public sector breeding programs emphasize more fundamental and longer term research such as genetic studies. As a result, germplasm enhancement in public programs has primarily targeted traits from more distant gene pools and has produced unique "landmark" cultivars or advanced germplasm with novel traits. Currently, public sector researchers primarily release enhanced germplasm that is further fine-tuned by increasingly sophisticated private sector breeders.

Germplasm enhancement can be accomplished using conventional plant breeding approaches (hybridization and selection) within the primary gene pool, where sexual barriers do not exist. Modifications to these strategies, including the use of bridging species, embryo rescue, and somatic hybridization, can facilitate gene transfer when some sexual incompatibility exists. Lastly, efficient transformation protocols permit access to genetic variability from very distantly related or unrelated organisms. All of these techniques may be most efficiently used in combination with well-developed genetic maps, BAC libraries, and DNA sequence data obtained from genomics studies.

9.6.2 Gene Pools: Evaluation and Utilization

One may define the gene pool as the sum of the available genes and gene combinations that can be used to improve the crop species. The inheritance and action of those genes may be important in selecting an appropriate breeding procedure. In some cases the genetic study and research on the disease, insect, and other problems will have already been done; otherwise the breeder may have to conduct these studies either before starting or simultaneously with the breeding work. When those studies and the breeding work are completed, field testing procedures must be carried out to ascertain the relative commercial worth of developed materials, especially if the ultimate goal is cultivar release.

9.6.2.1 Horticultural and Agronomic Characteristics

Many of the economic traits dependent upon the gene pool will be common for all the major lettuce types, including various resistances, earliness, slow-bolting, and seed germination characteristics. Others will be specific for type: heading characteristics, leaf texture, leaf color, leaf shape, and anthocyanin distribution. Each lettuce type has a specific array of traits that distinguish it from the other types. Butterhead lettuce, for example, has soft oily leaves, forming relatively small round heads, a few desirable interior and exterior color combinations, and a characteristic flavor. Crisphead lettuce forms a large, firm to very firm head of crisp textured leaves, preferably dark green on the exterior. Cos lettuce is elongated, with the leaves arranged mostly parallel to each other and relatively coarse textured, and the desired color may be dark green or yellow green, depending on the region where it is grown. Cutting lettuces vary immensely in color, leaf shape, size, and texture, and should not form a head, but an enlarged rosette. Superimposed on these differences are the common needs of disease, insect, and environmental resistances.

9.6.2.2 Biochemical and Quality Traits

As shown in Table 9.1, nutrient constituent values are lower in lettuce than in some other vegetables and iceberg lettuce has the least in many categories as compared to the other major lettuce types. Until recently little attention has been given to nutritional value of vegetables, including lettuce. This situation has been changing. A USDA program is exploring the nutrient and antioxidant values of lettuce (Mou and Ryder 2004). High performance liquid chromatography (HPLC) will examine leaf extracts for β-carotene, ascorbic acid, lutein, Vitamin E and phenolic compounds, while calcium, iron and other mineral constituents will be evaluated with inductively coupled atomic emission spectrometry. The results of the screening and establishment of genetic variability will determine likely candidates for breeding improvement (Mou 2005).

There are special needs for lettuce grown under protective cover that can be met through plant breeding. One of these is the reduction of nitrate content of lettuce grown in glasshouses during periods of low temperatures and low light intensity. Excess nitrate in winter grown lettuce, when consumed, may be converted to nitrite and may lead to two health problems. In infants, it may lead to a condition called methaemoglobinaemia, or blue baby. In adults the compound may be converted to nitrosamines, which may be carcinogenic. Nitrate accumulation is genetically controlled and it is possible to breed for low nitrate in lettuce cultivars (Reinink and Groenwold 1987; Reinink 1991, 1992). Screening of *L. sativa* and wild *Lactuca* species germplasm showed significant variation in nitrate content. In butterhead lettuce nitrate content was negatively correlated with dry matter content and positively with plant fresh weight. Within each of the cultivated morphotypes of lettuce, accessions with low nitrate content were identified that offered good prospects for breeding for low nitrate content (Reinink, Groenwold, and Bootsma 1987).

9.6.2.3 Reaction to Biotic and Abiotic Stresses

The interactions between plants and pathogens vary enormously at different levels. Wild and crop plants interacting with microorganisms are characterized by two main types of expression, basic incompatibility (nonhost plant resistance or nonspecific basic resistance) and basic compatibility (host plant resistance or parasite-specific resistance; Lebeda 1984; Heath 1991). Basic or nonhost resistance is maintained by a mixture of passive and active defences that exhibit interspecific and probably intraspecific differences and is characteristic for most of the plant–parasite interactions in nature. There are at least two types of pathogen-specific host plant resistance: race-specific and race-nonspecific.

Variation in host plant resistance is mostly mirrored by the diversity of pathogens. Genetic variation in pathogen populations is generated by the processes of spontaneous mutation, sexual recombination, and somatic hybridization. Variation in population structure can also occur through migration or a range of cytological and molecular changes (Burdon 1993).

Breeding for resistance against one or more of several lettuce diseases and pests is a major component of most lettuce breeding projects (Ryder 1986, 1999a; Pink and Keane 1993; Reinink 1999; Ryder et al. 2003). These include an array of viral, bacterial, and fungal diseases. Breeding and screening procedures vary, depending upon the nature of the disease organism, knowledge of its etiology and epidemiology, newness of the program, and the genetic basis of the resistance. There are also some important insect pests which must be seriously considered in lettuce breeding programs. Also, some abiotic factors and nutrient deficiencies are important from the breeding viewpoint. In all cases the availability of *Lactuca* germplasms with resistance or suitable character(s) is crucial to the breeding process.

In recent decades, the basic information about diseases and pests of lettuce have been summarized in various books and compendia (e.g., Dixon 1981; Sherf and Macnab 1986; Smith 1988; Davis et al. 1997; Barkai-Golan 2001; Capinera 2001). Similar summaries on resistance breeding have also been published (Kalloo 1988; Pink and Keane 1993; Ryder 1986, 1999a; Lebeda and Křístková 1999; Hintum et al. 2003).

Viral Pathogens

Lettuce Mosaic Virus (LMV). LMV, a member of the *Potyvirus* genus, causes one of the most destructive of lettuce diseases worldwide: lettuce mosaic (Dinant and Lot 1992). First identified as a virus disease in the United States (Jagger 1921), lettuce mosaic has been found wherever lettuce is grown, and can cause losses up to 100% (Ryder 1999a, 1999b). Plant loss is defined by severe symptom expression, including mottling of the leaves, yellowing, stunting, and often necrosis. The virus is a flexuous rod with a single positive-sense RNA strand of 10,080 nucleotides, with viral protein Vpg covalently linked at its $5'$ end, and a poly-A tail at its $3'$ end (Revers et al. 1997). The virus is vectored in the field by the aphid *Myzus persicae* Sulz.(green peach, peach-potato) (Grogan, Welch, and Bardin 1952). It is also transmitted through the seed of susceptible cultivars at an average rate of 1–3% (Couch 1955).

A number of virus variants have been described, varying in distribution, virulence, and seed transmission characteristics. Several were described in the United States but not thoroughly characterized (McLean and Kinsey 1962, 1963; Zink, Duffus, and Kimble 1973). A series of more recent studies identified additional variants, including those that broke the resistance conferred by one or both resistance alleles of a gene for disease reaction (Pink, Walkey, and McClement 1992a, 1992b; Krause-Sakate et al. 2002). Of special concern is the LMV-Most strain, which breaks resistance to both alleles and is also seed-transmitting (Krause-Sakate et al. 2002). This strain and several others have been partially or completely characterized in molecular studies. The LMV-Common (LMV-0) isolates are, by far, most frequently found worldwide. The complete nucleotide sequences of the genomic RNAs of two isolates, one resistance breaking, the other not, of the LMV have been determined (Revers et al. 1997).

Lettuce mosaic is controlled in field plantings in one of two ways: seed indexing, to minimize the seed-borne transmission trait (Grogan, Welch, and Bardin 1952), and genetic resistance (Von der Pahlen and Crnko 1965; Ryder 1968, 1970). There is one principal gene for disease reaction, *Mo-1mo-1,* which has two recessive resistance alleles, *mo-1^g* (*mo1^1*) and *mo-1^e* (*mo1^2*; Ryder 1970, 2002; Pink, Walkey, and McClement 1992a). Another gene, *mo-2*, confers resistance in only one known cultivar, Ithaca, an iceberg type (Pink, Walkey, and McClement 1992a). A third gene, *MiMi^1*, confers a mild reaction in its partially dominant form, *Mi* (Ryder 2002). The resistance reaction is manifested as a chlorotic lesion, although the infection is systemic as with susceptibility. Combining *mo* with *Mi* raises the level of resistance to a nearly complete state (Ryder 2002). An additional dominant allele, *Mo-3*, has been identified in IVT 1398, an accession of *L. virosa* (Maisonneuve et al. 1999).

The alleles of *mo-1* have been cloned and sequenced (Nicaise et al. 2003a, 2003b), showing that forms of the translation initiation factor 4E control the two types of resistance as well as the susceptibility reaction in various lettuce cultivars.

Two principal germplasm sources of resistance are the basis for most resistance studies and breeding. One is Gallega, a Latin type lettuce cultivar originating in Spain and grown there, in northern Africa, and in South America. The other is PI 251245, one of three accessions from Egypt; these are primitive oil seed type lettuces, wild in appearance but probably belonging to the species *L. sativa*. Gallega is the source of *mo-1^g*, while PI 251245 is the source of *mo-1^e*. Most lettuce cultivars developed in European countries used the Gallega source; in the U.S., most cultivars came from the Egyptian source (Ryder 1999). Other resistance sources include cv Ithaca (*mo-2*); IVT 1398 accession of *L. virosa* (*Mo-3*); Balady Aswan Green, a stem lettuce from Egypt (*Mi*) (Ryder 2002); and PI 226514, a cos-like landrace from Iran (Hayes and Ryder 2004), which has two recessive alleles, *mo* and another not yet named.

Screening for LMV resistance is primarily a greenhouse procedure. The virus is easily manipulated, allowing inoculation with the primary vector, the green peach aphid (*Myzus persicae*) or mechanically with the aid of a buffer and an abrasive material. Symptoms appear quickly, 10–14 days for susceptible plants and 20–30 days for tolerant plants. Symptoms are easily discerned, enabling early elimination of unwanted susceptible plants. Plants may be inoculated and evaluated at an early seedling stage, permitting screening of large numbers of plants.

Breeding for LMV resistance has been in progress since the 1960s. Iceberg, cos, and butterhead cultivars have been released from public institutions and seed companies (Reinink 1999).

Mirafiori Lettuce Virus (Lettuce Big Vein). Lettuce big vein virus [LBVV; recently Mirafiori lettuce virus (MiLV)] in lettuce is indicated by two principal symptoms. vein clearing (loss of chlorophyll in the immediate area surrounding the vascular system) and stiffening and crinkling of leaves. In later stages of growth there may be delay of heading or failure to form heads. The disease was first recorded in 1934 in California (Jagger and Chandler 1934) and has since been observed in most lettuce production areas around the world (Ryder 1999a). It is virus induced, vectored by a soil-borne root-feeding chytrid fungus, *Olpidium brassicae* (Wor.) Dang. (Grogan et al. 1958). Losses from big vein, particularly with crisphead lettuce, are often dependent upon market conditions (Zink and Grogan 1954). When prices are low, fields with smaller heads are less likely to be harvested than those with normal size heads. The vein clearing symptom itself does not usually affect harvestability.

Lettuce big vein (LBV) has been recognized as a virus disease for many years, but only recently has a specific culprit been identified. An Ophiovirus, MiLV, was implicated in the occurrence of the disease (Roggero et al. 2000) and subsequently confirmed as the cause (Lot et al. 2002; Roggero et al. 2003). A second virus, LBVV, is often associated with big vein infected plants and has been previously implicated as the cause (Kuwata et al. 1983; Vetten, Lesemamm, and Dalchow, 1987), but is no longer considered to be so. The virus is not seedborne in lettuce.

LBV occurs during cool weather, in soils that retain water well (Westerlund, Campbell, and Grogan 1978a, 1978b). It may therefore be most severe in the early spring in places like the coastal valleys of California and in January and February in winter production areas of the Northern Hemisphere. In the Southern Hemisphere, big vein would occur in similar circumstances during the July–October period (Latham and Jones 2004). LBV may reach 100% symptom expression in hydroponic systems under controlled conditions. The virus is transmitted by free-swimming zoospores of the fungus. It will remain for long periods in the soil residing in resting spores.

All resistance sources but one have contributed a form of resistance that may be characterized as delay of symptom expression and measured as percentage of plants showing vein clearing at a date near plant maturity (Ryder and Robinson 1995). Because of substantial environmental effect, it is most useful to measure reaction in populations rather than in individual plants. Given these constraints, significant reduction in the effect of the virus infection can be obtained with this form of resistance. All of the sources of resistance presently used in breeding programs are cultivars and breeding lines developed from crosses between cultivars. The one exceptional resistance source is IVT 280, a *L. virosa* accession from The Netherlands. This accession and several others of the species *L. virosa* are LBV asymptotic when inoculated with MiLV (Bos and Huijberts 1990; Hayes, Ryder, and Robinson 2004). It is not clear whether this level of resistance is transferable to cultivated lettuce.

The genetic basis for resistance to LBV, from either the cultivated sources or from *L. virosa*, has not been elucidated. Nevertheless, breeding for resistance as described above has been successful, primarily with the development of resistant iceberg type cultivars in the United States.

Screening for LBV resistance is difficult. The vector, *Olpidium brassicae*, is a root feeding fungus. In the greenhouse, infectious material, a slurry of ground infected roots in water, is applied to the roots in the seedling stage. Symptoms may take several weeks to appear, so plants must be transplanted to individual pots. This limits the total number of plants that can be handled. Symptoms are manifested only in cool temperatures, so the screening can be done only in the late-fall/winter/early-spring period. There is a high probability of escape for individual plants; therefore, resistance can only be confirmed in populations derived from single plants. Once resistance is established in the greenhouse, field testing is useful in confirming the resistance under commercial conditions.

All current LBV resistant cultivars and breeding materials from them are assumed to contain the virus or viruses responsible for LBV. Symptom expression is manifested in a varying fraction of inoculated plants when the temperature is relatively low and soil moisture is relatively high. This may be due to activation of the fungal vector producing the virus in the plant, to movement of the virus from the roots into the leaves, or from reaction of the virus and the plant to produce symptoms. Therefore resistance is measured as a proportion of plants showing symptoms at a given point in plant development, usually at harvest maturity. For a given population, this proportion will vary with season and soil type. Establishing a population group as resistant therefore requires repeated testing under varying conditions and only during periods of the year when LBV is expressed at all.

Beet Western Yellows Virus (BWYV). Several yellowing diseases of lettuce have been identified and described, caused by different viruses, vectored by different insect species, and occurred in different regions. Of the range of symptoms expressed, the common one is leaf yellowing, but with some variation in degree of expression and type of ancillary symptoms.

BWYV is the most widespread and most economically damaging of the yellowing viruses. First discovered in the United States and assigned the name radish yellows (Duffus 1960), it was subsequently shown to be mildly damaging to iceberg lettuce, which at that time was by far the dominant type grown in the United States. It is more severe on butterhead lettuce, and is considered to be the most economically damaging virus in the U.K. (Walkey and Pink 1990). It is severe in continental Europe as well, especially in Mediterranean countries (Maisonneuve, Chovelon, and Lot 1991). The most prominent symptom of the disease is interveinal yellowing, which appears first on older lower leaves. When restricted to these leaves, it is a relatively minor problem. The symptoms may progress to younger leaves, and their expression may become severe with this progression. With age, affected leaves become thick and brittle and the plant becomes stunted. The disease is caused by a luteovirus. The virus is transmitted in a persistent manner primarily by *Myzus persicae* (Sulz.) and remains confined to the phloem in the plant (Duffus 1973). Strains of the virus have been identified, which are differentiated by their effect on different species but not on cultivars within species (Govier 1985). The virus is not seedborne in lettuce.

Several sources of resistance to BWYV have been identified. Among cultivars, a few show substantially lower titer or symptom severity scores. Two of these, Bursc 17 and Crystal Heart, were used in inheritance and breeding studies (Walkey and Pink 1990; Pink, Lot, and Johnson 1991). The resistance is inherited as a single recessive (*bwy*). Walkey and Pink (1990) also identified high level resistance in accessions of *L. perennis* and *L. muralis*, which do not cross with lettuce. The *L. virosa* accession IVT 280, resistant also to LBV (see above), is resistant to BWYV (Maisonneuve, Chovelon, and Lot 1991). This resistance is inherited as a single dominant (*Bw*). Both forms of resistance have been used in breeding programs, in the U.K. and France, respectively, to transfer resistance into modern cultivars.

Other Yellowing Virus Diseases. Other yellowing diseases are of less importance. They may be restricted geographically, or to particular venues, or may occur sporadically. Beet pseudo-yellows virus (BPYV) is transmitted by the greenhouse whitefly (*Trialeurodes vaporariorum* Westwood) in a semi-persistent manner (Duffus 1965). The virus causes interveinal yellowing and stunting. It is not a field problem on lettuce nor is it a serious problem in greenhouse production.

Three other virus diseases with limited impact on lettuce production are: Sowthistle yellow vein virus (SYVV), vectored by the sowthistle aphid (*Hyperomyzus lactucae* L.; Duffus, Zink, and Bardin 1970); Beet yellow stunt virus (BYSV), transmitted by *Nasonovia lactucae* L., *Myzus persicae*, or *Macrosiphum euphorbiae* in a semi-persistent manner (Duffus 1972); and *Lettuce necrotic yellows virus* (LNYV), found in Australia and New Zealand, and also transmitted by *H. lactucae* (Stubbs and Grogan 1963).

Two additional whitefly-transmitted yellowing viruses severely affected iceberg type lettuces in the winter desert production areas of California and Arizona in the decade from 1981 to 1991. For the first few years, the disease, lettuce infectious yellows (caused by LIYV), was caused by a closterovirus, and transmitted by *Bemisia tabaci* Gennadius (sweetpotato whitefly; Duffus, Larsen, and Liu 1986). This virus occurred in very large numbers in summer and early fall and caused severe interveinal yellowing, stunting, and necrosis, resulting in losses as high as 100%. The insect and therefore the disease were effectively controlled by applications of imidacloprid. In 1990, a new whitefly species began to replace *B. tabaci*. It was a poor transmitter of the LIYV and that disease virtually disappeared. However, the new whitefly transmitted another closterovirus, which caused similar but less severe symptoms as LIYV. It was named lettuce chlorosis virus (Duffus et al. 1996). Imidacloprid continued to be effective and neither LIYV or lettuce chlorosis virus has occurred at detectable levels.

During the period that the diseases were a factor in production, resistant sources were identified. One was the iceberg cultivar Climax (McCreight, Kishaba, and Mayberry 1986); several accessions of *L. saligna* were also resistant (McCreight 1987). The genetic basis for resistance is not known. Breeding lines, but no cultivars, have been developed from both types of material.

Tomato Spotted Wilt Virus (TSWV). TSWV is a virus disease with an extremely wide host range that includes lettuce (Davis et al. 1997). It was first described on tomato in 1919. It is caused by a tospovirus and transmitted by several species of thrips. In Hawaii, where the disease is particularly damaging, the vector is *Frankliniella occidentalis* Pergande. Symptom expression includes yellowing, necrosis, wilting, and browning of the youngest leaves. The plant is affected primarily on one side, so that the leaves on the less affected side have curved midribs. The disease also occurs in Australia, California, South Africa, and Chile.

The disease is difficult to control by chemical or biological methods, although an integrated system has been worked out in Hawaii, which has had some success (Cho et al. 1989). Two sources of resistance have been identified. Butterhead cultivars Tinto and Ancora have shown a moderate level of resistance, and have been used in a breeding program in Hawaii (O'Malley and Hartmann 1989). Progenies from a cross between cultivated lettuce and *L. saligna* have been segregated for resistance to TSWV (Wang et al. 1992). The genetic basis for disease reaction is not known, although it is suggested that resistance is partially dominant (O'Malley and Hartmann 1989).

Cucumber Mosaic Virus (CMV). CMV also has a wide host range (Davis et al. 1997). On lettuce it resembles LMV in symptom expression, causing mottling, stunting, necrosis. The causal agent is a cucumovirus. It can be transmitted by over 60 aphid species, most important of which are *M. persicae* and *Aphis gossypii* Glover and is transmitted in a nonpersistent manner. It is a minor disease problem on lettuce in most areas, but is considered the most important lettuce virus problem in New York State (Provvidenti, Robinson, and Shail 1980).

Resistance has been identified in PI 261653, an accession of *Lactuca saligna*. It was transferred by pedigree selection to *L. sativa*, leading to the development of a crisphead cultivar, "Salad Crisp," suitable for late fall production in New York (Provvidenti, Robinson, and Shail 1980). The resistance was useful against one of two strains of CMV known at that time.

Tomato Bushy Stunt Virus (TBSV) and Lettuce Necrotic Stunt Virus (LNSV). TBSV and LNSV are closely related soilborne tombusviruses (Obermeier et al. 2001). This group of viruses causes the disease lettuce dieback, whose symptoms include stunting, yellowing, and chlorotic and necrotic spotting. A vector is not required for infection to occur, but may facilitate natural infections in the field. Screening for breeding and genetic studies is currently most reliably done in infested fields, but there has been some success in developing greenhouse evaluation protocols (Obermeier et al. 2001). Lettuce dieback can cause 100% losses in infested fields, but the disease is only regionally significant. Thus far, infested fields have been identified only within coastal California growing regions.

Not all lettuce types and cultivars are susceptible. Nearly all romaine and red leaf lettuce cultivars are susceptible, but all modern iceberg and many green leaf varieties do not show symptoms when planted in infested fields. Resistance has been identified in some romaine breeding lines and in wild and heirloom varieties of all types (Grube et al. 2005a, 2005b). It is not known whether resistant varieties are immune or merely fail to show symptoms.

Bacterial Pathogens

Corky Root. For many years the cause of corky root was debated; it was attributed by various proponents to *Pythium* species, decaying plant residues, or highly ammoniated nitrogen fertilizers (Grogan and Zink 1976; Hoff and Newhall 1960; Amin and Sequiera 1966). Finally it was shown to be caused by a Gram-negative soilborne bacterium, *Rhizomonas suberifaciens* gen. nova, sp. nova (Van Bruggen, Jochimsen, and Brown 1990). The bacterium has more recently been named *Sphingomonas suberifaciens* van Bruggen et al. (Yabuuchi et al. 1999). It is widespread, having been reported in many locations in North America, Europe, Australia, and Asia.

Corky root attacks the roots. Early symptoms are yellow-brown spots on the taproot, followed by severe greenish-brown discoloration and corky longitudinal ridges. The root may become constricted and break off near the crown. Many secondary and feeder roots may be destroyed.

Normal roots may grow up to 60 cm; corked roots may be 25 cm or less. Above ground symptoms, include yellowing, wilting, and stunting, are often sufficient to reduce or prevent harvest because of undersized or nonharvestable heads (Van Bruggen et al. 1988).

Resistance was first identified in several accessions of cos-like landraces, PI 171669, 174229, and 175739 (Dickson 1963). One line, PI 171669, was used in subsequent crosses to produce the first resistant cultivars (Sequeira 1970). A single gene governs plant reaction to the bacterium, with the recessive allele (*cor*) conferring resistance (Brown and Michelmore 1988). Four additional PI accessions of *L. serriola* (PI 491096, 491110, and 491239) and *L. virosa* (PI 273597c) showed high resistance to corky root in a series of tests (Mou and Bull 2004). None of the lines was linked to two molecular markers associated with *cor*, and they may represent one or more new genetic sources of resistance.

Bacterial Leaf Spot (BLS). BLS is incited by *Xanthomonas campestris* pv. *vitians* (Brown) Dye, a motile, aerobic Gram-negative rod (Burkholder 1954; Koike and Gilbertson 1997). It was first reported in 1918, and occurs sporadically in North America, Europe, and Asia.

The initial symptoms are water soaked lesions on leaf margins that later turn dark brown or black and often coalesce. On highly susceptible plants, the symptoms are likely to be severe enough to prevent harvest. BLS is not usually a serious problem. It may cause significant economic loss during cool wet seasons.

Relative susceptibility varied among cultivars tested by Carisse et al. (2000). In another study, Sahin and Miller (1997) found variation ranging from highly resistant to highly susceptible among a different set of cultivars. Goldman et al. (2003) identified nine heirloom cultivars as resistant. The genetics of resistance is unknown, as is the existence of resistance among noncultivated materials.

Fungal Pathogens

Downy Mildew. Lebeda, Pink, and Astley (2002) published a comprehensive review on lettuce downy mildew (*Bremia lactucae*) biology and ecology, sources of resistance, mechanisms, and genetic control of resistance in *Lactuca* species; including germplasm evaluations. In this part we summarize the most important information about the pathogen and host resistance.

The distribution of *B. lactucae* is worldwide, occurring on all continents except Antarctica (Marlatt 1974). *B. lactucae* occurs on cultivated lettuce wherever the crop is grown, especially in regions with a temperate climate (Crute and Dixon 1981). However, there is incomplete information about the distribution of the fungus on wild *Lactuca* species. Our knowledge of the geographic distribution of the fungus is based mainly on its occurrence on lettuce (*L. sativa*), chicory (*Cichorium intybus*), endive (*C. endivia*) and *Sonchus* spp. The economic impact of lettuce downy mildew can be very high (Crute 1992a). *B. lactucae* is known to infect more than two hundred species of Asteraceae (Compositae) from about 50 genera of the tribes Lactuceae, Cynareae, and Arctotideae (Crute and Dixon 1981; Lebeda, Pink, and Astley 2002). It is evident that *B. lactucae* is highly specific and from the viewpoint of parasitism mostly limited to the same genus of plants (Lebeda and Syrovátko 1988).

Of the 100 wild *Lactuca* species described (Lebeda et al. 2004b) only 14 are definitely known as a natural hosts of *B. lactucae* (Lebeda, Pink, and Astley 2002). *L. serriola*, as the most common wild *Lactuca* spp. occurring around the world (Lebeda et al. 2004b), could be considered as an important weedy host. However, except for the Czech Republic (Petrželová and Lebeda 2004a) there is no detailed information on the natural occurrence of lettuce downy mildew and its epidemiological impact on this species (Lebeda, Pink, and Astley 2002; Lebeda and Petrželová 2004).

In host plant–downy mildew interactions, including *Lactuca* spp.—*B. lactucae*, there is generally a very clear expression of compatibility or incompatibility. The characterization of virulence and classification of races is based on the pattern of compatible and incompatible reactions on differential host genotypes (Lebeda and Schwinn 1994).Variability in virulence of *B. lactucae* was first described in the United States. Similar situations were found subsequently in most countries where lettuce is widely grown (Crute 1987). Virulence variation was first categorized in terms of physiological races (Crute and Dixon 1981). Later it was described by specific virulence determinants (virulence phenotypes, virulence factors) based on the interpretation of the host–pathogen interaction in a gene-for-gene relationship (Crute 1987).

There is rather limited knowledge of virulence variation of *B. lactucae* in wild pathosystems (Lebeda 2002). Only isolates originating from natural populations of *L. serriola* have been investigated for specific virulence variation (Lebeda 2002; Lebeda, Pink, and Astley 2002; Lebeda and Petrželová 2004; Petrželová and Lebeda 2004b). In comparison with isolates obtained from the crop pathosystem (*L. sativa*), the isolates derived from the wild plant pathosystem (*L. serriola*) were characterized by rather simple virulence phenotypes. However, many different races were recognized (Lebeda and Petrželová 2004). Generally, *B. lactucae* isolates from the wild pathosystem are characterized in terms of v-factors mostly matching *Dm* genes or R-factors located or derived from *L. serriola* (Lebeda and Petrželová 2004).

The interaction between cultivars of *L. sativa* and *B. lactucae* is clearly race-specific (Crute and Johnson 1976; Lebeda 1984b; Farrara and Michelmore 1987). The first observations and experimental studies related to the specificity of interactions between wild *Lactuca* species and *B. lactucae* occurred in the early 1920s; more intensive research in this area started in the 1940s and is ongoing (Lebeda, Pink, and Astley 2002). A total of 15 wild *Lactuca* species and eight species of five related genera have been screened for resistance to *B. lactucae*. In most of these studies, *B. lactucae* isolates originating from *L. sativa* were used. However, there is also another gap in the information in these interactions, because most of the studies concerned interactions between *B. lactucae* and accessions of *L. serriola*, *L. saligna* and *L. virosa*, with only a few publications describing other interactions (Lebeda, Pink, and Astley 2002). Recently, searching for new sources of resistance and genes suitable for practical lettuce breeding (Lebeda and Zinkernagel 2003a; Beharav et al. 2006) is considered as very important.

Utilization of wild *Lactuca* germplasm in lettuce breeding was summarized by Lebeda, Pink, and Astley (2002). In commercial lettuce breeding programs only a limited number of race-specific resistance genes have been utilized and these have differed among countries (Crute 1987, 1992a, 1992c). However, the utilization of wild *Lactuca* species, as sources of resistance in practical breeding, started relatively early (Lebeda, Pink, and Astley 2002). Accessions of *L. serriola* [PI 91532 (correct number is PI 104584; Welch, Zink, and Grogan 1965) and PI 167150] originating from Russia and Turkey were used in the 1930s in the United States as sources of resistance against *B. lactucae*. These sources created the background for a new generation of lettuce cultivars (Imperial 410, Calmar, Valmaine) for outdoor cropping, which were introduced in the 1940s and 1950s (Whitaker et al. 1958). All of these cultivars have race-specific resistance (Table 9.14). In Europe, the utilization of wild *Lactuca* germplasm was based on two diverse strategies (the Netherlands and Great Britain). In the 1950s, genes originating from old German and French cultivars of *L. sativa* were mostly used (Crute 1992a, 1992c). At the end of the 1960 in the Netherlands an interspecific hybrid between *L. sativa* (cv. Hilde) and an accession of *L. serriola*, which is described in the literature as H×B, Hilde×*L. serriola* was released. Resistance derived from this material is assigned to the race-specific gene *Dm*11 (Crute 1992c; Table 9.14). In the 1970s and 1980s other sources of resistance to *B. lactucae*, derived from *L. serriola* with resistance genes (factors) described as *Dm*16 and R18 (Table 9.14), were used in the Netherlands. All of these genes have been used frequently in breeding programs in Europe during the last twenty years. However, resistance based on these genes is no longer effective against some *B. lactucae* isolates (Lebeda and Zinkernagel 1999, 2003b). From the end of the 1980s, there was increasing interest (esp. in the Netherlands and U.K.) for the utilization of resistance located in the hybrid line *L. serriola* (Swedish)×*L. sativa* (Brunhilde) and line CS–RL (Lebeda and Blok 1991) derived

Table 9.14 Examples of Race-Specific Resistance Genes (*Dm*) or Factors (R) Located or Derived from *Lactuca serriola*

Dm Gene (R-factor)	*L. serriola* Accession (line)	Origin	Occurrence in *L. sativa* Cultivars[a]	Linkage Group
*Dm*5	PI 167150	Turkey	Valmaine	2
*Dm*5/8+10	PI 104584	USSR	Sucrine	2
	PI 167150	Turkey		
*Dm*8	PI 104584	USSR	Avoncrisp Calmar Salinas	2
*Dm*6	PI 104584	USSR	Sabine	1
*Dm*7	LSE/57/15	U.K.	Great Lakes Mesa 659	3
*Dm*11	IVT Wageningen	?	Capitan	3
*Dm*15	PIVT 1309	Netherlands	—	1
*Dm*16	LSE/18	Czechoslovakia	Saffier Titania	1
*Dm*7+10+13	PI 114512	Sweden	Vanguard	
	PI 114535	U.K.		
	PI 125819	Afghanistan		
	(+*L.virosa* PI 125130)	Sweden		
R17	LS 102	France?	—	2
R18	LS 17	France?	Mariska	1
R19 (R18+?)	CS-RL LJ88356	Sweden	Libusa Miura	1
*Dm*7+R23	CGN 5153	USSR (Krym)	—	3,5?
R24+R25	CGN14255	Hungary	—	3,5
R24+R26	CGN14256	Hungary	—	3,4
R24+R27	CGN14270	Hungary	—	3,4?
R24+R28	CGN14280	Hungary	—	3,?
R24+R29	PI 491178	Turkey	—	3,?
R30	PI 491229	Greece	—	1
R? (+modifiers, probably RNS)	PI 281876	Iraq	—	?

Note: ?, Not known or unclear;—, *Dm* gene or R-factor not yet located in *L. sativa* cultivar(s); RNS, Race-nonspecific resistance.

[a] Only selected examples.

Source: From Lebeda, A., Pink, D. A. C., and Astley, D., *Advances in Downy Mildew Research*, Kulwer Academic Publishers, Dordrecht, 2002b, 85–117.

from this material. This line was highly resistant for a long time. However, recently a new race overcoming the resistance was described (Lebeda and Zinkernagel 2003a).

During the last two decades, researchers and breeders have been increasingly interested in the exploitation of new sources of resistance among other wild *Lactuca* species. The most important potential sources of resistance are considered to be *L. saligna* and *L. virosa*. Several studies showed that these wild species may possess a new and very interesting resistance to *B. lactucae* (Lebeda, Pink, and Astley 2002). As a result of studies in the 1990s a new lettuce cultivar Titan (Sluis & Groot) with the race-specific gene *Dm*6 plus resistance derived from *L. saligna* (pers. comm., K. Reinink, Rijk Zwaan, the Netherlands) was released in the Netherlands (Lebeda, Pink, and Astley 2002). However, this resistance is no longer effective (Lebeda and Zinkernagel 2003a, 2003b). Recently, a very intensive program of lettuce breeding based on introduction of newly located sources and genes of resistance from *L. serriola, L. saligna,* and *L. virosa* was developed in the United States (Michelmore 2002; Michelmore and Ochoa 2005). Also new sources of resistance were located in wild *Lactuca* spp. originating mostly from Middle East (Beharav et al. 2006).

The effectiveness of the expression of some *Dm* genes located in *L. serriola* can be dependent on environmental factors. Judelson and Michelmore (1992) showed that resistance (assessed as the absence of sporulation) based on *Dm*6, *Dm*7, *Dm*11, *Dm*15, and *Dm*16 became less effective or ineffective at temperatures below 10°C. The ecological and epidemiological consequences of this effect are not known (Lebeda, Pink, and Astley 2002).

The occurrence of race specificity in other wild *Lactuca* species and related genera has not been analysed in detail. However, recent analyses (Lebeda and Petrželová 2001; Lebeda, Pink, and Astley 2002) have shown that the occurrence of race-specific resistance in wild *Lactuca* species is a common phenomenon. In the section *Lactuca*, all of the species studied express race specificity after inoculation with isolates of *B. lactucae* from *L. sativa* and *L. serriola*. The presence of race-specific resistance in *L. saligna* was described as questionable because most of the screened accessions expressed complete or incomplete resistance at both the seedling and adult stage (Lebeda, Pink, and Astley 2002). However, recent results (Beharav et al. 2006) have shown that race specificity in *L. saligna* could be more common. A race-specific response has also been confirmed in some species from other sections of the genus *Lactuca* (*L. viminea*, *L. tatarica*, *L. quercina*, *L. indica*, *L. biennis*; Lebeda and Petrželová 2001). Moreover, there is clear evidence of the occurrence of a race-specific response in some species of related genera (e.g., *Cicerbita*, *Mycelis*; Lebeda, Pink, and Astley 2002).

Other types of resistance (race-nonspecific, field, nonhost; for detailed characterization see Lebeda, Pink, and Astley 2002) of *Lactuca* species against *B. lactucae* are not very well known. There is only limited information available about race-nonspecific resistance in wild *Lactuca* spp. germplasm. The presence of race-nonspecific resistance has only been reported in *L. serriola* (Lebeda, Pink, and Astley 2002). Currently only two *L. serriola* accessions can be considered as potential sources of this type of resistance. Lebeda (1986), Lebeda and Jendrůlek (1989), and Norwood, Crute, and Lebeda (1981) recognized that accessions PI 281876 and PI 281877 in the seedling stage were infected by some *B. lactucae* isolates. However, the intensity of sporulation was mostly very low, and in some interactions was followed by an expression of a necrotic response. Current thoughts are that this resistance is based on some major gene(s) and modifiers (Lebeda, Pink, and Astley 2002). *L. serriola* (PI 281876) has been used frequently in practical breeding programs (Lebeda and Pink 1998).

Research into sources of field resistance to *B. lactucae* has been focused on *L. sativa* (Lebeda, Pink, and Astley 2002). Recent results showed that some *L. sativa* germplasm (e.g., Grand Rapids) are underutilized from the viewpoint of breeding for field resistance and controlling downy mildew (Grube and Ochoa 2005). Wild *Lactuca* spp. germplasms have not been studied extensively with respect to their field resistance (Lebeda, Pink, and Astley 2002). Crute and Norwood (1981) were the first to point out that some wild *Lactuca* spp. could be potential sources of field resistance. Till now, the most comprehensive experiments were carried out by Lebeda (1990). In total, 31 accessions of four *Lactuca* species [*L. serriola*, *L. saligna*, *L. aculeata*, *L. indica* (syn. *L. squarrosa*)], and one *L. serriola*×*L. sativa* hybrid (line CS–RL) were studied in three years of field experiments. The disease incidence was significantly different across species and accessions. *L. saligna*, *L. aculeata* accessions and the *L. serriola*×*L. sativa* hybrid were free of infection during the observation period. This reaction implies the presence of effective unknown R-factors (Lebeda, Pink, and Astley 2002) in these genotypes. In the *L. serriola* accessions, significant differences in the level of field resistance were observed (Lebeda 1990). Some accessions were highly susceptible (e.g., PI 204753, PI 253468, PI 273596, PI 273617, PI 274359), in contrast, accessions PI 281876 and PI 253467 were free of disease symptoms (again implying the presence of effective unknown R-factors). However, the possible race-nonspecific resistance in PI 281876 is also likely to be expressed as field resistance (Lebeda 1990). Gustafsson (1989) also recognized in field experiments that *L. saligna* accessions were free of disease and a few *L. serriola* accessions displayed a high level of resistance.

Nonhost resistance is usually very effective, durable, and not influenced by changes of environmental conditions (Ride 1985). The only evidence for this in *Lactuca* is that some *L. saligna* accessions may possibly possess nonhost resistance (Lebeda and Reinink 1994). *L. saligna* is a common species in southern Europe and the Mediterranean area (Feráková 1977), but this species has never been recorded as a natural host of *B. lactucae*. Recent experimental results with new highly virulent isolates of *B. lactucae* originating from *L. sativa* have not confirmed the presence of race-specific resistance in *L. saligna* (Lebeda and Zinkernagel 2003a). However, recent findings indicate that, at least in some *L. saligna* accessions, an unknown race-specific resistance is located, apart from the expected nonhost resistance to *B. lactucae* (Beharav et al. 2006).

There is only limited information available on the histological, cytological, biochemical, and molecular background of resistance to lettuce downy mildew in *L. sativa* and wild *Lactuca* species. Some basic ideas and conclusions related to this subject were summarized by Mansfield, Bennett, and Bestwick (1997), Lebeda, Pink, Mieslerová (2001c, 2002a, 2002b), Jeuken and Lindhout (2002, 2004), and Lebeda and Sedlářová (2003). Data obtained in histological studies of resistance in wild *Lactuca* species suggest action of a wide range of resistance mechanisms in *Lactuca* species against *B. lactucae*.

Powdery Mildew. Powdery mildews are virulent pathogens. They affect and destroy primarily the leaves of their hosts, thus limiting photosynthesis and slowing plant growth. In extreme cases they may kill the host. In natural populations their destructive effect is usually not as severe; however, in commercial crops, they can cause severe problems (Bélanger et al. 2002). Lettuce powdery mildew is considered a disease of increasing importance (Turini et al. 2001). The economic importance of powdery mildew on commercial lettuce can be high, especially in some regions (e.g., in the United States) (Paulus 1997), and also in relation to seed production (Lebeda and Mieslerová 2003). However, our knowledge of its biology, taxonomy, host–pathogen interaction and epidemiology is rather poor (Lebeda 1999).

In the family Asteraceae, the best documented powdery mildew is *Erysiphe cichoracearum* DC. s.str. (Dixon 1981; Braun 1995), now assigned to *Golovinomyces cichoracearum* V.P. Gelyuta (Braun 1999). This species has been considered as the only species regularly described on *Lactuca* species (Lebeda 1985a, 1999; Braun 1995). Lebeda (1985a) clearly demonstrated that there is no significant variation in anamorph morphology of this species on *L. sativa* and wild *Lactuca* species. Occurrence of the sexual stage (cleistothecia) of *G. cichoracearum* on wild *Lactuca* species (*L. aculeata, L. perennis, L. saligna, L. serriola,* and *L. tatarica*) was well documented in the Czech Republic (Lebeda and Buczkowski 1986; Lebeda 1994). Recently, however, another powdery mildew species was collected and described on *Lactuca viminea* subsp. *chondrilliflora* found in southern France (Provence) (Lebeda, Pink, and Astley 2002). The comparison of this powdery mildew with *G. cichoracearum* showed substantial differences in morphology of the anamorph stage. Unfortunately, the teleomorph stage (cleistothecia) was not found on *L. viminea* subsp. *chondrilliflora*. Although the host range of *G. cichoracearum* included *L. viminea*, it can be excluded that both powdery mildew samples belong to the same species. This hypothesis, is supported by the new taxonomy of powdery mildews, where the anamorphs (mainly conidia arrangement) play an increasing important role (Braun and Takamatsu 2000), and have recently been confirmed by SEM examinations (Cook, Inman, and Billings 1997) and molecular data (Takamatsu, Hirata, and Sato 1998; Saenz and Taylor 1999). There is a strong probability that it represents a new species (assigned to *Erysiphe* emend.). However, the importance of this powdery mildew species for commercially grown lettuce is unknown (Lebeda, Pink, and Astley 2002).

There is rather limited information on geographic distribution of powdery mildew on *Lactuca* species (Lebeda 1999). Most reports deal with powdery mildew on *L. sativa* (Schnathorst et al. 1958; Paulus 1997) and in all cases the disease is attributed to *G. cichoracearum* (Lebeda 1985a). However, only limited knowledge is available on distribution of *G. cichoracearum* on naturally growing *Lactuca* species. Natural hosts of powdery mildew include *L. muralis, L. perennis, L. quercina, L. serriola, L. saligna, L. sibirica, L. viminea,* and *L. virosa* (Lebeda 1985a, 1999; Braun 1995). One of the most common species in Europe is *L. serriola* (prickly lettuce) which could be also considered as a common host of *G. cichoracearum*. In 1998 in the Czech Republic, research was started on the occurrence of powdery mildew in natural populations of *L. serriola*. A total of 32 natural populations of *L. serriola* were surveyed on 19 sites. On 16 sites (84.2%) and 21 populations (65.6%) the occurrence of natural infection was observed. Substantial variation in expression of the degree of infection between different sites or populations was recognized. It was concluded that *L. serriola* could be a reservoir of inoculum for lettuce infection (Lebeda 1999).

There is limited information about host resistance and genetics of resistance. The preliminary search for resistance of *L. sativa* germplasm showed that none of the crisphead and nonheading leafy types were resistant to *G. cichoracearum* (Dixon 1981). Lebeda (1985c) revealed substantial differences in disease severity in a set of 25 lettuce cultivars. Only two cultivars (Amanda Plus and Bremex) was free of the infection. There was a sporadic occurrence of powdery mildew pustules on cvs. Blondine and Bourguignonne. Heavy infection was recorded in cvs. Avoncrisp, Great Lakes, Larganda, Hilde, Type 57, Line 4/57/D, and Mesa 659 (crispheads and butterheads); nearly the whole or the whole head was infected with powdery mildew (Lebeda 1985c).

The screening of more than one hundred accessions of wild representatives of the genus *Lactuca* (*L. aculeata, L. dentata, L. perennis, L. saligna, L. serriola, L. tatarica, L. tenerrima, L. viminea,* and *L. virosa*) under conditions of natural infection by *G. cichoracearum* revealed high variability in resistance (Lebeda 1985b, 1994). The accessions of *L. serriola* were attacked most severely and *L. saligna* showed highly variable levels of resistance, while the lowest levels of infection were found in *L. virosa, L. viminea, L. tenerrima,* and *L. tatarica*. In some species (e.g., *L. saligna and L. serriola*) the interaction with the pathogen is probably based on race-specific resistance (Lebeda 1999; Lebeda and Mieslerová 2003; Table 9.15).

Reported results give important information on lettuce powdery mildew (*G. cichoracearum*), but open new questions for further investigation. From a breeding viewpoint, a search for resistance sources is essential. As nearly all lettuce cvs. and *L. serriola* accessions are susceptible, only limited possibilities appear with their use as resistant sources to powdery mildew. Some *L. saligna* accessions could be considered as a suitable sources, especially where they carry resistance to *Bremia lactucae* as well (Lebeda 1985b). *L. virosa* could be also considered as a suitable species; however, its resistance seems to be related to ontogenetic development (Lebeda 1985a). Thus, further investigations should be carried out to determine lettuce resistance to powdery mildew at various stages of ontogeny. There is little detailed information on fungus variation at the level of physiological races, however, Lebeda (1994) suggests their potential occurrence.

Anthracnose. Anthracnose is also known as shothole disease and as ringspot. It is caused by a fungus *Microdochium panattoniana* (Berl.) Sutton et al. It occurs in cool wet conditions and is found in various parts of the world, at low incidence, although it can be severe occasionally (Ochoa, Delp, and Michelmore 1987). Typical symptoms include small circular brown spots primarily on lower leaf blades. The centers dry and fall out, which accounts for the common name. Lesions on the midrib become necrotic, sunken, and elongated. The lesions are caused by splashes of rain or irrigation water carrying infective conidia. The initial infection may be soilborne or seedborne.

Table 9.15   Differential Reaction to Lettuce Powdery Mildew (*G. cichoracearum*) in Wild *Lactuca* Species

| *Lactuca* spp. | Accessions | |
	Resistant	Susceptible
L. saligna	CGN 05282, CGN 05306, CGN 05311, CGN 05330	CGN 05265, CGN 05267, CGN 05305, CGN 05796
L. serriola	PI 255665	CGN 05010, CGN 05019, CGN 05073, CGN 05153
L. virosa	CGN 04680, CGN 04954, CGN 05020, CGN 05869	PI 271938[a]

[a] Limited occurrence of powdery mildew mycelia was recorded.
Source: From Lebeda, A., *Acta Phytopath. Acad. Sci. Hung.,* 20, 149–162, 1985; Lebeda, A., *Euphytica,* 34, 521–523, 1985; Lebeda, A., *Genet. Resour. Crop Evol.,* 41, 55–57, 1994.

Sources of resistance have been identified in *L. sativa, L. serriola, L. saligna*, and *L. virosa* (Ochoa, Delp, and Michelmore 1987).

Stemphylium Leaf Spot. This fungus disease has been reported in many parts of the world (Raid 1997). It has relatively minor economic impact. Leaf spots are small, round, and brown, which may appear sunken because the tissue becomes necrotic. The disease is caused by *Stemphylium botryosum* f. *lactucum* Wallr. The only known source of resistance to the disease is an unidentified line of *L. saligna* found in Israel (Netzer et al. 1985). Reaction is controlled by two genes, with one allele dominant for resistance and the other recessive.

Sclerotinia Drop. Sclerotinia drop of lettuce can be caused by either of two fungal species, *Sclerotinia minor* Jagger or *Sclerotinia sclerotiorum* (Lib.) De Bary. In the Salinas valley of California, the former predominates, but the latter predominates in most other regions (Subbarao 1998). Because Sclerotinia is soilborne and persists for many years, the pathogen is difficult to control with cultural methods. Symptoms of the disease are complete collapse and death of the plant, and yield losses can be significant when environmental conditions favor the fungus.

The inoculation procedure for screening for resistance to *Sclerotinia minor* has been established for either greenhouse or field screening (Grube and Ryder 2004). It consists of placing inoculum, either fungal sclerotia or mycelia, close to the crown of the test plant. This is a fairly laborious process, especially in the field. However, at present, field screening is somewhat more reliable than greenhouse screening. Increasing the reliability of greenhouse screening is essential for efficient progress of a breeding program, as well as the elucidation of the genetic basis for resistance.

Extensive lettuce germplasm has been evaluated for resistance, but thus far no complete resistance has been identified. Partial resistance has been described among wild *Lactuca* species, PI accessions, breeding lines, and heirloom cultivars, using either *S. sclerotiorum* (Chupp and Sherf 1960; Elia and Piglionica 1964; Whipps et al. 2002; and others) or *S. minor* (Abawi et al. 1980; Subbarao 1998; Grube and Ryder 2004). It is unknown whether resistance to the two species is correlated. The most resistant accessions identified include the primitive *L. sativa* (PI251246). A RIL population has been developed from the cross between PI251246 and a susceptible variety for QTL mapping. Preliminary genetic analysis suggests that resistance is heritable, but that it may be associated with primitive growth habit (Grube 2004).

Verticillium Wilt. Verticillium wilt, incited by *Verticillium dahliae* Kleb, is a relatively new disease problem for lettuce, appearing for the first time in 1995 on a single farm in the Pajaro Valley, along the central coast of California (Subbarao et al. 1997). The organism has a broad host range among vegetables, fruits, fiber crops, ornamentals, and trees. Among the vegetables and other row crops, susceptible species include artichoke, strawberry, tomato, pepper, cauliflower, and mint (Bhat and Subbarao 1999). The earliest symptoms appear on the foliage with wilting of lower leaves. On the roots, vascular discoloration occurs, greenish black in color, that may extend from the tip of the taproot to the core area of the head. A cross section of the root shows a circular discolored pattern. The most severe symptoms occur at maturity, especially on iceberg type plants. The plant becomes completely yellow, the head becomes shriveled and it may become covered with large numbers of microsclerotia. The source of infection is the population of microsclerotia in the soil, which may occur in very large numbers. Losses of up to 100% may occur in head lettuce: smaller losses occur in other lettuce types.

Field resistance has been identified in various types of cultivated lettuce, except in the iceberg type (Grube et al. 2005a, 2005b). The Latin type had, overall, the highest level of resistance. The *L. virosa* accession IVT 280 and several other accessions of the species have shown high levels of resistance in field tests (Grube et al. 2005a, 2005b). The genetic basis for resistance is not known in any of the sources. Nevertheless, breeding work to transfer resistance, particularly to iceberg type lettuces, has been initiated.

Fusarium Wilt. Fusarium wilt was first reported on lettuce in Japan in 1967 (Matuo and Matahashi 1967), but it was not until many years later that its widespread occurrence and potential for economic damage was fully recognized (Hubbard and Gerik 1993; Fujinaga et al. 2003;

Garibaldi, Gilardi, and Gullino, 2004). The disease is incited by *Fusarium oxysporum* f. sp. *lactucae* n.f. (same as f. sp. *lactucum*; Hubbard and Gerik 1993; Fujinaga et al. 2003). Three races of the fungus have been identified (Fujinaga et al. 2003). It is a disease of the root vascular system. As one of several wilting diseases exhibiting yellowing and wilting of leaves and stunting and plant death, the principal diagnostic symptom is a reddish brown discoloration of the cortex and upper crown (Matheron and Koike 2003).

Resistance sources have been identified for Races 1 and/or 2, but not for Race 3 (Garibaldi, Gilardi, and Gullino 2004; Tsuchiya et al. 2004). All sources are cultivars of the various lettuce types. The genetic basis for resistance has not been established although preliminary work indicates that it may be recessive (Grube et al. 2005a, 2005b).

Nematodes, Insects, and Mites

Programs for insect resistance in lettuce are relatively uncommon. The two most active programs are for resistance to lettuce aphid and to leafminer.

The lettuce aphid (*Nasonovia ribisnigri* Mosley) colonizes the interior of the lettuce head and is therefore harder to control with chemical treatments than most aphid species, which tend to colonize outside leaves. It is a medium sized, olive-green aphid. Its primary hosts are *Ribes* species and secondary hosts are members of the Asteraceae. It transmits cucumber mosaic and beet western yellows virus but not lettuce mosaic virus. It has been a serious pest in Europe for many years but only recently became economically important in the western U.S.

Effective resistance, as manifested by noncolonization, was identified in *L. virosa* (accession IVT 280) and has been transferred to some cultivars by breeders of several Dutch seed companies (Eenink, Dieleman, and Groenwold 1982a, 1982b; Van der Arend, Ester, and van Schijndel 1999). Resistance is conferred by a partially dominant allele of a single gene. Transfer to *L. sativa* was accomplished through a bridge cross to *L. serriola*. It was also necessary to break a tight linkage to a deleterious gene causing slow growth and pale color.

The leafminer species *Liriomyza langei* Frick is a pest of lettuce. It is one of several species that cause damage to vegetables. Damage to the plant includes adult sting scars, mines and larvae in lower leaf tissue, and stunted growth. The insect has caused serious losses on lettuce and spinach.

Resistance has been identified in various types of lettuce and related species. Leafminer resistance can be measured in several ways: number of stings per leaf, length of adult survival, number of mines, number of pupae and number of emerging flies. Based on number of stings, resistance was identified in several *Lactuca* species: *L. sativa* (cvs. Merlot, Lolla Rossa, Salad Bowl, landraces PI 187238, and PI491212), *L. saligna* (PI 491099, and PI 509525), *L. serriola* (PI 274901, and PI 491178), and *L. virosa* (PI 273597, and PI 274375) (Mou and Liu 2004; Mou et al. 2004). The genetic basis for resistance is not known. It is not complete resistance and therefore is likely to be quantitative. A series of crosses and backcrosses is being employed to transfer resistance to various lettuce types.

Abiotic Stresses

Resistance screening procedures for other diseases in breeding programs fall generally within the scope of those described. Screening for resistance to physiological disorders, such as tipburn, are highly influenced by environment. They are also mature plant phenomena and, at present at least, cannot be implemented under controlled conditions. Selection for resistance to tipburn is done in the field when the disorder occurs and its incidence cannot easily be predicted.

Tipburn is the most common of the physiological disorders of lettuce, occurring wherever lettuce is grown. In certain areas it is a chronic problem, appearing year after year on a seasonal schedule. On field grown lettuce, symptoms appear at the time of maturity of the plant, thereby coinciding with the time of harvest. Where quality of product is of paramount importance, the crop will not be harvested. Thus, tipburn can have enormous economic consequences.

Usually, symptoms of tipburn appear near the margins of interior leaves in the form of small brown dots or lines, which coalesce to form larger brown lesions. The leaves affected are not at the center of the head but closer to the surface and are the most rapidly growing leaves in the head (Collier and Tibbitts 1982). The rapid growth is accompanied by failure of Ca ions to reach the marginal tissue, which becomes weakened, collapsing the tissue and causing necrosis. Tipburn is one of many disorders that are calcium related. Others include blossom end rot of tomato, blackheart of celery, and brownheart of endive (Shear 1975). The occurrence of tipburn is dependent upon environmental conditions. Essentially, any change that increases growth rate during the maturity phase of growth will affect leaf elongation and calcium transport, leading to tissue breakdown. These changes include: increase in temperature, increase in water, increase in light, or increase in nitrogen.

Cultivar differences in tipburn sensitivity were first identified by Cox and McKee (1976). Differences were not related to lettuce type. Sources of resistance for breeding programs have been cultivars and commercial type breeding lines. Several tipburn resistant cultivars have been released. The genetic basis for resistance is not known.

9.6.3 Cross-Compatible Breeding Strategies

These strategies apply to crosses within cultivated lettuce (*L. sativa*) and between *L. sativa*, *L. serriola* and sexually compatible *L. saligna* and *L. virosa*. The three main methods of breeding lettuce, typical for self-pollinated species, are the pure-line selection method, the pedigree method, and the backcross method.

9.6.3.1 Hybridization and Hybridity Testing

Lettuce flower heads (capitula) contains 12–20 florets, each consisting of a single petal (ligule) and a single carpel with two stigmas that is enclosed by a stamen tube. The anthers shed pollen from the inside surface of the tube as the style elongates (Figure 9.22), effectively insuring selfing before foreign pollen can reach the stigmatal surface (Thompson 1938). To make controlled pollinations, self pollen must be removed from the stigmatic surface before the pollen germinates. The parts are very delicate, making emasculation very difficult. Pollen is typically removed by spraying flowers with water or air or both prior to application of pollen from another parent. Regardless of the technique, some percentage of self-pollination can occur. Hybrids are usually identified by using at least one completely or incompletely dominant morphological marker in the pollen parent. Breeders typically grow parents alongside putative F_1 plants and use markers that are manifested in the seedling stage, such as leaf color, leaf shape, or downy mildew resistance. Molecular markers have been used for hybridity testing by some researchers (Chupeau et al. 1994), but their use in breeding programs has not become routine.

9.6.3.2 Pure-Line Selection

Pure-line selection does not involve crossing, but identification of a perceived difference between a single plant or several plants in a population and the bulk of the plants in the population. Progeny of the selections are then compared to the original cultivar to determine if the perceived difference was heritable and is sufficiently different and potentially useful to merit a new name and release as a new cultivar. Single plant selection is often supplemented by mass selection at the later stages of the cycle to build up populations of desirable and very similar types.

Pure-line selection has been a common practice in private sector breeding programs, as a result of which, groups of cultivars have been developed similar to an original cultivar. For example, after

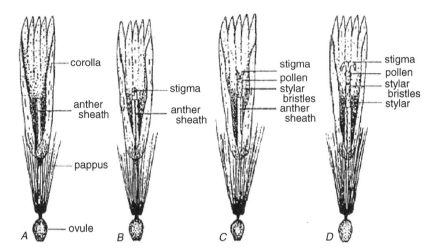

Figure 9.22   Opening of a lettuce floret. (*A*) Stigmas have not yet emerged; (*B*) Stigmas beginning to emerge; (*C*) Stigmas are separating, this is the preferred time for cross to be made; (*D*) Stigmas curved backward, self pollination complete, too late for a cross. (From Thompson, R. C., *Genetic Relations of Some Color Factors in Lettuce*, Tech. Bul. 620., USDA, Washington, DC, 1938. With permission.)

the crisphead cultivar Great Lakes was released in 1941 (Bohn and Whitaker 1951), several dozen Great Lakes types were selected and released by various companies over the next two decades. Similar activity occurred after the release of the crisphead cultivar Salinas in 1975 (Ryder 1986) and smaller "constellations" have appeared around several romaine and leaf cultivars. In many cases, pure-line selection was based on differences in single genes that had no economic significance, e.g., flower color and seed stalk color. For example, the brown-seeded cultivar Excell was selected from the black-seeded cultivar Empire. Selection is an ancient method of crop improvement, almost certainly antedating crossing procedures by centuries. Undoubtedly, both mutation and natural cross-pollination produced the variability enabling selection to be effective.

9.6.3.3 Pedigree Breeding

Pedigree breeding has been a staple method of improvement for many years. The basic premise is to hybridize two parent genotypes, and then to identify and select favorable combinations of genes that are superior to the parents. At its simplest, a cross is made between two parents, followed by selection of desirable plants in the F_2 generation and further selection of desirable families in the F_3 and subsequent inbreeding generations. Additional traits may be introduced by making crosses to one or more additional parents in early or late generations of selection.

An example of a cultivar developed from a single cross followed by selection is cv. Salinas. It was derived from a cross between Calmar and the breeding line 8830 (Ryder 1979). Calmar was a popular, elite Great Lakes type cultivar that made good head size, was widely adapted, and had some tolerance to big vein and tipburn. Breeding line 8830 had the color, texture, and internal quality of the cultivar Vanguard. Salinas was developed by inbreeding and selection to F_7. It had the appearance and quality of Vanguard, a high level of tipburn tolerance and was very uniform in maturity. Numerous lettuce cultivars have been bred this way, primarily when the parents were of the same general type; e.g., both crisphead.

Multiple cross procedures of various complexity have also been used. Thompson, a big vein tolerant cultivar, stemmed from an original cross between Merit and breeding line 2741, both of which had low-moderate big vein tolerance (Thompson and Ryder 1961). An F_5 line 9747-1, had higher big vein tolerance than either parent but an unsatisfactory head shape. It was crossed to Calmar to improve head shape and an F_6 line was released as Thompson, with big vein tolerance equivalent to 9747-1. Pedigrees can be considerably more complex. The pedigree of the leaf cultivar Salad Bowl includes eight crosses among nine different cultivars of several types (leaf, butterhead, crisphead, and romaine) (Thompson and Ryder 1961).

9.6.3.4 *Backcrossing*

Backcross breeding is typically used to transfer a small number of genes from a donor genotype into a horticulturally acceptable recipient (or recurrent) genotype. Most commonly, this method entails simultaneous selection for a few genes being transferred and for the phenotype of the recurrent parent. The result is typically a variety that is nearly isogenic to the recurrent parent and differs only for the trait(s) of interest. Because it can be difficult to differentiate hybrids from self-fertilized progeny of the recurrent parent in advanced generations, backcrossing is most often used for traits that are simply inherited and easily selected. With minor modifications, the backcross method can also be used for more complex traits such as those controlled by multiple genes.

Single genes have often been backcrossed into different lettuce types within *L. sativa*. For example, the cultivar Slobolt was derived by transferring the slow-bolting trait from the butterhead cultivar Giant Summer into the leaf lettuce Grand Rapids (Thompson and Ryder 1961). Back-crossing has also been used to introgress genes from more exotic germplasm into cultivated *L. sativa*. For example, the *Lettuce mosaic virus* (*LMV*) resistance allele *mo-1^e* was introgressed into several cultivars from the primitive *L. sativa* PI 251245 (Ryder 1979b). Downy mildew resistance genes from *L. serriola* have also been incorporated in several cultivars (Crute 1992b), and resistance to the lettuce aphid, *Nasonovia ribis-nigri*, was transferred from *L. virosa* into cultivated types (Eenink, Groenwold, and Dieleman 1982b).

Resistances to several other diseases and insect pests have been reported in exotic germplasm but have yet to be exploited. These are viewed as most promising for diseases where resistance within *L. sativa* is of an inadequate level or is genetically complex. The use of exotic germplasm is often hindered by the presence of undesirable horticultural characters that are incorporated alongside target traits. For wild *Lactuca* species, unfavorable traits are numerous and include having tough leaves and woody stems, often with spines, and the production of bitter compounds with sedative properties (Pink and Keane 1993). Ultimately, lettuce germplasm will only be useful if it has suitable plant morphology (e.g., leaf texture, plant size, and shape). Prior to initiating a long-term and laborious backcross program to transfer traits from wild species, it is important to consider whether the traits of interest are readily transferrable. For example, are secondary metabolites or unique physiological features of the wild species directly responsible for the resistance traits? If genes can be transferred, will they be as effective in a cultivated *L. sativa* genetic background as in the source species? For example, "nonhost" resistance to downy mildew from *L. saligna* was proposed as a durable alterna-tive to partial or race-specific resistance. Once transferred to *L. sativa* via backcrossing, however, resistance was shown to be due to race-specific and partial resistance genes like those already present in *L. sativa* (Jeuken and Lindhout 2002). For polygenic traits that may require a long-term breeding investment, it is wise to determine which alleles have the largest effects and whether some or all loci have undesirable epistatic or pleiotropic effects. Backcross inbred line (BIL) populations, such as the one that was used for simultaneously studying inheritance and transferring downy mildew resistance from *L. saligna* into *L. sativa*, are ideal for this purpose (Jeuken and Lindhout 2004).

An important consideration when backcrossing is the genomic location of favorable loci from the recurrent genotype in relation to those being introduced from the donor. This is particularly important when considering disease resistance genes. As with other plant species, resistance alleles with different specificities have been shown to cluster within the lettuce genome. Breeders may replace a resistance-gene rich segment of the genome with a segment from a wild species to introduce one useful gene, but could inadvertently replace one or more equally important genes.

9.6.3.5   Field Testing

Irrespective of breeding methods employed, field testing is vital for determining parameters either for release of advanced generation materials or cultivars. Field testing provides several types of information that may not have been learned during screening and selecting stages. For this to happen, it is necessary to test materials in a number of environments, which is achieved by growing them on different soil types, at different times of the year that cover most of the season, and for at least two and preferably three to five years. In this way, the breeder establishes adaptability; yield; time and uniformity of harvest; reaction to temperature extremes, particularly regarding tipburn and bolting tendency; and reaction to diseases that may not have been involved in the breeding goals. This information enables the breeder to recommend to growers the most suitable times and places to grow a new cultivar, and to inform recipients of advanced germplasm of the expected usefulness of the materials in further breeding.

9.6.4   Cross-Incompatible Breeding Strategies: Interspecific Hybridization

Thompson et al. (1941) placed *L. sativa*, *L. serriola*, *L. saligna*, and *L. virosa* in a compatibility group and follow-up cytogenetic studies have shown that those four species represent a group reproductively isolated from the other species of the genus (Koopman, de Jong, and de Vries, 1993; Koopman and de Jong 1996). *L. sativa* and *L. serriola* can easily be crossed in either direction producing fertile hybrids. From anatomical studies of some *L. sativa* × *L. serriola* hybrids, it is apparent there is a closer relationship between some genotypes of these species (Haque and Godward 1987). Crosses between *L. sativa* as one parent and other distant species are made with difficulty due to more complicated interspecific barriers. *L. saligna* is known to produce, as a female parent, hybrids with *L. sativa* and *L. serriola* (Pink and Keane 1993). However, the reciprocal crosses are possible and inability to obtain hybrids was based primarily on the choice of suitable accessions and not caused by unilateral incompatibility (de Vries 1990). Some *L. saligna* × *L. sativa* and reciprocal hybrids were phenotypically similar to *L. serriola*, which has resulted in a theory that *L. saligna* formed a part of the ancestral complex of cultivated lettuce (de Vries 1990). *L. saligna* was crossed to a cultivated iceberg type by R.W. Robinson (Provvidenti, Robinson, and Shail 1980) who developed a cultivar Salad Crisp from that cross. Crosses between *L. sativa* as one parent and *L. virosa* as the other parent are made with a great difficulty, resulting in low seed set, unviable seeds, stunted plants, or sterile hybrids (Lindqvist 1960a). Viable hybrid plants were obtained only when *L. serriola* was used as a bridging species (Thompson and Ryder 1961; Eenink, Groenwold, and Dieleman 1982b). In 1958, the cultivar Vanguard was developed from a cross between a *L. sativa* × *L. serriola* line, which was then crossed to *L. virosa*. With some manipulations, crosses have been made and have led to development of cultivars (Ryder 1999a).

Hybridization data on the species belonging to the different sections or groups of the genus *Lactuca* are limited to *L. viminea* and *L. tatarica*. Gorenwold (1983) reported partly fertile hybrids between *L. viminea* (section *Phoenixopus*) and *L. virosa* (section *Lactuca*). Also fertile interspecific hybrids have been produced between species *L. sativa* (section *Lactuca*) and *L. tatarica* (section *Mulgedium*) by using somatic hybridization (Chupeau et al. 1994; Maisonneuve et al. 1995).

In vitro techniques are also used to overcome the interspecific barriers in the genus. Plant regeneration from protoplasts has been reported in *L. sativa* (Berry et al. 1982; Engler and Grogan 1984; Brown et al. 1986) and *L. saligna* (Brown, Lucas, and Power 1987). *In vitro* rescue of immature embryos was used successfully for sexual hybridization between *L. sativa* and *L. virosa*. Protoplast fusion permitted the regeneration of somatic hybrids between *L. sativa* and either *L. tatarica* or *L. perennis* (Maisonneuve et al. 1995). Somatic hybrids between cultivated lettuce and *L. virosa* were produced by protoplast electrofusion (Matsumoto 1991). Hybrids had normal flower morphology, but all were sterile. *L. indica* (section *Tuberosae*) can be somatically hybridized with *L. sativa* to produce a viable callus (Mizutani et al. 1989).

9.6.5   Novel and Biotechnological Approaches

There are several novel tools that may play a more prominent role in increasing efficiency of lettuce breeding programs in the future. These include: (1) genic male sterility (GMS), (2) early flowering, (3) molecular marker-based hybridity testing and marker-assisted selection (MAS), (4) advances in phenotyping procedures, (5) cell and tissue culture, and (6) genomics-based approaches to create novel allelic variation.

9.6.5.1   *Genic Male Sterility (GMS)*

Several genes that confer GMS have been identified in lettuce, but they have not been used extensively in breeding programs (Robinson, McCreight, and Ryder 1983). GMS could increase the efficiency of backcross breeding in lettuce by eliminating the need for extensive hybridity testing. An efficient strategy could involve transferring dominant GMS to the female donor parent prior to backcrossing (Ryder 1971). When using distantly related donors, the initial crosses with the source of male sterility could serve a dual purpose as a bridging cross. If combined with an effective method for pollen transfer, GMS could also be used to produce hybrid lettuce seed on a commercial scale. Recent results suggest that insect-mediated pollination of lettuce may occur at a higher rate than was previously thought, which may make hybrid production feasible (Goubara and Takasaki 2004; Maisonneuve et al. 2004).

9.6.5.2   *Early Flowering*

Single early flowering genes that significantly reduce the generation time have been described in lettuce (Ryder 1983, 1985, 1988). Use of these genes could reduce the total time required to backcross a single gene into a recurrent cultivar by one-half (Ryder 1985).

9.6.5.3   *Molecular Marker-Based Hybridity Testing and Marker-Assisted Selection (MAS)*

Genetic markers can be used to rapidly identify hybrids or to track genes of interest in segregating populations. Several single genes that may be used as morphological markers have been described (Robinson, McCreight, and Ryder 1983). Inexpensive and easy to use morphological or disease resistance markers are routinely used for hybridity testing. Robust and inexpensive molecular markers could represent an advantage because of their insensitivity to plant age and environmental conditions. MAS of genes of interest can also aid in the elimination of undesired chromosome segments to reduce the total time required for the breeding process. The primary limitation to implementing MAS is the availability of suitable markers. Markers can include sequences that are tightly linked to genes of interest. Markers linked to the *cor* locus for corky root resistance and assays for high throughput use have been developed (Moreno-Vazquez et al. 2003; Dufresne et al. 2004). The availability of extensive molecular map and genome sequence data expands the possibilities for MAS in lettuce. For example, direct selection based on known sequences of LMV and downy mildew resistance alleles is now possible (Shen et al. 2002; Nicaise et al. 2003a, 2003b). Several agronomically important traits have been mapped in one or more of the published lettuce genetic maps. In the future, the ability to focus on candidate genes from available lettuce sequence databases (Michelmore et al. 2003) may speed marker development.

9.6.5.4   Advances in Phenotyping Procedures

Progress in solving many of the most significant problems faced by lettuce breeders has been hindered by difficulties in accurately defining and selecting the desirable genotypes. This has been the case for resistances to *Sclerotinia* drop, powdery mildew, and lettuce big vein, to name but a few. Any advances in developing more rapid or more reliable selection techniques will facilitate breeding. A good example is for the disease lettuce big vein. Cultivars with a moderate level of tolerance have been developed using field and greenhouse screening protocols that rely on phenotypic scoring of foliar symptoms on a family basis. The causative agents of this disease, primarily *Lettuce mirafiori virus* and also *Lettuce big vein virus*, were recently conclusively identified (Lot et al. 2002; Roggero et al. 2003). Using this information, sequence-based techniques for rapid identification of these viruses were developed (Navarro et al. 2004). The use of these tools has revealed that, while the most tolerant cultivars available support replication of one or both viruses, some wild species appear to be immune (Hayes, Ryder, and Robinson 2004). The ability to screen for immunity as opposed to tolerance may enable transfer of this trait into cultivated lettuce.

9.6.5.5   Cell and Tissue Culture

Techniques and approaches used for *in vitro* culture of lettuce have been reviewed thoroughly by Michelmore and Eash (1986), Alconero (1988), and Pink and Keane (1993). Those reviews summarize the results with lettuce and related species from the viewpoint of tissue regeneration, anther and pollen culture, embryo rescue technique, organogenesis in explants, protoplast culture and regeneration, somaclonal variation, genetic variability of *in vitro* regenerated plants, and genetic transformation. Some new achievements related to lettuce *in vitro* culture are summarized in Section 9.6.4. Since the 1990s there has been achieved substantial development only in the area of genetic transformation, and this approach will be reviewed in more detail.

9.6.5.6   Transformation

Transformation is generally thought of as a way to introduce genetic variation obtained from other organisms into the cultivated gene pool. An intriguing variation on this theme is the creation of novel allelic variation (Lassner and Bedbrook 2001). Using gene shuffling, knowledge about the effects of subtle sequence changes on a gene's function can be used to predict the optimum sequence for a certain desired function. Once generated, these optimum sequences could be introduced into cultivated gene pools via transformation. An example could be the design of custom resistance genes with very broad specificity (Michelmore 2003). Lettuce disease resistance gene sequences have been the focus of many studies, and could be a feasible target for this type of approach (Sicard et al. 1999; Chin et al. 2001; Plocik et al. 2004).

Initial attempts at producing transgenic lettuce with agronomically desirable traits were ineffective, and led researchers to postulate that transgene instability and low expression of transgenes in *Lactuca* might limit the potential use of this technology (Falk 1996; Gilbertson 1996; Dinant et al. 1997). Subsequent studies have revealed that stable expression of transgenes can be obtained in lettuce, but, as with other crops, success may depend on choice of promoter, host genotype, and limiting the number of insertion events for a given line (Curtis et al. 1994; Ampomah-Dwamena, Conner, Fautrier 1997; McCabe et al. 1999a, 1999b). Transformation of lettuce has been demonstrated by several research groups and can now be considered routine (Michelmore et al. 1987; Chupeau et al. 1989; Torres et al. 1993; Curtis et al. 1994; Davey et al. 2002).

Several genes that confer potentially useful traits have been introduced into lettuce via transformation. These include viral genes to confer LMV resistance (Dinant et al. 1993, 1997) and bacterial genes to confer herbicide resistance (McCabe et al. 1999; Mohapatra et al. 1999; Torres et al. 1999) or delay leaf senescence (McCabe et al. 2001). Genes from other plants have also been used. These include genes from soybean to increase iron uptake (Goto, Yoshihara, and Saiki 2000), from tobacco to decrease nitrate accumulation (Curtis et al. 1999), from cowpea to increase freezing tolerance (Pileggi et al. 2001), and from grape to enable production of resveratrol (Chia and Ng 2003). Lettuce was also transformed with the Maize Ac/Ds transposable elements to create a useful tool for fundamental research (Yang et al. 1993; Okubara et al. 1997). Despite the potential value of some of these traits, no transgenic lettuce has yet been commercialized. In some cases, this is because the transgenes did not have the expected desirable effects on the plant phenotype (Dinant et al. 1993; McCabe et al. 1999). For those that were effective, low commercial acceptance of transgenes from outside the crop genus in a fresh vegetable crop likely played a role in preventing their adoption.

Facile transformation expands the potential genetic variation available for germplasm enhancement by eliminating the barriers imposed by sexual incompatibility. There are two ways transgenic approaches could allow better access to the tertiary gene pool (*L. virosa* and others). One way would be through directly transforming *L. sativa* with characterized genes from wild species. The second method is less direct, and would be used in conjunction with somatic hybridization techniques. Transformation was used to create a universal hybridizer by combining dominant antibiotic resistance and recessive albinism markers in the same genotype (Chupeau et al. 1994). Another potentially valuable use of this technology is to increase or decrease the expression of endogenous genes, i.e., genes already present within cultivated lettuce. Several uncertainties remain, however. Regardless of the source of gene(s) used, the expense and time required to isolate and study genes of interest prior to their introduction is prohibitive. Further, the current emphasis on regulation of the process rather than the product means that any transformed crop will face more regulatory and public perception obstacles than a conventionally bred variety.

9.7 CONCLUSIONS

Lettuce (*Lactuca sativa*) belongs to the core group of crops and is one of the oldest domesticated plants (8,000 to 4,000 years before present) (Hancock 2004). The known artifacts indicate that lettuce was used as a food in Ancient Egypt and also in southwestern Asia in the Tigris–Euphrates region. Nevertheless our knowledge about the origin, process of domestication, diversification, and spread of lettuce around the globe is still very fragmentary and not fully understood. More historical, geographical, biological, and genetical evidence is needed for explanation of these processes.

Based on the available botanical literature, the genus *Lactuca* L. comprises approximately 100 wild species; however, detailed information about the biogeography and ecobiology of most of these species is not available (Lebeda et al. 2004b). Also the taxonomy of wild *Lactuca* species and related genera is currently unclear and much basic information is missing. A formal classification of *Lactuca* and its subgeneric divisions is suggested by Lebeda (1998), in which seven sections and two heterogeneous geographical groups are recognized. More complex and detailed elaboration of the taxonomy of the whole genus, involving all known and described species, is required. Information about ecology, biology, and variation of autochthonous African and Asian species is not available (Lebeda et al. 2004b). Relatively good knowledge is now available only on most of the European *Lactuca* species (Feráková 1977).

Also, the collection, conservation and evaluation of wild *Lactuca* germplasm are all rather incomplete. From approximately one hundred known *Lactuca* species, the world genebank collections of *Lactuca* species, and related genera, comprise relatively low numbers of accessions of only about 20 species. Most (about 90%) of these accessions belongs to the three species (*L. serriola*, *L. saligna*, and *L. virosa*) originating mostly from Europe and from the primary center of origin (Lebeda, Doležalová, and Astley 2004a). In some areas of the world local landraces are still grown.

The pressure of modern cultivars opens the possibility of genetic erosion. The available germplasms and some of the other wild *Lactuca* species may be the source of new and very valuable characters for future breeding progress (Lebeda and Boukema 2005). The development of internationally recognized descriptors have created a good background for future progress in characterization of *Lactuca* germplasm. Creation of the International *Lactuca* Database (Stavelikova, Boukema, and van Hintum 2002) have contributed to better communication in the international lettuce community. Recent studies have shown enormous variation in *L. serriola* germplasm for various parameters: morphology and phenology, resistance to pathogens and pests, nuclear DNA content, protein, and DNA markers. However, this variation is not sufficiently exploited in lettuce breeding. More intensive exploration, collection, characterization, and research of *Lactuca* germplasm is required to obtain a broader view about the variation and relationships between different gene pools and taxa of *Lactuca* species (Lebeda, Doležalová, and Astley 2004a, 2004b).

During the last sixty years, the genus *Lactuca* together with the species of related genera have been intensively studied with respect to exploitation of wild relatives in commercial lettuce breeding (Lebeda, Doležalová, and Astley 2004a, 2004b). These studies have enabled the development of the concept of lettuce genepools. However, the further development of this concept is dependent on the acquisition of new scientific data. The primary gene pool is represented mainly by *L. sativa*, with its enormous morphological and genetical variation, numerous cultivars and many landraces (Rodenburg 1960; Phillips and Rix 1993; Stickland 1998). Lettuce comprises seven main groups of cultivars (including oil seed lettuce), which differ phenotypically; they are usually described as morphotypes. All the various lettuce morphotypes are interfertile and form the primary gene pool together with the wild species *L. serriola* (de Vries and van Raamsdonk 1994). This species is considered as a primary progenitor of cultivated lettuce and is most closely related to *L. sativa*. For this reason, some taxonomists consider both species as conspecific. Koopman et al. (1998) regarded four *serriola*-like species (*L. sativa*, *L. serriola*, *L. altaica*, and *L. dregeana*) as conspecific. This conclusion was based only on the "molecular view" of this question. However, many other parameters (e.g., biogeography, and ecobiology) must be seriously considered in resolving this problem. The opinion expressed in this chapter, based on the broader ecobiological and taxonomical viewpoint, considers these taxa as independent species. *L. saligna* is included in the secondary gene pool. Because of very difficult crossing barriers, *L. virosa* belongs in the tertiary gene pool. Recently, the species from the secondary and tertiary gene pools have been frequently used as sources of different characters (e.g., resistance to pathogens) in lettuce breeding (Ryder 1999a, 1999b).

In the field of genetic mapping and genomics, substantial progress has been achieved in recent years. Molecular maps have now been constructed for intraspecific (*L. sativa* × *L. sativa*) and interspecific (*L. sativa* × *L. serriola* and *L. sativa* × *L. saligna*) populations in lettuce. Using common markers, several of the maps have now been integrated and combined.

More than 70 qualitative genes known to control disease resistances and morphological traits have been identified through classical genetic analyses of lettuce (Michelmore, Kesseli, and Ryder 1994). Some of these have been localized on one or more of the lettuce molecular maps (Grube et al. 2005a, 2005b). Most of the mapped genes in lettuce are disease resistance genes, the majority being *Dm* genes that control reaction to downy mildew (*Bremia lactucae*). The existence of detailed molecular maps for lettuce has permitted mapping of minor genes that control quantitative traits, i.e., quantitative trait loci (QTL). These include seed and seedling traits relating to germination and early plant growth (Argyris et al. 2005) and root architecture (Johnson et al. 2000). These techniques have also been applied to study disease resistance genes that confer partial or quantitative resistance, such as the downy mildew resistance QTL from *L. saligna* (Jeuken and Lindhout 2002, 2004) and *L. sativa* (Hand et al. 2003). Technologies of genomic manipulation (BAC library, transformation system) have become recently available for lettuce, (http://www.compgenomics. ucdavis.edu), including bioinformatics approaches (http://www.cgpdb.ucdavis.edu/).

During the last two decades, enormous progress has been made in germplasm enhancement and introduction of novel traits into cultivated lettuce. For lettuce, nearly all germplasm enhancement has targeted traits from the primary gene pool, but more recently the secondary and tertiary gene pools have been used. However, recent developments have provided access to more distant gene pools (e.g., *Lactuca* related genera and species). For lettuce, desirable traits especially include disease and insect resistances, early vegetative development, slow bolting, novel leaf colors and shapes, nutritional value, uniform development and harvest, and broad environmental adaptation. However, optimum horticultural and other characteristics differ greatly for different morphotypes of lettuce. Some characteristics have been in the center of interest for a long time (e.g., morphology, and disease resistance), while some others has been neglected (e.g., nutrient value) (Ryder 1999a). Recently this situation has been changing and a new USDA program is exploring the nutrient and antioxidant values of lettuce (Mou and Ryder 2004). There are some special qualitative traits, like nitrate content, for lettuce grown under protective cover and in glasshouses. In this area, good research progress has been made in germplasm evaluation and practical enhancement in lettuce breeding.

The reaction of lettuce to biotic and abiotic stresses has been in focus for a long time. Variation in host plant resistance is mostly mirrored by the diversity of pathogens. This phenomenon complicates the process of resistance breeding. Research on lettuce and wild *Lactuca* germplasm for resistance to virus diseases (e.g., LM, LBV, and BWY) yielded new valuable data, and breeding for virus resistance has been in progress in the last three decades. Many iceberg, cos, and butterhead cultivars have been released from public institutions and seed companies (Reinink 1999; Hayes and Ryder 2004).

There is increasing interest in lettuce bacterial diseases. At least two diseases are considered important. Corky root, a lettuce root disease, caused by the Gram-negative bacterium *Sphingomonas suberifaciens,* has been reported in many locations in North America, Europe, Australia, and Asia. New efficient sources of resistance have been recently located in *L. serriola* and *L. virosa* germplasm (Mou and Bull 2004). Breeding of new resistant lettuce cultivars is in progress. The knowledge of resistance sources and genetics of resistance against bacterial leaf spot, incited by *Xanthomonas campestris* pv. *vitians*, is rather limited.

In the group of fungal diseases there are numerous pathogens known; some of them could be very dangerous for lettuce crops. Probably the most important lettuce disease is downy mildew, caused by *Bremia lactucae*, which can be considered as a global pathogen of cultivated lettuce. However, there is very limited information available about its occurrence on wild *Lactuca* species in natural populations, which may be important sources of infection and virulence factors able to overcome new resistance genes originating from wild *Lactuca* species (Lebeda 2002; Lebeda and Zinkernagel 2003b). Interactions between wild *Lactuca* species and *B. lactucae* represent very diverse and complicated systems (Lebeda, Pink, and Astley 2002). The host range of this fungus is very broad, but only 14 *Lactuca* spp. and 6 species from related genera are known as natural hosts of *B. lactucae*. Variation in virulence of the fungus on *L. sativa* is known, but on wild *Lactuca* spp. this phenomenon has been studied only in isolates originating from prickly lettuce, *L. serriola* (Lebeda and Petrželová 2004).

Knowledge of the specificity of interactions between wild *Lactuca* spp. germplasm and *B. lactucae* is limited. Only a few wild *Lactuca* spp. (18) and related species (8) have been studied for host specificity. The wild *Lactuca* spp. studied are mostly characterized by a race-specific response. In some uncharacterized *L. saligna* accessions, very effective mechanisms of resistance have been discovered (Lebeda, Pink, and Astley 2002; Lebeda and Zinkernagel 2003a; Beharav et al. 2006). Very efficient sources of field resistance (rate-reducing) and possibly race-nonspecific resistance have been identified in some *L. serriola* accessions, which are being used in practical breeding (Lebeda and Pink 1998). Substantial variation in the response and the genetical background of resistance have also been recognized in *Lactuca* spp. germplasm. At least 16 race-specific *Dm* genes and/or R-factors have been identified in or derived from *L. serriola*. Some of these genes (e.g., *Dm*5/8, *Dm*6, *Dm*7, *Dm*11, *Dm*16, R18) are used frequently in commercial lettuce breeding. However, only limited information is available on the genetic control of resistance in other wild *Lactuca* spp. (e.g., *L. aculeata, L. saligna, L. virosa*) (Lebeda, Pink, and Astley 2002). Different mechanisms of resistance have been recognized in various wild *Lactuca* spp. based on histological and cytological studies (Lebeda and Pink 1998), but their background is still not well understood (Lebeda, Pink, Mieslerová 2001c).

Lettuce powdery mildew, caused by *Golovinomyces cichoracearum*, is the only species regularly described on *Lactuca* spp. (Lebeda 1985a, 1999). Powdery mildew is considered a disease of increasing importance (Turini et al. 2001). However, there is only limited information about host resistance and genetics of resistance. In some species (e.g., *L. saligna*, and *L. serriola*) the interaction with the pathogen is probably based on race-specific resistance (Lebeda and Mieslerová 2003). Breeding for resistance is still not well advanced.

There are some other fungal lettuce diseases (e.g., anthracnose, stemphylium leaf spot, sclerotinia drop, verticillium, and fusarium wilt) where there is still rather limited information about lettuce germplasm resistance and genetics of resistance. For this reason breeding for resistance is mostly in beginning stages (Ryder 1999a, 1999b).

Also, programs for insect resistance in lettuce are relatively uncommon. The two most developed programs are for resistance to lettuce aphid and to leafminer. An effective source of resistance against lettuce aphid (*Nasonovia ribisnigri*) was identified in *L. virosa* and has been transferred to some lettuce cultivars (Van der Arend, Ester, and van Schijndel 1999). Transfer to *L. sativa* was accomplished through a bridge cross to *L. serriola*. The resistance to leafminer (*Liriomyza langei*) was identified in several *L. sativa* cultivars and landraces, *L. saligna, L. serriola*, and *L. virosa* accessions (Mou and Liu 2004). The genetic basis for resistance is not known and may be quantitative. Transfer of resistance to various lettuce types is in progress.

Tipburn is the most important of the physiological disorders of lettuce and can have enormous economic consequences. Tipburn is a calcium related disorder and its occurrence is dependent upon environmental conditions. Sources of resistance have been located in some lettuce cultivars and several tipburn resistant cultivars have been released (Ryder 1999a).

Cross-compatible breeding strategies are based on crosses within cultivated lettuce (*L. sativa*) and between *L. sativa, L. serriola*, and sexually compatible germplasm from secondary (*L. saligna*) and tertiary (*L. virosa*) gene pools. Three main methods of breeding lettuce, typical for self-pollinated species, are currently used: the pure-line selection method, the pedigree method and the backcross method. Hybridization is the most common method in lettuce breeding for creation of new genetical material. Hybrids are often identified by using at least one completely or incompletely dominant morphological marker in the pollen parent. Also, molecular markers have been used for hybridity testing (Chupeau et al. 1994), but their use in breeding programs has not become routine.

Pure-line selection does not involve crossing, but identification of a perceived difference between a single plant or several plants and the main part of a population. This method has been a common practice in lettuce breeding programs, as a result of which, groups of cultivars have been developed similar to an original cultivar (Ryder 1986).

The basic premise of the pedigree method is to hybridize two parent genotypes, and then to identify and select favorable combinations of genes that are superior to the parents. This method has been used for lettuce breeding for many years (Ryder 1986).

Backcross breeding is used to transfer one or a few genes from a donor genotype into a horticulturally acceptable recipient (or recurrent) genotype. Backcrossing has been frequently and successfuly used to introgress genes from exotic germplasm (e.g., *L. serriola, L. saligna*, and *L. virosa*), with resistance to some diseases (downy mildew) or pests (lettuce aphid), into cultivated *L. sativa* (Pink and Keane 1993; Ryder 1999a).

Recently, cross-incompatible breeding strategies have been frequently used for interspecific hybridization between *L. sativa* and distant wild *Lactuca* species. However, these crosses are difficult due to complicated interspecific barriers. *L. saligna* is known to produce, as a female parent, hybrids with *L. sativa* and *L. serriola* (Pink and Keane 1993). However, the reciprocal crosses are possible and the ability to obtain hybrids is based primarily on the choice of suitable parents. Crosses with *L. virosa* as one parent are also difficult; viable hybrid plants have been obtained when *L. serriola* was used as a bridging species (Thompson and Ryder 1961). *In vitro* techniques are also used to overcome the interspecific barriers in the genus. Plant regeneration from protoplasts has been reported in *L. sativa* (Brown et al. 1986) and *L. saligna* (Brown, Lucas, and Power 1987). *In vitro* rescue of immature embryos was used successfully for sexual hybridization between *L. sativa* and *L. virosa*. Fertile interspecific hybrids have been produced between the species *L. sativa* and *L. tatarica* by using somatic hybridization (Maisonneuve et al. 1995). Somatic hybrids between cultivated lettuce and *L. virosa* were produced by protoplast electrofusion (Matsumoto 1991).

Currently there are several methodologies that may play a more prominent role in increasing efficiency of utilization of *Lactuca* germplasm in lettuce breeding programs. These include: (1) genetic male sterility (GMS), (2) early flowering, (3) molecular marker-based hybridity testing and marker-assisted selection (MAS), (4) advances in phenotyping procedures, (5) cell and tissue culture, and (6) genetic transformation and genomics-based approaches to transfer genes and create novel allelic variation. For future development in enhancement of *Lactuca* germplasm in lettuce breeding, the efficient combination of classic and novel approaches will be required.

ACKNOWLEDGMENTS

The work of A. Lebeda, I. Doležalová, and E. Křístková in this chapter was supported by the project MSM 6198959215 (Ministry of Education, Youth and Sports, Czech Republic).

The authors wish to thank the following colleagues for help and contributing information to this chapter: Prof. Dr. V. Feráková, Dr. M. Widrlechner, Mrs. D. Vondráková, Dr. V. Kryštof, A. Novotná, and Dr. R. Hayes. E. J. Ryder began work on this chapter as a Collaborator with U.S. Department of Agriculture, Agricultural Research Service and is now retired.

REFERENCES

Abawi, G. S., Robinson, R. W., Cobb, A. C., and Shail, J. W., Reaction of lettuce germplasm to artificial inoculation with *Sclerotinia minor* under greenhouse conditions, *Plant Dis.*, 64, 668–671, 1980.

Adler, W., Oswald, K., and Fischer, R., *Exkursionsflora von Österreich*, Stuttgart-Wien: Verlag Eugen Ulmer, 1994.

Aggrioni, P. P. and Mori, B., Numeri cromosomici per la flora Italiana, *Inf. Bot. Ital.*, 8, 269–276, 1976.

Alain, H., *Flora de Cuba. Tomo V*, La Habana: Asociacion de Estudiantes de Ciencias Biologicas Publicaciones, 1964.

Alconero, R., Lettuce (*Lactuca sativa* L.), In *Biotechnology in Agriculture and Forestry*, Bajaj, Y. P. S., Ed., Berlin: Springer-Verlag, pp. 351–369, 1988.

Amin, K. S. and Sequiera, L., Phytotoxic substances from decomposing lettuce residues in relation to the etiology of corky root rot of lettuce, *Phytopathology*, 56, 1054–1061, 1966.

Ampomah-Dwamena, C., Conner, A. J., and Fautrier, A. G., Genotypic response of lettuce cotyledons to regeneration *in vitro*, *Sci. Hort.*, 71, 137–145, 1997.

Anonymous, Convention on Biological Diversity, Text and Annexes, Switzerland: UNEP/CBD/94/1, 1994.

Argyris, J., Truco, M. J., Ochoa, O., Knapp, S. J., Still, D. W., Lenssen, G. M., Schut, J. W., Michelmore, R. W., and Bradford, K. J., Quantitative trait loci association with seed and seedling traits in *Lactuca*, *Theor. Appl. Genet.*, 111, 1365–1376, 2005.

Ayad, W. G., Hodgkin, T., Jaradat, A., and Rao, V. R., Eds., Molecular genetic techniques for plant genetic resources, *Report IPGRI Workshop*, 9–11 October 1995, Rome, Rome: IPGRI, pp. 1–137, 1997.

Babcock, E. B., *The Genus Crepis* I and II. University of California Publications in Botany, nos. 21 and 22. Berkeley, Los Angelos, CA, 1947.

Babcock, E. B., Stebbins, G. L., and Jenkins, J. A., Chromosomes and phylogeny in some Genera of the Crepidinae, *Cytologia*, 188–210, 1937.

Balick, M. J., Nee, M. H., and Atha, D. E., Checklist of the vascular plants of Belize with common names and uses, In *Memoirs of the New York Botanical Garden*, Vol. 85, New York: The New York Botanical Garden Press, 2000.

Barkai-Golan, R., *Postharvest Diseases of Fruits and Vegetables*, Amsterdam: Elsevier Academic Press, 2001.

Beharav, A., Lewinsohn, D., Lebeda, A., and Nevo, E., New wild *Lactuca* genetic resources with resistance against *Bremia lactucae*, *Genet. Resour. Crop Evol.*, 53, 467–474, 2006.

Bélanger, R. R., Bushnell, W. R., Dik, A. J., and Carver, T. L. W., Eds., *The Powdery Mildews. A Comprehensive Treatise*, St. Paul, MN: APS Press, 2002.

Bennett, M. D. and Leitch, I. J., Nuclear DNA amounts in angiosperms, *Ann. Bot.*, 76, 113–176, 1995.

Bennett, M. D. and Leitch, I. J., Plant genome size research: a field in focus, *Ann. Bot.*, 95, 1–6, 2005a.

Bennett, M. D. and Leitch, I. J., Nuclear DNA amounts in angiosperms: progress, problems and prospects, *Ann. Bot.*, 95, 45–90, 2005b.

Bentham, G., Notes on the classification, history, and geographical distribution of the Compositae, *Bot. J. Linn. Soc.*, 13, 335–577, 1873.

Berry, S. F., Lu, D. Y., Pental, D., and Cooking, E. C., Regeneration of plants from protoplasts of *Lactuca sativa* L., *Z. Pflanzenphysiol.*, 108, 31–38, 1982.

Bhat, R. G. and Subbarao, K. V., Host range specificity in *Verticillium dahliae*, *Phytopathology*, 89, 1218–1225, 1999.

Blanca López, G. and Cueto Romero, M., Números cromosomáticos de plantas occidentales, *Anal. Jard. Bot. Madrid*, 41, 185–189, 1984.

Bohm, B. A. and Stuessy, T. F., Eds., *Flavonoids of the Sunflower Family (Asteraceae)*, New York: Springer-Verlag, 2001.

Bohn, G. W. and Whitaker, T. W., Recently introduced varieties of head lettuce and methods used in their development, *U.S. Dept. Agri. Circ.*, No. 881, Washington, DC, 27 pp., 1951.

Bos, L. and Huijberts, N., Screening for resistance to big-vein disease of lettuce (*Lactuca sativa*), *Crop Protect.*, 9, 446–452, 1990.

Boukema, I. W., The International *Lactuca* Database (ILDB), In *Report of a Vegetables Network. Joint Meeting with an ad hoc group on Leafy Vegetables, 22–24 May 2003*, Thomas, G., Astley, D., Boukema, I. W., Daunay, M. C., Del Greco, A., Diez, M. J., van Dooijeweert, W., Keller, J., Kotlinska, T., Lebeda, A., Lipman, E., Maggioni, L., and Rosa, E., Eds., Rome: International Plant Genetic Resources Institute, p. 129, 2005.

Boukema, I. W. and de Groot, L., Status of the CGN leafy vegetables collection, In *Report of a Vegetables Network. Joint Meeting with an Ad Hoc Group on Leafy Vegetables, 22–24 May 2003, Skierniewice, Poland*, Thomas, G., Astley, D., Boukema, I. W., Daunay, M. C., Del Greco, A., Diez, M. J., van Dooijeweert, W., Keller, J., Kotlinska, T., Lebeda, A., Lipman, E., Maggioni, L., and Rosa, E., Eds., Rome: International Plant Genetic Resources Institute, pp. 113–116, 2005.

Boukema, I. W., Hazenkamp, Th., and van Hintum, Th. J. L., *The CGN Collection Reviews: The CGN Lettuce Collection*, Wageningen: Centre for Genetic Resources, 1990.

Boukema, I. W., Stavelíková, H., and van Hintum, Th. J. L., The international *Lactuca* database, In *Report of a Network Coordinating Group on Vegetables*, Maggioni, L. and Spellman, O., Eds., Rome: IPGRI, pp. 58–59, 2001.

Boulos, L., Revision systématique du genre Sonchus s.l, *Bot. Not.*, 125, 287–305, 1972.

Boulos, L., Revision systématique du genre Sonchus s.l, *Bot. Not.*, 126, 155–196, 1973.

Boulos, L., Revision systématique du genre Sonchus s.l, *Bot. Not.*, 127, 7–37, 1974a.

Boulos, L., Revision systématique du genre Sonchus s.l, *Bot. Not.*, 127, 402–451, 1974b.

Brako, L. and Zarucchi, J. L., *Catalogue of the Flowering Plants and Gymnosperms of Peru*, St. Louis: Missouri Botanical Garden Press, 1993.

Braun, U., *The Powdery Mildews (Erysiphales) of Europe*, Jena: Gustav Fischer Verlag, 1995.

Braun, U., Some critical notes on the classification and the generic concept of the Erysiphaceae, *Schlechtendalia*, 3, 48–54, 1999.

Braun, U. and Takamatsu, S., Phylogeny of the *Erysiphe*, *Microsphaera*, *Uncinula* (Erysipheae) and *Cystotheca*, *Podosphaera*, *Sphaerotheca* (Cystotheceae) inferred from rDNA ITS sequences—some taxonomic consequences, *Schlechtendalia*, 4, 1–33, 2000.

Bremer, K., Major clades and grades of the Asteraceae, In *Compositae: Systematics. Proceedings of the International Compositae Conference Kew (1994)*, Hind, D. J. N. and Beentje, H. J., Eds., Vol. 1, Kew: Royal Botanic Gardens, pp. 1–7, 1996.

Bremer, K., Anderberg, A. A., Karis, P. O., Nordenstam, B., Lundberg, J., and Ryding, O., *Asteraceae: Cladistic and Classification*, Portland: Timber Press, 1994.

Bretting, P. K. and Widrlechner, M. P., Genetic markers and plant genetic resource management, *Plant Breed. Rev.*, 13, 11–86, 1995.

Brickell, C. D., Baum, B. R., Hetterscheid, W. L. A., Leslie, A.C., McNeill, J., Trehane, P., Vrugtman, F., and Wiersema, J. H., *International code of Nomenclature for Cultivated Plants, 7th ed.*, Acta Hort, 647, i–xxi, Gent, Belgium, pp. 1 123, 2004.

Brown, P. R. and Michelmore, R. W., The genetics of corky root resistance in lettuce, *Phytopathology*, 78, 1145–1150, 1988.

Brown, C., Lucas, J. A., Crute, I. R., Walkey, G. D. A., and Power, J. B., An assessment of genetic variability in somacloned lettuce plants (*Lactuca sativa*) and their offspring, *Ann. Appl. Biol.*, 109, 391–407, 1986.

Brown, C., Lucas, J. A., and Power, J. B., Plant regeneration from protoplasts of a wild lettuce species (*Lactuca saligna* L.), *Plant Cell Rep.*, 6, 180–182, 1987.

Brullo, S. and Pavone, P., Numeri cromosomici per la flora italiana: 464–483, *Inf. Bot. Ital.*, 10, 248–265, 1978.

Burbinge, N. T. and Gray, M., *Flora of the Australian Capital Territory*, Canberra: Australian National University Press, 1970.

Burdon, J. J., The structure of pathogen populations in natural plant communities, *Annu. Rev. Phytopathol.*, 31, 305–323, 1993.

Burdon, J. J. and Jarosz, A. M., Wild relatives as source of disease resistance, In *The Use of Plant Genetic Resources*, Brown, A. H. D., Frankel, O. H., Marshall, D. R., and Williams, J. T., Eds., Cambridge: Cambridge University Press, pp. 281–296, 1989.

Burkholder, W. H., Three bacteria pathogenic on head lettuce in New York State, *Phytopathology*, 44, 592–596, 1954.

Capinera, J., *Handbook of Vegetable Pests*, Amsterdam: Elsevier Academic Press, 2001.

Carisse, O., Oiumet, A., Toussaint, V., and Philion, V., Evaluation of the effects of seed treatments, bactericides, and cultivars on bacterial leaf spot of lettuce caused by *Xanthomonas campestris* pv. *vitians*, *Plant Dis.*, 84, 295–299, 2000.

Chalkuziev, P., Species novae florae Schachimardanicae, *Bot. Mat. Gerb. Inst. Bot. Akad. Nauk Uzbek. SSR*, 19, 58–61, 1974.

Chia, T. F. and Ng, I., Production of resveratrol in transgenic red lettuce for chemoprevention of cancer and cardiovascular diseases, In *Agrobiotechnology and Plant Tissue Culture*, Enfield: Science Publishers, pp. 133–145, 2003.

Chin, D. B., Arroyo-Garcia, R., Ochoa, O. E., Kesseli, R. V., Lavelle, D. O., and Michelmore, R. W., Recombination and spontaneous mutation at the major cluster of resistance genes in lettuce (*Lactuca sativa*), *Genetics*, 157, 849–861, 2001.

Cho, J. J., Mau, R. F. L., German, T. L., Hartmann, R. W., Yudin, L. S., Gonsalves, D., and Provvidenti, R., A multidisciplinary approach to management of tomato spotted wilt virus in Hawaii, *Plant Dis.*, 73, 375–383, 1989.

Chupeau, M. C., Bellini, C., Guerche, P., Maisonneuve, B., Vastra, G., and Chupeau, Y., Transgenic plants of lettuce (*Lactuca sativa*) obtained through electroporation of protoplasts, *Bio/Technol.*, 7, 503–508, 1989.

Chupeau, M. C., Maisonneuve, B., Bellec, Y., and Chupeau, Y., A *Lactuca* universal hybridizer, and its use in creation of fertile interspecific somatic hybrids, *Mol. Gen. Genet.*, 245, 139–145, 1994.

Chupp, C. and Sherf, A. F., *Vegetable Diseases and Their Control*, New York: Ronald Press, 1960.

Chytilová, V., Křístková, E., Losík, J., Petříková, K., and Stavělíková, H., 7.9. Metodika řešení kolekce zelenin, aromatických a léčivých rostlin; 7.9.1. Speciální část: Zelenina (7.9. Methodology for collections of vegetables, aromatic and medicinal plants; 7.9.1. Special part: Vegetables), In *Rámcová metodika Národního programu konzervace a využívání genetických zdrojů rostlin a agro-biodiverzity; Genetické zdroje č. 90 (General Methodology of National Program of Conservation and Utilization of Genetic Resources of Plants and Agrobiodiversity)*, Dotlačil, L., Stehno, Z., Fáberová, I., and Holubec, V., Eds., Praha: Výzkumný ústav rostlinné výroby (Research Institute of Crop Production), pp. 70–75, 2004, http://www.vurv.cz

Cole, R. A., Sutherland, R. A., and Riggall, W. E., The use of polyacrylamide gradient gel electrophoresis to identify variation in isozymes as markers for *Lactuca* species and resistance to the lettuce root aphid *Pemphigus bursarius*, *Euphytica*, 56, 237–242, 1991.

Collier, G. F. and Tibbitts, T. W., Tipburn of lettuce, *Hort. Rev.*, 4, 49–65, 1982.

Cook, R. T. A., Inman, A. J., and Billings, C., Identification and classification of powdery mildew anamorphs using light and scanning electron microscopy and host range data, *Mycol. Res.*, 101, 975–1002, 1997.

Cox, E. F. and McKee, J. M. T., A comparison of tipburn susceptibility in lettuce under filed and glasshouse conditions, *J. Hort. Sci.*, 51, 117–122, 1976.

Couch, H. B., Studies on seed transmission of lettuce mosaic virus, *Phytopathology*, 45, 63–70, 1955.

Cronquist, A., Phylogeny and taxonomy of the Compositae, *Amer. Midl. Natur.*, 53, 478–511, 1955.

Cronquist, A., The Compositae revisited, *Brittonia*, 29, 137–153, 1977.

Cronquist, A., *Vascular Flora of the Southeastern United States. Vol. I, Asteraceae*, Chapel Hill: The University of North Carolina Press, 1980.

Cross, R. J., Review paper: global genetic resources of vegetables, *Plant Varieties Seeds*, 11, 39–60, 1998.

Crute, I. R., The geographical distribution and frequency of virulence determinants in *Bremia lactucae*: relationships between genetic control and host selection, In *Populations of Plant Pathogens: Their Dynamics and Genetics*, Wolfe, M. S. and Caten, C. E., Eds., Oxford: Blackwell Scientific Publications, pp. 193–212, 1987.

Crute, I. R., From breeding to cloning (and back again?): a case study with lettuce downy mildew, *Annu. Rev. Phytopathol.*, 30, 485–506, 1992a.

Crute, I. R., The contribution to successful integrated disease management of research on genetic resistance to pathogens of horticultural crops, *J. Royal Agric. Soc. Engl.*, 153, 132–144, 1992b.

Crute, I. R., The role of resistance breeding in the integrated control of downy mildew (*Bremia lactucae*) in protected lettuce, *Euphytica*, 63, 95–102, 1992c.

Crute, I. R. and Dixon, G. R., Downy mildew diseases caused by the genus *Bremia* Regel, In *The Downy Mildews*, Spencer, D. M., Ed., London: Academic Press, pp. 421–460, 1981.

Crute, I. R. and Johnson, A. G., The genetic relationship between races of Bremia lactucae and cultivars of *Lactuca sativa*, *Ann. Appl. Biol.*, 83, 125–137, 1976.

Curtis, I. S., Power, J. B., Blackhall, N. W., de Laat, A. M. M., and Davey, M. R., Genotype-independent transformation of lettuce using *Agrobacterium tumefaciens*, *J. Exp. Bot.*, 45, 1441–1449, 1994.

Curtis, I. S., Power, J. B., de Laat, A. M. M., Caboche, M., and Davey, M. R., Expression of a chimeric nitrate reductase gene in transgenic lettuce reduces nitrate in leaves, *Plant Cell Rep.*, 18, 889–896, 1999.

Davey, M., McCabe, M. S., Mohapatra, U., and Power, J. B., Genetic manipulation of lettuce, In *Transgenic Plants and Crops*, Khachatourians, G. C., McHughen, A., Scorza, R., Nip, W-K, and Hui, Y. H., Eds., New York: Marcel Dekker, pp. 613–635, 2002.

Davis, R. M., Subbarao, K. V., Raid, R. N., and Kurtz, E. A., Eds., *Compendium of Lettuce Diseases*, St. Paul: APS Press, 1997.

Dethier, D., Le genre *Lactuca* L. (*Asteraceae*) en Afrique centrale, *Bull. Jard. Bot. Nat. Belg.*, 52, 367–382, 1982.

Dickson, M. H., Resistance to corky root rot in head lettuce, *Proc. Amer. Soc. Hort. Sci.*, 82, 388–390, 1963.

Dinant, S., Maisonneuve, B., Albouy, J., Chupeau, Y., Chupeau, M. C., Bellec, Y., Gaudefroy, F. et al., Coat protein gene-mediated protection in *Lactuca sativa* against lettuce mosaic potyvirus strains, *Mol. Breed.*, 3, 75–86, 1997.

Dinant, S., Blaise, F., Kusiak, C., Astier-Manifacier, S., and Albouy, J., Heterologous resistance to potato virus Y in transgenic tobacco plants expressing the coat protein gene of lettuce mosaic potyvirus, *Phytopathology*, 83, 818–824, 1993.

Dinant, S. and Lot, H., Lettuce mosaic virus, a review, *Plant Pathol.*, 41, 528–542, 1992.

Dinoor, A. and Eshed, N., Plant conservation *in situ* for disease resistance, In *Plant Genetic Conservation, The In Situ Approach*, Maxted, N., Ford-Lloyd, B. V., and Hawkes, J. G., Eds., Dordrecht: Kluwer Academic Publishers, pp. 323–336, 1997.

Dixon, G. R., *Vegetable Crop Diseases*, London: MacMillan Publishers, 1981.

Doležalová, I., Lebeda, A., and Křístková, E., Prickly lettuce (*L. serriola* L.) collecting and distribution study in Slovenia and Sweden, *Plant Genet. Resour. Newsl.*, 128, 41–44, 2001.

Doležalová, I., Křístková, E., Lebeda, A., and Vinter, V., Description of morphological characters of wild *Lactuca* L. spp. genetic resources (English–Czech version), *Hort. Sci. Prague*, 29, 56–83, 2002a.

Doležalová, I., Lebeda, A., Janeček, J., Číhalíková, J., Křístková, E., and Vránová, O., Variation in chromosome numbers and nuclear DNA contents in genetic resources of *Lactuca* L. species (Asteraceae), *Genet. Resour. Crop Evol.*, 49, 383–395, 2002b.

Doležalová, I., Křístková, E., Lebeda, A., Vinter, V., Astley, D., and Boukema, I. W., Basic morphological descriptors for genetic resources of wild *Lactuca* spp., *Plant Genet. Resour. Newsl.*, 134, 1–9, 2003a.

Doležalová, I., Lebeda, A., Dziechciarková, M., Křístková, E., Astley, D., and van de Wiel, C. C. M., Relationships among morphological characters, isozymes polymorphism and DNA variability—the impact on *Lactuca* germplasm taxonomy, *Czech J. Genet. Plant Breed.*, 39, 59–67, 2003b.

Doležalová, I., Lebeda, A., Tiefenbachová, I., and Křístková, E., Taxonomic reconsideration of some *Lactuca* spp. germplasm maintained in world genebank collections, *Acta Hort.*, 634, 193–201, 2004.

Doležalová, I., Lebeda, A., Křístková, E., and Novotná, A., *Morphological variation of* Lactuca serriola *populations from some European countries*, XVII International Botanical Congress, Vienna, Austria, 17–23, July 2005, Abstracts, p. 458, Abstract P1393, 2005.

Doležel, J., Application of flow cytometry for the study of plant genomes, *J. Appl. Genet.*, 38, 285–302, 1997.

Dostál, J., *Nová květena ČSSR, 2. díl (New Flora of the Czechoslovakia Vol. 2)*, Prague: Academia, 1989.

Duffus, J. E., Radish yellows, a disease of radish, sugar beet, and other crops, *Phytopathology*, 50, 389–394, 1960.

Duffus, J. E., Beet pseudo-yellows virus, transmitted by the greenhouse whitefly (*Trialeurodes vaporariorum*), *Phytopathology*, 55, 450–453, 1965.

Duffus, J. E., Beet yellow stunt, a potentially destructive virus disease of sugar beet and lettuce, *Phytopathology*, 62, 161–165, 1972.

Duffus, J. E., The yellowing virus diseases of beet, *Adv. Virus Res.*, 18, 345–386, 1973.

Duffus, J. E., Zink, F. W., and Bardin, R., Natural occurrence of sowthistle yellow vein virus on lettuce, *Phytopathology*, 60, 1383–1384, 1970.

Duffus, J. E., Larsen, R. C., and Liu, H. Y., Lettuce infectious yellows virus—a new type of whitefly-transmitted virus, *Phytopathology*, 76, 97–100, 1986.

Duffus, J. E., Liu, H. Y., Wisler, G. C., and Li, R. H., Lettuce chlorosis virus—a new whitefly-transmitted closterovirus, *Eur. J. Plant Pathol.*, 102, 591–596, 1996.

Dufresne, P. J., Jenni, S., and Fortin, M. G., FRET hybridization probes for the rapid detection of disease resistance alleles in plants: detection of corky root resistance in lettuce, *Mol. Breed.*, 13, 323–332, 2004.

Dziechciarková, M., Lebeda, A., Doležalová, I., and Astley, D., Characterization of *Lactuca* spp. by protein and molecular markers—a review, *Plant Soil Environ.*, 50, 49–60, 2004a.

Dziechciarková, M., Lebeda, A., Doležalová, I., and Křístková, E., Isozyme variation in European *Lactuca serriola* germplasm, In *Genetic Variation for Plant Breeding, Proceedings of the 17th EUCARPIA General Congress, 8–11 September 2004, Tulln, Austria*, Vollmann, J., Grausgruber, H., and Ruckenbauer, P., Eds., Vienna: BOKU-University of Natural Resources and Applied Life Sciences, pp. 103–107, 2004b.

Eenink, A. H., Dieleman, F. L., and Groenwold, R., Resistance of lettuce (*Lactuca*) to the leaf aphid *Nasonovia ribis-nigri*. 2. Inheritance of the resistance., *Euphytica*, 31, 301–304, 1982a.

Eenink, A. H., Groenwold, R., and Dieleman, F. L., Resistance of lettuce (*Lactuca*) to the leaf aphid *Nasonovia ribis-nigri*. 1. Transfer of resistance from *L. virosa* to *L. sativa* by inrespecific crosses and selection of resistant breeding lines, *Euphytica*, 31, 291–300, 1982b.

Elia, M. and Piglionica, V., Preliminary observations on the resistance of some lettuce cultivars to collar rot caused by *Sclerotinia* spp., *Phytopathol. Medit.*, 3, 37–39, 1964.

Engler, D. E. and Grogan, R. G., Variation in lettuce regenerated from protoplasts, *J. Hered.*, 75, 426–430, 1984.

Falk, B. W., Basic approaches to lettuce virus disease control, Salinas: Iceberg Lettuce Advisory Board Annual Report, pp. 70–74, 1996.

FAO, *The State of the World's Plant Genetic Resources for Food and Agriculture*, Rome: Food and Agriculture Organization of the United Nations, 1998.

FAO (Food and Agriculture Organization of the United Nations), FAOSTAT 2005, http://faostat.fao.org (accessed April 2005).

Farrara, B. F. and Michelmore, R. W., Identification of new sources of resistance to downy mildew in *Lactuca spp.*, *HortScience*, 22, 647–649, 1987.

Farrara, B. F., Ilott, T. W., and Michelmore, R. W., Genetic analysis of factors for resistance to downy mildrew (*Bremia lactucae*) in species of lettuce (*Lactuca sativa* and *L. serriola*), *Plant Pathol.*, 36, 499–514, 1987.

Feráková, V., *Lactuca* L, In *Flora Europaea*, Tutin, T. G., Heywood, V. H., Burges, N. A., Moore, D. M., Valentine, D. H., Walters, S. M., and Webb, D. A., Eds., Vol. 4, Cambridge: Cambridge University Press, pp. 328–331, 1976.

Feráková, V., *The Genus Lactuca in Europe*, Bratislava: Komenský University Press, 1977.

Fernald, L. M., *GRAY's Manual of Botany, Eighth Edition-Illustrated*, New York: American Book Company, 1950.

Frietema, F. T., The systematic relationship of *Lactuca sativa* and *Lactuca serriola*, in relation to the distribution of prickly lettuce, *Acta Bot. Neerlandica*, 43, 79, 1994.

Frietema de Vries, F. T., Cultivated plants and the wild flora, In *Rijksherbarium/Hortus Botanicus*, The Netherlands: Leiden, 1996.

Frijters, A. C. J., Zhang, J., Damme, M. V., Wang, G-L. , Ronald, P. C., and Michelmore, R. W., Construction of a bacterial artificial chromosome library containing large EcoRI and HindIII genomic fragments of lettuce, *Theor. Appl. Genet.*, 94, 390–399, 1997.

Fujinaga, M., Ogiso, H., Tsuchiya, N., Saito, H., Yamanaka, S., Nozue, M., and Kojima, M., Race 3, a new race of *Fusarium oxysporum* f. sp. *lactucae* determined by a differential system with commercial cultivars, *J. Gen. Plant Pathol.*, 69, 23–28, 2003.

Funk, V. A., Bayer, R. J., Watson, L., Gemeinholzer, B., Oberprieler, C., Garcia-Jacas, N., and Susanna, A., Evolution of the Compositae: the big picture, *XVII International Botanical Congress, Vienna, Austria, 17–23 July 2005, Abstracts*, p. 103, Abstract 6.11.1, 2005.

Garibaldi, A., Gilardi, G., and Gullino, M. L., Varietal resistance of lettuce to *Fusarium oxysporum* f. sp. *lactucae*, *Crop Protect.*, 23, 845–851, 2004.

Gemeinholzer, B. and Kilian, N., Phylogeny and subtribal delimitation of the *Cichorieae* (Asteraceae), in *XVII International Botanical Congress, Vienna, Austria, 17–23 July 2005, Abstracts*, p. 103 Abstract 6.11.4, 2005.

George, R. A. T., *Vegetable Seed Production*, 2nd ed., Wallingford: CABI Publishing, 1999.

Gilbertson, R. L., Management and detection of LMV: production of LMV resistant lettuce and LMV coat protein antibodies, Salinas: Iceberg Lettuce Advisory Board Annual Report, pp. 87–91, 1996.

Goldman, P. H., Koike, S. T., Ryder, E., and Bull, C. T., Influence of bacterial populations on leaf spot development in resistant and susceptible lettuce cultivars, *Phytopathology*, 93, S29, 2003.

Goto, F., Yoshihara, T., and Saiki, H., Iron accumulation and enhanced growth in transgenic lettuce plants expressing the iron- binding protein ferritin, *Theor. Appl. Genet.*, 100, 658–664, 2000.

Goubara, M. and Takasaki, T., Pollination effects of the sweat bee *Lasioglossum villosulum* trichopse (Hymenoptera: Halictidae) on genic male sterile lettuce, *Appl. Entomol. Zool.*, 39, 163–169, 2004.

Govier, D. A., Purification and partial characterisation of beet mild yellowing virus and its serological detection in plants and aphids, *Ann. Appl. Biol.*, 107, 439–447, 1985.

Greuter, W., The Euro+Med treatment of *Cichorieae* (*Compositae*)—generic concepts and required new names, *Willdenovia*, 33, 229–238, 2003.

Greuter, W., McNeill, J., Barrie, F. R., Burdet, H. M., DeMoulin, V., Filgueiras, T. S., Nicolson, D. H., Silva, P. C., Skog, J. E., Trehane, P., Turland, N. J., and Hawksworth, D. L., *International Code of Botanical Nomenclature (St. Louis Code). Regnum Vegetabile*, Vol. 138, Königstein: Koeltz Scientific Books, 2000.

GRIN database, http://www.ars-grin.gov/npgs, (accessed April 2005).

Grogan, R. G. and Zink, F. W., Fertilizer injury and its relationship to previously described disease of lettuce, *Phytopathology*, 46, 416–422, 1976.

Grogan, R. G., Welch, J. E., and Bardin, R., Common lettuce mosaic and its control by the use of mosaic-free seed, *Phytopathology*, 42, 573–578, 1952.

Grogan, R. G., Zink, F. W., Hewitt, W. B., and Kimble, K. A., The association of *Olpidium* with the big-vein disease of lettuce, *Phytopathology*, 48, 292–296, 1958.

Grube, R. C., Genetic analysis of resistance to lettuce drop caused by *Sclerotinia minor*, *Acta Hort.*, 637, 49–53, 2004.

Grube, R. C. and Ochoa, O. E., Comparative genetic analysis of field resistance to downy mildew in the lettuce cultivars "Grand Rapids" and "Iceberg", *Euphytica*, 142, 205–215, 2005.

Grube, R. and Ryder, E., Identification of lettuce (*Lactuca sativa* L.) germplasm with genetic resistance to drop caused by *Sclerotinia minor*, *J. Amer. Soc. Hort. Sci.*, 129, 70–76, 2004.

Grube, R. C., Ryder, E. J., Koike, S. T., McCreight, J. D., and Wintermantel, W. M., Breeding for resistance to new and emerging lettuce diseases in California, In *Eucarpia Leafy Vegetables 2003, Proceedings of the Eucarpia Meeting on Leafy Vegetables Genetics and Breeding*, van Hintum, Th. J. L., Lebeda, A., Pink, D. A., and Schut, J. W., Eds., Wageningen: Centre for Genetic Resources, pp. 37–42, 2003.

Grube, R. C., Hayes, R., Mou, B., and McCreight, J. D., Lettuce breeding, *California Lettuce Research Board Annual Report, 2004–2005*. Salinas, CA: California Lettuce Research Board, 2005.

Grube, R. C., Wintermantel, W. M., Hand, P., Aburomia, R., Pink, D. A. C., and Ryder, E. J., Genetic analysis and mapping of resistance to lettuce dieback: a soilborne disease caused by tombusviruses, *Theor. Appl. Genet.*, 110, 259–268, 2005.

Grulich, V., *Lactuca*, L.,—locika, In *Květena České republiky 7* (Flora of the Czech Republic 7), Slavík, B. and Štěpánková, J., Eds., Praha: Academia, pp. 487–497, 2004.

Guarino, L., Ramanatha Rao, V., and Reid, R., *Collecting Plant Genetic Diversity, Technical Guidelines*, Wallingford: CAB International, 1995.

Gustafsson, I., Potential sources of resistance to lettuce downy mildew (*Bremia lactucae*) in different *Lactuca* species, *Euphytica*, 40, 227–232, 1989.

Hancock, J. F., *Plant Evolution and the Origin of Crop Species*, 2nd ed., Wallingford: CABI Publishing, 2004.

Hand, P., Kift, N., McClement, S., Lynn, J. R., Grube, R., Schut, J. W., van der Arend, A. J. M., and Pink, D. A. C., Progress towards mapping QTLs for pest and disease resistance in lettuce, In *Eucarpia Leafy Vegetables 2003, Proceedings of the Eucarpia Meeting on Leafy Vegetables Genetics and Breeding*, van Hintum, Th. J. L., Lebeda, A., Pink, D. A., and Schut, J. W., Eds., Wageningen: Centre for Genetic Resources, pp. 31–35, 2003.

Haque, M. Z. and Godward, M. B. E., Relationship between *Lactuca serriola* L. and *L. sativa* L. cultivars, *Genet. Agr.*, 41, 275–282, 1987.

Harborne, J. B., *Comparative Biochemistry of the Flavonoids*, New York: Academic Press, 1967.

Harlan, J., Lettuce and the sycamore: sex and romance in Ancient Egypt, *Econ. Bot.*, 40, 4–15, 1986.

Harlan, J. R., *Crops and Man*, 2nd ed., Madison: American Society of Agronomy Inc. Crop Science Society of America, 1992.

Harlan, J. R. and de Wet, J. M. J., Towards a rational classification of cultivated plants, *Taxon*, 20, 509–517, 1971.

Hayes, R. J. and Ryder, E., Breeding lettuce cultivars with high resistance to lettuce mosaic virus, Paper presented at *17th International Lettuce and Leafy Vegetables Conference*, Montreal, Quebec, Canada, 2004.

Hayes, R. J., Ryder, E., and Robinson, B., Introgression of big vein tolerance from *Lactuca virosa* L. into cultivated lettuce (*Lactuca sativa* L.), *HortScience*, 39, 881, 2004.

Heath, M. C., Evolution of resistance to fungal parasitism in natural ecosystems, *New Phytologist*, 119, 331–343, 1991.

Hedrick, U. P., *Sturtevant's Edible Plants of the World*, New York: Dover Press, 1972.

Helm, J., *Lactuca sativa* in morphologisch-systematischer Sicht, *Kulturpflanze*, 2, 72–129, 1954.

Henry, R. J., *Plant Diversity and Evolution: Genotypic and Phenotypic Variation in Higher Plants*, Wallingford: CABI Publishing, 2005.

Hickman, J. C., Ed., *The Jepson Manual Higher Plants of California*, Berkley: University of California Press, 1993.

Hill, M., Witsenboer, H., Zabeau, M., Vos, P., Kesseli, R., and Michelmore, R., PCR-based fingerprinting using AFLPs as a tool for studying genetic relationships in *Lactuca* spp., *Theor. Appl. Genet.*, 93, 1202–1210, 1996.

Hintum, Th. J. L. and Boukema, I. W., Genetic resources of leafy vegetables, In *Eucarpia Leafy Vegetables'99, Proceedings of the Eucarpia Meeting on Leafy Vegetables Genetics and Breeding*, Lebeda, A. and Křístková, E., Eds., Olomouc: Palacký University, pp. 59–72, 1999.

Hoff, J. K. and Newhall, A. G., Corky root rot of iceberg lettuce on the mucklands of New York., *Plant Dis. Rep.*, 44, 333–339, 1960.

Hubbard, J. C. and Gerik, J. S., A new wilt disease of lettuce incited by *Fusarium oxysporum* f. sp. *lactucum* forma specialis nov, *Plant Dis.*, 77, 750–754, 1993.

Huenneke, L. F., Ecological implications of genetic variation in plant populations, In *Genetics and Conserva-tion of Rare Plants*, Falk, D. A. and Holsinger, K. E., Eds., New York: Oxford University Press, pp. 31–44, 1991.

Hui-Lin, L., Tang-Shui, L., Tseng-Chieng, H., Tetsuo, K., and Vol de, Ch. E., *Flora of Taiwan*, Vol. 4, Taipei: Epoch Publishing Co, 1978.

Irwin, S. V., Kesseli, R. V., Waycott, W., Ryder, E. J., Cho, J. J., and Michelmore, R. W., Identification of PCR-based markers flanking the recessive LMV resistance gene *mo1* in an intraspecific cross in lettuce, *Genome*, 42, 982–986, 1999.

Iwatsuki, K., Yamazaki, T., Boufford, D. E., and Ohba, H., *Flora of Japan*, Vol. IIIb, Tokyo: Kodansha, 1995.

Jacobs, M., *The Unfinished Story of Lactuca Altaica*. M.Sc. Thesis, Wageningen University, 2001.

Jagger, I. C., A transmissible mosaic disease of lettuce, *J. Agri. Res.*, 20, 737–741, 1921.

Jagger, I. C. and Chandler, N., Big vein of lettuce, *Phytopathology*, 24, 1253–1256, 1934.

Jansen, R. C., New EU-funded cluster of PGR projects: GERMPLASM, *IPGRI Newlet. Eur.*, 20, 5, 2001.

Jeffrey, C., Notes on Compositae I. The Cichorieae in East tropical Africa, *Kew Bull*, 18, 427–486, 1966.

Jeffrey, C., *Lactuca* L, In *Flora of Turkey and the East Aegean Islands*, Davis, P. H., Ed., Edinburgh: Edinburgh University Press, pp. 776–782, 1975.

Jeffrey, C. and Beentje, H. J., Cichorieae, In *Flora of Tropical East Africa*, Beentje, H. J. and Smith, S. A. L., Eds., Rotterdam: Compositae I, pp. 63–108, 2000.

Jeuken, M. and Lindhout, P., *Lactuca saligna*, a non-host for lettuce downy mildew (*Bremia lactucae*), harbors a new race-specific *Dm* gene and three QTLs for resistance, *Theor. Appl. Genet.*, 105, 384–391, 2002.

Jeuken, M. J. W. and Lindhout, P., The development of lettuce backcross inbred lines (BILs) for exploitation of the *Lactuca saligna* (wild lettuce) germplasm, *Theor. Appl. Genet.*, 109, 394–401, 2004.

Jeuken, M., van Wijk, R., Peleman, J., and Lindhout, P., An integrated interspecific AFLP map of lettuce (*Lactuca*) based on two *L. sativa* × *L. saligna* F-2 populations, *Theor. Appl. Genet.*, 103, 638–647, 2001.

Johnson, W. C., Jackson, L. E., Ochoa, O., van Wijk, R., Peleman, J., St Clair, D. A., and Michelmore, R. W., Lettuce, a shallow-rooted crop, and *Lactuca serriola*, its wild progenitor, differ at QTL determining root architecture and deep soil water exploitation, *Theor. Appl. Genet.*, 101, 1066–1073, 2000.

Judelson, H. S. and Michelmore, R. W., Temperature and genotype interactions in the expression of host resistance in lettuce downy mildew, *Physiol. Mol. Plant Pathol.*, 40, 233–245, 1992.

Judd, W. S., Campbell, C. S., Kellogg, E. A., and Stevens, P. F., *Plant Systematics: A Phylogenetic Approach*, Sunderland: Sinauer Assoc., 1999.

Kalloo, G., *Vegetable Breeding*, Vol. II, Boca Raton: CRC Press, 1988.

Karp, A., Edwards, K. J., Bruford, M., Vosman, B., Morgante, M., Seberg, O., Kremer, A., Boursot, P., Arctander, P., Tautz, D., and Hewitt, G. M., Molecular technologies for biodiversity evaluation: oppor-tunities and challenges, *Nat. Biotechnol.*, 15, 625–628, 1997a.

Karp, A., Kresovich, S., Bhat, K. V., Ayad, W. G., and Hodgkin, T., *Molecular Tools in Plant Genetic Resources Conservation: A Guide to the Technologies IPGRI Technical Bulletin No. 2*, Rome: Inter-national Plant Genetic Resources Institute, 1997b.

Karp, A., Isaac, P. G., and Ingram, D. S., Eds., In Molecular Tools for Screening Biodiversity, London: Chapman & Hall, 1998.

Kartesz, T. J., *A Synonymized Checklist of the Vascular Flora of the United States, Canada and Greenland*, Vol. 1, 2nd ed., Checklist.Portland: Timber Press, 1994.

Keimer, L., *Die Gartenpflanzen in Alten Aegypten*, Hamburg: Hoffmann und Campe, 1924.

Kesseli, R. V. and Michelmore, R. W., Genetic variation and phylogenies detected from isozyme markers in species of *Lactuca*, *J. Hered.*, 77, 324–331, 1986.

Kesseli, R. V., Ochoa, O., and Michelmore, R. W., Variation at RFLP loci in *Lactuca* spp. and origin of cultivated lettuce (*L. sativa*), *Genome*, 34, 430–436, 1991.

Kesseli, R. V., Paran, I., and Michelmore, R. W., Analysis of a detailed genetic linkage map of *Lactuca sativa* (lettuce) constructed from RFLP and RAPD markers, *Genetics*, 136, 1435–1446, 1994.

Kilian, N., *Lactuca stebbinsii* (Lactuceae Compositae), a puzzling new species from Angola, *Willdenowia*, 31, 71–78, 2001.

Kirpicznikov, M. E., *Lactuca* L, In Flora of the USSR, Bobrov, E. G. and Cvelev, N. N., Eds., 29, Moscow: Nauka, pp. 274–324, 1964.

Kirpicznikov, M. E., *Lactuca* L., In *Flora of the USSR (2000),Vol. 29* (English Translation, unpublished), Komarov, V. L., Ed. pp. 274–324, 2000.

Koike, S. T. and Gilbertson, R. L., Bacterial leaf spot, In *Compendium of Lettuce Diseases*, Davis, R. M., Subbarao, K. V., Raid, R. N., and Kurtz, E. A., Eds., St. Paul: APS Press, pp. 27–28, 1997.

Koopman, W. J. M., Plant systematics as useful tool for plant breeders, examples from lettuce, In *Eucarpia Leafy Vegetables'99, Proceedings of the Eucarpia Meeting on Leafy Vegetables Genetics and Breeding*, Lebeda, A. and Křístková, E., Eds., Olomouc: Palacký University in Olomouc, pp. 95–105, 1999.

Koopman, W. J. M., Identifying lettuce species (*Lactuca* subs. *Lactuca*, Asteraceae). A practical application of flow cytometrysubs, *Euphytica*, 116, 151–159, 2000.

Koopman, W. J. M., Zooming in on the Lettuce Genome: Species Relationships in Lactuca s.l. Inferred from *Chromosomal and Molecular Characters*. Ph.D. diss., Wageningen University, 2002.

Koopman, W. J. M., Phylogenetic signal in AFLP data sets, *Syst. Biol.*, 54, 197–217, 2005.

Koopman, W. J. M. and de Jong, H. J., A numerical analysis of karyotypes and DNA amounts in lettuce cultivars and species (*Lactuca* subs. *Lactuca*, Compositae), *Acta Bot. Neerl.*, 45, 211–222, 1996.

Koopman, W. J. M., de Jong, H. J., and de Vries, I. M. D., Chromosome banding in lettuce species (*Lactuca* sect. *Lactuca*, Compositae), *Plant Syst. Evol.*, 185, 249–257, 1993.

Koopman, W. J. M., Guetta, E., Van de Wiel, C. C. M., Vosman, B., and Van den Berg, R. G., Phylogenetic relationships among *Lactuca* (Asteraceae) species and related genera based on ITS-1 DNA sequences, *Am. J. Bot.*, 85, 1517–1530, 1998.

Koopman, W. J. M., Zevenbergen, M. J., and van den Berg, R. G., Species relationships in *Lactuca* s.l (Lactuceae, Asteraceae) inferred from AFLP fingerprints, *Am. J. Bot.*, 88, 1881–1887, 2001.

Koopman, W. J. M., Hadam, J., and Doležel, J., Evolution of DNA content and base composition in *Lactuca* (Asteraceae) and related genera, in *Zooming in on the lettuce genome: Species relationships in Lactuca s.l. inferred from chromosomal and molecular characters*, Ph.D. Dissertation, Wageningen University, pp. 97–124, 2002.

Kotlińska, T., Genetic resources of leafy vegetables in Poland, In *Report of a Vegetables Network. Joint Meeting with an ad hoc Group on Leafy Vegetables*, 22–24 May 2003, Skierniewice, Poland, Thomas, G., Astley, D., Boukema, I. W., Daunay, M. C., Del Greco, A., Diez, M. J., van Dooijeweert, W., Keller, J., Kotlinska, T., Lebeda, A., Lipman, E., Maggioni, L., and Rosa, E., Eds., Rome: International Plant Genetic Resources Institute, pp. 119–121, 2005.

Krause-Sakate, R., Le Gall, H., Fakhfakh, O., Peypelut, M., Marrakchi, M., Varveri, C., Pavan, M. A., Souche, S., Lot, H., Zerbini, F. M., and Candresse, T., Molecular and biological characterization of *lettuce mosaic virus* (LMV) isolates reveals a distinct and widespread type of resistance-breaking isolate, LMV-Most, *Phytopathology*, 92, 563–572, 2002.

Křístková, E. and Chytilová, V., Genetic resources of leafy vegetables in the Czech Republic, In *Report of a Vegetables Network. Joint Meeting with an ad hoc Group on Leafy Vegetables*, 22–24 May 2003, Skierniewice, Poland, Thomas, G., Astley, D., Boukema, I. W., Daunay, M. C., Del Greco, A., Diez, M. J., van Dooijeweert, W., Keller, J., Kotlinska, T., Lebeda, A., Lipman, E., Maggioni, L., and Rosa, E., Eds., Rome: International Plant Genetic Resources Institute, pp. 96–102, 2005.

Křístková, E. and Lebeda, A., Collection of *Lactuca* spp. genetic resources in the Czech Republic, In *Eucarpia Leafy Vegetables'99, Proceedings of the Eucarpia Meeting on Leafy Vegetables Genetics and Breeding*, Lebeda, A. and Křístková, E., Eds., Olomouc: Palacký University in Olomouc, pp. 109–116, 1999.

Křístková, E., Lebeda, A., and Doležalová, I., Collecting and evaluating of *Lactuca serriola* germplasm in Europe, In *Broad variation and precise characterization—limitation for the future*, Swiecicki, W., Naganowska, B., and Wolko, B., Eds., Poznan: Eucarpia, Section Genetic Resources, pp. 49–52, 2001.

Křístková, E., Doležalová, I., Lebeda, A., and Vinter, V., Descriptors and codes for genetic resources of *Lactuca sativa* L.), unpublished manuscript, 2006.

Kuang, H., Woo, S-S., Meyers, B. C., Nevo, E., and Michelmore, R. W., Multiple genetic processes result in heterogeneous processes of evolution within the major cluster disease resistance genes in lettuce, *Plant Cell*, 16, 2870–2894, 2004.

Kubát, K., Hrouda, L., Chrtek jun, J., Kaplan, Z., Kirschner, J., and Štěpánek, J., Eds., *Klíč ke květeně České republiky* (Key to the flora of the Czech Republic), Praha: Academia, 2002.

Kumar, A. and Hirochika, H., Applications of retrotransposons as genetic tools in plant biology, *Trends Plant Sci.*, 6, 127–134, 2001.

Kuwata, S., Kubo, S., Yamashita, S., and Doi, Y., Rod-shaped particles, a probable entity of lettuce big vein virus, *Ann. Phytopathol. Soc. Japan*, 49, 246–251, 1983.

Ladizinsky, G., Ecological and genetic considerations in collecting and using wild relatives, In *The Use of Plant Genetic Resources*, Brown, A. H. D., Frankel, O. H., Marshall, D. R., and Williams, J. T., Eds., Cambridge: Cambridge University Press, pp. 297–305, 1989.

Ladizinsky, G., *Plant Evolution under Domestication*, Dordrecht: Kluwer Academic Publishers, 1998.

Landry, B. S., Kesseli, R. V., Farrara, B., and Michelmore, R. W., A genetic map of lettuce (*Lactuca sativa* L.) with restriction fragment length polymorphism, isozyme, disease resistance and morphological markers, *Genetics*, 116, 331–337, 1987.

Lasser, T., *Flora de Venezuela.Vol. X, Parte Segunda*, Caracas: Edicion Especial del Instituto Botanico, Ministerio de Agricultura y Cria, 1964.

Lassner, M. and Bedbrook, J., Directed molecular evolution in plant improvement, *Curr. Opin. Plant Biol.*, 4, 152–156, 2001.

Latham, L. J. and Jones, R. A. C., Deploying partially resistant genotypes and plastic mulch on the soil surface to depress spread of lettuce big-vein virus disease in lettuce, *Austral. J. Agric. Res.*, 55, 131–138, 2004.

Lebeda, A., A contribution to the general theory of host-parasite specificity, *Phytopath. Z.*, 110, 226–234, 1984a.

Lebeda, A., Race-specific factors of resistance to *Bremia lactucae* in the world assortment of lettuce, *Scientia Hort.*, 22, 23–32, 1984b.

Lebeda, A., Occurrence of natural infection of powdery mildew (*Erysiphe cichoracearum*) by the genus Lactuca in Czechoslovakia, *Acta Phytopath. Acad. Sci. Hung.*, 20, 149–162, 1985a.

Lebeda, A., Differences in resistance of wild *Lactuca* species to natural infection of lettuce powdery mildew (*Erysiphe cichoracearum*), *Euphytica*, 34, 521–523, 1985b.

Lebeda, A., Susceptibility of some lettuce cultivars to natural infection by powdery mildew, *Tests Agrochem. & Cult. No. 6 (Ann. Appl. Biol.*, 106(Suppl.), 158–159, 1985c.

Lebeda, A., Specificity of interactions between wild *Lactuca* spp. and *Bremia lactucae* isolates from *Lactuca serriola*, *J. Phytopathol.*, 117, 54–64, 1986.

Lebeda, A., The location of sources of field resistance to *Bremia lactucae* in wild *Lactuca* species, *Plant Breed.*, 105, 75–77, 1990.

Lebeda, A., Evaluation of wild *Lactuca* species for resistance of natural infection of powdery mildew (*Erysiphe cichoracearum*), *Genet. Resour. Crop Evol.*, 41, 55–57, 1994.

Lebeda, A., Biodiversity of the interactions between germplasm of wild *Lactuca* spp. and related genera and lettuce downy mildew (*Bremia lactucae*), Report on research programme OECD Biological Resource Management for Sustainable Agricultural Systems, Wellesbourne, U.K.: HRI, 1998.

Lebeda, A., Powdery mildew on lettuce and wild *Lactuca* species. Paper presented at *The First International Powdery Mildew Conference*, Avignon (France), 1999.

Lebeda, A., Occurrence and variation in virulence of *Bremia lactucae* in natural populations of *Lactuca serriola*, In *Advances in Downy Mildew Research*, Spencer-Phillips, P. T. N., Gisi, U., and Lebeda, A., Eds., Dordrecht: Kluwer Academic Publishers, pp. 179–183, 2002.

Lebeda, A. and Astley, D., World genetic resources of *Lactuca* spp., their taxonomy and biodiversity, In *Eucarpia Leafy Vegetables'99, Proceedings of the Eucarpia Meeting on Leafy Vegetables Genetics and Breeding*, Lebeda, A. and Křístková, E., Eds., Olomouc: Palacký University in Olomouc, pp. 81–94, 1999.

Lebeda, A. and Blok, I., Race-specific resistance genes to *Bremia lactucae* in new Czechoslovak lettuce cultivars and location of resistance in a *Lactuca serriola × Lactuca sativa* hybrid, *Archiv. Phytopathol. Pflanzenschutz*, 27, 65–72, 1991.

Lebeda, A. and Boukema, I. W., Leafy vegetables genetic resources, In *Report of a Network Coordinating Group on Vegatables; Ad hoc Meeting, 26–27 May 2000, Vila Real, Portugal*, Maggioni, L. and Spellman, O., Eds., Rome: IPGRI, pp. 48–57, 2001.

Lebeda, A. and Boukema, I. W., *Ad Hoc* meeting on leafy vegetables, In *Report of a Vegetables Network. Joint Meeting with an ad hoc Group on Leafy Vegetables*, 22–24 May 2003, Skierniewice, Poland, Thomas, G., Astley, D., Boukema, I. W., Daunay, M. C., Del Greco, A., Diez, M. J., van Dooijeweert, W., Keller, J., Kotlinska, T., Lebeda, A., Lipman, E., Maggioni, L., and Rosa, E., Eds., Rome: International Plant Genetic Resources Institute, pp. 82–94, 2005.

Lebeda, A. and Buczkowski, J., Occurrence of *Erysiphe cichoracearum* perithecia on wild *Lactuca* species, *J. Phytopathol.*, 115, 21–28, 1986.

Lebeda, A. and Jendrůlek, T., Application of multivariate analysis for characterizing the relationships between wild *Lactuca* spp. and *Bremia lactucae*, *Acta Phytopathol. Entomol. Hung.*, 24, 317–331, 1989.

Lebeda, A. and Křístková, E., Genetic resources of vegetable crops from the genus *Lactuca*, *Hort. Sci. (Prague)*, 22, 117–121, 1995.

Lebeda, A. and Křístková, E., Eds., *Eucarpia Leafy Vegetables '99, Proceedings of the Eucarpia Meeting on Leafy Vegetables Genetics and Breeding*, Olmouc: Palacký University in Olomouc, 1999.

Lebeda, A. and Mieslerová, B., Lettuce powdery mildew—an unknown disease of lettuce, In *Eucarpia Leafy Vegetables 2003, Proceedings of the Eucarpia Meeting on Leafy Vegetables Genetics and Breeding, Noordwijkerhout*, van Hintum, Th. J. L., Lebeda, A., Pink, D. A., and Schut, J. W., Eds., Wageningen: Centre for Genetic Resources, p. 164, 2003.

Lebeda, A. and Petrželová, I., Occurrence and characterization of race-specific resistance to *Bremia lactucae* in wild *Lactuca* spp., In *Broad Variation and Precise Characterization—Limitation for the Future, Eucarpia, Section Genetic Resources*, Swiecicki, W., Naganowska, B., and Wolko, B., Eds., Poznan: Prodruk, pp. 232–233, 2001.

Lebeda, A. and Petrželová, I., Variation and distribution of virulence phenotypes of *Bremia lactucae* in natural populations of *Lactuca serriola*, *Plant Pathol.*, 53, 316–324, 2004.

Lebeda, A. and Pink, D. A. C., Histological aspects of the response of wild *Lactuca* spp. and their hybrids, with *L. sativa* to lettuce downy mildew (*Bremia lactucae*), *Plant Pathol.*, 47, 723–736, 1998.

Lebeda, A. and Reinink, K., Histological characterization of resistance in *Lactuca saligna* to lettuce downy mildew (*Bremia lactucae*), *Physiol. Mol. Plant Pathol.*, 44, 125–139, 1994.

Lebeda, A. and Schwinn, F. J., The downy mildews-an overview of recent research progress, *J. Plant Dis. Protec.*, 101, 225–254, 1994.

Lebeda, A. and Sedlářová, M., Cellular mechanisms involved in the expression of specificity in *Lactuca* spp.—*Bremia lactucae* interactions, In *Eucarpia Leafy Vegetables 2003, Proceedings of the Eucarpia Meeting on Leafy Vegetables Genetics and Breeding*, Noordwijkerhout, van Hintum, Th. J. L., Lebeda, A., Pink, D. A., and Schut, J. W., Eds., Wageningen: Centre for Genetic Resources, pp. 55–60, 2003.

Lebeda, A. and Syrovátko, P., Specificity of Bremia lactucae isolates from Lactuca sativa and some Asteraceae Plants, *Acta Phytopathol. Entomol. Hung.*, 23, 39–48, 1988.

Lebeda, A. and Zinkernagel, V., Durability of race-specific resistance in lettuce against lettuce downy mildew (*Bremia lactucae*), In *Eucarpia Leafy Vegetables '99, Proceedings of the Eucarpia Meeting on Leafy Vegetables Genetics and Breeding*, Lebeda, A. and Křístková, E., Eds., Olomouc: Palacký University in Olomouc, pp. 183–189, 1999.

Lebeda, A. and Zinkernagel, V., Characterization of new highly virulent German isolates of *Bremia lactucae* and efficiency of resistance in wild *Lactuca* spp. germplasm, *J. Phytopathol.*, 151, 274–282, 2003a.

Lebeda, A. and Zinkernagel, V., Evolution and distribution of virulence in the German population of *Bremia lactucae*, *Plant Pathol.*, 52, 41–51, 2003b.

Lebeda, A., Doležalová, I., and Astley, D., Representation of wild *Lactuca* spp. (Asteraceae, Lactuceae) in world genebank collections, *Genet. Resour. Crop Evol.*, 51, 167–174, 2004a.

Lebeda, A., Doležalová, I., and Vondráková, D., The working collection of the genus *Lactuca* at the Palacký University, Czech Republic, In *Report of a Vegetables Network. Joint Meeting with an ad hoc group on Leafy Vegetables, 22–24 May 2003, Skierniewice, Poland*, Thomas, G., Astley, D., Boukema, I. W., Daunay, M. C., Del Greco, A., Diez, M. J., van Dooijeweert, W., Keller, J., Kotlinska T., Lebeda, A., Lipman, E., Maggioni, L., Rosa, E., Eds., International Plant Genetic Resources Institute, Rome, 103–107, 2005.

Lebeda, A., Pink, D. A. C., and Astley, D., Aspects of the interactions between wild *Lactuca* spp. and related genera and lettuce downy mildew (*Bremia lactucae*), In *Advances in downy mildew research*, Spencer-Phillips, P. T. N., Gisi, U., and Lebeda, A., Eds., Dordrecht: Kluwer Academic Publishers, pp. 85–117, 2002b.

Lebeda, A., Pink, D. A. C., and Mieslerová, B., Host-parasite specificity and defense variability in the *Lactuca* spp.—*Bremia lactucae* pathosystem, *J. Plant Pathol.*, 83, 25–35, 2001c.

Lebeda, A., Doležalová, I., Křístková, E., Vinter, V., Vránová, O., Doležal, K., Tarkowski, P. et al., Complex research of taxonomy and ecobiology of wild *Lactuca* spp. genetic resources, In *Eucarpia*

Leafy Vegetables '99, Proceedings of the Eucarpia Meeting on Leafy Vegetables Genetics and Breeding, Lebeda, A. and Křístková, E., Eds., Olomouc: Palacký University in Olomouc, pp. 117–131, 1999.

Lebeda, A., Doležalová, I., Křístková, E., Janeček, J., Vinter, V., Vránová, O., Doležal, K. et al., Biodiversity of genetic resources of wild *Lactuca* spp, In *Broad Variation and Precise Characterization—Limitation for the Future*, Święcicki, W., Naganowska, B., and Wolko, B., Eds., Poznań: Prodruk, pp. 53–56, 2001a.

Lebeda, A., Doležalová, I., Křístková, E., and Mieslerová, B., Biodiversity and ecogeography of wild *Lactuca* spp. in some European countries, *Genet. Resour. Crop Evol.*, 48, 153–164, 2001b.

Lebeda, A., Mieslerová, B., Doležalová, I., and Křístková, E., Occurrence of powdery mildew on *Lactuca viminea* subsp. *chondrilliflora* in South France, *Mycotaxon*, LXXXIV, 83–87, 2002a.

Lebeda, A., Doležalová, I., Feráková, V., and Astley, D., Geographical distribution of wild *Lactuca* species (Asteraceae Lactuceae), *Bot. Rev.*, 70, 328–356, 2004b.

Lebeda, A., Doležalová, I., Křístková, E., and Novotná, A., Morphological and developmental characteristics of *Lactuca serriola* germplasm originating from Europe, In *Summaries and Program of 17th International Lettuce and Leafy Vegetable Conference*, 28–31 August 2004, Sandman Hotel, Montreal-Longueuil, Agriculture and Agri-Food Canada, 28–29, 2004c, (Abstract).

Lebeda, A., Doležalová, I., Janeček, J., and Gasmanová, N., Differences in relative DNA content of *Lactuca serriola* germplasm collected in Europe, In *Summaries and Program of 17th International Lettuce and Leafy Vegetable Conference*, 29–30, Agriculture and Agri-Food Canada, 29–30, 2004d, (Abstract).

Lebeda, A., Doležalová, I., Křístková, E., and Novotná, A., Comparative study of variation of some morphological and developmental characteristics of *Lactuca serriola* germplasm collected in Central Europe (Czech Republic) and the British Isles (England, U.K.), *XVII Eucarpia Genetic Resources Section Meeting; "Plant Genetic Resources of Geographical and" other "islands (Conservation, Evaluation and Use for Plant Breeding),"* Bullitta, S., Ed., Castelsardo (Italy), 30 March–2 April 2005; Book of Abstracts, CNR-ISPAAM, sezione Sassari, Italy, 46, 2005c.

Lebeda, A., Gasmanová, N., Doležalová, I., and Janeček, J., Preliminary study of relative DNA content variation in some European populations of *Lactuca serriola*, In *XVII International Botanical Congress, Vienna, Austria, 17–23 July 2005, Abstracts*, p. 276, Abstract P0229, 2005d.

Lebeda, A., Widrlechner, M. P., Staub, J., Ezura, H., Zalapa, J., and Křístková, E., Cucurbits (Cucurbitaceae; *Cucumis* spp., *Cucurbita* spp., *Citrullus* spp.), In *Genetic resources, chromosome engineering, and crop improvement series*, Singh, R. J., Ed., Vol. 3, Boca Raton: CRC Press, 2006.

Lebeda, A., Doležalová, I., Křístková, E., Dehmer, K. J., Astley, D., van de Wiel, C. C. M., and van Treuren, R., Acquisition and ecological characterization of *Lactuca serriola* L. germplasm collected in the Czech Republic, Germany, The Netherlands and United Kingdom, *Genet. Resour. Crop Evol.*, 2006 (in press).

Lenné, J. M. and Wood, D., Plant disease and the use of wild germplasm, *Annu. Rev. Phytopathol.*, 29, 35–63, 1991.

Lindqvist, K., Cytogenetic studies in the *serriola* group of *Lactuca*, *Hereditas*, 46, 75–151, 1960a.

Lindqvist, K., On the origin of cultivated lettuce, *Hereditas*, 46, 319–350, 1960b.

Lot, H., Campbell, R. N., Souche, S., Milne, R. G., and Roggero, P., Transmission by *Olpidium brassicae* of Mirafiori lettuce virus and Lettuce big-vein virus, and their roles in lettuce big-vein etiology, *Phytopathology*, 92, 288–293, 2002.

Lundqvist, J. and Jäger, E. J., Eds., In Index Holmiensis, Vol. VIII Dicotyledoneae K-M, Stockholm: Swedish Museum of Natural History, 1995.

Maesen van der, L. J. G., Systematics of crop–weed complexes, *Acta Bot. Neerlandica*, 43, 78–79, 1994.

Maggioni, L., Conservation and use of vegetable genetic resources: a European perspective, *Acta Hort.*, 637, 13–30, 2004.

Maisonneuve, B., *Lactuca virosa*, a source of disease resistance genes for lettuce breeding: results and difficulties for gene introgression, In *Eucarpia Leafy Vegetables 2003, Proceedings of the Eucarpia Meeting on Leafy Vegetables Genetics and Breeding*, Noordwijkerhout, van Hintum, Th. J. L., Lebeda, A., Pink, D. A., and Schut, J. W., Eds., Wagenigen: Centre for Genetic Resources, pp. 61–67, 2003.

Maisonneuve, B., Chovelon, V., and Lot, H., Inheritance of resistance to beet western yellows virus in *Lactuca virosa* L, *HortScience*, 26, 1543–1545, 1991.

Maisonneuve, B., Chupeau, M. C., Bellec, Y., and Chupeau, Y., Sexual and somatic hybridization in the genus *Lactuca, Euphytica*, 85, 281–285, 1995.

Maisonneuve, B., Bellec, Y., Souche, S., and Lot, H., New resistances against downy mildew and lettuce mosaic potyvirus in wild *Lactuca* spp., In *Eucarpia Leafy Vegetables'99, Proceedings of the Eucarpia Meeting on Leafy Vegetables Genetics and Breeding*, Lebeda, A. and Křístková, E., Eds., Olomouc: Palacký University in Olomouc, pp. 191–197, 1999.

Maisonneuve, B., Valéry, S., Morison, N., and Vaissière, B. E., Pollen-mediated gene flow by insects in *Lactuca sativa*, In *17th International Lettuce and Leafy Vegetable Conference*, Jenni, S., Ed., Montreal: PQ, p. 27, 2004.

Mansfield, J. W., Bennett, M., Bestwick, Ch., and Woods-Tör, A., Phenotypic expression of gene-for-gene interaction involving fungal and bacterial pathogens: variation from recognition to response, In *The Gene-for-Gene Relationship in Plant–Parasite Interactions*, Crute, I. R., Holub, E. B., and Burdon, J. J., Eds., Wallingford: CAB International, pp. 265–291, 1997.

Marlatt, R. B., Biology, morphology, taxonomy and disease relations of the fungus *Bremia*, *Florida agricultural Experimental Station Technical Bulletin*, 764, 1974.

Matheron, M. E. and Koike, S. T., First report of fusarium wilt of lettuce caused by *Fusarium oxysporum* f. sp. *lactucae* in Arizona, *Plant Dis.*, 87, 1265, 2003.

Matsumoto, E., Interspecific somatic hybridization between lettuce (*Lactuca sativa*) and wild species *L. virosa*, *Plant Cell Rep.*, 9, 531–534, 1991.

Matuo, T. and Motohashi, S., On *Fusarium oxysporum* f. sp. *lactucae* n.f. causing root rot of lettuce, *Trans. Mycol. Soc. Japan*, 32, 13–15, 1967.

Mazier, M., Maisonneuve, B., Bellec, Y., Chupeau, M. C., Souche, S., and Chupeau, Y., Interest for protoplasts in lettuce breeding, In *Eucarpia Leafy Vegetables '99, Proceedings of the Eucarpia Meeting on Leafy Vegetables Genetics and Breeding*, Lebeda, A. and Křístková, E., Eds., Olomouc: Palacký University in Olomouc, pp. 239–244, 1999.

McCabe, M. S., Mohapatra, U. B., Debnath, S. C., Power, J. B., and Davey, M. R., Integration, expression and inheritance of two linked T-DNA marker genes in transgenic lettuce, *Mol. Breed.*, 5, 329–344, 1999a.

McCabe, M. S., Schepers, F., van der Arend, A., Mohapatra, U., de Laat, A. M. M., Power, J. B., and Davey, M. R., Increased stable inheritance of herbicide resistance in transgenic lettuce carrying a petE promoter-bar gene compared with a CaMV 35S-bar gene, *Theor. Appl. Genet.*, 99, 587–592, 1999b.

McCabe, M. S., Garratt, L. C., Schepers, F., Jordi, W., Stoopen, G. M., Davelaar, F., van Rhijn, J. H. A., Power, J. B., and Davey, M. R., Effects of P-SAG12-IPT gene expression on development and senescence in transgenic lettuce, *Plant Physiol.*, 127, 505–516, 2001.

McCreight, J. D., Resistance in wild lettuce to lettuce infectious yellows, *HortScience*, 22, 640–642, 1987.

McCreight, J. D., Kishaba, A. N., and Mayberry, K. S., Lettuce infectious yellows tolerance in lettuce, *J. Am. Soc. Hort. Sci.*, 111, 788–792, 1986.

McGregor, R. L., Barkley, T. M., Brooks, R. E., and Schofield, E. K., *Flora of the Great Plains*, Kansas: University Press of Kansas, 1986.

McGuire, P. E., Ryder, E. J., Michelmore, R. W., Clark, R. L., Antle, R., Emery, G., Hannan, R. M., Kesseli, R. V., Kurtz, E. A., Ochoa, O., Rubatzky, V. E., and Waycott, W., *Genetic resources of lettuce and Lactuca species in California. An assessment of the USDA and UC collections and recommendations for long-term security. Report No. 12*, University of California, Genetic Resources Conservation Program, Davis, CA, 1993.

McLean, D. L. and Kinsey, M. G., Three variants of lettuce mosaic virus and methods utilized for differentiation, *Phytopatholgy*, 52, 403–406, 1962.

McLean, D. L. and Kinsey, M. G., Transmission studies of a highly virulent variant of lettuce mosaic virus, *Plant Dis. Rep.*, 47, 474–476, 1963.

Mejia, L. and McDaniel, R. G., Electrophoretic characterization of lettuce cultivars, *Hort. Sci.*, 21, 278–280, 1986.

Mejías, J. A., Cytotaxonomic studies in the Iberian taxa of the genus *Lactuca* (Compositae), *Bot. Helvet.*, 103, 113–130, 1993.

Meusel, H. and Jäger, E. J., *Vergleichende Chorologie der Zentraleuropäischen Flora*, New York: Gustav Fischer Verlag, Jena-Stuttgart, 1992.

Meyers, B. C., Chin, D. B., Shen, K. A., Sivaramakrishnan, S., Lavelle, D. O., Zhang, Z., and Michelmore, R. W., The major resistance gene cluster in lettuce is highly duplicated and spans several megabases, *Plant Cell*, 10, 1817–1832, 1998a.

Meyers, B. C., Shen, K. A., Rohani, P., Gaut, B. S., and Michelmore, R. W., Receptor-like genes in the major resistance locus of lettuce are subject to divergent selection, *Plant Cell*, 10, 1833–1846, 1998b.

Michelmore, R. W., Genetic variation in lettuce, In *Annual Report 2001–2002*, Kurtz, E., Ed., Salinas: California Lettuce Research Board, pp. 77–86, 2002.

Michelmore, R. W., The impact zone: genomics and breeding for durable disease resistance, *Curr. Opin. Plant Biol.*, 6, 397–404, 2003.

Michelmore, R. W. and Eash, J. A., Lettuce, In *Handbook of Plant Cell Culture*, Evans, D. A., Sharp, W. R., and Ammirato, P. V., Eds., Vol. 4, New York: MacMillan, pp. 512–551, 1986.

Michelmore, R. W. and Meyers, B. C., Clusters of resistance genes in plants evolve by divergent selection and a birth-and-death process, *Genome Res.*, 8, 1113–1130, 1998.

Michelmore, R. W. and Ochoa, O. E., Breeding crisphead lettuce, California Lettuce Research Board, Annual Report, April 1, 2004 through March 31, 2005, Salinas: California Research Board, pp. 68–78, 2005.

Michelmore, R. W., Kozik, A., Truco, M. J., Matviencho, M., Ochoa, O., Damme, M. V., Lavelle, D., Lin, H., Pande, B., McHale, L., Sudarshana, P., Argyris, J., Ellison, P., Bradford, K., Jackson, L., and Kesseli, R., ESTs and candidate gene approaches in the Compositae Genome Project, van Hintum, Th. J. L., Lebeda, A., Pink, D., Schut, J. W., In *Eucarpia Leafy Vegetables 2003, Proceedings of the Eucarpia Meeting on Leafy Vegetables Genetics and Breeding*, 2003, Centre for Genetic Resources, Wageningen, 131–136.

Michelmore, R., Marsh, E., Seely, S., and Landry, B., Transformation of lettuce (*Lactuca sativa*) mediated by *Agrobacterium tumefaciens*, *Plant Cell Rep.*, 6, 439–442, 1987.

Michelmore, R. W., Kesseli, R. V., and Ryder, E. J., Genetic mapping in lettuce, In *DNA Based Markers in Plants*, Phillips, R. L. and Vasil, P. K., Eds., Dordrecht: Kluwer Academic Publishers, pp. 223–239, 1994.

Mizutani, T. and Tanaka, T., Genetic analyses of isozyme in lettuce, *Lactuca sativa*, and its relatives, *J. Jap. Soc. Hort. Sci.*, 72, 122–127, 2003.

Mizutani, T., Liu, X. J., Tashiro, Y., Miyazaki, S., and Shimazaki, K., Plant regeneration and cell fusion of protoplasts from lettuce cultivars and related wild species in Japan, *Bull. Fac. Agric., Saga University*, 67, 109–118, 1989.

Mohapatra, U., McCabe, M. S., Power, J. B., Schepers, F., Van der Arend, A., and Davey, M. R., Expression of the bar gene confers herbicide resistance in transgenic lettuce, *Transgen. Res.*, 8, 33–44, 1999.

Montmollin, B. de, Étude cytotaxonomique de la flore de la Crète, III. Nombres chromosomiques, *Candollea*, 41, 431–439, 1986.

Moreno-Vazquez, S., Ochoa, O. E., Faber, N., Chao, S. M., Jacobs, J. M. E., Maisonneuve, B., Kesseli, R. V., and Michelmore, R. W., SNP-based codominant markers for a recessive gene conferring resistance to corky root rot (*Rhizomonas suberifaciens*) in lettuce (*Lactuca sativa*), *Genome*, 46, 1059–1069, 2003.

Mou, B. Q., Genetic variation of beta-carotene and lutein contents in lettuce, *J. Amer. Soc. Hort. Sci.*, 130, 870–876, 2005.

Mou, B. Q. and Bull, C., Screening lettuce germplasm for new sources of resistance to corky root, *J. Amer. Soc. Hort. Sci.*, 129, 712–716, 2004.

Mou, B. Q. and Liu, Y. B., Host plant resistance to leafminers in lettuce, *J. Amer. Soc. Hort. Sci.*, 129, 383–388, 2004.

Mou, B. and Ryder, E. J., Screening and breeding for resistance to leafminer (*Liriomyza langei*) in lettuce and spinach, In *Eucarpia Leafy Vegetables 2003, Proceedings of the Eucarpia Meeting on Leafy Vegetables Genetics and Breeding*, van Hintum, Th. J. L., Lebeda, A., Pink, D. A., and Schut, J. W., Eds., Wageningen: Centre for Genetic Resources, pp. 43–47, 2003.

Mou, B. and Ryder, E. J., Relationship between the nutritional value and the head structure of lettuce, *Acta Hort.*, 637, 361–367, 2004.

Mou, B., Ryder, E. J., Tanaka, J., Liu, Y. B., and Chaney, W. E., Breeding for resistance to leafminer in lettuce, *Acta Hort.*, 637, 57–62, 2004.

Nash, D. L., Tribe XI, Cichorieae, In *Flora of Guatemala, Fieldiana, Botany 24 (12)*, Nash, D. L. and Williams, L. O., Eds., Chicago: Chicago Museum of Natural History, 1976.

Navarro, J. A., Botella, F., Maruhenda, A., Sastre, P., Sanchez-Pina, M. A., and Pallas, V., Comparative infection progress analysis of Lettuce big-vein virus and Mirafiori lettuce virus in lettuce crops by developed molecular diagnosis techniques, *Phytopathology*, 94, 470–477, 2004.

Nazarova, É. A., *Takhtajaniantha* Nazarova and *Lactucella* Nazarova: Two new genera of the tribe Lactuceae (family Asteraceae), *Biol. J. Armen.*, 43, 179–183, 1990.

Nessler, C. L., A systematic survey of the tribe Cichorieae in Virginia U.S.A., *Castanea*, 41, 226–248, 1976.

Netzer, D., Globerson, D., Weintal, Ch., and Elyassi, R., Sources and inheritance of resistance to Stemphylium leaf spot of lettuce, *Euphytica*, 34, 393–396, 1985.

Nicaise, V., German-Retana, S., Sanjuan, R., Dubrana, M. P., Mazier, M., Maisonneuve, B., Candresse, T., Caranta, C., and LeGall, O., The eukaryotic translation initiation factor 4E controls lettuce susceptibility to the potyvirus Lettuce mosaic virus, *Plant Physiol.*, 132, 1272–1282, 2003a.

Nicaise, V., German-Retana, S., Sanjuan, R., Dubrana, M. P., Mazier, M., Maisonneuve, B., Candresse, T., Caranta, C., and LeGall, O., Molecular characterization of *mo1*, a recessive gene associated with *Lettuce mosaic potyvirus* resistance in lettuce, In *Eucarpia Leafy Vegetables 2003, Proceedings of the Eucarpia Meeting on Leafy Vegetables Genetics and Breeding*, van Hintum, Th. J. L, Lebeda, A., Pink, D. A., and Schut, J. W., Eds., Wageningen: Centre for Genetic Resources, pp. 143–148, 2003b.

Norwood, J. M., Crute, I. R., and Lebeda, A., The location and characteristics of novel sources of resistance to *Bremia lactucae* Regel (downy mildew) in wild *Lactuca* L. species, *Euphytica*, 30, 659–668, 1981.

Obermeier, C., Sears, J. L., Liu, H. Y., Schlueter, K. O., Ryder, E. J., Duffus, J. E., Koike, S. T., and Wisler, G. C., Characterization of distinct tombusviruses that cause disease of lettuce and tomato in the western United States, *Phytopathology*, 91, 797–806, 2001.

Ochoa, O., Delp, B., and Michelmore, R. W., Resistance in *Lactuca* spp. to *Microdochium panattoniana* (lettuce anthracnose), *Euphytica*, 36, 609–614, 1987.

Ohwi, J., *Flora of Japan*, Washington DC: Smithsonian Institution, 1965.

Okubara, P. A., Arroyo-Garcia, R., Shen, K. A., Mazier, M., Meyers, B. C., Ochoa, O. E., Kim, S., Yang, C-H., and Michelmore, R. W., A transgenic mutant of *Lactuca sativa* (lettuce) with a T-DNA tightly linked to loss of downy mildew resistance, *Mol. Plant Microbe Interact.*, 10, 970–977, 1997.

O'Malley, P. J. and Hartmann, R. W., Resistance to tomato spotted wilt virus in lettuce, *HortScience*, 24, 360–362, 1989.

Ortega, J. and Navarro, B., Estudios en la flora de Macaronesia: Algunos números de cromosomas, *Bot. Macaron*, 3, 73–80, 1977.

Paulus, A. O., Powdery mildew, In *Compendium of Lettuce Diseases*, Davis, R. M., Subbarao, K. V., Raid, R. N., and Kurtz, E. A., Eds., St. Paul: APS Press, p. 23, 1997.

Pavlov, N. V., Ed, In Flora Kazakhstana IX, Alma-Ata: Izdavatelstvo Nauka, 1966.

Petrželová, I. and Lebeda, A., Occurrence of *Bremia lactucae* in natural populations of *Lactuca serriola*, *J. Phytopathol.*, 152, 391–398, 2004a.

Petrželová, I. and Lebeda, A., Temporal and spatial variation in virulence of natural populations of *Bremia lactucae* occurring on *Lactuca serriola*, In *Advances in downy mildew research*, Spencer-Phillips, P. T. N. and Jeger, M., Eds., Vol. 2, Dordrecht: Kluwer Academic Publishers, pp. 141–163, 2004b.

Phillips, R. and Rix, M., *Vegetables*, London: Pan Books Ltd, 1993.

Pickersgill, B., Biosystematics of crop-weed complexes, *Die Kulturpflanze*, 29, 377–388, 1981.

Pileggi, M., Pereiara, A. A. M., Silva, J. D., Pileggi, S. A. V., and Verma, D. P. S., An improved method for transformation of lettuce by Agrobacterium tumefaciens with a gene that confers freezing resistance, *Braz. Arch. Biol. Technol.*, 44, 191–196, 2001.

Pink, D. and Astley, D., Leafy vegetables in the United Kingdom, In *Report of a vegetables network. Joint Meeting with an ad hoc Group on Leafy Vegetables*, Thomas, G., Astley, D., Boukema, I. W., Daunay, M. C., Del Greco, A., Diez, M. J., van Dooijeweert, W., Keller, J., Kotlinska, T., Lebeda, A., Lipman, E., Maggioni, L., and Rosa, E., Eds., Rome: International Plant Genetic Resources Institute, pp. 127–128, 2005.

Pink, D. A. C. and Keane, E. M., Lettuce: *Lactuca sativa* L., In *Genetic improvement of vegetable crops*, Kalloo, G. and Bergh, B. O., Eds., Oxford: Pergamon Press, pp. 543–571, 1993.

Pink, D. A. C., Walkey, D. G. A., and McClement, S. J., Genetics of resistance to beet western yellows virus in lettuce, *Plant Pathol.*, 40, 542–545, 1991.

Pink, D. A. C., Kostova, D., and Walkey, D. G. A., Differentiation of pathotypes of lettuce mosaic virus, *Plant Pathol.*, 41, 5–12, 1992a.

Pink, D. A. C., Lot, H., and Johnson, R., Novel types of lettuce mosaic virus- breakdown of a durable resistance?, *Euphytica*, 63, 169–174, 1992b.

Pistrick, K. and Mal'cev, I. I., Expedition to the south-western Hissar Mountains (Southern Uzbekistan) for collecting plant genetic resources in 1995, *Genet. Resour. Crop Evol.*, 45, 225–233, 1998.

Plocik, A., Layden, J., and Kesseli, R., Comparative analysis of NBS domain sequences of NBS-LRR disease resistance genes from sunflower, lettuce, and chicory, *Mol. Phylogenet. Evol.*, 31, 153–163, 2004.

Pope, G. V., *Flora Zambesiaca 6 (1)*, Kew Botanical Gardens: Kew, 1992.

Provvidenti, R., Robinson, R. W., and Shail, W., A source of resistance to a strain of cucumber mosaic virus in *Lactuca saligna* L, *HortScience*, 15, 528–529, 1980.

Raid, R. N., Stemphylium leaf spot, In *Compendium of Lettuce Diseases*, Davis, R. M., Subbarao, K. V., Raid, R. N., and Kurtz, E. A., Eds., St. Paul: APS Press, pp. 25–26, 1997.

Rechinger, K. H., *Flora Iranica. Flora des Iranischen Hochlandes und der Umrahmenden Gebirge*, Graz: Academische Druck-u. Verlagsanstalt, 1977.

Reinink, K., Genetics of nitrate content of lettuce, 1: Analysis of generation means, *Euphytica*, 54, 83–92, 1991.

Reinink, K., Genetics of nitrate content of lettuce, 2: Components of variance, *Euphytica*, 60, 61–74, 1992.

Reinink, K., Lettuce resistance breeding, In *Eucarpia Leafy Vegetables '99, Proceedings of the Eucarpia Meeting on Leafy Vegetables Genetics and Breeding*, Lebeda, A. and Křístková, E., Eds., Olomouc: Palacký University in Olomouc, pp. 139–148, 1999.

Reinink, K. and Groenwold, R., The inheritance of nitrate content in lettuce (*Lactuca sativa* L.), *Euphytica*, 36, 733–744, 1987.

Reinink, K., Groenwold, R., and Bootsma, A., Genotypical differences in nitrate content in *Lactuca sativa* L. and related species and correlation with dry matter content, *Euphytica*, 36, 11–18, 1987.

Revers, F., Yang, S. J., Walter, J., Souche, S., Lot, H., LeGall, O., Candresse, T., and Dunez, J., Comparison of the complete nucleotide sequences of two isolates of lettuce mosaic virus differing in their biological properties, *Virus Res.*, 47, 167–177, 1997.

Ride, J. P., Non-host resistance to fungi, In *Mechanisms of Resistance to Plant Diseases*, Fraser, R. S. S., Ed., Dordrecht: Martinus Nijhoff/Dr. W. Junk Publishers, pp. 29–61, 1985.

Robinson, R. W., McCreight, J. D., and Ryder, E. J., The genes of lettuce and closely related species, *Plant Breed. Rev.*, 1, 267–293, 1983.

Rodenburg, C. M., *Varieties of Lettuce. An International Monograph*, Zwolle: W.E.J. Tjeenk Willink, 1960.

Roggero, P., Ciuffo, M., Vaira, A. M., Accotto, G. P., Masenga, V., and Milne, R. G., An *Ophiovirus* isolated from lettuce with big-vein symptoms, *Arch. Virol.*, 145, 2629–2642, 2000.

Roggero, P., Lot, H., Souche, S., Lenzi, R., and Milne, R. G., Occurrence of Mirafiori lettuce virus and Lettuce big-vein virus in relation to development of big-vein symptoms in lettuce crops, *Europ. J. Plant Pathol.*, 109, 261–267, 2003.

Rothmaler, W., Bassler, M., Jagger, E. J., and Werner, K., *Exkursionsflora von Deutschland, Bd. 2, Gefässpflanzen. Grundband*, 17th ed., Heidelberg: Spektrum Akademischer Verlag, 1999.

Roux, L., Chengjiu, Z., and Roux, Y., Characterization of *Lactuca sativa* L. and related species by electro-focusing of esterase, *Agronomie*, 5, 915–921, 1985.

Rubatzky, V. E. and Yamaguchi, M , *World Vegetables, Principles, Production, and Nutritive Values*, 2nd ed., New York: Chapman & Hall, 1997.

Rulkens, A. J. H., De CGN sla-collectie: inventarisatie, paspoortgegevens en enkele richtlijnen voor de toekomst *CGN report: CGN-T 1987-1*, Wageningen: CGN, 1987.

Ryder, E. J., Evaluation of lettuce varieties and breeding lines for resistance to common lettuce mosaic. U.S. Dept. Agri., ARS, Tech. Bull. 1391, Washington, DC, USA, 8 pp., 1968.

Ryder, E. J., Inheritance of resistance to common lettuce mosaic, *J. Amer. Soc. Hort. Sci.*, 95, 378–379, 1970.

Ryder, E. J., Genetic studies in lettuce (*Lactuca sativa* L.), *J. Amer. Soc. Hort. Sci.*, 96, 826–828, 1971.

Ryder, E. J., "Salinas" lettuce, *HortScience*, 14, 283–284, 1979a.

Ryder, E. J., Vanguard 75 Lettuce, *HortScience*, 14, 284–286, 1979b.

Ryder, E. J., Inheritance, linkage, and gene interaction studies in lettuce, *J. Amer. Soc. Hort. Sci.*, 108, 985–991, 1983.

Ryder, E. J., Use of early flowering genes to reduce generation time in backcrossing, with specific application to lettuce breeding, *J. Amer. Soc. Hort. Sci.*, 110, 570–573, 1985.

Ryder, E. J., Lettuce breeding, In *Breeding Vegetable Crops*, Bassett, M., Ed., Westport: AVI Publishing Co., 1986.

Ryder, E. J., Early flowering in lettuce as influenced by a second flowering time gene and seasonal variation, *J. Amer. Soc. Hort. Sci.*, 113, 456–460, 1988.

Ryder, E. J., *Lettuce, Endive and Cichory*, Wallingford: CABI Publishing, 1999.

Ryder, E. J., Genetics in lettuce breeding: past, present and future, In *Eucarpia Leafy Vegetables '99, Proceedings of the Eucarpia Meeting on Leafy Vegetables Genetics and Breeding*, Lebeda, A. and Křístková, E., Eds., Olomouc: Palacký University in Olomouc, pp. 81–94, 1999b.

Ryder, E. J., A mild systemic reaction to lettuce mosaic virus in lettuce (*Lactuca sativa*), Inheritance and interaction with an allele for resistance, *J. Amer. Soc. Hort. Sci.*, 127, 814–818, 2002.

Ryder, E. J. and Robinson, B. J., Big-vein resistance in lettuce, identifying, selecting, and testing resistant cultivars and breeding lines, *J. Amer. Soc. Hort. Sci.*, 120, 741–746, 1995.

Ryder, E. J. and Whitaker, T., Lettuce, *Lactuca sativa* (Compositae), In *Evolution of Crop Plants*, Smartt, J. and Simmonds, N. W., Eds. 2nd ed., Harlow: Longman Scientific & Technical, pp. 53–56, 1995.

Ryder, E. J., Grube, R. C., Subbarao, K. V., and Koike, S. T., Breeding for resistance to diseases in lettuce;-successes and challenges, In Proceedings of the Eucarpia Meeting on L:eafy Vegetables Genetics and Breeding, van Hintum, Th. J. L., Lebeda, A., Pink, D. A., and Schut, J. W., Eds., Wageningen: Centre for Genetic Resources, pp. 25–30, 2003.

Saenz, G. S. and Taylor, J. W., Phylogeny of the Erysiphales (powdery mildews) inferred from internal transcribed spacer (ITS) ribosomal DNA sequences, *Can. J. Bot.*, 77, 150–169, 1999.

Sahin, F. and Miller, S. A., Identification of the bacterial leaf spot pathogen of lettuce, *Xanthomonas campestris* pv. *vitians*, in Ohio, and assessment of cultivar resistance and seed treatment, *Plant. Dis.*, 81, 1443–1446, 1997.

Sanches-Monge, E. and Garcia-Olmego, F., Eds., In Interspecific Hybridization in Plant Breeding Proceedings of the Eighth Congress of Eucarpia, Madrid (Spain), May 23–25, 1977, Madrid: Universidad Politechnica de Madrid, 1978.

Schnathorst, W. C., Grogan, R. G., and Bardin, R., Distribution, host range, and origin of lettuce powdery mildew, *Phytopathology*, 48, 538–543, 1958.

Sequeira, L., Resistance to corky root rot in lettuce, *Plant Dis. Rptr.*, 54, 754–758, 1970.

Shear, C. B., Calcium-related disorders of fruits and vegetables, *J. Am. Soc. Hort. Sci.*, 18, 225–230, 1975.

Shen, K. A., Chin, D. B., Arroyo-Garcia, R., Ochoa, O. E., Lavelle, D. O., Wroblewsky, T., Meyers, B. C., and Michelmore, R. W., Dm3 is one member of a large constitutively expressed family of nucleotide binding site-leucine-rich repeat encoding genes, *Mol. Plant Microbe Interact.*, 15, 251–261, 2002.

Sherf, A. F. and Macnab, A. A., *Vegetable Diseases and Their Control*, New York: John Wiley & Sons, 1986.

Shih, C., On circumscription of the genus *Prenanthes* L. and *Notoseris* Shih: A new genus of Compositae from China, *Acta Phytotax. Sin.*, 25, 189–203, 1987.

Shih, C., Revision of *Lactuca* L. and two new genera of the tribe Lactuceae (Compositae) on the mainland of Asia, *Acta Phytotax. Sin.*, 26, 382–393, 1988a.

Shih, C., Revision of *Lactuca* L. and two new genera of the tribe Lactuceae (Compositae) on the mainland of Asia (cont.), *Acta Phytotax. Sin.*, 26, 418–428, 1988b.

Shih, C., On circumscription of the genus Cicerbita Wall., and two new genera of Compositae from Sino-Himalayan region, *Acta Phytotax. Sin.*, 29, 394–417, 1991.

Sicard, D., Woo, S. S., Arroyo-Garcia, R., Ochoa, O., Nguyen, D., Korol, A., Nevo, E., and Michelmore, R., Molecular diversity at the major cluster of disease resistance genes in cultivated and wild *Lactuca* spp., *Theor. Appl. Genet.*, 99, 405–418, 1999.

Smith, I. M., Ed, In European Handbook of Plant Diseases, Oxford: Blackwell Scientific Publishers, 1988.

Soest, L. J. M. van., Report of expedition to Uzbekistan in 1997, Itinerary, collected materials and data, *CPRO-DLO/CGN*. Wageningen, 1998.

Soest, L. J. M. and Boukema, I. W., Genetic resources conservation of wild relatives with a user's perspectives, *Bocconea*, 7, 305–316, 1997.

Soest, L. J. M., Baimatov, K. I., Chapurin, V. F., and Pimakhov, A. P., Multicrop collecting mission to Uzbekistan, *Plant Genet. Resour. Newslet.*, 116, 32–35, 1998.

Soják, J., Bemerkungen zu einigen *Compositen*, I, *Novitates Bot. Horti Bot. Pragensis*, 1961, 33–37, 1961.

Soják, J., Bemerkungen zu einigen *Compositen*, II, *Novitates Bot. Horti Bot. Pragensis*, 1962, 41–50, 1962.

Soltis, D. E. and Soltis, P. S., *Isozymes in Plant Biology*, Portland: Dioscorides Press, 1989.

Sretenović-Rajičić, T., van Hintum, Th. J. L., Lebeda, A., and Dehmer, K., Analysis of wild *Lactuca* genebank accessions: implication on taxonomic status, identification of redundancies and conservation. *Genet. Resour. Crop Evol.*, 2006 (submitted).

Stavelikova, H., Boukema, I. W., and van Hintum, Th. J. L., The International *Lactuca* database, *Plant Genet. Resour. Newslet.*, 130, 16–19, 2002.

Staub, J. E. and Serquen, F. C., Genetic markers, map construction, and their application in plant breeding, *Hort. Sci.*, 31, 729–741, 1996.

Staub, J. E., Serquen, F. C., and Gupta, M., Genetic markers, map contruction and their application in plant breeding, *Hort. Sci.*, 31, 729–741, 1996.

Stebbins, G. L., Critical notes on *Lactuca* and related genera, *J. Bot. (London)*, 75, 12–18, 1937.

Steyermark, J. A., *Flora of Missouri*, Ames, IA: The Iowa State University Press, 1963.

Stickland, S., *Vegetables. The Gardener's Guide to Cultivating Diversity*, London: Gaia Books Ltd., 1998.

Strausbaugh, P. D. and Core, E. L., *Flora of West Virginia*, 2nd ed., Grantsville: Seneca Books Inc., 1978.

Stebbins, G. L., *Chromosomal Variation in Higher Plants*, London: Arnold, 1971.

Stubbs, L. L. and Grogan, R. G., Necrotic yellows, a newly recognized virus disease of lettuce, *Austral. J. Agri. Res.*, 14, 439–459, 1963.

Subbarao, K. V., Progress towards integrated management of lettuce drop, *Plant Dis.*, 80, 28–33, 1998.

Subbarao, K. V., Hubbard, J. C., Greathead, A., and Spencer, G. A., Verticillium wilt, In *Compendium of Lettuce Diseases*, Davis, R. M., Subbarao, K. V., Raid, R. N., and Kurtz, E. A., Eds., St. Paul: American Phytopathological Society, pp. 26–37, 1997.

Takamatsu, S., Hirata, T., and Sato, Y., Phylogenetic analysis and predicted secondary structures of the rDNA internal transcribed spacers of the powdery mildew fungi (Erysiphaceae), *Mycoscience*, 39, 441–453, 1998.

Thomas, G., Astley, D., Boukema, I. W., Daunay, M. C., Del Greco, A., Diez, M. J., van Dooijeweert, W., Keller, J., Kotlinska, T., Lebeda, A., Lipman, E., Maggioni, L., and Rosa, E., Eds., *Report of a Vegetables Network. Joint Meeting with an ad hoc group on Leafy Vegetables, 22–24 May 2003, Skierniewice, Poland*, Rome: International Plant Genetic Resources Institute, 2005.

Thompson, R. C., Natural cross-pollination in lettuce, *Proc. Amer. Soc. Hort. Sci.*, 30, 545–547, 1933.

Thompson, R. C., *Genetic Relations of Some Color Factors in Lettuce*, Tech. Bul. 620, Washington DC: USDA, 1938.

Thompson, R. C., and Ryder, E. J., Descriptions and pedigrees of nine varieties of lettuce, U.S. Dept. Agri. Tech. Bul., 1244, 1961.

Thompson, R. C., Whitaker, T. W., and Kosar, W. F., Interspecific genetic relationships in *Lactuca*, *J. Agric. Res. (Washington D.C.)*, 63, 91–107, 1941.

Thompson, R. C., Whitaker, T. W., Bohn, G. W., and van Horn, C. W., Natural cross-pollination in lettuce, *Proc. Amer. Soc. Hort. Sci.*, 72, 403–409, 1958.

Tomb, A. S., Lactuceae—systematic review, In *The Biology and Chemistry of the Compositae, Vol. II*, Heywood, V. H., Harbone, J. B., and Turner, B. L., Eds., London: Academic Press, pp. 1067–1079, 1977.

Torres, A. C., Cantliffe, D. J., Laughner, B., Bieniek, B., Nagata, R., Ashraf, M., and Ferl, R. J., Stable transformation of lettuce cultivar South Bay from cotyledon explants, *Plant Cell Tissue Organ Cult.*, 34, 279–285, 1993.

Torres, A. C., Nagata, R. T., Ferl, R. J., Bewick, T. A., and Cantliffe, D. J., *In vitro* assay selection of glyphosate resistance in lettuce, *J. Amer. Soc. Hort. Sci.*, 124, 86–89, 1999.

Tracy, W. W., Jr., *American Varieties of Lettuce*, US Department of Agriculture, Washington DC., Bulletin No. 69, 1904.

Tsuchiya, N., Fujinaga, M., Ogiso, H., Usui, T., and Tsukada, M., Resistance tests and genetic resources for breeding fusarium root rot resistant lettuce, *J. Japan. Soc. Hort. Sci.*, 73, 105–113, 2004.

Tuisl, G., *Der Verwandtschaftskreis der Gattung Lactuca L. im iranischen Hochland und seinen Randgebieten*, Wien: Selbstverlag Naturhistorisches Museum Wien, 1968.

Turini, T., Koike, S., Ryder, E., and Grube, B., Powdery mildew management, In *California Lettuce Research Board, Annual Report 2000/2001*, Salinas, CA: California Lettuce Research Board, pp. 179–189, 2001.

Tzanoudakis, D., Chromosome studies in the Greek flora, I. Karyotypes of some Aegean angiosperms, *Bot. Helv.*, 96, 27–36, 1986.

UPOV, Guidelines for the conduct of tests for distinctness, homogeneity and stability, Lettuce (*Lactuca sativa* L.), Document No. TG/13/4 (26.10.1981), http://www.oecd.org/dataoecd/46/22/2741883.pdf, 1981.

USDA, *Vegetables 2003 Summary*, Washington DC: National Agricultural Statistics Service, 2004.

Valcárcel, J. V., José Díez, M., and Nuez, F., Status of leafy vegetable collections in Spain, In *Report of a Vegetables Network. Joint Meeting with an ad hoc group on Leafy Vegetables*, Thomas, G., Astley, D., Boukema, I. W., Daunay, M. C., Del Greco, A., Diez, M. J., van Dooijeweert, W., Keller, J., Kotlinska, T., Lebeda, A., Lipman, E., Maggioni, L., and Rosa, E., Eds., Rome: International Plant Genetic Resources Institute, pp. 122–126, 2005.

Van Bruggen, A. H. C., Grogan, R. G., Bogdanoff, C. P., and Waters, C. M., Corky root of lettuce in California caused by a gram-negative bacterium, *Phytopathology*, 78, 1139–1145, 1988.

Van Bruggen, A. H. C., Jochimsen, K. N., and Brown, P. R., *Rhizomonas suberifaciens* gen. nov., sp. nov., the causal agent of corky root of lettuce, *Inter. J. System. Bacteriol*, 40, 175–188, 1990.

Van der Arend, A. J. M., Ester, A., and van Schijndel, J. T., Developing an aphid resistant butterhead lettuce "Dynamite", In *Eucarpia Leafy Vegetables '99, Proceedings of the Eucarpia Meeting on Leafy Vegetables Genetics and Breeding*, Lebeda, A. and Křístková, E., Eds., Olomouc: Palacky University, pp. 149–157, 1999.

van Hintum, Th. J. L., Improving the management of genebanks: a solution from biotechnology, In *Eucarpia Leafy Vegetables '99, Proceedings of the Eucarpia Meeting on Leafy Vegetables Genetics and Breeding*, Lebeda, A. and Křístková, E., Eds., Olomouc: Palacký University in Olomouc, pp. 107–108, 1999.

van Hintum, Th. J. L., Molecular characterization of a lettuce germplasm collection, In *Eucarpia Leafy Vegetables 2003, Proceedings of the Eucarpia Meeting on Leafy Vegetables Genetics and Breeding*, van Hintum, Th. J. L., Lebeda, A., Pink, D., and Schut, J. W., Eds., Wageningen: Centre for Genetic Resources, pp. 99–104, 2003.

van Hintum, Th. J. L. and Knüpffer, H., Duplication within and between germplasm collections. I. Identifying duplication on the basis of passport data, *Genet. Resour. Crop Evol.*, 42, 127–133, 1995.

van Hintum, Th. J. L., Lebeda, A., Pink, D. A., and Schut, J. W., Eds., *Eucarpia Leafy Vegetables 2003, Proceedings of the Eucarpia Meeting on Leafy Vegetables Genetics and Breeding*, Wageningen: Centre for Genetic Resources, 2003.

Vavilov, N. I., *Five Continents*, Rome: International Plant Genetic Resources Institute, 1995.

Vermeulen, A., Desprez, B., Lancelin, D., and Bannerot, D. H., Relationship among *Cichorium* species and related genera as determined by analysis of mitochondrial RFLPs, *Theor. Appl. Genet.*, 88, 159–166, 1994.

Vetten, H. J., Lesemann, D. E., and Dalchow, J., Electron microscopical and serological detection of virus-like particles associated with lettuce big vein disease, *J. Phytopathol.*, 120, 53–59, 1987.

Von der Pahlen, A. and Crnko, J., Lettuce mosaic virus (*Marmor lactucae* Holmes) in Mendoza and Buenos Aires, *Rev. Invest. Agropecu., Ser. 5*, 4, 25–31, 1965.

Vosman, B., Molecular analysis of variation in *Lactuca* as a case study for the potential usages of molecular markers in the management of plant genetic resources, In *Molecular Genetic Techniques for Plant Genetic Resources*, Ayad, W. G., Hodgkin, T., Jaradat, A., and Rao, V. R., Eds., Rome: International Plant Genetic Resources, pp. 44–48, 1997.

Vries de, I. M., Crossing experiments of lettuce cultivars and species (*Lactuca* sect. *Lactuca, Compositae*), *Pl. Syst. Evol.*, 171, 233–248, 1990.

Vries de, I. M., Characterization and identification of L. sativa cultivars and wild relatives with SDS-electrophoresis (*Lactuca* sect. *Lactuca, Compositae*), *Genet. Resour. Crop Evol.*, 43, 193–202, 1996.

Vries de, I. M., Origin and domestication of *Lactuca sativa* L., *Genet. Resour. Crop Evol.*, 44, 165–174, 1997.

Vries de, I. M. and van Raamsdonk, L. W. D., Numerical morphological analysis of lettuce cultivars and species (*Lactuca* sect. *Lactuca*, Asteraceae), *Plant Syst. Evol.*, 193, 125–141, 1994.

Vuilleumier, B. S., The genera of *Lactuceae* (*Compositae*) in the southeastern United States, *J. Arnold. Arbor.*, 54, 42–93, 1973.

Walkey, D. G. A. and Pink, D. C. A., Studies on resistance to beet western yellows virus in lettuce (*Lactuca sativa*) and the occurrence of field sources of the virus, *Plant Pathol.*, 39, 141–155, 1990.

Wang, M., Chu, J. J., Provvidenti, R., and Hu, J. S., Identification of resistance to tomato spotted wilt virus in lettuce, *Plant Dis.*, 76, 642, 1992.

Waycott, W. and Fort, S. B., Differentiation of nearly identical germplasm accessions by a combination of molecular and morphological analyses, *Genome*, 37, 577–583, 1994.

Waycott, W., Fort, S. B., Ryder, E. J., and Michelmore, R. W., Mapping morphological genes relative to molecular markers in lettuce (*Lactuca sativa* L.), *Heredity*, 82, 245–251, 1999.

Welch, J. E., Zink, F. W., and Grogan, R. G., Calmar, *California Agric.*, 19, 3–4, 1965.

Westerlund, F. V., Campbell, R. N., and Grogan, R. G., Effect of temperature on transmission, translocation, and persistence of the lettuce big-vein agent and big-vein symptom expression, *Phytopathology*, 68, 921–926, 1978a.

Westerlund, F. V., Campbell, R. N., Grogan, R. G., and Duniway, J. M., Soil factors affecting the reproduction and survival of *Olpidium brassicae* and its transmission of big-vein agent to lettuce, *Phytopathology*, 68, 927–935, 1978b.

Whipps, J. M., Budge, S. P., McClement, S., and Pink, D. A. C., A glasshouse cropping method for screening lettuce lines for resistance to *Sclerotinia sclerotiorum*, *Europ. J. Plant Pathol.*, 108, 373–378, 2002.

Whitaker, T. W., Salads for everyone—a look at the lettuce plant, *Econ. Bot.*, 23, 261–264, 1969.

Whitaker, T. W., Bohn, G. W., Welch, J. E., and Grogan, R. G., History and development of head lettuce resistant to downy mildew, *Proc. Amer. Soc. Hort. Sci.*, 72, 410–416, 1958.

Wiel van de, C., Arens, P., and Vosman, B., Microsatellite fingerprinting in lettuce (*Lactuca sativa* L.) and wild relatives, *Plant Cell Rep.*, 17, 837–842, 1998.

Wiel van de, C., Arens, P., and Vosman, B., Microsatellite retrieval in lettuce (*Lactuca sativa* L.), *Genome*, 42, 139–149, 1999.

Wiel van de, C., Flavell, A., Syed, N., Antonise, R., van der Voort, J. R., and van der Linden, G., Analysis of gene flow in the lettuce crop-weed complex, In *Introgression from Genetically Modified Plants into Wild Relatives*, den Nijs, H. C. M., Bartsch, D., and Sweet, J., Eds., Wallingford: CABI Publishing, pp. 163–171, 2004.

Witsenboer, H., Kesseli, R. V., Fortin, M., Stanghellini, M., and Michelmore, R. W., Sources and genetic structure of a cluster of genes for resistance to three pathogens in lettuce, *Theor. Appl. Genet.*, 91, 178–188, 1995.

Witsenboer, H., Vogel, J., and Michelmore, R. W., Identification, genetic localization, and allelic diversity of selectively amplified microsatellite polymorphic loci in lettuce and wild relatives (*Lactuca* spp.), *Genome*, 40, 923–936, 1997.

Yabuuchi, E., Kosako, Y., Naka, T., Suzuki, S., Yano, I. et al., Proposal of *Sphingomonas suberifaciens* (van Bruggen, Yochimsen and Brown 1990) comb. nov., *Sphingomonas natatoria* (Sly 1985) comb. nov., *Sphingomonas ursincola* (Yurkov et al. 1997) comb. nov., and emendation of the Genus *Sphingomonas*, *Microbiol. Immunol.*, 43, 339–349, 1999.

Yamamoto, T., Nishikawa, A., and Oeda, K., D.N.A. polymorphisms in *Oryza sativa* L. and *Lactuca sativa* L. amplified by arbitrary primed PCR, *Euphytica*, 78, 143–148, 1994.

Yang, C-H., Carrol, B., Scofield, S., Jones, J., and Michelmore, R., Transactivation of Ds elements in plants of lettuce *Lactuca sativa*, *Mol. Gen. Genet.*, 241, 389–398, 1993.

Young, N. D., The genetic architecture of resistance, *Curr. Opinion Plant Biol.*, 3, 285–290, 2000.

Zeven, A. C. and Brandenburg, W. A., Use of paintings from the 16th to 19th centuries to study the history of domesticated plants, *Econom. Bot.*, 40, 397–408, 1986.

Zink, F. W. and Grogan, R. G., The inter-related effects of big-vein and market price on the yield of lettuce, *Plant Dis. Rep.*, 38, 844–846, 1954.

Zink, F. W., Duffus, J. E., and Kimble, K. A., Relationship of a non-lethal reaction to a virulent isolate of lettuce mosaic virus and turnip mosaic susceptibility in lettuce, *J. Amer. Soc. Hort. Sci.*, 98, 41–45, 1973.

Zohary, D., The wild genetic resources of cultivated lettuce (*L. sativa* L.), *Euphytica*, 53, 31–35, 1991.

Zohary, D. and Hopf, M., *Domestication of Plants in the Old World: The Origin and Spread of Cultivated Plants in West Asia, Europe and the Nile Valley*, 2nd ed., Oxford: Clarendon Press, 1993.

CHAPTER 10

Eggplant (*Solanum melongena* L.)

Major Singh and Rajesh Kumar

CONTENTS

10.1   INTRODUCTION

Eggplant [*Solanum melongena* L. ($2n=24$)], also known as aubergine or brinjal, is an important solanaceous vegetable crop in many countries. It is probably a native of India (Vavilov 1928). It is believed that eggplant may have originated in Indo-Burma region. China may be the secondary

Table 10.1 Nutritive Composition of 100 g Edible Portion of Fresh Eggplant

Nutrients	Value	Nutrients	Value
Moisture	92.7 g	Protein	1.2 g
Fat	0.2 g	Minerals	0.3 g
Fiber	1.3 g	Carbohydrate	5.5 g
Energy	24 kcal	Calcium	15 mg
Ash	0.5 g	Vitamin C	5 mg
Vitamin A (as β-carotene)	0.02 mg/33 I.U.	Niacin	0.6 mg
Thiamine	0.04 mg	Copper	0.07–0.1 mg
Sodium	0.9–2.5 mg	Potassium	190–238 mg
Phosphorous	37 mg	Iron	0.4 mg
Magnesium	10–15 mg	Manganese	0.11 mg

center of origin. The first record of eggplant in Europe was in the fifteenth century (Hedrick 1919); the name was probably derived from the white egg-like fruits.

Eggplant is an excellent source of minerals and vitamins, and in total nutritional value it can be compared with tomato. It is grown on 1,700,655 hectares, with a total production of 29,840,793 metric ton (FAO 2004; http://faostat.fao.org.) The important eggplant-growing countries are India, Japan, Indonesia, China, Bulgaria, Italy, France, the United States, and many African countries.

Eggplant contains small amounts of several important minerals and vitamins needed daily (Table 10.1). It is very low in sodium and calories. The raw eggplant contains only 24 kcal energy per 100 g. Even though it is a low calorie vegetable, the caloric value rises steeply when it is fried. It also supplies vitamin C, potassium, iron, niacin, calcium, phosphorous, fiber, and folic acid. One hundred grams of edible eggplant provides 19–38 mg of potassium. It is high in water content and has about 93% moisture. Due to its low calorie content and high potassium content, it is suitable to control diabetes, hypertension, and obesity. Eggplant is a rich source of bioflavonoids and clears stagnant blood in our bodies. Eggplant contains solanine (which inhibits calcium absorption), but when cooked the solanine is neutralized.

10.2 REPRODUCTIVE BIOLOGY

The reproductive biology of eggplant has been studied and discussed by several research workers (Frydrych 1964; Deshpande, Bankapus, and Nalawadi 1978; Nishi 1978). There are five stages of development between bud formation and full flowering (Popova 1958). The duration of the stages depends on the genotype and the temperature. According to Deshpande, Bankapus, and Nalawadi (1978), maximum anthesis occurred at 6:00 a.m. in "Arka Shirish" and at 8:00 a.m. in "Arka Sheel" and "Arka Kusmaker." After 8:00 a.m. there was a marked decline in anthesis in all varieties until noon. Stigma receptivity was highest at the time of flower opening. Popova (1958) reported that pollen remained viable for 7–10 days and stigma was receptive for 6–8 days.

Eggplant is not an obligate self-pollinator, but a facultative cross-pollinator. The number of seeds per fruit is higher in intravarietal crosses than in selfed plants but lower than in open-pollinated plants. Natural cross-pollination has been studied in different varieties in various environments (Sambandam 1964b), and varied from 0.7 to 15% with an average of 4.4% in India. According to Franceschetti and Lepori (1985), natural cross-pollination ranged from 10 to 29%. Purple-fruited attribute was used as a marker. Out-crossing is performed by insects.

10.3 CYTOGENETICS

10.3.1 Cytology

Cytological studies on eggplant species have been instrumental in classification of the plant. Although the basic chromosomal number ($x = 12$; n is the gametic chromosome number) is the same in all the varieties and species, the chiasma frequency during diplonema and diakinesis shows varied bivalents (Kalloo 1993). The cytology and chromosome numbers of many diploids, auto-triploids, autotetraploids, amphiploids, and androgenic haploids have been thoroughly studied in eggplant (Kirti, Moorty, and Rao 1984a or Kirti et al. 1984b; Rao and Baksh 1979; Siliak-Yakovlev and Isouard 1983). Imperfect homology between some chromosomes pairs in *S. melongena* suggested that hybridization and structural chromosome changes may have played a part in its evolution (Okoli 1988). The karyotype analysis showed that haploidy has no effect on the shape and size of the chromosomes (Venora, Russo, and Errico 1992). Karyotype comparisons between wild *S. sisymbriifolium* Lam., *S. torvum* Sw., and cultivated *S. melongena* show that *S. sisymbriifolium* is phylogenetically distant to *S. torvum* and *S. melongena*, which are closer to each other than the former (Russo et al. 1992).

10.3.2 Genetics

Genetics of various qualitative traits have been studied in eggplant (Table 10.2). Many such traits can be used as markers in the production of hybrid seeds and gene mapping. Data on genetic variability, heritability, correlation, path coefficient, gene effects, and stability parameters have been generated for several quantitative traits. Estimates of heritability were high for days to first flowering, number of flowers per cluster, fruit length, fruit girth, fruit number per plant, and fruit weight (Sidhu et al. 1980). Yield exhibited moderate heritability (Borikar, Makne, and Kulkarni 1981). High heritability and high genetic advance were exhibited by fruit per plant (Vadivel and Bapu 1989).

Fruit yield is positively correlated with the number and weight of fruits. According to Khurana et al. (1988), fruit yield showed positive correlation with fruit diameter and mean fruit weight. These attributes were positively correlated with number of branches, stem weight, leaf weight, leaf area, and leaf number.

Gene effect has been studied for a number of attributes such as yield, length of fruit, and number of flowers. Both additive and dominant gene effects were significant for days to flowering, the latter being larger. Additive gene effect was more important than dominant gene effect for number, weight, and girth of fruits (Sidhu et al. 1980).

10.3.3 Molecular Maps

Plant breeding programs worldwide are developing molecular linkage maps through Restriction fragment length polymorphism (RFLP), Random amplified polymorphic DNA (RAPD), and Amplified fragment length polymorphism (AFLP) (Tanksley et al. 1989; Williams et al. 1990; Vos et al. 1995). The RAPD, RFLP, and AFLP studies have also revealed distinction between wild and cultivated plants, and very effectively demonstrated genetic diversity and relatedness among various eggplants (Isshiki et al. 1998; Mace, Lester, and Gebhardt 1999; Kashyap et al. 1999; Kashyap 2002). Recently Nunome, Yoshida, and Hirai (1998) have constructed a linkage map of eggplant. Their aim was to identify molecular markers linked to important agronomic traits, especially those resistant to bacterial wilt caused by *Ralstonia solanacearum*.

Table 10.2 Genetics of Qualitative Traits in Eggplant

Traits	Dominance	Genetics	Reference
Chlorophyll deficient seedling		Monogenic recessive	Nuttall (1965)
Hypocotyl	Purple > green	Single gene	Sambandam (1964a)
Hypocotyl	Purple > green	Single gene (PhyPhy)	Wanjari and Khapre (1977)
Stem	Purple > green	(PstPst)	Wanjari and Khapre (1977)
Purple coloration of stem, petiole, and vein	Purple > green	All three by duplicate gene designated Pst_1 and Pst_2	Nimbalkar and More (1980)
Stem color	Purple > green	Single Pst	Khapre, Wanjari, and Deokar (1986)
Stem and leaf	Purple stem > green stem and leaf	Monogenic	Shariff and Habib (1977)
Presence of spine	> Absence of spine	Monogenic	
Purple and spine	> Green and spine	Digenic	
Plant height	Tall and dwarf	Monogenic	Choudhury (1972)
Deeply lobed leaf	Incompletely dominant over slightly lobed		Choudhury (1972)
Fruit color	Green (Gm) incompletely dominant over white (gm)		Choudhury (1972)
Leaf color and leaf vein	By two complementary genes Cla_1, Clb_1 and Plv_1, Plv_2, respectively		Patil and More (1983)
Stem color	By three complementary genes $Csta_1$, $Cstb_1$, and $Cstc_1$		Patil and More (1983)
Fruit color	By three complementary genes Pfa_1, Pfb_1, and Pfb_2; purple > green		Patil and More (1983)
Spines on stem, leaf, and petiole	Singe gene		Patil and More (1983)
Fruit color	Green streaked purple > greenish white		More, Patil, and Nimbalkar (1982)
Spined leaves and stem	Three complementary genes of which any two must be dominant		Khapre, Deokar, and Wanjari (1987)
Spined calyx	Two complementary genes and an inhibitor		Khapre, Deokar, and Wanjari (1987)
Male sterility	Two recessive nuclear genes ms_1, ms_2		Chauhan (1984)
Male sterility	Cytoplasmic control		Fang, Mao, and Xie (1985)
Erect versus spreading	Two complementary genes Egr_1 and Egr_2		Wanjari and Khapre (1981)
Fruit shape	Ofa, Ofb_1, Ofb_2, and Ofb_3		Nimbalkar and More (1981)
Fruit color	Purple > green and cream. Complex form of inheritance		Gopinath, Madalageri, and Somasekar (1986)

RFLP genetic linkage map for qualitative and quantitative traits has been constructed from an F_2 segregating population of eggplant derived from an interspecific cross between *S. melongena* and *S. linnaeanum* Hepper and Jaeger (Frary et al. 2000). More recently markers linked to fruit shape and color were identified in a molecular linkage map based on RAPD and AFLP (Nunome et al. 2001).

Therefore, the application of molecular-assisted selection (MAS) may also be applied in breeding eggplant. The markers can function as probes for the disease resistant trait. Further, this may facilitate rapid identification and isolation of the genes of interest. Genome mapping with probes and markers can help in studying both Mendelian and non-Mendelian genes. However, such studies in eggplant are in their infancy.

10.3.4 Phylogenetic Studies

The cytological studies carried out earlier are now verified through DNA fingerprinting. Sakata, Nishio, and Matthews (1991) initiated phylogenetic studies by comparing isozymes and pattern of cpDNA between wild and cultivated eggplants. They comprehend that scarlet eggplant should be treated as single species (*S. aethiopicum* L.) derived from single wild species (*S. anguivi* Lam.). In a similar study, Sakata and Lester (1994) showed that *S. melongena* and wild species *S. incanum* L. are closer in phylogeny than the wild species *S. marginatum* L. Allozyme patterns and RAPDs also showed cultivated *S. melongena*, the weedy form *S. insanum* L., and the wild species *S. incanum* are very close phylogenetically (Karihaloo and Gottlieb 1995; Karihaloo, Brauner, and Gottlieb 1995). Isshiki, Okubo, and Fujieda (1994, 2000) and Isshiki et al. (1998) conducted segregation studies for seven isozyme loci of five enzymes in cultivated and wild species of eggplant. Seed storage proteins have been used to characterize the wild and cultivated eggplant species (Menella et al. 1999).

10.4 GENE POOLS OF EGGPLANT

Bailey divided eggplant into three varieties: the round, egg-shaped *S. melongena* var. *esculentum*; the long, slender *S. melongena* var. *serpentinum*; and the dwarf variety known as *S. melongena* var. *depressum*. Results obtained from hybridization, electrophoresis of seed protein, morphological examination, and field observations indicated that all the African taxa of *Solanum* section Oliganthes series Aethiopica, namely *S. gilo* (including *S. olivare*), *S. zuccagnianum* (= *S. aethiopicum*) and *S. integrifolium*, comprise a single series.

Solanum melongena is crossable with *S. torvum* (McCammon and Honma 1983). *Solanum incanum* as female parent was crossable with *S. melongena* producing 201 seeds with 45% germination (Baksh and Iqbal 1979). The hybrid of S. *melongena* × *S. indica* was fertile and exhibited normal meiosis (Rao and Kumar 1980). *Solanum melongena* as female parent was crossable with *S. macrocarpon* as male. The fertility of F_1 progeny ranged from complete absence of flower to the production of fruits with few seeds (Schaff et al. 1980). However, according to Gowda, Shivashankar, and Joshi (1990), *S. melongena* × *S. macrocarpon* hybrid plants failed to set seed due to ovule abortion. F_1 hybrid showed resistance to *Leucinodes orbonalis*. *Solanum melongena* was crossed successfully with *S. incanum, S. indicum, S. indicum* var. *multiflorum*, and *S. integrifolium* in both directions, except in the case of *S. melongena* (female) × *S. incanum* (male). Meiosis of F_1 indicated that *S. incanum* and *S. melongena* are closely related species but dissimilar with respect to the degree of incompatibility among other *Solanum* species (Lakshmi, Moorty, and Rao 1981). *Solanum macrocarpon* and *S. melongena* were reciprocally intercrossed. The crossability and fertility of F_1 hybrids depended on the *S. melongena* lines used and the direction of the cross (Schaff, Boyer, and Pollack 1982).

Incompatibility in *S. melongena, S. melongena* var. *insanum*, and eight other species is attributed to a post-fertilization problem. However, crosses with *S. indicum* and *S. melongena* were incompatible as female parent owing to cytoplasmic action (Rao 1979).

In several cases, crosses were only successful by *in vitro* embryo rescue method. Sharma et al. (1980) were successful in obtaining fertile F_1 plants between *S. melongena* and *S. khasianum*

Clarke through embryo rescue, but only when *S. khasianum* was selected as the female parent and 25-day-old embryos were cultured on Nitsch and Nitsch (1969) medium. They also achieved partial success in obtaining hybrid plants between *S. melongena* × *S. sisymbriifolium* Lam., when the latter was taken as the male parent and the excised torpedo stage embryos were cultured. However, the plants could not survive for very long and died (Sharma, Sareen, and Chowdhury 1984). Daunay et al. (1998) were successful in producing 22 hybrids between *S. melongena* and wild species of eggplant. In another study, Blestsos et al. (1998) obtained hybrid plants between *S. torvum* and *S. sisymbriifolium* by culturing immature seeds on MS medium for 50 days. Thereafter, the embryos were dissected and cultured on MS medium in dark for 10 days and then transferred to light. They characterized hybrids morphologically and by isozymes. The successful interspecific crosses have been obtained with only a few wild species, but hybrids were sterile (Sharma et al. 1980; McCammon and Honma 1983; Sharma, Sareen, and Chowdhury 1984; Blestsos et al. 1998).

10.5   GENETIC RESOURCES

Germplasm collection, evaluation, maintenance, and conservation of eggplants have been extensively documented (Cuastero et al. 1985; Vadivel and Bapu 1989). India has major collections from India, Africa, Southeast Asia, Bangladesh and other countries. A collection of 475 varieties was classified into eleven groups on the basis of 18 attributes by numerical taxonomy (Martin and Rhodes 1979). Nuez et al. (1987) described collections of new eggplant accessions from various parts of the Europe. Studentsova (1980) examined 750 varieties and identified early, high-yielding eggplant suitable for mechanical harvesting and resistant to pests and pathogens.

A directory of germplasm collections of eggplant has been compiled by Toll and Sloten (1982). Use of a world-wide collection of eggplant varieties and species in breeding for various characters, particularly resistance breeding, has been outlined by Vorouina and Losputova (1982). Further, the Vavilov Institute of Plant Industry collection was studied over many years for valuable traits of economic importance. *Solanum melongena* var. "Black Beauty" and *S. integrifolium* "K 844" were identified as resistant to big bud mycoplasma. The Hungarian form K 2390 (*S. integrifolium*) and Ugandan K 919 (*S. sisymbriifolium*) were highly resistant to *Verticillium alboatrum* and *Tetranychus urticae* (Sharma et al. 1990).

Variation patterns in African scarlet eggplant have been studied, involving multivariate analysis based on 74 attributes in 97 accessions. It was suggested that a simple classification into the cultivar group and ornamental was sufficient (Lester et al. 1986). The use of genetic resources for improvement of eggplant has been discussed by Bannerot and Foury (1986). Germplasm lines of eggplant have been screened against several biotic stresses (Table 10.3).

10.6   BREEDING OBJECTIVES

The development of high-yielding, early, better quality and disease resistant varieties is the major objective of eggplant breeder. The color of the fruit and size and shape (Figure 10.1), the proportion of seeds to pulp, short cooking time, and lower solanine levels are important traits in assessing quality. Eggplant is susceptible to several pests and diseases such as wilt, *Phomopsis*, little leaf and root-knot nematodes, and to insects such as shoot and fruit borer, jassids, epilachna beetle, etc. Plants are susceptible to both low and high temperatures; therefore, attempts are being made at several centers to develop chilling or frost-tolerant and heat-tolerant varieties (Figure 10.2).

Table 10.3 Sources of Eggplant Genotypes Resistant to Pathogens/Pests

Pathogens/Pests	Genotypes	References
Fungal wilts		
Fusarium wilt	*S. indicum,* *S. integrifolium, S. incanum*	Yamakawa and Mochizuki (1979)
	F$_4$ plants from *S. melongena* × *S. indicum*	Rao and Kumar (1980)
Verticillium wilt	*S. caripense, S. periscum,* *S. sisymbriifolium, S. scabrum,* *S. torvum*	Sakata, Nishio, and Mon'ma (1989) and Petrov, Nakov, and Krasteva (1989)
Bacterial wilts	*S. integrifolium, S. torvum*	Yamakawa (1982) and Sheela, Gopalkrishnan, and Peter (1984)
	"Kopek", "Black Beauty"	Jenkins and Nesmith (1976)
Fruit rot		
Phomopsis vexans	*S. gilo, S. integrifolium*	Ahmad (1987)
Cercospora solani	*S. macrocarpon*	Madalageri et al. (1988)
Fruit and shoot borers		
L. orbonalis	*S. xanthocarpum, S. khasianum,* *S. integriifolium, S. sisymbriifolium*	Chelliah and Srinivasan (1983) and Sharma et al. (1980) and Khan, Rao, and Baksh (1978)
Root-knot nematodes		
Meloidogyne spp.	*S. sisymbriifolium, S. torvum,* *S. aethiopicum, S. warscewiczii*	Ahuja et al. (1987), Di Vito, Zaccheo, and Catalano (1992), Daunay and Dalmasso (1985) and Hebert (1985)
Spider mite	*S. macrocarpon, S. integriifolium,* *S. mamosum, S. pseudocapsium,* *S. Sisymbriifolium*	Schalk et al. (1975).
Other pests/pathogens		
Aphis gossypii	*S. sisymbriifolium, S. mammosum*	Sambandam and Chelliah (1983)
Tetranychuns urticae	*S. macrocarpon*	Schaff, Boyer, and Pollack (1982)
Egg plant mosaic virus (Tymovirus)	*S. hispidum*	Rao (1980)
Mycoplasma	*S. hispidum, S. Integrifolium*	Rao (1980) and Khan, Rao, and Baksh (1978)

10.6.1 Heterosis and Hybrid Seed Production

Basavaraja (1986) reported a hybrid seed production technique in eggplant. Flower biology, such as the peak anthesis at 5:50 a.m. anther dehiscence at 7:45 a.m. and stigma receptivity on the day of anthesis were important for hybridization. This produced about 1,300 seeds per fruit. Production of 1 kg of hybrid seed took about 17 hours. According to Nascimento, Torres, and Lima (2003), the quantity of hybrid seed production was increased by the use of pollen stored for 3 days at low temperature.

Male sterility can be used for the production of hybrid seeds. Functional male sterility is controlled by a single recessive gene (Nuttall 1963). Popova and Daskalov (1971) attempted to utilize functional male sterile lines originated from Canada. Male sterility was induced by spraying the flower buds with 10 ppm 2,4-dichlorophenoxyacetic acid (2,4-D) without causing female sterility, and hybrid seeds were obtained by hand pollination (Jyotishi and Hussain 1968). Hybrid seeds can be produced without prior emasculation of the flower. Pollination at bud coloration stage gave up to 97% hybrid seeds even without emasculation. Some exerted-stigma lines are available in eggplant and can be used for the production of hybrids. Eggplant has tremendous potential for heterosis for yield, quality, and resistance to biotic stresses. Heterosis for yield and color of fruits has been reported in long and round-fruited varieties of eggplant.

Figure 10.1 Variation in fruit shape and size in the collections of eggplant.

Figure 10.2 Giant of Banaras—a landrace from Varanasi with biggest fruit size in eggplant collections (mention place in the text).

10.6.2   Breeding for Resistance to Biotic and Abiotic Stresses

Eggplant is susceptible to a number of biotic and abiotic stresses. Among important biotic stresses are *Fusarium* wilt, *Verticillium* wilt, *Phomopsis* fruit rot, bacterial wilt, shoot and fruit borer, jassids, aphids, and root-knot nematodes. Sources for resistance to most of these stresses have been identified (Table 10.3). Nicklow (1983) proposed recurrent selection for improvement of *Verticillium* wilt resistance. The coefficient of genetic variation for resistance to *Verticillium* wilt was 20.5% and heritability was 85.8% (Melo and Costa 1985).

Resistance to *Phomopsis vexans* was recessive and polygenic. Dominance effect was usually more pronounced than additive effect, but additive × additive interactions were also significant in most cases (Kalda, Swarup, and Chowdhury 1977). An Indian isolate of *Pseudomonas solanacearum* was more virulent than the American isolates (Jenkins and Nesmith 1976). For screening germplasm against *P. solanacearum*, *Solanum* plants were grown in plastic pots with a hole at the bottom. The protruding roots were cut and soaked in a suspension of *P. solanacearum*. Inoculum concentrations greater than 10^8 colony forming units per ml and 20 ml per plant were required for adequate infection. *Solanum torvum* and "Taiwan-naga" were resistant and "Satoharu" was moderately resistant. *Solanum integrifolium*, a rootstock used for grafting, showed a specific resistance reaction to *P. solanacearum* (Ozaki and Kimura 1980). The bacterial wilt-resistant line of eggplant was used as rootstock for grafting tomato. The grafting was 100% successful in pot. This indicated that resistant *S. melongena* lines could be used as rootstock for tomato in infected soil (Sheela, Gopalkrishnan, and Peter 1984). The bacterial wilt resistance is controlled by a single dominant gene *Rps* (Gopinath, Madalageri, and Somasekar 1986). Four cycles of mass, single plant, pure line, and single seed descent selections for resistance to *P. solanacearum* were compared. Single seed descent was the most effective in raising the level of resistance, the disease incidence being reduced to 0.91% during the second cycle and 7.72% during the fourth selection cycle (Sankar, Jessykutty, and Peter 1987). Seven methods of fruit inoculation with *Colletotrichum gloeosporioides* (*Glomerella cingulata*) were compared, of which the most efficient, on the basis of necrotic area after 4 days, was the sub epidermal injection of 0.1 ml of a suspension of 10^4 conidia per ml (Madeira and Reifschneider 1986).

Higher levels of glycoalkaloids, peroxidase, and polyphenol oxidase activity were observed in a cultivar resistant to shoot and fruit borer (SM 17-4) than in the susceptible cultivar "Punjab Chamkila" (Bajaj, Singh, and G. Kaur 1989). Cultivars S 34 and S 250 showed multiple pest resistance, i.e., resistance to jassid and to shoot and fruit borer (Pawar et al. 1987). Resistance to greenhouse whitefly (*Trialeurodes vaporariorum*) in "Shinkuro" and *S. macrocarpon* was due to antibiotic activity (Malausa, Daunay, and Bourgoin 1988). Cultivars PBR 129-5 and SM 17-4 can be used as a part of integrated pest management depending on the complexity of the pest problem and the type of chemical protection given to the crop (Singh and Sidhu 1988).

Meloidogyne incognita reduced the resistance to *Pseudomonas solanacearum* in some lines. Plants of "Florida Market," a cultivar moderately resistant to *P. solanacearum*, inoculated with *M. incognita* 28 days prior to inoculation with *P. solanacearum*, were more susceptible to wilt than those inoculated with nematodes 14 days prior to inoculation with bacteria (Jenkins 1974). Higher concentrations of proline and phenol were detected in *M. incognita* infected roots of both susceptible ("Pusa Purple Long") and resistant (RHR 51) cultivars at 30, 60, and 90 days old, compared with uninfected roots of the same cultivars. The increase of phenolic content in resistant cultivars, evaluated after 60 days, was about six times (Sharma et al. 1990).

Breeding for tolerance to low temperature has three aims: (1) maintaining or improving tolerance of chilling and freezing, (2) maintaining vigor under low temperatures and avoiding reduction in growth rate, and (3) preventing poor fruit set, the development of poor-quality fruits and reduction in yield (Nishi 1978). Cultivars "Black Torpedo," "Long Tom 4," "Black Beauty," and BI proved most promising in appearance, texture, and yield under dry and wet seasons

(Lee 1979). *Solanum macrocarpon* K 2591 from Ghana and *S. melongena* cultivars "Supreme" and "Violette Round" were resistant under drought conditions (Gati 1983). Variety *Ralstonia* 34 was found to be resistant to high temperatures. Increased percentage of fruit set at high temperature was associated with less stigma exertions, and high percentages of pollen viability, pollen tube growth, and fertilization (Kalloo, Baswana, and Sharma 1990).

10.6.3   Breeding for Quality, Processing, and Physiological Traits

The quality of eggplant fruit mainly depends upon its physical appearance, namely, its color and shape. Anthocyanin controls the fruit color. Highest optical density value for purple color of fruit, which is a premium quality, was recorded in varieties H 7, H 8, and H 10 (Kalloo, Dixit, and Baswana 1991). Flavor quality is linked with the presence of glycoalkaloids in the solasonine structure concentrated in the placental pulp, and saponin, which occurs mainly in the seeds and pulp. These substances were studied in several varieties and their levels for acceptable flavor have been discussed (Aubert, Daunay, and Pochard 1989b). Saponin content was higher with lengthened harvesting period. It ranged from 1 to 3.5 mg per gram fresh weight in twelve varieties studied. Frequent harvesting produced better-quality fruits (Aubert, Daunay, and Pochard 1989a). Screening should be done at biological ripeness or before, and the seed or seed vessel should be tested in order to develop varieties with low solanin content (Andryush-chenko and Pilipenko 1989).

10.6.4   Biotechnology

10.6.4.1   Organogenesis

Organogenesis has been successfully achieved in cultivated and wild varieties as well as their hybrids. Fassuliotis (1975) was first to report regeneration in *S. sisymbriifolium*. Stem parenchyma cells were isolated and cultured on Linsmeier-Skoog (LS) medium supplemented with 6-(g,g-dimethylallylamino)-purine (2iP) and indole-3-acetic acid (IAA). Other wild species in which regeneration studies have been carried out include *S. xanthocarpum* Schrad and Wendl (Rao and Narayanaswami 1968), *S. aviculare* Forst, *S. gilo* Raddi (Gleddie, Keller, and Setter-field 1985; Kashyap and Rajam 1999), *S. khasianum* Clarke (Bhatt, Bhatt, and Sussex 1979; Kowalozyk, Mackenzie, and Cocking 1983), *S. indicum* and *S. torvum* Sw (Gleddie et al. 1985a; Kashyap and Rajam 1999). Kamat and Rao (1978) reported shoot regeneration from hypocotyl segments of *S. melongena* and F_1 hybrids in presence of cytokinins, kinetin, and zeatin. Cell suspensions culture raised from pith callus in eggplant have shown appearance of green nodules in presence of IAA and 2iP that fail to differentiate further (Fassuliotis, Nelson, and Bhatt 1981). Such cultures organized shoots when transferred to medium containing ascorbic acid or antiauxin *p*-chlorophenoxy isobutyric (PCIB) acid. Cytokinin is necessary for shoot differ-entiation in eggplant (Allicho, Del Grosso, and Boschieri 1982). Adventitious shoots have been formed by BAP, zeatin, and kinetin supplemented media (Gleddie et al. 1983; Mukherjee, Rathnasbapathi, and Gupta 1991; Sharma and Rajam 1995a). Sharma and Rajam (1995a) recorded that hypocotyl explants yield higher adventitious shoots than cotyledons or leaves. They have shown that morphogenetic response also varied within single explant and followed a basipetal pattern where the apex region was more responsive than the basal region. The use of thidiazuron (TDZ) enhanced shoot organogenesis; the leaves and the cotyledons respond best to TDZ (Magioli et al. 1998). Scoccianti et al. (2000) has reported the correlation between the organogenesis from the cotyledon explants of eggplant and hormone modulated enhancement of biosynthesis and conjugation of polyamines (PAs).

10.6.4.2 Somatic Embryogenesis

Somatic embryogenesis (SE) in eggplant has been reported from stems, hypocotyls, leaves, cell suspensions, isolated protoplasts, and roots (Matsuoka and Hinata 1979; Gleddie et al. 1983; Fobert and Webb 1988; Kalloo 1993; Sharma and Rajam 1995a, 1995b; Yadav 1997; Yadav and Rajam 1997, 1998). Yamada, Nakagawa, and Sinoto (1967) were first to report SE in eggplant in MS medium supplemented with IAA, whereas Matsuoka and Hinata (1979) obtained somatic embryos from hypocotyls on MS medium supplemented with 8 mg/l α-naphthaleneacetic acid (NAA). Studies on SE carried out so far reveal that NAA enhances somatic embryo differentiation. However, its optimal concentration varies with the initial explants. Leaf explants required 2–6 mg/l of NAA for SE differentiation while hypocotyl required 6–10 mg/l (Matsuoka and Hinata 1979; Gleddie et al. 1983; Sharma and Rajam 1995a).

Synthetic seeds have also been developed by encapsulating somatic embryos with sodium alginate and calcium chloride (Lakshmana Rao and Singh 1991; Mariani 1992). Somatic embryogenic response in eggplant was also genotype-dependent. Sharma and Rajam (1995a) compared genotype, explant types, and position\kern-2pton SE using various Indian cultivars. Leaves and cotyledons showed higher SE induction compared to hypocotyls. Interestingly, significant differences for morphogenetic potential were recorded within a single explant and terminal hypocotyl segment exhibited better embryogenetic potential than median segments (Sharma and Rajam 1995a, 1995b). Saito and Nishimura (1994) obtained non-vitrified somatic embryos from cell suspension cultures.

Momiyama et al. (1995) investigated expression of mRNA products during early stages of SE through differentiation. Eight new mRNA products were identified from 4-day-old tissue cultures. The expression of only one of the products remains uniform throughout early stages of initial 10 days of SE. They cloned the cDNA corresponding to this expression product. The product cloned has been found to share high similarity with *Arabidopsis* cDNA clone.

Hitomi, Amagai, and Ezura (1998) showed that auxin type used for SE could impart somaclonal variation. Using morphological features of somatic embryogenic-raised plants, they showed that the somaclonal variation was higher in plantlets obtained with NAA compared to 2,4-D. This is accounted from the variations in the plant habit (tall or dwarf) and leaf shape (narrow or thick leaves). Plants obtained with NAA expressed several somaclonal variants such as plants with larger leaves (4.5%), dwarf habit (4.5%), narrow leaves (3.5%), larger leaves and dwarf plants (6.6%), and narrow leaves with dwarf habit (1.4%), whereas the plants regenerated through 2,4-D had only one variation—narrow leaves with dwarf habit (11.8%).

PAs are important regulators for SE in eggplant. During SE in *S. melongena* cotyledon explants treated with NAA exhibited high free putrescine (Put) and spermidine (Spd) in comparison to those without NAA (Fobert and Webb 1988). The temporal changes in cellular PA concentrations were closely associated with SE process (Sharma and Rajam 1995a, 1995b; Yadav and Rajam 1998). Embryogenic potential of explants collected from different regions of leaf (apical and basal discs) was related to the spatial distribution of endogenous PA levels (Yadav and Rajam 1997). It has been shown that the apical region of leaf explants on eggplants with high SE potential have higher PAs than the basal region that yields lesser somatic embryos (Sharma and Rajam 1995b). Further, Put promoted somatic embryo formation by six-fold. By contrast, Spd and Spermine (Spm) have no stimulatory effect. When Put inhibitor difluoromethylornithine (DFMO) was added externally, there was a reduction in Put and Spd titres and a corresponding reduction in SE. The inhibitory effect of PA inhibitor difluoromethylarginine (DFMA) on SE could be restored by adding exogenous Put (Yadav and Rajam 1997). Yadav and Rajam (1998) studied the temporal regulation of SE by modulating PA biosynthesis in eggplant. Put levels increased during early stages of SE, and the judicious time and dosage of PA/PA biosynthesis inhibitor (DFMA) allowed the modulation of PA metabolism and further regulation of SE. Such studies may be helpful in promotion and induction of plant regeneration via SE in morphogenetically poor and recalcitrant species, respectively. This approach has been used successfully in rice for promotion of regeneration from poorly responding genotypes (Shoeb et al. 2001). The above studies clearly suggested that PAs have an intricate regulatory role in SE.

10.6.4.3 Anther Culture

Studies with anther culture have mostly been conducted for cultivated eggplant with the goal of obtaining double haploid parents for conventional breeding (Rotino 1996). The double haploid plants have been successfully used in conventional breeding programs to obtain pure lines faster than selfed inbreds. Double haploid plants are homozygous at all loci, and this may help to study the genetic basis of quantitative traits by overcoming the problems associated with the environmental variations. The double haploid parents have proven to be useful for breeding plants with useful agronomic traits, such as high yield, disease resistance, earliness, abiotic stress tolerance, and other characters in numerous crops (Jacobsen and Sopory 1978; Uhrig and Salamani 1987; Waari 1996).

Raina and Iyer (1973) were first to report plant regeneration from anther culture in eggplant. They regenerated homozygous diploid (double haploids) plants through callus developed from anthers cultured at uninucleate pollen stage that were previously treated with colchicine. Haploid plantlets were also obtained by Isouard, Raquin, and Demarly (1979). Dumas De Vaulx and Chambonnet (1982) studied extensively to improve the development of androgenic haploids. They showed that high temperature ($35 \pm 2°C$) incubation of anthers under dark conditions, first for 7–8 days, improved the efficiency of haploid plant formation. A combination of both auxin and cytokinin was essential during early stages of anther culture. Similarly, Rotino, Falavigna, and Restaino (1987) showed that haploid plant regeneration was affected by genotypes, temperatures, culture conditions, growth hormones, and anther stage. A higher temperature governs the shift of the microspores from gametophytic stage to sporophytic stage. The regeneration medium supplemented with the same cytokinin as in the induction medium would enhance the anther response (Rotino 1996).

The effect of somaclonal variations on agronomic traits of embryogenic and androgenic colchicine-treated double haploid lines was also investigated (Rotino et al. 1991). Plant height, fruit shape, and yield were some of the features affected by somaclonal variations. Miyoshi (1996) cultured isolated microspores on a medium of Litcher (1982) supplemented with NAA (0.5 mg/l) and BA (0.5 mg/l) initially for 4 weeks. The calli obtained were transferred to MS medium supplemented with zeatin (4 mg/l) and IAA (0.2 mg/l) for plant regeneration. The efficiency of haploid regeneration was additionally improved by pre-culturing anthers or microspores to induce callus formation prior to inducing plantlet regeneration (Gu 1979; Miyoshi 1996). Anther culture was successfully employed to reduce the ploidy of eggplant somatic hybrids of S. melongena, and wild species, S. integrifolium and S. aethiopicum (Rotino et al. 2001).

10.6.4.4 Protoplast Culture and Somatic Hybridization

Plant regeneration from protoplasts has been achieved in both cultivated and wild species of eggplant. Protoplast culture and somatic hybridization are useful tools for overcoming the pre- and post-fertilization breeding barriers encountered during conventional breeding. Protoplast culture is an excellent means for understanding cytological and ultrastructural changes during cell growth and differentiation, and behavioral patterns of plastids and mitochondria (Fournier, Lejeune, and Tourte 1995). For cultured eggplant of S. melongena, protoplasts isolated from mesophyll cells grew best using cytokinin and auxin (Sihachakr and Ducreux 1987). However, protoplasts isolated from petioles and stems showed better plant regeneration potential compared to the cells isolated from lamina (Sihachakr and Ducreux 1987). In addition to hormone, the regeneration of plantlets from protoplasts was highly genotype specific

(Gleddie et al. 1983). An alternate pathway for regeneration from protoplasts is via SE (Gleddie et al. 1983).

In wild species, regeneration response is varied between species. Gleddie et al. (1985a) compared plantlet regeneration of *S. sisymbriifolium*, *S. torvum*, *S. gilo*, *S. aviculare*, *S. aculeatissimum*, and *S. khasianum*. Plant regeneration was the best in *S. aviculare*. *Solanum sisymbriifolium* and *S. gilo* showed minimal response and plant regeneration was not observed. However, Guri, Volokita, and Sink (1987) were successful in establishing plantlets regenerated from *S. torvum* protoplasts. Protoplasts and protoplast obtained calli of *S. torvum* regenerated plants better on modified Kao Michayluk (KM) medium supplemented with cytokinins and auxin compared to MS medium Gleddie et al. (1985a).

Somatic hybridization using protoplast fusion has facilitated transfer of useful traits in eggplant between sexually incompatible species (Table 10.4). One of the main advantages of somatic hybridization over sexual hybridization is that hybrids with novel cytoplasmic and nuclear traits can be developed by hybridizing distantly related species of economic importance.

The first successful somatic hybrid was produced between *S. melongena* and *S. sisymbriifolium* through polyethylene glycol (PEG)-mediated protoplast fusion (Gleddie, Keller, and Selterfield 1986b). Flowers of somatic hybrid were light purple and displayed floral abnormalities like separate petals, which were absent in either of the parents. Since then, somatic hybrids have been developed by fusing protoplasts of cultivated varieties with *S. aethiopicum* group *aculeatum* (Daunay et al. 1993), *S. torvum* (Guri and Sink 1988a), and *S. khasianum* (Sihachakr et al. 1988). Interspecific somatic hybrids were also produced by fusing protoplasts of *S. melongena* with *S. nigrum* to transfer herbicide (atrazine) resistant trait from *S. nigrum* (Guri and Sink 1988b). Somatic hybrids obtained through organogenesis showed flower abscission. In another study, Samoylov and Sink (1996) showed that the regeneration ability of asymmetric somatic hybrids depends upon the irradiation dose. The asymmetric hybrids were developed by exposing the chromosomes of donor genome to different doses of gamma rays (100, 250, 500, 750, and 1,000 Gy). It was found that the hybrids exposed to 100 Gy gamma rays was a tetraploid ($4x$), whereas exposure to other doses formed hybrids between $5x$ and $9x$. The radiation dosage determined the ploidy level and the DNA content of these hybrids.

Protoplast fusion has been achieved through electrofusion and PEG. The protocols for both have been standardized (Sihachakr et al. 1993). The basal medium for protoplast culture and regeneration medium has been KM medium supplemented with 2,4-D, NAA, BAP or zeatin in combination or alone (Kao and Michayluk 1975; Sihachakr et al. 1993). Regenerated somatic hybrids were obtained through callus. Somatic hybrids have been characterized using isozyme patterns, morphology and cultural behavior, or molecular and physiological studies (Gleddie et al. 1986a; Gleddie, Keller, and Selterfield 1986b; Guri and Sink 1988b; Sakata, Nishio, and Matthews 1991; Filippone, Penza, and Romano 1992; Daunay et al. 1993; Sihachakr et al. 1993;). Ploidy was determined by flow cytometry (Filippone, Penza, and Romano 1992; Sihachakr et al. 1993).

Many useful traits from wild species like resistance to fungal and bacterial wilts and nematodes have been maintained in the somatic hybrids. Recently, somatic hybrids between *S. melongena*, *S. aethiopicum* group *aculeatum*, and *gilo* have been produced and found to be highly resistant to bacterial wilt caused by *Ralstonia solanacearum* (Collonnier et al. 2001).

10.6.4.5 Genetic Engineering

Successful eggplant genetic transformation was achieved as early as 1988 (Guri and Sink 1988a, 1988b) using *Agrobacterium*-mediated genetic transformation with the cointegrate vector— pMON 200 harboring *nptII* gene. This was followed by a series of successful attempts to

Table 10.4   Production of Somatic Hybrids from Protoplast Fusion in Eggplant

Parents Selected	Culture Response	Reference
S. melongena (Black Beauty) × *S. sisymbriifolium*	26 Hybrid lines were obtained mostly aneuploids having chromosome number close to 4,883 somatic hybrids obtained	Gleddie et al. (1985b), Gleddie, Keller, and Selterfield (1986b)
S. melongena (Dourga) × *S. khasianum*	Most hybrids were tetraploids (48) and few were aneuploids	Sihachakr et al. (1988)
S. melongena (Black Beauty) × *S. torvum*	10 Somatic hybrids were well established and chromosome number ranged from 46 to 48	Guri and Sink (1988a)
S. melongena (Black Beauty) × *S. nigrum*	Only two somatic hybrids could be obtained	Guri and Sink (1988b)
S. melongena (Dourga) × *S. torvum*	19 Somatic hybrids were obtained having chromosome number ranged from 46 to 48	Sihachakr et al. (1988)
S. melongena (Dourga) × *S. nigrum*	Only single somatic hybrid could be obtained which was aneuploid with 2*n* close to 96	Sihachakr et al. (1989)
S. melongena (Shironasu) × *N. tabacum*	Somatic hybrids were not successful since only green shoots could be obtained from two somatic hybrid colonies	Toki, Kameya, and Abe (1990)
S. melongena × *S. macrocarpon*	The developed hybrids were highly sterile and there was failure to set seed	Gowda, Shivashankar, and Joshi (1990)
S. melongena × *Lycopersicon* spp.	Only two hybrids with leaf like primordia were obtained	Guri and Sink (1991)
S. melongena (Dourga) × *S. aethiopicum*	35 Somatic hybrids were obtained of which 32 were tetraploids one was hexaploid and two mixoploids	Daunay et al. (1993)
S. melongena (Black Beauty) × hybrid of *L. esculentum* and *L. pennellii* (EP)	These were asymmetric hybrids. Only four hybrids could be obtained	Liu, Ly, and Sink (1995) and Samoylov and Sink (1996)
S. melongena × *S. torvum*	They selected 12 somatic hybrids to be incorporated in the breeding programs	Jarl, Rietveld, and de Haas (1999)
S. melongena × *S. aethiopicum*	30 Hybrid plants were obtained. All hybrids were fertile. Of these16 were tested tolerant against bacterial wilt	Collonnier et al. (2001)

improve the transformation protocol using gene constructs with nptII as a selection marker and several reporter genes like *gus* (*b-glucouronidase*), *cat* (*chloramphenicol acetyl transferase*), and *luciferase*. On the other hand, a number of reports describe transformation protocols, and a comprehensive study to optimize factors that dramatically affect transformation (explant type, genotype, and cultural conditions) is found lacking. A recent study has made some efforts to standardize transformation protocol, taking into account the influence of antibiotics and growth regulators (Billings et al. 1997).

The insecticidal protein gene *Bt* (*CryIIIb*) from *Bacillus thuringensis* was used to develop transgenic insect resistance in eggplant (Rotino et al. 1992b). Plants developed were resistant to Colorado potato beetle (Arpaia et al. 1997). Field trials showed that resistant transgenic lines had significantly higher fruit yield (Arpaia et al. 1998). There was no significant difference in colonies of non-target insects like green peach aphid (*Myzua persicae* Sulz.), flea beetle (*Altica* sp.), poatao tuber moth (*Phthorimaea operculella*), and lacewings (*Chrysoperla carnes* Stephens) in transgenic

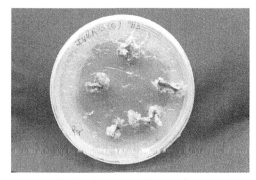
Explant in regeneration media

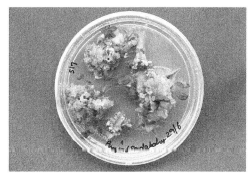
Shoot inititation in regeneration media

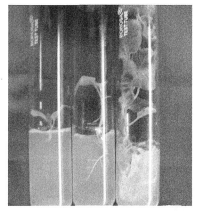
Root inititation in rooting media

Regenerated plantlet

Complete regenerated plant in pot

Hardening of plantlet in glasshouse

Figure 10.3 Genetic transformation of eggplant using *Cry 1Ac* gene.

and control fields. Introduction of a synthetic *Bt* (*CryIAb*) gene to eggplant provided protection against fruit borer, *Leucinodes orbonalis* (Kumar et al. 1998; Figure 10.3a–Figure 10.3f).

Parthenocarpic eggplant transgenics have been successfully produced by transferring *iaaM* gene from *Pseudomonas syringae* pv. *savatanoi*, driven by the *DefH9* promoter from *Antirrhinum majus*. Transgenic plants developed fruits from both pollinated and unpollinated flowers in comparison to the untransformed control plants, which formed fruits only when pollinated. In case of transgenic plants, seedless fruits were obtained only from unpollinated flowers (Rotino et al. 1997).

In greenhouse trials, transgenic parthenocarpic plants were more productive than both natural parthenocarpic hybrids and induced parthenocarpic plants using phytohormone sprays (Donzella, Spena, and Rotino 2000).

10.7 SUMMARY

Eggplant, an important vegetable crop, has been grown in India since ancient time due to its low caloric value and high potassium content. Supposedly originating in India, eggplant is consumed by most of the world. The plant is monocious, and its heaving hermaphrodite flower is predominantly self-pollinated. Almost all of the brinjal varieties are diploid. Chromosome changes might have played a significant role in its evolution. Interspecific crosses range from completely fertile to completely sterile. Sometime interspecific crosses are only successful by *in vitro* embryo rescue methods. Brinjal has been studied in detail for its genetic analysis, assessment of genetic diversity with conventional and non-conventional tools, and identification of molecular markers for many agronomically important traits. Molecular maps have been developed by many workers using different molecular techniques, viz., RAPD, RFLP, SSRs, and AFLP, etc. Plant breeder efforts of the past 60 years have improved brinjal for both qualitative and quantitative traits and developed many successful commercial hybrids. Many promising varieties resistant to biotic and abiotic stresses have been released for commercial cultivation. Using modern tissue culture techniques, plant breeding programs have been stimulated to develop new varieties. Transgenic brinjal for biotic stresses viz. bacterial wilt and Lepidopteron insects has developed and the experiments on the stable expression of the transgene is underway.

REFERENCES

Ahmad, Q., Sources of resistance in brinjal to *Phomopsis* fruit rot, *Ind. Phytopathol.*, 40, 98–99, 1987.

Ahuja, S. et al., Effects of infestation of eggplant (*Solanum melongena*) with root knot nematode (*Meloidogyne incognita*) on the oxidative enzymes and cell wall constituents in their roots, *Cap. Newsl.*, 6, 98–99, 1987.

Allicho, R., Del Grosso, E., and Boschieri, E., Tissue cultures and plant regeneration from different explants in six cultivars of *Solanum melongena*, *Experientia*, 38, 449–450, 1982.

Andryushchenko, V. K. and Pilipenko, A. D., Interrelation between water consumption by eggplants and the quality of the product obtained, *Capsicum Newsl.*, 7, 81, 1989.

Arpaia, S. et al., Production of transgenic eggplant (*Solanum melongena*) resistant to Colorado potato beetle (*Leptinotarsa decemlineata* Say), *Theor. Appl. Genet.*, 95, 329–334, 1997.

Arpaia, S. et al., *Field performance of Bt-transgenic eggplant lines resistant to Colorado potato beetle Proceedings of the 10th Eucarpia Meeting on Genetics and Breeding of Capsicum and Eggplant* pp. 191–194, Avignon, France, 1998.

Aubert, S., Daunay, M. C., and Pochard, E., Steroidal saponins of aubergine (*Solanum melongena* L.). 1. Food value, methods of analysis and localization in the fruit, *Agronomie*, 9, 641–651, 1989a.

Aubert, S., Daunay, M. C., and Pochard, E., Steroidal saponins from eggplant (*Solanum melongena* L.). 2. Effects of cultural conditions, genotypes and seeds per fruit, *Agronomie*, 9, 751–758, 1989b.

Bajaj, K. L., Singh, D., and Kaur, G., Biochemical basis of relative field resistance of eggplant (*Solanum melongena*) to the shoot and fruit borer (*Leucinodes orbonalis* Guen.), *Veg. Sci.*, 16, 145–149, 1989.

Baksh, S. and Iqbal, M., Compatibility relationships in some non-tuberous species of *Solanum*, *J. Hortic. Sci.*, 54, 163, 1979.

Bannerot, H. and Foury, C., Use of genetic resources and variety breeding, *Bull. Tech. d'Inform.*, 407, 93–106, 1986.

Basavaraja, N., Studies on hybrid seed production in brinjal (*Solanum melongena* L.), *Mysore J. Agric. Sci.*, 20, 249, 1986.

Bhatt, P. N., Bhatt, D. P., and Sussex, I. M., Organ regeneration from leaf discs of *Solanum nigrum*, *S. dulcamara* and *S. khasianum*, *Z. Pflanzenphysiol.*, 95, 355–362, 1979.

Billings, S. et al., The effect of growth regulators and antibiotics on eggplant transformation, *J. Am. Soc. Hort. Sci.*, 122(2), 158–162, 1997.

Blestsos, F. A. et al., Interspecific hybrids between three eggplant (*Solanum melongena* L.) cultivars and two wild species (*Solanum torvum* Sw. and *Solanum sisymbriifolium* Lam.), *Plant Breed.*, 117, 159–164, 1998.

Borikar, S. T., Makne, U. G., and Kulkarni, U. G., Note on diallel analysis in brinjal, *Indian J. Agric. Sci.*, 51, 51–52, 1981.

Chauhan, S. V. S., Studies in genic male-sterile *Solanum melongena* L, *Indian J. Genet. Plant Breed.*, 44, 367–371, 1984.

Chelliah, S. and Srinivasan, K., *Resistance in bhindi, brinjal and tomato to major insect and mite pests Proceedings of the National Seminar on Breeding Crop Plants for Resistance to Pests and Diseases* 1983 p. 47, Tamil Nadu, India.

Choudhury, H. C., Genetical studies in some West African: *Solanum melongena, Can. J. Genet. Cytol.*, 14, 446–449, 1972.

Collonnier, C. et al., Source of resistance against *Ralstonia solanaceraum* in fertile somatic hybrids of eggplant (*Solanum melongena* L.) with *Solanum aethiopicum* L, *Plant Sci.*, 301–313, 2001.

Cuastero, J. et al., Germplasm resources of *S. melongena* from Spain, *Capsicum Newsl.*, 4, 77–78, 1985.

Daunay, M. C. and Dalmasso, A., Multiplication of *Meloidogyne javanica, M. incognita* and *M. arenaria* on several *Solanum* species, *Revue de Nematologie*, 8, 31–45, 1985.

Daunay, M. C. et al., Production and characterization of fertile somatic hybrids of eggplant (*Solanum melongena* L.) with *Solanum aethiopicum* L, *Theor. Appl. Genet.*, 85, 841–850, 1993.

Daunay, M. C. et al., The use of wild genetic resources for eggplant *(Solanum melongena)* breeding. II. In Crossability and fertility of interspecific hybrids *Proceedings of the 10th Eucarpia Meeting on Genetics and Breeding of Capsicum and Eggplant*, pp. 19–24, Avignon, France, 1998

Deshpande, A. A., Bankapus, V. M., and Nalawadi, U. G., Some aspects of blossom biology in brinjal varieties (*Solanum melongena* L.), *Curr. Res. Univ. Agric. Sci. (Bangalore)*, 7(10), 174–175, 1978.

Di Vito, M., Zaccheo, G., and Catalano, F., Source for resistance to root knot nematodes *Meloidogyne spp.* eggplant In *Proceedings of the Eighth Meeting on Genetics and Breeding of Capsicum and Eggplant*, pp. 301–303, Rome, Italy, 1992.

Donzella, G., Spena, A., and Rotino, G. L., Transgenic parthenocarpic eggplants: superior germplasm for increased winter production, *Mol. Breed.*, 6, 79–86, 2000.

Dumas De Vaulx, R. and Chambonnet, D., Culture *in vitro* d0 anther d0 aubergine (*Solanum melongena* L.): stimulation de la production de plantes au moyen de traitements a þ 35 8C associes à de faibles teneures en substances de croissance, *Agronomie*, 2, 983–988, 1982. (in French with English summary).

Fang, M. R., Mao, R. C., and Xie, W. H., Breeding cytoplasmically male sterile lines of eggplant, *Acta Hortic. Sin.*, 12, 261–266, 1985.

Fassuliotis, G., Regeneration of whole plants from isolated stem parenchyma cells of *Solanum sismbriifolium*, *J. Am. Soc. Hort. Sci.*, 100, 636–638, 1975.

Fassuliotis, G., Nelson, B. V., and Bhatt, D. P., Organogenesis in tissue culture of *Solanum melongena* cv. Florida market, *Plant Sci. Lett.*, 22, 119–125, 1981.

Filippone, E., Penza, R., and Romano, R., Advanced biotechnologies applied to eggplant *(Solanum melongena* L.) breeding In *Proceedings of the Eighth Meeting on Genetics and Breeding of Capsicum and Eggplant*, pp. 260–265, Rome, Italy, 1992.

Fobert, P. R. and Webb, D. T., Effect of polyamines, polyamine precursors and polyamine inhibitors on somatic embryogenesis from eggplant (*Solanum melongena* L.) cotyledons, *Can. J. Bot.*, 66, 1734–1742, 1988.

Fournier, D., Lejeune, F., and Tourte, Y., Cytological events during the initiation of meristematic nodules in calli derived from eggplant protoplasts, *Biol. Cell*, 85, 93–100, 1995.

Franceschetti, U. and Lepori, G., Natural cross pollination in eggplant, *Sementi Elette*, 31(6), 25–28, 1985.

Frary, A. et al., A genetic linkage map of eggplant In *Proceedings of the Eighth Conference on Plant and Animal Genome*. January 9–12, 2000. (Abstr. No. P319)

Frydrych, J., Biology of flowering in the eggplant, Bull, *Vyzkumny Ustav Zelinarobs Olomone*, 8, 27–37, 1964.

Gati, G. K., Results of a field evaluation of resistance to drought in eggplant, *Nauchno-tekhnicheskii Bynlleteu Vsesoyuznogo Ordena, Lenina i Ordena Druzhby narodov Nauchno-issedovatel' skoko Instituta Rastenievodstva imeni*, Vavilova, N. I., Ed., Vol. 128, pp. 58–59.

Gleddie, S., Keller, W., and Setterfield, G., Somatic embryogenesis and plant regeneration from leaf explants and cell suspensions of *Solanum melongena* (eggplant), *Can. J. Bot.*, 61, 656–666, 1983.

Gleddie, S., Keller, W., and Setterfield, G., Plant regeneration from tissue, cell and protoplast cultures of several wild *Solanum* species, *J. Plant Physiol.*, 109, 405–418, 1985a.

Gleddie, S., Fassuliotis, G., Keller, W., and Setterfield, G., Somatic hybridisation as a potential method of transferring nematode and mite resistance into eggplant, *Z. Pflanzezüchtg.*, 94, 348–351, 1985b.

Gleddie, S., Keller, W., and Setterfield, G., Eggplant, In *Handbook of Plant Cell Culture, vol. 3, Techniques for Propagation and Breeding*, Evans, D. A. and Sharp, W. R., Eds., New York: MacMillan, pp. 500–511, 1986a.

Gleddie, S., Keller, W. A., and Selterfield, G., Production and characterization of somatic hybrids between *Solanum melongena* L. and *S. sisymbriifolium* Lam, *Theor. Appl. Genet.*, 71, 613–621, 1986b.

Gopinath, G., Madalageri, B. B., and Somasekar, C., A note on heredity of fruit colour in WCGR 112-8 brinjal, *Curr. Res. Univ. Agric. Sci. (Bangalore)*, 15, 17–18, 1986.

Gowda, P. H. R., Shivashankar, K. T., and Joshi, S., Interspecific hybridization between Solanum melongena and Solanum macrocarpon: study of F1 hybrid plants, *Euphytica*, 48, 59–61, 1990.

Gu, S. R., Plantlets from isolated pollen culture of eggplant (*Solanum melongena* L.), *Acta Bot. Sin.*, 21, 30–36, 1979.

Guri, A. and Sink, K. C., Interspecific somatic hybrid plants between eggplant *Solanum melongena* L. and *Solanum torvum*, *Theor. Appl. Genet.*, 76, 490–496, 1988a.

Guri, A. and Sink, K. C., Organelle composition in somatic hybrids between an atrazine resistant biotype of *Solanum nigrum* and *Solanum melongena*, *Plant Sci.*, 58, 51–58, 1988b.

Guri, A. and Sink, K. C., Somatic hybridization between selected *Lycopersicon* Nad. *Solanum* species, *Plant Cell. Rep.*, 10, 76–80, 1991.

Guri, A., Volokita, M., and Sink, K. C., Plant regeneration from leaf protoplasts of *Solanum torvum*, *Plant Cell. Rep.*, 6, 302–304, 1987.

Hebert, Y., Comparative resistance of mine species of the genes Solanum to bacterial wilt (*Psudomonas solanacearum*) and the nematode *Meloidogyne incognita*. Implications for the breeding of aubergine (*S. melongena*) in the humid tropical zone, *Agronomie*, 5, 27–32, 1985.

Hedrick, U. P., Sturtevant's notes on edible plants, *N.Y. Dept. Agric. Ann. Rep.*, 27, 212, 1919. 685

Hitomi, A., Amagai, H., and Ezura, H., The influence of auxin on the array of somaclonal variants generated from somatic embryogenesis of eggplant, *Solanum melongena* L, *Plant Breed.*, 117, 379–383, 1998.

Isouard, G., Raquin, C., and Demarly, Y., Obention de plantes haploids et diploids par culture *in vitro* d'aubergine (*Solanum melongena* L.), *CR Acad. Sci. Ser. D*, 288, 987–989, 1979. (in French, with English summary)

Isshiki, S., Okubo, H., and Fujieda, K., Genetic control of isozymes in eggplant and its wild species, *Euphytica*, 80, 145–150, 1994.

Isshiki, S., Okubo, H., and Fujieda, K., Segregation of isozymes in selfed progenies of a synthetic amphidiploid between *Solanum integrifolium* and *S. melongena*, *Euphytica*, 112, 9–14, 2000.

Isshiki, S. et al., RFLP analysis of a PCR amplified region of chloroplast DNA in eggplant and related *Solanum* species, *Euphytica*, 102, 295–299, 1998.

Jacobsen, E. and Sopory, S. K., The influence and possible recombination of genotypes on production of microspore embryoids in anther cultures of *Solanum tuberosum* and dihaploid hybrids, *Theor. Appl. Genet.*, 52, 119–123, 1978.

Jarl, C. I., Rietveld, M. E., and de Haas, J. M., Transfer of fungal tolerance through interspecific somatic hybridization between *Solanum melongena* and *S. torvum*, *Plant Cell Rep.*, 18, 791–796, 1999.

Jenkins, S. F., Interaction of *Pseudomonas solanacearum* and Meloidogyne incognita on bacterial wilt resistant and susceptible cultivars of eggplant, *Proc. Am. Phytopathol. Soc.*, 1, 69–70, 1974.

Jenkins, S. F. and Nesmith, W. C., Severity of southern bacterial wilt of tomato and eggplant as influenced by plant age, inoculation technique, isolate source, *Proc. Am. Phytopathol. Soc.*, 3, 337–338, 1976.

Jyotishi, R. P. and Hussain, S. M., Use of 2, 4-D as an aid in hybrid seed production in brinjal, JNKVV, *Res. J. Jabalpur*, 1, 20–22, 1968.

Kalda, T. S., Swarup, V., and Chowdhury, B., Resistance to Phomopsis blight in eggplant, *Veg. Sci.*, 4, 90–101, 1977.

Kalloo, G., Eggplant (*Solanum melongena*), *Genetic Improvement of Vegetable Crops*, Kalloo, G., Ed., Oxford: Pergamon Press, pp. 587–604, 1993.

Kalloo, G., Baswana, K. S., and Sharma, N. K., *Heat tolerance in eggplant (Solanum melongena* L.), XXIII *Int. Hortic. Congr.* Firenze, Italy, p. 1204, 1990.

Kalloo, G., Dixit, J., and Baswana, K. S., Hissar Shyamal (H8): is superior to BRl12 brinjal, *Indian Hortic.*, 36, 29–34, 1991.

Kamat, M. G. and Rao, P. S., Vegetative multiplication of eggplants using tissue culture techniques, *Plant Sci. Lett.*, 13, 57–65, 1978.

Kao, K. M. and Michayluk, M. R., Nutritional requirements for growth of *Vicia hajaatana* cells and protoplasts at very low population density in liquid media, *Planta*, 126, 105–110, 1975.

Karihaloo, J. L. and Gottlieb, L. D., Allozyme variation in the eggplant *Solanum melongena* L. (*Solanaceae*), *Theor. Appl. Genet.*, 90, 578–583, 1995.

Karihaloo, J. L., Brauner, S., and Gottlieb, L. D., Random amplified polymorphic DNA variation in the eggplant, *Solanum melongena* L. (*Solanaceae*), *Theor. Appl. Genet.*, 90, 767–770, 1995.

Kashyap, V. Morphological and molecular characterization of dihaploids derived from somatic hybrids between wild species and cultivated eggplant (*Solanum melongena* L.). Ph.D. thesis, University of Delhi (South Campus), New Delhi, India, 2002.

Kashyap, V. and Rajam, M. V., RAPD fingerprinting of eggplant *Solanum melongena* L. species for resistance to fungal wilts In *Proceedings of the International Conference on Life Sciences in Next Millennium.* Hyderabad, India, December 11–14, 1999.

Kashyap, V. et al., Plant regeneration in wild species of eggplant, In *Proceedings of the National Symposium on Role of Plant Tissue Culture in Biodiversity, Conservation and Economic Development* India: Almora pp. 14–15, June 7–9, 1999.

Khan, R., Rao, G. R., and Baksh, S., Cytogenetics of *Solanum* integrifolium and its possible use in eggplant breeding, *Indian J. Genet. Plant Breed.*, 38, 343–347, 1978.

Khapre, P. R., Wanjari, K. B., and Deokar, A. B., Inheritance of some colour characters in the interspecific cross *Solanum melongena* × *S. indicum*, *J. Maharashtra Agric. Univ.*, 11, 296–298, 1986.

Khapre, P. K., Deokar, A. B., and Wanjari, K. B., Inheritance of spineness in *Solanum melongena* × *Solanum indicum*, *J. Maharashtra Agric. Univ.*, 12, 107–108, 1987.

Khurana, S. C. et al., Correlation and path analysis in eggplant (*Solanum melongena*), *Indian J. Agric. Sci.*, 58, 799–800, 1988.

Kirti, P. B., Moorty, K. V., and Rao, B. G. S., Cytological studies on diploid, autotretraploid and autotriploid *Solanum sisymbriifolium*, *Curr. Sci. India*, 53, 12–13, 1984a.

Kirti, P. B. et al., Cytological observations on some autotetraploids and amphidiploids in spinous *Solanums* and their bearing on interrelationships, *Curr. Sci. India*, 53, 1256–1258, 1984b.

Nascimento, W. M., Torres, A. C., and Lima, L. B., Pollen viability in hybrid seed production of eggplant under tropical conditions, *Acta Hort. (ISHS)*, 607, 37–39, 2003.

Kowalozyk, T. P., Mackenzie, I. A., and Cocking, E. C., Plant regeneration from organ explants and protoplasts of medicinal plant *Solanum khasianum* CB. Clarke var. Chatterjeeanum Sengupta (Syn. *Solanum viarum* Dunal), *Z. Pflanzenphysiol.*, 11, 55–68, 1983.

Kumar, P. A. et al., Insect-resistant transgenic brinjal plants, *Mol. Breed.*, 4, 33–37, 1998.

Lakshmana Rao, P. V. and Singh, B., Plantlet regeneration from encapsulated somatic embryos of hybrid *Solanum melongena* L, *Plant Cell Rep.*, 10, 7–11, 1991.

Lakshmi, V. V. S., Moorty, K. V., and Rao, B. G. S., Studies on the incompatibilities among spinous solanum species. III. The degree of divergence of *S. incanum* L. and *S. melongena* L. in relation to other spinous *Solanums*, *Incompatibility Newsl.*, 13, 101–103, 1981.

Lee, C. T., Performance studies on eggplant cultivars during dry and wet seasons in the tropics (Abst.), *Hort. Sci.*, 14(3), 111–448, 1979.

Lester, R. N. et al., Variation patterns in the African scarlet eggplant, *Solanum aethiopicum* L., *Intraspecific Classification of Wild and Cultivated Plants*, Styles, B. T., Ed., Oxford: Clarendon Press, pp. 283–307, 1986.

Litcher, R., Induction of haploid plants from isolated pollen *Brasscia napus* rape, *Z. Pflanzenphysiol.*, 105, 427–434, 1982.

Liu, K. B., Ly, M., and Sink, K. C., Assymetric somatic hybrid plants between an interspecific *Lycopersicon* hybrid and *Solanum melongena*, *Plant Cell Rep.*, 14, 652–656, 1995.

Mace, E. S., Lester, R. N., and Gebhardt, C. G., AFLP analysis of genetic relationships among the cultivated eggplant, *Solanum melongena* L., and wild relatives (Solanaceae), *Theor. Appl. Genet.*, 99, 626–633, 1999.

Madalageri, B. B. et al., Reaction of eggplant genotypes to *Cercospora solani* and *Leucinodes orbonalis*, *Plant Pathol. Newsl.*, 6, 26–27, 1988.

Madeira, M. C. B. and Reifschneider, F. J. B., Evaluation of eggplant resistance to *Collectotrichum gleosporiodes*, *Capsicum Newsl.*, 5, 68–69, 1986.

Magioli, C. et al., Efficient shoot organogenesis of eggplant (*Solanum melongena* L.) induced by thidiazuron, *Plant Cell Rep.*, 17, 661–663, 1998.

Malausa, J. C., Daunay, M. C., and Bourgoin, T., Preliminary research on resistance of aubergine to the glasshouse whitefly, *Trialeurodes vaporariorum* Westwood (Homoptera: Aleyrodidae), *Agronomie*, 8, 693–699, 1988.

Mariani, P., Eggplant somatic embryogenesis combined with synthetic seed technology *Proceedings of the Eighth Meeting on Genetics and Breeding of Capsicum and Eggplant*, pp. 289–294, September 7–10.

Martin, F. W. and Rhodes, A. M., Subspecific grouping of eggplant cultivars, *Euphytica*, 28, 367–383, 1979.

Matsuoka, H. and Hinata, K., NAA induced organogenesis and embryogenesis in hypocotyl callus of *Solanum melongena* L, *J. Exp. Bot.*, 30, 363–370, 1979.

McCammon, K. R. and Honma, S., Morphological and cytogenetic analysis of an interspecific hybrid eggplant *S. melongena* × *S. torvum*, *Hort. Sci.*, 18, 894–895, 1983.

Melo, I. S. de and da Costa, C. D., Resistance reaction of aubergine progenies to Verticillium albo-atrum Reinkes Berth, *Summa. Phytopathol.*, 11, 180–185, 1985.

Menella, G. et al., Seed storage protein characterization of *Solanum* species and cultivars and androgenic lines of *S. melongena* L. by SDS-PAGE and AE-HPLC, *Seed Sci. Technol.*, 27, 23–35, 1999.

Miyoshi, K., Callus induction and plantlet formation through culture of isolated microspores of eggplant (*Solanum melongena* L.), *Plant Cell Rep.*, 15, 391–395, 1996.

Momiyama, Y. et al., Differential display identifies developmentally regulated genes during somatic embryogenesis in eggplant (*Solanum melongena* L.), *Biochem. Biophys. Res. Commun.*, 213, 376–382, 1995.

More, D. C., Patil, S. B., and Nimbalkar, V. S., Inheritance of some characters in brinjal cross SM2 × Nimbkar Green Round, *J. Maharashtra Agric. Univ.*, 7, 243–244, 1982.

Mukherjee, S. K., Rathnasbapathi, B., and Gupta, N., Low sugar and osmotic requirements for shoot regeneration from leaf pieces of *Solanum melongena* L, *Plant Cell Tiss. Org. Cult.*, 25, 12–16, 1991.

Nicklow, C. W., The use of recurrent selection in efforts to achieve *Verticillium* resistance in eggplant, *HortSci.*, 18(4), 111–600, 1983.

Nimbalkar, V. S. and More, D. C., Genetic studies in a brinjal (*S. melongena* L.) cross Mukta-Keshi × White Green, *J. Maharashtra Agric. Univ.*, 5, 208–210, 1980.

Nimbalkar, V. S. and More, D. C., Inheritance of fruit shape in a brinjal (*S. melongena* L.) cross White Green × Manjari Gota, *J. Maharashtra Agric. Univ.*, 6, 253–254, 1981.

Nishi, S., Saving energy in indoor cultivation of fruit and vegetables: breeding prospects, *Agric. Hortic. (Nogyi Oyobi Engei)*, 53, 1245–1248, 1978.

Nitsch, J. P. and Nitsch, C., Haploid plants from pollen grains, *Science*, 163, 85–87, 1969.

Nuez, F. et al., Germplasm resources of *S. melongena* from Spain, *Capsicum Newsl.*, 6, 87–88, 1987.

Nunome, T. et al., Mapping of fruit shape and color development traits in eggplant (*Solanum melongena* L.) based on RAPD and AFLP markers, *Breeding Sci.*, 51, 19–26, 2001.

Nunome, T., Yoshida, T., and Hirai, M., Genetic linkage map of eggplant In *Proceedings of the 10th Eucarpia Meeting on Genetics and Breeding of Capsicum and Eggplant*, Arignon, France pp. 239–242, 1998.

Nuttall, V. W., The inheritance and possible usefulness of functional male sterility in *S. melongena*, *Can. J. Genet. Cytol.*, 5, 197–199, 1963.

Nuttall, V. W., The inheritance of a chlorophyll deficient seedling mutant in *Solanum melongena*, *Can. J. Genet. Cytol.*, 7, 349–351, 1965.

Okoli, B. E., Cytotaxonomic study of five west African species of *Solanum* (*Solanaceae*), *Feddes Repertorium*, 99(5–6), 183–187, 1988.

Ozaki, K. and Kimura, T., Method for evaluating the resistance of Solanum plants to bacterial wilt caused by *Pseudomonas solanacearum* Bull, *Chugoku. Nat. Agric. Exp. Stn.*, 4, 103–107, 1980.

Patil, S. K. and More, D. C., Inheritance studies of some characters in brinjal, *J. Maharashtra Agric. Univ.*, 8, 47–49, 1983.

Pawar, D. B. et al., Promising resistant sources for jassid and fruit borer in brinjal, *Curro. Res. Rep. Mahatma Phule Agric. Univ.*, 3, 81–84, 1987.

Petrov, C., Nakov, B., and Krasteva, L., Studies on the resistance of eggplant introductions to Verticillium wilt, in Eucarpia In *VIIth Meeting on Genetics and Breeding on Capsicum and Eggplant*, Yugoslavia: Kragujevac, 1989 pp. 183–186 (27–30 June).

Popova, D., Some observations on the flowering, pollination and fertilization of the eggplant, I3v. Inst. Rasten. (News Inst. Plant Industr.), *Sofija*, 5, 211–214, 1958.

Popova, D. and Daskalov, S., Study of functionally male-sterile form of eggplant, *Grhdinarska (Lozarska Nauka)*, 8, 25–27, 1971.

Raina, S. K. and Iyer, R. D., Differentiation of diploid plants from pollen callus in anther cultures of *Solanum melongena* L, *Z. Pflanzenzüchtg*, 70, 275–280, 1973.

Rao, N. N., The barriers of hybridization between *Solanum melongena* and some other species of *Solanum*, In *The Biology and Taxonomy of the Solanaceae*, Hawkes, J. G., Lester, R. N., and Skelding, A. D., Eds., London: Academic Press, pp. 605–614, 1979.

Rao, G. R., Cytogenic relationship and barrier to gene exchange between *Solanum melongena* L. and *Solanum hispidum*, *Pers. Caryologia*, 33, 429–433, 1980.

Rao, G. R. and Baksh, S., Cytomorphological study of the amphidiploids derived from the hybrids of the crosses between *Solanum melongena* L. and *Solanum integrifolium* Poir, *Curr. Sci.*, 48, 316–317, 1979.

Rao, G. R. and Kumar, A., Some observations on interspecific hybrids of *Solanum melongena* L, *Proc. Indian Acad. Sci. (Plant Sci.)*, 89, 117–121, 1980.

Rao, P. S. and Narayanaswami, I. S., Induced morphogenesis in tissue cultures of *Solanum xanthocarpum*, *Planta*, 81, 372–375, 1968.

Rotino, G. L., Falavigna, A., and Restaino, F., Production of anther-derived plantlets of eggplant, *Capsicum Newsl.*, 6, 89–90, 1987.

Rotino, G. L., Haploidy in eggplant, *In vitro* Production in Higher Plants, Jain, S. M., Sopory, S. K., and Veilleux, R. E., Eds., vol. 3, Netherlands: Kluwer Academic Publishers, pp. 115–141, 1996.

Rotino, G. L. et al., Variation among androgenic and embryogenic lines of eggplant (*Solanum melongena* L.), *J. Genet. Breed.*, 45, 141–146, 1991.

Rotino, G. L. et al., In Agrobacterium mediated transformation of *Solanum spp.* using a Bacillus thuringiensis gene effective against coleopteran, In *Proceedings of the Eighth Meeting on Genetics and Breeding of Capsiscum and Eggplant*, pp. 295–300, 1992.

Rotino, G. L. et al., Genetic engineering of parthenocarpic plants, *Nature Biotech.*, 15(13), 1398–1401, 1997.

Rotino, G. L. et al., Towards introgression of resistance to *Fusarium oxysporum* F. sp. *melongenae* from *Solanum integrifolium* into eggplant In *Proceedings of the 11th Eucarpia Meeting on Genetics and Breeding of Capsicum and Eggplant*, pp. 303–307, 2001.

Russo, C. et al., Karyotype analysis in wild species of *Solanum* spp. In *Proceedings of the Eighth Meeting on Genetics and Breeding of Capsicum and Eggplant*, pp. 272–277, 1992.

Saito, T. and Nishimura, S., Improved culture conditions for somatic embryogensis using in aseptic ventilative filter in eggplant (*Solanum melongena* L.), *Plant Sci.*, 102, 205–211, 1994.

Sakata, Y. and Lester, R. N., Chloroplast DNA diversity in (*Solanum melongena*) and its related species *S. incanum* and *S. marginatum*, *Euphytica*, 80, 1–4, 1994.

Sakata, Y., Nishio, T., and Matthews, P. J., Chloroplast DNA analysis of eggplant (*Solanum melongena*) and related species for taxonomic affinity, *Euphytica*, 55, 21–26, 1991.

Sakata, Y., Nishio, T., and Mon'ma, S., *Resistance of Solanum* species to *Verticillium* wilt and bacterial wilt *Eucarpia VVth Meeting on Genetics and Breeding on Capsicum and Eggplant*, pp. 177–181, 1989.

Sambandam, C. N., Inheritance of hypocotyl colour in brinjal, *Indian J. Genet. Plant Breed.*, 24, 175–177, 1964a.

Sambandam, C. N., Natural cross pollination in eggplant, *Econ. Bot.*, 18, 128, 1964b.

Sambandam, C. N. and Chelliah, S., Breeding brinjal for resistance to *Aphis gossypii* G. In *Proceedings of the National Seminar on Breeding Crop Plants for Resistance to Pests and Diseases*, 15, Tamil Nadu Agricultural University, 1983.

Samoylov, V. M. and Sink, K. C., The role of irradiation dose and DNA content of somatic hybrid calli in producing asymmetric plants between an interspecific tomato hybrid and eggplant, *Theor. Appl. Genet.*, 92, 850–857, 1996.

Sankar, M. A., Jessykutty, P. C., and Peter, K. V., Efficiency of four selection methods to improve level of bacterial wilt resistance in eggplant, *Indian J. Agric. Sci.*, 57, 138–141, 1987.

Schaff, D. A., Boyer, C. H., and Pollack, B. L., Interspecific hybridization of *Solanum melongena* L.×*S. macrocarpon* L. (Abstr.), *Hort.Sci.*, 15, 419, 1980.

Schaff, D. A. et al., Hybridization and fertility of hybrid derivatives of *Solanum melongena* L. and *Solanum macrocarpon* L, *Theor. Appl. Genet.*, 62, 149–153, 1982.

Schalk, J. M. et al., Resistance in eggplant, *Solanum melongena* L. and nontuber-bearing *Solanum* species to carmine spider mite, *J. Am. Soc. Hort. Soc.*, 100, 479–481, 1975.

Scoccianti, V. et al., Organogenesis from *Solanum melongena* L. (eggplant) cotyledon explants is associated with hormone: modulation of polyamine biosynthesis and conjugation, *Protoplasma*, 211, 51–63, 2000.

Shariff, R. A. and Habib, A. F., Inheritance studies in interspecific crosses of *Solanum xanthocalpum* and *Solanum melongena* L, *Veg. Sci.*, 4, 22–24, 1977.

Sharma, P. and Rajam, M. V., Genotype, explant and position effects on organogenesis and somatic embryogenesis in eggplant (*Solanum melongena* L.), *J. Exp. Bot.*, 46, 135–141, 1995a.

Sharma, P. and Rajam, M. V., Spatial and temporal changes in endogenous polyamine levels associated with somatic embryogenesis from different hypocotyl segments of eggplant (*Solanum melongena* L.), *J. Plant Physiol.*, 146, 658–664, 1995b.

Sharma, D. R., Sareen, P. K., and Chowdhury, J. B., Crossability and pollination in some non-tuberous *Solanum* species, *Indian J. Agric. Sci.*, 54, 514–517, 1984.

Sharma, D. R. et al., Interspecific hybridization in the genus *Solanum*. A cross between *S. melongena* and *S. khasianum* throughout embryo culture, *Z. Pflanzenzücht.*, 85, 248–253, 1980.

Sharma, J. L. et al., Alterations in proline and phenol content of *Meloidogyne incognita* infected brinjal cultivars, *Pakistan J. Nematol.*, 8(1), 33–38, 1990.

Sheela, K. B., Gopalkrishnan, P. K., and Peter, K. V., Resistance to bacterial wilt in a set of eggplant breeding lines, *Indian J. Agric. Sci.*, 54, 457–460, 1984.

Shoeb, F. et al., Polyamines as biomarkers for plant regeneration capacity: improvement of regeneration by modulation of polyamine metabolism in different genotypes of indica rice, *Plant Sci.*, 160, 1229–1235, 2001.

Sidhu, A. S. et al., Genetics of yield components in brinjal (*Solanum melongena* L.), *Haryana J. Hort. Sci.*, 9, 160–164, 1980.

Sihachakr, D. and Ducreux, G., Cultural behaviour of protoplasts from different organs of eggplant (*Solanum melongena* L.) and plant regeneration, *Plant Tissue Org. Cult.*, 11, 179–188, 1987.

Sihachakr, D. et al., Electrofusion for the production of somatic hybrid plants of S*olanum melongena* L. and *Solanum khasianum* C.B, *Clark Plant Sci.*, 57, 215–223, 1988.

Sihachakr, D. et al., Somatic hybrid plants produced by electrofusion between *Solanum melongena* L. and *Solanum torvum*, *Theor. Appl. Genet.*, 77, 1–6, 1989.

Sihachakr, D. et al., Regeneration of plants from protoplasts of eggplant (*Solanum melongena* L.), In *Biotechnology in Agriculture and Forestry* In *Plant Protoplasts and Genetic Engineering IV*, Vol. 23, Balaji, Y. P.S., Ed. Berlin: Springer, pp. 108–121, 1993.

Siliak-Yakovlev, S. and Isouard, G., Contribution à l'étude caryologique de l'aubergine (*Solanum melongena* L.), *Agronomie*, 3, 81–86, 1983.

Singh, D. and Sidhu, A. S., Management of pest complex in brinjal, *Indian J. Entomol.*, 48, 305–311, 1988.

Studentsova, L. I., Promising breeding material of eggplant, *Trudy. Prikl. Bot. Genet. Sel.*, 66, 50–55, 1980.

Tanksley, S. D. et al., RFLP mapping in plant breeding: new tools for an old science, *Biotech*, 7, 257–264, 1989.

Toki, S., Kameya, T., and Abe, T., Production of a triple mutant, chlorophyll-deficient, streptomycin-, and kanamycin-resistant *Nicotiana tabacum*, and its use in intergeneric somatic hybrid formation with *Solanum melongena*, *Theor. Appl. Genet.*, 80, 588–592, 1990.

Toll, J. and van Sloten, D. H., Directory of Germplasm Collections 4, Vegetables, *Intern. Board Plant Genet. Resour.*, 187, 1982.

Uhrig, H. and Salamani, F., Diahploid plant production from 4X-genotypes of potato by use of efficient anther plants producing tetraploid strains (4X EAPR-clones): proposal of breeding methodology, *Plant Breed.*, 98, 228–235, 1987.

Vadivel, E. and Bapu, J. R. K., Evaluation and documentation of eggplant germplasm, *Capsicum Newsl.*, 7, 80, 1989.

Vavilov, N.I. (1928). Geographical centres of our cultivated plants. *Proc. Vth Int. Congo. Genet.*: 342. New York.

Venora, G., Russo, C., and Errico, A., Karyotype analysis in *Solanum melongena* L. In *Proceedings of the Eighth Eucarpia Meeting on Genetics and Breeding of Capsicum and Eggplant*, pp. 266–271, 1992.

Vorouina, M. V. and Losputova, Use of the world diversity of red peppers and eggplants for different trends in breeding, Byull, *Vsesofuznogo Ordena Lenina i Ordena Druzhby Narodoc nauchno-issledovatel' skogo Institute Rastenievodstva Imeni*, Vavilova, N. I., Ed., 120, P. 21.

Vos, P. et al., AFLP: a new technique for DNA fingerprinting, *Nuc. Acid Res.*, 23, 4407–4414, 1995.

Waari, S., The potentials of using dihaploid/diploid genotypes in breeding potato by somatic hybridization, *In Vitro* Production in Higher Plants, Jain, S. M., Sopory, S. K., and Veilleux, R. E., Eds., vol. 3, Netherlands: Kluwer Academic Publishers, 1996.

Wanjari, K. B. and Khapre, P. R., Inheritance of pigmentation in *S. melongena*×*S. indicum*, *Genet. Agr.*, 31, 327–332, 1977.

Wanjari, K.B. and Khapre, P.R. Study of branching and linkage in *S. melongena* L.×*S. indicum* L. Presented in 4th Int. SABRAO Congress, May 4–8, Universite Kebangsaan, Malaysia and Federal Hotel, Kuala Lumpur, 40, 1981.

Williams, J. G. K. et al., DNA polymorphisms amplified by arbitrary primers are useful as genetic markers, *Nuc. Acid Res.*, 18, 6531–6535, 1990.

Yadav, J.S., Polyamine-mediated regulation of somatic embryogenesis and lateral root differentiation in eggplant (*Solanum melongena*). PhD Thesis. University of Delhi South Campus, New Delhi, India, 1997.

Yadav, J. S. and Rajam, M. V., Spatial distribution of free and conjugated polyamines in leaves of *Solanum melongena* L. associated with differential morphogenetic capacity: efficient somatic embryogenesis with putrescine, *J. Exp. Bot.*, 48, 1537–1545, 1997.

Yadav, J. S. and Rajam, M. V., Temporal regulation of somatic embryogenesis by adjusting cellular polyamine content in eggplant, *Plant Physiol.*, 116, 617–625, 1998.

Yamada, T., Nakagawa, H., and Sinoto, Y., Studies on the differentiation in cultured cells. I. Embryogenesis in three strains of *Solanum* callus, *Bot. Mag.*, 80, 68–74, 1967.

Yamakawa, K. and Mochizuki, H., Nature and inheritance of *Fusarium* wilt resistance in eggplant cultivars and related wild *Solanum* species, Bull, *Vegetable and Ornamental Crops, Station A (Yasai Shibenjo Hokoku, A)*, 6, 19–27, 1979.

Yamakawa, K., Use of rootstocks in Solanaceous fruit-vegetable production in Japan, *Jpn. Agric. Res. Quart.*, 15, 175–179, 1982.

CHAPTER 11

Carrot

Philipp W. Simon and Irwin L. Goldman

CONTENTS

11.1   INTRODUCTION

Carrot is among the top ten most important vegetable crops in terms of area devoted to its production and tonnage of crop production, as presented in the introductory chapter of this volume. Like most of the other crops in the lower half of the top ten list, relatively little effort and resources have been devoted to carrot breeding, and even today, there are few breeding programs around the world with an effort predominantly on carrot and very few with an exclusive focus on carrot. Yet, in the last 50 years, there has been significant progress in carrot breeding.

World carrot production in 2003 was estimated at 23.3 million metric tons (ERS http://www. ers.usda.gov/data/sdp/view.asp?f=crops/carrots/), with China far out-producing the next two countries on the list, the United States and Russia. China's production of carrot today is nearly one-third of the world's total carrot production.

Carrot production in the United States is remarkably concentrated. California dominates the carrot production market in the United States. Fresh market carrot production in California grew from $250.4 million in 1993 to $522.2 million in 2003, while processing production value at the farm level grew from $9.7 million to $11.3 million during the same period. By comparison, from 1993 to 2003, fresh market carrot production in the United States grew in value from $337.7 million to $632.4 million. Processing carrot production in the United States during the same period did not increase, and instead fell by over $4 million from $37.6 million to $33.0 million (ERS, http://www. ers.usda.gov/data/sdp/view.asp?f=crops/carrots/). Clearly, the greatest opportunities for future growth in the United States are in fresh market production.

Though it produces the largest share of United States fresh market production, California also produces a sizable portion of the processing carrot crop, with 103,000 metric tons in 2003, compared to the United States total of 405,000 metric tons. By comparison, Washington and Wisconsin produced 139,000 and 87,000 metric tons in 2003, respectively.

11.2   TAXONOMY AND GERMPLASM RESOURCES

11.2.1   Taxonomy

Taxonomists today recognize approximately 20 species in the genus *Daucus*, where *Daucus carota* includes the cultivated carrot and several wild subspecies (Heywood 1983). Nearly all of the *Daucus* species are diploid, and the basic chromosome number varies from $x=9$ to $x=11$. Only carrot, *Daucus capillifolius* and a few other rare southern Mediterranean species have a basic chromosome number of $x=9$ (Sáenz Laín 1981). The only successful interspecific cross involving carrot is with *D. capillifolius* (McCollum 1975).

Carrot is a member of the Apiaceae or Umbelliferae family, and the diversity in chromosome numbers in *Daucus* is similar to that across the family, where a wide range in chromosome numbers occurs over primarily-diploid taxa. Chromosome size varies widely across *Daucus* and the Apiaceae, and almost nothing is known about chromosome evolution in the family. Major advances have been made in placing *Daucus* in the Apiaceae using classical and molecular tools, helping to clarify relationships in this diverse and complicated family (Plunkett et al. 2004).

11.2.2   Germplasm Resources

The center of diversity for carrot is in Central Asia, and the first cultivation of carrot for its storage root is reported to be in Afghanistan, approximately 1,100 years ago (Mackevic 1929). Long before carrot was domesticated, wild carrot had become widespread, as seeds were found in Europe dating back nearly 5,000 years. Today, wild carrot is found around the world in temperate regions with adequate precipitation on disturbed sites, such as vacant lots, roadsides, and agricultural lands.

Germplasm collections that include carrot have focused on open-pollinated cultivated carrots of local importance. Little germplasm of wild carrot has been collected. Most of the *Daucus* species other than carrot have a much more limited geographic range and are primarily located around the Mediterranean Sea, except for an Australian species (*Daucus glochidiatus*), a wide-spread American species (*Daucus pusillus*), and a more isolated American species (*Daucus montanus*) (Rubatzky, Quiros, and Simon 1999). The *Daucus* species held in germplasm collections are typically represented by only one or two accessions, and several are not held in any collection.

11.3 BREEDING SYSTEMS AND GENETICS

11.3.1 Reproductive Biology

Carrot is an outcrossing diploid species ($2n = 2x = 18$) typically described as a biennial. Most wild carrot might more accurately be described as "winter annual" because seedlings established in the autumn will usually flower and produce seed the following summer. Seedlings or storage roots typically are vernalized by exposure to cool temperatures for as little as 6–8 weeks, and many carrot breeding programs have taken advantage of this fact to complete the breeding cycle in one year, growing a vegetative crop in a warm climate winter nursery, harvesting and analyzing the crop in time for planting the vernalized roots for a summer seed crop in temperate climates. Other breeding programs grow both vegetative and seed crop in temperate summer climates over two years to handle carrot in a true biennial fashion.

Carrot flowers are insect-pollinated, perfect, protandrous and small, with several hundred to over one thousand in an inflorescence, the umbel (Peterson and Simon 1986; Rubatzky, Quiros, and Simon 1999). Hand emasculation is difficult, and each flower can produce only two seeds, so hand pollinations are only undertaken when absolutely necessary. Typically, male and female fertile F_1 hybrids are produced by moving pollen between plants with a brush or by placing receptive flowers from both parents in a cloth or mesh isolation cage containing houseflies or blowflies. Seed from both plants is then grown in adjacent rows, and hybrids selected based upon traits inherited from the pollen parent. There is no self-incompatibility system in carrots, so self-pollination is quite possible. Inbreeding depression, however, is often quite acute, usually making the distinction between hybrids and the products of self-pollination straightforward.

11.3.2 Genetics and Genetic Mapping

Approximately 40 monogenic traits with readily discernible phenotypes have been described for carrot. No linkage associations were noted until isozyme and DNA markers began being utilized in the 1990s, and today, over 1,000 carrot molecular and morphological markers have been mapped (Bradeen and Simon 2006). Several populations contributed to the mapping effort, but only a partially joined map has been constructed (Santos and Simon 2004). Most of the mapped carrot molecular markers are AFLPs, but RFLPs, isozymes, RAPDs, CAPs, and SCARs have also been mapped. Cytoplasm male sterility (CMS) is the only trait of carrot known to be conditioned by the cytoplasm.

Carrot germplasm is quite diverse, so that much plant-to-plant variation can be detected with molecular markers within wild populations and open-pollinated cultivars. This high level of variation has been utilized in several genetic studies where populations derived from self-pollination of heterozygous plants can be treated as F_2s (Westphal and Wricke 1991; Schulz, Westphal, and Wricke 1993). Crosses between cultivated and wild carrot have also been used for mapping studies to study the genes that distinguished these disparate partners (Santos and Simon 2002). There are no known crossing barriers between domesticated and wild carrots.

11.4 GENETIC IMPROVEMENT AND BREEDING

11.4.1 Germplasm Utilization

With the proximity of wild carrot to the cultivated crop during its domestication in central Asia, Middle East, Europe, and eastern Asia, it seems likely that gene flow and consequent introgression of wild carrot traits into the cultivated carrot occurred on a regular basis in many places over the 1,000-plus years of carrot domestication and cultivation. This likely continues in those increasingly rare places where local races of carrots are still being grown, and it has also continued in modern carrot-breeding programs. This fact is most abundantly evident in the 50-plus years of using wild carrot to breed for cytoplasmic male sterility (CMS). The successful introgression of wild carrot cytoplasm into cultivated germplasm, in fact, led to the discovery of most of the carrot genes discovered before 1980.

Wild carrot germplasm has also been important as a source of resistance to carrot fly (*Psila rosa*). In this case, the germplasm source was *D. capillifolius* and introgression of this trait led to the release of the cultivar "Flyaway," which bears that source of resistance (Ellis and Hardman 1981). Wild carrot subspecies *dentatus* and *hispanicus* have also been used as a source of genes for resistance to powdery mildew (*Erysiphe* sp.) (Bonnet 1983) and Northern root trait nematode (*Meloidogyne hapla*) (Frese 1983), respectively. Mildew resistance has been widely deployed in modern cultivars, while this source of nematode resistance has not been widely pursued.

European open-pollinated carrot cultivars have served as the germplasm source of several traits used for carrot improvement, including another source of *M. hapla* resistance, aster yellows mycoplasm resistance (Gabelman, Goldman, and Breitbach 1994), and brown anther CMS discussed below. Asian and Brazilian cultivars have been used as the source of resistance genes for alternaria leaf blight and tropical rootknot nematode (*M. javanica*; Vieira, Della Vecchia, and Ikuta 1983). To breed for elevated storage root carotene content, several studies have demonstrated that intercrosses between European and Asian orange open-pollinated cultivars yield transgressive segregants with higher carotene content than either parent (Simon et al. 1989).

More germplasm from landraces, old cultivars, wild carrot, and other *Daucus* species will likely be collected, opening up the opportunity for identifying more genetic variation of interest for introgression into modern cultivars. As new carrot products are developed, new quality components identified, and new disease and pest resistance evaluation techniques are devised, the incentive for utilizing carrot germplasm diversity will continue.

11.4.2 Origins of Carrot

Wild carrot appears in many temperate regions of the world, far beyond its Mediterranean and Asian centers of origin, where this plant displays great diversity. It is quite possible that ancient cultures in those regions used wild carrot as an herb, and it is also quite likely that the seeds were used medicinally in the Mediterranean region since antiquity (Banga 1957a, 1957b). Almost certainly, the wild and early forms of the domesticated carrot were used as an herb and as medicine before they were used as a root vegetable in the conventional sense of that term today. There is good evidence that wild carrot is the direct progenitor of the cultivated carrot (Simon 2000). Selection for a swollen-rooted type suitable for consumption undoubtedly took many centuries.

Carrot roots are colored due to pigments in the carotenoid biosynthetic pathway and pigments in the anthocyanin biosynthetic pathway. Orange-colored roots in cultivated carrot are due to α- and β-carotene, and the red color is due to an accumulation of lycopene (Umiel and Gabelman 1971). Purple-rooted carrot is due to anthocyanin pigmentation, and yellow-rooted carrot is due to xanthophylls, which are oxygenated derivatives of carotenoids (Simon 1996). When carrot roots are lacking in pigmentation, or low in relative amounts of these various pigments, they can appear white or very pale orange and yellow.

For several decades, researchers at several public institutions in the United States have focused on developing and improving pigmentation levels in carrot. These efforts began with the work of C.E. Peterson at Michigan State University and later USDA, ARS and the University of Wisconsin, and W.H. Gabelman at the University of Wisconsin in the 1950s and 1960s. In addition to selection for increased carotenoid pigmentation, the work of these two scientists and their students led to many discoveries on the genetic control of pigmentation in carrot roots. Many of these studies are reviewed in Simon (2000). Much of the publicly available carrot germplasm in the United States traces back to these two breeding programs and to the work of D. Franklin at the University of Idaho during the latter part of the twentieth century. The breeding programs of B. Michalik in Poland and A. Bonnet in France have had a similar impact in Europe.

The orange-rooted cultivated carrot popular in markets today originated as a purple-rooted domesticate of wild carrot in central Asia (Mackevic 1929; Banga 1963). Heywood (1983) has suggested that orange-rooted carrot may have been derived from crosses between wild carrot and cultivated purple-rooted carrot, though other explanations have been offered for the origin of the orange type (see below). Simon (1996) noted that orange-rooted types are present in Turkey in regions where primarily purple-rooted types appear, though it is not clear if these are contemporaneous. Simon (2000) and Koch (2005) suggest that it may have taken a considerable amount of time for the purple-rooted carrot to spread as a vegetable crop. One of the reasons for this could be the time it took to improve the eating quality of roots from the fibrous and woody roots of wild carrot.

The cultivated purple-rooted carrot appeared as a crop in the Mediterranean region in the thirteenth century, and it was here that it received its greatest attention as a crop. Banga (1957a, 1957b, 1963) has suggested that, while purple-rooted carrot was common in Europe during the fourteenth, fifteenth, and sixteenth centuries, a change toward other colors appeared in sixteenth and seventeenth century Dutch paintings (e.g., Figure 11.1). Banga's analysis, which is based on careful study of European paintings of that era, suggests that colored roots of a yellow or orange–yellow variety appear in these paintings beginning in the sixteenth century (Banga 1963). While the yellowish-colored carrot roots appeared to be quite popular in that period of time, they were supplanted by orange-rooted types shortly thereafter and their popularity diminished. Today, orange-rooted types dominate the marketplace, though purple, and in some cases, red- and yellow-rooted, types are making a comeback in specialty markets.

Banga and others have suggested that our modern orange-rooted carrot descended from populations that were known as "Long Yellow" or "Long Orange" and "Horn" in seventeenth century

Figure 11.1 *Greengrocer's Shop.* Painted by Aertsen, 1509–1575. Carrots in lower left hand corner of the painting are yellowish-orange, a departure from the yellow-rooted carrots that predominated in Europe at the time. (From Banga, O., *Main Types of the Western Carotene Carrot and Their Origin*, Zwolle, The Netherlands: W.E.J. Tjeenk Willink, 1963. With permission.)

Holland. These are the root types that Banga described from his analysis of oil paintings. The "Long Orange" population gave rise to full-season varieties that were meant to grow as long as possible and develop substantial root bulk. These included cultivars such as "Long Stump Winter," "Flanders," "Flakkee," "St. Valery," "Bauer's Kieler Rote," and "Meaux" or "Tendersweet." Interestingly, these latter two cultivars were the source of key germplasm that formed the basis of United States breeding programs in the 1940s (Goldman 1996). The fact that our modern carrot germplasm may be only one generation removed from the ancestral "Long Orange" has implications for genetic diversity discussed in this chapter.

Among the most important descendants of the "Long Orange" type was "Imperator," the primary cultivar used for United States fresh market production. "Imperator" results from a cross between "Chantenay" and "Nantes," and was first made by the Associated Seed Growers, Inc. of New Haven, Connecticut, in the United States. This company later became known as Asgrow Seed Company and was purchased in the 1990s by Seminis Vegetable Seeds of Woodland, California.

The "Horn" carrots gave rise to shorter, mid- and early-season varieties developed to meet the challenge of shorter growing seasons. Three primary types were found in this category: "Late Half Long Horn," "Early Half Long Horn," and "Early Short Horn" (Figure 11.2 through Figure 11.4). Examples of the former include "James Intermediate," "Danvers," and "Chantenay." These types had pointed tips and were considered easy to harvest. "Nantes" is an example of a derivative of the "Early Half Long Horn" type. "Parisienne" is an example of the smaller, round-rooted types descended from "Early Short Horn." These have become popular today for home gardeners, where its round radish-shaped roots are a novelty. It is also important to note that the "Early Half Long Horn" population gave rise to forcing cultivars such as "Amsterdam Forcing" and "Forcing Nantes." These are unique types suited to production in Northern Europe in a greenhouse production system. At the time of their development, royalty and other wealthy landowners experimented with forcing as a horticultural technique. Vegetable forcing, such as that done with carrots, was handled in greenhouses during the winter months.

The "Long Orange" and "Horn" carrots were brought to the United Sates by early European settlers and served as the core cultivar types listed in United States seed catalogues from their inception in the 1700s up through the early 1900s (Table 11.1) (Burr 1865; Magruder et al. 1940; Babb, Kraus, and Magruder 1950). Of particular interest to modern carrot production is the fact that the majority of carrot populations available to the carrot industry in the early part of the twentieth century in the United States all descended from these two populations. More specifically, the full-season varieties likely descended primarily from "Long Orange," thereby suggesting what is possibly a very significant population

Figure 11.2 "Late Half Long Horn," a carrot population described by Banga. (From Banga, O., *Main Types of the Western Carotene Carrot and Their Origin*, Zwolle, The Netherlands: W.E.J. Tjeenk Willink, 1963. With permission.)

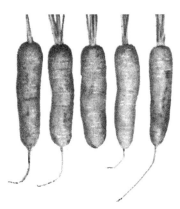

Figure 11.3 "Early Half Long Horn," a carrot population described by Banga. (From Banga, O., *Main Types of the Western Carotene Carrot and Their Origin*, Zwolle, The Netherlands: W.E.J. Tjeenk Willink, 1963. With permission.)

bottleneck for modern carrot germplasm. The primary founding populations available to the carrot breeder and carrot industry in 1940 in the United States are provided in Figure 11.5. These illustrate populations that descend from "Long Orange" and "Horn." In fact, the "Long Orange" population was available and in use in 1940, though it does not appear in production today.

Despite this relatively narrow bottleneck through which passed modern carrot germplasm, a variety of shapes exist in today's marketplace. The primary type used in the United States fresh market, for both standard cello-pack production and for cut-and-peel production, is the "Imperator" type. This type resulted from a cross of parents derived from "Long Orange" (Banga 1963). This type is very long and thin, with a small core and the potential to be grown at very high densities. The "Nantes" type is the preferred type in European fresh market production, and is characterized by a blunt tip and a length less than "Imperator." It has very high eating quality and is also very popular in Asia. "Kuroda" is a large, thick-rooted type produced in Asia and South America. "Kuroda" has a wedge-shaped root and exhibits a great deal of taper towards the tip. Its carotenoid pigmentation is typically not as high as "Imperator" or "Nantes," in part because of color preferences in regions where it is grown. Processing cultivar types include "Danvers," "Chantenay," and in some cases "Nantes," though the first two predominate. "Danvers" and "Chantenay" are large, bulky-rooted types that have the potential to grow to large diameter. "Chantenay" in particular has a larger core and does not

Figure 11.4 "Early Short Horn," a carrot population described by Banga. (From Banga, O., *Main Types of the Western Carotene Carrot and Their Origin*, Zwolle, The Netherlands: W.E.J. Tjeenk Willink, 1963. With permission.)

Table 11.1   Chronology of Carrot Cultivars Sold in Seed Catalogues

		Root Size and Color Category						
Year	Company and Location	Short Orange	Intermediate Orange	Long Orange	Long Yellow	Short Yellow	White	Purple
1793	Garden & Graf's seeds, Richmond, VA			Large Orange Large Red				
1810	William Booth, Baltimore, MD		Early Orange Early Horn	Large Orange	Large Lemon			
1834	William Prince & Sons, Flushing, NY	Short Orange	Early Scarlet Horn*	Large Orange Long Red Surrey/Studley Long Red Croisy Altringham* Long Orange* Blood Red	French Pale Yellow*	Early Yellow Horn* French Early Short	Long White (Green top)* Large Long White*	French Large Purple/Violet*
1834–35	William Prince & Sons, Flushing, NY	9 marked with * for 1834 above plus:					Large Field	
1835	William Prince & Sons, Flushing, NY	9 marked with * for 1834 above plus:			Large Yellow Field			
1839	Ellis & Bosson, Boston, MA		Early Horn	Long Orange Altringham				Blood Red/Purple
1849 and 1852	Wethersfield Seed Garden CT		Early Horn	Long Orange Large Altringham	Large Lemon		Large White Belgian	Blood Red/Purple
1853	G.H. Barr & Co., New York, NY		Early Horn New intermediate	Long Orange Altringham				
1856 (additional items listed in 1861)	G.U. Dreer's, Philadelphia, PA		Early Horn (Early Scarlet Short Horn)	Long Orange Altringham (Long Surrey)				
1860	William R. Prince & Company		Early Short Horn Early Half-long Red	Long Orange Altringham			Large White Field	Long Blood or Violet
1865	Fearing Burr, Jr. (not a catalogue, but rather a book)	Early frame	New intermediate Early Half Long Scarlet Early Horn	Long Orange Altringham Flanders Large Pale Scarlet Long Red Belgian Long Surrey Studley	Long Yellow		Long White Short White White Belgian White Belgian Horn	Purple/Blood Red
1890	Peter Henderson, New York, NY			Improved Long Orange Altringham	Yellow Belgian		White Belgian	
1890, Autumn	Peter Henderson, New York, NY	Early French forcing	Henderson's intermediate Danvers Early Scarlet Horn Half Long Red Chantenay Early Half Long Scarlet Carentan	Long Orange improved				

Year	Source					
1890	E.J. Bowen, San Fransisco, CA	Early oxheart	Scarlet Horn Half Long Stump-rooted Danvers	Long Orange Finest improved Long Orange		Large White Belgian
1890	W. Atlee Burpee, Philadelphia, PA	Oxheart/Guerande Early very short scarlet/Golden ball	Saint Valery/New intermediate Danvers Half Long Orange Chantenay/Model Short Horn/Early Scarlet Horn Half Long Scarlet	Burpee's Long improved Orange Coreless Long Red Long Red Altringham	Large Yellow Belgian	Large White Vosges Large White Belgian Large White Belgian
1890	Henry A. Dreer, Philadelphia, PA	Early Short Horn Scarlet Guerande/Oxheart	Early Scarlet Horn Early Half Long Scarlet Early Half Long Carentan Danver's Half Long Scarlet Saint Valery	Improved Long Orange Long Scarlet Altringham		Large White Belgian
1890	James M. Thorburn, New York, NY	Extra early forcing New French Bellot Half Long Stump-rooted Guerande Strain	Early Scarlet Horn Half Long Pointed James Intermediate Half Long Stump-rooted Half Long Nantes Strain Half Long Chantenay Strain Half Long Carentan Half Long Danvers	Best Long Orange Altringham		Long White
1890	J. Seulberger, Oakland, CA	Early French Forcing	Early Scarlet Horn	Long Orange		Long White Belgian
1891	Fred P. Burr & Co., Middleton, CT	Early French forcing Guerande Stump-rooted	Danvers Half Long	Long Orange Improved		
1891	William C. Beckert, Allegheny, PA	Guerande/Oxheart	Half Long Stump-rooted Half Long Point-rooted Early Scarlet Horn Intermediate Red Danvers Half Long Half Long Nantes Chantenay	Improved Long Orange	Large Yellow Belgian	Large White Belgian
1920	C.C. Morse, San Francisco, CA	Oxheart French Forcing	Chantenay Danvers Half Long Nantes Scarlet Horn James Intermediate	Long Orange Red St. Valery		Short White

*Cultivars marked with asterisks in the 1834 entry are referred to in the 1834–35 and 1835 entries.

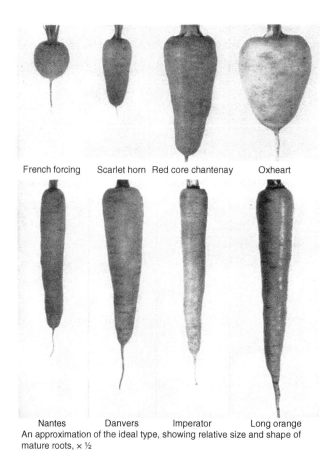

French forcing   Scarlet horn Red core chantenay   Oxheart

Nantes   Danvers   Imperator   Long orange
An approximation of the ideal type, showing relative size and shape of
mature roots, × ½

Figure 11.5 Primary types of carrot roots in United States, 1940. Note that "Long Orange" and "Horn," the two ancestral orange rooted types, were present in the middle of the twentieth century in the United States. (From Magruder, R., et al., *United States Department of Agriculture Publication*, 361, 48, 1940. With permission.)

possess the high eating quality found in "Imperator" or "Nantes." It is common to find both of these types grown for slicing and dicing markets in the United States, though "Chantenay" types are predominant in the dicing market and "Danvers" and "Imperator" types in the slicing market. To some extent, processing carrot types should technically include the United States "baby" or "cut-and-peel" type product, though processing typically refers to heavy processing, such as canning and freezing. The bulk of this is done on the two coasts and in the midwest, while the bulk of cut-and-peel production takes place in California.

11.4.3   F₁ Hybrid Development

Until the 1960s, it was not possible to find F_1 hybrid carrot, though this type of cultivar had been envisioned in the late 1940s and early 1950s by carrot breeders. The only cultivars that were available were open-pollinated, and attempts to develop hybrids were difficult due to the suscep-tibility of carrot to inbreeding depression. The trick to developing an F_1 hybrid cultivar was the development of a pollination control system whereby pollen could be transferred reliably from male fertile plants to male sterile plants. The existence of a cytoplasmic-genic sterility system in carrot was first reported by Welch and Grimball in 1947, and this type, which was discovered in cultivated carrots such as "Tendersweet," was referred to as "Brown Anther." In this type, anthers degenerate, leaving nonfunctional pollen and a resulting male sterile plant. Welch and Grimball determined that this condition was caused by the interaction of genes in the nucleus and the cytoplasm, a model first reported by Henry Jones for sterility in onion some eleven years earlier. While many early attempts to develop F_1 hybrid carrot involved the use of the brown anther type, the discovery of petaloid sterility, itself also a CMS system, enabled improvements in hybrid production. The petaloid system worked by a homeotic allele that removed anthers completely and replaced them with a whorl of petals (discovered by Munger in 1953, reviewed in Peterson and Simon 1986). This obviously eliminated any possibility of pollen production and enabled a more efficient system for F_1 hybrid production. Several other germplasm sources of petaloidy have been discovered, all in wild carrot. In addition to petaloidy, several other types of male, and sometimes female, sterility discovered in wild carrot are being characterized (Nothnagel, Straka, and Linke 2000).

In the petaloid CMS system, maintainer lines carrying normal, fertile cytoplasm and recessive alleles at the nuclear restorer locus are used to propagate the male sterile lines, which carry a sterile cytoplasm and recessive alleles at the nuclear restorer locus (Thompson 1962; Wolyn and Chahal 1998). Several rounds of backcrossing between the maintainer and sterile lines assure greater genetic similarity between these two "A" and "B" lines, thereby making them into a usable pair. Once a stable sterile "A" line is established, it can be pollinated with a fertile line of the breeder's choosing. F_1 hybrids in carrot may be three-way hybrids, where the female plant is actually an F_1 hybrid itself, or two-way hybrids where the female is a cytoplasmically sterile "A" inbred line. Seed production can be accomplished by allowing roots to become vernalized in the field in their first season, with pollination and seed production in the subsequent season. A more-typical method would be to harvest roots of both sterile and fertile lines at the end of the first season of production, vernalize these roots in a controlled environment, and re-plant them in the second season for seed production.

Brown anther CMS continues to be used to some extent today, but petaloidy is the major CMS source used by carrot breeders today. Nuclear genes restoring fertility are found in numerous open-pollinated cultivars. While the incidence of restorers varies across root types, that incidence is typically low enough to assure success in discovering primary sterile and maintainer breeding stocks.

11.4.4 Major Breeding Objectives

The successful production of a marketable carrot crop includes several stages of progressive plant growth to a harvestable size, followed by varying types and duration of post-harvest storage, movement of the crop to markets, and in some cases, processing. With success in these endeavors by growers, shippers, and processors, consumers have a range of products to purchase and consume. This process can take from 75 to 180 days or more, with crop production typically accounting for the vast majority of that time.

Carrot domestication transformed the relatively small, white, heavily-divided (forked or sprangled) strong-flavored taproot of a plant with annual or biennial flowering habit to the large, orange, smooth, good-flavored storage root of a uniformly biennial or "winter" annual crop we know today. Modern carrot breeders have further refined the carrot, improving flavor, sweetness, reducing bitterness, and improving texture and color, to name a few.

11.4.4.1 Production

The reliable production of a marketable carrot crop is the prime mandate for breeders. Before the sixteenth century, the only distinction noted by historians among carrots was root color—yellow or purple. By the seventeenth century, orange carrots were most common, and in Europe, an array of root shapes was beginning to be described—first categorizing them as long or short, as described above. Eventually, carrots included a wide array of shapes and sizes, ranging from the small, spherical "French Forcing" (Figure 11.5) to large, long tapered "Altringham" types. The diversity of Eastern carrot types through recent centuries is not so well-documented, but based upon modern cultivars, it was bred to similar extremes. The first carrots described in North America included a subset of the European cultivars, predominantly those in the less extreme range of root size. The most prevalent root shape of carrots grown around the world today is the "Nantes" type, but "Chantenay" and "Danvers" types are widely grown for canning, freezing, and storage (Goldman 1996; Simon, Matthews, and Roberts 2000). The "Imperator" type is most widely grown in North America, "Kuroda" type in Asia, and "Brasilia" type in South America. It is surmised that relatively few genes control root shape, and breeders routinely intercross among root type classes with a goal of breeding back to one of the parental types, or to select for other root shapes. Regardless of what root type is the target of carrot breeders, uniformity of the shape, size, and color are paramount breeding goals. Smooth roots and vigorous top (leaf) growth are also important. Simple phenotypic recurrent mass selection has proven to be a successful breeding strategy. Like many vegetable crops, carrot yield is measured as the weight of the fraction of the crop that is marketable. All of these aforementioned most-important carrot breeding goals are achieved more readily in hybrid cultivars than in open-pollination, and this realization was the stimulus for the heyday of modern carrot breeding in the 1950s through the 1970s, when carrot and other vegetable breeding research was more actively pursued by universities and research institutes around the world.

Another crucial trait in carrot breeding is resistance to bolting or premature flowering. When carrots are grown in cold temperatures, such as those in northern latitudes, bolting or premature flowering can occur. This renders the crop unmarketable, and is therefore a highly-undesirable characteristic. Selection for nonbolting habit can be conducted during the breeding process, and in fact, such selection pressure must be continuously applied to limit the appearance of bolting types in segregating populations (Peterson and Simon 1986).

Following a typical vernalization treatment over a minimum of 8–10 weeks at approximately 5°C, most roots in a population should flower. Such a treatment should insure that only roots that are capable of flowering (i.e., those not resistant to bolting) remain in the population. Peterson and Simon (1986) recommend early spring planting in northern latitudes to insure that cold-temperature exposure occurs. A second option is to evaluate segregating populations in regions where cold temperatures during the winter might not be severe enough to completely vernalize carrot plants. In such a case, the breeder can select nonbolting types to insure that there is a degree of bolting resistance in the population.

11.4.4.2 Disease and Pest Resistance

Diseases and pests account for a reduction in the fraction of marketable carrot storage roots in all world production areas. Alternaria leaf blight caused by *Alternaria dauci* is a widespread foliar disease, and often noted as the most significant in its impact on global yield reduction. Immunity to leaf blight has not been reported, but several sources of resistance have been incorporated into cultivars. The genetics of resistant to this disease has not been well-described, but several studies suggest that resistance is conditioned by several genes with varying levels of dominance (Strandberg et al. 1972; Vieira et al. 1991; Boiteux, DellaVecchia, and Reifschneider 1993; Simon and Strandberg 1998).

Several species of nematodes attack carrots. Northern root-knot nematode, *Meloidogyne hapla*, is prevalent in most cool temperate production areas of Northern Europe and North and South America, while the tropical and Southern root-knot nematodes *M. javanica* and *M. incognita* are widespread in warmer carrot production areas of southern Europe, southern United States, northern South America, and Asia. A two-gene model for *M. hapla* resistance has been described by Wang and Goldman (1996) while a single gene confers resistance to *M. javanica* and *M. incognita* (Simon, Matthews, and Roberts 2000). Cultivars resistant to the latter nematode species have been developed for Brazil and North America (Huang, Della Vecchia, and Ferreira 1986).

Cavity spot caused by *Pythium* sp. is among the most destructive diseases of carrots in post-harvest storage. Identification of reliable sources of genetic resistance has been difficult, and consequently, breeding for resistance has not been possible.

Aster yellows, an insect-vectored disease caused by a mycoplasma-like organism (MLO) and carried by the aster leafhopper, is a destructive carrot disease in the upper midwestern United States, Canada, and other parts of the temperate world. Symptoms of aster yellows-infected plants include a profusion of crown shoots and excessive growth of root hairs, rendering roots unmarketable. Selection in naturally and artificially infested field sites was carried out in Wisconsin from 1982 to 1989. Sites were established where lettuce plants, which also attract the leaf hopper, were interspersed with rows of carrot plants. Twenty-three carrot germplasm lines inbred for a minimum of five generations and three hybrids were developed from a synthetic population known as AYSYN (Gabelman, Goldman, and Breitbach 1994). Significant reductions in aster yellows infection were observed in many selected lines compared to standard cultivars, and several inbred lines with substantial field resistance were released to the seed industry in the mid-1990s.

Sources of resistance to carrot fly and powdery mildew have come from wild carrot mentioned above. Cultivars resistant to both diseases have been developed. Prospects for advancing strong levels of resistance to adapted carrot cultivars vary with disease. Because disease pressure varies across environments and some level of genetic resistance is available for the most widespread diseases, low pesticide application and organic production is feasible and underway in several production areas.

11.4.4.3 Consumer Quality

Production of marketable carrots is essential to assure availability of the crop for consumers, and carrot breeders have dedicated much of their effort to this end. Consumers have long discerned differences in quality among carrots available to them. Consequently, modern carrot breeders include flavor, nutritional quality, and processing quality as major breeding goals.

Even before organized breeding efforts for carrot improvement, historians noted a preference for the flavor of purple carrot over yellow. Among modern Western orange carrots, the Nantes types invariably are noted as better-tasting than other types. Modern breeders have taken advantage of the genetic variation for flavor noted in diverse carrot germplasm. Carrot breeding for flavor improvement has primarily focused on raw carrot flavor, which can most simply be broken down into sweetness and harsh flavor.

Harsh flavor of carrots is caused by volatile monoterpenoids and sesquiterpenoids. More than 20 of these 10- and 15-carbon compounds have been in carrot roots, but the monoterpenoid terpinolene, the sesquiterpenes caryophyllene, and E-gamma-bisabolene tend to the most abundant (Buttery, Seifert, and Ling 1968). Volatile terpenoids accumulate in oil ducts and collectively account for much of the distinctive flavor of raw carrot. Carrots vary widely in volatile terpenoid content, from less than 10 to over 500 ppm (Senalik and Simon 1987), and above levels of approximately 50 ppm, most consumers note a harsh, turpentine-like chemical flavor. Harsh flavor is sometimes referred to as bitter; in fact, this latter term should be reserved for the truly bitter flavor that develops in stored carrots exposed to ethylene or similar gases and has a very different chemical basis. Carrot breeders have taken advantage of the wide variation in volatile terpenoid content in carrot germplasm to exercise selection for low harsh, or mild, flavor. Mild flavor, along with sweetness, have become major breeding goals for North American fresh market carrots because these two flavor attributes are crucial in the consumer appeal of whole and "baby" carrots. Heritability estimates of volatile terpenoid content and harsh flavor have not been reported as such, but are intermediate to relatively high. Scores for harsh flavor for a given group of diverse carrots across growing locations and years tends to have similar relative ranking for harsh flavor

(Simon, Peterson, and Lindsay 1980). Stressful growing conditions, such as extremes in heat, intermittent drought, or excess moisture, tend to increase volatile terpenoid content and consequent perception of harsh flavor. One interesting aspect of harsh flavor is that F_1 hybrids between mild and harsh parents tend to have reliably mild flavor. This incomplete dominance for mild flavor can be used to some advantage in hybrid-breeding programs, although intercrosses where both parents have mild flavor will more reliably yield mild-flavored progeny.

Most consumers would prefer sweeter carrots, which not surprisingly is correlated to sugar content. The carbohydrate profile of carrots is relatively simple in that it includes structural carbo-hydrates (cellulose, pectin, etc.) and the free sugars glucose, fructose, and sucrose with little or no starch or fructans. A single major gene, *Rs*, conditions the predominant type of sugar—sucrose verses reducing sugars (Freeman and Simon 1983), while sugar quantity is a quantitative trait with a heritability estimated as 0.45 (Stommel and Simon 1989).

The texture of raw carrots varies widely across diverse germplasm. Succulent, brittle, juicy texture is clearly preferred by most consumers and is typical, most notably, of the Nantes type. However, other root shapes can also have a succulent brittle texture, and not all carrots with a Nantes shape do. While preferred by consumers, brittle texture is the bane of carrot growers and shippers because roots with this quality break and crack before reaching consumers, consequently ending up discarded. Most carrots grown today have a more durable texture, making them tough enough to reach market. Consumers accept a less-brittle texture as long as it is not fibrous. Fibrous carrot texture is invariably an indication that the flowering process has begun, even if flower stalk elongation has not yet begun.

Carrot breeders exercise strict selection against premature flowering, so fibrous quality should be a defect that never reaches the commercial market. The balance between brittle and more tough (but not fibrous) texture presents an ongoing challenge for fresh market carrot breeders, attempting to develop the most succulent cultivars that still can reliably reach market without breakage. Phenotypic selection is often exercised as part of the evaluation of flavor and dependent upon skill and experience of the selector. Laboratory methods for evaluation of texture have been reported, but are not typically used as part of the routine breeding process. Patterns of inheritance and heritability have not been reported for texture, although it tends to be a trait that is reliably expressed across environments in genetically-uniform breeding stocks.

The tendency for carrot roots to "crack" longitudinally or vertically is a trait that carrot breeders have fastidiously selected against. If cracking occurs before harvest and growth continues, this is referred to as a "growth crack" or "split," while cracking during or after harvest is a "shatter crack." Gentle handling can reduce incidence of shatter cracks, and methods for breeders to select for that trait have been described. Dickson (1966) suggested a single dominant gene predisposes roots to shatter cracking.

Cracking or splitting of carrot roots may occur during root growth, harvest, or post-harvest handling. Cracks or splits may be of minimal length and therefore not affect the marketability of the root, or they may extend to the entire root length and cause severe problems in product acceptance by processors or consumers. Roots contain both xylem and phloem tissues. These tissues can accumulate parenchyma cells, which are a contributing factor to cracking and splitting. When parenchyma cells are abundant, crack proliferation may occur because the root lacks lignified tissue or air spaces that can stop cracks from forming.

During early root growth, water potential in the root determines the degree to which root cracking will occur. As maturity nears, water potential in the root is of far less importance to cracking, and may be more influenced by cell wall composition. Cultivar differences have been reported for cracking resistance, though definitive mechanisms for this resistance have not been reported. Production factors that may promote cracking include irrigation or heavy rainfall late in crop development, harvest machinery that lifts carrot plants by their leaves instead of digging roots from under the soil, and harvesting prior to crop maturity.

Root texture interacts with sugar content in influencing the perception of sweet flavor, in that carrots with more succulent texture tend to be perceived as more sweet. Harsh flavor also influences the perception of sweetness in masking it in high-terpenoid breeding stocks. These interactions between texture, harshness, and sweetness mean that carrot breeders cannot accurately predict sweet flavor based upon sugar content alone in a given breeding population, unless neither texture nor harsh flavor vary significantly. Because of these interactions, carrot breeders must rely heavily on tasting of breeding stock. Single plant selection is relatively successful, suggesting relatively high heritability of these components of raw carrot organoleptic quality. Flavor evaluation of breeding stock grown in environments of the targeted production regime that tend to stimulate less desirable flavor (more harsh, less sweet) is a breeding strategy that is useful in developing germplasm with desirable flavor with wide adaptation.

Genetic improvement of carrots for use in canning and freezing has been underway since these processing methods were first used. Volatile terpenoids are driven off or altered during canning or cooking, so harsh flavor of raw carrots is irrelevant after processing. Sweet flavor, on the other hand, tends to carry through from raw carrots to the processed product. The intensity and uniformity of orange color in raw carrots is often changed during canning. Consequently, a serious carrot breeding program targeting carrot improvement for canning must process breeding stocks to gain a reasonable assessment of their performance.

As carrot has become more popular ingredient in juices, evaluation of breeding stocks and cultivars has been initiated for juice quality. Color, flavor, and turbidity varies widely among orange carrots, and with the entry of purple, yellow, and red carrots into the marketplace, these "new" colors (Nicolle et al. 2004; Surles et al. 2004) are also being evaluated for juice quality. Preferred cultivars have been identified by processors, but no genetic evaluation characteristics important for other processed product have been reported.

The most important nutrient of modern carrots are the provitamin A carotenoids, α- and β-carotene. Carrots are the richest source of these compounds in the United States diet, accounting for 50% of the total dietary carotene supply and 20–30% of the United States vitamin A supply (Simon 1992). Vitamin A is an essential nutrient that contributes significantly to immune function and eye health. It also may offer significant potential to inhibit carcinogenesis and improve cardiovascular health.

Carrot growers and breeders have noted and selected for orange color, often including selection targets in cultivar names such as "Long Orange," "Red Cored Chantenay," and "Early Scarlet Orange." While monikers other than orange are often used, all of these names refer to orange provitamin A carotenes, not lycopene.

Orange color is noticeable with carotene content as low as 5 ppm, and visual selection for intensity and distribution of orange color is successful up to 150 or 200 ppm. Above these levels, differences are difficult to visually discern, and laboratory methods involving extraction and spectrophotometricanalysis are required to reliably make progress in selection. Values up to 1000 ppm total carotenes (fresh weight basis) have been recorded, making carrot the richest whole food source of provitamin A carotenoids (Simon et al. 1989). Among orange carrots, heritability of 0.40 and approximately 20 major QTL have been reported to control carotenoid content (Santos and Simon 2002). Underlying these quantitative genetics factors, two major qualitative genes, Y and Y_2, conditioning orange color have been described (Buishand and Gabelman 1979). Orange carrots are yyy_2y_2, while white carrots are YYY_2Y_2, with yellow and pale orange color in other genotypes. Most modern carrot breeding effort has exclusively involved intercrosses among orange carrots and the numerous QTL involved in that color class. A major exception to this generalization has been the use of white wild carrot as a source of CMS. As yellow, red, and even white cultivated carrots become more popular, the major genes and eventually QTL conditioning these colors will be better described.

Little investigation of carotenoid biosynthesis has been conducted with carrot, in part because of its limitations as an experimental organism. All previous studies on the genetic control of β-carotene synthesis in carrot root tissue have indicated the presence of pigment (i.e., orange roots) is recessive to white or nonpigmented roots. However, Goldman and Breitbach (1996) identified and characterized a recessive gene that causes a 93% reduction in carotenoid content, suggesting a new interpretation of carotenoid biosynthesis in carrot roots. This gene, designated *rp*, likely causes a lesion in the carotenoid biosynthetic pathway, and may provide new clues as to the details of this important process in carrot. The first several leaves of *rprp* plants are white and speckled, suggesting an effect of the *rp* allele on chlorophyll through the reduction of carotenoid pigmentation. Leaves appear identical to wild type appearance by the sixth leaf, suggesting developmental effects of the *rp* allele.

Through analysis of this mutant, Koch and Goldman (2004, 2005) determined that carrots produce tocopherols, particularly α-tocopherol or provitamin E. This has led to projects designed to screen carrot germplasm for both provitamins A and E, and to a breeding program focusing on increasing both compounds in carrot. In experiments designed to assess the impact of the *rp* allele (Koch and Goldman 2005), the reduced pigment (*rp*) mutation of carrot exhibited a 96% reduction in levels of α- and β-carotene and a 25–43% reduction in α-tocopherol when compared to a near-isogenic line. In plants homozygous for *rp*, a substantial increase was observed in phytoene, a precursor to carotenoids, suggesting the location of the *rp* lesion in the carotenoid synthesis pathway.

Koch and Goldman (2004) developed a one-pass method for extraction and analysis of carotenoids and tocopherols in carrot. The method allows for ease of detection of these compounds, facilitating a more rapid increase through selecting them simultaneously. The technique, which is based on Grela, Jensen, and Jakobsen (1999), allows for detection of both classes of compounds in high-moisture plant tissue such as fruits and vegetables. Koch and Goldman (2005) conducted a two-year field experiment that was carried out at two locations to assess levels of major carotenoids and tocopherols in carrot (*Daucus carota*) root and leaf tissue. Levels of compounds in root tissue reported on a dry weight basis were: α-tocopherol, 0.04–0.18 ppm; lycopene, 0.00–52.94 ppm; α-carotene, 10.63–1504.76 ppm; and β-carotene, 26.69–1673.76 ppm. Higher levels of all carotenoids were measured in phloem tissue than xylem. Leaf tissue levels of tocopherols measured on a dry weight basis ranged from 0.02 to 0.85 ppm, while levels of carotenoids ranged from 12.81 to 411.66 ppm. In xylem tissue, α-tocopherol was significantly ($P \leq 0.001$) positively correlated with α-carotene ($r = 0.65$) and with β-carotene ($r = 0.52$). This positive correlation indicates it may be possible to select for both increased α-tocopherol and carotenoids in carrot. Purple carrots bring a different category of nutrients to consumers, the anthocyanins. A single dominant gene, P_1, conditions purple root color, while a second gene, P_2, is hypostatic to P_1 and conditions purple pigmentations in aerial plant parts (Simon 1996). The wide range of genetic variation in purple carrot pigment type, amount, and distribution awaits future genetic analysis and application in breeding programs.

11.5   BIOTECHNOLOGY

11.5.1   Tissue Culture and Genetic Transformation

The production of callus and regeneration of carrot tissue culture is readily achieved. In fact, carrot has been used as a model organism for research on somatic embryogenesis and totipotency since F.C. Steward's seminal studies in the 1950s and 1960s (Steward et al. 1964). While no comprehensive evaluations have been carried out of variation across diverse carrot germplasm for success in initiation of callus and suspension cultures and regeneration through embryogenesis and organogenesis, a relatively wide range of carrot germplasm has been used in tissue culture studies, including wild carrot and both European and Asian cultivars.

Agrobacterium-mediated genetic transformation of carrot is also readily achieved (Hardegger and Sturm 1998), as well as transformants targeting resistance to herbicides, alternaria leaf blight toxin, and increased carotene content with varying success in achieving desired phenotype. While the development of transgenic carrots with traits of interest for growers and consumers is feasible, trangene escape to wild carrot seems likely with the widespread distribution of wild carrot to many seed production areas. This possibility may be a significant issue in considering deployment strategies of transgenes in carrot.

11.5.2 Marker-Assisted Selection

Few molecular markers in or linked to carrot major genes or QTL have been developed. Examples have been reported for carotene QTL (Santos and Simon 2002) and the Y_2 gene (Bradeen and Simon 1998), the *Mj-1* nematode resistance gene (Boiteux et al. 2000, 2004), cytoplasmic male sterile (Bach, Olesen, and Simon 2002), and the *Rs* sugar type gene (Yau and Simon 2003; Yau, Santos, and Simon 2005), with marker-assisted selection exercised successfully in the latter case. Yau, Santos, and Simon (2005) suggest that these identifications can be done on leaf tissue from early growth, thereby removing the need for growing mature roots and analyzing sugar content. The identification of markers for resistance to alternaria leaf blight, aster yellows, and powdery mildew, as well as for soluble solids, carotenoids, and tocopherol, should be feasible and useful as the genetics of these important traits are studied. As codominant markers are more widely developed and maps are joined, the application of these genomic tools can have immediate application in marker-based breeding.

11.6 CONCLUSIONS AND FUTURE DIRECTIONS

Carrot is a relatively recently domesticated crop, being cultivated for only 1000 years, and carrot breeding has received serious attention for little more than 50 years. Significant strides have been made in meeting the demands of growers and shippers, and improvements in consumer quality have also been realized. The importance of carrots as a source of income and nutrients will hopefully point to a sustained carrot-breeding effort in the future. New products from orange carrots, as well as a rebirth of purple, yellow, red, and white carrots may foretell expanded consumer markets for carrots, and its efficient biomass production could point to industrial uses. As carrot genetics is better described, new biotechnological tools are applied to carrot improvement, and new germplasm is introgressed, a heightened efficiency of the breeding process seems likely. A great expansion of tools for marker-assisted selection will be of particular interest, and the development of efficient systems for haploidization and transposon-mediated genetic change would further accelerate the generation and identification of useful variation for carrot improvement. One example of where such technologies would benefit carrot breeders is in improving the identification of sterile lines and maintainer lines during the inbred-hybrid breeding process. Another area in which such technologies would benefit carrot breeders is in developing disease and pest resistance in their populations. Screening germplasm collections via seedling assays or field tests for particular pests is difficult and time-consuming, but identification of useful allelic variation through the use of such markers might make germplasm screening very efficient. Despite these possibilities, a broad understanding of the crop and classical field breeding will nevertheless continue to be the most essential keys to successful carrot breeding.

In their chapter on carrot breeding, Peterson and Simon (1986) mention that hybrid carrot development was of primary importance to carrot breeders up through the 1970s. Today, that goal has largely been achieved, and hybrid carrot is a reality worldwide. However, elements of this system have received little attention from breeders, and will need attention by scientists in order for carrot to continue its rise in popularity and productivity.

One of these factors is seed production, particularly in the inbred-hybrid production system. Seed-yielding ability is still limited, as selection for improved inbreds often does not focus on seed yield during the inbreeding process. Improved seed production ability would increase the efficiency and profitability of carrot seed production, particularly for hybrid seed. Selection for seed-yielding ability *per se* could be practiced during inbred development and hybrid testing. Preferences for individual inbreds by pollinators have received little research attention, and improving this element could greatly enhance hybrid seed production. The absence of improvement in these characteristics is not due to a lack of interest, but rather to a very limited number of carrot breeding programs. Further compounding this problem is that the majority of United States programs have been based in the midwestern United States, where seed production is not practiced. To improve seed production characteristics, selection and evaluation must take place in the area where seed production occurs, which is the western United States.

Fifty years ago, processing carrot production held a much larger share of the carrot acreage than it does today, excluding, of course, the cut-and-peel carrot products, which may properly be referred to as minimally processed. Given the rise in fresh-market carrot production and the relative decrease of processing production, carrot breeders are in danger of losing valuable processing carrot germplasm. Many of the fresh market hybrids are hybrids, and a substantial number of high-quality processing cultivars are hybrid cultivars as well. The few remaining open-pollinated cultivars that are typically used for processing production, particularly for canning and freezing, are not widely maintained by carrot breeders. The primary types used for this purpose are "Chantenay" and "Danvers," with the former being devoted mostly to dicing and the latter to slicing. In the last decade, efforts have begun to preserve and maintain this germplasm so that it can be of use for future breeding efforts. Maintenance of open-pollinated carrot cultivars could also be important for preserving valuable landrace material and other germplasm preserved in seed banks.

Recently, Simon and colleagues have conducted collecting expeditions for *Daucus* germplasm in the Mediterranean, the Middle East, and parts of Central Asia. These expeditions have provided a large number of unique genetic resources, including open-pollinated landrace populations, that could benefit carrot-breeding efforts in the future. These materials will need proper maintenance and evaluation to be of use to breeders. It is likely that a focus on maintaining populations will be of great value to the carrot breeding community in the next fifty years.

Finally, improvement efforts in carrot have not yet focused on developing germplasm suited for organic and sustainable production systems, despite the popularity of the crop for such systems worldwide. Indeed, organic growers have developed carrot production systems that enhance the ability of the crop to do well without inputs of synthetic fertilizer, herbicides, and pesticides, yet breeders have not made this a focus of their work. Given the importance of this crop in the rotation for organic vegetable production systems, and its high quality under such systems, carrot breeders may further enhance carrot production and profitability by directing their efforts toward cultivar development for these environments.

REFERENCES

Babb, M. F., Kraus, J. E., and Magruder, R., Synonymy of orange-fleshed varieties of carrots, *U.S.D.A. Circular*, 833, 1–100, 1950.

Bach, I. C., Olesen, A., and Simon, P. W., PCR-based markers to differentiate the mitochondrial genomes of petaloid and male fertile carrot (*Daucus carota* L.), *Euphytica*, 127, 353–365, 2002.

Banga, O., Origin of the European cultivated carrot, *Euphytica*, 6, 54–63, 1957a.

Banga, O., The development of the original European carrot material, *Euphytica*, 6, 64–76, 1957b.

Banga, O., *Main Types of the Western Carotene Carrot and their Origin*, Zwolle, The Netherlands: W.E.J. Tjeenk Willink, 1963.

Boiteux, L. S., Belter, J. G. D., Roberts, P. A. D., and Simon, P. W. D., RAPD linkage map of the genomic region encompassing the root knot nematode (*Meloidogyne javanica*) resistance locus in carrot, *Theoretical and Applied Genetics*, 100, 439–446, 2000.

Boiteux, L. S., DellaVecchia, P. T., and Reifschneider, F. J. B., Heritability estimate for resistance to *Alternaria dauci* in carrot, *Plant Breeding*, 110, 165–167, 1993.

Boiteux, L. S., Hyman, J. R., Bach, I. C., Fonseca, M. N., Matthews, W. C., Roberts, P. A., and Simon, P. W., Employment of flanking codominant STS markers to estimate allelic substitution effects of a nematode resistance locus in carrot, *Euphytica*, 136, 37–44, 2004.

Bonnet, A., Source of resistance to powdery mildew for breeding cultivated carrots, *Agronomie*, 3, 33–37, 1983.

Bradeen, J. M. and Simon, P. W., Conversion of an AFLP fragment linked to the carrot Y_2 locus to a simple, codominant, PCR-based marker form, *Theoretical and Applied Genetics*, 97, 960–967, 1998.

Bradeen, J. M. and Simon, P. W., Carrot, In *Genome Mapping & Molecular Breeding*, Kole, C., Ed., *Vegitables*, vol. 5, Heidelberg, Berlin, New York, and Tokyo: Springer, 2006 (in press).

Buishand, J. G. and Gabelman, W. H., Investigations on the inheritance of color and carotenoid content in phloem and xylem of carrot roots (*Daucus carota* L.), *Euphytica*, 28, 611–632, 1979.

Burr, F., *The Field and Garden Vegetables of America*, 19–20, 1988 reprint, Chilicothe, IL: American Botanist Booksellers, 1865.

Buttery, R. G., Seifert, R. M., and Ling, L. C., Characterization of some volatile constituents of carrots, *Journal of Agricultural and Food Chemistry*, 16, 1009–1015, 1968.

Dickson, M. H., The inheritance of longitudinal cracking in carrots, *Euphytica*, 15, 99–101, 1966.

Ellis, P. R. and Hardman, J. A., The consistency of the resistance to carrot fly attack at several centers in Europe, *Annals of Applied Biology*, 98, 491–497, 1981.

Freeman, R. E. and Simon, P. W., Evidence for simple genetic control of sugar type in carrot (*Daucus carota* L.), *Journal of the American Society for Horticultural Science*, 108, 50–54, 1983.

Frese, L., Resistance of the wild carrot *Daucus carota* ssp. *hispanicus* to the root-knot nematode, *Meloidogyne hapla*, *Journal of Plant Diseases and Protection*, 81, 396–403, 1983.

Gabelman, W. H., Goldman, I. L., and Breitbach, D. W., Evaluation and selection for resistance to aster yellows in carrot (*Daucus carota* L.), *Journal of the American Society for Horticultural Science*, 119, 1293–1297, 1994.

Goldman, I. L., F_1 hybrid, inbred line, and open-pollinated population releases from the University of Wisconsin carrot breeding program, *HortScience*, 31, 882–883, 1996.

Goldman, I. L. and Breitbach, D. N., Inheritance of a recessive character controlling reduced carotenoid pigmentation in carrot (*Daucus carota* L.), *Journal of Heredity*, 87, 380–382, 1996.

Grela, E. R., Jensen, S. K., and Jakobsen, K., Fatty acid composition and content of tocopherols and carotenoids in raw and extruded grass pea (*Lathyrus sativus* L.), *Journal of the Science of Food and Agriculture*, 79, 2075–2078, 1999.

Hardegger, M. and Sturm, A., Transformation and regeneration of carrot (*Daucus carota* L.), *Molecular Breeding*, 4, 119–129, 1998.

Heywood, V. H., Relationships and evolution in the *Daucus carota* complex, *Israel Journal of Botany*, 32, 51–65, 1983.

Huang, S. P., Della Vecchia, P. T., and Ferreira, P. E., Varietal response and estimates of heritability of resistance to *Meloidogyne javanica* in carrots, *Journal of Nematology*, 18, 496–501, 1986.

Koch, T., Development of carrot (*Daucus carota*) germplasm with elevated levels of carotenoids and tocopherols, and effects of the *Rp* locus on phytochemical composition, University of Wisconsin-Madison, Ph.D. discussion, 163, 2005.

Koch, T. and Goldman, I. L., A one-pass semi-quantitative method for extraction and analysis of carotenoids and tocopherols in carrot, *HortScience*, 39, 1260–1261, 2004.

Koch, T. and Goldman, I. L., Relationship of carotenoids and tocopherols in a sample of carrot root-color accessions and carrot germplasm carrying *Rp* and *rp* alleles, *Journal of Agricultural and Food Chemistry*, 53, 325–331, 2005.

Mackevic, V. I., The carrot of Afghanistan, *Bulletin of Applied Botany, Genetics and Plant Breeding*, 20, 517–556, 1929.

Magruder, R., Boswell, V. R., Elmsweller, S. L., Miller, J. C., Hutchins, A. E., Wood, J. F., Parker, M. M., and Zimmerley, H. H., Descriptions of types of principal American varieties orange-fleshed carrots, *US Department of Agriculture Publication*, 361, 48, 1940.

McCollum, G. D., Interspecific hybrid of *Daucus carota* × *D. capillifolius*, *Botanical Gazette*, 136, 201–206, 1975.

Nicolle, C., Simon, G., Rock, E., Amouroux, P., and Remesy, C., Genetic variability influences carotenoid, vitamin, phenolic, and mineral content in white, yellow, purple, orange, and dark-orange carrot cultivars, *Journal of the American Society for Horticultural Science*, 129, 523–529, 2004.

Nothnagel, T., Straka, P., and Linke, B., Male sterility in populations of *Daucus* and the development of alloplasmic male-sterile lines of carrot, *Plant Breeding*, 119, 145–152, 2000.

Peterson, C. E. and Simon, P. W., Carrot breeding, In *Breeding Vegetable Crops*, Bassett, M. J., Ed., Westport, CT: AVI, pp. 321–356, 1986.

Plunkett, G. M., Chandler, G. T., Lowry, II, P. P., Pinney, S. M., and Sprenkle, P. S., Recent advances in understanding *Apiales* and a revised classification, *South African Journal of Botany*, 70, 371–381, 2004.

Rubatzky, V. E., Quiros, C. F., and Simon, P. W., *Carrots and Related Vegetable Umbelliferae*, New York: CABI Publishing, 1999.

Sáenz Laín, C., Research on *Daucus* L. (Umbelliferae), *Anal Jardin Botánico Madrid*, 37, 481–533, 1981.

Santos, C. A. F. and Simon, P. W., QTL analyses reveal clustered loci for accumulation of major provitamin A carotenes and lycopene in carrot roots, *Molecular Genetics and Genomics*, 268, 122–129, 2002.

Santos, C. A. F. and Simon, P. W., Merging carrot linkage groups based on conserved dominant AFLP markers in F_2 populations, *Journal of the American Society for Horticultural Science*, 129, 211–217, 2004.

Schulz, B., Westphal, L., and Wricke, G., Linkage groups of isozymes, RFLP and RAPD markers in carrot (*Daucus carota* L. *sativus*), *Euphytica*, 74, 67–76, 1993.

Senalik, D. and Simon, P. W., Quantifying intra-plant variation of volatile terpenoids in carrot, *Phytochemistry*, 26, 1975–1979, 1987.

Simon, P. W., Genetic improvement of vegetable carotene content, In *Biotechnology and Nutrition: Proceedings of the Third International Symposium*, Bills, D. D. and Kung, S.-D., Eds., London: Butterworth-Heinemann, pp. 291–300, 1992.

Simon, P. W., Inheritance and expression of purple and yellow storage root color in carrot, *Journal of Heredity*, 87, 63–66, 1996.

Simon, P. W., Domestication, historical development, and modern breeding of carrot, *Plant Breeding Reviews*, 19, 157–190, 2000.

Simon, P. W. and Strandberg, J. O., Diallel analysis of resistance in carrot to *Alternaria* leaf blight, *Journal of the American Society for Horticultural Science*, 123, 412–415, 1998.

Simon, P. W., Peterson, C. E., and Lindsay, R. C., Genetic and environmental influences on carrot flavor, *Journal of the American Society for Horticultural Science*, 105, 416–420, 1980.

Simon, P. W., Wolff, X. Y., Peterson, C. E., Kammerlohr, D. S., Rubatzky, V. E., Strandberg, J. O., Bassett, M. J., and White, J. M., High carotene mass carrot population, *HortScience*, 24, 174, 1989.

Simon, P. W., Matthews, W. C., and Roberts, P. A., Evidence for simply inherited dominant resistance to *Meloidogyne javanica* in carrot, *Theoretical and Applied Genetics*, 100, 735–742, 2000.

Steward, F. C., Mapes, M. O., Kent, A. E., and Holsten, R. D., Growth and development of cultured plant cells, *Science*, 143, 20–27, 1964.

Stommel, J. R. and Simon, P. W., Phenotypic recurrent selection and heritability estimates for total dissolved solids and sugar type in carrot, *Journal of the American Society for Horticultural Science*, 114, 695–699, 1989.

Strandberg, J. O., Bassett, M. J., Peterson, C. E., and Berger, R. D., Sources of resistance to *Alternaria dauci*, *HortScience*, 7, 345, 1972.

Surles, R. L., Weng, N., Simon, P. W., and Tanumihardjo, S., Carotenoid profiles and consumer sensory evaluation of specialty carrots (*Daucus carota*, L.) of various colors, *Journal of Agricultural and Food Chemistry*, 52, 3417–3421, 2004.

Thompson, D. J., Studies on the inheritance of male-sterility in the carrot, *Daucus carota* L, *Proceedings of the American Society for Horticultural Science*, 78, 332–338, 1962.

Umiel, N. and Gabelman, W. H., Analytical procedures for detecting carotenoids of carrot roots and tomato fruits, *Journal of the American Society for Horticultural Science*, 96, 702–704, 1971.

Vieira, J. V., Casali, V. W. D., Milagres, J. C., Cardoso, A. A., and Regazzi, A. J., Heritability and genetic gain for resistance to leaf blight in carrot (*Daucus carota* L.) populations evaluated at different times after sowing, *Revista Brasileira de Genética*, 14, 501–508, 1991.

Vieira, J. V., Della Vecchia, P. T., and Ikuta, H., Cenoura 'Brasilia', *Horticultural Brasileira*, 1, 42, 1983.

Wang, M. and Goldman, I., Resistance to root knot nematode (*Meloidogyne hapla* Chitwood) in carrot is controlled by two recessive genes, *Journal of Heredity*, 87, 119–123, 1996.

Welch, J. E. and Grimball, E. L., Male sterility in carrot, *Science*, 106, 594, 1947.

Westphal, L. and Wricke, G., Genetic and linkage analysis of isozyme loci in *Daucus carota* L, *Euphytica*, 56, 259–267, 1991.

Wolyn, D. J. and Chahal, A., Nuclear and cytoplasmic interactions for petaloid male-sterile accessions of wild carrot (*Daucus carota* L.), *Journal of the American Society for Horticultural Science*, 123, 849–853, 1998.

Yau, Y. Y. and Simon, P. W., A 2.5-kb insert eliminates acid soluble invertase isozyme II transcript in carrot (*Daucus carota* L.) roots, causing high sucrose accumulation, *Plant Molecular Biology*, 53, 151–162, 2003.

Yau, Y. Y., Santos, K., and Simon, P. W., Molecular tagging and selection for sugar type in carrot roots using co-dominant, PCR-based markers, *Molecular Breeding*, 16, 1–10, 2005.

Index

Printed and bound by CPI Group (UK) Ltd, Croydon, CR0 4YY

23/10/2024

01778257-0005